Springer Collected Works in Mathematics

More information about this series at http://www.springer.com/series/11104

Kôsaku Yosida

Kôsaku Yosida

Collected Papers

Editor
Kiyosi Itô

Reprint of the 1992 Edition

 Springer

Author
Kôsaku Yosida (1909-1990)
The University of Tokyo
Tokyo, Japan

Editor
Kiyosi Itô (1915-2008)
Research Institute for Mathematical Sciences
Kyoto University
Kyoto, Japan

ISSN 2194-9875
ISBN 978-4-431-55050-1
Springer Tokyo Heidelberg New York Dordrecht London

Library of Congress Control Number: 2014957322

Preface

Kôsaku Yosida, born on February 7, 1909, was brought up in Tokyo. Having majored in Mathematics at University of Tokyo, he was appointed to Assistant at Osaka University in 1933 and promoted to Associate Professor in 1934. He received the title of Doctor of Science from Osaka University in 1939. In 1942 he was appointed to Professor at Nagoya University, where he worked very hard with his colleagues to promote and expand the newly established Department of Mathematics. He was appointed to Professor at Osaka University in 1953 and then to Professor at University of Tokyo in 1955. After retiring from University of Tokyo in 1969, he was appointed to Professor at Kyoto University, where he also acted as Director of the Research Institute for Mathematical Sciences. He retired from Kyoto University in 1972 and worked as Professor at Gakushuin University until 1979.

Yosida acted as President of the Mathematical Society of Japan, as Member of the Science Council of Japan, and as Member of the Executive Committee of the International Mathematical Union.

In 1967 he received the Japan Academy Prize and the Imperial Prize for his famous work on the theory of semigroups and its applications. In 1971 he was elected Member of the Japan Academy.

Yosida went abroad many times to give series of lectures at mathematical institutions and to deliver invited lectures at international mathematical symposia. In 1969 he organized a large scale international symposium on Functional Analysis and Related Topics in Tokyo, which contributed very much to promote the international activities of mathematicians in Japan.

To our great regret, Yosida passed away on June 20, 1990 at the age of 81, just before the International Congress of Mathematicians, Kyoto, 1990, whose realization owes much to his tremendous pioneering effort.

This volume contains almost all of his mathematical papers written in English. Since his work ranges over many fields of analysis, it is classified in ten sections. At the beginning of each section we included some comments which, we hope, will be helpful to the reader.

Everyone who reads this volume will be able to picture the image of an outstanding mathematician, Kôsaku Yosida, who made a tremendous contribution to promote functional analysis in Japan up to the international level.

Preface

As a student at University of Tokyo he wrote some interesting papers on meromorphic functions and ordinary differential equations (Section I). When he took his position at Osaka University in 1933, he became interested in functional analysis through discussions with Mitio Nagumo and Shizuo Kakutani. Yosida published many interesting papers (Sections II and IV) and, together with Nagumo and Kakutani, initiated research in functional analysis in Japan. Yosida's passion toward functional analysis continued throughout his life and culminated in the publication of a beautiful book: Functional Analysis (Springer-Verlag, 1964) in which his own important results are included. The most original work during his stay in Osaka (1933–42) is probably a series of papers on the operator-theoretical treatment of ergodic theory, some written jointly with Kakutani (Section III).

The work that made Yosida's name world-famous must be the theory of semigroups and its applications. This theory was published in 1948 when he was at Nagoya University. It is now called the Hille-Yosida theory, as is fully explained in Kato's comments (Section V). The influence of this theory was far-reaching: Yosida and many others applied it to numerous areas in analysis such as diffusion equations (Section VI), Markov processes (Section VII), hyperbolic equations (Section VIII), and potential theory (Section IX).

After 1980 he was mainly interested in modernizing operational calculus from Mikusiński's viewpoint (Section X).

This volume was compiled with the eager cooperation of the following mathematicians, most of whom were Yosida's students: Hiroshi Fujita, Daisuke Fujiwara, Takeyuki Hida, Teruo Ikebe, Seizô Itô, Yuji Ito, Tosio Kato, Tosihusa Kimura, Hikosaburo Komatsu, S. T. Kuroda, Shigetake Matsuura, Sigeru Mizohata, Shinzo Watanabe. It should also be mentioned that most reprints of Yosida's papers included in this volume were offered by Mrs. Fumiko Yosida. I would like to express my sincere gratitude to all who helped in the production of this collection.

June 1992   Kiyosi Itô

Contents

[†] Numbers in brackets refer to the Bibliography at the end of this volume.

Contents

Contents

V. Semigroups and Evolution Equations

VI. Diffusion Equations

Contents

I. Meromorphic Functions and Ordinary Differential Equations

Comments (Tosihusa Kimura)

In the paper [10], the following two conditions are considered for a meromorphic function $f(z)$ in $\mathbf{C}$.

(A) Given any sequence $\{a_i\}$, $a_i \in \mathbf{C}$, the family $f(z + a_i)$ is normal in any bounded domain.

(B) For any sequence $\{a_i\}$, there exists no subsequence of $f(z + a_i)$ which converges to a constant function.

Yosida proves two necessary and sufficient conditions that $f(z)$ satisfies (A), and a necessary and sufficient condition that $f(z)$ satisfies both (A) and (B). In addition several theorems are proved.

In the paper [12], Yosida proves an inequality for a meromorphic function $f(x)$ and its derivative $f'(x)$ and gives three applications.

The paper [3] deals with a linear differential equation

$$P_0(x)y^{(n)} + P_1(x)y^{(n-1)} + \cdots + P_n(x)y = 0,$$

where $P_i(x)$ are polynomials. It is shown that under certain hypotheses the general solution can be written as

$$y = \prod_{i=1}^{p} (x - \alpha_i)^{\nu_i} \sum_{j=1}^{n} C_j e^{\lambda_j x} R_j(x).$$

This result is a generalization of a theorem due to Halphen.

The paper [4] is not so long but monumental. "A Malmquist's theorem" in the title of the paper is stated as follows. Consider the differential equation

$$(1) \qquad \frac{dy}{dx} = R(x, y),$$

where $R(x, y)$ is a rational function of x and y. If the equation is not of the Riccati type, then its single-valued solution is a rational function, that is to say, if the equation admits at least one transcendental meromorphic solution, then the equation is of the Riccati type.

In this paper two theorems are proved. The first one is stated as follows. Let $R(x, z)$ and $R_1(x, z)$ be rational functions of x and z and let d and d_1 be the degrees

1

of R and R_1 with respect to z. If the differential equation

$$R\left(x, \frac{d^n y}{dx^n}\right) = R_1(x, y)$$

admits at least one transcendental meromorphic solution, then $(n + 1)d \geq d_1$ holds. The second one says that if the differential equation

$$\left(\frac{dy}{dx}\right)^m = R(x, y) \quad (R:\text{ rational in } x \text{ and } y)$$

admits at least one transcendental meromorphic solution, then R is a polynomial in y of at most $2m$ degrees.

The proof is based upon the Nevanlinna theory, which found in this paper, the first application to a field other than the theory of functions. The idea of the proof is simple, as is explained for the original Malmquist theorem as follows. Let $f(x)$ be a transcendental meromorphic solution of (1) and let $T(r, f(x))$ denote the characteristic function of f in the sense of Nevanlinna. It suffices to derive from the equality

$$T(r, f'(x)) = T(r, R(x, f(x)))$$

that R is a polynomial in y of degree at most 2. To this end, the following two estimations are used. For any positive ε,

$$T(r, f'(x)) \leq 2(1 + \varepsilon)T(r, f(x))$$

except for a sequence of intervals whose total measure is finite. The second estimation, due to Valiron, is a key lemma: For a rational function $R(x, z)$ in x and z,

$$T(r, R(x, f(x))) \leq dT(r, f(x)) + 0(\log r),$$

where d is the degree of R with respect to z.

This method was adopted later in the book of Bieberbach: Theorie der gewöhulichen Differentialgleichungen, 1953 and has been developed for more general differential equations by several mathematicians (See, Hille, Ordinary differential equations in the complex domain, 1976).

In the paper [6] a differential equation

$$y'' = (F(x) + iG(x)y)$$

is considered, where F and G are real continuous functions on an interval $0 \leq x \leq c$. Given a complex number a, let $y(x)$ be a solution satisfying $y(0) = a$ and $|y(x)|'_{x=0} > 0$ and let x_1 be the a-point next to $x = 0$ of $y(x)$. Yosida gives two estimates for x_1 from below.

The following theorem is due to iversen. The asymptotic values of a transcendental meromorphic function $y = f(x)$ can be identified with the transcendental singularities of the (infinitely many-valued) inverse functions $x = \rho(y)$ of $f(x)$. In the paper [7] Yosida gives a method of obtaining an asymptotic value η of a transcendental meromorphic solution of the equation

$$F(x, y, y') = 0 \quad (F:\text{ polynomial in } x, y \text{ and } y').$$

Next he considers the equation

$$y' = R(x, y) \qquad (R: \text{rational in } x \text{ and } y)$$

and supposes that the equation admits a transcendental meromorphic solution $y(x)$. Then the equation must be of the Riccati type by the Malmquist theorem:

$$(2) \qquad \frac{dy}{dx} = \frac{P_0(x) + P_1(x)y + P_2(x)y^2}{Q(x)},$$

where $P_i(x)$ and $Q(x)$ are polynomials of degrees p_i and q. Some results are proved for the asymptotic values (finite or infinite) of $y(x)$. For example, the solution can have asymptotic values only in case

$$(3) \qquad p = \max\{p_i\} \geqq q - 1.$$

In the paper [8], it is proved that for a transcendental meromorphic solution $y(x)$ of (2) we have

$$T(r, y(x)) = \begin{cases} 0(r^{2(p-q)+2}) & \text{if} \quad 2(p - q) \neq -1, \\ 0((\log r)^2) & \text{if} \quad 2(p - q) = -1. \end{cases}$$

This result implies that the following conjecture stated in [7] is true: If (2) admits at least one transcendental meromorphic solution, then (3) holds.

The theory of Nevanlinna was extended to (finite-valued) algebroid functions by several mathematicians. The theorem of Malmquist stated in [4] was generalized by Yosida to differential equations with transcendental meromorphic solutions by the use of the theory of Nevanlinna. Therefore it is natural to try generalizations of results obtained in the papers [4], [7] and [8] to differential equations with transcendental algebroid solutions. Yosida succeeded in extending his results in this direction.

In the paper [13], the Picard-Vessiot theory of homogeneous linear ordinary differential equations is discussed. One of his aims is to treat the theory from an abstract point of view. But this point of view is not complete. For example, differential fields are supposed to be sets of functions meromorphic in a domain of **C**. It should be noted, however, that he was the first one to state the Galois correspondence between a Galois subgroup and a differential subfield.

In the paper [62], Yosida gives another proof of the expansion theorem due to Titchmarsh and Kodaira for singular second-order differential operator of the form $-d^2/dx^2 + q(x)$. His method is based upon approximation by compact intervals of the open interval where $q(x)$ is defined. This idea is adopted independently by N. Levinson at the same time, but their proofs are slightly different.

The theorem of Malmquist stated in [4] is proved by Wittich and Hille in alternate methods. In the paper [102], Yosida gives a proof of the theorem of Malmquist, using neither the theorem of Valiron stated in [4] nor the estimations of Wittich and Hille given in their proofs.

A remark to a theorem due to Halphen

Japan. J. Math. **9** (1932) 231–232

Consider the linear differential equation with polynomial coefficients

$$(1) \qquad P_0(x)\frac{d^n y}{dx^n} + P_1(x)\frac{d^{n-1}y}{dx^{n-1}} + \ldots + P_n(x)y = 0.$$

We assume

α) that the degree p of $P_0(x)$ is not less than that of any other coefficient;

β) that the singular points $a_i(i=1, 2, \ldots, p)$ of (1) which lie in the finite part of the x-plane are regular in the sense of Fuchs;

γ) that the quotient of any two solutions of (1) is one-valued.

Under these three conditions we can prove the following generalisation of a theorem due to Halphen.[1]

Theorem 1.[2] The general solution of (1) must be of the form

$$(2) \qquad y(x) = \prod_{i=1}^{p}(x-a_i)^{\nu_i}\left\{\sum_{j=1}^{n}C_j e^{\lambda_j x}R_j(x)\right\}$$

where C_j are constants of integration, ν_i and λ_j are some definite constants, and $R_i(x)$ are rational functions.

Proof. By the condition γ) we conclude that the exponents relative to the singular points a_i must differ by integers from each other, and that no logarithmic terms can occur in the solutions. Let

$$\nu_i, \nu_i + e_{i1}, \nu_i + e_{i2}, \ldots, \nu_i + e_{i,n-1}$$

be the exponents relative to a_i, where $e_{i1}, e_{i2}, \ldots, e_{i,n-1}$ denote positive integers.

If we put

$$(3) \qquad y(x) = Y(x)f(x) \; ; \; f(x) = \prod_{i=1}^{p}(x-a_i)^{\nu_i}$$

the equation (1) will be transformed into

$$(4) \qquad Q_0(x)\frac{d^n Y}{dx^n} + Q_1(x)\frac{d^{n-1}Y}{dx^{n-1}} + \ldots + Q_n(x)Y = 0,$$

where

---

[1] See Picard; Traite d'Analyse, III, 3 ième éd. p. 431.

[2] Compare the analogous treatment of Halphen when $P_j(x)$ are doubly periodic functions. Picard, loc. cit., p. 451.

$$(5) \quad \begin{cases} Q_0(x) = P_0(x), \quad Q_1(x) = P_1(x) + {}_nC_1\dfrac{f'(x)}{f(x)}P_0(x), \ldots \\[2ex] Q_l(x) = P_l(x) + {}_{n-l+1}C_1\dfrac{f'(x)}{f(x)}P_{l-1}(x) + {}_{n-l+2}C_2\dfrac{f''(x)}{f(x)}P_{l-2}(x) \\[2ex] \qquad + \ldots + {}_nC_l\dfrac{f^{(l)}(x)}{f(x)}P_0(x), \\[2ex] \qquad \ldots\ldots\ldots\ldots\ldots\ldots\ldots\ldots\ldots \end{cases}$$

Thus $Q_i(x)(i=0,1,2,\ldots.n)$ are rational and $Q_i(x)/Q_0(x)$ are finite at infinity. By virtue of (3) the condition β) is satisfied for the equation (4), whose general solution $Y(x)$ is one-valued.

Therefore, by the above mentioned theorem of Halphen, the general solution of (4) must be of the form

$$Y(x) = \sum_{j=1}^{n} C_j e^{\lambda_j x} R_j(x).$$

Hence, by (3), our theorem is proved.

Theorem 2. n linearly independent functions of the form

$$(6) \qquad y_i(x) = \prod_{j=1}^{p} (x-a_j)^{\nu_j}\{e^{\lambda_i x}R_i(x)\}, \quad (i=1,2,\ldots,n),$$

a, ν and λ denoting some constants and $R(x)$ being rational functions, satisfy a linear differential equation with rational coefficients

$$(7) \qquad \frac{d^n y}{dx^n} + P_1(x)\frac{d^{n-1}y}{dx^{n-1}} + \ldots + P_n(x)y = 0 \,(^3)$$

such that $\qquad \lim_{x\to\infty} P_i(x) = \text{finite} \quad (i=1,2,\ldots.n)$

Proof. By the converse $(^4)$ of Halphen's theorem n linearly independent one-valued functions

$$1, \; y_2(x)/y_1(x), \; y_3(x)/y_1(x), \; \ldots, \; y_n(x)/y_1(x)$$

satisfy the differential equation of the type

$$(8) \qquad \frac{d^n z}{dx^n} + Q_1(x)\frac{d^{n-1}z}{dx^{n-1}} + \ldots + Q_{n-1}(x)\frac{dz}{dx} = 0$$

such that the rational functions $Q_i(x)$ $(i=1,2,\ldots,n-1)$ are finite at infinity.

Thus putting

$$z(x) = y(x)/y_1(x)$$

and transforming (8) into (7), we can easily prove our theorem, remembering the transformation formulae (1), (3), (4) and (5).

Mathematical Institute,
Tokyo Imperial University.

$(^3)$ Evidently the conditions β) and γ) are satisfied for this equation.

$(^4)$ cf. Picard, loc. cit., p. 435.

A generalisation of a Malmquist's theorem

Japan. J. Math. **9** (1932) 253–256

J. Malmquist([1]) has proved the following important theorem :
If the differential equation

$$\frac{dy}{dx} = R(x, y), \quad R \text{ rational function},$$

is not a Riccati's equation([2]), all its one-valued solutions must be rational.

In the present paper, I intend to give a generalisation of it (see the Theorem 1 and 2 below), as an application of the Nevanlinna's([3]) theory of meromorphic functions.

1. Let $R(x, z)$ be an irreducible rational function whose degree in z is d. If we denote by $z(x)$ any transcendental meromorphic function we shall have, in Nevanlinna's notation :

Lemma 1. (of G. Valiron)([4]),

$$(1) \qquad T(r, R(x, z(x))) = d\, T(r, z(x)) + O(\log r),$$

and

Lemma 2. For any positive ε,

$$(2) \qquad T(r, z^{(n)}(x)) < (1 + \varepsilon)(n + 1) T(r, z(x)),$$

for sufficiently large r, except possibly in a sequence of intervals where the total variation of $\log r$ is finite.

Proof. Evidently we have

$$T(r, z^{(k)}(x)) = N(r, z^{(k)}(x)) + m(r, z^{(k)}(x))$$

$$< (k + 1)N(r, z(x)) + m(r, z^{(k-1)}(x)) + m\left(r, \frac{z^{(k)}(x)}{z^{(k-1)}(x)}\right)$$

([1]) J. Malmquist ; Acta Mathematica, 36, p. 297.

([2]) The differential equation $\frac{dy}{dx} = R(x, y)$ is called to be of the Riccati's type if $R(x, y)$ is a polynomial in y of the degree≤ 2.

([3]) R. Nevanlinna; Le théorème de Picard-Borel et la théorie des fonctions méromorphes.

([4]) G. Valiron; Bull. de la Soc. Math de France, 1931.

$$= N(r, z(x)) + \{kN(r, z(x)) + m(r, z^{(k-1)}(x))\} + m\left(r, \frac{z^{(k)}(x)}{z^{(k-1)}(x)}\right)$$

$$< T(r, z(x)) + T(r, z^{(k-1)}(x)) + m\left(r, \frac{z^{(k)}(x)}{z^{(k-1)}(x)}\right).$$

On the other hand, Nevanlinna has proved that([5])

$$m\left(r, \frac{z^{(k)}(x)}{z^{(k-1)}(x)}\right) = O(\log r, \, T(r, z^{(k-1)}(x))),$$

except possibly in the sequence of intervals where the total variation of $\log r$ is finite. Thus assuming the inequality (2) to be true when $n = k-1$ it is also true when $n = k$. And as (2) is evidently true when $n = 1$([6]), by the above reasoning, our proof by mathematical induction is completed.

Remark. If $z(x)$ has only a finite number of poles, we have, instead of (2),

$$(2)' \qquad\qquad T(r, z^{(n)}(x)) < (1+\varepsilon)\, T(r, z(x)),$$

because in this case

$$N(r, z^{(n)}(x)) \leqq (n+1)N(r, z(x)) = O(\log r).$$

2. Besides $R(x, z)$, consider an irreducible rational fuuction $R_1(x, z)$ whose degree in z is d_1.

Then by the above two lemmas we may prove

Theorem 1. If the differential equation

$$(3) \qquad\qquad R\left(x, \frac{d^n y}{d x^n}\right) = R_1(x, y)$$

admits at least one transcendental meromorphic solution $y(x)$, we must have

$$(4) \qquad\qquad (n+1)d \geqq d_1.$$

Proof. From

$$T(r, R(x, y^{(n)}(x))) = T(r, R_1(x, y(x)))$$

follows, by the lemma 1,

$$d\, T(r, y^{(n)}(x)) = d_1 T(r, y(x)) + O(\log r).$$

Thus, for any positive ε,

---

(5) R. Nevanlinna, loc. cit., p. 63.

(6) R. Nevanlinna; loc. cit., p. 104.

$$(5) \qquad \begin{cases} d(n+1)(1+\varepsilon)T(r, y(x)) > d_1 T(r, y(x)) + O(\log r) \\ \qquad\qquad \text{(by the Lemma 2),} \end{cases}$$

for sufficiently large r, except possibly in the sequence of intervals where the total variation of $\log r$ is finite. As $y(x)$ is supposed to be transcendental, we must not have

$$\lim_{r \to +\infty} \frac{T(r, y(x))}{\log r} = \text{finite} \, (^7).$$

Therefore from (5) follows the demanded inequality (4). C. Q. F. D.
By the remark of the Lemma 2 we have

Remark. If $y(x)$ admits only a finite number of poles, we must have

$$(4)' \qquad\qquad\qquad d \geqq d_1.$$

Corollary 1. If $(n+1)d < d_1$, any meromorphic solution of (3) is rational.

Corollary 2. If $d < d_1$ any meromorphic solution of (3) with a finite number of poles is rational.

Theorem 2. When the differential equation

$$(6) \qquad\qquad \left(\frac{dy}{dx}\right)^m = R_1(x, y), \; m \text{ a positive integer,}$$

admits at least one transcendental meromorphic solution $y(x)$, $R_1(x, z)$ must be a polynomial in z of the degree at most equal to $2m$.

Proof. Let the degree in z of the numerator and that of the denominator of $R(x, z)$ be p and q respectively. By the transformation

$$Y = \frac{1}{y - \alpha}, \; \alpha \text{ being a properly chosen constant,}$$

the equation (6) will be transformed into

$$\left(\frac{dY}{dx}\right)^m = R_2(x, Y),$$

the degree in z of the numerator and that of the denominator of the irreducible rational function $R_2(x, z)$ are either

$$p \text{ and } p - 2m \text{ respectively, when } p - 2m \geqq q,$$
or
$$q + 2m \text{ and } q \text{ respectively, when } q > p - 2m.$$

Thus by the Theorem 1 we must have

$(^7)$ R. Nevanlinna; loc. cit., p. 40.

either $2m \geqq p \geqq q + 2m$ or $2m \geqq q + 2m > p$.

This proves our theorem.

Corollary 1. When $q \neq 0$ or when $p > 2m$, any meromorphic solution of (6) is rational.

In particular we have

Corollary 2([8]). If $m = 1$, any meromorphic solution of (6) is rational, so long as (6) is not a Riccati's equation.

Remark 1 ([9]). When $y(x)$ is one-valued and has only a finite number of essential singular points, our Theorem 1 and 2 will still hold since the Nevanlinna's theory may easily be extended to this class of functions. Therefore Malmquist's theorem is newly demonstrated and extended in the light of the modern functiontheoretical method.

Remark 2. While this paper was in the press, Prof. T. Shimizu kindly pointed out that the Lemma 2 is in A. Bloch; Les fonctions holomorphes et méromorphes dans le cercle-unité, p. 30.

---

(8) This is just the above mentioned Malmquist's theorem. See the remark 1.
(9) This remark I owe to Prof. T. Shimizu.

On the distribution of α-points of solutions for linear differential equation of the second order

Proc. Imp. Acad. Tokyo **8** (1932) 335–336

(Comm. by T. YOSIE, M.I.A., Oct. 12, 1932.)

Consider two differential equations

$$(1) \qquad y'' = \{F(x) + iG(x)\}y \,,$$

$$(2) \qquad Z'' = H(x)Z \,,$$

where $F(x)$, $G(x)$ and $H(x)$ are real and continuous functions in the domain $D: 0 \leq x \leq c$, such that $H(x) \leq F(x)$ in D.

We put $|y(x)| = R(x)$ and $\arg y(x) = \theta(x)$, then $R(x)$ will satisfy the differential equation

$$(3) \qquad R'' = \{F(x) + \theta'(x)^2\}R \,.$$

Lemma. Let the initial values of $R(x)$, $Z(x)$ be given by

$$(4) \qquad \begin{cases} R(0) = Z(0) = |a| \,, \text{ a being a constant, real or imaginary,} \\ R'(0) = Z'(0) > 0 \,. \end{cases}$$

If we denote by x_1 and x_2 the next $|a|$—points to $x=0$ in D of $R(x)$ and $Z(x)$ respectively, then we must have

$$(5) \qquad x_1 \geq x_2 \,.$$

Proof. By (4), $R(x)$ and $Z(x)$ are both $> |a|$ for $0 < x < \min(x_1, x_2)$. From (2) and (3) we obtain $\dfrac{d}{dx}(R'Z - RZ') = (\theta'^2 + F - H)RZ$. Integrating from 0 to x and remembering (4), we have

$$(6) \qquad \begin{cases} R'(x)Z(x) - R(x)Z'(x) = \int_0^x (\theta'^2 + F - H)RZ\,dx \\ \qquad\qquad\qquad\qquad \geq \int_0^x (\theta'^2 + F - H)a^2\,dx \geq 0 \,, \\ 0 \leq x \leq \min(x_1, x_2) \,. \end{cases}$$

Therefore we shall have $R(x) \geq Z(x) \geq |a|$ for $0 \leq x \leq \min(x_1, x_2)$. Hence, if it be possible that $x_1 < x_2$, we should have $|a| = R(x_1) \geq Z(x_1) \geq |a|$, that is $Z(x_1) = |a|$ or $x_1 = x_2$. This is a contradiction, and so we must have $x_1 \geq x_2$.

Remark. The equality sign of (5) holds if and only if $\theta'(x) \equiv 0$ and $F(x) \equiv H(x)$ in $[0, x_2]$.

Corollary. If we take $H(x) = -\max\limits_{0 \leq x \leq x_1} |F(x)| = F_{x_1}$ we may obtain

$$x_1 \geq \frac{\pi - 2t_g^{-1}\left(\dfrac{R(0)}{R'(0)}\sqrt{F_{x_1}}\right)}{\sqrt{F_{x_1}}} \quad {}^{1)}.$$

Because in this case

$$Z(x) = R(0)\sqrt{1 + \frac{R'(0)^2}{R(0)^2 F_{x_1}}} \; \sin\left(\sqrt{F_{x_1}}\,x + t_g^{-1}\left(\frac{R(0)}{R'(0)}\sqrt{F_{x_1}}\right)\right)$$

and hence

$$x_2 = \frac{\pi - 2t_g^{-1}\left(\dfrac{R(0)}{R'(0)}\sqrt{F_{x_1}}\right)}{\sqrt{F_{x_1}}} \; .$$

From these results we deduce the

Theorem. Let $y(x)$ be an integral of (1) such that $y(0)=a$, $|y(x)|'_{x=0}=R'(0)>0$. If the next a-points to $x=0$ of $y(x)$ in D be denoted by x_3, then we must have

$$(8) \qquad \begin{cases} x_3 \geq x_2 \,, \\[2mm] x_3 \geq \dfrac{\pi - 2t_g^{-1}\left(\dfrac{|a|}{R'(0)}\sqrt{F_{x_1}}\right)}{\sqrt{F_{x_1}}} \; . \end{cases}$$

Proof. Since $x_3 \geq x_1$, (8) follows immediately from (5) and (7).

Corollary. In particular, when $a=0$, (8) gives

$$(9) \qquad x_3 = x_1 \geq \frac{\pi}{\sqrt{F_{x_1}}} \quad {}^{2)}$$

where the equality sign holds if and only if $F(x) \equiv -F_{x_1}$, $G(x) \equiv 0$ in $[0, x_3]$.

Remark. Combined with the method of zero-free regions of E. Hille[3] the formula (9) will play some useful rôle in zero-point problems for linear differential equations of the 2nd order with imaginary coefficients.

---

1) The branch of tg^{-1} is to be taken such that $0 \leq tg^{-1}\alpha \leq \dfrac{\pi}{2}$ if $\alpha \geq 0$.

2) This idea of the generalisation of Sturm's classical theorem was proposed to Mr. H. Nakano, then the result was obtained by either of us independently, almost at the same time. Compare the succeeding paper of Mr. Nakano.

3) See E. Hille; Zero point broblems for linear differential equations of the second order, Mathematisk Tidsskrift, B, 2 (1927).

A note on Riccati's equation

Proc. Phys.-Math. Soc. Japan **15** (1933) 227–232

(Read April 2, 1933)

§ 1. We owe to F. Iversen[1] the following beautiful theorem:

The asymptotic values of a transcendental meromorphic function $f(x)$ can be identified with the transcendental singularities (of the infinite determination) of its inverse function $\rho(y)$.

By virtue of this theorem we can determine the possible asymptotic values η of any transcendental meromorphic solution $y(x)$ of the differetial equation

$$(1) \qquad \begin{cases} F(x, y, y')=0, \\ F \text{ an irreducible polynomial} \end{cases}$$

as follows.

Putting $x=\dfrac{1}{\chi}$ and transforming (1) into

$$(2) \qquad \begin{cases} G\left(y, \chi, \dfrac{d\chi}{dy}\right)=0, \\[2mm] G \text{ an irreducible polynomial} \end{cases}$$

we have only to determine such values η of y that some solution $\chi(y)$ of (2) (tending to zero when y tends to η) has η as its transcendental singular points.

Denoting by $\chi'(\chi, y)$ the function $\dfrac{d\chi}{dy}$ defined by (2) and following the researches of P. Painlevé[2], we see that such values η must belong to one of the three categories:

α) One of the determination of $\chi'(\chi, y)$ is indeterminate for

$$\chi=0, \quad y=\eta.$$

β) At least two critical points $\chi=g_i(y)$ and $\chi=g_j(y)$ of one of the determination of $\chi'(\chi, y)$ are equal to zero for $y=\eta$: $g_i(\eta)=g_j(\eta)=0$.

γ) The developpement of one of the determination $\chi'(\chi, y)$ in a vicinity of the critical point $\chi=g(y)$

---

(1) .F. Iversen: Recherches sur les fonctions inverses des fonctions méromorphes. p. 13.

(2) P. Painlevé: Théorie anılytiques de l'equation differentielles.

$$\chi' - g'(y) = \mu_\lambda(y)[\chi - g(y)]^{\frac{\lambda}{\nu}} + \mu_{\gamma+1}(y)[\chi - g(y)]^{\frac{\lambda+1}{\nu}} + \quad \cdots$$

is such that

$$g(\eta) = 0, \quad \mu_\lambda(\eta) = 0, \quad \lambda < \nu - 1.$$

These values η can algebraically be determined.

§2. Consider the case when the equation (1) is of the first degree

$$(3) \qquad \frac{dy}{dx} = R(x, y), \quad R \text{ a rational function.}$$

If (3) admits at least one transcendental meromorphic solution, then (3) must be of the Riccati's type[1] :

$$(4) \qquad \frac{dy}{dx} = \frac{P_0(x) + P_1(x)y + P_2(x)y^2}{Q(x)}$$

where $P_i(x)$ and $Q(x)$ are polynomials of degree p_i and q respectively and the right hand member of (4) is irreducible with respect to its arguments. Thus the equation corresponding to (2) is

$$(5) \qquad \left[P_0\!\left(\frac{1}{\chi}\right) + P_1\!\left(\frac{1}{\chi}\right)y + P_2\!\left(\frac{1}{\chi}\right)y^2\right]\frac{d\chi}{dy} + \chi^2 Q\!\left(\frac{1}{\chi}\right) = 0.$$

Therefore the values η should, in this case, belong to the category α). We distinguish two cases:

1st case. $q - 2 \geqq \max(p_i)$. The required values η do not exist, because $\dfrac{d\chi}{dy}$ does not become indeterminate for any finite value η when $\chi = 0$. Concerning the the case $\eta = 0$ see the Remark below.

2nd case. $q - 2 < \max(p_i)$. Let the terms of the highest degree in $P_i(x)$ be $a_i x^{r_i}$. Then the values η are given as the roots of the algebraic equation

$$(6) \qquad a_0 \varepsilon_0 + a_1 \varepsilon_1 y + a_2 \varepsilon_2 y^2 = 0$$

where

$$\left\{\begin{array}{lll}
\varepsilon_0 = \varepsilon_1 = 1, \ \varepsilon_2 = 0 & \text{when} & p_0 = p_1 > p_2 \\[4pt]
\varepsilon_0 = \varepsilon_2 = 0, \ \varepsilon_1 = 1 & \text{,,} & p_1 > \max(p_0, p_2) \\[4pt]
\varepsilon_0 = \varepsilon_2 = 1, \ \varepsilon_1 = 0 & \text{,,} & p_0 = p_2 > p_1
\end{array}\right.$$

$$\left\{\begin{array}{lll}
\varepsilon_0 = \varepsilon_1 = \varepsilon_2 = 1 & \text{,,} & p_0 = p_1 = p_2 \\[4pt]
\varepsilon_0 = 0, \ \varepsilon_1 = \varepsilon_2 = 1 & \text{,,} & p_1 = p_2 > p_0 \\[4pt]
\varepsilon_0 = \varepsilon_1 = 0, \ \varepsilon_2 = 1 & \text{,,} & p_2 > \max(p_0, p_1)^{(2)}
\end{array}\right.$$

---

(1) See J. Malmquist: Acta Mathematica 36. Or K. Yosida : Japanese Journal of Mathematics (1933) p. 253.

(2) $\varepsilon_1 = \varepsilon_2 = 0$, $\varepsilon_0 = 1$ when $p_0 > \max(p_1, p_2)$. In this case $y(x)$ may admit ∞ as its asymptotic value. See the Remark below.

Examples. Examples corresponding to the 1st two cases and to the last 4 cases are given respectively by

$$\begin{cases} y' = x^l(y-a) \\ y(x) = \exp.\left[\dfrac{x^{l+1}}{l+1}+c\right]+a, \end{cases}$$

$$\begin{cases} y' = x^{m+l}y^2 - \{(a+b)x^l + mx^{-1}\}y + abx^{l-m} \\ y(x) = \dfrac{1}{x^m}b\,\exp.\left[(a-b)\left(\dfrac{x^{l+1}}{l+1}+c\right)\right] - a\,\exp.\left[(a-b)\left(\dfrac{x^{l+1}}{l+1}+c\right)\right]-1, \end{cases}$$

where a, b, c are constants and l, m non-negative integers.

Remark. Putting $Y=\dfrac{1}{y}$ and transforming (5) into

$$(7) \qquad \left[P_0\left(\frac{1}{X}\right)Y^2 + P_1\left(\frac{1}{X}\right)Y + P_2\left(\frac{1}{X}\right)\right]\frac{dX}{dY} - X^2 Q\left(\frac{1}{X}\right)=0.$$

we see, by the same reasoning as above, that any transcendental meromorphic solution $y(x)$ may (or cannot) admit ∞ as its asymptotic value in case $q-2<\max(p_l)$ (or case $q-2\geqq\max(p_l)$).

Therefore the possible asymptotic values (finite or infinite) of $y(x)$ may be given as the roots (finite or infinite) of the equation (6) in case $q-2<\max(p.)$ if any, and there are no asymptotic values (finite or infinite) of $y(x)$ in case $q-2\geqq\max(p_l)$[1].

It seems probable to me that there will exist at least one asymptotic value (finite or infinite) for any transcendental meromorphic solution $y(x)$ of (4).

§ 3. When the equation (4) admits a rational solution $R(x)$ we may transform the equation (4) into

$$Y' + \frac{[P_1(x)+2R(x)P_2(x)]Y+P_2(x)}{Q(x)}=0$$

putting $\dfrac{1}{y-R(x)}=Y$, so that the general solution of (4) may be given as

$$y(x) = \exp.\left(\int\frac{P_1+2RP_2}{Q}\,dx\right)\left(c-\int\frac{P_2}{Q}\exp.\left(\int\frac{P_1+2RP_2}{Q}\,dx\right)dx\right)+R(x).$$

If, therefore, $y(x)$ is equal to a transcendental meromorphic function for a certain value of the constant of integration c, we see easily that it must admit at least one asymptotic value (finite or infinite).

---

(1) The case $q-2\geqq\max(p_l)$ cannot occur, because otherwise the point $x=\infty$ must be an algebraic singular doint for $y(x)$.

Now the question whether (4) admits a rational solution or not is a difficult one. The following result may be of some interest.

By a rational transformation

$$y = \frac{Y - \frac{1}{2}\left(\frac{P_1}{Q} + \frac{P_0'}{P_0} - \frac{Q'}{Q}\right)}{P_2} Q.$$

the equation (4) is transformed into

(8)
$$Y' = Y^2 + R(x)$$

where $R(x)$ denotes rational function.

Let the poles of $R(x)$ be all simple and $R(\infty) = $ finite. By this assumption the poles of any rational solution $Y(x)$ of (8) all lie in the finite part of the x-plane and are all simple with the residue -1, so that we have, for the contour of integration c which contains all the poles of $Y(x)$

$$-P = \frac{1}{2\pi i}\int_c Y(x)\,dx$$

denoting by P the number of poles of $Y(x)$. Thus $P = -a_{-1}$ if we denote the expansion of $Y(x)$ in the vicinity of $x = \infty$ by

$$Y(x) = a_0 + \frac{a_{-1}}{x} + \frac{a_{-2}}{x^2} + \ldots + \frac{a_{-n}}{x^n} + \ldots$$

To determine $a_{-1} = -P$, we substitute the expansions

$$Y(x) = a_0 + \frac{a_{-1}}{x} + \ldots + \frac{a_{-n}}{x^n} + \ldots \quad \text{and} \quad R(x) = R(\infty) + \frac{c_{-1}}{x} + \ldots$$

$$+ \frac{c_{-n}}{x^n} + \ldots$$

in (8):

$$\frac{-a_{-1}}{x^2} + \frac{-2a_{-2}}{x^3} + \ldots + \frac{-na_{-n}}{x^{n+1}} + \ldots$$

$$= a_0^2 + \frac{2a_0 a_{-1}}{x} + \frac{a_{-1}^2 + 2a_0 a_{-2}}{x^2} + \ldots + \frac{\sum\limits_{0 \leq \nu \leq n} a_{-\nu}a_{-n+\nu+1}}{x^{n+1}} + \ldots$$

$$+ R(\infty) + \frac{c_{-1}}{x} + \ldots + \frac{c_{-n-1}}{x^{n+1}}$$

Thus

$$(9) \quad \begin{cases} a_0^2 + R(\infty) = 0 \\ 2a_0 a_{-1} + c_{-1} = 0 \\ a_{-1}^2 + 2a_0 a_{-2} + a_{-1} + c_{-2} = 0 \\ \sum_{0 \leq \nu \leq n} a_{-\nu} a_{-n+\nu-1} + n a_{-n} + c_{-n-1} = 0 \end{cases}$$

From (9) we obtain, for example, the following formulae

$$(10) \quad \begin{cases} -a_{-1} = P = \dfrac{c_{-1}}{2\sqrt{-R(\infty)}}, \ \text{if } R(\infty) \neq 0, \\[3mm] -a_{-1} = P = \dfrac{1 \pm \sqrt{1 - 4c_{-2}}}{2}, \ \text{if } R(\infty) = 0. \end{cases}$$

Therefore we have the result:

Under the hypothesis mentioned above concerning the rational function $R(x)$, the right hand member of (10) must be a non-negative integer not smaller than the degree of $R(x)$[1].

§4. As the asymptotic values η of a transcendental meromorphic function $y(x)$ may probably have some intimate relations with the defects[2] with respect to η of $y(x)$, the following results may be of some interest:

All the defect and the index of multiplicity of any transcendental meromorphic solution $y(x)$ of (4) are equal to zero in case $P_2(x) \not\equiv 0$ and $P_0(r) + P_1(x)\alpha + P_2(x)\alpha^2 \not\equiv 0$ for any finite constant α[3]. *In particular the equation (4) can not admit solution which is a transcendental integral function when $P_2(x) \not\equiv 0$ and $P_0(x) + P_1(x)\alpha + P_2(x)\alpha_2 \not\equiv 0$ for any finite constant α.*

Proof. Let $y(x)$ be a transcendental meromorphic solution of (4). If we put

$$Y(x) = \frac{1}{y(x) - \alpha}, \ \alpha \text{ a finite constant,}$$

$Y(x)$ will satisfy the differential equation

$$Y' = -\frac{P_2(x) + (P_1(x) + 2\alpha P_2(x))Y + (P_0(x) + \alpha P_1(x) + \alpha^2 P_2(x))Y^2}{Q(x)}.$$

As the coefficient of Y^2 is $\not\equiv 0$, we have

---

(1) As P greater than the number of poles of $R(x)$.

(2) For the theory and the terminology used in Nevanlinnas theory of meromorphic functions see R. Nevanlinna: Le théoreme de Picard-Borel et la théorie des fonction méromorphes

(3) When $P_2(x) \equiv 0$ or when $P_0(x) + \alpha P_1(x) + \alpha^2 P_2(x) \equiv 0$ the differential equation is formally integrated so that there is no theoretical interest.

$$(11) \qquad T(r, Y'(x)) = 2T(r, Y(x)) + O(\log r)^{(1)},$$

where $T(r, f(x))$ denotes Nevanlinna's characteristic function. On the other hand we have, for any transcendental meromorphic function $Y(x)$

$$\varlimsup_{r \to \infty} \frac{T(r, Y'(x))}{T(r, Y(x))} \leqq 2 - \delta(\infty) - \mu(\infty)^{(2)}$$

denoting by $\delta(\infty)$ and $\mu(\infty)$ the defect and the index of multiplicity with respect to the value ∞ of $Y(x)$ respectively, whence follow the required formulas

$$\delta(\infty) = 0, \quad \mu(\infty) = 0.^{(3)}$$

Mathematical Institute,

Osaka Imperial University.

(Received May 8. 1933)

Remark. During the press of this paper, Prof. M. Fukuhara kindly gave me a letter communicating that the proposition enunciated at the end of the § 2 may be proved to be true by the method of asymptotic integration.

---

(1) To obtain the formula (11) the following theorem due to G, Valiron is used: Let $R(x, y)$ be an irreducible rational function whose degree with respect to y is d. If we substitute for y any transcendental meromorphic function $y(x)$, the characteristic function of $R(x, y(x))$ is equal to

$$dT(r, y(x)) + O(\log r)$$

See G. Valiron: Bullt. des Soc. Math. de France 1930.

(2) See R. Nevanlinna. loc. cit. p. 104.

(3) For the defect and the index of multiplicity relative to ∞ of $y(x)$, the result may be obtained by the same reasoning as above.

On the characteristic function of a transcendental meromorphic solution of an algebraic differential equation of the first order and of the first degree

Proc. Phys.-Math. Soc. Japan **15** (1933) 337–338

(Read July 1, 1933)

Consider the differential equation

$$(1) \qquad \frac{dy}{dx} = R(x, y), \quad R \text{ being a rational function of } x \text{ and } y.$$

If (1) admits at least one transcendental meromorphic solution $y(x)$, $R(x, y)$ must be a polynomial with respect to y of degree at most equal to 2.[1] Let it be

$$(2) \qquad \frac{dy}{dx} = \frac{P_0(x) + P_1(x)y + P_2(x)y^2}{Q(x)},$$

where $Q(x)$ and $P_i(x)$ are polynomials of degree q and p_i respectively and the right-hand member of (2) is irreducible.

In the preceding paper[2] I enunciated the following proposition:

If (2) admits at least one transcendental meromorphic solution $y(x)$, we must have

$$(3) \qquad p = \max(p_i) \geqq q - 1.$$

In the present paper I will give a generalisation of the above proposition by showing that we must have

$$(4) \qquad \left\{ \begin{aligned} T(r, y(x)) &= O(r^{2(p-q)+2}), \text{ if } 2(p-q)+1 \neq -1, \\ &= O((\log r)^2), \text{ if } 2(p-q)+1 = -1.^{[3]} \end{aligned} \right.$$

For let (4) be true, then we cannot have $2(p-q)+3 \leqq 0$ if $y(x)$ is meromorphic (*not necessarily transcendental*) $\not\equiv$ const.

Proof of the formula (4). By the researches on meromorphic functions due to Prof. T. Shimizu[4] we have the following equality:

---

(1) See K. Yosida: A generalisation of a Malmquist's theorem (Japanese Journal of Mathematics, Vol. 9, 1933)

(2) A note on Riccati's equation (This Proceeding, June, 1933)

(3) $T(r, y(x))$ denotes Nevanlinna's characteristic function for $y(x)$.

(4) See T. Shimizu: On the theory of meromorphic functions (Japanese Journal of Mathematics, Vol. 6, 1929).

$$\frac{d\,T(r,f(x))}{d\log r} = A(r,f(x)) = \frac{1}{\pi\left(1-\dfrac{1}{1+r^2}\right)} \int_0^r\int_0^{2\pi} \frac{|f'(\rho e^{i\theta})|^2}{(1+|f(\rho e^{i\theta})|^2)^2}\,\rho\,d\rho\,d\theta$$

for any meromorphic function $f(x)$. Substituting for $f(x)$ the above mentioned transcendental meromorphic solution $y(x)$ of (2),

$$\begin{cases} \dfrac{d\,T(r,y(x))}{d\log r} = \dfrac{1}{\pi\left(1-\dfrac{1}{1+r^2}\right)} \displaystyle\int_0^r\int_0^{2\pi} \dfrac{|\,P_0(x)+P_1(r)y(x)+P_2(x)\,y(x)^2\,|^2}{|\,Q(x)\{1+|y(x)|^2\}\,|^2}\,\rho\,d\rho\,d\theta \\[2ex] x = \rho e^{i\theta} \text{ in the integrand.} \end{cases}$$

Hence we have

$$\frac{d\,T(r,y(x))}{d\log r} = O(r^{2(p-q)+2}), \text{ if } 2(p-q)+1 \neq -1,$$
$$= O(\log r), \text{ if } 2(p-q)+1 = -1. \qquad \text{q. e. d.}$$

Remark. In the paper[1] quoted above I have also proved that when the differential equation

$$\left(\frac{dy}{dx}\right)^m = R_1(x,y),\ R_1 \text{ being a rational function of } x \text{ and } y \text{ and } m \text{ a}$$

positive intéger,

admits at least one transcendental meromorphic solution $y(x)$, $R_1(x,y)$ must be a polynomial in y of degree at most equal to $2m$:

$$(5) \qquad \left(\frac{dy}{dx}\right)^m = \frac{P_0(x)+P_1(x)\,y+\ldots+P_{2m}(x)\,y^{2m}}{Q(x)}$$

where $Q(x)$ and $P_i(x)$ are polynomials of degree q and p_i respectively. Similarly we can prove that we must have

$$T(r,y(x)) = O(r^{\frac{2}{m}(p-q)+2}), \text{ if } \frac{2}{m}(p-q)+1 \neq -1,$$
$$(4)'$$
$$= O((\log r)^2), \text{ if } \frac{2}{m}(p-q)+1 = -1.$$

When $Q(x)$ and $P_i(x)$ are constants, (5) is Briot-Bouquet's differential equation whose one-valued solutions $y(x)$ are elliptic functions or their degenerated functions. In this case we have, by (4)', $T(r,y(x)) = O(r^2)$. Since the order of characteristic function for elliptic function $y(x)$ is 2 we conclude that (4)' gives an exact result at least in the case $p=q=0$. Mathematical Institute,

Osaka Imperial University.

(Received July 10, 1933)

---

[1] K. Yosida: A generalisation etc.

On algebroid solutions of ordinary differential equations

Japan. J. Math. **10** (1933) 199–208

(Received, October 20, 1933.)

A classical theorem of J. Malmquist([1]) may be stated as follows:
If the differential equation

(E) $\dfrac{dy}{dx} = R(x, y)$, *R being a rational function of its asguments,*
is not of Riccati's type([2]), all its one valued solutions must be rational functions.

In a paper([3]) titled as " A Generalisation of a Malmquist's theorem" I tried to generalise the above theorem as an application of the Nevanlinna's theory on meromorphic functions. Recently([4]) I also gave the order of the characteristic function $T(r, y(x))$ for the transcendental meromorphic solution $y(x)$ of (E).

But by H. Selberg([5]), G. Valiron([6]), E. Ullrich([7]) and others the Nevanlinna's theory has been extended to the finite-valued algebroid-functions([8]).

By virtue of their theory we may generalise my results quoted above to the algebroid-solutions of (E), and the results will be regarded as

---

([1]) J. Malmquist: Acta Mathematica, **36**.

([2]) The equation (E) is called to be of Riccati's type if $R(x, y)$ is a polynomial with respect to y of degree ≤ 2.

([3]) K. Yosida: Jap. J. of Math., **9**, 1933, p. 253–

([4]) K. Yosida: On the characteristic function etc., Proceed. Physico-Math. Soc. of Japan, 1933, September.

([5]) H. Selberg: Über die Wertverteilung der algebroiden Funktionen, Math. Zschr., 1930, p. 709–

([6]) G. Valiron: Sur la derivée des fonct. algébroides, Bul. Soc. math. France, **59**, 1931, p. 17–

([7]) E. Ullrich: Über den Einfluss der Verzweigtheit usw., J. reine und angewandte Math., **167**, 1931, p. 198–

([8]) $y(x)$ is called a k-valued algebroid-function, if it is defined by the irreducible equation of the type

$$\sum_{i=0}^{k} A_i(x) y^i = 0, \quad A_k(x) \not\equiv 0$$

$A_i(x)$ being meromorphic functions.

some extensions of the well-known researches of P. Painlevé([9]) and others. In the course of this paper I will give two theorems (Theorem 1 and Theorem 6) on algebroid-functions which are of some interest.

I—A theorem concerning algebroid-functions.([10])

Theorem 1. For the characteristic functions of a k-valued algebroid-function $y(x)$ and of its m-th derivative $y^{(m)}(x)$ there exists the inequality

$$(1) \quad T(r, y^{(m)}) \leq \{2k(2m-1)-3(m-1)\} T(r, y) + O(\log r T(r, y))$$

except possibly in a sequence of intervals where the total variation of $\log r$ is finite.

Proof. We have, in a vicinity of $x=\alpha$, the following expansions

$$(2) \qquad y(x) = a + (x-\alpha)^{\pm \frac{\tau_i}{\lambda_i}} P\left((x-\alpha)^{\frac{1}{\lambda_i}}\right),$$

where $i=1, 2, \ldots, j$; $j \leq k$; $1 \leq \lambda_i$; $\sum \lambda_i = k$; $\tau_i \geq 1$; $P(0) \neq 0$, and $P(t)$ is a regular power series of t. We call α a τ-ple a-point of the algebroid $y(x)$, if $y(x) - a$ (or $\dfrac{1}{y(x)}$ when $a=\infty$) is equal to

$$(x-\alpha)^{\frac{\tau}{\lambda}} P\left((x-\alpha)^{\frac{1}{\lambda}}\right)$$

with positive τ for a properly chosen branch.

The poles of $y^{(m)}(x)$ occur certainly from the branch of $y(x)$ which is $=\infty$ at $x=\alpha_1$:

$$(2)' \qquad y(x) = (x-\alpha_1)^{-\frac{\tau}{\lambda}} P\left((x-\alpha_1)^{\frac{1}{\lambda}}\right),$$

and possibly from the branch which is $\neq \infty$ at $x=\alpha_2$ and for which $\lambda_i \geq 2$:

$$(2)'' \qquad y(x) = a + (x-\alpha_2)^{\frac{\tau}{\lambda}} P\left((x-\alpha_2)^{\frac{1}{\lambda}}\right).$$

Differentiating m-times the formula (2) we know that the contribution to $n(r, y^{(m)})$ which arises from the points α_1 is equal to

$$\sum_{|\alpha_1| \leq r} \tau + m \sum_{|\alpha_1| \leq r} \lambda.$$

([9]) See for example P. Boutroux's book : Leçons sur les fonct. défini. par les eq. dif. etc., p. 141-

([10]) Throughout this paper I use the notation $n(r, f)$, $\bar{n}(r, f)$, $N(r, f)$, $\overline{N}(r, f)$, $m(r, f)$, $T(r, f)$, $T(r, f')$ etc. in tae sense as in E. Ullrich's paper, loc. cit.

Similarly we know, by differentiation of (2), that the contribution to $n(r, y^{(m)})$ arising from the points α_2 is not greater than

$$\sum_{|\alpha_2|\leq r}(m\lambda-\tau)^+$$

where $u^+=\max(0, u)$.

Hence we have

$$n(r, y^{(m)})$$
$$\leq\sum_{|\alpha_1|\leq r}\tau+m\sum_{|\alpha_1|\leq r}\lambda+\sum_{|\alpha_2|\leq r}^{\lambda\geq 2,\tau\geq 1}(m\lambda-\tau)^+,$$

$$\leq\sum_{|\alpha_1|\leq r}(\tau+\lambda)+(m-1)\sum_{|\alpha_1|\leq r}\lambda+(m-1)\sum_{|\alpha_2|\leq r}^{\lambda\geq 2}\lambda+\sum_{|\alpha_2|\leq r}^{\lambda\geq 2}(\lambda-\tau)^+.$$

Since

$$\sum_{|\alpha_1|\leq r}(\tau+\lambda)+\sum_{|\alpha_2|\leq r}^{\lambda\geq 2,\tau\geq 1}(\lambda-\tau)^+\leq\sum_{|\alpha_2|\leq r}(\tau+1)+\sum_{|\alpha_1|,|\alpha_2|\leq r}(\lambda-1)$$

$$\leq 2n(r, y)+\sum_{|\alpha_1|,|\alpha_2|\leq r}(\lambda-1),$$

$$\sum_{|\alpha_1|\leq r}\lambda=\sum_{|\alpha_1|\leq r}(\lambda-1)+\bar{n}(r, y),$$

$$\sum_{|\alpha_2|\leq r}^{\lambda\geq 2}\lambda\leq\sum_{|\alpha_2|\leq r}(\lambda-1)+\sum_{|\alpha_2|\leq r}(\lambda-1),$$

the above inequality may be written as follows:

$$n(r, y^{(m)})\leq 2n(r, y)+\sum_{|\alpha_1|,|\alpha_2|\leq r}(\lambda-1)+(m-1)\left\{\sum_{|\alpha_1|\leq r}(\lambda-1)+\bar{n}(r, y)\right\}$$

$$+(m-1)\left\{\sum_{|\alpha_2|\leq r}(\lambda-1)+\sum_{|\alpha_2|\leq r}(\lambda-1)\right\}$$

$$\leq(m+1)n(r, y)+(2m-1)\sum_{|\alpha_1|,|\alpha_2|\leq r}(\lambda-1).$$

Thus integrating logarithmically we have

$$(3)\qquad N(r, y^{(m)})\leq(m+1)N(r, y)+k(2m-1)N_{\mathfrak{k}}(r),$$

where

$$\left\{\begin{array}{l}N_{\mathfrak{k}}(r)=\dfrac{1}{k}\displaystyle\int_0^r\dfrac{n_{\mathfrak{k}}(t)-n_{\mathfrak{k}}(0)}{t}dt+\dfrac{1}{k}n_{\mathfrak{k}}(0)\log r,\\[3mm] n_{\mathfrak{k}}(r)=\displaystyle\sum_{|\alpha_1|,|\alpha_2|\leq r}(\lambda-1).\end{array}\right.$$

Since the number $N_{\xi}(r)$ is proved to be $\leq 2(k-1)T(r, y)+O(1)$ by E. Ullrich([11]), (3) may be written as

$$(3)' \quad N(r, y^{(m)}) \leq (m+1)N(r, y)+2k(k-1)(2m-1)T(r, y)+O(1).$$

On the other hand, we have the following theorem due to H. Selberg([12]): *For any finite-valued algebroid-function $y(x)$ we must have*

$$(4) \qquad m\left(r, \frac{y'}{y}\right)=O(\log r \, T(r, y))$$

except possibly in a sequence of intervals where the total variation of $\log r$ is finite.

Thus combining (3)$'$ and (4) we obtain finally, by mathematical induction with respect to m, the formula (1).([13])

II—Generalisations of Malmquist's theorem.

1. As an application of the theorem 1 we may generalise the results which I gave in the above mentioned paper as follows.

Lemma (of G. Valiron)([14]). Let $R(x, z)$ be an irreducible rational function whose degree in z is d. If we denote by $z(x)$ any finite-valued algebroid-function we shall have

$$T(r, R(x, z(x)))= dT(r, z(x))+O(\log r).$$

Besides $R(x, z)$, consider an irreducible rational function $R_1(x, z)$ whose degree in z is d_1. Then we may prove

Theorem 2. If the differential equation

$$R(x, y^{(m)})= R_1(x, y)$$

admits at least one transcendental k-valued algebroid-solution $y(x)$, we must have

$$(5) \qquad k \geq \frac{\dfrac{d_1}{d}+3(m-1)}{2(2m-1)}$$

Proof. From

$$T(r, R(x, y^{(m)}(x)))= T(r, R_1(x, y(x)))$$

it follows, by the lemma and the theorem 1, that

([11]) E. Ullrich : loc. cit., p. 210.

([12]) H. Selberg: Über eine Eigenschaft der log. Ableitung usw., Avhandlingar Oslo Akademi, Math-Naturw. Kl., 14, 1930.

([13]) We have. by (4), $m(r, y^{(k)})=m\left(r, \frac{y^{(k)}}{y^{(k-1)}}y^{(k-1)}\right) \leq m\left(r, \frac{y^{(k)}}{y^{(k-1)}}\right)+m(r, y^{(k-1)})= O(\log r T(r, y^{(k-1)}))+m(r, y^{(k-1)})$.

([14]) G. Valiron : loc. cit., p. 18.

$$d[2k(2m-1)-3(m-1)]T(r, y(x))+O(\log r\,T(r, y(x)))$$

$$\geq d_1 T(r, y(x)), \text{ for the ordinary value of } r.$$

Since $y(x)$ is supposed to be transcendental, we must have

$$\lim_{r\to\infty}\frac{\log(r\,T(r, y(x)))}{T(r, y(x))}=0,(\,^{15}\,)$$

so that we obtain (5) by letting r tend to ∞.

Corollary. *If*

$$k'<\frac{\dfrac{d_1}{d}+3(m-1)}{2(2m-1)},$$

any k'-valued algebroid-solution of the above equation is an algebraic function.

Theorem 3. *(1st generalisation of Malmquist's theorem). When the differential equation*

(6) $$\left(\frac{dy}{dx}\right)^m=R(x, y), \text{ m being a positive integer,}$$

admits at least one transcendental k-valued solution $y(x)$, we must have

(7) $$k\geq\max\left(\frac{p}{2m},\ \frac{q}{2m}+1\right)$$

where p and q denote the degree in z of the numerator and that of the denominator of $R(x, z)$ respectively.

Proof. By the transformation

$$Y=\frac{1}{y-\alpha},\ \alpha \text{ a properly chosen constant,}$$

the equation (6) will be transformed into

(8) $$\left(\frac{dY}{dx}\right)^m=R_2(x, Y),$$

the degree in z of the numerator and that of the denominator of the irreducible rational function $R_2(x, z)$ are either

$$p \text{ and } p-2\,m \text{ respectively, when } p-2\,m\geq q$$

or $\qquad q+2\,m \text{ and } q \text{ respectively, when } q>p-2\,m.$

Thus. by the theorem 2 we must have $2\,km\geq\max\,(p, q+2m)$.

Corollary 1. *If the equation (6) admits at least one transcendental meromorphic solution, $R(x, y)$ must be a polynomial in y of degree at most equal to 2m.*

([15]) H. Selberg: Über die Wertverteilung usw.; loc. cit.

Proof. The formula (7) gives $p \leqq 2\,m$ and $q = 0$ when $k = 1$.

Corollary 2. If $k' < \max\left(\dfrac{p}{2m}, \dfrac{q}{2m} + 1\right)$, any k'-valued algebroid-solution $y(x)$ of (6) is an algebraic function.

Remark. Malmquist's theorem is contained in the above corollary 1, because the theorem 1 and the lemma will still hold, by a slight modification, for the finite-valued function which behaves like the algebraic function in the whole x-plane, except at a finite number of points x_i. And, as Painlevé showed us([16]), any solution $y(x)$ of (6) has only a finite number of transcendental singular points.

2. According to E. Ullrich([17]), we have

$$(9) \quad \begin{cases} \varlimsup_{r \to \infty} \dfrac{T'(r, f')}{T'(r, f)} \leqq 2 - \vartheta(\infty) + \xi, \\[2ex] \xi = \varlimsup_{r \to \infty} \dfrac{N_\xi(r)}{T(r, f)} \leqq 2(k - 1), \\[2ex] 0 \leqq \vartheta(\infty) = 1 - \varlimsup_{r \to \infty} \dfrac{\overline{N}(r, f)}{T'(r, f)} \leqq 1, \end{cases}$$

for any k-valued algebroid-function $f(x)$.

By definition, ξ gives a measure of *Verzweigtheit* of $f(x)$ over x-plane, and (in particular) $\xi = 0$ when $f(x)$ is one-valued (meromorphic). And $\vartheta(\infty)$, called as the defect with respect to the value ∞ of $f(x)$, gives a measure in what order $f(x)$ does not take the value ∞ in comparison to general values a.

Now let the equation (6) admit at least one transcendental k-valued algebroid-solution $y(x)$, then we may suppose the equation (6) to be of the form

$$\left(\frac{dy}{dx}\right)^m = \frac{\displaystyle\sum_{i=0}^{2km-j} P_i(x)y^i}{\displaystyle\sum_{i=0}^{2(k-1)m-j} Q_i(x)y^i}, \quad j \geqq 0,$$

by applying the transformation $Y = \dfrac{1}{y - \alpha}$, if necessary (see the proof of the theorem 3).

Therefore we have, by the lemma,

$$m\,T(r, y'(x)) = (2km - j)\,T(r, y(x)) + O(\log r),$$

so that (9) may be written as

<hr>

([16]) P. Boutroux: loc. cit., p. 141.

([17]) E. Ullrich: loc. cit. p. 217.

$$\frac{2km-j}{m}\leqq 2-\vartheta(\infty)+\xi, \quad \xi\leqq 2(k-1)$$

(10) or $$2k-\frac{j}{m}-2+\vartheta(\infty)\leqq\xi\leqq 2(k-1)$$

This gives a more precise result than the formura (9) which holds for any k-valued algebroid-function $f(x)$.

Considering the case $\xi=0$ (which is surely the case if $y(x)$ is meromorphic) we obtain,

Theorem 4. (2nd generalisation of Malmquist's theorem). If the equation (6) admits at least one finite-valued algebroid-solution $y(x)$ for which $\xi=0$, this equation (6) must be of the form

(11) $$\left(\frac{dy}{dx}\right)^m=\frac{\displaystyle\sum_{i=0}^{2m}P_i(x)y^i}{Q_0(x)}.$$

Proof. It is $2\,km-j-2\,m\geqq 0$ and $\vartheta(\infty)\geqq 0$, so that we must have

$$2\,km-j-2\,m=0, \quad \vartheta(\infty)=0,$$

if $\xi=0$. Hence $2\,km-j=2\,m$.

Remark. The above equation (11) is that of the Riccati's type when $m=1$.

The above proof shows that $\vartheta(\infty)=0$ is a consequence of the two equalities $2\,km-j=2\,m$ (that is $P_{2m}(x)\not\equiv 0$ and $P_{2m+i}(x)\equiv 0$) and $\xi=0$. The ξ which corresponds to $Y(x)=\dfrac{1}{y(x)-\alpha}$ (α a constant) is equal to 0 if the ξ which corresponds to $y(x)$ is 0. Therefore by applying the transformation

$$Y=\frac{1}{y-\alpha}$$

to the equation (11) we have the following.

Theorem 5. All the defects of the above mentioned solution of (11) are equal to zero, if $P_{2m}(x)\not\equiv 0$ and

$$\sum_{i=0}^{2m}P_i(x)\alpha^i\not\equiv 0$$

for any finite constant α.

For the transformed equation is of the form

$$\left(\frac{dY}{dx}\right)^m=(-1)^m\frac{\left\{\displaystyle\sum_{i=0}^{2m}P_i(x)\alpha^i\right\}Y^{2m}+\text{polynomial in } y \text{ of degree } (2\,m-1)}{Q_0(x)}$$

III—Properties of the algebroid-solutions of the equation

$$\left(\frac{dy}{dx}\right)^{m}=R(x,\,y).$$

1. Following the researches of Prof. T. Shimizu([18]) on meromorphic functions we will give a method which enables us to determine the order of the characteristic functinn $T(r,\,y(x))$ of the finite-valued algebroid-solution $y(x)$ of the equation (6). To this purpose we define the quantity $A_0(r,f)$ for the k-valued algebroid-function $f(x)$:

$$A_0(r,f)=\iint\limits_{|x|\leq r}\frac{|f'(x)|^2}{(1+|f(x)|^2)^2}\,dx_1dx_2,\quad x=x_1+ix_2.$$

$A_0(r,f)$ is equal to the area of the domain mapped by $w=f(x)$ for $|x|\leq r$, and projected on the Riemann sphere of radius $\dfrac{1}{2}$ touching the w-plane at the origin of this plane.

Dividing $A_0(r,f)$ by the k-ply counted area of the domain mapped by $f(x)\equiv x$ for $|x|\leq r$ and projected on the Riemann sphere, we obtain $A(r,f)$:

$$A(r,\,f)=\frac{A_0(r,\,f)}{k\pi\left(1-\dfrac{1}{1+r^2}\right)}.$$

$A(r,f)$ is thus $\dfrac{1}{k}$ of the mean number of sheets (calculated by the measure on the spherical area) of the Riemann surface of the inverse function of $f(x)$ in $|x|\leq r$.

Thus we have

$$A(r,\,f)=\frac{1}{k\pi\left(1-\dfrac{1}{1+r^2}\right)}\iint n\,(r,\,a)\frac{da_1da_2}{(1+|a|^2)^2},\quad a=a_1+ia_2$$

Therefore, by logarithmic integration, we have

$$\int_{\varepsilon}^{r}\frac{A(t,f)}{t}\,dt\approx T'(r,f),\quad \varepsilon>0,$$

since

$$T'(r,\,f)\approx\int_{\varepsilon}^{r}\frac{n(t,\,a)}{t}\,dt,\quad \varepsilon>0$$

for all a, except possibly for the set of values a whose linear measure

---

([18]) T. Shimizu: On the theory of meromorphic functions, Jap. J. of Math., 6, 1929.

on the Riemann sphere is equal to zero.([19]) Therefore we have

$$A(r,f) \approx \frac{1}{\text{spherical area of } A'} \iint_{a \text{ in } A'} n(r,a) \frac{da_1 da_2}{(1+|a|^2)^2}, \quad a = a_1 + ia_2,$$

where A' denotes any set of values a which form an areal set (in a-plane) whose spherical area is positive, Thus we have

Theorem 6. For any finite-valued algebroid function $f(x)$ the order of the characteristic function $T(r,f)$ is given by the following formula,

$$(12) \quad \begin{cases} T(r,f) \approx \displaystyle\int_\varepsilon^r \frac{A_1(t,f)}{t}\, dt, \quad \varepsilon > 0, \\[2ex] A_1(t,f) = \dfrac{1}{\text{spherical area of } A'} \displaystyle\iint_{x \text{ in } A'_t} \frac{|f'(x)|^2}{(1+|f(x)|^2)^2}\, dx_1 dx_2, \quad x = x_1 + ix_2, \end{cases}$$

where A'_t denotes the values of $x(|x| \le t)$ for which the corresponding points $y(x)$ are contained in the point set A', A' having the property mentioned above.

2. Now if the equation (6) admits at least one transcendental k-valued algebroid-solution $y(x)$, (6) may be considered to be of the form

$$(6) \qquad \left(\frac{dy}{dx}\right)^m = \frac{\displaystyle\sum_{i=0}^{2km-j} P_i(x) y^i}{\displaystyle\sum_{i=0}^{2(k-1)m-j} Q_i(x) y^i}, \quad j \ge 0,$$

where the polynomials $P_i(x)$ and $Q_i(x)$ are of degree p_i and q_i respectively, by making a suitable transformation $Y = \dfrac{1}{y-\alpha}$ if necessary. Then substituting $y(x)$ for $f(x)$ in (12) we obtain

$$\begin{cases} T(\rho, y) \approx \dfrac{1}{\text{spherical area of } A'} \displaystyle\int_\varepsilon^\rho \frac{1}{r} \\[3ex] \qquad \left\{ \displaystyle\iint_{x \text{ in } A'_r} \frac{\left|\displaystyle\sum_{i=0}^{2km-i} P_i(x) y^i(x)\right|^{\frac{2}{m}} dx_1 dx_2}{\left|\displaystyle\sum_{i=0}^{2(k-1)m-j} Q_i(x) y^i(x)\right|^{\frac{2}{m}}(1+|y(x)|^2)^2} \right\} dr, \\[3ex] x = x_1 + ix_2 \end{cases}$$

([19]) This can easily be demonstrated after the proof given by G. Valiron for meromorphic functions. See R. Nevanlinna's book: Le théoreme de Picard-Borel et la theorie des fonctions méromorphes, p. 82.

Therefore we have, by properly chosing A',

Theorem 7. The above mentioned solution $y(x)$ must satisfy the following condition:

$$(13) \quad \begin{cases} T(r, y) = O(r^{\frac{2}{m}(p-q)+2}), \text{ if } \dfrac{2}{m}(p-q)+1 \neq -1, \\[2mm] \quad\quad = O((\log r)^2), \text{ if } \dfrac{2}{m}(p-q)+1 = -1, \\[2mm] p = \max(p_i), \quad q = \min(q_i). \end{cases}$$

Since $T(r, y) \to \infty$ for $r \to \infty$, if $y(x) \neq \text{const.}$, we must have

$$\frac{2}{m}(p-q)+3 > 0.$$

Hence we obtain the

Theorem 8. If the differential equation (6) admits at least one transcendental finite-valued solution, (6) must be of the form (6) such that

$$p - q > \frac{-3m}{2}.$$

Corollary. If $p - q \leqq \dfrac{-3m}{2}$, any finite-valued algebroid-solution of (6)′ is an algebraic function.

Remark. All the defects of the above mentioned solution $y(x)$ are equal to zero under the simple condition as given in the theorem 5.

Example. The k-valued algebroid-function

$$Y(x) = \sqrt[k]{\frac{\wp(x)-a}{\wp(x)-b}}, \quad a \text{ and } b \text{ being constants, } b \neq a,$$

and $\wp(x)$ denoting Weierstrass' elliptic function, satisfies the differential equation

$$\left(\frac{dy}{dx}\right)^2 = \frac{(4b^3 - bg_2 - g_3)y^{4k} + \text{polynomial of } y \text{ of degree } 3\,k}{y^{2k-2}k^2(a-b)^2}.$$

We may easily verify that the equal-sign in the formula (7) is attained by this example, and that the order of $T(r, Y(x))$ is equal to 2 as the formula (13) shows us.

Mathematical Institute,
Osaka Imperial University.

On a class of meromorphic functions

Proc. Phys.-Math. Soc. Japan **16** (1934) 227–235

(Read April 5, 1934.)

It is easy to see that elliptic functions and their degenerated functions—simply periodic functions and rational functions—$y(z)$ have the following property (A):

Given any sequence of complex numbers $\{a_i\}$, *the family of meromorphic functions*

$$(1) \qquad y_i(z) = y(z + a_i), \quad i = 1, 2, \ldots$$

is a normal family in any closed and finite domain.

Among the functions cited above, elliptic functions $y(z)$ satisfy the additional condition (B):

For any sequence $\{a_i\}$ *there exists no partial sequence among the family* (1) *having constant as a limiting function.*

The class of meromorphic functions having the property (A) may be called the class (A) and this class is divided into two categories; the functions of the 1st category are those which satisfy the condition (B), the rest being of the 2nd category.

In § 1 we shall give some characteristic properties of the functions of the class (A) and some of their consequences. § 2 is devoted to the study of the functions of the 1st category. There we see that these functions $y(z)$ are of order 2 such that $\lim\limits_{r \to \infty} \dfrac{N_1(r)}{T(r,y)} = 2$, having no asymptotic values, like elliptic functions.[1]

§ 1.

1. *Theorem 1. In order that* $y(z)$ *may belong to the class* (A), *it is necessary and sufficient that we have the differential inequality*

$$(2) \qquad \frac{|y'(z)|}{1 + |y(z)|^2} < K,$$

for all z, with a constant K.

---

(1) Throughout this paper I use Nevanlinna's notation for meromorphic functions. See R. Nevanlinna: Le théorème de Picard-Borel et la théorie des fonction méromorphes.

Proof. This theorem is an immediate cosequence from Marty's criterion[1] for a normal family, which may easily be derived from the consideration of spherical oscillation of meromorphic functions. Marty's criterion may be stated as follows.

Let $\{y_i(z)\}$ be a family of functions which are meromorphic in an open domain D. In order that this family may be normal in D, it is necessary and sufficient that we have

$$\frac{|y_i{}'(z)|}{1+|y_i(z)|^2} < K(\overline{D})$$

in any closed domain $\overline{D}$ contained in D, where $K(\overline{D})$ is a constant independent of the suffix i.

Remark. The function $\dfrac{z+1}{z}\,\wp(z)$ satisfy (2), so that the class (A) is essentially more general than the class of elliptic functions and the degenerated functions of elliptic functions. An important class of functions—Julia's exceptional functions—belongs to the class (A), because Marty[2] has proved that a meromorphic function $y(z)$ is a Julia's exceptional function when and only when $\dfrac{|y'(z)|}{1+|y(z)|^2}=O\left(\dfrac{1}{|z|}\right)$.

Corollary 1. For the function $y(z)$ of the class (A) we have

$$T(r,\,y)=O(r^2)$$

For, by the formula of Prof. T. Shimizu[3] we have

$$(3) \qquad T(r,\,y)+O(1)=\int_0^r \frac{dr}{r}\iint_{|z|\leq r}\frac{|y'(z)|^2}{(1+|y(z)|^2)^2}\,dx\,dy, \quad z=x+iy,$$

for any meromorphic function $y(z)$.

Corollary 2. If a meromorphic function $y(z)$ satisfies the condition

$$\varlimsup_{r\to\infty}\frac{T(r,\,y)}{r^2}=\infty$$

then there must exist a sequence of numbers $\{a_i\}$ with the following property:

For any positive constant ε and for any partial sequences $\{a_{n_i}\}$ of $\{a_i\}$ $y(z)$ assumes every value, except possibly two, in an infinite number of times in the sequence of circles $\{|z-a_{n_i}|\leq\varepsilon\}$.

---

(1) M. Marty: Recherches sur la répartition des valeurs d'une fonction meromorphe, Ann. Fac. Univ. Toulouse 3. s. 23 (1931).

(2) M. Marty: loc. cit.

(3) T. Shimizu: On the theory of meromorphic functions, Jap. Journal of Math. Vol. 6, (1929).

This corollary has some relations with the results of Julia[1] and Milloux.[2]

2. Besides theorem 1, there is another characteristic property of the class (A). To this purpose, consider four equations

$$(4) \qquad y(z) = a, \ y(z) = b, \ y(z) = c, \ y(z) = d,$$

where a, b, c, d are four arbitrary constants, different from each other; and denote by $\{a_\lambda\}$, $\{b_\mu\}$, $\{c_\nu\}$, $\{d_\rho\}$ the roots of these equations respectively. Then we have the

Theorem 2. The function $y(z)$ of the class (A) is characterised by the condition that the lower limit of the difference of any two roots, taken respectively from different root-groups, is not equal to zero.

Proof. The condition is necessary. Assume, for example, there be two sequences of roots $\{a_{\lambda_i}\}$, $\{b_{\mu_i}\}$ such that $\lim_{i\to\infty} (a_{\lambda_i} - b_{\mu_i}) = 0$, then the sequence $y_i(z) = y(z + a_{\lambda_i})$, normal in any vicinity of the origin by the hypothesis, would admit, as a limiting function, a meromorphic function which assumes a and b simultaneously at the origin. And this is a contradiction.

The condition is sufficient. For if there be a sequence of numbers $\{\alpha_i\}$ such that the sequence $y_i(z) = y(z + \alpha_i)$ is not normal in a domain D, then there should exist a point z_0 (in D) in any vicinity of which the sequence of functions $y_i(z)$ takes all values, except possibly two. So that $y(z)$ assumes at least two of the values a, b, c, d, at two points whose distance can be made as small as possible. This contradicts the condition of the theorem.

Combining theorem 2 with the corollary 1 of the theorem 1, we obtain

Corollary. If a meromorphic function $y(z)$ satisfies the condition

$$\overline{\lim_{r\to\infty}} \frac{T(r, y)}{r^2} = \infty$$

there exist two sequences of the roots of the equations (4), $\{a_{\lambda_i}\}$ and $\{b_{\mu_i}\}$ for example, such that $\lim_{i\to\infty} (a_{\lambda_i} - b_{\mu_i}) = 0$, for any constants a, b, c and d, different from each other.

§ 2.

1. *Theorem 3. If $y(z)$ is a function of the class (A), the necessary*

---

(1) See, for exampl, P. Montel's book: Leçons sur les familles normales etc.

(2) H. Milloux: Le théorème de M. Picard, suites de fonctions holomorphes, fonctions méromorphes et fonctions entières, Journal de Math. (1924).

and sufficient condition that $y(z)$ may be of the 1st category is given by

$$\text{Lower limit}_{\text{for all } a}\left\{\iint\limits_{|z-a|\leqq\delta} \frac{|y'(z)|^2}{(1+|y(z)|^2)^2}\, dx\, dy\right\} = \varepsilon(\delta)>0, \quad z=x+iy,$$

where δ denotes any positive constant.

Proof. The condition is necessary. Assume that there be a sequence $\{a_i\}$ such that

$$\lim_{i\to\infty}\left\{\iint\limits_{|z-a_i|\leqq\delta} \frac{|y'(z)|^2}{(1+|y(z)|^2)^2}\, dx\, dy\right\} = 0,$$

and consider the sequence of functions $y_i(z)=y(z+a_i)$ for such a sequence $\{a_i\}$. Since $y(z)$ is of the 1st category by the hypothesis, we can choose a partial sequence $\{y_{n_i}(z)\}$ from $\{y_i(z)\}$ such that

$$\lim_{i\to\infty} y_{n_i}(z) = Y(z) \not\equiv \text{constant}$$

uniformely for $|z|\leqq\delta$. $Y(z)$ being $\not\equiv$ constant, we have

$$\iint\limits_{|z|\leqq\delta} \frac{|Y'(z)|^2}{(1+|Y(z)|^2)^2}\, dx\, dy>0.$$

On the other hand by the uniform convergence, we must have

$$\iint\limits_{|z|\leqq\delta} \frac{|Y'(z)|^2}{(1+|Y(z)|^2)^2}\, dx\, dy = \lim_{i\to\infty}\iint\limits_{|z-a_{n_i}|\leqq\delta} \frac{|y'(z)|^2}{(1+|y(z)|^2)^2}\, dx\, dy = 0$$

and this is a contradiction.

The condition is sufficient. If $y(z)$ is of the 2nd category, there must exist at least one sequence of numbers $\{a_i\}$ such that the sequence $y_i(z)=y(z+a_i)$ converges uniformly to a constant for $|z|\leqq\delta$, Then we must have $\lim\limits_{i\to\infty}\dfrac{|y'_i(z)|}{1+|y_i(z)|^2}=0$ uniformely for $|z|\leqq\delta$. Hence we shall have

$$\lim_{i\to\infty}\left\{\iint\limits_{|z-a_i|\leqq\delta} \frac{|y'(z)|^2}{(1+|y(z)|^2)^2}\, dx\, dy\right\} = 0$$

contrary to the hypothesis.

Combining this theorem with the formula (2) and (3) we have

Corollary. *The function $y(z)$ of the 1st category is of order 2, that is*

$$\varlimsup_{r\to\infty} \frac{T(r,y)}{r^2} = \text{constant} \ (\neq 0, \infty).$$

By the reasoning analogus to the proof of the above theorem we obtain

Theorem 4.   The necessary and sufficient condition that $y(z)$ of the class (A) may belong to the 1st category is given by

$$\text{Lower } \lim_{\substack{\text{for all } a}} \left\{ \max_{|z-a|\leqq\delta} \left[\frac{|y'(z)|}{1+|y(z)|^2} \right] \right\} = \varepsilon(\delta) > 0,$$

where δ denotes any positive constant.

Combining with the theorem 1, we obtain

Corollary.   Let $y(z)$ be a function of the 1st category.   Then for any sequence $\{a_i\}$, a limiting function of any convergent sequences chosen among the family $\{y_i(z)\}$, $y_i(z)=y(z+a_i)$, is of the 1st category.

2.   The sum of the indices of multiplicity of the function $\wp(z)$, which is of the 1st category, is equal to 2.   And $\wp(z)$ admits no asymptotic values.   Analogous theorem holds for all the functions of the 1st category as follows.

Theorem 5.   If $y(z)$ is of the 1st category, we must have

$$2T(r, y) - N_1(r) = O(r)$$

so that, since $y(z)$ is of order 2 by the corollary of the theorem 3, we have

$$(5) \qquad\qquad \lim_{r\to\infty} \frac{N_1(r)}{T(r, y)} = 2$$

The proof may be obtained by the following two lemmas.

Lemma 1.   To any meromorphic function $y(z)$, there corresponds a real function $D(r, y)$ defined by

$$D(r, y) = \frac{-1}{2\pi} \int_{\substack{0 \\ |z|=r}}^{2\pi} \log \frac{|y'(z)|}{1+|y(z)|^2}\, d\theta, \quad z=re^{i\theta},$$

with the following properties :

(α) $\qquad\qquad\qquad D(r, y)$ *is convex in* $\log r$,

(β) $\qquad\qquad\qquad D(r, y) = 2T(r, y) - N_1(r) + O(1).$

Proof. (α):   If we put $v(z)=\log(1+|y(z)|^2)-\log|y'(z)|$, $v(z)$ satisfies Liouville's differential equation[1]

---

(1)   This may be verified as follows.

$$\Delta = \frac{\partial^2}{\partial x^2} + \frac{\partial^2}{\partial y^2} = 4\frac{\partial^2}{\partial z\partial\bar{z}}. \quad \Delta\log\frac{|f'(z)|}{1+|f(z)|^2} = 2\frac{\partial^2}{\partial z\partial\bar{z}}\left(\log\frac{f'\bar{f}'}{(1+f\bar{f})^2}\right)$$

$$= 2\frac{\partial}{\partial z}\left\{ \frac{\frac{\partial}{\partial\bar{z}}f'}{f'} + \frac{\frac{\partial}{\partial\bar{z}}\bar{f}'}{\bar{f}'} - 2\frac{\frac{\partial f}{\partial\bar{z}}\bar{f}+f\frac{\partial\bar{f}}{\partial\bar{z}}}{1+f\bar{f}} \right\} = 2\frac{\partial}{\partial z}\left\{ \frac{\bar{f}''}{\bar{f}'} - 2\frac{f\bar{f}'}{1+f\bar{f}} \right\} = -4\frac{\partial}{\partial z}\left(\frac{f\bar{f}'}{1+f\bar{f}}\right)$$

$$\frac{\partial^2 v}{\partial x^2}+\frac{\partial^2 v}{\partial y^2}=4e^{-2v}, \quad z=x+iy.$$

Hence $v(z)$ is a subharmonic function, and therefore $D(r, y)$ is convex in $\log r$ by the theorem of F. Riesz.[1]

$$(\beta) \quad D(r, y)=\frac{1}{2\pi}\int_0^{2\pi}\log(1+|y(re^{i\theta})|^2)d\theta-\frac{1}{2\pi}\int_0^{2\pi}\log|y'(re^{i\theta})|d\theta$$

$$=2m(r, y)+O(1)-m(r, y')+m\left(r, \frac{1}{y'}\right)$$

$$=2m(r, y)+O(1)+N(r, y')-N\left(r, \frac{1}{y'}\right)$$

$$=2m(r, y)+2N(r, y)-\left\{2N(r, y)-N(r, y')+N\left(r, \frac{1}{y'}\right)\right\}+O(1)$$

$$=2T(r, y)-N_1(r)+O(1).$$

Lemma 2. Let C be any circular arc of (fixed length) 1, whose centre is origin and whose curvature $\leqq$ a fixed positive constant k. Then if $y(z)$ is of the 1st category, we must have

(6) $\qquad$ Upper limit for all possible arc C $\quad |\{M(y, C)\}|=$ finite

where $M(y, C)$ denotes the mean value of $-\log\dfrac{|y'(z)|}{1+|y(z)|^2}$ on C.

Proof. There exists a positive constant K such that a circle of radius K contains entirely any circular arc of length 1 if the middle point of this arc coincides wiih the centre of the circle. If a function $Y(z)$ is meromorphic $\not\equiv$ constant, we have

$\qquad$ Upper limit for all possible $\overline{C}$ $\quad |\{M(Y, \overline{C})\}|=$ finite constant $M(Y)$

$\overline{C}$ denoting any circular arc of length 1 and of curvature $\leqq k$ whose

---

$$=-4\left\{\frac{\frac{\partial f}{\partial z}\overline{f}'+f\frac{\partial \overline{f}}{\partial z}}{1+f\overline{f}}-\frac{f\overline{f}'\left(\frac{\partial f}{\partial z}\overline{f}+f\frac{\partial \overline{f}}{\partial z}\right)}{(1+f\overline{f})^2}\right\}=-4\left\{\frac{f'\overline{f}'}{1+f\overline{f}}-\frac{f\overline{f}'f'\overline{f}}{(1+f\overline{f})^2}\right\}$$

$$=-4\frac{f'\overline{f}'}{(1+f\overline{f})^2}=-4\ \exp.\left(\log\frac{|f'(z)|^2}{(1+|f(z)|^2)^2}\right).$$

(1) F. Riesz: Über subharmonische Funktion und ihre Rolle in der Funktionentheorie und in der Potentialtheorie, Acta Szeged, Tomus 2 (1924-1926).

middle point lies at the origin.

Now let (6) be not true, then there must exist a sequence of arcs C, $\{C_i\}$, such that $\lim\limits_{i\to\infty} |M(y, C_i)| = \infty$. Denoting by a_i the middle point of an arc C_i, we consider the normal family of functions $\{y_i(z)\}$, $y_i(z) = y(z+a_i)$, in $|z| \leq K$. Since $y(z)$ is assumed to be of the 1st category, any limiting function $Y(z)$ of $\{y_i(z)\}$ is $\not\equiv$ constant. So that the lemma is evident by the above arguments.

Corollary. All the defects of $y(z)$ are equal to zero, so that the function of the 1st category does not admit exceptional values.

Proof. We have, by Nevanlinna's 2nd fundamental theorem[1]

$$2T(r, y) - N_1(r) + S(r) > \sum_{a_i} (T(r, y) - N(r, a_i))$$

thus dividing this inequality by $T(r, y)$ and remembering (5), we obtain

$$\sum_{a_i} \delta(a_i) = 0.$$

Theorem 6. If $y(z)$ is of the 1st category, $y(z)$ admits no asymptotic values.

Proof. Let there be a path C tending to ∞, on which $y(z)$ tends to its asymptotic value a. Taking a sequence of numbers $\{a_i\}$ on C such that $\lim\limits_{i\to\infty} a_i = \infty$ we consider the family of functions $\{y_i(z)\}$, $y_i(z) = y(z+a_i)$.

Since constant a is the only one limiting function of this family, $y(z)$ must be of the 2nd category, contrary to the hypothesis.

3. *Theorem 6. Let $y(z)$ be a function of the 1st category. Then there exists a positive constant R such that $y(z)$ assumes every value at least once in any circle of radius R.*

Proof.[2] For, if not, there must exist a sequence of number $\{a_i\}$ and a sequence of positive numbers $\{R_i\}$ with the following conditions:

    (i) $\lim\limits_{i\to\infty} R_i = +\infty$

    (ii)   In the circle $|Z - a_i| \leq R_i$, $y(z)$ omitts to assume at least one value b_i.

Now consider the family of functions $\{y_i(z)\}$; $y_i(z) = y(z+a_i)$. $y(z)$

---

(1) R. Nevanlinna : loc. cit.

(2) The auther's proof was originally somewhat more complicated. The proof given here ows to Mr. S. Kakutani's suggestion. I express my cordial thanks to him.

being of the 1st category, there is a partial sequence $\{y_{n_i}(z)\}$ such that when $i\to\infty\,y_{n_i}(z)$ tends to a meromorphic function $Y(z)\neq$ const uniformly in any finite and closed domain. Let b be one of the limiting value of the sequence $\{b_{n_i}\}$. We assume that the partial sequence $\{y_{n_i}(z)\}$ is already so closen that $\lim_{i\to\infty} b_{n_i}=b$.

By the corollaries of the theorem 4 and 5 $Y(z)$ assumes every value. Suppose z_0 be one of the roots of the equation $Y(z)=b$. Then there exists two positive constants ε and δ such that $Y(z)$ assumes every value Y whose spherical distance from $b\leqq\varepsilon$ for $|z-z_0|\leqq\delta$. According to the uniforme convergence $y_{n_i}(z)$ assumes every value Y whose spherical distance from $b\leqq\dfrac{\varepsilon}{2}$ for $|z-z_0|\leqq\delta$, if i is sufficiently large. Hence $y(z)$ assumes every value Y whose spherical distance from $b\leqq\dfrac{\varepsilon}{2}$ in the circle C_i, $|z-(z_0+a_{n_i})|\leqq\delta$, if i is sufficiently large.

As $R_i\to\infty$, we may choose i so large that the circle $|z-a_{n_i}|\leqq R_{n_i}$ contains entirely the circle C_i. On the other hand $y(z)$ omitts to assume b_{n_i} in $|z-a_{n_i}|\leqq R_{n_i}$ by the assumption. This is a contradiction, since we have assumed that

$$\lim_{i\to\infty}\{\text{spherical distance }(b_{n_i},\ b)\}=0.$$

Remark. From the above theorem we see that the fundamental domains[1] for the function of the 1st category are similar to the parallelograms of periodicity for elliptic functions.

Prof. T. Shimizu kindly read through this manuscript and remarked me the result of G. Julia and J. Favard. My results have some relations with theirs.

In his book—Leçons sur les fonctions uniformes etc.—Julia considered the family

$$f_{p,\,q}(z)=f(z+p\omega_1+q\omega_2)$$

for meromorphic function $f(z)$, where ω_1, ω_2 denotes two constants such that $\omega_1/\omega_2\neq$ real, and $p,\,q$ being integers. The class of functions $f(z)$ for which the above family is normal coincides with our class (A), by our theorem 1. Julia has proved the corollary 1 and 2 of our theorem 1.

Favard has proved that the class of functions which we defined as of the 1st category coincides with the class of fast-periodic functions,

---

(1) See T. Shimizu: On the fundamental domains and the groups for meromorphic functions, 1, Jap. Journal of Math. Vol. 8 (1931).

defined by Bessonoff. Cf. J. Favard: Sur les fonctions méromorphes normales du groupe des translations, C. R. t. 185, p, 1434. Bessonoff: Sur les fonctions presque périodiques d'une variable complexe définies dans tout le plan, C. R. t. 182, p. 1011.

Mathematical Institute,
Osaka Imperial University.

(Received May 2, 1934)

A theorem concerning the derivatives of meromorphic functions

Proc. Phys.-Math. Soc. Japan **17** (1935) 170–173

(Read April 3, 1935.)

Given a meromorphic function $y=f(x)$ we let $x=g(y)$ be its inverse function defined on the Riemann surface R.

We assume that the branches $x=g_i(y)$, $(i=1, 2, \ldots.^{[1]})$ of the function $x=g(y)$ be expansible in regular power series $P_i(t)$, $(i=1, 2, \ldots.,)$:

$$(1) \qquad x=g_i(y)=P_i(t), \; t=y^{\frac{1}{\tau_i}} \text{ in } |t|<\delta^{\frac{1}{\tau_i}},$$

where τ_i denote integers $\geqq 1$ and δ a fixed positive constant not larger than $1.^{[2]}$

For any positive constant $\delta_1<\delta$ the circles $|t|<\delta_1^{\frac{1}{\tau_i}}$ are thus mapped upon the *schlicht* finite domains B_i of the x-plane by $x=P_i(t)$.

We denote by $\overline{B}_i$ the mapped images of the circles $|t|<\delta^{\frac{1}{\tau_i}}$ by $x=P_i(t)$. If the origin $x=0$ belongs to any one of the domains $\overline{B}_i$ or on its boundary, we choose it as $\overline{B}_1$. Then we have

Theorem. If we choose arbitrarily one point x within B_i $(i=2, 3, \ldots.)$ then

$$(2) \qquad \sum_{i=2}^{\infty} \frac{|f(x_i)|^{2\frac{\tau_i-1}{\tau_i}}}{|x_i|^2 \{1+(\log|2x_i|)^2+\pi^2\} |f'(x_i)|^2} < \frac{128\pi}{(\delta-\delta_1)^2}$$

Proof. The functions $x=P_i(t)$ are regular and *schlicht* in the circles $|t-t_i|<\delta^{\frac{1}{\tau_i}}-\delta_1^{\frac{1}{\tau_i}}$, where $x_i=P_i(t_i)$. By "*Verzerrungssatz*" the mapped images of the circles $|t-t_i|<\delta^{\frac{1}{\tau_i}}-\delta_1^{\frac{1}{\tau_i}}$, by $x=P_i(t)$ contain entirely the circles C_i:

$$(3) \qquad |x-x_i|<\frac{1}{4}(\delta^{\frac{1}{\tau_i}}-\delta_1^{\frac{1}{\tau_i}})|P_i'(t_i)|.$$

By the assumption the circles C_i do not contain the origin $x=0$ in their interior, if $i=2, 3, \ldots.$. Hence we have, by (3),

---

(1) These functions $x=g_i(y)$ need not exhaust all the branches of $x=g(y)$ at the point $y=0$.

(2) The assumption $\delta\leqq 1$ does not diminuate the generality of the problem.

$$|x_l| \geqq \frac{1}{4}(\delta^{\frac{1}{\tau_i}} - \delta_1^{\frac{1}{\tau_i}})|P_l'(t_l)|, \quad i \geqq 2,$$

and therefore

$$(4) \qquad |x| < |x_l| + \frac{1}{4}(\delta^{\frac{1}{\tau_i}} - \delta_1^{\frac{1}{\tau_i}})|P_l'(t_l)| < 2|x_l| \quad \text{in } C_l, \quad i = 2, 3, \ldots .$$

By the transformation $z = \log x$ the x-plane cut along the positive real axis is monovaluedly (*schlicht*) mapped upon the strip $-\pi < \Im(z) \leqq \pi$ of the z-plane. The mapped image D_i of the circle C_l, by this transformation, is not necessarily a connected domain. But these D_i are all contained in the strip and do not overlap each other.

Now consider a circular cylinder $\mathfrak{C}$ of diameter 1, which touches the z-plane along the imaginary axis; and denote by G the generating line of $\mathfrak{C}$ whose distance from the z-plane is 1. If we project the strip $-\pi < \Im(z) \leqq \pi$ upon $\mathfrak{C}$ by the straight lines orthogonal to the line G, the mapped image will have the area $= 2\pi^2$:

$$\iint_{-\pi < \Im(x) \leqq \pi} \frac{d\Omega}{1 + |z|^2} = 2\pi^2, \quad d\Omega = \text{areal element on the } z\text{-plane.}$$

Hence we have

$$\sum_{i=2}^{\infty} \iint_{D_l} \frac{d\Omega}{1 + |z|^2} < 2\pi^2,$$

So that

$$(5) \qquad \left\{ \begin{aligned} &\sum_{i=2}^{\infty} \iint_{C_l} \frac{\dfrac{d\omega}{|x|^2}}{1 + (\log|x|)^2 + \pi^2} < 2\pi^2 \\ &d\omega = \text{areal element on the } x\text{-plane.} \end{aligned} \right.$$

From (4) and (5), we obtain

$$\sum_{i=2}^{\infty} \frac{\iint_{C_i} d\omega}{4|x_i|^2(1 + (\log|2x_i|)^2 + \pi^2)} < 2\pi^2.$$

Therefore by (3),

$$\sum_{i=2}^{\infty} \frac{\pi \left\{ \dfrac{1}{4}(\delta^{\frac{1}{\tau_i}} - \delta_1^{\frac{1}{\tau_i}})|P_l'(t_l)| \right\}^2}{4|x_i|^2 \{ 1 + (\log|2x_i|)^2 + \pi^2 \}} < 2\pi^2.$$

This proves the theorem, since we have

$$(\delta - \delta_1) < \tau_i(\delta^{\frac{1}{\tau_i}} - \delta_1^{\frac{1}{\tau_i}}) < \log \frac{\delta}{\delta}, \quad (0 < \delta_1 < \delta \leqq 1, \tau_i \geqq 1),$$

$$|P_i'(t_i)| = \left| \frac{dg_i(y)}{dy} \frac{dy}{dt} \right|_{t_i} = \frac{\tau_i |f(x_i)|^{\frac{\tau_i-1}{\tau_i}}}{|f'(x_i)|}.$$

§2. Applications.

1. Ullrich's theorem. Let $\tau_i = 1$ and $|f'(x_i)| > a$ positive constant α, for $i = 2, 3, \ldots$, the above theorem shows that

$$(6) \qquad \sum_{i=2}^{\infty} \frac{1}{|x_i|^{2+\varepsilon}} \text{ convergent}, \ \varepsilon > 0.$$

This is the theorem 3 of E. Ullrich's paper[2]. The theorem 4[3] of his paper is also contained in our theorem.

2. Selberg's theorem. Let the regular power series $x = P_i(t)$, $i = 1, 2, \ldots$, $|t| < \delta^{\frac{1}{\tau_i}}_i$, given in (1), represent all the branches $x = g_i(x)$, $i = 1, 2, \ldots$ of the inverse function $x = g(y)$ which correspond to the point $y = 0$ on R. We denote by B_i' the mapped images of the circles $|t| < \delta_1^{\frac{1}{\tau_i}}$ by $x = P_i(t)$.

Let $\tau(r)$ be the maximum of the integers τ_i which correspond to the domains B'_i cut or touched by the circle $|x| = r$. Then we have, since $\delta_1 \leqq 1$, by (2),

$$(7) \qquad \frac{1}{|f'(x)|} \leqq C(x) \frac{\sqrt{128\pi}}{(\delta - \delta_1)} \frac{|x|\sqrt{1 + (\log|2x|)^2 + \pi^2}}{|f(x)|^{\frac{\tau(r)-1}{\tau(r)}}}, \ r = |x|,$$

at the point x where $|f(x)| \leqq \delta_1$, $C(x)$ denoting a positive function $(\leqq 1)$ which tends to zero as x tends to ∞.

This result is essentially the same as that obtained by H. Selberg[4], by using an asymptotic estimation concerning modular functions.

3. A Generalisation of Collingwood-Cartan theorem[5]. The hypothesis is the same as above. We obtain

$$(8) \qquad m\left(r, \frac{1}{f}\right) \leqq \tau(r)\left\{ m\left(r, \frac{f'}{f}\right) + O(\log r)\right\},$$

---

(3) E. Ullrich: Über eine Anwendung des Verzerrungssatzes auf meromorphe Funktionen, Crelle's Journal, 166, 4 (1932).

(4) H. Selberg: Algebroide Funktionen und Umkehrfunktionen Abelscher Integrale, Avh. Norske Vid. Akad. i Oslo, No. 4, p. 58 (1934). Our theorem (2) above may also be extended to algebroid functions as in Selberg's paper.

(5) E. Collingwood: Sur les valeurs exceptionelles des fonctions entières d'ordre fini, C.R., 179, p. 1125 (1924).

H. Cartan: Sur les valeurs exceptionelles d'une fonction dans tout le plan, C.R., 190, p. 1003 (1930).

by the inequality

$$m\left(r, \frac{1}{f}\right) \leqq m\left(r, \frac{f'}{f}\right) + \frac{1}{2\pi}\int_{|f|<1} \overset{+}{\log} \frac{1}{|f'(re^{i\theta})|}\, d\theta,$$

and the formula (7) as in Selberg's paper[6].

The formula (8) shows that whenever $\tau(r) = O(T(r, f)^{1-\varepsilon})$, $\varepsilon > 0$, the defect $\delta(0)$ of the function $f(x)$ is 0[7]. Mr. Shizuo Kakutani obtained elegantly a result more precise in the following paper.[8]

Mathematical Institute,
Osaka Imperial University.

(Received 18. April, 1935.)

---

(6) H. Selberg: loc. cit., p. 60.

(7) Of course the function $f(z)$ is supposed to be transcendental, not rational. Hence $\varlimsup_{r\to\infty} \dfrac{T(r,f)}{\log r} = \infty$.

(8) S. Kakutani: On the exceptional value of meromorphic functions, this Proceedings, June, p. 174 (1935).

On the groups of rationality for linear differential equations

Proc. Phys.-Math. Soc. Japan **17** (1935) 498–510

(Read September 28, 1935.)

The analogy between the linear differential equations and the algebraic equations was first pointed out by E. Picard[1] and was continuated by E. Vissiot[2] in his thesis.

Picard's method for the construction of the group of rationality for linear differential equation depends upon the use of the so-called *resolvent*. In the same direction there are important papers of M. Beke[3], A. Loewy[4] and Picard[5] himself. Without the use of the resolvent Vessiot defined the group of rationality rather directly by virtue of the characteristic property of this group.

Though Vessiot's method is very plane, his results are somewhat restricted and formal than those of Picard, as was emphasized by Beke[6]. According to Vessiot, the group of rationality is the group of all linear homogeneous substitutions which leave *formally invariant* all the rational functions of the solutions (and their derivatives) of the given linear differential equation with rational coefficients, which are *numerically rational* in the independent variable x. By Picard's definition of the group of rationality the *invariance* means *numerical invariance*, so that the analogy to the theory of algebraic equations is c'oser in Pieard's treatment than in Vessiot's thesis[7].

In the present paper the author's aim is to modify Vessiot's direct method in order to fulfil this gap between their theories, and at the same time to treat the whole theory from somewhat abstract point of view, so as to define the notions employed more precisely and then to give the style of the modern Galois theory in algebra to the theory. So

---

(1) Picard 1. (Citations in this form refer to the Literature.)

(2) Vessiot 1.

(3) Beke 1.

(4) Loewy 1.

(5) Picard 2. p. 524-599.

(6) Beke 1.

(7) Cf. Vessiot 2, where the point of view of numerical invariance is adop'ed, without any comment how to modify the arguments given in his thesis.

far as I know, the theory has never been presented explicitely in the duality form (3) and (4) given below. The classical results correspond to the formula (3).

The notion of the *irreducibility* of algebraic differential equations is necessary in the theory. However I did not make use of Königsberger's theorem directly in the proofs, because in the enunciation of Königsberger's theorem there are exceptional solutions and so the application of Königsberger's theorem without appropriate cares seems to be somewhat insufficient.

§1. The domain of rationality and its automorphism.

Definition 1. A set $\Re$ of functions $a(x)$ which are meromorphic in the given open region D is called a *domain of rationality* (defined in the region D) if it satisfies the conditions:

1) if $a \in \Re$ then $a' = \dfrac{da}{dx} \in \Re,$

2) if $a, b \in \Re$ then $a \pm b, ab, \dfrac{1}{a} \, (a \not\equiv 0) \in \Re$

3) $\Re \ni a(\equiv x)$ and hence $\Re$ contains all rational functions of x, by 2).

Thus a domain of rationality is *closed with respect to* rational operation and differentiation.

Let the coefficients of the differential equation

$$(1) \qquad \frac{d^n y}{dx^n} + p_1(x) \frac{d^{n-1} y}{dx^{n-1}} + \cdots + p_n(x) y = 0$$

be the elements $\in \Re$.

We assume that the solutions of (1) are all meromorphic in D (if not, we have only to take a suitable open sub-region of D as the initial region). Let $y_1, y_2, \ldots, y_n$ be a fundamental system of solutions of (1). The totality of functions meromorphic in D which can be expressed rationally in terms of $y_1, y_2, \ldots, y_n$ (and their derivatives) and the elements of $\Re$, clearly satisfies the conditions of Definition 1. We denote it by $\Re(y)$. $\Re(y)$ is said to be obtained from the domain of rationality $\Re$ by the *adjunction* of all the solution of (1). $\Re(y) \supseteq \Re$ by construction.

Remark. Two elements of $\Re(y)$ are to be regarded as the same element when and only when they are numerically identical as functions of x. Hence the same element of $\Re(y)$ may be expressed rationally in terms of $y_1, y_2, \ldots, y_n$ (and their derivatives) and the element of $\Re$ in many different ways. For example, we may express the element $y^{(n)}$ of $\Re(y)$ as $-(p_1(x)y_1^{(n-1)} + p_2(x)y_1^{(n-2)} + \ldots + p_n(x)y_1)$. Thus the element of $\Re(y)$ has only *one numerical value* but may have *various expressions*. We denote by the same letter α the numerical value of the element $\alpha \in \Re(y)$ and by $\alpha(y_1, y_2, \ldots, y_n)$ one of its expressions.

Definition 2. An operation A by which to every element of $\Re(y)$ corresponds at least one element of $\Re(y)$ is called an *automorphism* of $\Re(y)$ if it satisfies the conditions:

1) for any element β of $\Re(y)$ there exists at least one element α of $\Re(y)$ such that

$$\alpha \rightarrow \beta \quad {}^{8)}$$

2) if $\alpha \in \Re$ then $\qquad \alpha \leftarrow\!\!-\!\!\rightarrow \alpha^{(9)},$

3) if $\alpha \rightarrow \beta$ then $\dfrac{d\alpha}{dx} = \alpha' \rightarrow \beta' = \dfrac{d\beta}{dx}$ and hence $\alpha^{(m)} \rightarrow \beta^{(m)}$ for any positive integer m,

4) if $\alpha \rightarrow \beta$, $\gamma \rightarrow \delta$, then

$$\alpha \pm \gamma \rightarrow \beta \pm \delta, \quad \alpha\gamma \rightarrow \beta\delta \quad \text{and} \quad \frac{1}{\alpha} \rightarrow \frac{1}{\beta} \, (\alpha \not\equiv 0).$$

Theorem 1. *The automorphism A of $\Re(y)$ sets up a one-to-one correspondence between all the elements of $\Re(y)$ and all the elements of $\Re(y)$ itself.*

Proof. Let $\alpha \rightarrow \beta$, $\alpha \rightarrow \gamma$ by A, then as $\alpha - \alpha = 0 \in \Re$ we have $\beta - \gamma = 0$ by 2) and 4) of Definition 2. In the same way if $\alpha \rightarrow \gamma$, $\beta \rightarrow \gamma$ by A, we have $\alpha - \beta = 0$ since $\gamma - \gamma = 0 \in \Re$.

Definition 3. Let α be any element of $\Re(y)$ and $\alpha(y_1, y_2, \ldots, y_n)$ be any one of the expressions of α. Regarding the functions $y_1, y_2, \ldots, y_n$ and their derivatives as *indeterminate functions* of x connected by

$$y_i^{(k)} = \frac{dy_i^{(k-1)}}{dx}$$

we may effect differentiations

$$\mathfrak{d} \cdot \alpha(y_1, y_2, \ldots, y_n) = \frac{\partial \alpha(y_1, y_2, \ldots, y_n)}{\partial x} + \sum_{i=1}^{n} \frac{\partial \alpha(y_1, y_2, \ldots, y_n)}{\partial y_i} y_i' + \cdots$$

$$+ \sum_{i=1}^{n} \frac{\partial \alpha(y_1, y_2, \ldots, y_n)}{\partial y_i^{(k)}} y_i^{(k+1)} + \cdots$$

---

(8) β corresponds to α by A.

(9) If α be $\in \Re$, the equation $A \cdot \beta = \alpha$, $\beta \in \Re(y)$, is satisfied by $\beta = \alpha$ only.

This operation $\mathfrak{b}$ is called *formal differentiation*. The operation L which transforms $y_i^{(k)}$ $(i=1, 2, \ldots, n; \ k=0, 1, \ldots)$ involved in $\alpha(y_1, y_2, \ldots, y_n)$ by

$$\sum_{j=1}^{n} a_{ij} y_j^{(k)}, \qquad a_{ij} = \text{constant}$$

is called *linear substitution*. If $\det. \ \|a_{ij}\| \neq 0$ such a substitution is said to be *non-singular*.

Lemma 1. Two operations, the formal differentiation $\mathfrak{b}$ and the linear substitution L are commutative :

$$\mathfrak{b} \cdot (L \cdot \alpha(y_1, y_2, \ldots, y_n)) = L \cdot (\mathfrak{b} \cdot \alpha(y_1, y_2, \ldots, y_n)).$$

The proof is easy.

Theorem 2. To any automorphism A (of $\mathfrak{R}(y)$) there corresponds a uniquely determined non-singular linear substitutions L which leaves any element α (of $\mathfrak{R}(y)$) $\in \mathfrak{R}$ <u>numerically invariant</u>; that is $L \cdot \alpha(y_1, y_2, \ldots, y_n) = \alpha$ whatever expression $\alpha(y_1, y_2, \ldots, y_n)$ of α we take, if $\alpha \in \mathfrak{R}$. Conversely to any such non-singular linear substitution L there corresponds a uniquely determined automorphism A of $\mathfrak{R}(y)$.

Proof. If we put $A \cdot y_i = Y_i$ $(i=1,2,\ldots n)$, Y_i satisfies (1) if y_i satisfies (1), by the condition 2), 3) and 4) of Definition 2. As A sets up a one-to-one correspondence, by Theorem 1, this linear substitution L corresponding to A must be non-singular. As to the converse part of the theorem we first prove that L sets up a one-to-one correspondence between all the elements of $\mathfrak{R}(y)$ and all the elements of $\mathfrak{R}(y)$, as in the proof of Theorem 1. Then we verify the condition of Definition 2. 2) is satisfied by the hypothesis, 3) by Lemma 1, and 4) is satisfied immediately. C. Q. F. D.

By these two theorems we obtain easily

Theorem 3. All the automorphisms of $\mathfrak{R}(y)$ constitute a gronp $\mathfrak{G}$, so that the corresponding non-singular linear substitutions form a group $\mathfrak{L}$ isomorphic to $\mathfrak{G}$.

Remark 1. The corresponding elements of these two groups $\mathfrak{G}$ and $\mathfrak{L}$ effect the same operations upon the elements $\in \mathfrak{R}(y)$. The group of linear substitutions is a topological group as usual, hence we have obtained a topology in the group $\mathfrak{G}$ of automorphisms.

Remark 2. We started with a definite system $y_1, y_2, \ldots, y_n$ of fundamental solutions of (1). If another fundamental system of solutions

$$\bar{y}_i = \sum_{j=1}^{n} a_{ij} y_j, \quad (i=1, 2, \ldots, n)$$

were taken at first, we would obtain a group $\mathfrak{L}'$ isomorphic to $\mathfrak{L}$, instead of $\mathfrak{L}$ (the group $\mathfrak{G}$ being unaltered); that is $A\,\mathfrak{L}\,A^{-1}$, where A denotes the non-singular linear substitution given by the matrix $\|a_{ij}\|$. In what follows we restrict our arguments to the fundamental system $y_1, y_2, \ldots, y_n$ so that we need not distinguish $\mathfrak{G}$ and $\mathfrak{L}$.

§ 2. Two Lemmas from the theory of continuous groups.[10]

It is well known that a topological group is homogeneous, that is its topological structure is locally same everywhere over the group. A connected topological group is said to be a Lie group if the vicinity of the identical element is generated by a finite number of the linearly independent infinitesimal transformations (operations). This number is the *dimension* (or parameter number) of the Lie group.

Lemma 2 (of Cartan-Neumann[11]). Let a connected topological subgroup $\mathfrak{H}$ of a connected Lie group $\mathfrak{G}$ be <u>closed in</u> $\mathfrak{G}$, then H is also a connected Lie group.

Lemma 3. Let a connected Lie group $\mathfrak{H}$ be a subgroup of a connected Lie group $\mathfrak{G}$. If $\mathfrak{H} \neq \mathfrak{G}$, we must have

$$\text{dim. } \mathfrak{H} < \text{dim. } \mathfrak{G}.$$

Proof. Assume dim. $\mathfrak{H} =$ dim. $\mathfrak{G}$ then these two groups have the same vicinity of the identical element, since the numbers of the linearly independent infinitesimal transformations are the same for $\mathfrak{H}$ and $\mathfrak{G}$. Now, by a well known theorem of Schreier[12], any connected topological group $\mathfrak{G}$ is completely generated by its vicinity $V(e)$ of the identical element e. That is, any element g of $\mathfrak{G}$ may be expressed as the result of composition of a certain finite number of elements $\in V(e)$. Hence we would have $\mathfrak{G} = \mathfrak{H}$ contrary to the hypothesis $\mathfrak{G} \neq \mathfrak{H}$.

§ 3. The group of rationality.

Let $\mathfrak{G}$ be a topological group, then the set of all elements $\in \mathfrak{G}$ connected with its identical element constitutes the greatest connected subgroup $\overline{\mathfrak{G}}$ of $\mathfrak{G}$. $\overline{\mathfrak{G}}$ is called the *component* of $\mathfrak{G}$.

Theorem 4. The component $\overline{\mathfrak{G}}$ of the group $\mathfrak{G}$ of automorphisms of $\mathfrak{R}(y)$ is a connected Lie group.

Proof. It is easy to see that the group $\overline{\mathfrak{G}}$ is closed in the component

---

(10) In the following paragraphs "group" means "group of linear substitutions".

(11) Cartan 1. p. 24.

(12) Schreier 1.

of the group of all non-singular linear substitutions in $y_1, y_2, \ldots, y_n$; this component being a Lie group, we have the theorem by Lemma 2.

Definition 4. The component $\overline{\mathfrak{G}}$ of the group $\mathfrak{G}$ of automorphisms of $\mathfrak{R}(y)$ is called the *group of rationality* of $\mathfrak{R}$ with respect to the equation (1).

Definition 5. A subset $\mathfrak{R}_1 \supseteqq \mathfrak{R}$ of $\mathfrak{R}(y)$ is called a *sub-domain of rationality* of $\mathfrak{R}(y)$ over $\mathfrak{R}$ if it is closed with respect to differentiation and rational operation.

Let $\mathfrak{R}_1$ be a given sub-domain of rationality of $\mathfrak{R}(y)$ and let α be any one element of $\mathfrak{R}_1$, then the set $\mathfrak{G}_\alpha$ of all elements of the group $\mathfrak{G}$ which leave α *numerically invariant* forms a group closed in $\mathfrak{G}$. The set $\mathfrak{G}_\infty$ which consists of all the elements common to all the groups $\mathfrak{G}_\alpha$, $\alpha \in \mathfrak{R}_1$ also forms a group closed in $\mathfrak{G}$. Hence the component $\overline{\mathfrak{G}}_\infty$ of $\mathfrak{G}_\infty$ is closed in the component $\overline{\mathfrak{G}}$ of $\mathfrak{G}$, so that $\overline{\mathfrak{G}}_\infty$ is a connected Lie group by Lemma 2. In the same way the component $\overline{\mathfrak{G}}_\alpha$ of the group $\mathfrak{G}_\alpha$ is a connected Lie group, closed in $\overline{\mathfrak{G}}$.

From the above arguments we have

Theorem 5. (*The maximal property of the group* $\overline{\mathfrak{G}}_\alpha$ *and* $\overline{\mathfrak{G}}_\infty$). *If a connected subgroup* $\mathfrak{G}'$ *of* $\mathfrak{G}$ *leaves the element* $\alpha \in \mathfrak{R}_1$ (*or all the elements* $\alpha \in \mathfrak{R}_1$) *numerically invariant, we have*

$$\mathfrak{G}' \subseteqq \overline{\mathfrak{G}}_\alpha \quad (or \quad \mathfrak{G}' \subseteqq \overline{\mathfrak{G}}_\infty).$$

Moreover the groups $\overline{\mathfrak{G}}_\alpha$ *and* $\overline{\mathfrak{G}}_\infty$ *are closed in the group of rationality* $\overline{\mathfrak{G}}$.

Thus the group $\overline{\mathfrak{G}}_\alpha$ (or $\overline{\mathfrak{G}}_\infty$) is uniquely determined by α (or $\mathfrak{R}_1$). Hence we may write $\overline{\mathfrak{G}}_\alpha = \overline{\mathfrak{G}}(\alpha)$ (or $\overline{\mathfrak{G}}_\infty = \overline{\mathfrak{G}}(\mathfrak{R}_1)$). In the same way we may write $\overline{\mathfrak{G}} = \overline{\mathfrak{G}}(\mathfrak{R})$.

Theorem 6. *There exists at least one element* $\alpha \in \mathfrak{R}_1$ *such that* $\overline{\mathfrak{G}}(\mathfrak{R}_1) = \overline{\mathfrak{G}}(\alpha)$.

Proof. For every $\beta \in \mathfrak{R}_1$, $\overline{\mathfrak{G}}(\beta)$ is a connected Lie group of dimension $\leqq n^2$. Hence there exists at least one element $\alpha \in \mathfrak{R}_1$ such that

$$\text{dim. } \overline{\mathfrak{G}}(\alpha) \leqq \text{dim. } \overline{\mathfrak{G}}(\beta), \text{ for every } \beta \in \mathfrak{R}_1.$$

To prove that this $\overline{\mathfrak{G}}(\alpha)$ is the desired group we have only to show that $\overline{\mathfrak{G}}(\alpha) \subseteqq \overline{\mathfrak{G}}(\beta)$ for any $\beta \in \mathfrak{R}_1$, because of the inequality $\overline{\mathfrak{G}}(\alpha) \supseteqq \overline{\mathfrak{G}}(\mathfrak{R}_1)$ and of the maximal property of the group $\overline{\mathfrak{G}}(\mathfrak{R}_1)$.

Let g be a general substitution ($\in \overline{\mathfrak{G}}(\mathfrak{R})$) not belonging to $\overline{\mathfrak{G}}(\beta)$, and we put

$$g \cdot \beta = \beta_1 \neq \beta, \quad g \cdot \alpha = \alpha_1.$$

As $\dfrac{\alpha_1-\alpha}{\beta_1-\beta}$ depends upon at most n^2 parameters, there exists at least one rational function $r(x)$ which cannot be expressible in the form $\dfrac{\alpha_1-\alpha}{\beta_1-\beta}$ for any $g(\in \overline{\mathfrak{G}}(\mathfrak{R}))$ not belonging to $\overline{\mathfrak{G}}(\beta)$. Hence we must have

$$\overline{\mathfrak{G}}(\alpha-r(x)\beta)\leqq\overline{\mathfrak{G}}(\beta), \quad \alpha-r(x)\beta\in\mathfrak{R}_1$$

and so $\overline{\mathfrak{G}}(\alpha-r(x)\beta)\leqq\overline{\mathfrak{G}}(\alpha)$, whence $\overline{\mathfrak{G}}(\alpha-r(x)\beta)\leqq\overline{\mathfrak{G}}(\alpha)\cdot\overline{\mathfrak{G}}(\beta)$.

Therefore if we assume $\overline{\mathfrak{G}}(\alpha)\lneqq\overline{\mathfrak{G}}(\beta)$, we would obtain $\overline{\mathfrak{G}}(\alpha-r(x)\beta)\neq\overline{\mathfrak{G}}(\alpha)$, $\overline{\mathfrak{G}}(\alpha-r(x)\beta)\leqq\overline{\mathfrak{G}}(\alpha)$. Hence Lemma 3 gives dim. $\overline{\mathfrak{G}}(\alpha-r(x)\beta)<$ dim. $\overline{\mathfrak{G}}(\alpha)$, contrary to the definition of $\overline{\mathfrak{G}}(\alpha)$. Thus we must have $\overline{\mathfrak{G}}(\alpha)\leqq\overline{\mathfrak{G}}(\beta)$.

Theorem 7. For any element $\alpha\in\mathfrak{R}(y)$, the substitutions of Lie group $\overline{\mathfrak{G}}(\alpha)$ may topologically be regarded as points on an algebraic manifold M, defined by a finite number of algebraic relations among the n^2 coefficients of the linear substitutions of this group $\overline{\mathfrak{G}}(\alpha)$, such that the vicinity (on M) of the point of M which corresponds to the identical substitution is univalently covered by the points of M, corresponding to the substitution of $\overline{\mathfrak{G}}(\alpha)$ contained in a certain vicinity of the identical substitution.

Proof. Let dim. $\overline{\mathfrak{G}}(\alpha)=m$. We take α in one of its expressions, say $\alpha(y_1, y_2, \ldots, y_n)$. The condition that a non-singular linear substitution L leaves $\alpha(y_1, y_2, \ldots, y_n)$ *numerically invariant* is given by a finite number of mutually independent algebraic relations among the n^2 coefficients of L. For, this condition is given by

$$(2) \qquad \alpha(L\cdot y_1, L\cdot y_2, \ldots, L\cdot y_n)_{x=x_i}=\alpha_{x=x_i}, \quad (i=1, 2, \ldots),$$

where $x_1, x_2, \ldots, x_k, \ldots$ denotes any sequence of points of the region D such that,

$$x_j\neq x_k, \quad \text{if} \quad j\neq k$$

and x_k tending to an inner point x of D. The equation (2) is an algebraic relation among the n^2 coefficients of L for any i. Since the group $\overline{\mathfrak{G}}(\alpha)$ is a connected Lie group, the infinite number of relations must be equivalent to a finite number of mutually independent algebraic relations.

If we take α in another expression $\alpha_0(y_1, y_2, \ldots, y_n)$, we may have to adjoin new algebraic relations (independent of the former relations) in order that L leaves $\alpha_0(y_1, y_2, \ldots, y_n)$ *numerically invariant*; and so on. Moreover L must leave every element of $\mathfrak{R}$ numerically invariant, since $\overline{\mathfrak{G}}(\alpha)\leqq\overline{\mathfrak{G}}(\mathfrak{R})$. If new algebraic relations (independent of the former relations) are adjoined, the dimension of the algebraic manifold defined by the obtained algebraic relations must be diminished at least by one, compared with that of the manifold defined by the former relations only. Hence we obtain, after a finite step, an algebraic manifold M of dimension m, with the desired property.　　　　　C. Q. F. D.

Thus we say that the group $\overline{\mathfrak{G}}(\alpha)$ is an *algebraic group*. By theorem 6 the groups $\overline{\mathfrak{G}}(\mathfrak{R}_1)$ (and $\overline{\mathfrak{G}}(\mathfrak{R})$) are also algebraic groups. *The essential parameters of such Lie group may be so chosen that all the coefficients of the linear substitution of this group are the rational functions of these parameters.*[13]

§4. Galois theory for linear differential equations.

We found that to any sub-domain of rationality $\mathfrak{R}_1$ there corresponds uniquely a subgroup $\overline{\mathfrak{G}}(\mathfrak{R}_1)$ of $\overline{\mathfrak{G}}(\mathfrak{R})$ defined as the component of the totality of the elements of $\overline{\mathfrak{G}}(\mathfrak{R})$ which leave all the elements of $\mathfrak{R}_1$ *numerically invariant.* This group $\overline{\mathfrak{G}}(\mathfrak{R}_1)$ was a *connected Lie group, algebraic* and *closed in* the group of rationality $\overline{\mathfrak{G}}(\mathfrak{R})$. Conversely to any such subgroup $\overline{\mathfrak{G}}$ there corresponds uniquely a sub-domain of rationality $\mathfrak{S}$ defined as the totality of the elements of $\mathfrak{R}(y)$ which are *numerically invariant* by $\overline{\mathfrak{G}}$. The proof is easy. As $\mathfrak{S}$ is closed with respect to rational operation, we have only to show that the condition 1) of Definition 1 is satisfied. This may be effected by Lemma 1. We denote this sub-domain $\mathfrak{S}$ by $\mathfrak{S}(\overline{\mathfrak{G}})$.

Now the fundamental result of Picard-Vessiot may be generalized so as to take the duality form as follows.

Fundamental theorem :

$$(3)\qquad\qquad \mathfrak{S}(\overline{\mathfrak{G}}(\mathfrak{R}_1))=\mathfrak{R}_1,$$

$$(4)\qquad\qquad \overline{\mathfrak{G}}(\mathfrak{S}(\overline{\mathfrak{G}}))=\overline{\mathfrak{G}}.$$

Proof of (3). Since $\mathfrak{S}\supseteq\mathfrak{R}_1$ by definition, we have only to show that $\mathfrak{S}\subseteq\mathfrak{R}_1$.

Now, by Theorem 7, $\overline{\mathfrak{G}}(\mathfrak{R}_1)=\overline{\mathfrak{G}}(\alpha)$, for an element $\alpha\in\mathfrak{R}_1$. Hence $\mathfrak{S}(\overline{\mathfrak{G}}(\mathfrak{R}_1))=\mathfrak{S}(\overline{\mathfrak{G}}(\alpha))$. We then prove that any element β of $\mathfrak{S}(\overline{\mathfrak{G}}(\mathfrak{R}_1))$ can be expressed rationally in terms of α, $\dfrac{d\alpha}{dx}$, $\dfrac{d^2\alpha}{dx^2}$, $\ldots$, and the elements of $\mathfrak{R}$, so that we shall have $\mathfrak{S}(\overline{\mathfrak{G}}(\alpha))\subseteq\mathfrak{R}_1$. The proof proceeds as follows.

Let $\alpha(y_1, y_2, \ldots, y_n)$ and $\beta(y_1, y_2, \ldots, y_n)$ be one of the expressions

---

(13)　Picard 2. p. 549.

of α and β respectively. Let g and h be substitutions of $\overline{\overline{\mathfrak{G}}}$, where $\overline{\overline{\mathfrak{G}}}$ denotes the component of group of all non-singular linear substitutions:

$$g \cdot y_i^{(k)} = \bar{y}_i^{(k)} = \sum_{j=1}^{n} a_{ij} y_j^{(k)}$$

$$h \cdot y_j^{(l)} = Y_j^{(l)} = \sum_{k=1}^{n} b_{jk} y_k^{(l)}$$

we have

$$g \cdot \alpha = \alpha(\bar{y}_1, \bar{y}_2, \ldots, \bar{y}_n)$$
$$g \cdot \beta = \beta(\bar{y}_1, \bar{y}_2, \ldots, \bar{y}_n)$$
$$h \cdot \alpha = \alpha(Y_1, Y_2, \ldots, Y_n)$$
$$h \cdot \beta = \beta(Y_1, Y_2, \ldots, Y_n).$$

We can show, as in the proof of Theorem 6, that there exists at least one rational function $r(x)$ such that

$$(5) \qquad\qquad r(x) \neq \frac{h \cdot \beta - g \cdot \beta}{g \cdot \alpha - \alpha}$$

for any substitutions g, h of $\overline{\overline{\mathfrak{G}}}$, if $g \overline{\in} \overline{\mathfrak{G}}(\mathfrak{R}_1)$, that is $g \in \overline{\mathfrak{G}}(\alpha)$. $\beta(Y_1, Y_2, \ldots, Y_n)$ depends upon the parameters b_{jk} of the Lie group $\mathfrak{G}$, if h denotes the general element of $\overline{\overline{\mathfrak{G}}}$. We first fix the expression $\beta(y_1, y_2, \ldots, y_n)$ and then we may eliminate these parameters b_{jk} from $Z = \beta(Y_1, Y_2, \ldots, Y_n)$ and the results of its formal differentiation $Z' = \mathfrak{d} \cdot \beta(Y_1, Y_2, \ldots, Y_n)$, $Z'' = \mathfrak{d} \cdot (\mathfrak{d} \cdot \beta(Y_1, Y_2, \ldots, Y_n))$, $\ldots$, by algebraic operations, since the group $\overline{\overline{\mathfrak{G}}}$ is algebraic.

The result of elimination does not vanish identically, for every equation above introduces one new derivative of Z.

Thus we obtain an algebraic differential equation, whose general solutions are $\beta(Y_1, Y_2, \ldots, Y_n) = h \cdot \beta(Y_1, Y_2, \ldots, Y_n)$, h = general element of $\overline{\overline{\mathfrak{G}}}$. Let $\Phi(z) = 0$ be this equation, then we may assume the coefficients of $\Phi = 0$ be $\in \mathfrak{R}$. For we shall obtain the same differential equation $\Phi(z) = 0$ as the result of elimination from whatever $h \cdot \beta (h \in \overline{\overline{\mathfrak{G}}})$ we start, and hence the coefficients of $\Phi = 0$ may be considered as numerical invariants of the group $\overline{\overline{\mathfrak{G}}}$, and such invariants must be $\in \mathfrak{R}$ by a Lemma due to P. Appell[14].

Thus if we put

$$V = r(x)\alpha + h \cdot \beta, \quad h = \text{general element of } \overline{\overline{\mathfrak{G}}}.$$

---

(14) By Appell's lemma, such invariant may be expressed rationally in terms of the coefficients $p_i(x)$, $i = 1, 2, \ldots, n$ (and their derivatives) of the eqnation (1). See Picard 2. p. 541.

we have, for V the differential equation

$$\Phi(V - r(x)\alpha) = 0.$$

In the same way, we obtain an algebraic differential equation

$$\Psi(V) = 0$$

with coefficients $\in \mathfrak{R}$, whose general solutions are

$$V = r(x)(g \cdot \alpha) + g \cdot \beta, \quad g = \text{general element of } \overline{\overline{\mathfrak{G}}}.$$

These two equations $\Phi = 0$ and $\Psi = 0$ have only one particular solution in common, that is

$$V_0 = r(x)\alpha + \beta.$$

For, if

$$r(x)\alpha + h \cdot \beta = r(x)(g \cdot \alpha) + g \cdot \beta,$$

we must have, by (5), $g \in \overline{\mathfrak{G}}(\alpha)$ and hence

$$(6) \qquad\qquad h \cdot \beta = g \cdot \beta$$

so that we must have $h \cdot \beta = \beta$, as $\beta \in \mathfrak{S}(\overline{\mathfrak{G}}(\alpha))$ by definition.

This common solution $V = V_0$ satisfies an algebraic differential equation with coefficients $\in \mathfrak{R}(\alpha)$, denoting by $\mathfrak{R}(\alpha)$ the sub-domain of rationality obtained from $\mathfrak{R}$ by the adjunction of the element α (and its derivatives); for example, the equation

$$\Phi(V - r(x)\alpha) = 0.$$

Let $H(V) = 0$ be one of these algebraic differential equations of the smallest order with coefficients $\in \mathfrak{R}(\alpha)$, satisfied by $V = V_0$. We assume, moreover, $H(V) = 0$ is algebraically irreducible in $\mathfrak{R}(\alpha)$; that is, indecomposable into factors with coefficients $\in \mathfrak{R}(\alpha)$. Let this order be λ and let the degree in $\dfrac{d^\lambda V}{dx^\lambda}$ of $H(V)$ be d.

Then we must have $\lambda = 0$ and $d = 1$. For, if we assume $\lambda \geq 1$ (or $\lambda = 0$, $d \geq 2$), then not all the general solutions of $H(V) = 0$ are the particular solution common to $\Phi = 0$ and $\Psi = 0$, that is V_0, a one-valued (meromorphic) function of x in the region D. Thus, if $\lambda \geq 1$ (or $\lambda = 0$, $d \geq 2$), we may obtain from $H = 0$ and $\Phi = 0$ (or from $H = 0$ and $\Psi = 0$) an algebraic differential equation $H_1(V) = 0$ with coefficients $\in \mathfrak{R}(\alpha)$, satisfied by $V = V_0$ whose order is either $< \lambda$ or $= \lambda$ and of degree in $\dfrac{d^\lambda V}{dx^\lambda} < d$, contrary to the hypothesis concerning the equation $H(V) = 0$. (Cf. the analogus arguments given in Loewy 1, p. 131-134)

Thus we have $\lambda = 0$ and $d = 1$. Therefore $V_0 = r(x)\alpha + \beta$ and hence $\beta \in \mathfrak{R}(\alpha)$. This proves the formula (3). C. Q. F. D.

Proof of (4). If we put $\overline{\mathfrak{G}}(\mathfrak{S}(\overline{\mathfrak{G}})) = \overline{\mathfrak{G}}_1$, we have $\mathfrak{S}(\overline{\mathfrak{G}}_1) = \mathfrak{S}(\overline{\mathfrak{G}})$ by (3). Hence we have only to derive a contradiction from

$$\overline{\mathfrak{G}} \subseteqq \overline{\mathfrak{G}}_1, \quad \overline{\mathfrak{G}} \neq \overline{\mathfrak{G}}_1, \quad \mathfrak{S}(\overline{\mathfrak{G}}) = \mathfrak{S}(\overline{\mathfrak{G}}_1).$$

This may be effected as follows.

By Lemma 3, we have dim. $\overline{\mathfrak{G}} <$ dim. $\overline{\mathfrak{G}}_1$, since $\overline{\mathfrak{G}} \subseteqq \overline{\mathfrak{G}}_1$, $\overline{\mathfrak{G}} \neq \overline{\mathfrak{G}}_1$. Hence the respective essential parameters $(a_1, a_2, \ldots, a_r)$ and $(a_1, a_2, \ldots, a_r, a_{r+1}, \ldots, a_s)$, $r < s$, of these two Lie groups $\overline{\mathfrak{G}}$ and $\overline{\mathfrak{G}}_1$ may be so chosen that the element of $\overline{\mathfrak{G}}_1$ whose parameters are $(a_1, a_2, \ldots, a_r, a_{r+1}^0, a_{r+2}^0, \ldots, a_s^0)$ is the element of $\overline{\mathfrak{G}}$ whose parameters are $(a_1, a_2, \ldots, a_r)$, denoting by $(a_{r+1}^0, \ldots, a_s^0)$ a fixed system of values. Hence if the elements g and g_1 stand for the parameters $(a_1, a_2, \ldots, a_r)$ and $(a_1, a_2, \ldots, a_r, a_{r+1}, \ldots, a_s)$ respectively, there must exist at least one element $\alpha \in \mathfrak{R}(y)$ such that the function $g_1 \cdot \alpha$ depends on parameters at least one more than those of $g \cdot \alpha$, for $\overline{\mathfrak{G}} \neq \overline{\mathfrak{G}}_1$ means that $\overline{\mathfrak{G}}_1$ contains at least one automorphism $A \overline{\in \mathfrak{G}}$.

The group $\overline{\mathfrak{G}}$ being algebraic, we may obtain, as in the *proof of (3)*, an algebraic differential equation $E(z) = 0$ with coefficients $\in \mathfrak{S}(\overline{\mathfrak{G}})$, whose general solutions are $z = g \cdot \alpha$, $g =$ general element of $\overline{\mathfrak{G}}$. Hence all the functions $g_1 \cdot \alpha$, $g_1 \in \overline{\mathfrak{G}}_1$, does not satisfy $E(z) = 0$. On the other hand, the coefficients of $E = 0$ being invariant by the group $\overline{\mathfrak{G}}_1 (\mathfrak{S}(\overline{\mathfrak{G}}) = \mathfrak{S}(\overline{\mathfrak{G}}_1)$ by hypothesis), we should have

$$E(g_1 \cdot \alpha) = 0,$$

by applying the automorphism $g_1 \in \overline{\mathfrak{G}}_1$ to the equality $E(\alpha) = 0$. Thus we arrived at a contradiction.   C. Q. F. D.

§5. Reduction of the group of rationality.

Let α be any element $\in \mathfrak{R}(y)$. We denote by $\mathfrak{S}(\overline{\mathfrak{G}}(\mathfrak{R}) \cdot \alpha)$ the subdomain of rationality constructed from $\mathfrak{R}$, by adjoining all the elements $g \cdot \alpha$, $g \in \overline{\mathfrak{G}}(\mathfrak{R})$, and their derivatives.

Theorem 8[15]. $\overline{\mathfrak{G}}(\mathfrak{S}(\overline{\mathfrak{G}}(\mathfrak{R}) \cdot \alpha))$ *is an invariant subgroup of* $\overline{\mathfrak{G}}(\mathfrak{R})$.

Proof. By theorem 6, there exists an element $\beta \in \mathfrak{S}(\overline{\mathfrak{G}}(\mathfrak{R}) \cdot \alpha)$ such that

$$\overline{\mathfrak{G}}(\mathfrak{S}(\overline{\mathfrak{G}}(\mathfrak{R}) \cdot \alpha)) = \overline{\mathfrak{G}}(\beta).$$

Let g be any one automorphism of $\overline{\mathfrak{G}}(\mathfrak{R})$ and we put $\beta_1 = g \cdot \beta$. Then it is obvious that $\overline{\mathfrak{G}}(\beta_1) = g \overline{\mathfrak{G}}(\beta) g^{-1}$. Since $\beta_1 \in \mathfrak{S}(\overline{\mathfrak{G}}(\mathfrak{R}) \cdot \alpha)$ by definition, we have $\overline{\mathfrak{G}}(\mathfrak{S}(\overline{\mathfrak{G}}(\mathfrak{R}) \cdot \alpha)) \subseteqq \overline{\mathfrak{G}}(\beta_1)$ and hence

$$\overline{\mathfrak{G}}(\mathfrak{S}(\overline{\mathfrak{G}}(\mathfrak{R}) \cdot \alpha)) \subseteqq g \overline{\mathfrak{G}}(\mathfrak{S}(\overline{\mathfrak{G}}(\mathfrak{R}) \cdot \alpha)) g^{-1}.$$

---

(15) Cf. Picard 2. p. 593–595.

These two connected Lie groups are of the same dimension, we must have $\overline{\mathfrak{G}}(\mathfrak{S}(\overline{\mathfrak{G}}(\mathfrak{R})\cdot\alpha))=g\overline{\mathfrak{G}}(\mathfrak{S}(\overline{\mathfrak{G}}(\mathfrak{R})\cdot\alpha))g^{-1}$ by Lemma 3. As g was any element of $\overline{\mathfrak{G}}(\mathfrak{R})$, the group $\overline{\mathfrak{G}}(\mathfrak{S}(\overline{\mathfrak{G}}(\mathfrak{R})\cdot\alpha))$ is thus invariant in the group $\overline{\mathfrak{G}}(\mathfrak{R})$.

Theorem 9 [15]. *If α satisfies an algebraic differential equation of order λ and of degree 1 in $\dfrac{d^\lambda z}{dx^\lambda}$ with coefficients $\in\mathfrak{R}$, we have*

$$dim.\ \overline{\mathfrak{G}}(\mathfrak{R})-dim.\ \overline{\mathfrak{G}}(\mathfrak{S}(\overline{\mathfrak{G}}(\mathfrak{R})\cdot\alpha))\leqq\lambda,$$

or more precisely dim. (Lie factorgroup $\overline{\mathfrak{G}}(\mathfrak{R})/\overline{\mathfrak{G}}(\mathfrak{S}(\overline{\mathfrak{G}}(\mathfrak{R})\cdot\alpha))$ [16].

Proof. Let the equation be

$$(7)\qquad \frac{d^\lambda z}{dx^\lambda}-f\left(\frac{d^{\lambda-1}z}{dx^{\lambda-1}},\ \frac{d^{\lambda-2}z}{dx^{\lambda-2}},\ \ldots,z\right)=0$$

By applying the automorphisms $g\in\overline{\mathfrak{G}}(\mathfrak{R})$ to the equality

$$\frac{d^\lambda\alpha}{dx^\lambda}-f\left(\frac{d^{\lambda-1}\alpha}{dx^{\lambda-1}},\ \frac{d^{\lambda-2}\alpha}{dx^{\lambda-2}},\ \ldots,\alpha\right)=0$$

we know that, for any $g\in\overline{\mathfrak{G}}(\mathfrak{R})$, $g\cdot\alpha$ satisfies (7) since the coefficients of (7) are $\in\mathfrak{R}$. Hence if g denotes *a* general substitution of the group $\overline{\mathfrak{G}}(\mathfrak{R})$, $g\cdot\alpha$ depends essentially upon at most λ parameters.

Now as the element β introduced in the proof of Theorem 8 being $\in\mathfrak{S}(\overline{\mathfrak{G}}(\mathfrak{R})\cdot\alpha)$, β is an rational function of α, $\dfrac{d\alpha}{dx},\ldots\ldots$, (and their transforms by some finite number of $g\in\overline{\mathfrak{G}}(\mathfrak{R})$) with coefficients $\in\mathfrak{R}$. Hence if g denotes *a* general transformation of the group $\overline{\mathfrak{G}}(\mathfrak{R})$, $g\cdot\beta$ depends essentially upon at most λ parameters as $g\cdot\alpha$ does. The dependence is necessarily analytical. Hence the condition that the group $\overline{\mathfrak{G}}(\mathfrak{R})$ leaves β numerically invariant is given by at most λ independent analytical relations among the parameters of this group $\overline{\mathfrak{G}}(\mathfrak{R})$. Thus we have

$$dim.\ \overline{\mathfrak{G}}(\mathfrak{R})-dim.\ \overline{\mathfrak{G}}(\beta)=dim.\ \overline{\mathfrak{G}}(\mathfrak{R})-dim.\ \overline{\mathfrak{G}}(\mathfrak{S}(\overline{\mathfrak{G}}(\mathfrak{R})\cdot\alpha))\leqq\lambda.$$

Corollary [17]. *If $\lambda=1$ and if $dim.\ \overline{\mathfrak{G}}(\mathfrak{R})-dim.\ \overline{\mathfrak{G}}(\mathfrak{S}(\overline{\mathfrak{G}}(\mathfrak{R})\cdot\alpha))>0$, the group $\overline{\mathfrak{G}}(\mathfrak{S}(\overline{\mathfrak{G}}(\mathfrak{R})\cdot\alpha))$ is an invariant subgroup of $\overline{\mathfrak{G}}(\mathfrak{R})$ such that dim. $\overline{\mathfrak{G}}(\mathfrak{R})-dim.\ \overline{\mathfrak{G}}(\mathfrak{S}(\overline{\mathfrak{G}}(\mathfrak{R})\cdot\alpha))=1$, that is dim. (Lie factorgroup $\overline{\mathfrak{G}}(\mathfrak{R})/\overline{\mathfrak{G}}(\mathfrak{S}(\overline{\mathfrak{G}}(\mathfrak{R})\cdot\alpha)))=1$.*

---

(16) Let a connected Lie group $\mathfrak{B}$ be an invariant subgroup of a connected Lie group $\mathfrak{A}$. If $\mathfrak{B}$ be closed in $\mathfrak{A}$, then the component $(\overline{\mathfrak{A}/\mathfrak{B}})$ of the factor group $(\mathfrak{A}/\mathfrak{B})$ is a connected Lie group and

$$dim.\ (\overline{\mathfrak{A}/\mathfrak{B}})=dim.\ \mathfrak{A}-dim.\ \mathfrak{B}$$

Cf. van der Waerden 1.

(17) Cf. Vessiot 1. p. 241.

Remark. This corollary, combined with the formula (3) of our fundamental theorem, enables us to deduce Vessiot's theorem. This theorem states that a necessary and sufficient condition for the integrability of the equation (1) by quadrature is the integrability (in Lie's sense) of the group $\overline{\mathfrak{G}}(\mathfrak{R})^{(18)}$.

Mathematical Institute,
Osaka Imperial University.

(Received Oct. 5, 1935.)

LITERATURE.

M. Beke 1. "Zur Gruppentheorie der homogenen linearen Differentialgleichungen" Math. Ann. **49** (1897), 573-580.

E. Cartan 1. La théorie des Groupes Finis et Continus et l'Analysis Situs. Paris, 1930.

A. Loewy 1. "Die Rationalitätsgruppe einer linearen homogenen Differential-gleichungen', Math. Ann. **65** (1908), 129-160.

E. Picard 1. "Sur les groupes de transformations des équations linéaires". C. R. Avril (1883).

E. Picard 2. Traité d'Analyse, III. (3e déition), Paris, 1928.

C. Schreier 1. "Die Verwandschaft stetiger Gruppen im Grossen". Abh. math. Seminar Hamburg, **5** (1927), 233-244.

E. Vessiot 1. "Sur l'intégration des équations differentielles linéaires". Ann. Ecole Norm. 3e série, 9 (1892), 197-280.

E. Vessiot 2. "Rationelle Integrationstheorieen". Encyklop. Math. Wiss. II, A.1, 288-293.

van der Waerden 1. Vorlesungen über kontinuierliche Gruppen. Göttingen, 1929.

---

(18) See Picard 2. p. 597. A new proof of Vessiot's theorem from our point of view will be published in the next paper of the same title:

On Titchmarsh-Kodaira's formula concerning Weyl-Stone's eigenfunction expansion

Nagoya Math. J. 1 (1950) 49–58

1. Introduction. Let $q(x)$ be real and continuous in the infinite open interval $(-\infty, \infty)$ and let $y_1(x, \lambda)$, $y_2(x, \lambda)$ be the solutions of

$$(1.1) \qquad y'' + \{\lambda - q(x)\}y = 0 \text{ [1]}$$

with the initial conditions

$$(1.2) \qquad y_1(0, \lambda) = 1, \quad y_1'(0, \lambda) = 0, \quad y_2(0, \lambda) = 0, \quad y_2'(0, \lambda) = 1.$$

For appropriate homogeneous real boundary conditions at $x = -\infty$, $x = \infty$ of the differential operator

$$(1.3) \qquad L_x = q(x) - \frac{d^2}{dx^2},$$

there corresponds real symmetric positive definite matrix

$$(1.4) \qquad P(u_2) - P(u_1) = (p_{jk}(u_2) - p_{jk}(u_1)), \quad (j, k = 1, 2),$$
$$-\infty < u_1 < u_2 < \infty,$$

such that we have Weyl [2]-Stone's [3] expansion (in the sense of L_2-convergence):

$$(1.5) \qquad \text{for real-valued } f(x) \in L_2(-\infty, \infty),$$

$$f(x) = \lim_{n \to \infty} \int_{-\infty}^{\infty} du \left\{ \sum_{j,k=1}^{2} \int_0^u y_j(x, u) dp_{jk}(u) \int_{-n}^{n} f(s) y_k(s, u) ds \right\}.$$

Recently and independently of each other, E. C. Titchmarsh [4] and K. Kodaira [5]

Received December 20, 1949. (Added March 5, 1950). The result was communicated to Prof. K. Kodaira at Princeton, who informed to the author that a similar treatment may be carried on by Prof. N. Levinson. So a copy of the manuscript was sent to Prof. Levinson, who, in his letter of February 25, informed to the author that his work was submitted to the Duke Math. Journal in May, 1949. He says that his method is different from the pressent note; he proceeds in his proof from the Parseval relation of the Sturm-Liouville orthonormal functions.

[1] The case of finite or half finite open interval may be treated exactly in the same manner. Moreover (apparently) general equation $(p(\xi)z_\xi)_\xi + \{\lambda r(\xi) - s(\xi)\}z = 0$ may be reduced to (1.1) by the Liouville Transformation $x = \int_0^\xi (p^{-1}r)^{1/2}d\xi$, $y = (pr)^{1/4}z$.

[2] Über gewöhnlichc Differentialgleichungen mit Singularitäten und die zugehörigen Entwicklungen willkürlichen Funktionen, Math. Ann., **68** (1910), 220–269.

[3] Linear transformations in Hilbert space, Amer. Math. Soc. Coll. Publ. XV (1932).

[4] Eigenfunction expansisons associated with second order differential equations, Oxford (1946).

[5] The eigenvalue problem for ordinary differential equations of the second order and Heisenberg's theory of S-matrices, Amer. J. of Math., **71** (1949), 921–945.

gave the explicite formula for the density matrix (1.4). The result is much important, since it enables us to unify the classical expansion theorem such as Fourier series and integrals, the Bessel-, Hermite- or Laguerre-function expansions etc. Titchmarsh's method makes use of Cauchy's calculus of residues and does not rely upon the theory of integral equations nor the theory of Hilbert space. Kodaira's method is a modernization, by the theory of Hilbert space, of Weyl's original method. The purpose of the present paper is to show that W-S-T-K's theory may be obtained, by making use of Weyl's analysis (see 2 below), as a natural limiting case of the classical expansion theorem due to Hilbert-Schmidt which concerns with the case of finite closed interval.

2. Preliminaries (Wey'ls analysis). For the sake of exposition and completeness we first develop, with or without proof, Weyl's analysis.[6]

Let, for $0 \leqq x \leqq b < \infty$,

(2.1) $\qquad L_x F(x, \lambda) = \lambda F(x, \lambda), \quad L_x G(x, \lambda') = \lambda' G(x, \lambda')$.

We have Green's formula

(2.2) $\qquad (\lambda' - \lambda) \int_0^\infty FG\,dx = \int_0^\infty \{FL_x G - GL_x F\}dx = \int_0^\infty \{-FG'' + F''G\}dx$
$$= W_0(F, G) - W_x(F, G), \text{ where}$$

$$W_x(H, K) = H(x)K'(x) - H'(x)K(x).$$

We see, by putting $\lambda' = \lambda$, that $W_x(F(x, \lambda), G(x, \lambda))$ is independent of x. Let

(2.3) $\qquad W_x(F(x, \lambda), \quad G(x, \lambda)) = \omega(\lambda)$.

Next let

(2.4) $\qquad F(x, \lambda) = y_2(x, \lambda), \quad G(b, \lambda) = -\sin\beta, \quad G'(b, \lambda) = \cos\beta,$

then

(2.5) $\qquad \omega(\lambda) = W_b(F(x, \lambda), G(x, \lambda)) = F(b, \lambda)\cos\beta + F'(b, \lambda)\sin\beta$.

Thus the condition $\omega(\lambda_0) = 0$ is equivalent to the condition that λ_0 is an eigenvalue of

$$L_x y = \lambda y, \quad 1\cdot y(0) + 0\cdot y'(0) = 0, \quad y(b)\cos\beta + y'(b)\sin\beta = 0,$$

the corresponding eigenfunction being $y_2(x, \lambda)$. Hence the roots λ_0 of $\omega(\lambda) = 0$ must be real.

The homogeneous real boundary condition at $x = b$ of the solution of $L_x y = \lambda y$:

(2.6) $\qquad \{y_1(b, \lambda) + h_b(\lambda)y_2(b, \lambda)\}\cos\beta + \{y_1'(b, \lambda) + h_b(\lambda)y_2'(b, \lambda)\sin\beta = 0$

defines

(2.7) $\qquad h_b(\lambda) = -\dfrac{y_1(b, \lambda)\cos\beta + y_1'(b, \lambda)\sin\beta}{y_2(b, \lambda)\cos\beta + y_2'(b, \lambda)\sin\beta}.$

---

[6] H. Weyl or E. C. Titchmarsh, loc. cit.

Since the denominator is $\omega(\lambda)$, $h_b(\lambda)$ is a meromorphic function in λ whose poles are all real. The numerator is also an $\omega(\lambda)$ and thus the zeros of $h_b(\lambda)$ are also all real. Thus, if $\Im(\lambda) = v \neq 0$,

$$(2.8) \qquad h_b(\lambda, z) = -\frac{y_1(b, \lambda)z + y_1'(b, \lambda)}{y_2(b, \lambda)z + y_2'(b, \lambda)}$$

describes a finite circle $C_b(\lambda)$ in the complex plane when z describes the real axis. Since

$$(2.9) \qquad h_b\left(\lambda, -\frac{y_2'(b, \lambda)}{y_2(b, \lambda)}\right) = \infty,$$

the centre of the circle $C_b(\lambda)$ is given by

$$h_b\left(\lambda, -\frac{\overline{y_2'(b, \lambda)}}{\overline{y_2(b, \lambda)}}\right) = -\frac{W_b(y_1, \bar{y}_2)}{W_b(y_2, \bar{y}_2)}.$$

The radius $r_b(\lambda)$ of the circle $C_b(\lambda)$ is given by

$$\left| h_b(\lambda, 0) + \frac{W_b(y_1, \bar{y}_2)}{W_b(y_2, \bar{y}_2)} \right| = \left| \frac{y_1'(b, \lambda)}{y_1(b, \lambda)} - \frac{W_b(y_1, \bar{y}_2)}{W_b(y_2, \bar{y}_2)} \right|$$

$$= \left| \frac{W_b(y_1, y_2)}{W_b(y_2, \bar{y}_2)} \right| = \left(2|v| \int_0^b |y_2(x, \lambda)|^2 dx\right)^{-1}, \quad (\lambda = u + iv),$$

since $W_b(y_1, y_2) = W_0(y_1, y_2) = 1$,

$$(2.10) \qquad y_j(x, \bar{\lambda}) = \overline{y_j(x, \lambda)}, \quad y_j'(x, \bar{\lambda}) = \overline{y_j'(x, \lambda)}$$

and hence, by Green's formula,

$$(2.11) \qquad 2v \int_0^b |y_2(x, \lambda)|^2 dx = 2v \int_0^b y_2(x, \lambda) y_2(x, \bar{\lambda}) dx$$

$$= iW_0(y_2(x, \lambda), y_2(x, \bar{\lambda})) - iW_b(y_2(x, \lambda), y_2(x, \bar{\lambda}))$$

$$= -iW_b(y_2, \bar{y}_2).$$

Thus, by (2.9) and

$$\Im\left(-\frac{y_2'(b, \lambda)}{y_2(b, \lambda)}\right) = \frac{i}{2}\left(\frac{y_2'(b, \lambda)}{y_2(b, \lambda)} - \frac{\overline{y_2'(b, \lambda)}}{\overline{y_2(b, \lambda)}}\right) = \frac{-i}{2}\frac{W_b(y_2, \bar{y}_2)}{|y_2(b, \lambda)|^2}$$

$$= \frac{v \int_0^b |y_2(x, \lambda)|^2 dx}{|y_2(b, \lambda)|^2},$$

we see that the interior of $C_b(\lambda)$ corresponds to the lower (upper) z-plane if $v > 0$ $(v < 0)$. Thus, since

$$z = -\frac{y_2'(b, \lambda)h_b + y_1'(b, \lambda)}{y_2(b, \lambda)h_b + y_1(b, \lambda)},$$

the two conditions $v > 0$, $h \in$ the interior of the circle $C_b(\lambda)$ is equivalent to

$$i\left(-\frac{y_2'(b, \lambda)h + y_1'(b, \lambda)}{y_2(b, \lambda)h + y_1(b, \lambda)} + \frac{\overline{y_2'(b, \lambda)\bar{h} + y_1'(b, \lambda)}}{\overline{y_2(b, \lambda)\bar{h} + y_1(b, \lambda)}}\right) > 0.$$

By (2.10) and by a simple calculation we see that it is equivalent to

$$iW_b(y_1 + hy_2, \bar{y}_1 + \bar{h}\bar{y}_2) > 0.$$

Hence, if $v = \mathfrak{J}(\lambda) > 0$, we have, by Green's formula and $W_0(y_2, \bar{y}_1) = -1$, $W_0(y_2, \bar{y}_2) = 0$, $W_0(y_1, \bar{y}_1) = 0$, $W_0(y_1, \bar{y}_2) = 1$,

$$2 v \int_0^b |y_1 + hy_2|^2 dx = i(W_0(y_1 + hy_2, \bar{y}_1 + \bar{h}\bar{y})$$
$$- W_b(y_1 + hy_2, \bar{y}_1 + \bar{h}\bar{y}_2)) < iW_0(y_1 + hy_2, \bar{y}_1 + \bar{h}\bar{y}_2) = 2\,\mathfrak{J}(h).$$

Therefore if $v = \mathfrak{J}(\lambda) > 0$, the two conditions $h \in$ the interior of $C_b(\lambda)$ and $h \in$ the perimeter of $C_b(\lambda)$ are respectively equivalent to

$$(2.12) \qquad \int_0^b |y_1(x, \lambda) + hy_2(x, \lambda)|^2 dx < v^{-1}\mathfrak{J}(h) \text{ and } = v^{-1}\mathfrak{J}(h).$$

Thus, if $\mathfrak{J}(\lambda) \neq 0$,

$$(2.13) \qquad b < b' \text{ implies } C_{b'}(\lambda) \subseteqq C_b(\lambda),$$

and hence

$$(2.14) \qquad C_\infty(\lambda) = \bigcap_{b>0} C_b(\lambda)$$

is either i) a circle with a positive radius

$$r_\infty(\lambda) = \lim_{b\to\infty} r_b(\lambda) = (2|v| \int_0^\infty |y_2(x, \lambda)|^2 dx)^{-1} \text{ (the limit circle case)},$$

or ii) a point = a circle with radius 0:

$$r_\infty(\lambda) = \lim_{b\to\infty} r_b(\lambda) = (2|v| \int_0^\infty |y_2(x, \lambda)|^2 dx)^{-1} = 0 \text{ (the limit point case)}.$$

We see from (2.12), that in the limit circle case (for λ_0, $\mathfrak{J}(\lambda_0) \neq 0$), all the solutions of $L_x y = \lambda_0 y$ belong to $L_2(0, \infty)$. And this implies, as will be seen easily,[7] that all the solution of $L_x y = \lambda y$ belong to $L_2(0, \infty)$ for every (real or complex) λ. Conversely, if all the solutions of $L_x y = \lambda_0 y$ (for $\mathfrak{J}(\lambda_0) \neq 0$) are

---

[7] Let $K(x, s) = y_1(x, \lambda_0)y_2(s, \lambda) - y_1(s, \lambda_0)y_2(x, \lambda)$, then $u(x) = (Kv)(x) = \int_0^x K(x, s)v(s)ds$ satisfies $L_x u - \lambda_0 u = v$, $u(0) = u'(0) = 0$. Hence the solution of $L_x \hat{y} = \lambda \hat{y}$, $\hat{y}(0, \lambda) = \gamma$, $\hat{y}'(0, \lambda) = \delta$, when expanded as

$$\hat{y}(x, \lambda) = u_0(x) + (\lambda - \lambda_0)u_1(x) + (\lambda - \lambda_0)^2 u_2(x) + \cdots,$$

gives us

$$u_n(x) = (Ku_{n-1})(x), \quad n = 1, 2, \ldots.$$

For, we must have

$$(L_x - \lambda_0)u_0 = 0, \quad u_0(0) = \gamma, \quad u_0'(0) = \delta, \quad (L_x - \lambda_0)u_n = u_{n-1}, \quad u_n(0) = u_n'(0) = 0$$

by comparing the coefficients of $(\lambda - \lambda_0)^n$ on both sides of $L_x \hat{y} = \lambda \hat{y}$.

Thus we obtain, by induction with respect to n,

$$|u_n(x)|^2 \leqq \frac{1}{(n-1)!} \int_0^\infty |u_0(x)|^2 dx \times \frac{d}{dx}\left(\int_0^x k(s)ds\right)^n, \quad k(x) = \int_0^x |K(x, s)|^2 ds.$$

Hence, by $k(x) \in L_1(0, \infty)$, we obtain $\hat{y}(x, \lambda) \in L_2(0, \infty)$ easily. This proof is due to Kodaira.

$\in L_2(0, \infty)$, then we are in the limit circle case. This we see from

$$r_\infty(\lambda) = (2|\Im|(\lambda_0)|\int_0^\infty |y_2(x, \lambda_0)|^2 dx)^{-1} \neq 0.$$

Therefore the limit point case is characterised, independently of the parameter λ, by the fact that, for at least one λ (real or complex), $L_x y = \lambda y$ admits solution $\overline{\in} L_2(0, \infty)$. That, in this case, the equation $L_x y = \lambda y$ for $\Im(\lambda) \neq 0$ admits solution $\in L_2(0, \infty)$ will be seen from (2.12):

$$(2.15) \qquad \int_0^\infty |y_1(x, \lambda) + C_\infty(\lambda) y_2(x, \lambda)|^2 dx \leq v^{-1} \Im(C_\infty(\lambda)).$$

Since the centre and the radius of the circle $C_b(\lambda)$ depend continuously on b, λ we see, from (2.13), that in the limit circle case there exists a sequence $\{b_n\}$ with $b_n \uparrow \infty$ such that $h_{b_n}(\lambda)$ converges to a function $m_2(\lambda)$ regular for $\Im(\lambda) \neq 0$, uniformly in any bounded closed λ-domain not containing the real numbers. In the limit point case we may replace $\lim_{n \to \infty} h_{b_n}(\lambda)$ by $\lim_{l \to \infty} h_b(\lambda)$ $(= m_2(\lambda) = C_\infty(\lambda))$.

Similarly we may define, for $-\infty < a \leq 0$,

$$(2.16) \qquad h_a(\lambda) = -\frac{y_1(a, \lambda) \cos \alpha + y_1'(a, \lambda) \sin \alpha}{y_2(a, \lambda) \cos \alpha + y_2'(a, \lambda) \sin \alpha}, \qquad (\alpha = \text{real})$$

and the finite circle $C_a(\lambda)$. We have, if $v = \Im(\lambda) \neq 0$,

$$(2.17) \qquad \int_a^0 |y_1(x, \lambda) + h y_2(x, \lambda)|^2 dx < -v^{-1}\Im(h) \text{ or } = -v^{-1}\Im(h)$$

according as $h \in$ the interior of $C_a(\lambda)$ or $\in$ the perimeter of $C_a(\lambda)$. There exists, as above, a sequence $\{a_n\}$ with $a_n \downarrow -\infty$ such that $h_{a_n}(\lambda)$ converges to a function $m_1(\lambda)$ regular for $\Im(\lambda) \neq 0$, uniformly in any bounded closed λ-domain not containing real numbers. In the limit point case we may replace $\lim_{n \to \infty} h_{a_n}(\lambda)$ by $\lim_{a \to -\infty} h_a(\lambda)$ $(= m_1(\lambda) = C_{-\infty}(\lambda) = \bigcap_{a<0} C_a(\lambda))$.

3. Weyl-Stone's expansion for the finite closed interval [a, b]. Let a real-valued function $f(x)$ in $(-\infty, \infty)$ be such that $f''(x)$ is continuous and $f(x) \equiv 0$ for $-\infty < x \leq a'$, $b' \leq x < \infty$, where $a' > a$, $b' < b$. By Hilbert-Schmidt's expansion theorem we have absolutely and uniformly convergent expansion:

$$(3.1) \qquad f(x) = \sum_n f_{n,a,b} y_{n,a,b}(x), \qquad a \leq x \leq b,$$

$$f_{n,a,b} = (f, y_{n,a,b}) = \int_{-\infty}^\infty f(x)\overline{y_{n,a,b}(x)}dx = \int_a^b f(x)\overline{y_{n,a,b})x}dx,$$

where $\{y_{n,a,b}(x)\}$ is a complete system of normed orthogonal eigen-functions of

$$(3.2) \qquad L_x y_{n,a,b} = \lambda_{n,a,b} y_{n,a,b},$$

$$y_{n,a,b}(a) \cos \alpha + y_{n,a,b}'(a) \sin \alpha = 0,$$

$$y_{n,a,b}(b) \cos \beta + y_{n,a,b}'(b) \sin\beta = 0.$$

Also, by Hilbert-Schmidt's theorem, the unique solution $y(x, \lambda)$ of

(3.3)
$$L_x y - \lambda y = f(x), \quad \Im(\lambda) \neq 0,$$
$$y(a, \lambda) \cos \alpha + y'(a, \lambda) \sin \alpha = 0, \quad y(b, \lambda) \cos \beta + y'(b, \lambda) \sin \beta = 0$$

may be expanded in absolutely and uniformly convergent series

(3.4)
$$y(x, \lambda) = \sum_n - (\lambda - \lambda_{n,a,b})^{-1} f_{n,a,b} y_{n,a,b}(x), \quad a \leqq x \leqq b.$$

It is well-known (and it may be verified easily), that we have another expression for $y(x, \lambda)$:

(3.5)
$$y(x, \lambda) = \int_a^b G_{a,b}(x, s, \lambda) f(s) ds, \quad a \leqq x \leqq b,$$

where the Green's function $G_{a,b}(x, s, \lambda)$ is given by

(3.6)
$$G_{a,b}(x, s, \lambda) = - W_x(y_a, y_b)^{-1} y_b(x, \lambda) y_a(s, \lambda), \quad x \geqq s,$$
$$- W_x(y_a, y_b)^{-1} y_a(x, \lambda) y_b(s, \lambda), \quad x < s,$$
$$y_a(x, \lambda) = y_1(x, \lambda) + h_a(\lambda) y_2(x, \lambda),$$
$$y_b(x, \lambda) = y_1(x, \lambda) + h_b(\lambda) y_2(x, \lambda).$$

We have, by (1.2),

(3.7)
$$W_x(y_a, y_b) = W_0(y_a, y_b) = - h_a(\lambda) + h_b(\lambda).$$

Hence we have, by (3.4) and (3.5),

(3.8)
$$\sum_n (\lambda - \lambda_{n,a,b})^{-1} f_{n,a,b} y_{n,a,b}(x)$$
$$= (h_b(\lambda) - h_a(\lambda))^{-1} y_b(x, \lambda) \int_a^x y_a(s, \lambda) f(s) ds$$
$$+ (h_a(\lambda) - h_a(\lambda))^{-1} y_a(x, \lambda) \int_x^b y_b(s, \lambda) f(s) ds.$$

Thus we have, from (3.1),

(3.9)
$$f(x) = \sum_n f_{n,a,b} y_{n,a,b}(x) = \text{residue sum of } \sum_n (\lambda - \lambda_{n,a,b})^{-1} f_{n,a,b} y_{n,a,b}(x)$$
$$= \sum_n (- \nu_n)^{-1} (y_1(x, \lambda_n) + \mu_n y_2(x, \lambda_n))(y_1(s, \lambda_n) + \mu_n y_2(s, \lambda_n), f(s))$$
$$+ \sum_m (\mu'_{2m} - \mu'_{1m})^{-1} y_1(x, \lambda'_m)(y_1(s, \lambda'_m), f(s))$$
$$+ \sum_k \mu''_{1k} \mu''_{2k} (\mu''_{2k} - \mu''_{1k})^{-1} y_2(x, \lambda''_k)(y_2(s, \lambda''_k), f(s)),$$

where

(3.10)
$$h_a(\lambda_n) = h_b(\lambda_n) = \mu_n \neq 0, \quad h_a(\lambda) - h_b(\lambda) \sim (\lambda - \lambda_n) \nu_n;$$
$$h_a(\lambda) \sim \mu'_{1m}(\lambda - \lambda'_m), \quad h_b(\lambda) \sim \mu'_{2m}(\lambda - \lambda'_m);$$
$$h_a(\lambda) \sim \mu''_{1k}(\lambda - \lambda''_k)^{-1}, \quad h_b(\lambda) \sim \mu''_{2k}(\lambda - \lambda''_k)^{-1}.$$

That the zeros and poles of $h_a(\lambda)$, $h_b(\lambda)$ are all simple may be proved as follows. Let, for example, $h_b(\lambda)$ have multiple zero u_0. Then we obtain a contradiction from (2.12), by putting $h = h_b(\lambda)$, $\lambda = u_0 + iv$ and letting $v \downarrow 0$.

Hence (3.9) may be written as

$$(3.11) \qquad f(x) = \int_{-\infty}^{\infty} du \Big(\sum_{j,k=1}^{2} \int_{0}^{u} y_j(x, u) dp_{jk}^{(a,b)}(u) (f(s), y_k(s, u)) \Big)$$

where

$$(3.12) \qquad p_{11}^{(a,b)}(u_2) - p_{11}^{(a,b)}(u_1) = (2\pi i)^{-1} \int_{C(u_1, u_2)} (h_b(\lambda) - h_a(\lambda))^{-1} d\lambda,$$

$$p_{12}^{(a,b)}(u_2) - p_{12}^{(a,b)}(u_1) = p_{21}^{(a,b)}(u_2) - p_{21}^{(a,b)}(u_1) = (2\pi i)^{-1}$$
$$\frac{1}{2} \int_{C(u_1, u_2)} (h_a(\lambda) + h_b(\lambda))(h_b(\lambda) - h_a(\lambda))^{-1} d\lambda,$$

$$p_{22}^{(a,b)}(u_2) - p_{22}^{(a,b)}(u_1) = (2\pi i)^{-1} \int_{C(u_1, u_2)} h_a(\lambda) h_b(\lambda)(h_b(\lambda) - h_a(\lambda))^{-1} d\lambda.$$

Here the path of integration $C(u_1, u_2)$ is a polygonal line connecting $u_1 - iv$, $u_2 - iv$, $u_2 + iv$, $u_1 + iv$, $u_1 - iv$ in this order, v being any positive number. Since

$$(3.13) \qquad h_a(\bar{\lambda}) = \overline{h_a(\lambda)}, \quad h_b(\bar{\lambda}) = \overline{h_b(\lambda)}$$

by (2.10), we see that

$$(3.14) \qquad p_{jk}^{(a,b)}(u_2) - p_{jk}^{(a,b)}(u_1) = \lim_{v \downarrow 0} \pi^{-1} \int_{u_1}^{u_2} f_{jk}^{(a,b)}(u + iv) du,$$

$$f_{11}^{(a,b)}(\lambda) = \Im(h_a(\lambda) - h_b(\lambda))^{-1},$$

$$f_{12}^{(a,b)}(\lambda) = f_{21}^{(a,b)}(\lambda) = \frac{1}{2} \Im(h_a(\lambda) + h_b(\lambda))(h_a(\lambda) - h_b(\lambda))^{-1},$$

$$f_{22}^{(a,b)}(\lambda) = \Im h_a(\lambda) h_b(\lambda)(h_a(\lambda) - h_b(\lambda))^{-1}.$$

4. Titchmarsh-Kodaira's formula.

LEMMA 1. Let a harmonic function $h(z)$ in $|z| < 1$ be such that

$$(4.1) \qquad (2\pi)^{-1} \int_{-\pi}^{\psi} |h(re^{i\theta})| d\theta = q_r(\psi) \leqq C < \infty \text{ for } -\pi \leqq \psi \leqq \pi, 0 \leqq r < 1.$$

Then we have Poisson's representation:

$$(4.2) \qquad h(re^{i\theta}) = \frac{1}{2\pi} \int_{-\pi}^{\pi} \Re \frac{e^{i\psi} + re^{i\theta}}{e^{i\psi} - re^{i\theta}} dq(\psi), \quad \frac{1}{2\pi} \int_{-\pi}^{\pi} |dq(\psi)| \leqq C.$$

Proof. See, for exmple, Stone's book, loc, cit., p. 570.

LEMMA 2. Let $v = \Im(\lambda) > 0$. Then we have

$$(4.3) \qquad f_{11}^{(a,b)}(\lambda) = v \int_{a}^{b} |G_{a;b}(0, s, \lambda)|^2 ds,$$

$$f_{12}^{(a,b)}(\lambda) = f_{21}^{(a,b)}(\lambda) = v \int_{a}^{b} G_{a,b}(0, s, \lambda) \overline{G'_{a,b}(0, s, \lambda)} ds,$$

$$f_{22}^{(a,b)}(\lambda) = v \int_{a}^{b} |G'_{a,b}(0, s, \lambda)|^2 ds.$$

In particular we have, for $\Im(\lambda) > 0$,

$$(4.4) \qquad f_{11}^{(a,b)}(\lambda) \geqq 0, \quad f_{22}^{(a,b)}(\lambda) \geqq 0, \quad |f_{12}^{(a,b)}(\lambda)| \leqq (f_{11}^{(a,b)}(\lambda)f_{22}^{(a,b)}(\lambda))^{1/2}.$$

Proof is easy from (1.2), (2.12), (2.17), (3.6) and (3.7).

Therefore, by Lemma 1 and the transformation

$$(4.5) \qquad \lambda = u + iv = i(1-z)(1+z)^{-1}, \quad z = re^{i\theta}, \quad (0 \leqq r < 1)$$

we obtain

$$(4.6) \qquad f_{jk}^{(a,b)}(\lambda) = f_{jk}^{(a,b)}\left(i\frac{1-z}{1+z}\right) = \frac{1}{2\pi} \int_{-\pi}^{\pi} \Re \frac{e^{i\psi} + re^{i\theta}}{e^{i\psi} - re^{i\theta}} \, dq_{jk}^{(a,b)}(\psi),$$

where

$$(4.7) \qquad q_{jj}^{(a,b)}(\psi) \text{ is monotone increasing and } q_{jj}^{(a,b)}(-\pi) = 0, \quad (2\pi)^{-1}$$

$$q_{jj}^{(a,b)}(\pi) = f_{jj}^{(a,b)}(i),$$

$$q_{12}^{(a,b)}(\psi) = q_{21}^{(a,b)}(\psi) \text{ is of bounded variation and } q_{12}^{(a,b)}(-\pi) = 0,$$

$$(2\pi)^{-1} \int_{-\pi}^{\pi} |dq_{12}^{(a,b)}(\pi)| \leqq (f_{11}^{(a,b)}(i)f_{22}^{(a,b)}(i))^{1/2}.$$

Putting

$$(4.5)' \qquad s = i(1-e^{i\psi})(1+e^{i\psi})^{-1} = \tan(\psi/2), \quad \lambda = u + iv = i(1-z)(1+z)^{-1}$$

we obtain

$$(4.8) \qquad \Re \frac{e^{i\psi} + re^{i\theta}}{e^{i\psi} - re^{i\theta}} = \Re \frac{(i-s)(i+s)^{-1} + (i-\lambda)(i+\lambda)^{-1}}{(i-s)(i+s)^{-1} - (i-\lambda)(i+\lambda)^{-1}} = \Re \frac{i(\lambda s + 1)}{\lambda - s}$$

$$= \Re \frac{i|\lambda|^2 s - i\lambda s + i(\bar{\lambda} - s)}{|\lambda - s|^2} = \frac{v(s^2 + 1)}{(u-s)^2 + v^2}$$

and thus

$$(4.9) \qquad f_{jk}^{(a,b)}(u + iv) = \int_{-\infty}^{\infty} v((u-s)^2 + v^2)^{-1}(s^2 + 1) \, dq_{jk}^{(a,b)}(2\tan^{-1} s).$$

Now $f_{jk}^{(a,b)}(\lambda)$ converges, when $a = a_n \downarrow -\infty$, $b = b_n \uparrow \infty$, to $f_{jk}(\lambda)$ uniformly in any bounded closed λ-doamin not containing the real numbers. Here

$$(4.10) \qquad f_{11}(\lambda) = \Im(m_1(\lambda) - m_2(\lambda))^{-1},$$

$$f_{12}(\lambda) = f_{21}(\lambda) = \frac{1}{2} \Im(m_1(\lambda) + m_2(\lambda))(m_1(\lambda) - m_2(\lambda))^{-1},$$

$$f_{22}(\lambda) = \Im m_1(\lambda)m_2(\lambda)(m_1(\lambda) - m_2(\lambda))^{-1}.$$

For the proof we must show that $m_1(\lambda) \neq m_2(\lambda)$ if $\Im(\lambda) \neq 0$. This may be proved as follows. Firstly $m_1(\lambda) \not\equiv m_2(\lambda)$. If, otherwise, we would obtain a contradiction

$$\int_{-\infty}^{\infty} |y_1(x, \lambda) + m_1(\lambda)y_2(x, \lambda)|^2 dx \leqq v^{-1}\Im(m_1(\lambda) - v^{-1}\Im(m_1(\lambda)) = 0$$

from (2.12) and (2.17). Secondly let $m_1(\lambda_0) - m_2(\lambda_0) = 0$ for $\Im(\lambda_0) \neq 0$. Then, by Hurwitz's theorem on the sequence of uniformly convergent regular functions, $h_{a_n}(\lambda) - h_{b_n}(\lambda)$ has zero in any vicinity of λ_0, if n is taken sufficiently large. Such (non-real) zero is an eigenvalue of the boundary value problem corresponding to the finite closed interval (a_n, b_n). This is a contradiction.

Hence, by applying Helly's theorem, we see that, when $a = a_{n'} \downarrow - \infty$, $b = b_{n'} \uparrow \infty$,

$$(4.11) \qquad \text{finite limit } q_{jk}^{(a,b)}(2\tan^{-1} s) \text{ existe for all } s, \text{ and}$$

$$(4.12) \qquad f_{jk}(u + iv) = \int_{-\infty}^{\infty} v((u - s)^2 + v^2)^{-1}(s^2 + 1)dq_{jk}(2\tan^{-1} s) .$$

By, (4.9),

$$p_{jk}^{(a,b)}(u_2) - p_{jk}^{(a,b)}(u_1) = \lim_{v \downarrow 0} \pi^{-1}\int_{-\infty}^{\infty} dq_{jk}^{(a,b)}(2\tan^{-1} s)\int_{u_1}^{u_2} v((u - s)^2 + v^2)^{-1}$$
$$(s^2 + 1)du$$
$$= \int_{u_1}^{u_2}(s^2 + 1)dq_{jk}^{(a,b)}(2\tan^{-1} s) ,$$

if $q_{jk}^{(a,b)}(2\tan^{-1} s)$ is continuous at $s = u_1, u_2$.

Hence, by (4.11) and (4.12), we see that, when $a = a_{n'} \downarrow - \infty$ and $b = b_{n'} \uparrow \infty$

$$(4.13) \qquad \lim (p_{jk}^{(a,b)}(u_2) - p_{jk}^{(a,b)}(u_1)) = \int_{u_1}^{u_2}(s^2 + 1)dq_{jk}(2\tan^{-1} s)$$
$$= \lim_{v \downarrow 0} \pi^{-1}\int_{-\infty}^{\infty} dq_{jk}(2\tan^{-1} s)\int_{u_1}^{u_2} v((u - s)^2 + v^2)^{-1}(s^2 + 1)du$$
$$= \lim_{v \downarrow 0} \pi^{-1}\int_{u_1}^{u_2} f_{jk}(u + iv)du = p_{jk}(u_2) - p_{jk}(u_1)$$

if $q_{jk}(2\tan^{-1} s)$ is continuous at $s = u_1, u_2$. Here we again make use of Helly's theorem, and assume that $s = u_1, u_2$ are not discontinuous points of the functions $q_{jk}^{(a,b)}(2\tan^{-1} s)$, $(j, k = 1, 2 ; a = a_{n'}, b = b_{n'}, n = 1, 2, \ldots)$.

From (3.11) we obtain

$$(4.14) \qquad (g, f) = \int_{-\infty}^{\infty} du\Big\{ \sum_{j,k=1}^{2} \int_{0}^{u}(g(x), y_j(x, u))dp_{jk}^{(a,b)}(u)(f(s), y_k(s, u))\Big\}$$

if the real-valued function $g(x)$ is, like $f(x)$, such that $g''(x)$ is continuous and $g(x) \equiv 0$ for $x \in [a', b']$. By the positive definiteness of the density matrix

$$(p_{jk}^{(a,b)}(u_2) - p_{jk}^{(a,b)}(u_1)) , \quad j, k = 1, 2, \quad (u_2 > u_1) ,$$

we easily obtain

$$\sum_{m}\Big| \sum_{j,k}\int_{u_{m-1}}^{u_m}(g(x), y_j(x, u))dp_{jk}^{(a,b)}(u)(f(s), y_k(s, u))\Big| \leqq$$
$$\sum_{m}\Big[\sum_{j,k}\int_{u_{m-1}}^{u_m}(g(x), y_j(x, u))dp_{jk}^{(a,b)}(u)(g(s), y_k(s, u))\Big]^{1/2} \times$$
$$\Big[\int_{u_{m-1}}^{u_m}(f(x), y_j(x, u)dp_{jk}^{(a,b)}(u)(f(s), y_k(s, u))\Big]^{1/2} \leqq (g, g)^{1/2}(f, f)^{1/2} .$$

Thus, letting $a = a_{n'} \downarrow -\infty$ and $b = b_{n'} \uparrow \infty$ in (4.14) we obtain

$$(4.15) \qquad (g,f) = \int_{-\infty}^{\infty} du \left\{ \sum_{j,k=1}^{2} \int_{0}^{u} (g(x), y_j(x, u)) \, dp_{jk}(u) \, (f(s), y_k(s, u)) \right\}.$$

This is W-S-T-K's *theorem* for real-valued functions $f(x)$, $g(x)$ if $f''(x)$, $g''(x)$ are continuous and $f(x) \equiv g(x) \equiv 0$ for sufficiently large $|x|$. The general case: $f(x)$, $g(x) \in L_2(-\infty, \infty)$, may be obtained from this special case by the customary limiting process.

Remark. As was shown in 2, the condition that the boundary point $x = \infty$ ($x = -\infty$) is in the limit point case is independent of λ and $m_2(\lambda)$ is independent of β (λ and $m_1(\lambda)$ is independent of α). Thus, we may obtain Titchmarsh-Kodaira's formula

$$(4.17) \qquad m_2(\lambda) = \lim_{b \to \infty} -y_1(b, \lambda)/y_2(b, \lambda), \quad (m_1(\lambda) = \lim_{a \to -\infty} -y_1(a, \lambda)/y_2(a, \lambda)).$$

In the limit point case the follwing formula may also be of use:

$$(4.18) \qquad m_2(\lambda) = y'(0)/y(0) \text{ for any solution of}$$

$$y'' + \{\lambda - q(x)\}y = 0, \quad 0 < \int_0^{\infty} |y(x)|^2 dx < \infty,$$

and similarly for $m_1(\lambda)$. The proof is easy, since, in the limit point case of $x = \infty$, the above solution y must be a constant multiple of

$$(4.19) \qquad y_\infty(x, \lambda) = y_1(x, \lambda) + m_2(\lambda)y_2(x, \lambda)$$

Mathematical Institute, Nagoya University

Correction to my paper "On Titchmarsh-Kodaira's formula concerning Weyl-Stone's eigenfunction expansion" in Nagoya Mathematical Journal, 1 (1950) 49–58

Nagoya Math. J. **6** (1953) 187–188

Recently Mr. Seizô Itô kindly called the author's attention to the fact that the derivation of (4.15) in the above referred paper is insufficient since the paper does not contain the proof of the compactness condition:

$$(1) \quad \lim_{T \uparrow \infty} \left\{ \int_{-\infty}^{-T} + \int_{T}^{\infty} \right\} \left| d_u \left[\sum_{j,\,k=1}^{2} \int_{0}^{u} (g(x),\, y_j(x,\,u))\, dp_{jk}^{(a,\,b)}(u)(f(s),\, y_k(s,\,u)) \right] \right| = 0$$

uniformly in a and b

for C^2 functions $g(x)$ and $f(x)$ vanishing outside the interval (a', b'), $-\infty < a < a' < b' < b < \infty$. The purpose of the present note is to give a proof to (1) as follows.

We have, by definition (page 49),

$$L_x y_j(x,\, u) = \left(q(x) - \frac{d^2}{dx^2} \right) y_j(x,\, u) = u y_j(x,\, u).$$

Thus, since $g(x) \in C^2$ vanishes outside the interval (a', b'), we have, by partial integration,

$$(g(x),\, y_j(x,\, u)) = (g(x),\, u^{-1} L_x y_j(x,\, u)) = u^{-1}(L_x g(x),\, y_j(x,\, u)).$$

And similarly for $f(x)$. We have also the completeness relation

$$(g(x),\, f(x)) = \int_{-\infty}^{\infty} d_u \left[\sum_{j,\,k=1}^{2} \int_{0}^{u} (g(x),\, y_j(x,\, u))\, dp_{jk}^{(a,\,b)}(u)(f(s),\, y_k(s,\, u)) \right]$$

for continuous functions $g(x)$ and $f(x)$ vanishing outside the interval (a', b'). The proof was given on page 57 for C^2 functions $g(x)$ and $f(x)$. The extension to continuous functions may be obtained by customary limiting process. Hence, by the positive definiteness of the density matrix $(dp_{jk}^{(a,\,b)}(u))$, we have, for C^2 function $g(x)$ vanishing outside the interval (a', b'),

$$\left\{ \int_{-\infty}^{-T} + \int_{T}^{\infty} \right\} d_u \left[\sum_{j,\,k=1}^{2} \int_{0}^{u} (g(x),\, y_j(x,\, u))\, dp_{jk}^{(a,\,b)}(u)(g(s),\, y_k(s,\, u)) \right]$$

Received June 12, 1953.

$$\leq T^{-2}\left\{\int_{-\infty}^{-T} + \int_{T}^{\infty}\right\} du \left[\sum_{j,\,k=1}^{2} \int_{0}^{u} (L_x g(x),\, y_j(x,\, u)) dp_{jk}^{(a,\,b)}(u)(L_s g(s),\, y_j(s,\, u)) \right]$$

$$\leq T^{-2}\int_{-\infty}^{\infty} du [\quad] = T^{-2}(L_x g(x),\, L_x g(x)).$$

We have also the similar inequality for C^2 function $f(x)$ vanishing outside $(a',\, b')$. Therefore we have proved (1) by the Schwarz's inequality

$$\left|\left\{\int_{-\infty}^{-T} + \int_{T}^{\infty}\right\} du \left[\sum_{j,\,k=1}^{2} \int_{0}^{u} (g(x),\, y_j(x,\, u)) dp_{jk}^{(a,\,b)}(u)(f(s),\, y_k(s,\, u)) \right]\right|^{2}$$

$$\leq \left\{\int_{-\infty}^{-T} + \int_{T}^{\infty}\right\} du \left[\sum_{j,\,k=1}^{2} \int_{0}^{u} (g(x),\, y_j(x,\, u)) dp_{jk}^{(a,\,b)}(u)(g(s),\, y_k(s,\, u)) \right]$$

$$\times \left\{\int_{-\infty}^{-T} + \int_{T}^{\infty}\right\} du \left[\sum_{j,\,k=1}^{2} \int_{0}^{u} (f(x),\, y_j(x,\, u)) dp_{j,\,k}^{(a,\,b)}(u)(f(s),\, y_k(s,\, u)) \right].$$

Mathematical Institute,
Osaka University

A note on malmquist's theorem on first order algebraic differential equations

Proc. Japan Acad. **53** (1977) 120–123

(Communicated by Kôsaku Yosida, M. J. A., Feb. 12, 1977)

The entitled theorem reads as follows: *If the differential equation*
(1) $dw/dz = R(z, w)$ (*R is a rational function of z and w*)
has a transcendental meromorphic solution $w(z)$, then the equation must be of the Riccati type, i.e., $R(z, w)$ must be a polynomial of the second degree in w.

In 1933 the present author gave, as an application of the Nevanlinna theory ([5]) of meromorphic functions, another proof of this striking theorem of J. Malmquist [4] dating 1913. In this proof (Yosida [9] and [10]), a decisive role was played by a theorem of G. Valiron [6]:
(2) $T(r, R(z, w(z)))^{1)} = d \cdot T(r, w(z)) + O(\log r),$
where d is the degree in w of $R(z, w)$. In 1950, H. Wittich ([7] and [8]) gave an alternate proof which is based upon the fact that the order of the meromorphic function $w(z)$ is finite and that its proximity function $m(r, w(z))$ is $O(\log r)$. Recently in 1974, E. Hille ([2] and [3]) gave another approach proposing a geometric argument instead of Wittich's estimation via the calculus of residues. It is to be noted here that, for the finiteness of the order of the meromorphic solution $w(z)$, the author gave in 1934 a straightforward proof ([10], Theorem 7) relying upon the T. Shimizu-L. Ahlfors-H. Cartan interpretation (see, e.g., [5], 165–) of the Nevanlinna characteristic $T(r, w(z))$.

In view of the above, I should like to show that my original idea in [9] and [10] can be pursued to the result without appealing to the theorem of Valiron nor to the Wittich-Hille type estimation.

We may assume that
(3) $R(z, w) = P(z, w)/Q(z, w) = (\sum_{j=0}^{p} p_j(z)w^j)/(\sum_{k=0}^{q} q_k(z)w^k)$
with polynomial coefficients p_j's and q_k's such that $p_p(z) \cdot q_q(z) \not\equiv 0$ and w-polynomials $P(z, w)$ and $Q(z, w)$ have no factos in common. By virtue of the *defect relation* in the Nevanlinna theory, we have

---

1) We shall follow notations in [1]:

$T(r, f(z)) = m(r, f(z)) + N(r, f(z)),$ $\quad m(r, f(z)) = (2\pi)^{-1} \int_0^{2\pi} \log^+ |f(re^{i\theta})| \, d\theta, \; N(r, f(z))$

$= \int_0^t t^{-1}(n(t, f(z)) - n(0, f(z)))dt + n(0, f(z)) \cdot \log r,$ where $n(r, (f(z))$ denotes the number of poles of $f(z)$ for $|z| \leq r$, multiple poles being counted with the multiplicity.

$$(4) \qquad \varliminf_{r \uparrow \infty} N\!\left(r, \frac{1}{w(z)-C}\right) \Big/ T(r, w(z)) = 1$$

except possibly for countable sequence of complex numbers C's. Hence there exist a complex number C_0 and an increasing sequence of positive numbers r_j's with $r_j \uparrow \infty$ such that

$$(4)' \qquad \lim_{j \to \infty} N\!\left(r_j, \frac{1}{w(z)-C_0}\right) \Big/ T(r_j, w(z)) = 1$$

and

$$(5) \qquad P(z, C_0) \not\equiv 0 \quad \text{and} \quad Q(z, C_0) \not\equiv 0.$$

Thus, by the first fundamental theorem of the Nevanlinna theory, we have

$$(6) \qquad \lim_{j \to \infty} m\!\left(r_j, \frac{1}{w(z)-C_0}\right) \Big/ T(r_j, w(z)) = 0.$$

Next, by the transformation

$$(7) \qquad W(z) = \frac{1}{w(z)-C_0}$$

we obtain, from (1) and (3),

$$(8) \qquad dW/dz = P_1(z, W)/Q_1(z, W)$$

where the degrees in W of $P_1(z, W)$ and of $Q_1(z, W)$ are

$$(9) \qquad \begin{cases} p \text{ and } p-2, \text{ respectively when } p-2 \geq q \\ \text{or} \\ q+2 \text{ and } q, \text{ respectively when } q > p-2. \end{cases}$$

Hence we have

$$(10) \qquad dW/dz = a_2(z)W^2 + a_1(z)W + a_0(z) + (P_2(z, W)/Q_1(z, W))$$

where $a_j(z)$'s are rational functions with $a_2(z) \not\equiv 0$ and the degree $\hat{p}_2$ in W of $P_2(z, W)$ satisfies

$$(11) \qquad \begin{cases} \hat{p}_2 \leq (p-2)-1 \qquad \text{when } (p-2) \geq q, \\ \text{or} \\ \hat{p}_2 \leq q-1 \qquad \text{when } q > (p-2). \end{cases}$$

Therefore the theorem of Malmquist is proved if we can show that

$$(12) \qquad P_2(z, W)/Q_1(z, W) \text{ does not contain } W.$$

The Proof of (12). Assume the contrary. Then (11) implies that

$$(13) \qquad \begin{cases} P_2(z, W(z))/Q_1(z, W(z)) = 0 \text{ at every pole of } W(z) \text{ except} \\ \text{possibly for finite number of } z\text{'s.} \end{cases}$$

Moreover,

$$(14) \qquad P_2(z, W(z))/Q_1(z, W(z)) = W'(z) - a_2(z)W(z)^2 - a_1(z)W(z) - a_0(z)$$

takes finite value for all z at which $W(z)$ is finite, except possibly for finite number of z's. Therefore (as in Wittich [8] and Hille [3])

$$(15) \qquad N(r, P_2(z, W(z))/Q_1(z, W(z))) = O(\log r).$$

Again by (14), we obtain (as in Wittich [8] and Hille [3])

$$(16) \qquad \begin{cases} m(r, P_2(z, W(z))/Q_1(z, W(z))) \\ \qquad \leq m(r, W'(z)) + 2m(r, W(z)) + m(r, W(z)) + O(\log r) \\ \qquad \leq m(r, W(z)) + m(r, W'(z)/W(z)) + 3m(r, W(z)) + O(\log r). \end{cases}$$

On the other hand, as was stated in the beginning of the present note, we obtain

(17) $\qquad T(r, W(z)) = O(r^k) \qquad$ with a positive integer k

by using to *the Shimizu-Ahlfors-Cartan interpretation of the Nevanlinna characteristic* $T(r, W(z))$:

(17)′ $\qquad \begin{cases} \text{for any } r_0 > 0 \text{ and any non-void open set } G \text{ of the complex plane,} \\[4pt] T(r, W(z)) = O\left(\displaystyle\int_{r_0}^{r} s^{-1} ds \iint_{D(G, r_0 \leq t \leq s)} \frac{|W'(te^{i\theta})|^2}{(1 + |W(te^{i\theta})|^2)^2} \, t \, dt \, d\theta \right), \\[6pt] \text{where} \\[2pt] D(G, r_0 \leq t \leq s) \text{ denotes the set } \{te^{i\theta} ; W(te^{i\theta}) \in G, r_0 \leq t \leq s\}. \end{cases}$

In fact, substituting (8) in (17)′ and taking $r_0 > 0$ and G appropriately, we obtain (17). The finiteness of the order of $W(z)$, expressed in (17), implies $m(r, W'(z)/W(z)) = O(\log (r \cdot T(r, W(z))))$ ([1], Theorem 2.2) and so

(18) $\quad T(r, P_2(z, W(z))/Q_1(z, W(z))) = O(\log (r \cdot T(r, W(z)))) + O(m(r, W(z))$

by (15) and (16). Moreover, we have, by (13) and the first fundamental theorem of the Nevanlinna theory,

$$T(r_j, P_2(z, W(z))/Q_1(z, W(z)) \geq N(r_j, Q_1(z, W(z))/P_2(z, W(z))) + O(1)$$
$$\geq N(r_j, W(z)) + O(\log r_j) \qquad (\text{as } j \to \infty).$$

This contradicts to (18) because of (4)′, (6), $T(r, W(z)) = T(r, w(z)) + O(1)$ and the fact that $\log (r \cdot T(r, W(z))) = o(T(r, W(z))$ which is implied by the transcendental meromorphic assumption of $W(z)$.

We have thus proved (12) and so the proof of the theorem of Malmquist is completed.

References

[1] W. K. Haymann: Meromorphic Functions. Oxford at the Clarendon Press (1954).

[2] E. Hille: Finiteness of the order of meromorphic solutions of some non-linear ordinary differential equations. Proc. Roy. Soc. Edinburgh, (A) **72-29**, 331–336 (1973–74).

[3] ——: Ordinary Differential Equations in the Complex Domain. John Wiley and Sons (1976).

[4] J. Malmquist: Sur les fonctions à un nombre fini des branches définies par les équations différentielles du premier ordre. Acta Math., **36**, 297–343 (1913).

[5] R. Nevanlinna: Eindeutige Analytische Fuktionen. Springer-Verlag (1936).

[6] G. Valiron: Sur la dérivée des fonctions algebroides. Bull. Soc. Math. de France, **49**, 17–39 (1931).

[7] H. Wittich: Ganze tranzendente Lösungen algebraischer Differential-gleichungen. Math. Ann., **122**, 221–234 (1950).

[8] ——: Neuere Untersuchungen über Eindeutige Analytische Funktionen. zweite Auflage. Springer-Verlag (1955).

[9] K. Yosida: A generalization of a Malmquist's theorem. Jap. J. Math., **9**, 253–256 (1933).

[10] ——: On algebroid solutions of ordinary differential equations. Jap. J. Math., **10**, 253–256 (1933).

II. Topological Groups and Lie Groups

Comments (Seizô Itô)

As may be seen in his early works ([1] ~ [9]), Kôsaku Yosida was interested in ordinary differential equations in the first half of 1930's, and also in Lie transformation groups in connection with S. Lie's theory on the integration of differential equations. Since then, he wrote several papers on (local) Lie groups and general topological groups. Though formulations and results in some of those papers are somewhat classical from the present-day view-point, most of them contain results interesting at the time. Some of those results are mentioned below.

In the paper [14], it is shown that, if a group G embedded in a metrically complete ring R is locally compact and connected, then G is a Lie group; this result is an extension of a theorem by J. von Neumann (Math. Z. Bd. 30, 1929) which treated the case where G is a closed subgroup of all non-singular matrices whose entries are complex numbers.

Based on the result of [14], several facts on representations of Lie groups and Lie algebras are shown in [15] and subsequent papers. Among others, the result of [18] stated below seems to be particularly interesting: any irreducible Lie algebra of matrices is the Lie algebra of a Lie group of matrices. In [17] it is proved that "irreducible Lie algebra" in the statement above is replaced by "semi-simple Lie algebra." These results are closely related to Weyl's theory of determining all irreducible representations of semi-simple Lie groups and to early works concerning global consideration of Lie groups.

In the paper [46], a simple proof of the Tannaka duality theorem for non-commutative compact groups is given; the theorem reads as follows. Let G be a compact group and let R be the ring of Fourier polynomials in the element of a set of mutually inequivalent irreducible unitary representations of G. Let T be the set of linear homomorphisms t of R into the field of complex numbers such that $t \cdot e = 1$ and $t \cdot \bar{x} = \overline{t \cdot x}$. Then T is isomorphic to G. The proof relies on the use of the Stone-Gelfand abstraction of the Weierstrass polynomial approximation theorem.

The content of [52] is a proof of the equivalence of two topologies on the character group G^* of a locally compact, separable abelian group G, one being the usual compact-open topology and the other being induced by the weak topology on the linear functionals on $L^1(G)$ (with respect to Haar's measure); here "separable" means "satisfying the second axiom of countability." The proof is essentially based on the fact that G^* is σ-compact with respect to the compact-open topology (accordingly G^* is a Baire space since the local-compactness of G^* is well known). It is also proved that G^* is a topological group with respect to the weak topology even when G is not separable.

On the group embedded in the metrical complete ring

Japan. J. Math. **13** (1936) 7–26

(Received February 10, 1936)

We owe to J. von Neumann([1]) the following important theorem:

Let $\mathfrak{G}_n$ be a group of n-dimensional matrices $A = \|a_{ij}\|$ whose coefficients a_{ij} are complex numbers. If $\mathfrak{G}_n$ be closed in the group of all non-singular n-dimensional matrices, then $\mathfrak{G}_n$ is a Lie group.

In the above theorem the topology is defined by the absolute value

$$|A| = \sqrt[2]{\sum_{i,j=1}^{n} |a_{ij}|^2}.$$

Neumann's paper depends implicitely upon the following facts. Let $\mathfrak{M}_n$ denote the set of all n-dimensional matrices $A = \|a_{ij}\|$ with complex numbers as its coefficients a_{ij}. As usual, $\mathfrak{M}_n$ is a ring with respect to the matrix addition $A+B$ and the matrix multiplication AB, having a unit element E, the unit matrix. With regard to the scalar-multiplication aA, the set of all complex numbers a is an *Operatorenbereich* of $\mathfrak{M}_n$. By the absolute value $|A| = \sqrt[2]{\sum_{i,j=1}^{n} |a_{ij}|^2}$, defined above, $\mathfrak{M}_n$ becomes a *metrical ring*. In fact, $\mathfrak{M}_n$ is a *metrical complete ring* and is *locally compact*. Hence, by virtue of this absolute value, we may introduce an *analysis* into $\mathfrak{M}_n$.

Neumann defined exp A and ln A in $\mathfrak{M}_n$ and then, by virtue of these functions, derived the theorem quoted above.

In the present paper the author's aim is to generalise this theorem to the group embedded in a metrical complete ring. The result is:

If a group $\mathfrak{G}$ embedded in a metrical complete ring $\mathfrak{R}$ be locally compact (in itself) and connected, then $\mathfrak{G}$ is a Lie group.

(The terminologies will be explained later in the following paragraphs.)

Now, in its extended form, the famous problem of Hilbert concerning continuous groups may read as follows:

Is it true that an n-parameter continuous group be topologically isomorphic with an n-parameter Lie group of transformations?

Here two topological group $\mathfrak{G}_1$ and $\mathfrak{G}_2$ are called to be topologically isomorphic with each other, if there exists an topological transformation of

---

$\mathfrak{G}_1$ on $\mathfrak{G}_2$ which is also an isomorphic transformation of the group $\mathfrak{G}_1$ on the group $\mathfrak{G}_2$.

Thus our result may be considered as an answer to a special case of Hilbert's problem. It is to be noted, however, that our result is not concerned with the (topological) isomorphism. It gives directly the analytical representation of the group in question.([2])

From our point of view, Neumann's theorem corresponds to the case where the metrical complete ring $\mathfrak{R}$ is locally compact (in itself).

The plan of this paper is as follows. In §1 is defined the metrical complete ring $\mathfrak{R}$. §1 deals with the exp-, ln- and Inv-functions in $\mathfrak{R}$. Exp-and ln- function corresponds to Neumann's exp-and ln-functions in $\mathfrak{M}_n$, and Inv-function corresponds to the inverse element of the group $\mathfrak{G}$ embedded in $\mathfrak{R}$.([3]) As we are concerned with the abstract group, we define, in §3 and §4, the *differentiability* of the group $\mathfrak{G}$ embedded in $\mathfrak{R}$. Such group $\mathfrak{G}$ is differentiable, if it be locally compact in itself. In §5 is defined the *Lie-ring* of the differentiable group $\mathfrak{G}$ embedded in $\mathfrak{R}$. It is the set of all the *differential quotients* of $\mathfrak{G}$. In §6 we give the proof of our result quoted above, by virtue of the differentiability of the group $\mathfrak{G}$.

As an important and new example of our theory, we give, in §7, a theorem of M. Nagumo concerning one-parameter continuous group $\mathfrak{G}$ embedded in $\mathfrak{R}$.

§1. Metrical complete ring $\mathfrak{R}$.([4])

Let a set $\mathfrak{R}$ of elements $A, B, C,\ldots\ldots$ constitutes a *ring* with respect to an addition $A+B$ and a multiplication AB; that is, $\mathfrak{R}$ is an abelian group (additive group or modul in the terminologies of algebras) with regard to the addition and satisfies

$$(AB)C = A(BC), \qquad \text{(Associative law)}$$

$$\left.\begin{array}{l}(A+B)C = AC+BC, \\ A(B+C) = AB+AC.\end{array}\right\} \qquad \text{(Distributive law)}$$

The ring $\mathfrak{R}$ is said to have complex numbers as its *Operatorenbereich*, if there is defined a scalar-multiplication aA which satisfies the conditions:

([2]) J. von Neumann solved affirmatively the Hilbert's problem for the case where the n-parameter group is compact by virtue of the Haar's invariant-mass in topological group and the theorem quoted at the beginning of this paper. See J. von Neumann: Die Einführung analytischer parameter in topologischen Gruppen, Annals of Mathematics (1933).

([3]) Cf. Hilb: Uber die Auflösung von Gleichungen mit unendlichen vielen Unbekannten, Sitzungsberichte d. phys. med. Sozietät., Erlangen (1908).

([4]) For this notion I owe to M. Nagumo.

For any complex number a and for any element $A \in \mathfrak{R}$ aA is a uniquely determined element $\in \mathfrak{R}$ such that,

$$aA = Aa, \qquad 1A = A,$$
$$(a+b)A = aA+bA,$$
$$a(A+B) = aA+aB,$$
$$a(bA) = (ab)A,$$
$$(aA)(bB) = (ab)(AB)(^5)$$

In what follows we deal only with the ring $\mathfrak{R}$ which has complex numbers as its *Operatorenbereich*, and having the *unit-element E*:

$$AE = EA = A \qquad \text{for any } A \in \mathfrak{R}.$$

Hence we do not distinguish the number 0 and the zero-element $0(^6)$ of $\mathfrak{R}$. In the same way $(-a)A$ is denoted by $-aA$, and especially $(-1)A$ by $-A(^7)$.

The ring $\mathfrak{R}$ is called *metrical*, if to each element $A \in \mathfrak{R}$, there corresponds a uniqualy determined real number $|A|$, the *absolute value* of A, which satisfies the postulates:

(1) $\quad |A| \geq 0, \quad$ and $\quad |A| = 0 \quad$ when and only when $A = 0(^8)$

(2) $\quad |A+B| \leq |A|+|B|,$

(3) $\quad |AB| \leq |A||B|,$

(4) $\quad |aA| = |a||A|.(^9)$

Hence a metrical ring $\mathfrak{R}$ is a *linear metrical space*(^{10}) (with complex numbers as its coefficients) with the *distance* $|A-B|$. Thus a topology is introduced into $\mathfrak{R}$. We write

$$\lim_{n \to \infty} A_n = A \qquad \text{or more simply } A_n \to A,$$

when $\lim_{n \to \infty} |A_n - A| = 0$.

In order to introduce analysis into $\mathfrak{R}$, we assume the metrical ring $\mathfrak{R}$ to be *complete*; that is, any fundamental sequence $\{A_n\}$ of elements $\in \mathfrak{R}$,

$$\lim_{n \to \infty} |A_{n+m} - A_n| = 0 \qquad \text{for any } m \geq 1,$$

(^{5}) The complex numbers are denoted by the letters a, b, c, . . .

(^{6}) The zero-element 0 of a ring $\mathfrak{R}$ satisfies $A+0 = 0+A = A$ for all $A \in \mathfrak{R}$. The number 0 and the zero-element 0 are connected by $0E = 0$.

(^{7}) $A+(-A) = (-A)+A = A-A = 0$, zero-element of $\mathfrak{R}$.

(^{8}) $|E|$ need not equal to 1.

(^{9}) $|a|$ denotes the ordinary absolute value of the complex number a.

(^{10}) A linear metrical space is a set which satisfies the same conditions as the metrical ring $\mathfrak{R}$ defined above, except the postulates concerning the multiplication AB. If the space is complete in the above sense, it is called a linear metrical complete space.

determines uniquely an element $A \in \mathfrak{R}$ such that

$$A_n \to A .$$

Such a ring $\mathfrak{R}$ is called a *metrical complete ring*. An important example of such a ring is the set of all bounded (continuous) linear transformations T of a linear metrical complete space (or complex Banach space) $\mathfrak{C}$ into $\mathfrak{C}$, the absolute value $|T|$ being defined by

$$|T| = \text{lowest upper bound} \ \frac{|Tx|}{|x|},$$
$$x \in \mathfrak{C}$$

where $|x|$, $|Tx|$ denote the absoulte value of the elements x, Tx in $\mathfrak{C}$.

A special case of the above metrical complete ring is the set of all n-dimensional matrices $A = \|a_{jj}\|$ whose coefficients a_{ij} are complex numbers. The three operations in this ring, $A+B$, AB and aA have the usual meanings, the absolute value being defined by

$$|A| = \sqrt[2]{\sum_{i,j=1}^{n} |a_{ij}|^2} .\text{(11)}$$

In the following paragraphs $\mathfrak{R}$ denotes a metrical complete ring.

§2. The functions exp, ln and Inv.

We define three functions in $\mathfrak{R}$:

$$\exp A = E + \sum_{\nu=1}^{\infty} \frac{A^\nu}{\nu!}$$

$$\ln A = \sum_{\nu=1}^{\infty} \frac{(-1)^{\nu-1}}{\nu} (A-E)^\nu$$

$$\text{Inv} \ A = E + \sum_{\nu=1}^{\infty} (E-A)^\nu ,$$

(A^ν denotes the results of ν-times multiplication of A).

As $\mathfrak{R}$ is complete, the inequalities

$$\left| \sum_{\nu=n+1}^{n+m} \frac{A^\nu}{\nu!} \right| \leqq \sum_{\nu=n+1}^{n+m} \frac{|A|^\nu}{\nu!} ,$$

$$\left| \sum_{\nu=n+1}^{n+m} \frac{(-1)^{\nu-1}}{\nu} (B-E)^\nu \right| \leqq \sum_{\nu=n+1}^{n+m} \frac{|B-E|^\nu}{\nu} ,$$

---

(11) This is the case where the linear metrical space $\mathfrak{C}$ is of finite dimension. The topological structures are the same whether we take the absolute value $|A| = \sqrt[2]{\sum_{i,j=1}^{n} |a_{ij}|^2}$ or $|A| = \text{lowest upper bound} \ \frac{|Ax|}{|x|}.$
$$x \in \mathfrak{C}$$

$$\left| \sum_{\nu=n+1}^{n+m} (E-C)^\nu \right| \leqq \sum_{\nu=n+1}^{n+m} |E-C|^\nu,$$

obtained from (2), (3) and (4) show that exp A, ln B, and Inv C, respectively define a uniquely determined element $\in \Re$ for

$$A \in \Re, \quad |B-E|<1, \quad \text{and} \quad |C-E|<1.$$

Concerning these functions we have the formulae:

(5) $\qquad \exp(\ln A) = A \qquad$ for $\qquad |A-E|<1.$

(6) $\qquad \ln(\exp A) = A \qquad$ for $\qquad |A|<\ln 2.$

(7) $\qquad (\text{Inv } A)A = A(\text{Inv } A) = E \qquad$ for $\qquad |A-E|<1.$

(8) $\qquad (\exp A)(\exp B) = \exp(A+B) \qquad$ if $\qquad AB = BA.$

(9) $\qquad \ln(AB) = \ln A + \ln B \qquad$ if $\quad AB = BA \qquad$ and

$$|A-E|<1, \qquad |B-E|<1, \qquad |AB-E|<1.$$

(10) $\qquad \exp A = E+A+O(|A|^2)$

(11) $\qquad \ln A = A-E+O(|A-E|^2)$

(12) $\qquad |\exp(A+B)-(\exp A)(\exp B)| = O(|A||B|).$

The notation $O(|A|)$ (or $o(|A|)$) denotes a complex number or element $\in \Re$ whose absolute value is of order $O(|A|)$ (or $o(|A|)$).

Moreover the O-terms in (10), (11) and (12) are uniform, that is,

$$\frac{|O(|A|^2)|}{|A|^2}, \qquad \frac{|O(|A-E|^2)|}{|A-E|^2} \qquad \text{and} \qquad \frac{|O(|A||B|)|}{|A||B|},$$

are respectively limited from above by positive numbers determined by

$$|A|, \qquad |A-E| \qquad \text{and} \qquad |A||B|,$$

such that these numbers are limited when $|A|$, $|A-E|$ and $|A||B|$ tend to zero respectively.

The above formulae may easily be obtained by the comparison with the ordinary power series of exp a, ln a and $\dfrac{1}{a}$. Neumann[12] proved (5), (6), (8), (9), (10), (11) and (12) for the case where $\Re$ is the ring of all n-dimensional matrices $A = \|a_{ij}\|$, with the absolute value $|A| = \sqrt[2]{\sum_{i,j=1}^{n} |a_{ij}|^2}$. Since the proofs are exactly the same as given by Neumann we do not here reproduce them. We shall only prove (7).

The proof of (7).

By definition we have

$$A(\text{Inv } A) = \left\{E-(E-A)\right\}\left\{E+\sum_{\nu=1}^{\infty}(E-A)^\nu\right\}$$

---

(12) J. I. p. 7–14.

$$= E + \sum_{\nu=1}^{\infty} (E-A)^{\nu} - \left\{ (E-A) + \sum_{\nu=2}^{\infty} (E-A)^{\nu} \right\}$$

$$= E.$$

Similarly we have $(\operatorname{Inv} A)A = E$. C. Q. F. D.

Thus by (7) the simultaneous equations

$$XA = AX = E, \qquad |A-E| < 1.$$

admits a unique solution in $\Re$, that is, $X = \operatorname{Inv} A$. Hence we way write $\operatorname{Inv} A = A^{-1}$. If there exists a unique solution X in $\Re$ of the simultaneous equations

$$AX = XA = E,$$

where $|A-E| \geq 1$ we also write $X = A^{-1}$. We put $(A^{-1})^p = A^{-p}$ for positive integer p. Thus we have

$$(13) \qquad A^m A^l = A^l A^m = A^{l+m}; \ l,\ m \ \text{integers} \gtreqless 0,$$

defining $A^0 = E$ for any $A \in \Re$.

We have also $(k \gtreqless 0)$

$$(14) \quad \ln A^k = k \ln A, \qquad \text{if} \qquad |A-E| < 1, \qquad |A^2-E| < 1, \ldots\ldots,$$

$$|A^{|k|}-E| < 1 \qquad \text{and} \qquad |A^k-E| < 1.$$

Proof. By virtue of (13), we have

$$\ln A^{|k|} = |k| \ln A,$$

by applying (9) successively. Hence when $k < 0$, we have by (13) and (9)

$$\ln(A^{|k|} A^k) = \ln A^{|k|} + \ln A^k = |k| \ln A + \ln A^k$$

$$= \ln E = 0,$$

Thus $\ln A^k = -|k| \ln A = k \ln A$. C. Q. F. D.

The following three lemmas will be frequently used in this paper.

Lemma 1. $\exp A$, $\ln B$ and $\operatorname{Inv} C$ are continuous for $A \in \Re$, $|B-E| < 1$ and $|C-E| < 1$ respectively; that is, if

$$A_n \to A, \qquad B_n \to B, \qquad C_n \to C,$$

$$|B_n-E| < 1, \qquad |B-E| < 1, \qquad |C_n-E| < 1, \qquad |C-E| < 1,$$

we have

$$\exp A_n \to \exp A, \qquad \ln B_n \to \ln B, \qquad \operatorname{Inv} C_n \to \operatorname{Inv} C.$$

Proof.([13]) It is sufficient to prove the inequalities

$$|A^p - B^p| \leq p\, a^{p-1} |A-B|, \qquad p = 0, 1, 2 \ldots\ldots$$

from $|A| \leq a$, $|B| \leq a$. This may be effected by mathematical induction.

The inequality is evident for $p = 0, 1$. Let it be satisfied for $p = 0, 1, 2, \ldots\ldots, n$. Then by (2) and (3)

([13]) Cf. J. I. p. 8.

$$|A^{n+1}-B^{n+1}| = |(A^n-B^n)A+B^n(A-B)| \leqq |A^n-B^n|\,|A|+|B^n|\,|A-B|$$

$$\leqq n\,a^{n-1}\,|A-B|\,a+a^n\,|A-B| = (n+1)a^n\,|A-B|.$$

C. Q. F. D.

Lemma 2.[14] *Let* $A_i \in \mathfrak{R}$, $A_i \to E$ *for* $i \to \infty$. *Let* ε_i *be real numbers such that* $\varepsilon_i \to 0$ *for* $i \to \infty$. *Then*

$$\lim_{i\to\infty}\frac{A_i-E}{\varepsilon_i} = \lim_{i\to\infty}\frac{\ln A_i}{\varepsilon_i},$$

in the sense that, if the limit on either side exists (and hence defines a uniquely determined element $\in \mathfrak{R}$, *since* $\mathfrak{R}$ *is complete), the other limit also exists and the above equality holds.*

Proof. As $A_i \to E$, $\ln A_i$ has a sense for large i, we have by (11)

$$\frac{\ln A_i}{\varepsilon_i} = \frac{(A-E)}{\varepsilon_i}+\frac{1}{\varepsilon_i}O(|A_i-E|^2)$$

$$= \frac{(A-E)}{\varepsilon_i}+\varepsilon_i O\left(\left|\frac{A_i-E}{\varepsilon_i}\right|^2\right).$$

If $\lim_{i\to\infty}\dfrac{A_i-E}{\varepsilon_i}$ exists then $\left|\dfrac{A_i-E}{\varepsilon_i}\right|$ is limited in i, so that $\lim_{i\to\infty}\dfrac{\ln A_i}{\varepsilon_i}$ exists and is equal to $\lim_{i\to\infty}\dfrac{A_i-E}{\varepsilon_i}$.

In the same way, the equality

$$\frac{A_i-E}{\varepsilon_i} = \frac{1}{\varepsilon_i}(\exp \ln A_i - E) = \frac{1}{\varepsilon_i}\left(\ln A_i + O(|\ln A_i|^2)\right)$$

$$= \frac{1}{\varepsilon_i}\ln A_i + \varepsilon_i O\left(\left|\frac{\ln A_i}{\varepsilon_i}\right|^2\right),$$

obtained from (5) and (10) proves the converse part of the lemma.

C. Q. F. D.

Lemma 3. Let δ_1, δ_2 *be two positive numbers.* $\delta_1 \leqq \delta_2$. *If an element* $A \in \mathfrak{R}$, $|A-E| < 1$, *satisfies*

$$\delta_1 \leqq |k \ln A| \leqq \delta_2$$

for an integer $k \geqq 0$, *then we have*

$$\delta_1\left(1-\frac{\delta_2}{2(1-\delta_2)}\right) < |A^{\pm k}-E|; \qquad \delta_2 < 1,$$

$$|A^m-E| < \exp(\delta_2)-1; \qquad m = 0, \pm 1, \qquad \pm 2, \ldots\ldots, \pm k.[15]$$

---

(14) Cf. J. I. p. 16, where this lemma is proved for positive ε_i.

(15) This inequality holds good if A satisfies $|A-E| < 1$ and $|k \ln A| \leqq \delta_i$ only.

Proof. By (2) and (3), we have

$$| \exp (m \ln A) - E | \leq \sum_{\nu=1}^{\infty} \frac{| m \ln A |^{\nu}}{\nu !} \leq \exp (\delta_2) - 1 ,$$

since $| m \ln A | = | m | \, | \ln A | \leq k | \ln A | = | k \ln A | \leq \delta_2$ by (4).
In the same way, we have

$$| \exp (\pm k \ln A) - E | \geq | k \ln A | - \sum_{\nu=2}^{\infty} \frac{| k \ln A |^{\nu}}{\nu !}$$

$$\geq \delta_1 \Big(1 - \sum_{\nu=2}^{\infty} \frac{\delta_2^{\nu-1}}{\nu !} \Big) > \delta_1 \Big(1 - \frac{1}{2} \sum_{\nu=1}^{\infty} \delta_2^{\nu} \Big) = \delta_1 \Big(1 - \frac{\delta_2}{2(1-\delta_2)} \Big)$$

On the other hand, we have, for integer $p \geq 0$.

$$\exp (p \ln A) = (\exp \ln A)^p = A^p$$

by (8) and (5). Hence

$$\exp (n \ln A) = A^n$$

for any integer $n \gtrless 0$, as we have

$$\big\{ \exp (\pm n \ln A) \big\} \big\{ \exp (\mp n \ln A) \big\} = E$$

by (8). This proves the lemma. C. Q. F. D.

§ 3. The group $\mathfrak{G}$ embedded in $\mathfrak{R}$.

Let a subset $\mathfrak{G}$ of $\mathfrak{R}$ be a group with respect to the multiplication AB in $\mathfrak{R}$. The identity element E_0 of the group $\mathfrak{R}$ does not necessarily coincides with the unit-element E of $\mathfrak{R}$. $\mathfrak{G}$ is called a *group embedded in* $\mathfrak{R}$, if $E_0 = E$. In this case, the inverse element of $A \in \mathfrak{G}$ is A^{-1}, as defined above.

When $E_0 \neq E$, we consider the sub-ring $\mathfrak{R}'$ of $\mathfrak{R}$ which consists of all the elements $\in \mathfrak{R}$ of the form $E_0 A E_0$, $A \in \mathfrak{R}$. $\mathfrak{R}'$ is again a metrical complete ring with complex numbers as its *Operatorenbereich*, the unit-element of $\mathfrak{R}'$ being $= E_0$. Hence, without loss of generality, we deal only with the group $\mathfrak{G}$ embedded in $\mathfrak{R}$.

Theorem 1. The group $\mathfrak{G}$ is a topological group by the topology determined by the distance $| A - B |$; that is if A_n, B_n, A, $B \in \mathfrak{G}$, then

$$A_n B_n \to AB \qquad and \qquad A_n^{-1} \to A^{-1} ,$$

when $A_n \to A$, $B_n \to B$.

Proof.

$$| AB - A_n B_n | = | AB - A_n B + A_n B - A_n B_n |$$

$$\leq | A - A_n | \, | B | + | A_n | \, | B - B_n | ,$$

by (2) and (3). Since $A_n \to A$ we have $| A_n | = | A | + O(1)$, thus $A_n B_n \to AB$.

Therefore if we denote $A^{-1}A_n = C_n$, we have $C_n \to E$. Hence $C_n^{-1} \to E$ by lemma 1. Thus $A_n^{-1} = C_n^{-1}A^{-1} \to A^{-1}$.

C. Q. F. D.

The above theorem enables us to introduce the analysis, defined in $\mathfrak{R}$, into the group $\mathfrak{G}$.

§ 4. The differentiability of the group $\mathfrak{G}$.

The group $\mathfrak{G}$ embedded in $\mathfrak{R}$ is said to be *differentiable at* $A \in \mathfrak{G}$, if it satisfies the condition:

From any sequence $\{A_n\}$ of elements $\in \mathfrak{G}$ such that

$$A_n \neq A, \qquad A_n \to A,$$

we may choose a subsequence $\{A_{n'}\}$ and a sequence of real numbers $\{\varepsilon_{n'}\}$, $\varepsilon_{n'} \neq 0$, $\varepsilon_{n'} \to 0$ with the property

$$\frac{A_{n'} - A}{\varepsilon_{n'}} \to \text{an element } U(\neq 0) \in \mathfrak{R}.$$

U need not belong to $\mathfrak{G}$; it is called the *differential quotient of* $\mathfrak{G}$ at A. By definition aU is also the differential quotient of $\mathfrak{G}$ at A for any real a, if $a \neq 0$.

The terms, differentiable and differential quotient are justified by the following two theorems.

Theorem 2.   The group $\mathfrak{G}$ is either differentiable everywhere on $\mathfrak{G}$ or is differentiable nowhere on $\mathfrak{G}$.

Proof.   Let $\mathfrak{G}$ be differentiable at A. We shall prove that $\mathfrak{G}$ is differentiable at B. Let $\{B_n\}$ be a sequence of elements $\in \mathfrak{G}$, $B_n \neq B$, $B_n \to B$. Then the sequence $\{AB^{-1}B_n\}$ of elements $\in \mathfrak{G}$ satisfies

$$AB^{-1}B_n \neq A, \qquad AB^{-1}B_n \to A \qquad \text{(by Theorem 1).}(^{16})$$

As $\mathfrak{G}$ is supposed to be differentiable at A, we may choose a subsequence $\{AB^{-1}B_{n'}\}$ and a sequence $\{\varepsilon_{n'}\}$ of real numbers, $\varepsilon_{n'} \neq 0$, $\varepsilon_{n'} \to 0$, such that

$$\frac{AB^{-1}B_{n'} - A}{\varepsilon_{n'}} \to U(\neq 0) \in \mathfrak{R}.$$

Therefore

$$\frac{B_{n'} - B}{\varepsilon_{n'}} \to BA^{-1}U \in \mathfrak{R}$$

by the same reasoning as in the proof of theorem 1. If $BA^{-1}U = 0$, we have $U = AB^{-1}(BA^{-1}U) = 0$, contrary to the hypothesis $U \neq 0$. Thus $\mathfrak{G}$ is differentiable at B.

C. Q. F. D.

(^{16}) A^{-1} and B^{-1} exist, since $\mathfrak{G}$ is a group.

Theorem 3.[17] *Let $\mathfrak{G}$ be differentiable at E and let U be one of the differential quotients of $\mathfrak{G}$ at E. Then for any real $\varepsilon, -1 \leqq \varepsilon \leqq 1$, $(\varepsilon \neq 0)$, there exists at least one element A_ε in $\mathfrak{G}$ such that*

$$\lim_{\varepsilon \to 0} \frac{A_\varepsilon - E}{\varepsilon} = U.$$

Proof. By definition, there exists a sequence $\{A_n\}$ of elements $\epsilon \, \mathfrak{G}$ and a sequence of real numbers $\{\varepsilon_n\}$ such that

$$A_n \neq E, \qquad A_n \to E; \qquad \varepsilon_n \neq 0, \qquad \varepsilon_n \to 0,$$

$$\frac{A_n - E}{\varepsilon_n} \to U,$$

For any real $\varepsilon, -1 \leqq \varepsilon \leqq 1 (\varepsilon \neq 0)$, we take a positive integer $p(\varepsilon)$ such that

$$|\varepsilon_{p(\varepsilon)}| < \varepsilon^2,$$

and choose a positive integer $q(\varepsilon)$ satisfying the inequalities

$$\big(q(\varepsilon) - 1\big) \, |\varepsilon_{p(\varepsilon)}| < \varepsilon \leqq q(\varepsilon) \, |\varepsilon_{p(\varepsilon)}|.$$

$\lim_{\varepsilon \to 0} \varepsilon_{p(\varepsilon)} = 0$, since $|\varepsilon_{p(\varepsilon)}| < \varepsilon^2$, so that $\lim_{\varepsilon \to 0} p(\varepsilon) = \infty$ and hence

$$q(\varepsilon) \geqq \frac{\varepsilon}{|\varepsilon_{p(\varepsilon)}|} > \frac{1}{\varepsilon}$$

tends to ∞ if $\varepsilon \to 0$.

We define

$$A\varepsilon = \{A_{p(\varepsilon)}\}^{q(\varepsilon)\,\text{sign}\,(\varepsilon_{p(\varepsilon)})} \, (18)$$

$A\varepsilon \in \mathfrak{G}$, since $\mathfrak{G}$ is a group.

By lemma 2 $\dfrac{\ln A_{p(\varepsilon)}}{\varepsilon_{p(\varepsilon)}} \to U$ as $\varepsilon \to 0$, and hence

$$\frac{\ln A_{p(\varepsilon)}}{\varepsilon_{p(\varepsilon)}} = O(1), \qquad \ln A_{p(\varepsilon)} = O(|\varepsilon_{p(\varepsilon)}|)$$

$$q(\varepsilon) \ln A_{p(\varepsilon)} = O(q(\varepsilon) \, |\varepsilon_{p(\varepsilon)}|).$$

Hence for sufficiently small ε, by the lemma 3

$$|A_{p(\varepsilon)}^r - E| < 1; \qquad r = 0, \pm 1, \pm 2, \ldots\ldots, \pm q(\varepsilon).$$

Therefore by (14)

$$\ln A_\varepsilon = \ln \{A_{p(\varepsilon)}^{q(\varepsilon)\,\text{sign}\,(\varepsilon_{p(\varepsilon)})}\} = q(\varepsilon)\,\text{sign}\,(\varepsilon_{p(\varepsilon)}) \ln A_{p(\varepsilon)}$$

and hence

---

(17) For the case of matrices, Neumann proved this theorem for $0 < \varepsilon \leqq 1$. Cf. J. I. 17–18.

$$\frac{\ln A_\varepsilon}{\varepsilon} = \frac{q(\varepsilon)\,|\,\varepsilon_{p(\varepsilon)}\,|}{\varepsilon}\;\frac{\ln A_{p(\varepsilon)}}{\varepsilon_{p(\varepsilon)}}$$

$$\rightarrow 1\cdot U = U, \qquad \text{as} \qquad \varepsilon \rightarrow 0\,.$$

Thus we have

$$\lim_{\varepsilon\to 0}\frac{A_\varepsilon - E}{\varepsilon} = U$$

by Lemma 2. C. Q. F. D.

Now we can give a sufficient condition for the differentiability of the group $\mathfrak{G}$; that is, the local compactness (in itself) of $\mathfrak{G}$.

The group $\mathfrak{G}$ embedded in $\mathfrak{R}$ is called to be *locally compact (in itself)* at $A \in \mathfrak{G}$, if it satisfies the condition:

From any sequence $\{A_i\}$ of elements $\in \mathfrak{G}$, $|A_i - A| \leqq \delta$, we may choose a subsequence $\{A_{i'}\}$ such that

$$\lim_{i'\to\infty} A_{i'} = \text{an element } A' \in \mathfrak{G},$$

if δ is sufficiently small; that is, the set $\mathfrak{G}_{A,\delta}$ of all the elements $B \in \mathfrak{G}$, $|B - A| \leqq \delta$, is *compact (in itself)* for sufficiently small δ.

As the transformation $X \rightarrow X'(X' = AX, A \in \mathfrak{G})$ defined in $\mathfrak{G}$ is a topological transformation[19] of $\mathfrak{G}$ on $\mathfrak{G}$, $\mathfrak{G}$ is locally compact (in itself) everywhere on $\mathfrak{G}$, if $\mathfrak{G}$ is locally compact (in itself) at E. Hence we deal only with the local compactness (in itself) of the group $\mathfrak{G}$ at E.

Fundamental theorem 1. If a group $\mathfrak{G}$ embedded in $\mathfrak{G}$ be locally compact (in itself) at E, $\mathfrak{G}$ is differentiable everywhere on $\mathfrak{G}$.

Proof. Let $\{A_i\}$ be any sequence of elements $\in \mathfrak{G}$ such that $A_i \neq E$ $A_i \rightarrow E$. Let δ be a positive number such that

$$1 - \frac{\delta}{1 - 2\delta} > 0, \qquad 1 > \exp(2\delta) - 1\,.$$

As $A_i \rightarrow E$, $\ln A_i$ has a sense and $|\ln A_i| < \delta$ for $i \geqq i(\delta)$ by lemma 1. $|\ln A_i| \neq 0$, since otherwise $\ln A_i = 0$ by (1) and hence we would obtain $\exp \ln A_i = A_i = E$ by (5).

Thus by (4), there exists a positive integer n_i such that

$$\begin{cases} |\,k_i \ln A_i\,| < \delta, & k_i = 0, \pm 1, \pm 2, \ldots\ldots, \pm(n_i - 1), \\ \delta \leqq |\,n_i \ln A_i\,| < 2\delta, \end{cases}$$

for $i \geqq i(\delta)$.

Hence by lemma 3

---

(18) sign $(\varepsilon_{p(\varepsilon)}) = 1$ or -1, according as $\varepsilon_{p(\varepsilon)} > 0$ or < 0.

(19) For $\mathfrak{G}$ is a topological group by theorem 1.

85

K. Yosida

$$(15) \quad \begin{cases} |\,A_i^{k_i}-E\,| < \exp{(2\delta)}-1 < 1, \qquad k_i = 0,\ \pm 1,\ \pm 2,\ \ldots\ldots,\ \pm n_i\,, \\[2mm] 0 < \delta - \dfrac{\delta^2}{1-2\delta} < |\,A_i^{\pm n_i}-E\,| < \exp{(2\delta)}-1\,. \end{cases}$$

As $A_i^{\pm n_i} \in \mathfrak{G}$ and the group $\mathfrak{G}$ is locally compact (in itself) at E, we know, by taking δ sufficiently small, that there exists at least one partial sequence $\{A_{i'}^{n_{i'}}\}$ of $\{A_i^{n_i}\}$ such that

$$A_{i'}^{n_{i'}} \to \text{an element } A \in \mathfrak{G}$$

$$0 < \delta - \frac{\delta^2}{1-2\delta} \leqq |\,A-E\,| \leqq \exp{(2\delta)}-1\,.$$

Thus by (14), (15) and lemma 1

$$n_{i'} \ln A_{i'} = \ln{(A_{i'}^{n_{i'}})} \to \ln A \neq 0\,.$$

(If $\ln A = 0$, we would obtain $A = \exp \ln A = E$ by (5)).

Hence, by lemma 2.

$$n_{i'}(A_{i'}-E) \to \ln A \neq 0\,.$$

Thus we have our theorem, by theorem 2.

Remark. Formula (15) shows that the group $\mathfrak{G}$ does not contain an arbitrarily small cyclic group as its subgroup; that is, if $\delta > 0$ be sufficiently small, $\mathfrak{G}$ has no element $A \neq E$ which satisfies the inequalities

$$|\,A^n-E\,| \leqq \delta\,, \qquad n = 0,\ \pm 1,\ \pm 2,\ \ldots\ldots.$$

By taking

$$(4)' \qquad\qquad |\,aA\,| \leqq L\,|\,a\,|\,|\,A\,| \qquad (L \text{ denotes a definite constant})$$

instead of (4), in the definition of the *absolute value*, we may define a *generalised metrical complete ring* $\overline{\mathfrak{R}}$. Then we may prove

Fundamental Theorem 1'.

Let a group $\mathfrak{G}$ embedded in a generalised metrical complete ring $\overline{\mathfrak{R}}$, be locally compact (in itself) at E. Then $\mathfrak{G}$ is differentiable everywhere on $\mathfrak{G}$, if $\mathfrak{G}$ does not contain an arbitrarily small cyclic group as its subgroup.

The proof will be obtained by slightly modifying the above proof, remembering the continuity property of the composition-rule of the group (theorem 1).

§5. The Lie-ring $\mathfrak{J}$ of the group $\mathfrak{G}$.

Let a group $\mathfrak{G}$ embedded in $\mathfrak{R}$ be differentiable. We denote by $\mathfrak{J}$ the set of all the differential quotient of $\mathfrak{G}$ at E. For convention,[20] we let 0 belong to $\mathfrak{J}$. Then we have

---

[20] If $\mathfrak{G}$ is not differentiable, we say that $\mathfrak{J}$ of $\mathfrak{G}$ consists of 0 alone.

Theorem 4.[21] *Let* U, V *be* $\in \mathfrak{F}$, *then.*

$$\begin{cases} aU+\beta V \in \mathfrak{F} \text{ for any real } a, \beta; \text{ and} \\ UV-VU \in \mathfrak{F}. \end{cases}$$

Proof. Let $U \neq 0$, $V \neq 0$. By theorem 3, there exist two sequences $\{A_n\}$, $\{B_n\}$ of elements $\in \mathfrak{G}$ such that

$$A_n \neq E, \qquad B_n \neq E,$$
$$n(A_n-E) \to U, \qquad n(B_n-E) \to V.$$

Hence $an(A_n-E) \to aU$, thus $aU \in \mathfrak{F}$.

By

$$n(A_nB_n-E) = n(A_n-E)+n(B_n-E)+n(A_n-E)(B_n-E)$$
$$\to U+V+0(UV) = U+V,$$

we know that $U+V \in \mathfrak{F}$ when $U+V \neq 0$, since $A_nB_n \in \mathfrak{G}$ and $A_nB_n \neq E$ for large n in this case. If $U+V = 0$, $U+V \in \mathfrak{F}$ by definition.

Next we consider

$$n^2(A_nB_nA_n^{-1}B_n^{-1}-E) = n^2(A_nB_n-B_nA_n)A_n^{-1}B_n^{-1}.$$

As $A_n^{-1}B_n^{-1} \to E$ by lemma 1 and

$$n^2(A_nB_n-B_nA_n) = n(A_n-E)n(B_n-E)-n(B_n-E)n(A_n-E)$$
$$\to UV-VU,$$

we have

$$n^2(A_nB_nA_n^{-1}B_n^{-1}-E) \to (UV-VU)E = UV-VU.$$

Hence $UV-VU \in \mathfrak{F}$ when $UV-VU \neq 0$, since $A_nB_nA_n^{-1}B_n^{-1} \in \mathfrak{G}$ and $A_nB_nA_n^{-1}B_n^{-1} \neq E$ for large n in this case. If $UV-VU = 0$, $UV-VU \in \mathfrak{F}$, by definition.

Thus with respect to the addition $U+V$ and the *non-associative multiplication* $U \times V = UV-VU$, $\mathfrak{F}$ is a ring with real numbers as its Operatorenbereich. Except the existence of a finite number of linearly independent basis[22], $\mathfrak{F}$ obeys the customary rule of the ordinary Lie-ring:[23]

$$U+V = V+U,$$
$$U \times V = -V \times U,$$
$$U \times (V+W) = U \times V+U \times W,$$

---

(21) Cf. J. I., p. 18–19.

(22) The linear independence means with real coefficients. $\mathfrak{F}$ is called to have a linearly independent base $(U_1, U_2, \ldots\ldots, \ldots\ldots, U_n)$, if any element U of $\mathfrak{F}$ is uniquely represented as $U = \sum_{i=1}^{n} a_i U_i$, with real a.

(23) The set of infinitesimal transformations (Operations) of an ordinary Lie group of transformations is usually termed as a Lie-ring.

$$(V+W) \times U = V \times U + W \times U,$$

$$U \times (V \times W) + V \times (W \times U) + W \times (U \times V) = 0.$$

Hence it is justified to call $\mathfrak{J}$ the *Lie-ring* of the group $\mathfrak{G}$.

§ 6. Generation of the group $\mathfrak{G}$ by its Lie-ring.

A group $\mathfrak{G}$ embedded in $\mathfrak{R}$ is called to be *locally closed in $\mathfrak{R}$ at E*, when it satisfies the condition :

From $A_n \in \mathfrak{G}$, $A_n \to A \in \mathfrak{R}$, we conclude that $A \in \mathfrak{G}$, if $|A-E|$ is sufficiently small.

$\mathfrak{G}$ is thus locally closed in $\mathfrak{R}$ at E, if $\mathfrak{G}$ is locally compact at E.

Now we have the

Theorem 5. Let a group $\mathfrak{G}$ embedded in a metrical complete ring $\mathfrak{R}$ (or in a generalised metrical complete ring $\overline{\mathfrak{R}}$) be differentiable and locally closed in $\mathfrak{R}$ ($\overline{\mathfrak{R}}$) at E. Let U be any element of the Lie-ring $\mathfrak{J}$ of $\mathfrak{G}$, then $\exp(U) \in \mathfrak{G}$.

Proof. Let $U \neq 0$, and let a be a real number. By theorem 3 and 4, there exists a sequence $\{A_n\}$ of elements $\in \mathfrak{G}$ such that $n(A_n - E) \to aU$.

Hence by lemma 2,

$$n \ln A_n \to aU,$$

thus by (5), (8) and lemma 1,

$$A_n^n = \exp(n \ln A_n) = (\exp \ln A_n)^n \to \exp(aU).$$

By lemma 1, $\lim_{a \to 0} \exp aU = E$, and hence

$$\exp(aU) \in \mathfrak{G},$$

for a sufficiently small a.

When $U = 0$, $aU = 0$ and hence $\exp(aU) = E \in \mathfrak{G}$.

Thus $\exp\left(\dfrac{U}{n}\right) \in \mathfrak{G}$ for a sufficiently large positive integer n and $\left(\exp\left(\dfrac{U}{n}\right)\right)^n \in \mathfrak{G}$, since $\mathfrak{G}$ is a group.

This proves the theorem, for by (8) we have

$$\left(\exp\left(\frac{U}{n}\right)\right)^n = \exp U.$$

C. Q. F. D.

Corollary 1. Let a group $\mathfrak{G}$ embedded in a metrical complete ring $\mathfrak{R}$ be locally compact at E. Let U be any element of the Lie-ring $\mathfrak{J}$ of $\mathfrak{G}$. Then

$$\exp U \in \mathfrak{G}.$$

Proof. Since $\mathfrak{G}$ is differentiable by the fundamental theorem 1.

Corollary 2. Let a group $\mathfrak{G}$ embedded in a metrical complete ring $\mathfrak{R}$ be locally compact (in itself) at E (or let a group $\mathfrak{G}$ embedded in a generalised metrical complete ring $\overline{\mathfrak{R}}$ be differentiable and locally closed in $\overline{\mathfrak{R}}$ at E). Then the Lie ring $\mathfrak{I}$ of $\mathfrak{G}$ is closed in $\mathfrak{R}$.

Proof. Let $U_1 U_2, \ldots\ldots \in \mathfrak{I}$ and $\lim_{i \to 0} U_i = U(\neq 0) \in \mathfrak{R}$.

Then by (4) or by (4)' $\lim_{a \to 0} aU_i = 0$ uniformly in i. As

$$\exp aU_i \in \mathfrak{G} \quad \text{for any real } a,$$

and

$$\lim_{i \to \infty} \exp aU_i = \exp aU,$$

by lemma 1, $\exp aU \in \mathfrak{G}$ for a sufficiently small a, since $\mathfrak{G}$ is locally closed in $\overline{\mathfrak{R}}$ $(\mathfrak{R})$ at E. By lemma 2 we have

$$\lim_{a \to 0} \frac{\exp aU - E}{a} = \lim_{a \to 0} \frac{aU}{a} = U$$

and hence $U \in \mathfrak{I}$.

Corollary 3. Let a gronp $\mathfrak{G}$ embedded in a metrical complete ring $\mathfrak{R}$ be locally compact at E. Then the Lie-ring $\mathfrak{I}$ of $\mathfrak{G}$ is of finite dimension; that is, $\mathfrak{I}$ has a finite number, say n, of linearly independent base $(U_1, U_2, \ldots\ldots, U_n)$.

Prcof. Denote by $\mathfrak{G}_\delta$ the set of elements $A \in \mathfrak{G}$ satisfying $|A - E| \leq \delta$. If δ is sufficiently small, $\ln A$ has a sense for $|A - E| \leq \delta$, and the set $\overline{\mathfrak{G}}_\delta$ of elements $\ln A$, $A \in \mathfrak{G}_\delta$ is compact (in itself), since $\mathfrak{G}_\delta$ is compact (in itself) and the function $\ln$ is continuous (Lemma 1).

Now by (6) and the above theorem, the set $\mathfrak{I}_\varepsilon$ of elements $U \in \mathfrak{I}$ satisfying $|U| \leq \varepsilon$ is $\subseteq \overline{\mathfrak{G}}_\delta$ for a sufficiently small ε, and $\mathfrak{I}_\varepsilon$ is closed in $\overline{\mathfrak{G}}_\delta$ becauce of the above corollary. Hence $\mathfrak{I}_\varepsilon$ is compact (in itself). As $\mathfrak{I}$ is a linear manifold closed in a linear metrical complete space $\mathfrak{R}$, $\mathfrak{I}$ must be of finite dimension by a theorem due to F. Riesz. (See, for example, Banach; Théorie des Opérations Linéaires, Warsaw 1932, p. 84)

Lemma 4. Let a group $\mathfrak{G}$ embedded in $\mathfrak{R}$ be locally compact (in itself) at E (or be differentiable and locally closed in $\overline{\mathfrak{R}}$ at E). If $\mathfrak{G}$ be of dimension n, then the Lie-ring $\mathfrak{I}$ of $\mathfrak{G}$ is of dimension $\leq n$: that is, $\mathfrak{I}$ has at most n linearly independent base $(U_1, U_2, \ldots\ldots, U_k)$, $k \leq n$.

Proof. Assume $k \geq n+1$, then

$$\exp \left(\sum_{i=1}^{n+1} a_i U_i \right), \quad a \text{ real,}$$

belongs to $\mathfrak{G}$ by the above theorem. Because of the linear independence of $U_1, U_2, \ldots\ldots, U_{n+1}$ the above set of elements $\exp \left(\sum_{i=1}^{n+1} a_i U_i \right)$ is of dimension

$n+1$ at E for sufficiently small $\sum_{i=1}^{n+1} |a_i|$, contrary to the hypothesis.

C. Q. F. D.

Let $\mathfrak{G}$ be a group embedded in a metrical complete ring $\mathfrak{R}$ (or in a generalised metrical complete ring $\overline{\mathfrak{R}}$), and let $\mathfrak{J}$ be the Lie-ring of $\mathfrak{G}$. $\mathfrak{G}$ is called a *Lie group*, if it satisfies the conditions:

α) $\exp U \in \mathfrak{G}$ for any $U \in \mathfrak{J}$.

β) there exists a positive number ε such that any element $A \in \mathfrak{G}$ for which $|A - E| \leq \varepsilon$, is represented uniquely in the form

$$A = \exp U, \qquad U \in \mathfrak{J} . \qquad (\lim_{A \to E} U = 0)$$

γ) any element $A \in \mathfrak{G}$ can be represented as the result of composition

$$A = (\exp V_1)(\exp V_2)\cdots\cdots(\exp V_m), \qquad V \in \mathfrak{J},$$

where the number m may depend upon A.

Now we may prove the following

Fundamental Theorem 2.

> *Let a group $\mathfrak{G}$ embedded in a metrical complete ring $\mathfrak{R}$ be*
> i) *locally compact (in itself) at E,*
> ii) *connected,*

then $\mathfrak{G}$ is a Lie group.

(Here, a topological group is called to be *connected* when it cannot be decomposed into two open sets without common elements.)

Proof. By the corollary 3 of the theorem 5 the Lie-ring $\mathfrak{J}$ of $\mathfrak{G}$ contains a finite number n of linearly independent base $(U_1, U_2, \ldots\ldots, U_n)$. Then, by theorem 5, the set $\mathfrak{R}_\delta$ of elements $\in \mathfrak{R}$ defined by

$$\exp\left(\sum_{i=1}^{n} a_i U_i \right), \qquad a \text{ real and } \sum_{i=1}^{n} |a_i| \leq \delta$$

belongs to $\mathfrak{G}$ for any $\delta > 0$.

Then $\mathfrak{R}_\delta$ is compact (in itself). For the function $\exp$ is continuous by lemma 1 and the linear manifold $\sum_{i=1}^{n} a_i U_i$, a real and $\sum_{i=1}^{n} |a_i| \leq \delta$, is compact (in itself).

We next prove that a sufficiently small neighbourhood of the identical element of the group $\mathfrak{G}$ is contained in $\mathfrak{R}_\delta$; that is, if $A_i \in \mathfrak{G}$ and $A_i \to E$, then $A_i \in \mathfrak{R}_\delta$ for a sufficiently large i.

Assume the contrary and let

(16) $$A_i \,\bar{\in}\, \mathfrak{R}_\delta, \qquad i = 1, 2, \ldots\ldots.$$

Since $\mathfrak{R}_\delta$ contains the inverse of any one of its element $\exp\left(\sum_{i=1}^{n} a_i U_i \right)$,

---

(^{24}) Cf. our proof with the arguments given by Neumann for the case of matrix group. J. I., p. 22-23.

that is, $\exp\left(-\sum_{i=1}^{n} a_i U_i\right)$ by (8), the assumption (16) means that

$$|TA_i - E| > 0, \qquad i = 1, 2, \ldots\ldots,$$

for any $T \in \mathfrak{N}_\delta$. Since $\mathfrak{N}_\delta$ is compact (in itself) and the distance $|TA_i - E|$ is continuous in T, there must exist at least one element $T_i \in \mathfrak{N}_\delta$ such that

$$|TA_i - E| \geqq |T_i A_i - E| = \delta_i > 0, \qquad T \in \mathfrak{N}_\delta,$$

for any i. As $E \in \mathfrak{N}_\delta$ and $A_i \to E$, we have $\delta_i \to 0$, and so $T_i \to E$.

$\mathfrak{G}$ is differentiable by the hypothesis, hence there exist a subsequence $\{T_{i'} A_{i'}\}$ of $\{T_i A_i\}$ and a sequence $\{\eta_{i'}\}$ of real numbers, $\eta_{i'} \to 0$, $\eta_{i'} \neq 0$, such that

$$\text{(17)} \qquad \lim_{i' \to \infty} \frac{T_{i'} A_{i'} - E}{\eta_{i'}} = W(\neq 0) \in \mathfrak{F}.$$

Hence $T_{i'} A_{i'} = E + \eta_{i'} W + \mathrm{o}(|\eta_{i'}|^2)$, and $\exp(-\eta_{i'} W) = E - \eta_{i'} + \mathrm{O}(|\eta_{i'}|^2)$ by (10). We have thus

$$\exp(-\eta_{i'} W) T_{i'} A_{i'} - E = \big(E - \eta_{i'} W + \mathrm{o}(|\eta_{i'}|)\big)\big(E + \eta_{i'} W + \mathrm{O}(|\eta_{i'}|^2)\big)$$
$$= E + \eta_{i'} W + \mathrm{o}(|\eta_{i'}|) - \eta_{i'} W + \mathrm{o}(|\eta_{i'}|) + \mathrm{o}(|\eta_{i'}|) = \mathrm{o}(|\eta_{i'}|).$$

Now since $T_{i'} \in \mathfrak{N}_\delta$ we have

$$T_{i'} = \exp V_{i'}, \qquad V_{i'} \in \mathfrak{F},$$

where $V_{i'} = \ln \exp V_{i'} = \ln T_{i'} \to 0$ by (6) and lemma 1.

Thus

$$\text{(18)} \qquad \big|\{\exp(-\eta_{i'} W)\}(\exp V_{i'})A_{i'} - E\big| = \mathrm{o}(|\eta_{i'}|).$$

Hence by (2) and (3)

$$\big|\{\exp(-\eta_{i'} W + V_{i'})\}A_{i'} - E\big|$$
$$\leqq \big|\{\mathrm{ex}(-\eta_{i'} W + V_{i'})\}A_{i'} - \{\exp(-\eta_{i'} W)\}(\exp V_{i'})A_{i'}\big| .$$
$$+ \big|\{\exp(+\eta_{i'} W)\}(\exp V_{i'})A_{i'} - E\big| ,$$

and so by (3), (12) and (18)

$$\big|\{\exp(-\eta_{i'} W + V_{i'})\}A_{i'} - E\big| \leqq \mathrm{O}(|\eta_{i'} W| \, |V_{i'}|)|A_{i'}| + \mathrm{o}(|\eta_{i'}|)$$
$$= \mathrm{o}(|\eta_{i'}|),$$

since $A_{i'} \to E$ and $V_{i'} \to 0$.

As $\exp V_{i'} \in \mathfrak{N}_\delta$ and $-\eta_{i'} W + V_{i'} \to 0$, we have, for a sufficiently large i',

$$\exp(-\eta_{i'} W + V_{i'}) \in \mathfrak{N}_\delta \quad \text{and hence}$$
$$\mathrm{o}(|\eta_{i'}|) \geqq \delta_{i'}$$

for a sufficiently large i'. This is a contradiction, since we have

K. Yosida

$$\frac{\delta_{i'}}{|\eta_{i'}|} \to |W| \neq 0$$

by (1) and (17).

Thus a sufficiently small neighbourhood of the identical element of the group $\mathfrak{G}$ is contained in $\mathfrak{N}_\delta$. Hence the condition $a)$, $\beta)$ of the Lie group is satisfied for $\mathfrak{G}$, by (6).

This proves our theorem, since by a well-know theorem of 0. Schreier[25], any connected topological group $\mathfrak{G}$ is completely generated by any of its neighbourhood $V(E)$ of the identical element E; that is, any element A of $\mathfrak{G}$ can be represented as the result of composition:

$$A = E_1 E_2 \cdots\cdots E_m, \qquad E_i \in V(E).$$

C. Q. F. D.

By lemma 4 and theorem 5, we have also

Fundamental Theorem 2'. Let a group $\mathfrak{G}$ embedded in a generalised metrical complete ring $\overline{\mathfrak{R}}$ be

i)   *of finite dimension,*

ii) *differentiable at E,*

iii) *locally closed in $\mathfrak{R}$ at E,*

iv) *connected.*

Then $\mathfrak{G}$ is a Lie group.

Remark. If $\mathfrak{G}$ is a Lie group, then the equation

$$A = \exp X; \qquad A \in \mathfrak{G}$$

admits a unique solution X in the Lie ring $\mathfrak{J}$ of the group $\mathfrak{G}$, if $|A - E|$ is sufficiently small.

M. Nagumo[26] proved that the equation

$$A = \exp X, \qquad A \in \mathfrak{R}$$

is solvable in $\mathfrak{R}$, if and only if there exists a connected abelian group $\mathfrak{G}$ embedded in $\mathfrak{R}$ such that $A \in \mathfrak{G}$.

§ 7. Nagumo's theorem.

Let a one-parameter family of elements $G(t), -\infty < t < +\infty$, constitute a continuous group $\mathfrak{G}$ embedded in $\mathfrak{R}$:

$$\begin{cases} G(t)G(s) = G(t+s), \\ G(0) = E, \\ \lim_{s \to t} G(s) = G(t). \end{cases}$$

---

(25) O. Schreier: Abstrakte kontinuierliche Gruppen. Hamburg, Abh. math. Sem. **4** (1925).

(26) M. Nagumo: This Journal.

M. Nagumo[27] proved that $G(t)$ is differentiable in the ordinary sense:

$$\lim_{\Delta t \to 0} \frac{G(t+\Delta t) - G(t)}{\Delta t} = H(t) \in \Re,$$

and then proved the analytical representation of the group $\mathfrak{G}$ as

$$G(t) = \exp(tH(0)),$$

by integrating the abstract differential equation

$$\frac{dG(t)}{dt} = H(0)G(t).$$

This theorem may easily be obtained by slightly modifying the proof of the fundamental theorem 2.

I will here give a direct proof of this theorem depending upon the same idea as in the arguments developed in the preceding paragraphs.

Proof. As $\lim_{t \to 0} G(t) = E$, we may suppose that $|G(t) - E| < 1$ for $|t| \le 2$.[28]

Let $t_1, t_2, \cdots\cdots$ be any sequence of positive numbers, $t_i \to 0$. We choose positive integer p_i such that

$$(p_i - 1)t_i < 1 \le p_i t_i,$$

then

$$G(t_i)^{p_i} = G(p_i t_i) \to G(1),$$

and hence

$$\ln G(t_i)^{p_i} = p_i \ln G(t_i) \to \ln G(1),$$

by lemma 1 and (14) since $|p_i t_i| < 2$ for a sufficiently large i.

Thus by lemma 2.

$$p_i\big(G(t_i) - E\big) \to \ln G(1),$$

and hence

$$(19) \qquad \frac{G(t_i) - E}{t_i} = \frac{p_i(G(t_i) - E)}{p_i t_i} \to \ln G(1).$$

Next let t be a positive number and let n denote positive integers, then by (19)

$$\lim_{n \to \infty} \frac{G\left(\dfrac{t}{n}\right) - E}{\dfrac{t}{n}} = \ln G(1),$$

---

[27]. M. Nagumo: loc. cit.
[28] If not, we have only to consider the group $G(at)$ for sufficiently small positive a.

$$\lim_{n\to\infty} n\left(G\left(\frac{t}{n}\right)-E\right) = t \ln G(1) .$$

Hence by lemma 2

$$\lim_{n\to\infty} n \ln G\left(\frac{t}{n}\right) = t \ln G(1) .$$

Thus if $\dfrac{t}{n} < 2$, we have by (5), (8) and lemma 1,

$$\exp\left(n \ln G\left(\frac{t}{n}\right)\right) = \left\{\exp\left(\ln G\left(\frac{t}{n}\right)\right)\right\}^n = G\left(\frac{t}{n}\right)^n = G(t) \to \mathrm{ext}\left(t \ln G(1)\right).$$

Hence by the functional equation $G(t)\ G(s) = G(t+s)$ and (8)

$$G(t) = \exp\left(t \ln G(1)\right)$$

for all real t. From this equation we also know that $G(t)$ is differentiable in the ordinary sense :

$$\lim_{\varDelta t\to 0} \frac{G(t+\varDelta t)-G(t)}{\varDelta t} = \left\{\ln G(1)\right\}G(t)$$

C. Q. F. D.

Remark. Nagumo's theorem may be applied to the one-parameter continuous group of linear continuous (bounded) transformations of a linear metrical complete space $\mathfrak{C}$ into $\mathfrak{C}$. In a particular case where the space $\mathfrak{C}$ is a Hilbert space $\mathfrak{H}$, it is to be compared with the famous theorem of M. H. Stone,[29] which concerns with the one-parameter group $U(t)$ $(-\infty < t < +\infty)$ of the unitary transformations in $\mathfrak{H}$.

Mathematical Institute,
Ôsaka Imperial University.

---

(29) M. H. Stone; Proc. Nat. Acad. 16 (1930), p. 172–175.

During the proof correction the author found that Nagumo's theorem was also obtained by D. S. Nathan (Duke Math. J. 1).

By the use of this Nagumo-Nathan's theorem we may show that any continuous representation $\mathfrak{D}$ (embedded in $\mathfrak{R}$) of the locally compact topological group $\mathfrak{G}$ may be generated by infinitesimal operators, if $\mathfrak{G}$ satisfies the first axiom of countability and if the group $\mathfrak{D}$ be of finite dimention. This result will be published soon in this Journal.

On the group embedded in the
metrical complete ring. II

Japan. J. Math. **13** (1936) 459–472

(Received July, 8, 1936)

According to M. Nagumo and D. S. Nathan[1] any continuous representation $\mathfrak{D}$ (embedded in the metrical complete ring[2] $\mathfrak{R}$) of the additive group $\mathfrak{G}$ of real numbers may be generated by an infinitesimal operator $\epsilon\mathfrak{R}$. By this theorem we may prove, after B. L. van der Waerden[3], that any continuous representation $\mathfrak{D}$ (embedded in $\mathfrak{R}$) of the Lie group $\mathfrak{G}$ is an analytical representation. This constitutes a generalisation of the result due to J. von Neumann[4] and B. L. van der Waerden[5] concerning the matrix-representation of the Lie group.

By modifying the method used in my preceding paper[6] (to be cited as [I] below) I intend to extend the above result to the case where the group $\mathfrak{G}$ is a topological group, not necessarily a Lie group.

In §1 of this paper are given the necessary definitions and our main theorem, the proof of it will be given in §2.

Formulae and the properties concerning exp.- and ln-functions as developped in [I] will be frequently used in this paper with citations.

§ 1. Definitions and the theorem.

Let $\mathfrak{R}$ be the metrical complete ring as defined in [I] and let $\mathfrak{G}$ denote a topological group[7].

A *representation* $\mathfrak{D}$ of $\mathfrak{G}$ in $\mathfrak{R}$ is a function D, defined for all the elements $\epsilon\mathfrak{G}$, which takes the values from $\mathfrak{R}$, such that we have the functional equation

---

(1) *M. Nagumo:* Einige analytische Untersuchungen in linearen metrischen Ringen. Jap. J. of Math. 13 (1936): *D. S. Nathan:* One-parameter groups of transformations in abstract vector spaces. Duke Math. J. 1. Another proof is given in *K. Yosida:* On the group embedded in the metrical complete ring. Jap. J. of Math. 13 (1936).

(2) *M. Nagumo* or *K. Yosida:* loc. cit.

(3) *B. L. van der Waerden:* Stetigkeitssätze für halbeinfache Liesche Gruppen. Math. Zeits. 36.

(4) *J. von Neumann:* Über die analytischen Eigenschaften von Gruppen linearer transformationen und ihrer Darstellungen. Math. Zeits. 30.

(5) *B. L. van der Waerden:* loc. cit.

(6) *K. Yosida:* loc. cit.

(7) For the definition of the topological group see, for example, *E. R. van Kampen:* Locally Bicompact Abelian Groups and their Chracter Groups. Ann. of Math. 36, No. 2.

$$\begin{cases} D(a)D(b) = D(ab)\,, & (a,\,b \,\epsilon\, \mathfrak{G}) \\ D(e) = E\,, \end{cases}$$

where e and E denote the unit-element of $\mathfrak{G}$ and $\mathfrak{R}$ respectively.

The representation $\mathfrak{D}$ is called *contiuuons* if $D(a)$ is a continuous function of a, the topology in $\mathfrak{R}$ being defined by the metric $|A-B|$ in $\mathfrak{R}$[8].

The continuous representation $\mathfrak{D}$ is called a *Lie representation* if there exists a finite number, say n, of elements $U_1,\, U_2,\,\ldots,\, U_n \,\epsilon\, \mathfrak{R}$ which are linearly independent (with real numbers as coefficients) such that the following conditions are satisfied :

i) If we denote by $\mathfrak{J}$ the set of all elements $\epsilon\, \mathfrak{R}$ of the form $\sum_{i=1}^{n} a_i U_i$, a real, we have $(UV-VU)\,\epsilon\,\mathfrak{J}$ for $U,\, V\,\epsilon\,\mathfrak{J}$ and

$$\exp U \epsilon \mathfrak{D}\,, \quad \text{when } U \epsilon \mathfrak{J}\,.$$

ii) $D(a) = \exp U,\; U = \sum_{i=1}^{n} a_i U_i$, if $a\,\epsilon\,\mathfrak{G}$ be sufficiently near e and $\lim \sum_{i=1}^{n} |\,a_i\,| = 0$ if $\lim a = e$.

iii) Any element of $\mathfrak{D}$ can be expressible as the result of composition $\exp V_1 \exp V_2 \cdots \exp V_m$, $V_k = \sum_{i=1}^{n} a_{ik} U_i \epsilon \mathfrak{J}$ where $\sum_{i=1}^{n} |\,a_{ik}\,|$ may be taken as small as we please.

It is to be noted that a Lie reqresentation $\mathfrak{D}$ is *not* necessarily a *Lie group* in the sense as defined in [I], for we cannot conclude that $\lim_{j \to \infty} \sum_{i=1}^{n} |\,a_{ij}\,| = 0$ from $\lim_{j \to \infty} \exp V_j = E$, $V_j = \sum_{i=1}^{n} a_{ij} U_i$. Such an example is given by $D(t) = \exp tU$, $-\infty < t < \infty$, $U = \left\| \begin{matrix} i\pi & o \\ o & i\tau\pi \end{matrix} \right\|$, $i = \sqrt{-1}$, $\tau =$ irrational.

Now our main theorem reads as follows.

Theorem. *Let $\mathfrak{G}$ be a locally bicompact, connected topological group which satisfies the first axiom of countability*[9]. *Then the continuous representation $\mathfrak{D}$ of $\mathfrak{G}$ in $\mathfrak{R}$ is a Lie representation, if the group $\mathfrak{D}$ be of finite dimension.*

---

(8) *See* [I]

(9) A topological group $\mathfrak{G}$ is called locally bicompact, if there exists a vicinity $U(e)$ of the unit-element e such that the closure $\bar{U}(e)$ of $U(e)$ is bicompact (for the sake of brevity the term *in itself* is omitted throughout this paper). A bicompact space is defined by the property that each covering by open sets of the space contains a finite covering. Any continuous image of a bicompact space is bicompact. Any infinite set of points of a bicompact space admits at least one accumulation point in the space. The first axiam of countability asserts that any accumulation point a of the sequence $\{a_n\}$ of points is a limiting point of at least one partial sequence $\{a_{n_i}\}$ of $\{a_n\}$:

$$\lim_{i \to \infty} a_{n_i} = a$$

Remark added during the proof. The two assumption, i) the group $\mathfrak{G}$ satisfies the first axion of countability and ii) the group $\mathfrak{D}$ is of finite dimension, are not necessary for the proof of the theorem, as is shown in the **Appendix**.

§ 2. Proof of the theorem.

We first assume the theorem is true when $\mathfrak{D}$ is an isomorphic representation of $\mathfrak{G}$:

$$D(a) = D(b) \text{ when and only when } a = b\,;$$

and then deduce the theorem from this special case.

This may be done as follows.

Since $\mathfrak{D}$ is continuously homomorphic to $\mathfrak{G}$, $\mathfrak{D}$ is isomorphic to a quotient group $\mathfrak{G}/\mathfrak{N}$, where $\mathfrak{N}$ denotes an invariant subgroup closed in $\mathfrak{G}$. We can introduce a topology in $\mathfrak{G}/\mathfrak{N}$ by considering any set of $\mathfrak{G}/\mathfrak{N}$, open in $\mathfrak{G}/\mathfrak{N}$, if and only if it corresponds to an open set in $\mathfrak{G}$, by the homomorphism $\mathfrak{G} \rightarrow \mathfrak{G}/\mathfrak{N}$.

The homomorphic mapping $\mathfrak{G} \rightarrow \mathfrak{G}/\mathfrak{N}$ is thus a continuous mapping which maps any open set in $\mathfrak{G}$ to an open set in $\mathfrak{G}/\mathfrak{N}$. Hence the topological group $\mathfrak{G}/\mathfrak{N}$ is locally bicompact and connected with $\mathfrak{G}$[10].

Moreover, by the isomorphic mapping $\mathfrak{G}/\mathfrak{N} \rightarrow \mathfrak{D}$, $\mathfrak{D}$ becomes a continuous representation of $\mathfrak{G}/\mathfrak{N}$, since to any open set in $\mathfrak{D}$ there corresponds an open set in $\mathfrak{G}/\mathfrak{N}$. And $\mathfrak{G}/\mathfrak{N}$ satisfies the f.a. of c. with $\mathfrak{G}$.

Thus $\mathfrak{D}$ is a Lie representation of $\mathfrak{G}/\mathfrak{N}$ by our hypothesis. Hence $\mathfrak{D}$ is also a Lie representation of $\mathfrak{G}$, for to any element $\epsilon\,\mathfrak{G}$ near e corresponds an element $\epsilon\,\mathfrak{G}/\mathfrak{N}$ near the unit-element of $\mathfrak{G}/\mathfrak{N}$ and any element $\epsilon\,\mathfrak{G}$ can be written as the result of composition of a finite number of elements $\epsilon\,\mathfrak{G}$ arbitrarily near e, by a theorem due to O. Schreier[11].

Now the proof of our special case may be effected by the following series of lemmas.

Lemma 1. *Let $U(e)$ be a vicinity of the unit-element e such that the closure $\overline{U}(e)$ of $U(e)$ be bicompact. The common part $V(e)$ of $U(e)$ and $U(e)^{-1}$ ($U(e)^{-1}$ is the set of all elements a^{-1}, $a\epsilon U(e)$) has the same property and $V(e) = V(e)^{-1}$. If $U(e)$ be so small that we have*

$$| D(a) - E | < a \text{ certain positive } a < 1 \text{ for } a\epsilon\overline{V}(e),$$

$\mathfrak{G}$ *does not contain a cyclic subgroup ($\neq$ identical group) $\subseteq \overline{V}(e)$.*

Proof. We assume the contrary, and let $a \neq e$, $a\epsilon\overline{V}(e)$, be such that

---

(10) For the proof see *E. R. van Kampen*, loc. cit., where the general properties of locally bicompact $\mathfrak{G}$ is treated.

(11) *O. Schreier:* Abstracte kontinuierliche Gruppen. Abh, Math. Seminar Hamburg. 4.

$$a^n \epsilon \overline{V}(e), \qquad n = 0,\ \pm 1,\ \pm 2, \ldots\ .$$

Then $D(a)^n = D(a^n)$ and hence $|D(a)^n - E| < \alpha$ for $n = o,\ \pm 1,\ \pm 2, \ldots$. Thus we have, for any integer n,

$$|n|\,|lnD(a)| = |n\,lnD(a)| = |ln\{D(a)^n\}| \leq \sum_{m=1}^{\infty} \frac{|D(a)^n - E|^m}{m} \leq \sum_{m=1}^{\infty} \frac{\alpha^m}{m},$$

by (2), (3), (4) and (14) of [I].
This is a contradiction, since we have

$$|\,lnD(a)\,| \neq 0$$

by $D(a) \neq E$ and (1), (5) of [I]. $\hspace{3cm}$ Q.E.D.

Lemma 2. *Let there exist a sequence $\{a_i\}$ of elements $\epsilon \mathfrak{G}$ such that*

$$a_i \neq e, \qquad \lim a_i = e.$$

Then $\mathfrak{G}$ contains a one-parameter continuons subgroup $a(t)$, $-\infty < t < +\infty$:

$$\begin{cases} a(t)a(s) = a(t+s),\ a(o) = e, \\ \lim_{t \to t_0} a(t) = a(t_0),\ a(t)\epsilon \overline{V}(e) \quad \text{for} \quad -1 \leq t \leq +1, \end{cases}$$

where $a(1) = \lim\limits_{i_1 \to \infty} a_{i_1}^{n_{i_1}} \neq e$ for a certain partial sequence $\{a_{i_1}\}$ of $\{a_i\}$ and a certain sequence $\{n_{i_1}\}$ of positive integees.

Proof. Let $V_1(e)$ be a vicinity of e such that $V_1(e) = V_1(e)^{-1}$ and $V_1(e)^2 \leq V(e)$ ($V_1(e)^2$ denotes the set of all the elements of the form ab where a and b $\epsilon V_1(e)$). We may assume $a_i \epsilon V_1(e)$ for $i = 1, 2, \ldots$, since $\lim\limits_{i \to \infty} a_i = e$.

By Lemma 2, there exists a sequence $\{n_i\}$ of positive integers such that

$$\begin{cases} a_i^n \epsilon \overline{V}(e), \quad n = 0,\ \pm 1, \ldots,\ \pm n_i, \\ a_i^{\pm(n_i+1)}\ \bar{\epsilon}\,\overline{V}(e). \end{cases}$$

We have

$$a_i^{\pm n_i}\ \bar{\epsilon}\, V_1(e), \qquad i = 1, 2, \ldots,$$

since otherwise $a_i^{\pm(n_i+1)} \epsilon V_1(e)^2 \leq V(e)$.

As $\overline{V}(e)$ is bicompact we may choose a subsequence $\{a_{i_1}^{n_{i_1}}\}$ of $\{a_i^{n_i}\}$ such that $\lim\limits_{i_1 \to \infty} a_{i_1}^{n_{i_1}} = a$. $a \neq e$, since $a_{i_1}^{n_{i_1}} \bar{\epsilon} V_1(e)$.

If we define the positive integer m_{i_1} by

$$2m_{i_1} - 1 < n_{i_1} \leq 2m_{i_1}$$

the sequence $\{a_{i_1}^{m_{i_1}}\}$ is a convergent sequence. For, if not, we may choose two convergent subsequences $\{a_{i_2}^{m_{i_2}}\}$, $\{a_{i_3}^{m_{i_3}}\}$ of $\{a_{i_1}^{m_{i_1}}\}$ such that

$$\lim_{i_2 \to \infty} a_{i_2}^{m_{i_2}} = a' \neq a'' = \lim_{i_3 \to \infty} a_{i_3}^{m_{i_3}},$$

since $\overline{V}(e)$ is bicompact. Clearly a', $a'' \epsilon \overline{V}(e)$ and

$$a'^2 = a''^2 = a \epsilon \overline{V}(e) .$$

Thus we would have

$$D(a') \neq D(a'') , \ D(a')^2 = D(a'')^2.$$

We have thus $2lnD(a') = 2lnD(a'')$ by (14) of [I], and hence $D(a') = D(a'')$ by (5) of [I], contrary to the inequality $D(a') \neq D(a'')$.

Thus $\lim_{i_1 \to \infty} a_{i_1}^{m_{i_1}}$ exists. We denote this limit by $a^{\frac{1}{2}}$. In the same way we define, successively, $a^{\frac{1}{4}} = (a^{\frac{1}{2}})^{\frac{1}{2}}, \ldots, a^{\frac{1}{2n}}$. We have

$$a^{\frac{1}{2n}} \neq e \text{ with } a \neq e,$$

$$(a^{\frac{1}{2n}})^{2l} = a^{\frac{1}{2n-l}} \quad \text{for} \quad 0 \leq l \leq n,$$

and $a^{\frac{m}{2n}} = (a^{\frac{1}{2n}})^m \epsilon \overline{V}(e)$, $0 \leq m \leq 2^n$, since $\lim_{i_1 \to \infty} a_{i_1} = e$ and $a_{i_1}^k \epsilon \overline{V}(e)$ for $k = 0, \pm 1, \ldots, \pm n_{i_1}$.

Moreover we have

$$\lim_{n \to \infty} a^{\frac{1}{2n}} = e .$$

For, if not, we would obtain a partial sequence $\{a^{\frac{1}{2n'}}\}$ of $\{a^{\frac{1}{2n}}\}$ such that

$$\lim_{n' \to \infty} a^{\frac{1}{2n'}} = b \neq e .$$

Then we have

$$\lim_{n' \to \infty} (a^{\frac{1}{2n'}})^l = b^l$$

for any integer l and as $(a^{\frac{1}{2n'}})^l \epsilon \overline{V}(e)$ for $0 \leq l \leq 2^{n'}$, the following inclusions must hold

$$b^l \epsilon \overline{V}(e) , \qquad l = 1, 2, \ldots .$$

Thus we must have $b = e$, by the same reasoning as used in the proof of Lemma 1, contrary to the inequality $b \neq e$.

Hence, in the same way, we may deduce easily the following fact:
Let r_1, r_2, ..., r_n .... be any sequence of rational numbers of the form $\dfrac{k}{2^h}$, then we have

$$\lim_{n\to\infty} a^{r_n} = e \quad \text{if} \quad \lim_{n\to\infty} r_n = 0 .$$

Now we define $a\left(\dfrac{m}{2^n}\right)$ by $(a^{\frac{1}{2^n}})^m$ for any integer $m \gtrless 0$, then

$$a\left(\frac{m}{2^n}\right) a\left(\frac{m'}{2^{n'}}\right) = a\left(\frac{2^{n'}m}{2^{n+n'}}\right) a\left(\frac{2^n m'}{2^{n+n'}}\right)$$

$$= a\left(\frac{1}{2^{n+n'}}\right)^{2^{n'}m+2^n m'} = a\left(\frac{m}{2^n}+\frac{m'}{2^{n'}}\right) .$$

Hence by continuity we easily obtain the desired group $a(t)$, such that $a(1) = a$, $a(t)\epsilon \overline{V}(e)$ for $-1 \leqq t \leqq +1$. $\qquad$ Q.E.D.

If we put $D(a(t)) = D_t$, D_t is continuous in t and

$$D_t D_s = D_{t+s} , \quad D_0 = E .$$

Thus by Nagumo-Nathan's[12] theorem

$$D_t = \exp (tU) , \quad U\epsilon\Re .$$

We call such an element $U\epsilon\Re$ the *infinitesimal operator* of the representation $\mathfrak{D}$.

Lemma 3. *The set $\mathfrak{J}$ of all the infinitesimal operators of $\mathfrak{D}$, added 0 of $\Re$, consitutes a linear manifold with real coeffieients, that is, we have*
   i) $U+ V\epsilon\mathfrak{J}$, *if* U, $V\epsilon\mathfrak{J}$,
   ii) $aU\epsilon\mathfrak{J}$ *for real a if* $U\epsilon\mathfrak{J}$.

Proof. ii) Let $a \neq 0$. If we put $a(at) = c(t)$, $c(t)$ is also a one-parameter continuous subgroup ($\neq e$) of $\mathfrak{G}$, and we have

$$D(c(t)) = \exp \{t(aU)\} , \quad aU\epsilon\Re$$

Hence $aU\epsilon\mathfrak{J}$. If $a = 0$ we have $aU = 0\epsilon\mathfrak{J}$ by definition.

   i) Let $b(t)$ ($\neq e$) be another one-parameter continuous subgroup of $\mathfrak{G}$ such that

$$|D(b(t))-E| < 1 \quad \text{for} \quad -1 \leqq t \leqq 1 ,$$

and let

$$D(b(t)) = \exp (tV) , \quad V\epsilon\mathfrak{J} .$$

---

(12) *M. Nagnmo or K. Yosida :* loc. cit.

We assume $U + V \neq 0$. Then, if we put $c_n = a\left(\dfrac{1}{n}\right) b\left(\dfrac{1}{n}\right)$, $c_n \neq e$ for sufficiently large positive interger n. For, if otherwise, $a\left(\dfrac{1}{n}\right) = b\left(\dfrac{1}{n}\right)^{-1} = b\left(\dfrac{-1}{n}\right)$, and hence $\exp\left(\dfrac{U}{n}\right) = \exp\left(\dfrac{-V}{n}\right)$ for sufficiently large n and hence $U = -V$ by (6) of $[\mathrm{I}]$, contrary to the hypothesis.

Thus $c_n \neq e$ for $n \geq n_0$. Let n_0 be so large that we have $c_n \in V_1(e)$ for $n \geq n_0$. Then by Lemma 1 there exists, for $n \geq n_0$, a sequence $\{l_n\}$ of positive integers such that

$$\begin{cases} c_n^m \in \overline{V}(e), \quad m = 0, \pm 1, \ldots, \pm l_n, \\ c_n^{\pm(l_n+1)} \bar{\in} \overline{V}(e). \end{cases}$$

We put $m_n = \min.(l_n, n)$. As $0 < \dfrac{m_n}{n} \leq 1$ we may choose a subsequence $\left\{\dfrac{m_{n'}}{n'}\right\}$ of $\left\{\dfrac{m_n}{n}\right\}$ such that

$$\begin{cases} 0 \leq \lim \dfrac{m_{n'}}{n'} = t_0 \leq 1, \\ \lim\limits_{n' \to \infty} c_{n'}^{m_{n'}} = c \in \overline{V}(e). \end{cases}$$

Hence

$$\lim_{n' \to \infty} D(c_{n'})^{m_{n'}} = \lim_{n' \to \infty} D(c_{n'}^{m_{n'}}) = D(c).$$

Thus, by (14) and Lemma 2 of $[\mathrm{I}]$, we have

$$\lim_{n' \to \infty} m_{n'}\{D(c_{n'}) - E\} = \lim_{n' \to \infty} m_{n'} \ln D(c_{n'}) = \lim_{n' \to \infty} \ln\{D(c_{n'})^{m_{n'}}\} = \ln D(c).$$

As

$$m_{n'}\{D(c_{n'}) - E\} = m_{n'}\left\{ D\left(a\left(\tfrac{1}{n'}\right)\right) D\left(b\left(\tfrac{1}{n'}\right)\right) - E\right\}$$

$$= m_{n'}\left\{ \left(D\left(a\left(\tfrac{1}{n'}\right)\right) - E\right) + \left(D\left(b\left(\tfrac{1}{n'}\right)\right) - E\right) \right.$$

$$\left. + \left[D\left(a\left(\tfrac{1}{n'}\right)\right) - E\right]\left[D\left(b\left(\tfrac{1}{n'}\right)\right) - E\right]\right\},$$

and, by virtue of (6) and Lemma 2 of $[\mathrm{I}]$,

---

(13) To apply (6) of $[\mathrm{I}]$ we must have $|U| < \ln 2$, $|V| < \ln 2$. This assumption may be made without the loss of generality, because of ii) of our Lemma.

$$\lim_{n'\to\infty} m_{n'}\left\{D\left(a\left(\frac{1}{n'}\right)\right)-E\right\} = \lim_{n'\to\infty} \frac{m_{n'}}{n'}\, n'\left\{D\left(a\left(\frac{1}{n'}\right)\right)-E\right\}$$

$$= t_0 \lim_{n'\to\infty} n'\, lnD\left(a\left(\frac{1}{n'}\right)\right) = t_0 \lim_{n'\to\infty} ln\left\{D\left(a\left(\frac{1}{n'}\right)\right)^{n'}\right\} = t_0\, lnD\left(a\left(\frac{n'}{n'}\right)\right)$$

$$= t_0\, lnD(a(1)) = t_0\, ln\, \exp U^{(13)} = t_0 U,$$

$$\lim_{n'\to\infty} m_{n'}\left\{D\left(b\left(\frac{1}{n'}\right)\right)-E\right\} = t_0 V^{(13)},$$

$$\lim_{n'\to\infty} m_{n'}\left\{D\left(a\left(\frac{1}{n'}\right)\right)-E\right\}\left\{D\left(b\left(\frac{1}{n'}\right)-E\right\} = t_0 U\cdot 0 = 0,$$

we have

$$lnD(c) = t_0(U+V).$$

The number t_0 is $\neq 0$. For, if $m_{n'} = n'$ for infinitely many n', we have $t_0 = 1$ by taking such subsequence $\{n'\}$.

If $m_{n'} < n'$ for $n' \geq n_0$, we have

$$c_{n'}^{m_{n'}}\, \bar\epsilon\, V_1(e), \qquad n' \geq n_0,$$

since, otherwise

$$c_{n'}^{l_{n'}+1} = c_{n'}^{m_{n'}+1} = c_{n'}^{m_{n'}} c_{n'}\, \epsilon\, V_1(e)^2 \leq V(e),$$

contrary to the definition of $l_{n'}$. Thus $c\,\bar\epsilon\, V_1(e)$ and hence $c \neq e$. Therefore $D(c) = \exp\{t_0(U+V)\} \neq E$ and hence $t_0 \neq 0$ by (6) of [I].

Thus, by the Lemma 2, there exists a one-parameter continuous subgroup $c(t)$ of $\mathfrak{G}$ such that

$$c(1) = c \neq e, \quad D(c(t)) = \exp\{t\,[t_0(U+V)]\}^{(14)}.$$

This proves i) of our Lemma by ii). $\qquad\qquad$ Q.E.D.

Now, by (5) and Lemma 1 of [I] the mapping $A \to lnA$, $A\epsilon\mathfrak{R}$, is a homeomorphic mapping if $|A-E|$ is <1. Hence by Lemma 3 and the dimensionality hypothesis concerning $\mathfrak{D}$, there exists a finite number, say n, of linearly linearly independent (with real coefficients) base $U_1, U_2,\ldots\ldots U_n$ of $\mathfrak{F}$ such that any element of $\mathfrak{F}$ can be written uniquely in the form $\sum_{i=1}^{n} a_i U_i$. a real.

---

(14) For the above reasoning shows that

$$D\left(c^{\frac{k}{2^h}}\right) = \exp\left\{\frac{k}{2^h}\, t_0(U+V)\right\}$$

for any integers k, h.

Lemma 4. $\exp(\sum_{i=1}^{n} a_i U_i) \epsilon \mathfrak{D}$ *for any system of real numbers a, and if* $\sum_{i=1}^{n} |a_i|$ *tends to 0, the element $a \epsilon \mathfrak{G}$ for which*

$$D(a) = \exp\left(\sum_{i=1}^{n} a_i U_i\right)$$

also tends to e.

Proof. Let $U = \sum_{i=1}^{n} a_i U_i$ be any element $\epsilon \mathfrak{J}$ such that $\sum_{i=1}^{n} |a_i| = 1$. Then by definition, there exists a one-parameter continuous subgroup $a(t) \leq \mathfrak{G}$ for which $D(a(t)) = \exp(t \sum_{i=1}^{n} a_i U_i)$. As $\lim_{t \to 0} a(t) = e$ and $V(e)$ is open, there exists a positive t_0 such that

$$\begin{cases} a(t) \epsilon V(e), & -t_0 < t < t_0, \\ a(t_0) \bar{\epsilon} V(e), & a(t_0) \epsilon \overline{V}(e), \end{cases}$$

because of the Lemma 1.

The positive number t_0 may depend upon the system of real numbers $(a_1, a_2, \ldots, a_n)$, $\sum_{i=1}^{n} |a_i| = 1$. We write, for this reason,

$$t_0 = t_0(a_1, a_2, \ldots, a_n).$$

Then we have, for $\sum_{i=1}^{n} |a_i| = 1$,

$$\text{largest lower bound } t_0(a_1, a_2, \ldots, a_n) = t_1 > 0.$$

For, if otherwise, there exists a sequence $\{(a_1^{(k)}, a_2^{(k)}, \ldots, a_n^{(k)})\}$ of system of real numbers such that

$$\begin{cases} \lim_{k \to \infty} t_0(a_1^{(k)}, a_2^{(k)}, \ldots, a_n^{(k)}) = 0, \\ \lim_{k \to \infty} a_i^{(k)} = a_i, \sum_{i=1}^{n} |a_i| = 1. \end{cases}$$

Thus we must have

$$\lim_{k \to \infty} \exp\left\{t_0(a_1^{(k)}, a_2^{(k)}, \ldots, a_n^{(k)}) \sum_{i=1}^{n} a_i^{(k)} U_i\right\} = E$$

This is a contradiction, since any limiting element $\epsilon \mathfrak{G}$ of the set of elements a_k defined by

$$D(a_k) = \exp\left\{t_0(a_1^{(k)}, a_2^{(k)}, \ldots, a_n^{(k)}) \sum_{i=1}^{n} a_i^{(k)} U_i\right\}$$

is not equal to e by $a_k \bar{\epsilon} V(e)$.

Thus, if $\sum_{i=1}^{n} |a_i| \leq t_1$, $t_1 > 0$, the element $a \epsilon \mathfrak{G}$ for which

$D(a) = \exp\left(\sum_{i=1}^{n} a_i U_i\right)$ must satisfy $a \epsilon \overline{V}(e)$. As the function $D(a)$ defines a continuous isomorphic mapping of $\mathfrak{G}$ on $\mathfrak{D}$ and $\overline{V}(e)$ being bicompact, we must have $\lim a = e$ for $\lim \sum_{i=1}^{n} |a_1| = 0$. Q.E.D.

Lemma 5. *Let $\{a_i\}$ be any sequence of elements $\epsilon \mathfrak{G}$ for which $a_i \neq e$, $\lim_{i \to \infty} a_i = e$. Then, if i is sufficiently large, we have*

$$D(a_i) = \exp U^{(i)}$$

where $U^{(i)} = \sum_{k=1}^{n} a_{ik} U_k$, a_{ik} real, such that $\lim_{i \to \infty} \sum_{k=1}^{n} |a_{ik}| = 0$.

Proof. Let $\mathfrak{J}'$ be the set of all the elements $U = \sum_{i=1}^{n} a_i U_i$ for which $\sum_{i=1}^{n} |a_i| \leq \delta$. By Lemma 4, we may choose $\delta > 0$ so small that any element $a \epsilon \mathfrak{G}$, which satisfies $D(a) = \exp U$, $U \epsilon \mathfrak{J}'$, belongs to $\overline{V}(e)$. We denote by $\mathfrak{D}'$ the set of all elements $\epsilon \mathfrak{D}$ of the form $\exp U$, $U \epsilon \mathfrak{J}'$. $\mathfrak{D}'$ is surely bicompact with $\mathfrak{J}'$ and it contains the inverse of any of its element $\exp U$, that is $\exp(-U)$.

Now let us assume that $D(a_i) \epsilon \mathfrak{D}'$ for $i = 1, 2, \ldots$. Then

$$|D(a_i)T - E| \neq 0 \quad \text{for any } T \epsilon \mathfrak{D}',$$

since otherwise $D(a_i)T' = E$ for an element $T' \epsilon \mathfrak{D}'$ by (1) of [I], and hence $D(a_i) = T'^{-1} \epsilon \mathfrak{D}'$, contrary to the hypothesis.

Since $\mathfrak{D}'$ is bicompact, there exists, for $i = 1, 2, \ldots$ at least one element $T_i \epsilon \mathfrak{D}'$ such that

$$0 < \eta_i = |D(a_i)T_i - E| \leq |D(a_i)T - E| \quad \text{for all } T \epsilon \mathfrak{D}'.$$

$\lim_{i \to \infty} \eta_i = 0$, since $E = \exp 0 \epsilon \mathfrak{D}'$ and hence $\lim_{i \to \infty} T_i = E$ by $\lim_{i \to \infty} D(a_i) = E$.

Thus if we put $T_i = \exp V_i$, $V_i \epsilon \mathfrak{J}'$, $\lim_{i \to \infty} V_i = 0$ by (6) of [I] and as $V_i \epsilon \mathfrak{J}'$ the element $b_i \epsilon \mathfrak{G}$ defined by $T_i = D(b_i)$ must satisfy

$$\lim_{i \to \infty} b_i = e,$$

by Lemma 4.

Hence $D(a_i b_i) = D(a_i)T_i \neq E$, $a_i b_i \neq e$. As $\lim_{i \to \infty} a_i b_i = e$ we may choose a subsequence $\{a_{i'} b_{i'}\}$ of $\{a_i b_i\}$ and a sequence $\{p_{i'}\}$ of positive integers such that, for $i' \geq i_0'$,

$$(a_{i'} b_{i'})^{p_{i'}} \epsilon \overline{V}(e), \quad \lim (a_{i'} b_{i'})^{p_{i'}} = d \neq e,$$

by Lemma 2. Thus, as in the proof of Lemma 3, we have
$$\lim_{i' \to \infty} p_{i'} \{D(a_{i'})T_{i'} - E\} = \ln D(d) \neq 0,$$
and so
$$\lim_{i' \to \infty} p_{i'} \eta_{i'} \neq 0,$$
by (4) of [I].

If we put $lnD(d) = W \epsilon \mathfrak{J}$, we have

$$D(a_{i'})T_{i'} = E + \frac{W}{p_{i'}} + o\left(\frac{1}{p_{i'}}\right),$$

and

$$\exp\left(\frac{-W}{p_{i'}}\right) = E - \frac{W}{p_{i'}} + o\left(\frac{1}{p_{i'}}\right)$$

by (10) of [I].   Therefore, by $\lim_{i'\to\infty} p_{i'} \eta_{i'} \neq 0$,

$$D(a_{i'})T_{i'}\exp\left(\frac{-W}{p_{i'}}\right) - E = o\left(\frac{1}{p_{i'}}\right) = o(\eta_{i'}).$$

On the other hand, we have by (2) of [I],

$$\left| D(a_{i'})\exp\left(V_{i'} - \frac{W}{p_{i'}}\right) - E \right|$$

$$\leq \left| D(a_{i'})\exp\left(V_{i'} - \frac{W}{p_{i'}}\right) - D(a_{i'})\exp V_{i'}\exp\left(\frac{-W}{p_{i'}}\right) \right|$$

$$+ \left| D(a_{i'})\exp V_{i'}\exp\left(\frac{-W}{p_{i'}}\right) - E \right|.$$

The first term of the right hand member is

$$\leq \left| D(a_{i'}) \right| O\left(\left| V_{i'} \right| \left| \frac{W}{p_{i'}} \right| \right)$$

by (2), (12) of [I].   The second term we proved above to be $o(\eta_{i'})$.   Thus we have,

$$\left| D(a_{i'})\exp\left(V_{i'} - \frac{W}{p_{i'}}\right) - E \right| = o\left(\frac{1}{p_{i'}}\right) = o(\eta_{i'})$$

since $\lim_{i'\to\infty} | V_{i'} | = 0$.

This is a contradiction, since

$$V_{i'} - \frac{W}{p_{i'}} \epsilon \mathfrak{J}',$$

for sufficiently large i', by

$$V_{i'} \epsilon \mathfrak{J}', \quad W \epsilon \mathfrak{J} \text{ and } \lim V_{i'} = 0, \quad \lim_{i'\to\infty} \frac{W}{p_{i'}} = 0. \quad \text{Q.E.D.}$$

Lemma 6. *Any element $a \epsilon \mathfrak{G}$ can be expressed as the result of composition $a_1 a_2 \ldots a_m$, where a_i may be chosen from any vicinity of e, by a theorem of O. Schreier[15] for connected group.*

(15)   *O. Schreier :*   loc. cit.
(16)   *See J. von Neumann :*   loc. cit.

These six Lemmas with the following remark complete the proof of our theorem for the case where $\mathfrak{D}$ is continuously isomorphic to $\mathfrak{G}$.

Remark. *By the method as used in the proof of Lemma 3, we see that $\mathfrak{F}$ contains the* commutatorproduct *(Poissons parenthesis)* $UV - VU$ *of any of its elements* U, V.

We have only to put $c_n = a\left(\frac{1}{n}\right)b\left(\frac{1}{n}\right)a\left(\frac{-1}{n}\right)b\left(\frac{-1}{n}\right)$ and $m_n = \min.$

(l_n, n^2) and $t_0 = \lim\frac{m_{n'}}{(n')^2}$ in the proof of Lemma 3.

Mathematical Institute,
Osaka Imperial University.

APPENDIX (added August, 1936).

(Received August, 6, 1936)

In the above theorem the two assumptions, i) the group $\mathfrak{G}$ satisfies the first axiom of countability and ii) the group $\mathfrak{D}$ is of finite dimension, are not necessary for the proof. Thus we may improve the theorem as follows.

Theorem. *Let $\mathfrak{G}$ be a locally bicompact, connected topological group. Then the continuous repesentation $\mathfrak{D}$ of $\mathfrak{G}$ in $\mathfrak{R}$ is a Lie representation.*

Proof. As above we deal with the case of isomorphic representation.

i) By Lemma 1, $\mathfrak{G}$ is a locally bicompact topological group which does not contain an arbitrarily small cyclic subgroup ($\neq$ identical group). According to S. Kakutani and A. Komatu such a group $\mathfrak{G}$ necessarily satisfies the first axiom of countability. Their proof runs as follows.

Let $U(e)$ be a vicinity of the unit-element e such that its closure $\bar{U}(e)$ is bicompact. We assume that $\bar{U}(e)$ does not contain cyclic subgroup ($\neq$ identical group) of $\mathfrak{G}$. Then we may prove that the sequence $\{U_k(e)\}$ of the vicinity of e defined by

$$U_k^{-1}(e) = U_k(e), \quad U_k(e)^2 \leq U_{k-1}(e), \quad U_0(e) = U(e) \cdot U^{-1}(e),$$

constitutes a complete system of the vicinity of e. To this purpose we have only to prove that for any open set $V \ni e$ of $\mathfrak{G}$

$$V \geq U_k(e)$$

for sufficiently large k. If this inclusions do not hold we must have a sequence $\{a_k\}$ of elements $\epsilon\,\mathfrak{G}$ such that

$$a_k \epsilon U_k(e), \qquad a_k \bar{\epsilon}\, V.$$

106

Since $\bar{U}_0(e) \leq \bar{U}(e)$ is bicompact and $\bar{U}_k(e) \leq U_k^2(e) \leq U_0(e)$, $k = 0, 1, 2, \ldots$, the sequence $\{a_k\}$ has at least one accumulation point a. Then

$$\begin{cases} a \epsilon \bar{U}_k(e), & k = 0, 1, 2, \ldots, \\ a \bar{\epsilon} V \end{cases}$$

Hence $a \neq e$. From $a \epsilon \bar{U}_k(e)$ $(k = 0, 1, 2, \ldots)$ and $U_k(e) = U_k^{-1}(e)$, $U_k^2(e) \leq U_{k-1}(e)$ we conclude that

$$a^n \epsilon \bar{U}_0(e) \leq \bar{U}(e), \quad n = 0, \pm 1, \pm 2, \ldots.$$

contrary to the hypothesis.

ii) The hypothesis of the finite dimensionally concerning the group $\mathfrak{D}$ is only used to prove the existence of a finite number of a linearly independent (with real coefficients) base for $\mathfrak{J}$.

This existence may be proved, without this hypothesis, as follows.

Let $V_1(e)$ be the vicinity of e as defined in Lemma 1. Then the set $\mathfrak{K}$ of elements $\epsilon \mathfrak{R}$ defined by

$$lnD(a), \quad a \epsilon \bar{V}_1(e)$$

is bicompact, since it is the continuous image of the bicompact set $\bar{V}_1(e)$.

Any element $U(\neq 0) \epsilon \mathfrak{J}$ satisfies, by definition,

$$D(a(t)) = \exp(tU)$$

where $a(t)$ denotes a one-parameter continuous subgroup $(\neq$ identical group) of $\mathfrak{G}$:

$$a(t)a(s) = a(t+s), \quad \lim_{t \to t_0} a(t) = a(t_0), \quad a(o) = e.$$

Since $V_1(e)$ is open and $V_1(e) = V_1^{-1}(e)$, there must exist, by Lemma 1, a positive t_U such that

$$\begin{cases} a(t) \epsilon V_1(e) \quad \text{for} \quad -t_U < t < t_U, \\ a(t_U) \epsilon \bar{V}_1(e) - V_1(e). \end{cases}$$

For all $U \epsilon \mathfrak{J}$ which satisfies $|U| = 1$ we have (greatest lower bound of t_U) > 0, since otherwise there must exist at least one sequence $\{U(i)\}$ of elements $\epsilon \mathfrak{J}$ such that

$$|U(i)| = 1, \quad \lim_{i \to \infty} t_{U(i)} = 0,$$

and hence

$$\lim_{i \to \infty} \exp\{t_{U(i)}U(i)\} = \exp(0) = E.$$

This is a contradiction, as the element $a_i \epsilon \mathfrak{G}$ defined by

$$D(a_i) = \exp \{t_{U(i)} U(i)\}$$

satisfies

$$a_i \, \bar{\epsilon} \, V_1(e) , \quad a_i \, \epsilon \overline{V}_1(e)$$

and hence any limiting point of the sequence $\{a_i\}$ is not e.

Thus, remembering (4), (6) of [I], there exists a positive t_0 for which the set $\mathfrak{J}'$ of elements $U \epsilon \mathfrak{J}$ defined by

$$|U| \leq t_0$$

all belong to $\mathfrak{R}$.

Hence the linear metrical (real) space $\mathfrak{J}$ is locally compact and thus it must be of finite dimension by a theorem due to F. Riesz (Cf. S. Banach: Théorie des Operations Linéaries, p. 84).

Remark added during the proof correction. Meanwhile I wrote three papers which are closely related to the present paper;

K. Yosida: A note on the continuous representations of the topological group. Proc. of the Imp. Acad. 12 (1936),

K. Yosida: A problem concerning the second fundamental theorem of Lie. Proc. of the Imp. Acad. 13 (1937).

K. Yosida: A theorem concerning the semi-simple Lie groups. To appear soon in the Tôhoku Math. Journal.

A note on the continuous representation of topological groups

Proc. Imp. Acad. Tokyo **12** (1936) 329 331

(Comm. by T. Yosie, m.i.a., Dec. 12, 1936.)

§ I. In a recent paper,[1] the author treated the group $\mathfrak{D}$ embedded in the *metrical complete ring* $\mathfrak{R}$. Such a group $\mathfrak{D}$ is, by [I], a *Lie group* (as defined in [I][2]) if and only if $\mathfrak{D}$ is locally compact.

In another paper,[3] I obtained a result saying that, if $\mathfrak{D}$ is continuously homomorphic to a connected and locally bicompact topological group $\mathfrak{G}$ (*without any countability axiom*), then $\mathfrak{D}$ is a *Lie representation* (defined in [II]).

A Lie representation $\mathfrak{D}$ is not necessarily a Lie group as defined in [I] (see [II]), though the *infinitesimal operators* of $\mathfrak{D}$ obey the customary rule of the ordinary *Lie-ring*, when $\mathfrak{G}$ is a Lie group (see [II]).

However we may add the following remark :

The representation $\mathfrak{D}$ of $\mathfrak{G}$ is a Lie group (as defined in [I]) if and only if the homomorphic mapping $\mathfrak{G} \rightarrow \mathfrak{D}$ is open.

Here a continuous mapping is called *open* if the mapped image of any open set is an open set.

Proof. $\mathfrak{D}$ is isomorphic to the quotient group $\mathfrak{G}/\mathfrak{N}$, where $\mathfrak{N}$ is an invariant subgroup closed in $\mathfrak{G}$. We call any set in $\mathfrak{G}/\mathfrak{N}$ open if and only if it corresponds to an open set in $\mathfrak{G}$ by the homomorphic mapping $\mathfrak{G} \rightarrow \mathfrak{G}/\mathfrak{N}$. Then $\mathfrak{G}/\mathfrak{N}$ is connected and locally bicompact with $\mathfrak{G}$. Thus $\mathfrak{D}$ is continuously isomorphic to $\mathfrak{G}/\mathfrak{N}$ and the mapping $\mathfrak{G} \rightarrow \mathfrak{D}$ is open if and only if the mapping $\mathfrak{G}/\mathfrak{N} \rightarrow \mathfrak{D}$ is open.

Hence we may- and shall- assume that $\mathfrak{D}$ is continuously isomorphic to $\mathfrak{G}$.

Thus if the mapping $\mathfrak{G} \rightarrow \mathfrak{D}$ is open, $\mathfrak{G}$ and $\mathfrak{D}$ are homeomorphic with each other, and hence $\mathfrak{D}$ is locally bicompact and connected with $\mathfrak{G}$. This proves the sufficiency of the condition of the remark.

As the group $\mathfrak{D}$ is embedded in $\mathfrak{R}$, $\mathfrak{D}$ does not contain an arbitrarily small cyclic subgroup ($\neq$ identical group[4]). $\mathfrak{G}$ enjoys the same property, for $\mathfrak{D}$ is continuousy isomorphic to $\mathfrak{G}$. By a theorem due to A. Komatu and S. Kakutani (see [II]) $\mathfrak{G}$ satisfies the first axiom of countability, since $\mathfrak{G}$ is locally bicompact. Hence $\mathfrak{G}$ is metrisable by a result of S. Kakutani.[5]

---

1) K. Yosida: On the group embedded in the metrical complete ring, Jap. J. of Math. **13** (1936). This paper will be cited as [I].

2) The condition γ) in the definition of a *Lie group* in [I] is not an essential one, it states that the group is connected. Hence it may be omitted out of the definition.

3) K. Yosida: On the group embedded in the metrical complete ring, II, to appear soon in Jap. J. of Math. This paper will be cited as [II].

4) For the proof see [I].

5) S. Kakutani: Über die Metrisation der topologischen Gruppen, Proc. **12** (1936). Cf. also G. Birkhoff: A note on topological groups, Comp. Math. **3** (1936).

Thus $\mathfrak{G}$ is locally separable and connected. We see that such a group is separable, by applying Schreier's theorem[1] on connected groups.

Hence by H. Freudenthal's[2] result, the mapping $\mathfrak{G} \to \mathfrak{D}$ is open if $\mathfrak{D}$ is locally compact, for then $\mathfrak{D}$ is a Lie group by [I].

§ II. With regards to the dimension relation by the continuous homomorphic mapping $\mathfrak{G} \to \mathfrak{D}$ we may prove the following remark:

If $\mathfrak{G}$ is compact and separable,[3] we have

$$(d) \qquad \dim \mathfrak{D} \leqq \dim \mathfrak{G},$$

without assuming the group $\mathfrak{D}$ to be embedded in $\mathfrak{R}$. It may be any topological group.

Proof. $\mathfrak{D}$ is compact and separable with $\mathfrak{G}$, and hence the mapping $\mathfrak{G} \to \mathfrak{D}$ is open.[4] Thus $\mathfrak{D}$ is topologically isomorphic to $\mathfrak{G}/\mathfrak{R}$, where $\mathfrak{R}$ denotes an invariant subgroup closed in $\mathfrak{G}$.[5]

By a theorem of H. Freudenthal,[6] there exists a decreasing sequence $\{\mathfrak{H}_m\}$ of closed invariant subgroups in $\mathfrak{G}$, $\lim_{m \to \infty} \mathfrak{H}_m = e$ (unit element of $\mathfrak{G}$), such that $\mathfrak{G}/\mathfrak{H}_m$ $(m = 1, 2, \ldots)$ is topologically isomorphic to a compact matrix group (Lie group by [I]). $\mathfrak{G}$ is thus G_n-*adic generated*[7] by the sequence $\{\mathfrak{G}/\mathfrak{H}_m\}$, and $\dim \mathfrak{G} = \lim_{m \to \infty} \dim \mathfrak{G}/\mathfrak{H}_m$.[8]

As $\mathfrak{H}_m \supseteqq \mathfrak{H}_{m+1}$, $\lim \mathfrak{H}_m = e$, it is easy to see that the group $\mathfrak{G}/\mathfrak{R}$ and $\mathfrak{R}$ are G_n-adic generated by the sequences

$$\left\{ \mathfrak{G}/\mathfrak{R} \Big/ \mathfrak{R}\mathfrak{H}_m/\mathfrak{R} \right\} \qquad \text{and} \qquad \left\{ \mathfrak{R}/\mathfrak{R}V\mathfrak{H}_m \right\}^{9)}$$

respectively. $\mathfrak{R}\mathfrak{H}_m/\mathfrak{H}_m$ is a compact matrix group (Lie group by [I]), since it is a closed invariant subgroup in the compact matrix group $\mathfrak{G}/\mathfrak{H}_m$. Thus the topological isomorphisms

$$\begin{cases} \mathfrak{G}/\mathfrak{R}\mathfrak{H}_m \cong \mathfrak{G}/\mathfrak{H}_m \Big/ \mathfrak{R}\mathfrak{H}_m/\mathfrak{H}_m \cong \mathfrak{G}/\mathfrak{R} \Big/ \mathfrak{R}\mathfrak{H}_m/\mathfrak{R}, \\[2ex] \mathfrak{R}\mathfrak{H}_m/\mathfrak{H}_m \cong \mathfrak{R}/\mathfrak{R}V\mathfrak{H}_m{}^{10)} \end{cases}$$

---

1) O. Schreier: Abstrakte kontinuierliche Gruppen, Hamburg Abh. Math. Sem. **4** (1925).

2) H. Freudenthal: Einige Sätze über topologische Gruppen, Ann. of Math. **37** (1936), p. 47. This paper will be cited as FI.

3) The separability hypothesis may be replaced by the first axiom of countability, for then $\mathfrak{G}$ is metrisable by Kakutani's theorem, loc. cit. When $\mathfrak{G}$ is locally compact, connected, separable and zero-dimensional, (d) is obtained by H. Freudenthal, loc. cit. p. 51.

4) FI, p. 47.

5) FI, p, 49. The topology in quotient group is defined as in § I.

6) H. Freudenthal: Topologische Gruppen mit genügend vielen fastperiodischen Funktionen, Ann. of Math. **37** (1936). This paper will be cited as FII. Cf. also E. R. van Kampen: Almost periodic functions and compact groups, Ann. of Math. **37** (1936).

7) FII, p. 69.

8) FII, p. 71.

9) $\mathfrak{R}\mathfrak{H}$ denotes the set of all the products nh, where $n \in \mathfrak{R}$, $h \in \mathfrak{H}$. $\mathfrak{R}V\mathfrak{H}$ denotes the set of all the elements common to $\mathfrak{R}$ and $\mathfrak{H}$.

10) FI, p. 50.

show that $\mathfrak{N}/\mathfrak{N}V\mathfrak{H}_m$ and $\mathfrak{G}/\mathfrak{N}\mathfrak{H}_m$ are compact Lie groups, and hence we have[1]

$$\dim \mathfrak{G}/\mathfrak{N} = \lim_{m \to \infty} \dim \mathfrak{G}/\mathfrak{N}\mathfrak{H}_m , \qquad \dim \mathfrak{N} = \lim_{m \to \infty} \mathfrak{N}/\mathfrak{N}V\mathfrak{H}_m .$$

By considering the *canonical parameters* we obtain

$$\dim \mathfrak{G}/\mathfrak{N}\mathfrak{H}_m = \dim \mathfrak{G}/\mathfrak{H}_m \Big/ \mathfrak{N}\mathfrak{H}_m/\mathfrak{H}_m = \dim \mathfrak{G}/\mathfrak{H}_m - \dim \mathfrak{N}\mathfrak{H}_m/\mathfrak{H}_m ,$$

and thus[2]

$$(d') \quad \dim \mathfrak{G}/\mathfrak{N} = \lim_{m \to \infty} \dim \mathfrak{G}/\mathfrak{H}_m - \lim_{m \to \infty} \dim \mathfrak{N}/\mathfrak{N}V\mathfrak{H}_m = \dim \mathfrak{G} - \dim \mathfrak{N} .$$

This proves (d) by $\mathfrak{D} \cong \mathfrak{G}/\mathfrak{N}$.

Q. E. D.

---

1) FII, p. 71.

2) After this paper is completed the author found that (d') was also obtained by van Kampen in somewhat another way. E. R. van Kampen: A note on a theorem by Pontrjagin, Amer. J. of Math. 51, **1** (1936) p. 178.

A theorem concerning the semi-simple Lie groups

Tohoku Math. J. **43** (1937) 81–84

1. The theorem and its applications.

Let $\mathfrak{R}$ denote the set of all the matrices of a fixed degree, say n, with complex numbers as coefficients. We define the topology in $\mathfrak{R}$ by the *absolute value* $|A|=\sqrt[2]{\sum_{i,j=1}^{n}|a_{ij}|^{2}}$, $A=\|a_{ij}\|$. A real linear subspace $\overline{\mathfrak{J}}\subseteq\mathfrak{R}$ is called a *Lie-ring* if it satisfies the condition: $[X,Y]=XY-YX\in\overline{\mathfrak{J}}$ for $X,Y\in\overline{\mathfrak{J}}$. The two ring-operations are the matrix-addition and the *commutator-multiplication* $[X,Y]$. $\overline{\mathfrak{J}}$ surely has a linearly independent base with real coefficients. Let it be X_1, $\cdots$, X_m. We write $\overline{\mathfrak{J}}=(X_1,\cdots,X_m)$.

Then the set $\overline{\mathfrak{G}}$ of all the matrices of the form:

$$\exp\!\left(\sum_{i=1}^{m}t_iX_i\right)^{(1)}, \quad t \text{ real and } \sum_{i=1}^{m}|t_i|<\varepsilon, \quad \varepsilon>0,$$

constitutes a Lie *group-germ*(^{2}), that is, if $X,Y\in\overline{\mathfrak{G}}$ are sufficiently near the unit-matrix E, X^{-1} and XY also $\in\overline{\mathfrak{G}}$. Next we denote by $\mathfrak{G}$ the set of all the products of a finite number of elements $\in\overline{\mathfrak{G}}$ and of the limit matrices of such products, so long as they are non-singular. Then $\mathfrak{G}$ is a locally compact topological group, and hence it is a Lie group(^{3}). The Lie-ring $\mathfrak{J}$ of this Lie group $\mathfrak{G}$ is the set of all the *differential quotients* of $\mathfrak{G}$ at E(^{4}). Such differential quotient is defined by the limit of the form $\lim_{i\to\infty}((A_i-E)/\varepsilon_i)$, where $A_i(\neq E)\in\mathfrak{G}$ and real $\varepsilon_i(\neq0)$ are such that $\lim_{i\to\infty}A_i=E$, $\lim_{i\to\infty}\varepsilon_i=0$. Thus $\mathfrak{J}\supseteq\overline{\mathfrak{J}}$.

In a previous paper(^{5}) the author showed that $\mathfrak{J}$ does not necessarily coincide with $\overline{\mathfrak{J}}$, and proved the theorem:

(^{1}) $\exp(X)=\sum_{n=0}^{\infty}(X^n/n!)$.

(^{2}) The second fundamental theorem of Lie.

(^{3}) K. Yosida: Jap. J. of Math. **13** (1936), 7. Cf. also J. von Neumann: Math. Zeitsch. **30** (1929), 3.

(^{4}) K. Yosida: loc. cit.

(^{5}) K. Yosida: Proc. of the Imp. Acad. **5** (1927), 152.

$\overline{\mathfrak{G}}$ *is a vicinity of the identity of the topological group* $\mathfrak{G}$ *(that is* $\mathfrak{J}=\overline{\mathfrak{J}}$*), if the Lie-ring* $\overline{\mathfrak{J}}$ *is irreducible (that is if the matrix group* $\mathfrak{G}$ *is irreducible).*

In the present note I intend to prove the following

Theorem. $\overline{\mathfrak{G}}$ *is a vicinity of the identity of the topological group* $\mathfrak{G}$ *(that is* $\mathfrak{J}=\overline{\mathfrak{J}}$*), if the Lie-ring* $\overline{\mathfrak{J}}$ *is semi-simple.*

As an application of the theorem we obtain the result:

Any continuous matrix-representation $\mathfrak{G}_2$ *of a connected Lie group* $\mathfrak{G}_1$ *is an open (gebietstreu) representation, if either* $\mathfrak{G}_1$ *is semi-simple or if* $\mathfrak{G}_2$ *is irreducible.* For a continuous representation of a connected Lie group may be obtained by the customary method of the *infinitesimal representation*([1]).

2. The proof of the theorem.

We require two lemmas.

Lemma 1. $\overline{\mathfrak{G}}$ is a Lie invariant subgroup-germ of $\mathfrak{G}$, that is, $\overline{\mathfrak{J}}$ is an ideal of $\mathfrak{J}$: $[X, Y] \in \overline{\mathfrak{J}}$ if $X \in \overline{\mathfrak{J}}$, $Y \in \mathfrak{J}$.

Proof. See Lemma 1 of the paper cited in the footnote (5) of p. 81.

Lemma 2. Let a matrix $Y \in \mathfrak{J}$ be commutative with every matrix $\in \overline{\mathfrak{J}}$: $[Y, X] = 0$ for $X \in \overline{\mathfrak{J}}$. Then $Y = 0$, if $\overline{\mathfrak{J}}$ is semi-simple.

Proof. Being semi-simple, the Lie-ring $\overline{\mathfrak{J}}$ is completely reducible([2]). Thus every matrix $X \in \overline{\mathfrak{J}}$ may be considered to be of the form

$$\left\|\begin{array}{cccc} X^{(1)} & 0 & \cdots & 0 \\ 0 & X^{(2)} & \cdots & 0 \\ \vdots & & \ddots & \vdots \\ 0 & & \cdots & X^{(s)} \end{array}\right\|,$$

where $X^{(t)}(1 \leq t \leq s)$ constitutes an irreducible Lie-ring $\overline{\mathfrak{J}}^{(t)}(1 \leq t \leq s)$ homomorphic to $\overline{\mathfrak{J}}$. $\overline{\mathfrak{J}}^{(t)}$ is semi-simple with $\overline{\mathfrak{J}}$ and hence([3]) we have

---

([1]) See, for example, K. Yosida: Jap. J. of Math. **13**(1927), 459. By virtue of this paper we see that the above result is also true, when $\mathfrak{G}_1$ is a locally bicompact connected topological group (not necessarily a Lie group) and the matrix representation $\mathfrak{G}_2$ is irreducible.

([2]) H. Weyl: Math. Zeitsch. **24**(1925), 381. The complete reducibility also holds for the semi-simple Lie-ring of matrices with real parameters.

([3]) A semi-simple Lie-ring $\overline{\mathfrak{J}}$ coincides with its commutator-ring, that is any element $\in \overline{\mathfrak{J}}$ may be obtained as the linear combination of the commutator-products $[X, Y]$, where X, Y both $\in \overline{\mathfrak{J}}$.

$$\text{trace} (X^{(t)}) = 0, \quad \text{for any} \quad X^{(t)} \in \overline{\mathfrak{J}}^{(t)}.$$

Thus by the formula

$$\det (\exp (A)) = \exp (\text{trace} (A)),$$

every matrix $Y \in \mathfrak{J}$ must be of the form

$$\begin{Vmatrix} Y^{(1)} & 0 & \cdots & 0 \\ 0 & Y^{(2)} & \cdots & 0 \\ \vdots & & \ddots & \vdots \\ 0 & & \cdots & Y^{(s)} \end{Vmatrix}, \quad \text{trace} (Y^{(t)}) = 0,$$

where $Y^{(t)}$ is of the same degree as that of $X^{(t)}$. By the assumption $Y^{(t)}$ is commutative with the irreducible Lie group-germ $\overline{\mathfrak{G}}^{(t)}$ whose Lie-ring is $\overline{\mathfrak{J}}^{(t)}$. Hence we have $Y = 0$, by applying Schur's lemma.

Proof of the theorem. By Lemma 1, $\overline{\mathfrak{J}}$ is an ideal of $\mathfrak{J}$. We assume $\mathfrak{J} \neq \overline{\mathfrak{J}}$, for otherwise the proof is unnecessary. Thus let $(X_1, \cdots, X_m)$ and $(X_1, \cdots, X_m, Y_1, \cdots, Y_k)$ be the bases of $\overline{\mathfrak{J}}$ and $\mathfrak{J}$ respectively.

We denote by Y any one of $Y_1, \cdots, Y_k$. As $\overline{\mathfrak{J}}$ is an ideal of $\mathfrak{J}$, $\mathfrak{J}_1 = (X_1, \cdots, X_m, Y)$ is a Lie-ring and $\overline{\mathfrak{J}}$ is an ideal of $\mathfrak{J}_1$. $\mathfrak{J}_1$ is not semi-simple. For if not the ideal $\overline{\mathfrak{J}}$ is a direct summand[1] of $\mathfrak{J}_1$ and hence there exists an element $Y' \equiv Y$ (mod. $\overline{\mathfrak{J}}$) commutative with every matrix $\in \overline{\mathfrak{J}}$, proving $Y \in \overline{\mathfrak{J}}$ by Lemma 2, contrary to the hypothesis.

Hence the *radical* (the maximal *nilpotent* ideal) of $\mathfrak{J}_1$ is $\neq 0$. Let it be denoted by $\mathfrak{R}_1$. As $\overline{\mathfrak{J}}$ is semi-simple, $\mathfrak{R}_1$ is not contained in $\overline{\mathfrak{J}}$. Thus we may assume $\mathfrak{R}_1 = (X_{l+1}, \cdots, X_m, Y')$, where $Y' \equiv Y$ (mod. $\overline{\mathfrak{J}}$) and $\mathfrak{J}_1 = (X_1, \cdots, X_m, Y) = (X_1, \cdots, X_m, Y')$, $\overline{\mathfrak{J}} = (X_1, X_2, \cdots, X_m)$.

There are two cases.

Case 1. $l + 1 \leq m$. As $\overline{\mathfrak{J}}$ and $\mathfrak{R}_1$ are ideals of $\mathfrak{J}_1$, $\overline{\mathfrak{J}}_2 = (X_{l+1}, \cdots, X_m)$ is an ideal of $\overline{\mathfrak{J}}$. $\overline{\mathfrak{J}}_2$ is nilpotent, for it is a subring of the nilpotent Lie-ring $\mathfrak{R}_1$. This is a contradiction, for $\overline{\mathfrak{J}}$ is semi-simple.

Case 2. $l + 1 > m$ or $\mathfrak{R}_1 = (Y')$. As $\mathfrak{R}_1$ and $\overline{\mathfrak{J}}$ are ideals of $\mathfrak{J}_1$ we

---

[1] É. Cartan: Thèses (1894), p. 53. Cartan's theorem is stated for the semi-simple Lie-ring with complex parameters. It may be proved that a semi-simple Lie-ring with real parameters is a direct sum of simple and semi-simple ideals with real parameters.

must have

$$[X, Y'] \in \text{the intersection } (\mathfrak{N}_1 \cdot \overline{\mathfrak{J}}) \text{ for every } X \in \overline{\overline{\mathfrak{J}}}.$$

Hence $[X, Y'] = 0$ for every $X \in \overline{\mathfrak{J}}$. Thus $Y' = 0$ by Lemma 2. This proves $Y \in \overline{\mathfrak{J}}$, contrary to the hypothesis.

Hence, in any case, we must have $\overline{\mathfrak{J}} = \mathfrak{J}$.

Mathematical Institute,
Osaka Imperial University.

(Received August 24, 1937.)

A problem concerning the second fundamental theorem of Lie

Proc. Imp. Acad. Tokyo **13** (1937) 152–155

(Comm. by T. Yosie, M.I.A., May 12, 1937.)

§ 1. The problem and the theorem.

Let $\mathfrak{R}$ denote the set of all the matrices of a fixed degree, say n, with complex numbers as coefficients. We introduce a topology in $\mathfrak{R}$ by the *absolute value*

$$| A | = \sqrt{\sum_{i, j=1}^{n} | a_{ij} |^2}, \qquad A = \| a_{ij} \| .$$

If $\mathfrak{G}$, a subset of non-singular matrices $\in \mathfrak{R}$, is a group with respect to the matrix-multiplication, it is a topological group by the distance $| A - B |$.

The topological group $\mathfrak{G}$ is called a *Lie group*, if there exist a finite number, say m, of elements $X_1, X_2, \ldots\ldots, X_m \in \mathfrak{R}$ which satisfy the conditions:

1). $X_1, X_2, \ldots\ldots, X_m$ are linearly independent with real coefficients.

2). $\exp \left(\sum_{i=1}^{m} t_i X_i\right) \in \mathfrak{G}$, t real.[1]

3). There exists a positive ε such that any element $A \in \mathfrak{G}$ may be represented uniquely in the form

$$A = \exp \left(\sum_{i=1}^{m} t_i X_i\right), \; t \text{ real},$$

if $| A - E | \leqq \varepsilon$ (E the unit-matrix of $\mathfrak{R}$).

By a theorem of J. von Neumann[2] $\mathfrak{G}$ is a Lie group if and only if it is locally compact. Here, for convention, a discrete group is also called a Lie group. If $\mathfrak{G}$ is a Lie group, the set $\mathfrak{J}$ of all the elements $\sum_{i=1}^{m} t_i X_i$, t real, satisfies:

(α). $\mathfrak{J}$ is a real linear space which has a finite base with real coefficients, viz, $X_1, X_2, \ldots\ldots, X_m$.

(β). $[X, Y] = XY - YX \in \mathfrak{J}$ with $X, Y \in \mathfrak{J}$.

$\mathfrak{J}$ is called the *Lie ring* of the Lie group $\mathfrak{G}$, the two ring-operations being the vector-addition and the *commutator-multiplication* $[X, Y]$. It is the set of all the *differential quotients* of $\mathfrak{G}$ at E.[3] The differential quotient of $\mathfrak{G}$ at E is defined by $\lim_{i \to \infty} \left((A_i - E)/\varepsilon_i\right)$, where $A_i (\neq E) \in \mathfrak{G}$ and real $\varepsilon_i (\neq 0)$ are such that $\lim_{i \to \infty} A_i = E$, $\lim_{i \to \infty} \varepsilon_i = 0$.

---

1) $\exp (X) = \sum_{n=0}^{\infty} (X^n / n !)$.

2) See K. Yosida: Jap. J. of Math. **13** (1936), p. 7. Neumann's original statement (M. Z. **30** (1929), p. 3) reads as follows:

$\mathfrak{G}$ is a Lie group if $\mathfrak{G}$ is closed in the group of all the non-singular matrices $\in \mathfrak{R}$.

3) Cf. K. Yosida: loc. cit.

Conversely let $\mathfrak{F}$ denote a subset of $\mathfrak{R}$ which satisfies (α) and (β). Then, by the second fundamental theorem of Lie, the set $\overline{\mathfrak{G}}$ of all the elements of the form

$$\exp\left(\sum_{i=1}^{m} t_i X_i\right),\ t\ \text{real and}\ \sum_{i=1}^{m}|t_i| < \epsilon,\qquad \epsilon > 0,$$

constitutes a *Lie group-germ*. That is, if $X, Y \in \overline{\mathfrak{G}}$ are sufficiently near E, X^{-1} and YX also $\in \overline{\mathfrak{G}}$. $\mathfrak{F}$ is called the Lie ring of the Lie group-germ $\overline{\mathfrak{G}}$.

Then the set $\widetilde{\mathfrak{G}}$ of all the products of a finite number of elements $\in \overline{\mathfrak{G}}$ and of the limit matrices of such products, so long as they are non-singular, forms a locally compact group. Hence $\widetilde{\mathfrak{G}}$ is a Lie group. Let $\widetilde{\mathfrak{F}}$ be the Lie ring of this Lie group $\widetilde{\mathfrak{G}}$, then $\widetilde{\mathfrak{F}} \supseteq \mathfrak{F}$. However, $\widetilde{\mathfrak{F}}$ does not necessarily coincide with $\mathfrak{F}$, as the following example shows us :

$$\text{the base of}\ \overline{\mathfrak{F}} = \begin{matrix} \sqrt{-1} & 0 \\ 0 & \tau\sqrt{-1} \end{matrix},\ \ \tau/2\pi\ \text{irrational}.$$

Hence the Lie group-germ $\overline{\mathfrak{G}}$ is not necessarily a vicinity of the identity of the topological group $\widetilde{\mathfrak{G}}$.

Thus it may be of some interest to obtain the conditions by which $\widetilde{\mathfrak{F}}$ coincides with $\mathfrak{F}$. As an answer to this problem, I intend to prove the following

Theorem. The Lie group-germ $\overline{\mathfrak{G}}$ is a vicinity of the identity of the Lie group $\widetilde{\mathfrak{G}}$, if the ring $\mathfrak{F}$ is irreducible.

Here $\mathfrak{F}$ is called *irreducible* if the group $\widetilde{\mathfrak{G}}$ is irreducible, that is, if all the matrices of $\mathfrak{F}$ are not simultaneously similar to the matrices of the form

$$\begin{matrix} A & 0 \\ * & B \end{matrix}\ .$$

§ 2. The proof of the theorem.

Lemma 1. $\overline{\mathfrak{G}}$ is a Lie invariant subgroup-germ of $\widetilde{\mathfrak{G}}$, viz. $BAB^{-1} \in \overline{\mathfrak{G}}$ for any $B \in \widetilde{\mathfrak{G}}$ if $A \in \overline{\mathfrak{G}}$ is sufficiently near E.

Proof. Let $A = \exp(X)$, $X \in \mathfrak{F}$. Then $BAB^{-1} = \exp(BXB^{-1})$ and BXB^{-1} tends to 0 as X tends to 0. Thus it is sufficient to prove

$$(*)\qquad BXB^{-1} \in \mathfrak{F}\ \text{with}\ X \in \mathfrak{F},\quad \text{if}\ B \in \widetilde{\mathfrak{G}}.$$

$(*)$ is evident in the special case $B \in \overline{\mathfrak{G}}$, for then the transformation $X \to BXB^{-1}$ is induced by the so-called linear adjoint Lie group-germ of $\overline{\mathfrak{G}}$. The general case $B \in \widetilde{\mathfrak{G}}$ may be obtained from this special case, by limiting process.

Lemma 2 (due to E. Cartan[1]). The vicinity of the identity of the irreducible Lie group $\widetilde{\mathfrak{G}}$ is a direct product of a *semi-simple* Lie

---

1) E. Cartan : Ann. Ec. Norm. Sup. (3) **26** (1909), p. 148. For the proof see H. Freudenthal : Ann. of Math. 37, **1** (1936), p. 63. In the course of the proof of our theorem, $\overline{\mathfrak{G}}_i$ ($i=1, 2$) will be proved to be not only Lie group-germ but also a vicinity of the identity of the Lie group.

group-germ $\bar{\mathfrak{G}}_1$ and an abelian Lie group-germ $\bar{\mathfrak{G}}_2$, where det. $(A)=1$ for any $A \in \bar{\mathfrak{G}}_1$ and the matrices of $\bar{\mathfrak{G}}_2$ are all of the form aE, a denoting complex numbers.

As a special case of this Lemma we have

Lemma 2′. $\mathfrak{G}$ is a semi-simple Lie group if $\bar{\mathfrak{J}}$ is irreducible and

$$(**) \qquad \text{trace } (X)=0 \qquad \text{for } X \in \bar{\mathfrak{J}}.$$

Proof. For then the matrices of $\mathfrak{G}$ and hence of $\tilde{\mathfrak{G}}$ are all of determinant 1.[1]

The above condition $(**)$ is surely satisfied if the Lie ring $\bar{\mathfrak{J}}$ is semi-simple. For a semi-simple Lie ring $\bar{\mathfrak{J}}$ coincides with its *commutator-ring*,[2] that is, any element of $\bar{\mathfrak{J}}$ may be obtained as the commutator-product $[X, Y]$, where X and $Y \in \bar{\mathfrak{J}}$.

Proof of the theorem. By Lemma 1 the sub-ring $\mathfrak{J}$ is an *ideal* in $\tilde{\mathfrak{J}}$, viz. $[X, Y] \in \mathfrak{J}$ for $X \in \mathfrak{J}$, $Y \in \tilde{\mathfrak{J}}$. We will prove that this ideal $\mathfrak{J}$ is a direct summand of the Lie ring $\tilde{\mathfrak{J}}$.

By Lemma 2 the Lie ring $\bar{\mathfrak{J}}$ is a direct sum of the semi-simple Lie ring $\bar{\mathfrak{J}}_1$ of the Lie group-germ $\bar{\mathfrak{G}}_1$ and the abelian Lie ring $\bar{\mathfrak{J}}_2$ of the Lie group-germ $\bar{\mathfrak{G}}_2$. Thus $\bar{\mathfrak{J}}_1$ is commutative with $\bar{\mathfrak{J}}_2 : [X, Y]=0$ for $X \in \bar{\mathfrak{J}}_1$, $Y \in \bar{\mathfrak{J}}_2$.

The semi-simple Lie ring $\bar{\mathfrak{J}}_1$ is a direct sum of simple and semi-simple ideals, by a theorem of E. Cartan.[3] Hence any ideal of $\bar{\mathfrak{J}}_1$ is semi-simple. As $\bar{\mathfrak{G}}_2$ consists of the matrices of the form aE, the base of the abelian Lie ring $\bar{\mathfrak{J}}_2$ is either

i). aE, where a denotes a real or complex number ($a=0$ if $\bar{\mathfrak{J}}_2=0$), or

ii). E and $\sqrt{-1}E$.

Thus, in any case, $\tilde{\mathfrak{J}}$ is a direct sum of simple ideals. Hence the ideal $\mathfrak{J}$ is a direct summand of $\tilde{\mathfrak{J}}$. We next prove that $\mathfrak{J} \supseteq \bar{\mathfrak{J}}_1$.

Let $\tilde{\mathfrak{J}}=\mathfrak{J}+\mathfrak{J}'$ be a direct decomposition of $\tilde{\mathfrak{J}}$. Then, as $\mathfrak{J}$ and $\mathfrak{J}'$ are ideals in $\tilde{\mathfrak{J}}$, $\mathfrak{J}$ is commutative with $\mathfrak{J}'$:

$$(***) \qquad [X, Y]=0 \qquad \text{for } X \in \mathfrak{J}, \qquad Y \in \mathfrak{J}'.$$

Hence, if $\mathfrak{J}$ does not contain $\bar{\mathfrak{J}}_1$, there must exist a semi-simple ideal $\mathfrak{J}'_1 \subseteq \bar{\mathfrak{J}}_1$, commutative with $\mathfrak{J}$ by $(***)$. Thus the matrices $\in \tilde{\mathfrak{G}}$ of the form $\exp(X)$, $X \in \mathfrak{J}'_1$, are permutable with every matrix of the irreducible group-germ $\mathfrak{G}$. Hence, by Schur's Lemma, $\exp(X)$ $(X \in \mathfrak{J}'_1)$ and consequently every matrix $\in \mathfrak{J}'_1$ must be of the form aE. $\mathfrak{J}'_1$ is thus an abelian Lie ring and hence is not semi-simple. This is a contradiction, and so we must have $\mathfrak{J} \supseteq \bar{\mathfrak{J}}_1$.

The same reasoning shows that, if $\bar{\mathfrak{J}}$ is irreducible and semi-simple, we must have $\mathfrak{J}=\bar{\mathfrak{J}}$. For, then $\mathfrak{J}$ is semi-simple by Lemma 2′. Hence, in the above Lemma 2, $\bar{\mathfrak{G}}_1$ *and* $\bar{\mathfrak{G}}_2$ *are not only Lie group-germ but also the vicinities of the identities of Lie groups.*

Next we will prove that $\mathfrak{J} \supseteq \bar{\mathfrak{J}}_2$. There are two cases.

---

1) det. $(\exp(X))=\exp(\text{trace }(X))$.

2) See, for example, H. Freudenthal: loc. cit.

3) E. Cartan: Thèses (1894), p. 53.

Case 1.   Base of $\bar{\bar{\mathfrak{J}}}_2 = aE\,(a=0$ if $\mathfrak{J}_2=0)$.

Assume that $\bar{\bar{\mathfrak{J}}}_2 \neq 0$ and $\bar{\mathfrak{J}}=\bar{\mathfrak{J}}_1$. Then the group-germ $\bar{\mathfrak{G}}$ is a vicinity of the identity of the Lie group $\bar{\mathfrak{G}}_1$ whose Lie ring are $\bar{\mathfrak{J}}_1 = \bar{\mathfrak{J}}$. Thus $\bar{\bar{\mathfrak{J}}}_2 = 0$, contrary to the hypothesis. This proves $\bar{\mathfrak{J}} \supseteqq \bar{\bar{\mathfrak{J}}}_2$.

Case 2.   Base of $\bar{\bar{\mathfrak{J}}}_2 = E$ and $\sqrt{-1}E$.

If both E and $\sqrt{-1}E$ do not belong to $\bar{\mathfrak{J}}$, we obtain $\bar{\bar{\mathfrak{J}}}_2 = 0$ as above, contrary to the hypothesis $\bar{\bar{\mathfrak{J}}}_2 \neq 0$. Next let either one of E and $\sqrt{-1}E$, E for example, belong to $\bar{\mathfrak{J}}$. Then, as E is permutable with every matrix, any matrix $\epsilon\,\bar{\mathfrak{G}}$ must be of the form

$$A_1 A_2 \ldots\ldots A_k Y, \quad \text{where}\begin{cases} A_i \,\epsilon\, \text{the intersection } (\bar{\mathfrak{G}}\cdot\bar{\mathfrak{G}}_1)\,, \\ Y = \exp\,(tE),\ t\ \text{real}, \end{cases}$$

or the limit matrix of such matrices. Thus, by Lemma 2, $\det.\,(X) = \exp\,(t)$, t real, for $X \,\epsilon\, \bar{\mathfrak{G}}$, and hence X is not of the form $\exp\,(s\cdot\sqrt{-1}E)$, s real. Then $\sqrt{-1}E$ does not belong to the Lie ring $\bar{\mathfrak{J}}$. This is a contradiction, and so we must have $\bar{\mathfrak{J}} \supseteqq \bar{\bar{\mathfrak{J}}}_2$.

Thus, in any case, $\bar{\mathfrak{J}} = \tilde{\mathfrak{J}}$.

Q. E. D.

———

A remark on a theorem of B. L. van der Waerden

Tohoku Math. J. **43** (1937) 411–413

A L ie group is called L-simple (simple in the sense of L ie) if it does not contain L ie invariant subgroups other than itself and eventually the identity group. An abstract group is called g-simple if it is simple purely group-theoretically. An L-simple group $\mathfrak{G}$ with discrete center $\mathfrak{Z}$ is locally topologically isomorphic to a g-simple L ie group, viz. $\mathfrak{G}/\mathfrak{Z}$.

B. L. v an d er W aerd en([1]) proved that for any connected, compact g-simple L ie group the topology may be defined purely algebraically.

In the present note the author's aim is to give a converse to this theorem. The result reads as follows.

The Komplettierung (in the sense of D. v an Dan tzig([2])) of any g-simple group topologised in the van d er Waerd en's way is locally topologically isomorphic to a connected, compact g-simple L ie group.

It is to be remarked that the group in question thus becomes finite-dimensional necessarily.

1. The hypothesis.

L et $\mathfrak{G}$ be a g-simple abstract group. We denote by $\mathfrak{M}(a)$ the set of all the elements $\epsilon\ \mathfrak{G}$ of the form

$$\prod_{i=1}^{n} [c_i (b_i ab_i^{-1} a^{-1}) c_i^{-1}] \quad \text{(group composition)},$$

where $a, b_i, c_i \in \mathfrak{G}$ and n denotes any fixed integer $\geqq 1$. $\mathfrak{M}(a) \ni e$ (the unit-element of $\mathfrak{G}$) for any a, and $\mathfrak{M}(a) \neq e$ if $a \neq e$, as $\mathfrak{G}$ is simple.

We assume that the following four conditions are satisfied.

1).   For any $a \neq e$ there exist $a' \neq e$ and $a'' \neq e$ such that([3])
$$\mathfrak{M}(a) \supseteqq [\mathfrak{M}(a')]^{-1}, \quad \mathfrak{M}(a) \supseteqq [\mathfrak{M}(a'')]^2.$$

([1])   B. L. v an d er W aerd en: M.Z. **36** (1933), 780.

([2])   D. v an Dan tzig: M.A. **107** (1932), 612.

([3]) $[\mathfrak{M}(a)]^{-1}$ denotes the set of all the elements of the form x^{-1}, where $x \in \mathfrak{M}(a)$. Similarly $[\mathfrak{M}(a)]^2$ denotes the set of all the elements of the form xy, where $x, y \in \mathfrak{M}(a)$.

2). There exists a sequence $\{a_i\}$ of elements $(\neq e)$ which satisfies the two conditions:

(α). $\mathfrak{M}(a_i) \supseteq \mathfrak{M}(a_{i+1})$, and the intersection of all $\mathfrak{M}(a_i)$ is e.

(β). For any $a \neq e$,
$$\mathfrak{M}(a) \supseteq \mathfrak{M}(a_i) \text{ for sufficiently large } i.$$

3). For any $\mathfrak{M}(a_i)$ there exists a finite number of elements $a_{i1}, a_{i2}, \ldots, a_{ik(i)}$ such that

$$\mathfrak{G} = \sum_{l=1}^{k(i)} [a_{il} \cdot \mathfrak{M}(a_i)] \quad \text{(set-theoretical sum)}.$$

4). $\mathfrak{G}$ is generated by any $\mathfrak{M}(a)$, that is, any element of $\mathfrak{G}$ may be obtained as the result of composition of a finite number of elements ϵ $\mathfrak{M}(a)$.

According to 1) and 2) $\mathfrak{G}$ is a topological group by the system $\{\mathfrak{M}(a)\}$ of the vicinity of e.

It is easy to see that the topology of van der Waerden mentioned above satisfies these four conditions.

In II, I will prove that the *Komplettierung* of such topologised group $\mathfrak{G}$ is locally topologically isomorphic to a connected, g-simple Lie group.

II. The proof.

As the topological group $\mathfrak{G}$ satisfies the first axiom of countability by 2), $\mathfrak{G}$ is metrisable (S. Kakutani's theorem([1])). Since $b \cdot \mathfrak{M}(a) \cdot b^{-1} = \mathfrak{M}(a)$ for any b, the metric $d(a, b)$ may be taken so as to be left- and right-invariant:

$$(*) \qquad\qquad d(a, b) = d(caf, cbf).$$

From 3) we see that the metrical group $\mathfrak{G}$ is totally bounded. Thus $\mathfrak{G}$ is separable and hence it is *komplettierbar*([2]) by (*). The *Komplettierung* $\overline{\mathfrak{G}}$ of $\mathfrak{G}$ is a compact metrical group.

$\overline{\mathfrak{G}}$ is generated by any of its vicinity of the unit-element e by 4). It is easy to see that such a compact group is connected([3]).

Hence $\overline{\mathfrak{G}}$ is a connected, separable and compact group, and $\mathfrak{G}$ is a subgroup dense in $\overline{\mathfrak{G}}$.

Thus $\overline{\mathfrak{G}}$ is G_n-*adic generated*([4]) by a sequence $\{\mathfrak{G}_i\}$ of connected,

---

([1]) S. Kakutani: Proc. of Imp. Acad., **12**(1936), 82.

([2]) D. van Dantzig: loc. cit.

([3]) D. van Dantzig: Comp. Math., **3** (1936), 57.

([4]) H. Freudenthal: Ann. of Math., **37** (1936), 57.

compact Lie groups, by H. Freudenthal's theorem. Here, by definition, $\mathfrak{G}_i$ is continuously homomorphic to $\overline{\mathfrak{G}}$ and the invariant subgroup $\mathfrak{N}_i$ closed in $\overline{\mathfrak{G}}$ defined by the topological isomorphism

$$\overline{\mathfrak{G}}/\mathfrak{N}_i \approx \mathfrak{G}_i$$

tends to e with $i \to \infty$.

Let $\mathfrak{G} \to \mathfrak{G}_i'$ by the homomorphic mapping $\mathfrak{G} \to \mathfrak{G}_i$. As $\overline{\mathfrak{G}}$ and $\mathfrak{G}_i$ are compact metrical groups, the continuous mapping $\overline{\mathfrak{G}} \to \mathfrak{G}_i$ is *open (gebietstreu)*. Thus the subgroup $\mathfrak{G}_i'$ is dense in $\mathfrak{G}_i$, since $\mathfrak{G}$ is dense in $\overline{\mathfrak{G}}$. So the closure of $\mathfrak{G}_i'$ coincides with $\mathfrak{G}_i$. $\mathfrak{G}_i'$ is g-simple with $\mathfrak{G}$. Thus $\mathfrak{G}_i'$ is topologically isomorphic with $\mathfrak{G}$ for large i as $\mathfrak{N}_i \to e$. Consequently $\overline{\mathfrak{G}}$ is topologically isomorphic with $\mathfrak{G}_i$ for large i. Hence $\overline{\mathfrak{G}}$ is a connected, compact Lie group.

The same reasoning shows that $\overline{\mathfrak{G}}$ is L-simple. The center $\mathfrak{Z}$ of this L-simple compact group $\overline{\mathfrak{G}}$ is a finite group. For otherwise e is not isolated in $\mathfrak{Z}$ and hence $\mathfrak{Z}$ must contain[1] at least one one-parameter subgroup $(\neq e)$, and consequently $\overline{\mathfrak{G}}$ does not coincide with its Lie commutatorgroup, contrary to 4). Q.E.D.

Mathematical Institute,
Ôsaka Imperial University.

(Received April 6, 1937.)

---

[1] E. Cartan: Mémorial des Sc. Math., **17**, p. 22. Cf. also K. Yosida: Jap. J. of Math., **13** (1916), 7.

On the exponential-formula in the
metrical complete ring

Proc. Imp. Acad. Tokyo **13** (1937) 301–304

(Comm. by T. Yosie, M.I.A., Oct. 12, 1937.)

In this note we shall solve the functional equation $\exp(X)\cdot\exp(Y)=$ $\exp\big(Z(X, Y)\big)$ in the *Lie-ring* embedded in the *metrical complete ring*,[1] following after a paper due to F. Hausdorff.[2]

We may replace his *symbolical differentiation* by the differentiation with respect to the *canonical parameters*. The (formal) power series employed in our proof are convergent by the topology defined in the metrical complete ring. In this way the deduction of the final result is much simplified than that of Hausdorff.

The formula $Z(X, Y)$ obtained is expressed in terms of the canonical parameters and the structure-constants of the Lie-ring. It is easy to see that[3] this formula also applies to the ordinary Lie-ring of (analytical) linear differential operators of the first order, for our proof is carried out formally. This constitutes *the converse of the second fundamental theorem of Lie.*

§ 1. Let $\mathfrak{J}$ be a Lie-ring embedded in the metrical complete ring $\mathfrak{R}$. By definition, $\mathfrak{J}$ is a real linear subspace $\leq \mathfrak{R}$ of finite dimension such that we have

$$(1) \qquad [X, Y]=XY-YX \in \mathfrak{J} \qquad \text{with } X, Y \in \mathfrak{J}.$$

Consider the functional equation

$$(2) \qquad \exp(X)\cdot\exp(Y)=\exp\big(Z((X, Y)\big) \qquad \text{for } X, Y \in \mathfrak{J},$$

where

$$\exp(A)=\sum_{n=0}^{\infty}(A^n/n!), \qquad A^0=E.$$

It admits a unique solution $Z(X, Y) \in \mathfrak{R}$, if $|X|$, $|Y|$ are sufficiently small, viz.

---

1) K. Yosida: On the group embedded in the metrical complete ring. Jap. J. of Math. **13** (1936), p. 7. For the sake of comprehension we will here reproduce the definition of the metrical complete ring.

Let the field of complex numbers be the *Operatorenbereich* of a (non-commutative) ring $\mathfrak{R}$ with the unit E, such that $aA=Aa$ for any $A \in \mathfrak{R}$ and for any complex number a. $\mathfrak{R}$ is called metrical if there is defined a absolute value $|A|$ satisfying the conditions: i) $|A| \geq 0$, and $|A|=0$ if and only if $A=0$, ii) $|A+B| \leq |A|+|B|$, $|AB| \leq |A||B|$ and $|aA|=|a||A|$. The metrical ring $\mathfrak{R}$ is called complete if it is complete in the topology defined by the metric $|A-B|$.

2) F. Hausdorff: Die symbolische Exponentialformel in der Gruppentheorie. Leipziger Berichte, Bd. **58** (1906), p. 19.

3) See Hausdorff: loc. cit.

$$(3) \qquad Z(X, Y) = \ln\big(\exp(X) \cdot \exp(Y)\big) = \sum_{n=1}^{\infty} (-1)^{n-1}$$

$$\big(\exp(X) \cdot \exp(Y) - E\big)^n / n .$$

In reality, we may prove that

$$Z(X, Y) \in \mathfrak{J} .$$

As the infinite series which occur in the following paragraphs are all convergent for sufficiently small $|X|$, $|Y|$, etc. by the topology in $\mathfrak{R}$, the proof will be carried out formally without mentioning of their convergences.

§ 2. *Lemma.*[1] For $X, U \in \mathfrak{R}$ and for a small real number a we have

$$(4) \qquad \begin{cases} \exp(X + aU) = \big(\exp(X)\big)\big(E + a\varphi(U, X) + 0(a^2)\big) \\ \qquad\qquad = \big(E + a\psi(U, X) + 0(a^2)\big)\big(\exp(X)\big) \end{cases}$$

where

$$(5) \quad \varphi(U, X) = \psi(U, -X) = U/1! + [U, X]/2! + \big[[U, X], X\big]/3! + \cdots\cdots$$

Proof. Let $F(X) \in \mathfrak{R}$ be any power series in X. We denote by $\overline{F(X+U)}$ the sum of all terms in $F(X+U)$ which have U as one-times factor. Then the first part of (4) is equivalent to

$$\sum_{n=0}^{\infty} \overline{(X+U)^n} / n! = \big(\exp(X)\big) \cdot \varphi(U, X) .$$

Thus we have to show

$$(6) \qquad \begin{cases} \dfrac{\overline{(X+U)^n}}{n!} = \dfrac{(UX^{n-1})}{0! \, n!} + \dfrac{X(UX^{n-2})}{1! \, (n-1)!} + \cdots\cdots + \dfrac{X^{n-2}(UX)}{(n-2)! \, 2!} \\[2mm] \qquad\qquad + \dfrac{X^{n-1}U}{(n-1)! \, 1!} , \qquad n = 0, 1, 2, \cdots\cdots , \end{cases}$$

where $\qquad (UX) = [U, X], \qquad (UX^2) = \big[[U, X], X\big], \cdots\cdots$

(6) is easily be proved by mathematical induction, remembering the identity

$$\overline{(X+U)^n} / n! = \big(\overline{(X+U)^{n-1}} / (n-1)!\big) \cdot (X/n) + X^{n-1}U/n! .$$

§ 3. Let $V(X, Y) \in \mathfrak{R}$ be defined by the equation

$$\psi(V, Y) = X, \qquad \text{for } X, Y \in \mathfrak{J} .$$

By comparing $\psi(V, Y) = X$ with the numerical equation $v\big(\exp(-y) -1\big)/(-y) = x$, we obtain

---

1) Hausdorff obtained this Lemma by introducing an unknown symbol W such that $X = [W, X]$.

$$(7) \qquad V(X, Y) = c_0 X + c_1 (XY) + c_2 (XY^2) + \cdots\cdots,$$

where the coefficients c_i are given by the ordinary power series

$$y/\big(1 - \exp(-y)\big) = c_0 + c_1 y + c_2 y^2 + \cdots\cdots$$

Now we have, by (4)

$$\exp\Big(Z\big(X + aX, \, Y - aV(X, Y)\big)\Big) = \big(\exp(X)\big)\big(E + 0\,(a^2)\big)\big(\exp(Y)\big),$$

and hence

$$\frac{\partial \exp\Big(Z\big(X + aX, \, Y - aV(X, Y)\big)\Big)}{\partial a} = 0 \qquad \text{for } a = 0.$$

Therefore, by termwise differentiating $\ln\big(\exp(Z)\big) = Z$, we obtain

$$(8) \qquad \frac{\partial Z\big(X + aX, \, Y - aV(X, Y)\big)}{\partial a} = 0 \qquad \text{for } a = 0.$$

§ 4. *We will solve (8) with the initial condition $Z(0, Y) = Y$.*

To this purpose, let $X_1, X_2, \cdots\cdots, X_n$ be the linearly independent base of $\mathfrak{F}$. We put $X = \sum_{i=1}^{n} t_i X_i$, $Y = \sum_{i=1}^{n} s_i X_i$. Then by (1), (7)

$$V(X, Y) = \sum_{i=1}^{n} v_i(t_1, t_2, \cdots\cdots, t_n, s_1, s_2, \cdots\cdots, s_n) X_i \in \mathfrak{F},$$

where $v_i(t, s) = v_i(t_1, t_2, \cdots\cdots, t_n, s_1, s_2, \cdots\cdots, s_n)$ is linear homogeneous in $t_1,$ $t_2, \cdots\cdots, t_n$.[1] Thus we have

$$Z\big(X + aX, \, Y - aV(X, Y)\big) = Z\Big((1 + a)\sum_{i=1}^{n} t_i X_i, \, \sum_{i=1}^{n}\big(s_i - av_i(t, s)\big)X_i\Big),$$

and hence by (8)

$$(9) \qquad \sum_{i=1}^{n} t_i \frac{\partial Z(t, s)}{\partial t_i} = \sum_{i=1}^{n} v_i(t, s)\frac{\partial Z(t, s)}{\partial s_i}, \qquad Z(t, s) = Z(X, Y).$$

In $Z(t, s)$ let $Z_k(t, s)$ be the term which is homogeneous of k-th degree in $t_1, t_2, \cdots\cdots, t_n$: $Z(t, s) = Z_0(t, s) + Z_1(t, s) + \cdots\cdots$

By the initial condition $Z(0, Y) = Y$ we have $Z_0(t, s) = \sum_{i=1}^{n} s_i X_i$. As $v_i(t, s)$ is linear homogeneous in $t_1, t_2, \cdots\cdots, t_n$ we must have

$$kZ_k(t, s) = \sum_{i=1}^{n} v_i(t, s)\frac{\partial Z_{k-1}(t, s)}{\partial s_i},$$

by (9). Hence[2]

---

1) See the Remark below.
2) Essentially this formula coincides with that obtained by Hausdorff.

$$(10) \quad Z(X, Y) = Z(t, s) = \left(E + A/1! + A^2/2! + A^3/3! + \cdots\cdots\right) . \sum_{i=1}^{n} s_i X_i ,$$

where A denotes the differential operator

$$A = \sum_{i=1}^{n} v_i(t, s) \frac{\partial}{\partial s_i} .$$

Thus $Z(X, Y) \in \mathfrak{F}$ with $X, Y \in \mathfrak{F}$, if $|X|, |Y|$ are sufficiently small.

Remark. Let the structure-constants of $\mathfrak{F}$ be given by

$$[X_i, X_j] = \sum_{k=1}^{n} c_{ijk} X_k, \qquad (i, j = 1, 2, \ldots\ldots, n) .$$

Then, by (5), it is easy to see that we have

$$\left\| \begin{matrix} v_1(t, s) \\ v_2(t, s) \\ \vdots \\ v_n(t, s) \end{matrix} \right\| = \left\| \frac{\exp\left(-V(s)\right) - E}{-V(s)} \right\|^{-1} \cdot \left\| \begin{matrix} t_1 \\ t_2 \\ \vdots \\ t_n \end{matrix} \right\| ,$$

where

$$\begin{cases} V(s) = \| V_{ij}(s) \| , \\ V_{ij}(s) = \sum_{k=1}^{n} s_k c_{jki} , \qquad (i, j = 1, 2, \ldots\ldots, n) . \end{cases}$$

A note on the differentiability of the topological group

Proc. Phys.-Math. Soc. Japan **20** (1938) 6–10

(Read September 25, 1937.)

We owe to J. von Neumann[1] the notion of the "differentiability" of the matrix group. With the aide of this notion he obtained a necessary and sufficient condition that a matrix group be a Lie group.

We may extend the "differentiability" to some other classes of the topological groups: i) the group $\mathfrak{G}$ embedded in the metrical complete ring[2], and ii) the topological group-germ $\overline{\mathfrak{G}}$ of analytical transformations in the euclidian space[3]. Concerning these "differentiabilities" we recall the following results from [I] and [II].

$\mathfrak{G}(\overline{\mathfrak{G}})$ is a Lie gronp (Lie group-germ) if and only if $\mathfrak{G}(\overline{\mathfrak{G}})$ satisfies the conditions:

α) it is *komplete*,

β) it is *differentiable*,

γ) its *Lie-ring* is of finite dimension.

(The terminologies will be explained later.)

It is to be remarked that these three conditions for $\mathfrak{G}$ are equivalent to the local compactness of $\mathfrak{G}$[4].

In the present note the author intend to show that, for both classes, the *dimensional hypothesis* is not necessary. In fact, we may prove that γ) is a consequence from α) and β). (For $\mathfrak{G}$ only β) suffices, see Theorem I below.) Thus we see that the two conditions α) and β) amount to a kind of "compactness".

§ 1. The first class.

Def. 1.[5] Let the field of complex numbers be the *Operatorenbereich* of a ring $\mathfrak{R}$ with the unit E, such that $1A=A$ for any $A \in \mathfrak{R}$. $\mathfrak{R}$ is called *metrical* if there is defined an *absolute value* $|A|$ with the conditions:

---

(1) J. von Neumann: Math. Zeitsch. **30** (1929), p. 3. Neumann did not use the term "differentiable".

(2) K. Yosida: Jap. J. of Math. **13** (1936). p. 7 This paper will be cited as [I].

(3) H. Cartan: Sur les groupes de transformations analytiques. Paris (1935). This book will be referred to as [II].

(4) [I], p. 22.

(5) [I], p. 8.

$$|A| \geqq 0 \text{ and } |A| = 0 \text{ if and only if } A = 0. \quad |A+B| \leqq |A| + |B|.$$

$$|AB| \leqq |A|\,|B|. \quad |aA| = |a|\,|A|.$$

The metrical ring $\Re$ is called *complete* if it is complete in the topology defined by the distance $|A-B|$.

In what follows $\Re$ denotes *metrical complete ring*.

Def. 2. A subset $\mathfrak{G}$ of $\Re$ is called a *group embedded in* $\Re$, if it is a group with respect to the ring-multiplication, such that the unit-element of $\mathfrak{G}$ coincides with E.

Theorem 1.[6] $\mathfrak{G}$ is a topological group by the topology in $\Re$.

Def. 3.[7] The group $\mathfrak{G}$ is called *differentiable* (at E), if from any sequence $\{T_i\}$ of $\mathfrak{G}$, $T_i \neq E$, $\lim\limits_{i \to \infty} T_i = E$, we may choose a subsequence $\{T_{i'}\}$ and a sequence $\{\varepsilon_{i'}\}$ of real numbers such that

$$(1) \qquad \lim_{i \to \infty} \left((T_{i'} - E)/\varepsilon_{i'} \right) = U (\neq 0) \in \Re.$$

Theorem 2.[8] The set $\mathfrak{T}$ of all such *differential quotients* of $\mathfrak{G}$ (at E) constitutes the *Lie-ring* (of $\mathfrak{G}$), that is, for $U, V \in \mathfrak{T}$ we have

$$\begin{cases} aU + bV \in \mathfrak{T}, & (a, b \text{ real numbers}), \\ [U, V] = UV - VU \in \mathfrak{T}. \end{cases}$$

We obtain $\lim\limits_{i \to \infty} |T_{i'} - E|/|\varepsilon_{i'}| = |U| \neq 0, \infty$ by (1). Hence we have the

Theorem 3. $\mathfrak{G}$ is differentiable (at E), if and only if for any sequence $\{T_i\}$ of $\mathfrak{G}$, $T_i \neq E$, $\lim\limits_{i \to \infty} T_i = E$, the sequence

$$(2) \qquad \left\{ \frac{\pm (T_i - E)}{|T_i - E|} \right\}$$

is compact.

Def. 4. $\mathfrak{G}$ is called *komplete* if it satisfies the condition: any fundamental sequence $\{T_i\}$ of $\mathfrak{G}$, $\lim\limits_{i, j \to \infty} T_i T_j^{-1} = E$, defines an element $T \in \mathfrak{G}$ such that $\lim\limits_{i \to \infty} T_i = T$.

As the topological group is homogeneous, $\mathfrak{G}$ is komplete if and only if it is *locally komplete* (at E). By Theorem 1 we see that $\mathfrak{G}$ is locally komplete (at E) if and only if $\mathfrak{G}$ is *locally closed* (at E). Here $\mathfrak{G}$ is called locally closed (at E) if there exist a positive ε such that we obtain $T \in \mathfrak{G}$ from $T_i \in \mathfrak{G}$, $\lim\limits_{i \to \infty} T_i = T$ and $|T - E| \leqq \varepsilon$.

Hence we have the

---

(6) [1], p. 14.

(7) [1], p. 15.

(8) [1], p. 19. For convention, we let 0 belong to $\mathfrak{T}$.

(9) [1], p. 20.

Theorem 4.[10] $\mathfrak{G}$ is a Lie group if and only if it satisfies α, β) and γ). (For convention, a discrete group is also called a Lie group).

After these preliminaries we now prove our

Theorem I. The Lie-ring $\mathfrak{T}$ of $\mathfrak{G}$ is of finite dimension if $\mathfrak{G}$ is differentiable.

Proof. We will show that $\mathfrak{T}$ has a finite base. To this purpose assume the contrary. Then there exists an infinite sequence $\{U_i\}$ of $\mathfrak{T}$ such that

$$(3) \qquad |U_i|=1, \quad |U_i-U_j|\geqq\frac{1}{2} \quad \text{for } i\neq j.$$

This may be proved by induction. Suppose $U_1, U_2, \ldots, U_m \in \mathfrak{T}$ be already chosen so that they satisfy (3). As $\mathfrak{T}$ does not have a finite base there exists an element $U \in \mathfrak{T}$ such that $\left|U-\sum_{i=1}^{m}a_iU_i\right|\geqq\delta>0$ for any system $(a_1, a_2, \ldots, a_m)$ of real numbers. Let $\sum_{i=1}^{m}b_iU_i \in \mathfrak{T}$ be such that $2\delta\geqq\left|U-\sum_{i=1}^{m}b_iU_i\right|\geqq\delta$. Then the system

$$U_1, U_2, \ldots, U_m, U_{m+1}=\left(U-\sum_{i=1}^{m}b_iU_i\right)\bigg/\left|U-\sum_{i=1}^{m}b_iU_i\right|$$

surely satisfies (3).

U_i being the differential quotient of $\mathfrak{G}$ with $|U_i|=1$, there exists a sequence $\{T_i\}$ of $\mathfrak{G}$, $T_i\neq E$, $\lim_{i\to\infty} T_i=E$, such that

$$\left|\frac{\eta_i(T_i-E)}{|T_i-E|}-U_i\right|\leqq\frac{1}{i}, \quad \eta_i=+1 \text{ or } -1, \quad i=1, 2, \ldots,$$

by (1), (2).

By (3) the sequence $\{\eta_i(T_i-E)/|T_i-E|\}$ is not compact, contrary to the differentiability of $\mathfrak{G}$.

Thus $\mathfrak{T}$ must be of finite dimension.

§ 2. The second class.

Let $\mathfrak{E}$ denote an n-dimensional euclidian space and let $\mathfrak{D}$ be a domain in $\mathfrak{E}$. A transformation $M\to M'$ $(M'=\varphi(M), M\in\mathfrak{D}, M'\in\mathfrak{E})$ is called *analytical*, if the coordinates $\varphi^1(M), \varphi^2(M), \ldots, \varphi^n(M)$ of the point M' are analytical[11] functions in the coordinate $M^1, M^2, \ldots, M^n$ of the point M. Given two analytical transformations T, S

$$T: M'=\varphi_T(M), \quad S: M'=\varphi_S(M),$$

---

(10) [I], p. 22.

(11) analytical=developpable in taylor series at each point of the domain.

we denote by $d(\mathfrak{D}', T, S)$ the number $\max\limits_{M\in\mathfrak{D}'} |\varphi_T(M)-\varphi_S(M)|$, where $|\varphi_T(M)-\varphi_S(M)|$ is the maximum of the coordinates $\varphi_T{}^1(M)-\varphi_S{}^1(M)$, $\varphi_T{}^2(M)-\varphi_S{}^2(M),\ldots,\varphi_T{}^n(M)-\varphi_S{}^n(M)$ of the point $\varphi_T(M)-\varphi_S(M)$.

Let a set $\overline{\mathfrak{G}}$ of analytical transformation defined in $\mathfrak{D}$ be a *topological group-germ* by the topology defined by the distance $d(\mathfrak{D}_1, T, S)$, where $\mathfrak{D}_1$ denotes a fixed closed domain interior to $\mathfrak{D}$. The identical transformation will be denoted by E.

Def. 5.[12] $\overline{\mathfrak{G}}$ is called *differentiable* (at E) if it satisfies the condition: From any sequence $\{T_i\}$ of $\overline{\mathfrak{G}}$, $d(\mathfrak{D}_1, T_i, E)\neq0$, $\lim\limits_{i\to\infty} d(\mathfrak{D}_1, T_i, E)=0$, we may choose a subsequence $\{T_{i'}\}$ and a sequence $\{\varepsilon_{i'}\}$ of real numbers such that

$$(4)\qquad \lim_{i'\to\infty} ((T_{i'}(M)-M)/\varepsilon_{i'})=\psi(M)\not\equiv0,$$

uniformly in any closed domain $\mathfrak{D}_2$ completely interior to $\mathfrak{D}_1$, where $\psi(M)$ is analytic in $\mathfrak{D}_1$.

Such $\psi(M)$ (a system of n analytical functions) is called the *differential quotient* of $\mathfrak{G}$.

Theorem 5.   As we obtain

$$\lim_{i\to\infty} \max_{M\in\mathfrak{D}_2} |T_{i'}(M)-M|/\varepsilon_{i'}| =\max_{M\in\mathfrak{D}_2} |\psi(M)|\neq0,$$

we may replace (4) by

$$(5)\qquad \lim_{i'\to\infty}\frac{\pm(T_{i'}(M)-M)}{d(\mathfrak{D}_2, T_{i'}, E)}=\widetilde{\psi}(M)\quad\text{uniformly for } M\in\mathfrak{D}_2.$$

Def. 6. $\overline{\mathfrak{G}}$ is called *komplete* if any fundamental sequence $\{T_i\}$ of $\overline{\mathfrak{G}}$, $\lim\limits_{i,j\to\infty} d(\mathfrak{D}_1, T_i, T_j)=0$, determines a transformation $T\in\overline{\mathfrak{G}}$ such that $\lim\limits_{i\to\infty} d(\mathfrak{D}_1, T_i, T)=0$.

Theorem 6.[13]   Let $\overline{\mathfrak{G}}$ be komplete, then the set $\mathfrak{T}$ of all the differential quotients of $\overline{\mathfrak{G}}$ constitutes the *Lie-ring* (of $\overline{\mathfrak{G}}$), that is, for $\psi_1(M), \psi_2(M)\in\mathfrak{T}$ we have

$$\begin{cases} a\,\psi_1(M)+b\,\psi_2(M)\in\mathfrak{T}, & (a, b \text{ real numbers}), \\ [\psi_1(M), \psi_2(M)]\in\mathfrak{T}^{(14)}. \end{cases}$$

---

(12) [II], p. 38. Cartan used the term "$\overline{\mathfrak{G}}$ jouit de la propriété [P]". The convergence means the convergences of the coordinates. Such vectorial notations will be used in the following lines.

(13) [II], p. 34. For convention, we let $\psi(M)\equiv0$ belong to $\mathfrak{T}$.

(14) The coordinates of this Poisson's parenthesis are given by

$$[\psi_1(M), \psi_2(M)]^k=\sum_{i=0}^{n}\left(\psi_1{}^i(M)\frac{\partial\psi_2{}^k(M)}{\partial M^i}-\psi_2{}^i(M)\frac{\partial\psi_1{}^k(M)}{\partial M^i}\right)$$

Theorem 7.[15] $\overline{\mathfrak{G}}$ is a Lie group-germ if and only if it satisfies α), β) and γ). (For convention, a discrete group-germ is also called a Lie group-germ.)

Now we come to the proof of

Theorem II. The Lie-ring $\mathfrak{T}$ of $\overline{\mathfrak{G}}$ is of finite dimension if $\overline{\mathfrak{G}}$ is komplete and differentiable.

Proof. We will prove that $\mathfrak{T}$ has a finite base. Let us assume the contrary. Then, as in the proof of Theorem I, we see that there exists an infinite sequence $\{\psi_i(M)\}$ of $\mathfrak{T}$ satisfying

$$(6) \qquad d(\mathfrak{D}_2', \psi_i, 0) = 1, \ d(\mathfrak{D}_2', \psi_i, \psi_j) \geqq \frac{1}{2} \quad \text{for } i \neq j,$$

where $\mathfrak{D}_2'$ denotes a closed domain completely interior to $\mathfrak{D}_1$.

By $d(\mathfrak{D}_2', \psi_i, 0) = 1$ and Theorem 5, there exists a sequence $\{T_i\}$ of $\overline{\mathfrak{G}}$ such that

$$\left| \frac{\eta_i(T_i(M) - M)}{d(\mathfrak{D}_2', T_i, E)} - \psi_i(M) \right| \leqq \frac{1}{i} \quad (\eta_i = +1 \text{ or } -1) \text{ for } M \in \mathfrak{D}_2'.$$

This, combined with (6), is contradictory to the differentiability of $\mathfrak{G}$. Hence $\mathfrak{T}$ must be of finite dimension.

Mathematical Institute,
Osaka Imperial University.

(Received November 4, 1937.)

---

(15) [II], p. 38. Cartan's statement reads as follows.

Let $\overline{\mathfrak{G}}$ be locally euclidian of finite dimension. Then $\overline{\mathfrak{G}}$ is a Lie group-germ if and only if $\overline{\mathfrak{G}}$ " jouit de la propriété [P]".

Of this " euclidian character " he made use of the condition α) and γ) only.

A characterisation of the adjoint representations of the semi-simple Lie-rings

Japan. J. Math. **14** (1938) 169–173

(Received January 12th, 1938.)

According to E. Cartan, the group of the continuous automorphisms of a semi-simple or a compact Lie group $\mathfrak{G}$ coincides with the inner automorphism group of $\mathfrak{G}$, in the vicinity of the identical automorphism. I intend to give a proof of this result, starting from a characterisation of the adjoint representation of a semi-simple Lie-ring. This constitutes the contents of §1, 2 and 3 of the present note. In §4 we obtain applications of this characterisation to the semi-simple Lie-rings and the completely reducible matric Lie-rings.

§1. A complex [1] vector space $\mathfrak{R}$ is called a *Lie-ring* if there is defined a *commutator-multiplication* $[x, y]$:

$$(1) \qquad [x, y] \in \mathfrak{R} \ ,$$

$$(2) \qquad [x+y, z+w] = [x, z]+[x, w]+[y, z]+[y, w] \ ,$$

$$(3) \qquad [x, [y, z]]+[y, [z, x]]+[z, [x, y]] = 0 \ ,$$

$$(4) \qquad [x, x] = 0 \ ,$$

$$(5) \qquad [ax, y] = [x, ay] = a[x, y] \quad \text{for scalar } a \ .$$

If $\mathfrak{R}$ is a finite dimensional vector space, it is called a Lie-ring of finite rank: rank $\mathfrak{R} = $ dimension $\mathfrak{R}$.

We have, by (2) and (4)

$$(6) \qquad [x, y] = -[y, x] \ .$$

By (1), (2) and (5) any element $x \in \mathfrak{R}$ induces a linear mapping T_x of $\mathfrak{R}$ *in* $\mathfrak{R}$: $T_x \cdot y = [x, y]$. We obtain, by (3) and (6), $T_x T_y \cdot z - T_y T_x \cdot z = [x, [y, z]]-[y, [x, z]] = [[x, y], z] = T_{[x,y]} \cdot z$, viz.

$$(7) \qquad [T_x, T_y] = T_x T_y - T_y T_x = T_{[x, y]}$$

Hence we have the

---

[1] or real. Except the theorems 2 and 3, the underlying field may be any field which is algebraically closed and of characteristic zero.

Lemma 1. The totality $\mathfrak{A}_\mathfrak{R}$ of T_x is a Lie-ring with the commutator-multiplication $[T_x, T_y]$. $\mathfrak{A}_\mathfrak{R}$ is isomorphic to $\mathfrak{R}/\mathfrak{Z}$, where $\mathfrak{Z}$ denotes the *central* of $\mathfrak{R}$.

$\mathfrak{A}_\mathfrak{R}$ is called the *adjoint representation* of $\mathfrak{R}$.

By (7) we have $T_x \cdot [y, z] = T_x T_y \cdot z = T_{[x, y]} \cdot z + T_y T_x \cdot z = [[x, y], z] + [y, [x, z]] = [T_x \cdot y, z] + [y, T_x \cdot z]$. A linear mapping T of $\mathfrak{R}$ *in* $\mathfrak{R}$ is called a *derivation*([2]) (of $\mathfrak{R}$) if it satisfies

$$(8) \qquad T \cdot [y, z] = [T \cdot y, z] + [y, T \cdot z] .$$

Let T, S be two derivations, then $TS \cdot [y, z] = T \cdot \{ [S \cdot y, z] + [y, S \cdot z] \} = [TS \cdot y, z] + [S \cdot y, T \cdot z] + [T \cdot y, S \cdot z] + [y, TS \cdot z]$ and hence $[T, S] = TS - ST$ is also a derivation. Thus

Lemma 2. The set $\mathfrak{D}_\mathfrak{R}$ of all the derivations T of R constitutes a Lie-ring with the commutator-multiplication $[T, S] = TS - ST$. $\mathfrak{A}_\mathfrak{R}$ is an *ideal* of $\mathfrak{D}_\mathfrak{R}$.

Proof. Let $T \in \mathfrak{D}_\mathfrak{R}$, $T_x \in \mathfrak{A}_\mathfrak{R}$. Then $TT_x \cdot y = T \cdot [x, y] = [T \cdot x, y] + [x, T \cdot y] = T_{T \cdot x} \cdot y + T_x T \cdot y$, viz.

$$(9) \qquad [T, T_x] = TT_x - T_x T = T_{T \cdot x} .$$

This proves that $\mathfrak{A}_\mathfrak{R}$ is an ideal of $\mathfrak{D}_\mathfrak{R}$.

§2. A Lie-ring $\mathfrak{R}$ of finite rank is called *semi-simple*, if and only if it does not contain a *solvable (integrable)* ideal $\neq 0$. Such ring $\mathfrak{R}$ is a *direct sum* of *simple* and semi-simple ideals, and the central $\mathfrak{Z}$ of $\mathfrak{R}$ is $= 0$.([3])

Theorem 1. $\mathfrak{D}_\mathfrak{R} = \mathfrak{A}_\mathfrak{R}$, if $\mathfrak{R}$ is semi-simple.

Proof. As the central $\mathfrak{Z}$ of $\mathfrak{R}$ is $= 0$, $\mathfrak{A}_\mathfrak{R}$, is semi-simple by the lemma 1. There may occur two cases.

Case 1. $\mathfrak{D}_\mathfrak{R}$ is semi-simple. By the lemma 2, $\mathfrak{A}_\mathfrak{R}$ is an ideal of $\mathfrak{D}_\mathfrak{R}$, and hence we have the direct decomposition $\mathfrak{D}_\mathfrak{R} = \mathfrak{A}_\mathfrak{R} + \mathfrak{D}_1$. As the *intersection* $(\mathfrak{A}_\mathfrak{R} \cdot \mathfrak{D}_1)$ is $= 0$ we have $[T, T_x] = 0$ for $T \in \mathfrak{D}_1$, $T_x \in \mathfrak{A}_\mathfrak{R}$. Hence, by (9), $T_{T \cdot x} = 0$. This means that $T \cdot x \in \mathfrak{Z}$ for any $x \in \mathfrak{R}$. As $\mathfrak{Z} = 0$ we have $T = 0$. Thus $\mathfrak{D}_1 = 0$, $\mathfrak{D}_\mathfrak{R} = \mathfrak{A}_\mathfrak{R}$.

Case 2. $\mathfrak{D}_\mathfrak{R}$ is not semi-simple, viz. there exists a solvable ideal $\mathfrak{D}_1 \neq 0$ of $\mathfrak{D}_\mathfrak{R}$. $\mathfrak{A}_\mathfrak{R}$ and $\mathfrak{D}_1$ being the ideals of $\mathfrak{D}_\mathfrak{R}$, we have $[T, T_x] \in (\mathfrak{A}_\mathfrak{R} \cdot \mathfrak{D}_1)$ for $T \in \mathfrak{D}_1$, $T_x \in \mathfrak{A}_\mathfrak{R}$. $(\mathfrak{A}_\mathfrak{R} \cdot \mathfrak{D}_1)$ is an ideal of $\mathfrak{A}_\mathfrak{R}$ and of $\mathfrak{D}_\mathfrak{R}$. Hence $(\mathfrak{A}_\mathfrak{R} \cdot \mathfrak{D}_1)$ is a solvable ideal of $\mathfrak{A}_\mathfrak{R}$. Thus we must have $(\mathfrak{A}_\mathfrak{R} \cdot \mathfrak{D}_1) = 0$ by the semi-simplicity of $\mathfrak{A}_\mathfrak{R}$. Then we obtain $\mathfrak{D}_1 = 0$ as in the case 1, contrary to the hypothesis. Therefore the case 2 does not occur. Q. E. D.

---

(2) This terminology is due to N. Jacobson : Trans. Amer. Math. Soc. **42** (1937), p. 206.

(3) E. Cartan : Thèse (1894), p. 53.

Let $\mathfrak{R}$ be a Lie-ring of finite rank. A linear mapping $\overline{T}$ of $\mathfrak{R}$ *on* $\mathfrak{R}$ is called an *automorphism* of $\mathfrak{R}$ if it satisfies

$$(10) \qquad \overline{T} \cdot [x, y] = [\overline{T} \cdot x, \overline{T} \cdot y] .$$

The totality $\overline{\mathfrak{A}}_{\mathfrak{R}}$ of the automorphisms $\overline{T}$ constitutes a group, the automorphism group of $\mathfrak{R}$. We may topologise $\overline{\mathfrak{A}}_{\mathfrak{R}}$ as follows. By choosing a linearly independent base of $\mathfrak{R}$, any $\overline{T} \in \overline{\mathfrak{A}}_{\mathfrak{R}}$ is given by a matrix $\overline{T} = || t_{ij} ||$. Then $\overline{\mathfrak{A}}_{\mathfrak{R}}$ is a *locally compact* topological group by the distance $| T - T' | = \sqrt{\sum_{i,j} | t_{ij} - t'_{ij} |^2}$.

Hence $\overline{\mathfrak{A}}_{\mathfrak{R}}$ is a Lie group [4]. The infinitesimal operators T of $\overline{\mathfrak{A}}_{\mathfrak{R}}$ are the linear mappings of $\mathfrak{R}$ *in* $\mathfrak{R}$ of the form

$$(11) \qquad T = \lim_{i \to \infty} (\overline{T}_i - \overline{E})/\varepsilon_i ,$$

where $\overline{T}_i \in \overline{\mathfrak{A}}_{\mathfrak{R}}$, $\overline{E}$ the identical mapping and $\{\varepsilon_i\}$ is a sequence of real numbers such that $\lim_{i \to \infty} \varepsilon_i = 0$. By (10) and (11) we see that the infinitesimal operator T of $\overline{\mathfrak{A}}_{\mathfrak{R}}$ is a derivation of $\mathfrak{R}$. Conversely, if T is a derivation of $\mathfrak{R}$, we obtain by (8)

$$[\exp T \cdot x, \exp T \cdot y] = \exp T \cdot [x, y], \ \exp T = \overline{E} + \sum_{n=1}^{\infty} (T^n/n !) ,$$

viz. $\overline{T} = \exp T$ is an automorphism of $\mathfrak{R}$. Hence

Theorem 2. The Lie-ring of the infinitesimal operators of the automorphism group of a Lie-ring (of finite rank) $\mathfrak{R}$ coincides with $\mathfrak{D}_{\mathfrak{R}}$.

§ 3. Let $\overline{T}$ be a *continuous* automorphism of a Lie group $\mathfrak{G}$. Then by $\overline{T}$ a one-parameter Lie subgroup of $\mathfrak{G}$ is mapped on a one-parameter Lie subgroup of $\mathfrak{G}$. Thus it is easy to see that $\overline{T}$ induces an automorphism T of the Lie-ring $\mathfrak{R}$ of $\mathfrak{G}$. Conversely any automorphism T of $\mathfrak{R}$ defines a continuous automorphism $\overline{T}$ of $\mathfrak{G}$. It it well known that, for the *inner* automorphisms $\{\overline{T}\}$ of $\mathfrak{G}$ the Lie-ring of the infinitesimal operators of the corresponding $\{T\}$ is $= \mathfrak{A}_{\mathfrak{R}}$.

Next let $\mathfrak{G}$ be a *compact* Lie group. Then $\mathfrak{G}$ is locally a *direct product* [5] of a compact semi-simple Lie group $\mathfrak{G}_1$ and a compact abelian Lie group $\mathfrak{G}_2$, which is the central of $\mathfrak{G}$. As a compact abelian Lie group is *locally isomorphic* to a *toroidal group* [6], any continuous automorphism of $\mathfrak{G}_2$ *connected* to the identical automorphism is the identical automorphism.

---

[4] K. Yosida: Jap. J. of Math. **13** (1936), p. 7.

[5] E. Cartan: Mém. des Sc. Math. **42** (1930), p. 37.

[6] A direct product of a finite number of groups, each of which is an additive group of real numbers modulo 1.

Thus, by the theorems 1 and 2, we obtain

Theorem 3([7]) (of E. Cartan).   The group of the continuous auto-morphisms of a semi-simple or a compact Lie group $\mathfrak{G}$ coincides with the inner automorphism group of $\mathfrak{G}$, in the vicinity of the identical auto-morphism.

§4.   Let a Lie-ring $\mathfrak{R}$ of finite rank be a direct sum of simple ideals: $\mathfrak{R} = \mathfrak{R}_1 + \mathfrak{R}_2 + \cdots + \mathfrak{R}_k$.   Then some of the $\mathfrak{R}_i$ are abelian and the rests are semi-simple, so that $\mathfrak{R}$ is a direct sum of a semi-simple ideal $\mathfrak{H}$ and the central $\mathfrak{Z} : \mathfrak{R} = \mathfrak{H} + \mathfrak{Z}$.   We have $\mathfrak{H} = \mathfrak{H}^2 = [\mathfrak{H}, \mathfrak{H}]$, viz. any element of $\mathfrak{H}$ may be obtained as a linear combination of the elements of the form $[y, z]$, where $y, z \in \mathfrak{H}$.   For, if otherwise, $\mathfrak{H}$ would contain an abelian ideal $\neq 0$ isomorphic to $\mathfrak{H}/\mathfrak{H}^2$.   Thus we have $\mathfrak{H} = \mathfrak{H}^2 = \mathfrak{R}^2$.

Lemma 3.   We assume $\mathfrak{R} = \mathfrak{H} + \mathfrak{Z}$ be a subring of a Lie-ring $\mathfrak{S}$.   Let an element $x \in \mathfrak{S}$ satisfy $[x, \mathfrak{R}] \in \mathfrak{R}$.   Then there exists a uniquely deter-mined $x_1 \in \mathfrak{R}^2 (= \mathfrak{H})$ such that $[x - x_1, \mathfrak{R}] = 0$.

Proof. x and $\mathfrak{R}$ generates a subring $\mathfrak{R}'$ of $\mathfrak{S}$ and $\mathfrak{R}$ is an ideal of $\mathfrak{R}'$. Then, by (3), the linear mapping T_x defined by $T_x \cdot y = [x, y], y \in \mathfrak{R}$ is a derivation of $\mathfrak{R}$.   Let $y, z \in \mathfrak{H}$, then by (8), $T_x [y, z] = [T_x \cdot y, z] + [y, T_x \cdot z]$ and hence $T_x \cdot \mathfrak{H}^2 \subseteq \mathfrak{H}$.   As $\mathfrak{H}^2 = \mathfrak{H} = \mathfrak{R}^2$, T_x is a derivation of $\mathfrak{R}^2$.   There-fore, by theorem 1, there exists an element $x_1 \in \mathfrak{R}^2$ which satisfies

$$(12) \qquad [x, y] = [x_1, y] \qquad \text{for all} \quad y \in \mathfrak{R}^2$$

x_1 is uniquely determined by (12), for the central of the semi-simple $\mathfrak{R}^2$ is $= 0$. Q. E. D.

By this lemma we obtain

Theorem 4([8]).   Let a subring $\mathfrak{R}$ of a Lie-ring $\mathfrak{S}$ be semi-simple.   Then the maximal Lie-ring $\mathfrak{R}'$ $(\subseteq \mathfrak{S})$ which contains $\mathfrak{R}$ as an ideal is given by

$$\mathfrak{R}' = \mathfrak{R} + \mathfrak{R}_1 \quad \text{(direct sum)},$$

where $\mathfrak{R}_1$ is the set of all the elements $z \in \mathfrak{S}$ which satisfy $[z, \mathfrak{R}] = 0$.

Next let $\mathfrak{S}$ be the set of all the matrices of n-th degree with complex numbers as coefficients.   Then $\mathfrak{S}$ is a Lie-ring with the commutator-multi-plication $[X, Y] = XY - YX$.   Let a subring $\mathfrak{R}$ of $\mathfrak{S}$ be *completely reducible*. We assume that $\mathfrak{R}$ is given in completely reduced form

$$(13) \qquad \mathfrak{R} = \left\|\begin{matrix} \mathfrak{R}_1 & & & 0 \\ & \mathfrak{R}_2 & & \\ & & \ddots & \\ 0 & & & \mathfrak{R}_k \end{matrix}\right\|$$

---

[7]   Bull. des Sc. Math. France **49** (1925), p. 361 and Mém. des Sc. Math **42** (1930), p. 38.
[8]   Cf. E. Cartan: Thèse (1894), p. 113.

where $\Re_1 = \Re_2 = \cdots = \Re_{k_1}$, $\Re_{k_1+1} = \Re_{k_1+2} = \cdots = \Re_{k_2}$, $\ldots$, $\Re_{k_m+1} = \Re_{k_m+2}$ $= \cdots = \Re_{k_{m+1}} = \Re_k$, and $\Re_i$ is not *equivalent* to $\Re_j$ if $i \neq j$. Since $\Re_i$ is irreducible, we have the direct decomposition([9]) $\Re_i = \mathfrak{H}_i = \mathfrak{Z}_i$, where $\mathfrak{H}_i = \Re_i^2$ is semi-simple and $\mathfrak{Z}_i$ consists of the matrices of the form aE_i (a = scalar and E_i = the unit matrix). Therefore we have the direct decomposition

$$(14) \qquad\qquad \Re = \mathfrak{H} + \mathfrak{Z} \,,$$

where $\mathfrak{H} = \Re^2$ is semi-simple and the central $\mathfrak{Z}$ consists of matrices

$$(15) \qquad\qquad \left\|\begin{array}{cccc} a_1E_1 & & & 0 \\ & a_2E_2 & & \\ & & \ddots & \\ 0 & & & a_kE_k \end{array}\right\|$$

where $a_1 = a_2 = \cdots = a_{k_1}$, $a_{k_1+1} = a_{k_1+2} = \cdots = a_{k_2}$, $\ldots$, $a_{k_m+1} = a_{k_m+2} = \cdots$ $= a_{k_{m+1}} = a_k$.

Let a matrix $X \in \mathfrak{S}$ satisfy $[X, \Re] \in \Re$. Then by the lemma 3,

$$\begin{cases} X = X_1 + Z, \; X_1 \in \Re^2, \; Z \in \mathfrak{S} \,, \\ [Z, \Re^2] = 0 \,. \end{cases}$$

As $\Re_i$ ($i = 1, 2, \ldots, k$) is irreducible we see, by (13), (14), (15) and a theorem([10]) from the matric algebras, that $[Z, \mathfrak{Z}] = 0$. Thus $[Z, \Re] = 0$. Therefore we obtain

Theorem 5. Let $\Re$ be a completely reducible matric Lie-ring, and let $\Re'$ be the maximal matric Lie-ring which contains $\Re$ as an ideal. Then we have

$$\Re' = \Re^2 + \Re_1 \qquad \text{(direct sum),}$$

where $\Re_1$ is the set of all the matrices commutative with every matrix $\in \Re$.

Mathematical Institute,
Osaka Imperial University.

([9]) E. Cartan : Ann. Éc. Norm. Sup. **26** (1909), p. 147.

([10]) B. L. van der Waerden : Grupp. von linear. Transf. (1935), p. 48.

On the fundamental theorem of the tensor calculus

Proc. Imp. Acad. Tokyo **14** (1938) 211–213

(Comm. by T. Yosie, m.i.a., June 13, 1938.)

Let $\mathfrak{G}$ be a connected semi-simple Lie group with s real parameters (r. p.).   Consider the

Problem of the construction of all matrix groups continuously homomorphic to the universal covering group $\widetilde{\mathfrak{G}}$ of $\mathfrak{G}$.
This Problem was treated by H. Weyl in his famous memoirs.[1]   His method consists in the combined use of the "unitarian trick" and the theory of "Gewichts."

In the present note I intend to show that the Problem can be solved, in principle, without the theory of "Gewichts."   We make use of the *theory of almost periodic functions* in place of the theory of "Gewichts."   Then we easily obtain the fundamental theorem of the tensor calculus, as was developed in Weyl's paper loc. cit.

The Lemma below will make clear a critical point which, so far as I know, has never been considered in the literature.   It asserts that any matrix Lie ring $\mathfrak{D}$ homomorphic to the Lie ring of $\mathfrak{G}$ coincides with the Lie ring of the Lie group *generated* by $\mathfrak{D}$.

I. *A Lemma.*   Let a matrix group $\mathfrak{D}$ be continuously homomorphic to $\widetilde{\mathfrak{G}}$. $\widetilde{\mathfrak{G}}$ being semi-simple with $\mathfrak{G}$, the homomorphic mapping $\widetilde{\mathfrak{G}} \to \mathfrak{D}$ is *open*[2] (*Gebietstreu*).   Thus $\mathfrak{D}$ is locally compact with $\widetilde{\mathfrak{G}}$ and hence $\mathfrak{D}$ is a Lie group.[3]   The Lie ring with r. p. of $\mathfrak{D}$ and $\mathfrak{G}$ be denoted by $\mathfrak{R}_{\mathfrak{D}}$ and $\mathfrak{R}_{\mathfrak{G}}$ respectively.   As $\mathfrak{D}$ is *open* homomorphic to $\widetilde{\mathfrak{G}}$, $\mathfrak{D}$ is locally *open* homomorphic to $\mathfrak{G}$.   Thus $\mathfrak{R}_{\mathfrak{D}}$ is homomorphic to $\mathfrak{R}_{\mathfrak{G}}$ with real numbers as *Operatorenbereich*.

Next let a matrix Lie ring $\mathfrak{R}_{\mathfrak{D}'}$ with r. p. be homomorphic to $\mathfrak{R}_{\mathfrak{G}}$ with real numbers as Operatorenbereich.   We denote by $\overline{\mathfrak{D}}'$ the Lie *group germ* whose Lie ring is $\mathfrak{R}_{\mathfrak{D}'}$.   Let $\mathfrak{D}'$ be the set of all the products of the form $A_1 A_2 \cdots A_k$, $A_i \in \overline{\mathfrak{D}}'$, and the limit matrices of such products, so long as they are non-singular. $\mathfrak{D}'$ is called the group *generated* by $\mathfrak{R}_{\mathfrak{D}'}$.   It is a connected locally compact group, and hence it is a Lie group.[3]   The Lie ring with r. p. of $\mathfrak{D}'$ coincides with $\mathfrak{R}_{\mathfrak{D}'}$, as $\mathfrak{R}_{\mathfrak{D}'}$ is semi-simple with $\mathfrak{R}_{\mathfrak{G}}$.[4]   Therefore the connected matrix group $\mathfrak{D}'$ is continuously (in reality, *open*) homomorphic to $\widetilde{\mathfrak{G}}$.

Summing up we have the

---

1) H. Weyl: Math. Zeitsch. **23** (1925), **24** (1926).

2) K. Yosida: Tohoku Math. J. **43**, 2 (1937). For the general properties of the continuous (not necessarily open) representation of the topological group Cf. K. Yosida: Jap. J. of Math. **13** (1937) and K. Yosida: Proc. **12** (1936).

3) J. von Neumann: Math. Zeitsch. **30** (1929). See also K. Yosida: Jap. J. of Math. **13** (1936).

4) K. Yosida: Tôhoku Math. J. loc. cit.

Lemma. A connected matrix group $\mathfrak{D}$ continuously homomorphic to $\mathfrak{G}$ is a Lie group whose Lie ring $\mathfrak{R}_{\mathfrak{D}}$ with r. p. is homomorphic to the Lie ring $\mathfrak{R}_{\mathfrak{G}}$ with r. p. of $\mathfrak{G}$. Conversely the group $\mathfrak{D}$ generated by a matrix Lie ring $\mathfrak{R}_{\mathfrak{D}}$ with r. p. homomorphic to $\mathfrak{R}_{\mathfrak{G}}$ with r. p. is a connected Lie group continuously homomorphic to $\mathfrak{G}$, such that the Lie ring with r. p. of $\mathfrak{D}$ coincides with $\mathfrak{R}_{\mathfrak{D}}$. In these statements the term " continuously " can be replaced by the term " open."

Remark. The above lemma is not always true if $\mathfrak{G}$ is not semi-simple, as the following example shows us:

$$\mathfrak{G} = \text{additive group of real numbers,}$$

$$\mathfrak{R}_{\mathfrak{D}} = \left\{ \left\| \begin{array}{cc} \sqrt{-1} & 0 \\ 0 & \sqrt{-1}\,a \end{array} \right\| \right\} \text{where} \quad a/\pi \neq \text{rational.}$$

II. *Reduction of the Problem.* $\mathfrak{G}$ being semi-simple, the adjoint representation $\mathfrak{A}$ of $\mathfrak{G}$ is locally topologically isomorphic to $\mathfrak{G}$. Thus we may replace $\mathfrak{G}$ by the universal covering group $\widetilde{\mathfrak{A}}$ of $\mathfrak{A}$ in the Problem.

Let the Lie ring with r. p. of $\mathfrak{A}$ be denoted by $\mathfrak{R}_{\mathfrak{A}}$. For any Lie ring $\mathfrak{R}$ with r. p. we denote by $\{\mathfrak{R}\}$ the Lie ring obtained from $\mathfrak{R}$ by regarding these r. p. as complex parameters (c. p.). Surely[1] $\{\mathfrak{R}_{\mathfrak{A}}\}$ is semi-simple with $s(2s)$ c. p. (r. p.).

If a matrix Lie ring $\mathfrak{R}_{\mathfrak{D}}$ with r. p. is homomorphic to $\mathfrak{R}_{\mathfrak{A}}$ with r. p., $\{\mathfrak{R}_{\mathfrak{D}}\}$ is homomorphic to $\{\mathfrak{R}_{\mathfrak{A}}\}$ with complex numbers as Operatorenbereich. Conversely let a matrix Lie ring $\mathfrak{R}_{\mathfrak{D}}'$ with c. p. be homomorphic to $\{\mathfrak{R}_{\mathfrak{A}}\}$ with c. p., then the homomorphic image $\mathfrak{R}_{\mathfrak{D}}'$ of $\mathfrak{R}_{\mathfrak{A}}$ is a Lie ring with r. p.

Hence, by the Lemma, the Problem is reduced to the

Problem I of the construction of all matrix Lie rings with c. p. homomorphic to $\{\mathfrak{R}_{\mathfrak{A}}\}$ with complex numbers as Operatorenbereich.

The Lie ring $\{\mathfrak{R}_{\mathfrak{A}}\}$ with $2s$ r. p. has a base $(x_1, x_2, \ldots, x_s, \sqrt{-1}\,x_1, \sqrt{-1}\,x_2, \ldots, \sqrt{-1}\,x_s)$ such that $(x_1, x_2, \ldots, x_s)$ with s r. p. constitutes the Lie ring of a semi-simple Lie group germ $\overline{\mathfrak{A}}_{\mathfrak{u}}$ of unitary matrices with s r. p. This results from the " unitarian trick " of H. Weyl.[2] By the Lemma, the semi-simple $\overline{\mathfrak{A}}_{\mathfrak{u}}$ is a vicinity of the connected compact semi-simple Lie group $\mathfrak{A}_{\mathfrak{u}}$ generated by $(x_1, x_2, \ldots, x_s)$ with s r. p.

Thus the Problem I is reduced to the

Problem II of the construction of all matrix Lie rings with r. p. homomorphic to the Lie ring $(x_1, x_2, \ldots, x_s)$ with s r. p. of $\mathfrak{A}_{\mathfrak{u}}$.

As $\mathfrak{A}_{\mathfrak{u}}$ is compact semi-simple, the universal covering group $\widetilde{\mathfrak{A}}_{\mathfrak{u}}$ of $\mathfrak{A}_{\mathfrak{u}}$ is compact by Weyl's theorem.[3] By Schreier's theorem,[4] any connected matrix group locally *open* homomorphic to $\mathfrak{A}_{\mathfrak{u}}$ is open homomorphic to $\widetilde{\mathfrak{A}}_{\mathfrak{u}}$ and vice versa.

---

1) E. Cartan: Ann. Ecol. Norm. Sup. **31** (1914). The elements of $\mathfrak{R}_{\mathfrak{A}}$ are real matrices.

2) H. Weyl: loc. cit.

3) H. Weyl: loc. cit.

4) O. Schreier: Abh. Math. Seminar Hamburg, **4** (1926).

$\widetilde{\mathfrak{A}}_{\mathfrak{u}}$ being semi-simple we see, by the Lemma, that the Problem II is equivalent to the

Problem III of the construction of all continuous representations of $\widetilde{\mathfrak{A}}_{\mathfrak{u}}$.

III. *Solution of the Problem III.* $\widetilde{\mathfrak{A}}_{\mathfrak{u}}$ is a connected compact Lie group. Hence, by von Neumann-Pontrjagin-Freudenthal's theorem,[1] there exists a matrix group $\mathfrak{M}$ topologically isomorphic to $\widetilde{\mathfrak{A}}_{\mathfrak{u}}$.

As $\mathfrak{M}$ is a compact matrix group, all continuous representations of $\mathfrak{M}$ can be obtained by taking complex conjugates and decomposing Kronecker's products from $\mathfrak{M}$, by van Kampen's theorem.[2]

Therefore the Problem III and hence the original Problem is completely solved.

Remark. A compact matrix group is *completely reducible.*[3] Thus the above $\mathfrak{M}$ is obtained as the *sum representation* of a finite number of suitably chosen *irreducible* continuous representations of $\widetilde{\mathfrak{A}}_{\mathfrak{u}}$. Hence, by the Lemma and the arguments in II, we see that the Lie ring with r. p. of such matrix group $\mathfrak{M}$ with r. p. can be obtained in the following way. There exists a finite number of *irreducible* matrix Lie rings $\mathfrak{R}_1, \mathfrak{R}_2, \ldots, \mathfrak{R}_l$ with r. p. homomorphic to $(x_1, x_2, \ldots, x_s)$ with r. p. such that i) the *sum Lie ring*

$$\mathfrak{R} = \left\| \begin{matrix} \mathfrak{R}_1 & & & \\ & \mathfrak{R}_2 & & \\ & & \ddots & \\ & & & \mathfrak{R}_l \end{matrix} \right\|$$

with r. p. is isomorphic to $(x_1, x_2, \ldots, x_s)$ with s r. p. and ii) the Lie group generated by $\mathfrak{R}$ is simply connected. Such $\mathfrak{R}$ is precisely the Lie ring with r. p. of the desired group $\mathfrak{M}$.

IV. *The fundamental theorem of the tensor calculus.* The Lie ring $\{\mathfrak{R}_{\mathfrak{G}}\}$ with c. p. is isomorphic to the matrix Lie ring $\{\mathfrak{R}_{\mathfrak{A}}\}$ with c. p. Hence any matrix Lie ring with c. p. homomorphic to $\{\mathfrak{R}_{\mathfrak{G}}\}$ with c. p. is *completely reducible* and can be obtained by decomposing the *infinitesimal Kronecker's product* from the matrix Lie ring with c. p.

$$\left\| \begin{matrix} \{\mathfrak{R}_1\} & & & \\ & \{\mathfrak{R}_2\} & & \\ & & \ddots & \\ & & & \{\mathfrak{R}_l\} \end{matrix} \right\|$$

Here $\mathfrak{R}_i$ are the Lie rings with r. p. obtained in III. Hence $\{\mathfrak{R}_i\}$ with c. p. are *irreducible* matrix Lie rings homomorphic to $\{\mathfrak{R}_{\mathfrak{G}}\}$. This constitutes a generalisation of the fundamental theorem of the tensor calculus.[4]

---

1) J. von Neumann: Ann. of Math. **34** (1933). L. Pontrjagin: C. R. **198** (1934). H. Freudenthal: Ann. of Math. **37** (1936).

2) E. R. van Kampen: Ann. of Math. **37** (1936).

3) J. von Neumann: Trans, American Math. Soc. **36, 3** (1934).

4) Cf. H. Weyl: loc. cit.

On the duality theorem of non-commutative compact groups

Proc. Imp. Acad. Tokyo **19** (1943) 181–183

(Comm. by T. TAKAGI, M.I.A., April 12, 1943.)

1. The duality theorem of L. Pontrjagin[1] concerning the compact commutative group was, by T. Tannaka[2], ingeneously extended to arbitrary compact group. To this extension, S. Bochner[3] and M. Krein[4] respectively gave new proofs recently. They three all start with the proof of the positivity or the continuity of certain homomorphisms. The proof given below is a direct one and will be shorter than theirs. It may be considered as a simplification of Tannaka's original proof. Our method lies in the use of Gelfand-Silov's abstraction[5] of Weierstrass' polynomial approximation theorem.

2. Let $\mathfrak{G}$ be a compact (=bicompact) Hausdorff group and let $\mathfrak{U}$ be a complete set of mutually inequivalent, continuous, unitary, irreducible representations $U(s)=\big(u_{ij}(s)\big)$ of $\mathfrak{G}$. The completeness (Peter-Weyl-Neumann's theory of almost periodic functions) implies:
1) for any pair of distinct points $s, t \in \mathfrak{G}$, there exists an $U(s) \in \mathfrak{U}$ such that $U(s) \neq U(t)$, 2) if $U_1(s)$, $U_2(s) \in \mathfrak{U}$ then the *product (complex conjugate) representation* $U_1(s) \times U_2(s)\big(\bar{U}_1(s)\big)$ is, as a unitary representation of $\mathfrak{G}$, completely reducible to a sum of a finite number of representations $\in \mathfrak{U}$. Let $\mathfrak{R}$ be the totality of *Fourier polynomials*:

$$x(s)=\sum a_{ij}^{(\eta)}u_{ij}^{(\eta)}(s),$$

viz. finite linear combinations of $u_{ij}^{(\eta)}(s)$, where $\big(u_{ij}^{(\eta)}(s)\big) \in \mathfrak{U}$ and $a_{ij}^{(\eta)}$ denote complex numbers. By 2) $\mathfrak{R}$ is a ring with unit e $\big(e(s)\equiv 1$ on $\mathfrak{G}\big)$ and complex multipliers. Here the sum and the multiplication in $\mathfrak{R}$ is the ordinary function sum and function multiplication. Let $\mathfrak{T}$ be the totality of the linear homomorphisms T of $\mathfrak{R}$ onto the field $\mathfrak{K}$ of complex numbers such that

$$(1) \quad \begin{cases} T \cdot e = 1, \\ T \cdot \bar{x} = \overline{T \cdot x} \quad \big(\text{bar indicates complex conjugates}: \ \bar{x}(s)=\overline{x(s)}\big). \end{cases}$$

$\mathfrak{T}$ is not void since each $s \in \mathfrak{G}$ induces such a homomorphism T_s:

---

1) Topological groups, Princeton (1939).

2) Über den Dualitätssatz der nichtkommutativen topologischen Gruppen, Tôhoku Math. J., **45** (1938).

3) 位相數學, 第 **4** 卷, 第 **1** 號 (昭和 17 年).

4) On positive functionals on almost periodic functions, C. R. URSS, **30** (1941).

5) Über verschiedene Methoden der Einführung der Topologie in die Menge der maximalen Ideale eines normierten Ringes, Rec. Math., **9, 7** (1941). Cf. H. Nakano: 連續函數ノ ring 及ビ vector lattice, 全國紙上數學談話會 **218** (1941).

$$(2) \qquad T_s \cdot x = x(s) \, .$$

We have, by 1), $\qquad T_s \neq T_t \quad \text{if} \quad s \neq t \, .$

$\mathfrak{T}$ may be considered as a group which contains $\mathfrak{G}$ as a subgroup.

Proof: We define the product $T = T_1 T_2$ in $\mathfrak{T}$ as follows. Let $U(s) = \big(u_{ij}(s) \big)_1^n$ be any member of $\mathfrak{U}$, then we put

$$(3) \qquad T \cdot u_{ij} = \sum_{k=1}^n (T_1 \cdot u_{ik})(T_2 \cdot u_{kj}) \qquad (i, j = 1, 2, \ldots, n) \, .$$

Because of the linear independence of $u_{ij}^{(\eta)}(s)$'s, $\big(u_{ij}^{(\eta)}(s) \big) \in \mathfrak{U}$, we may extend T linearly on the whole $\mathfrak{R}$. It is easy to see that the extension T is also a member of $\mathfrak{T}$ and that $\mathfrak{T}$ is a group with the unit T_{s_0} ($s_0 =$ the identity of $\mathfrak{G}$). $\mathfrak{G}$ is thus isomorphically embedded in the group $\mathfrak{T}$ by the correspondence $s \leftrightarrow T_s$.

Next introduce a *weak topology* in $\mathfrak{T}$ by taking for a neighbourhood of the unity T_{s_0} every set

$$\mathop{\mathfrak{E}}_T \{ | T \cdot x_i - T_{s_0} \cdot x_i | < \varepsilon, \quad x_i \in R, \quad i = 1, 2, \ldots, n \}$$

$\mathfrak{T}$ is a compact Hausdorff space as a closed subset of the infinite dimensional torus. It is easily seen that the isomorphic embedding $s \leftrightarrow T_s$ is also a topological one. Hence $\mathfrak{G}$ may be considered as a closed subgroup of the compact Hausdorff group $\mathfrak{T}$. In the truth we have the

Theorem (of T. Tannaka). $\mathfrak{T} = \mathfrak{G}$, *viz. every* $T \in \mathfrak{T}$ *is equal to a certain* T_s:

$$T \cdot x = x(s) \, , \qquad x \in \mathfrak{R} \, .$$

Proof. By the weak topology each $x(s) \in \mathfrak{R}$ may be considered as a continuous function $x(T)$ on the compact Hausdorff space $\mathfrak{T}$ such that $x(T_s) = x(s)$. The ring $\mathfrak{R}(\mathfrak{T})$ of continuous functions $x(T)$, $x \in \mathfrak{R}$, satisfies the following three conditions: i) $1 = e(T) \in \mathfrak{R}(\mathfrak{T})$, ii) for any two distinct points $T_1 \neq T_2$ of $\mathfrak{T}$ there exists $x \in \mathfrak{R}$ such that $x(T_1) \neq (T_2)$, iii) for any $x(T) \in \mathfrak{R}(\mathfrak{T})$ there exists the complex conjugate function $\bar{x}(T) = \overline{x(T)}$ in $\mathfrak{R}(\mathfrak{T})$.

Now let $\mathfrak{T} - \mathfrak{G}$ be not void. Then there exists a point $T_0 \in \mathfrak{T} - \mathfrak{G}$ and a continuous function $y(T)$ on $\mathfrak{T}$ such that

$$(4) \qquad y(T) \geq 0 \quad \text{on} \quad \mathfrak{T} \, , \qquad y(s) = 0 \quad \text{on} \quad \mathfrak{G} \quad \text{and} \quad y(T_0) = 1 \, .$$

By i)–iii) and the Gelfand-Silov's theorem referred to above, there exists, for any $\varepsilon > 0$, $x(s) = \sum a_{ij}^{(\eta)} u_{ij}^{(\eta)}(s) \in \mathfrak{R}$ such that

$$| y(T) - \sum a_{ij}^{(\eta)} u_{ij}^{(\eta)}(T) | \leq \varepsilon \quad \text{on} \quad \mathfrak{T}$$

and, in particular,

$$| y(s) - \sum a_{ij}^{(\eta)} u_{ij}^{(\eta)}(s) | \leq \varepsilon \quad \text{on} \quad \mathfrak{G} \, .$$

Let $u_{11}^{(\eta_0)}(s) = e(s) \equiv 1$, then by taking *Haar-Neumann's mean* we obtain

$$\left| \underset{T}{M}\big(y(T)\big) - a_{11}^{(\eta_0)} \right| \leqq \varepsilon, \qquad \left| \underset{s}{M}\big(y(s)\big) - a_{11}^{(\eta_0)} \right| \leqq \varepsilon.$$

This is a contradiction, since we have $\underset{T}{M}\big(y(T)\big) > 0$, $\underset{s}{M}\big(y(s)\big) = 0$ from (4).

Q. E. D.

Remark 1. The above proof also gives a new proof of a theorem of E. R. van Kampen[1] which is an extension of W. Burnside's theorem[2]. This was kindly pointed out to me by T. Tannaka. It is to be noted that Kampen's theorem plays an important rôle in Tannaka's proof.

Remark 2. In another note the author will give a proof of Pontrjagin's duality theorem for locally compact commutative group, by combining the generalized Plancherel's theorem[3] and the theory of Haar's measure[4]. It intends to be a simplification of Raikov's proof[5], published recently.

---

1) Almost periodic functions and compact groups, Ann. of Math., **37** (1936).

2) Theory of groups of finite order, 2nd ed., Cambridge (1911), 299.

3) M. Krein: Sur une généralisation du théorème de Plancherel au cas des intégrales de Fourier sur les groupes topologiques commutatifs, C, R. URSS, **30** (1941).

4) A. Weil: La réciproque du théorème de Haar dans "L'integration dans les groupes topologiques et ses applications," Paris (1940). Cf. also K. Kodaira: Über die Beziehung zwischen den Massen und Topologien in einer Gruppe, Proc. Phys.-Math. Soc. Japan, **23** (1941).

5) Generalized duality theorem for commutative groups with an invariant measure, C. R. URSS, **30** (1941).

Equivalence of two topologies of Abelian groups

(with T. Iwamura)

Proc. Imp. Acad. Tokyo **20** (1944) 451–453

(Comm. by T. Takagi, m.i.a., July 12, 1944.)

Let G be a locally compact (=bicompact), separable abelian group and let X be the totality of continuous characters[1] $\chi(g)$ of G. It is well known[2] that X is also a locally compact, separable abelian group by the multiplication

$$\chi_1\chi_2(g)=\chi_1(g)\chi_2(g)$$

and by Pontrjagin's topology induced from the (closed) neighbourhood:

$$U(\chi_1)=\{\chi\,;\,\sup_{g\in G_0}|\,\chi(g)-\chi_1(g)\,|\leqq\varepsilon,\qquad G_0=\text{compact subset of }G\}\,.$$

X also constitutes a locally compact, separable topological space $\widetilde{X}$ by the topology induced from the (closed) neighbourhood:

$$\widetilde{V}(\chi_1)=\Big\{\chi\,;\,\Big|\int_G x_i(g)\chi(g)dg-\int_G x_i(g)\chi_1(g)dg\,\Big|\leqq\varepsilon,\quad i=1,2,\ldots,n\Big\}$$

where $x_i(g)\in L_1(G)$ viz. $x_i(g)$ denote measurable functions integrable over G with respect to Haar's invariant measure dg on G. The latter topology is introduced by I. Gelfand and D. Raikov[3], and its equivalence to Pontrjagin's topology plays a fundamental rôle in the ring-theoretic treatment and extension of the classical Fourier analysis based upon the theory of normed ring[4]. However the proof of the equivalence is, so far as we know, not published by the Russian school, though stated and used by them repeatedly[5].

The purpose of the present note is i): to give it a proof and ii) to show that the character group is a topological group in Gelfand-Raikov's topology even when G is not separable. For the purpose we make use of the following

Lemma. For any χ_2, the mapping

$$\chi\to\chi_2\chi$$

---

1) A continuous character of G is a continuous homomorphic mapping of G in the topological group of rotations of a circle.

2) L. Pontrjagin: Topological group, Princeton (1939), 127.

3) C. R. URSS, **28**, 3 (1940).

4) D. Raikov: C. R. URSS, **28**, 4 (1940). M. Krein: C. R. URSS, **30**, 6 (1941). D. Raikov: C. R. URSS, **30**, 7 (1941). K. Yosida: Proc. **20** (1944), 269. The author (Yosida) wishes to withdraw the §3 of this note, since the Lemma 2 is valid for $z\in L_1(G)$ only and thus the arguments in §3 is insufficient. A complete proof and the fact that Bochner-Raikov's theorem may be derived from Plancherel's theorem will be published elsewhere.—During the proof, Y. Kawada kindly communicated that 3° may be obtained from Bochner-Raikov's theorem.

5) H. Anzai kindly communicated M. Fukamiya's unpublished proof of the equivalence, which is entirely different from ours.

of $\widetilde{X}$ on $\widetilde{X}$ is a topological one.

Proof. The inclusion $\chi \in \widetilde{U}(\chi_1)$ is equivalent to the inclusion $\chi_2\chi \in \widetilde{U}(\chi_2\chi_1)$ where

$$\begin{cases} U(\chi_2\chi_1) = \left\{\chi'\; ; \left|\int_G x_i'(g)\chi'(g)dg - \int_G x_i'(g)\chi_2\chi_1(g)dg\right| \leqq \varepsilon, \quad i=1,2,\ldots,n\right\} \\ x_i'(g) = x_i(g)\chi_2^{-1}(g), \qquad \chi_2^{-1}(g) = \chi_2(-g) \end{cases}$$

Q. E. D.

Proof of i). It is evident that the mapping

$$(1) \qquad\qquad X \ni \chi \to \chi \in \widetilde{X}$$

is continuous. Hence, by the lemma, we have only to show that for any neighbourhood $U(\chi_0)$ in X of the unit-character $\chi_0\big(\chi_0(g)\equiv 1\big)$ there exists a neighbourhood $\widetilde{V}(\chi_0)$ in $\widetilde{X}$ of χ_0 satisfying $U(\chi_0) \supseteqq \widetilde{V}(\chi_0)$ as subset (without topology) of X.

Let $W(\chi_0)$ be a compact and symmetric neighbourhood in X of χ_0:

$$(2) \qquad\qquad W(\chi_0) = W(\chi_0)^{-1} = \{\chi^{-1}; \chi \in W(\chi_0)\}$$

such that

$$(3) \qquad\qquad U(\chi_0) \supseteqq W(\chi_0)^2 = \{\chi_a\chi_\beta; \chi_a, \chi_\beta \in W(\chi_0)\}.$$

Since X is separable there exists an enumerable sequence $\{\chi_i\} \subseteqq X$ such that

$$X = \bigcup_{i=1}^{\infty} \chi_i W(\chi_0),$$

where

$$(4) \qquad\qquad \chi_i W(\chi_0) = \{\chi; \chi = \chi_i\chi', \chi' \in W(\chi_0)\}.$$

By the continuity of the mapping (1), the image $\widetilde{\chi_i W}(\chi_0)$ in $\widetilde{X}$ of the compact set $\chi_i W(\chi_0)$ in X is also a compact set of $\widetilde{X}$. $\widetilde{X}$ being complete as a locally compact space, at least one compact set $\widetilde{\chi_i W}(\chi_0)$ contains a neighbourhood $\widetilde{V}_1(\chi_a)$. This results from the fact that a complete space is not of Baire's first category. Hence, by the lemma, the neighbourhood $\chi_a^{-1}\widetilde{\chi_i W}(\chi_0)$ in $\widetilde{X}$ contains a neighbourhood $\widetilde{V}_1(\chi_0)$ in $\widetilde{X}$ of χ_0. Thus there exists χ such that the neighbourhood $\chi\widetilde{W}(\chi_0)$ in $\widetilde{X}$ contains a neighbourhood in $\widetilde{X}$ of χ_0. Hence $\chi\widetilde{W}(\chi_0) \ni \chi_0$ and thus $\widetilde{W}(\chi_0) \ni \chi^{-1}$, $\widetilde{W}(\chi_0) \ni \chi$ by (2). Therefore, by (3), there exists a neighbourhood $\widetilde{V}(\chi_0)$ in $\widetilde{X}$ of χ_0 which satisfies

$$U(\chi_0) \supseteqq W(\chi_0)^2 \supseteqq \widetilde{V}(\chi_0), \qquad \text{as subsets (without topology) of } X.$$

Q. E. D.

Remark. The separability of G is only used in (4). Hence i) holds good if, for example, X is compact or connected.

Proof of ii). Since $x(g) \in L_1(G)$ implies $y(g) = x(-g) \in L_1(G)$ and v. v., a subset $\widetilde{V}$ of $\widetilde{X}$ is a neighbourhood $\chi \in \widetilde{X}$ if and only if V^{-1} is a neighbourhood of χ^{-1}. Thus χ^{-1} is a continuous function of χ.

It remains to show that $\chi_1\chi_2$ is a continuous function of two variables χ_1, χ_2. By the lemma and by the commutativity of $\widetilde{X}$, it is sufficient to prove that for every neighbourhood $\widetilde{V}$ of the unit-character χ_0 there exists a neighbourhood $\widetilde{W}$ of χ_0 such that $\widetilde{W}^2 \subseteq \widetilde{V}$.

A generic neighbourhood of χ_0 is the intersection of a finite number of sets of the form

$$\widetilde{U}(x,\varepsilon)=\left\{\chi\,;\,\left|\int_G x(g)\chi_0(g)dg - \int_G x(g)\chi(g)dg\right|<\varepsilon\right\},$$

where $\varepsilon>0$ and $x(g)\in L_1(G)$. For any such ε, x there exists a step function $y(g)\in L_1(G)$ satisfying $\|x-y\|<\varepsilon/3$. Since this inequality implies that for every $\chi\in\widetilde{X}$ $\left|\int_G x(g)\chi(g)dg - \int_G y(g)\chi(g)dg\right|<\varepsilon/3$, we have $\widetilde{U}(y,\varepsilon/3)\subseteq\widetilde{U}(x,\varepsilon)$. We may therefore take as $x(g)$ only step functions from $L_1(G)$, and so only characteristic functions x_E of measurable subsets E of G with $0<m(E)<\infty$ (m indicates Haar's measure).

Put

$$\widetilde{U}(E,\varepsilon)=\left\{\chi\,;\,\left|\int_G x_E(g)\chi_0(g)dg - \int_G x_E(g)\chi(g)dg\right|<\varepsilon\right\}$$

$$=\left\{\chi\,;\,\left|\int_E (1-\chi(g))dg\right|<\varepsilon\right\}.$$

Let $\varepsilon'>0$ and put $D(\chi,E,\varepsilon')=\{g\,;\,g\in E,\,|1-\chi(g)|>\varepsilon'\}$; this is a measurable set with finite measure. It is easily seen that, if $\varepsilon'<\varepsilon$ and if ε', $\delta>0$ are sufficiently small, the set $V(E,\varepsilon',\delta)=\left\{\chi\,;\,m\big(D(\chi,E,\varepsilon')\big)<\delta\right\}$ is contained in $\widetilde{U}(E,\varepsilon)$. Moreover, we have always $V(E,\varepsilon'/2,\delta)^2\subseteq V(E,\varepsilon',\delta)$. Hence it is sufficient to show that for any ε', $\delta>0$ there exists an $\varepsilon>0$ such that $\widetilde{U}(E,\varepsilon)\subseteq V(E,\varepsilon',\delta')$. Such an ε may be given, as will be shown below, by $\varepsilon=\varepsilon''\delta$, where $\varepsilon''=\inf_{|1-\chi(g)|>\varepsilon'}\cdot\big(1-\Re\chi(g)\big)$, $\Re$ indicating the real part of a complex number. Since $|\chi(g)|=1$, we have $1-\Re\chi(g)\geq 0$ everywhere and $\varepsilon''>0$.

Suppose $\chi\in\widetilde{U}(E,\varepsilon)$, i.e. $\left|\int_E(1-\chi(g))dg\right|<\varepsilon$. Then $D(\chi,E,\varepsilon')$ coincides with $C=\{g\,;\,g\in E,\,1-\Re\chi(g)>\varepsilon''\}$. Since $\left|\int_E(1-\chi(g)dg\right|^2=\left\{\int_E(1-\Re\chi(g))dg\right\}^2+\left\{\int_E\Im\chi(g)dg\right\}^2$, where $\Im$ indicates the imaginary part of a complex number, we have

$$\varepsilon>\int_E\big(1-\Re\chi(g)\big)dg\geq\int_C\big(1-\Re\chi(g)\big)dg\geq\varepsilon''m(C)=\varepsilon''m\big(D(\chi,E,\varepsilon')\big).$$

Hence $m\big(D(\chi,E,\varepsilon')\big)<\varepsilon/\varepsilon''=\delta$, i.e. $\chi\in V(E,\varepsilon',\delta)$. Thus we have shown $\widetilde{U}(E,\varepsilon)\subseteq V(E,\varepsilon',\delta)$, and the proof is complete.

（東京 八五四）

III. Ergodic Theory

Comments (Yuji Ito)

The mean ergodic theorem proved by J. von Neumann and the individual (point-wise) ergodic theorem proved by G. D. Birkhoff in the years 1931–32 started the development of ergodic theory. Soon after their appearance, a number of mathematicians including F. Riesz, C. Visser, E. Hopf, A. Khintchine, A. Kolmogoroff, N. Wiener, N. Dunford and others tried to simplify the original proofs of these theorems and to generalize them in various directions. Early development of the subject was reported in the excellent monograph *Ergodentheorie* written by Hopf which appeared in 1937. Inspired, no doubt, by this monograph, Yosida got very much interested in ergodic theory and from 1938 to 1941 he worked intensively in this area and obtained, in collaboration with S. Kakutani, a number of significant results generalizing both mean and individual ergodic theorems and applied them to give a unified treatment of the then developing theory of Markov processes. We will describe below in some detail Yosida's contributions to ergodic theory and related topics.

1. On the mean and the uniform ergodic theorems and their applications to the theory of Markov processes.

The original mean ergodic theorem of von Neumann stated that if T is a unitary operator on a Hilbert space $\mathcal{H}$, then for any $x \in \mathcal{H}$, the sequence $x_n = \dfrac{1}{n} \sum_{k=1}^{n} T^k x$, $n = 1, 2, \ldots$ converges strongly to a point in $\mathcal{H}$.

This theorem was subsequently generalized in two directions: (1) to more general Banach spaces and (2) to more general classes of linear operators. Visser was the first one to extend this theorem to the case of more general bounded linear operators T on a Hilbert space by showing that if the norms of the iterates T^n, $n = 1, 2$, ... are uniformly bounded, then the sequence x_n converges weakly. Kakutani then noticed that this theorem of Visser can be proved in any Banach space if the operator T is weakly compact, an assumption which is automatically satisfied in Visser's case. Then, Yosida in paper [25] made a striking observation that in a general Banach space the existence of a weakly convergent subsequence of x_n, defined as above for operators whose iterates are uniformly bounded, implies the strong convergence of the sequence x_n itself, and thus, combining Kakutani's result he was able to prove a very general and definitive extension of the mean ergodic theorem. In a joint work with Kakutani [26], it was noted that if T is a linear operator on a Banach space which is quasi-weakly compact (i.e., there is a weakly compact operator V and an integer $m \geq 1$ such that $\|T^m - V\| < 1$), then T satisfies the key hypothesis made for the mean ergodic theorem proved by Yosida in [25] mentioned above, and therefore, if such a T has uniformly bounded iterates, then the mean ergodic theorem is valid for T.

Yosida and Kakutani realized that the mean ergodic theorem extended as above can be applied to operators T which do not necessarily possess an inverse, and therefore, can be utilized where irreversible processes are involved. They showed, in particular, that many of the results in the theory of Markov processes previously obtained by W. Doeblin, J. Doob, M. Fréchet and N. Kryloff and N. Bogoliouboff can be obtained operator theoretically by using their mean ergodic theorem and the following theorem, which they called the uniform ergodic theorem:

If T is a linear operator on a Banach space which is quasi-compact (i.e., there is a compact operator V and an integer $m \geq 1$ such that $\|T^m - V\| < 1$), then T has only a finite number of eigenvalues $\{\lambda_1, \ldots \lambda_N\}$ of absolute value 1, each of finite multiplicity, and there exists a constant M such that for $i = 1, \ldots, N$

$$\left\| \frac{1}{n} \sum_{k=0}^{n-1} \left(\frac{T}{\lambda_i} \right)^k - P_i \right\| \leq \frac{M}{n} \ (n = 1, 2, \ldots),$$ where P_i is the projection onto the subspace $\{x \,|\, Tx = \lambda_i x\}$.

Both Yosida and Kakutani published their own proofs of this theorem (which are considerably different) in separate papers. Yosida's proof appeared in [24]. Ideas and methods introduced in these papers are clarified and polished subsequently and their efforts led to the now classic paper [38] that appeared in the Annals of Mathematics in 1941. Results obtained by them in a series of papers (among them the papers [24], [25], [26], [28], [30], [32]), including both the mean and the uniform ergodic theorems mentioned above, are reorganized with strengthening of some of the results and simplification of some proofs and are presented in this paper in a beautiful exposition. Markov processes are treated by the consideration of two linear operators induced by them:

$$x \to T(x) = y : y(E) = \int_\Omega x(dt) P(t, E)$$

defined on the Banach space (M) of complex-valued countably additive set functions, and

$$x \to \overline{T}(x) = y : y(t) = \int_\Omega P(t, ds) x(ds)$$

defined on the Banach space (M^*) of complex-valued bounded Borel measurable functions. The authors assume that the Markov process they consider satisfies the condition imposed by Kryloff and Bogoliouboff stating that the operator T defined above is quasi-compact. With this assumption they can apply their uniform ergodic theorem and with further analysis they obtain the decomposition of the state space into ergodic kernels and the dissipative part, and further decomposition of each ergodic part into cyclic subergodic kernels. These results are more precise than those obtained earlier by Doeblin in his investigation of Markov processes satisfying the so-called Doeblin condition, which is weaker than the conditions imposed by Doob, Fréchet and others. Yosida and Kakutani then go on to show that the Doeblin condition actually implies the Kryloff-Bogoliouboff condition, and therefore, they are able to obtain all the results of Doeblin in a more precise form with less hypothesis.

Later in paper [35], Yosida proves a mean ergodic theorem in the space $L^1(\phi)$ for an operator arising from Markov process $P(x, E)$ with a stable distribution (i.e., finite invariant measure) ϕ. For this operator, the hypothesis of the general mean ergodic theorem mentioned earlier is not satisfied. In [35], Yosida investigates further the question of the existence of finite invariant measure and extends the results of Kryloff-Bogoliouboff given for homeomophisms to the case of Markov processes when the state space is compact and the operator induced from the process maps the space of continuous functions into itself. The results are extended further in [56] to the case where the state space of the Markov process is locally compact, and in addition, the decomposition of the state space into ergodic and dissipative parts were obtained. In an earlier paper with Kakutani [33], the same type of argument was used to obtain finite invariant measure (called the mean sojourn in that paper) and the decomposition of the state space into dissipative and ergodic parts in case of a countably infinite state space.

2. On the individual ergodic theorems.

The original ergodic theorem of Birkhoff stated that if ϕ is a measure preserving transformation of a measure space $(S, \mathscr{B}, m)$, then for any function $x(s) \in L^1(S)$, the sequence $x_n(s) = \dfrac{1}{n} \sum_{k=0}^{n-1} x(\phi^k(s))$, $n = 1, 2, \ldots$, converges almost everywhere on S to a function $\bar{x}(s) \in L^1(S)$.

Birkhoff actually discussed the case where S is a manifold and ϕ is a measure preserving homeomorphism of S, but the general case where S is a space with measure but no topology was discussed by Khintchine and A. Stepanov. A number of people tried to simplify the original proof of this theorem by Birkhoff, and to extend the result to more general classes of transformations and to operators defined on $L^p(S)$. Kolmogoroff in 1937 gave a proof simpler than the original one by proving and using a maximal inequality. Wiener proved in 1935 the so-called dominated ergodic theorem, and showed that the individual ergodic theorem can be deduced from the mean ergodic theorem of von Neumann with the use of this dominated ergodic theorem.

Inspired by these papers of Kolmogoroff and Wiener, Yosida, together with Kakutani, introduced in paper [31] what they called the maximal ergodic theorem and showed that both the Birkhoff ergodic theorem and the dominated ergodic theorem can be derived directly (without appealing to the mean ergodic theorem) from this maximal ergodic theorem. Their proof of the maximal ergodic theorem itself is quite elegant and elementary, and this paper inspired a host of susequent work by a number of people including F. Riesz, Y. Dowker, B. Marcus and K. Petersen.

Yosida also made a significant contribution to the subject of individual ergodic theorems by showing in paper [34] that the mean ergodic theorem together with a theorem of Banach and Saks will give the individual ergodic theorem if for any

$x \in L^1(S)$ $\varlimsup\limits_{n \to \infty} |x_n(s)| < \infty$ almost everywhere on S. As the mean ergodic theorem of

Yosida and Kakutani mentioned earlier applies to operators T which are much more general than the unitary operator $x \to Vx$: $Vx(s) = x(\phi s)$, induced from a measure preserving transformation ϕ, which was considered for the Birkhoff theorem, Yosida's idea of utilizing the Banach-Saks theorem in addition to the mean ergodic theorem attracted attention of a number of people who tried to extend the individual ergodic theorem to more general operators on $L^p(S)$. Yosida himself has formulated an operator theoretic version of individual ergodic theorems in a general setting in papers [24] and [36] and showed among other things that he can deduce from his general theorem individual ergodic theorems for N-parameter flows which were obtained earlier by Wiener. Then in 1954 Hopf was able to extend by exploiting Yosida's idea the individual ergodic theorem (in all $L^p(S)$ $(1 \leq p < \infty)$) to positive operators T which map $L^1(S)$ into $L^1(S)$ and $L^\infty(S)$ into $L^\infty(S)$ and satisfies $\|T\|_1 \leq 1$ and $\|T\|_\infty \leq 1$. Later, Dunford and Schwartz showed that one can even remove the assumption of positivity in Hopf's theorem. Although the subsequent work by A. Garcia, A. Brunel and others made it unnecessary to appeal to Yosida's method in proving Hopf-Dunford-Schwartz type individual ergodic theorems, it provided one of the key elements in the development of the theory surrounding individual ergodic theorems.

3. On other related limit theorems.

For semi-groups of linear operators T_t on a Banach space X, ergodic theorems ask for the existence of the limit $\lim\limits_{t \to \infty} \dfrac{1}{t} \displaystyle\int_0^t T_s ds$ taken in various sense (mean convergence, almost everywhere convergence and so on). E. Hille considered instead the existence of the limits of the form

$$\lim_{\lambda \downarrow 0} \int_0^\infty e^{-\lambda t} T_t dt \qquad \text{and} \qquad \lim_{\lambda \uparrow \infty} \int_0^\infty e^{-\lambda t} T_{tdt}$$

and called the results abelian ergodic theorems. These problems can be stated in terms of the resolvent operators associated with the given semi-group, and Yosida in paper [86] showed that one can prove theorems more general than those of Hille by proving limit theorems for pseudo-resolvents on a complete locally convex linear topological space.

Abstract integral equations and the homogeneous stochastic process

Proc. Imp. Acad. Tokyo **14** (1938) 286–291

(Comm. by T. TAKAGI, M.I.A., Oct. 12, 1938.)

§1. *Introduction.* Let each point x of a complex Banach space $\mathfrak{B}$ represent a state (x) of a physical or a mathematical system. Consider a *temporally homogeneous stochastic process* by which the state (x) is transferred to the state (y) after the elapse of a unit time. We assume that this transition is realised by a *linear mapping* T in $\mathfrak{B}$: $y = T \cdot x$. Under their respective restrictions on T and on $\mathfrak{B}$, A. Markov, B. Hostinsky, M. Fréchet, N. Kryloff-N. Bogoliouboff and other authors investigated the asymptotic behaviour of the n-th iterate T^n of T for large n. In the present note I intend to treat the problem by the *abstract integral equations* due to F. Riesz[1] and the theory of *resolvents* due to M. Nagumo.[2] The theorem below is a generalisation of Fréchet-Kryloff-Bogoliouboff's theorem.[3] The lemma 1 and the lemma 3 respectively generalise the theorem of Riesz and that of Nagumo. I express my hearty thanks to S. Kakutani who kindly collaborated with me in the discussion of the present note.[4] In the next paper[5] the *mean ergodic theorem* of J. von Neumann is extended to $\mathfrak{B}$, in a way as to be applied to the problem of the homogeneous stochastic process.

§2. *The theorem.* A linear mapping T of a complex Banach space $\mathfrak{B}$ in $\mathfrak{B}$ is called a (linear) *operator* in $\mathfrak{B}$. T is called *continuous* if its *norm* (absolute value) $\|T\| = \text{l.u.b.}_{\|x\|\leq 1} \|T \cdot x\|$ is finite. A continuous operator T is called *completely continuous* if it maps the unit sphere $\|x\| \leq 1$ of $\mathfrak{B}$ on a compact point set in $\mathfrak{B}$.

Let T satisfy the following two conditions:

(1) there exists a completely continuous operator V such that $\|T - V\| < 1$,

(2) there exists a constant a such that $\|T^n\| \leq a$ for $n = 1, 2, \ldots$.

Then we obtain the

Theorem. The proper values of T with modulus. 1 are isolated proper values of finite multiplicities. Let these proper values be $\lambda_1, \lambda_2, \ldots, \lambda_k$. Then there exist completely continuous operators $T_1, T_2 \ldots, T_k$, a continuous operator S and positive constants β, ε such that

1) Acta Math. **41** (1918), 71–98.

2) Jap. J. of Math. **13** (1936), 75–80.

3) M. Fréchet: Quart. J. of Math. **5** (1934), 106–144. N. Kryloff and N. Bogoliouboff: C. R. Paris, **204** (1937), 1386–1388.

4) He also obtained another proof of our theorem, by virtue of the mean ergodic theorem in $\mathfrak{B}$. See the following paper of Kakutani.

5) Proc. **14** (1938), 292.

$$(3) \quad \begin{cases} T = \sum_{i=1}^{k} \lambda_i T_i + S, \ \ T_i^2 = T_i, \ \ T_i T_j = 0 (i \neq j), \ \ T_i S = S T_i = 0 \\ \qquad\qquad\qquad\qquad\qquad (i, j = 1, 2, \ldots, k), \\ \| S^n \| \leq \beta/(1+\varepsilon)^n \qquad (n = 1, 2, \ldots\ldots). \end{cases}$$

Corollary 1. *There exist positive constants γ_λ such that, if $|\lambda| = 1$,*

$$(4) \quad \begin{cases} \left\| \dfrac{(T/\lambda) + (T/\lambda)^2 + \cdots + (T/\lambda)^n}{n} - T_\infty(\lambda) \right\| \leq \gamma_\lambda/n \qquad (n = 1, 2, \ldots), \\ T_\infty(\lambda) = T_i \ \text{if} \ \lambda = \lambda_i, \ \ T_\infty(\lambda) = 0 \ \text{if} \ \lambda \neq \lambda_1, \lambda_2, \ldots, \lambda_k. \end{cases}$$

Corollary 2. *$(T/\lambda)^n$ converges (necessarily to $T_\infty(\lambda)$) if and only if there are no proper values of T with modulus 1 other than λ.*

Corollary 3. *We replace the condition (1) by*

$$(5) \quad \begin{cases} \text{there exist positive integer } m \text{ and a completely continuous} \\ \text{operator } V \text{ such that } \| T^m - V \| < 1. \end{cases}$$

Then there exist positive constants γ'_λ such that, if $|\lambda| = 1$,

$$\left\| \frac{(T/\lambda) + (T/\lambda)^2 + \cdots + (T/\lambda)^n}{n} - T'_\infty(\lambda) \right\| \leq \gamma'_\lambda/n \qquad (n = 1, 2, \ldots),$$

$$T'_\infty(\lambda) = \frac{(T/\lambda) + (T/\lambda)^2 + \cdots + (T/\lambda)^{m-1}}{m-1} \lim_{n \to \infty} \frac{(T/\lambda)^m + (T/\lambda)^{2m} + \cdots + (T/\lambda)^{nm}}{n}.$$

Remark.[1] Put $T_0 = E - \sum_{i=1}^{k} T_i$, where E denotes the identical mapping of $\mathfrak{B}$. Then, by (3), $T_0^2 = T_0$, $T_0 T_i = T_i T_0 = 0$ $(i \geq 1)$. Hence, if $\mathfrak{B}_j$ denotes the image of $\mathfrak{B}$ by T_j, we have the direct decomposition $\mathfrak{B} = \mathfrak{B}_0 + \mathfrak{B}_1 + \cdots + \mathfrak{B}_k$. Each point of $\mathfrak{B}_j$ is invariant by T_j, as $T_j^2 = T_j$. $\mathfrak{B}_i \ (i \geq 1)$ is of finite dimension by Riesz's theorem since $T_i^2 = T_i$ and $T_i \ (i \geq 1)$ is completely continuous. Let $x \in \mathfrak{B}_0$, then $T \cdot x = T T_0 \cdot x = S \cdot x$, $\ldots, T^n \cdot x = S^n x$. Let $x \in \mathfrak{B}_i \ (i \geq 1)$, then $T \cdot x = T T_i \cdot x = \lambda_i T_i \cdot x = \lambda_i x, \ldots,$ $T^n \cdot x = \lambda_i^n \cdot x$. Hence $\lim_{n \to \infty} T^n \cdot x = 0$ uniformly for $x \in \mathfrak{B}_0$, and $T^n \cdot x$ $(x \in \mathfrak{B}_i, i \geq 1)$ moves in $\mathfrak{B}_i$ almost periodically with respect to n. $\mathfrak{B}_0$ and $\mathfrak{B}_i \ (i \geq 1)$ may respectively be called the *dissipative* part and the *ergodic* part of $\mathfrak{B}$.

§ 3.   *Three lemmas for the proof of the theorem.*

Lemma 1.[2] *Let T satisfy the condition (1). Then the proper values of T do not accumulate to the point not interior of the unit circle in the complex plane.*

Proof. Put $T = V + U$, then $\| U \| = \delta < 1$. We have to derive a contradiction from

$$(6) \quad T \cdot x_i = \lambda_i \cdot x_i, \ x_i \in \mathfrak{B}, \ x_i \neq 0, \ \lambda_i \neq \lambda_j \ (i \neq j,) \ \lim_{i \to \infty} \lambda_i = \lambda, \ |\lambda| \geq 1.$$

---

1) Cf. N. Kryloff and N. Bogoliouboff: Bult. Soc. Math. France, **64** (1936), 49–56.

2) If T is completely continuous this lemma reduces to the Satz 12 in Riesz, loc. cit. p. 90: the only accumulation point of the proper values of T is the point zero. For, in this case, λT satisfies (1) for any λ.

We have $T^n = T^n - (T-V)^n + (T-V)^n$. $T^n - (T-V)^n$ is completely continuous with V, and $\|(T-V)^n\| \leq \delta^n$, $T^n \cdot x_i = \lambda_i^n \cdot x_i$. Therefore it suffices to derive a contradiction from (6) when $\delta < (1/4)$. This may be carried out as follows.

$x_1, x_2, \ldots, x_n$ are linearly independent for any n. The proof is obtained by induction with respect to n. Let $x_1, x_2, \ldots, x_{n-1}$ be linearly independent and let x_n be linearly dependent with $x_1, x_2, \ldots, x_{n-1}$:
$x_n = \sum_{i=1}^{n-1} a_i x_i$. Then we obtain $\sum_{i=1}^{n-1} a_i (\lambda_n - \lambda_i) x_i = 0$ from $T \cdot x_n = \lambda_n x_n$, $T \cdot x_n = \sum_{i=1}^{n-1} a_i T \cdot x_i$, contrary to the hypothesis of the induction.

Thus the linear space $\Re_{n-1}$ spanned by $x_1, x_2, \ldots, x_{n-1}$ is a proper subspace of the linear space $\Re_n$ spanned by $x_1, x_2, \ldots, x_n$. By Riesz's theorem there exists a sequence $\{y_i\}$ such that $y_i \in \Re_i, \|y_i\| = 1, \|y_i - x\| > (1/2)$ for all $x \in \Re_{i-1}$. We have $T(y_i/\lambda_i) - T(y_j/\lambda_j) = y_i - \{y_i - T(y_i/\lambda_i) + T(y_j/\lambda_j)\}$. $y_i - T(y_i/\lambda_i) \in \Re_{i-1}$ as $y_i \in \Re_i$. Hence

(7) $\qquad \| V(y_i/\lambda_i) - V(y_j/\lambda_j)\| + \delta \|(y_i/\lambda_i) - (y_j/\lambda_j)\| > (1/2) \quad$ for $\quad j < i$.

V being completely continuous and $\|y_i\| = 1$, $\lim_{i \to \infty} \lambda_i = \lambda$, $|\lambda| \geq 1$, there exists a partial sequence $\{i'\}$ of $\{i\}$ such that $\lim \| V(y_{i'}/\lambda_{i'}) - V(y_{j'}/\lambda_{i'})\| = 0$. Thus, by (7), $\delta < (1/4)$, $\|y_{i'}\| = 1$, $\lim \lambda_{i'} = \lambda$ and $|\lambda| \geq 1$, we obtain a contradiction.

Lemma 2. Let $\mathfrak{D}$ be a domain in the complex λ-plane. A family $V(\lambda)$ of completely continuous operators in $\mathfrak{B}$ be regular in $\lambda \in \mathfrak{D}$. Let $\mathfrak{F}$ denotes the set of points (in $\mathfrak{D}$) at each point of which the equation $\big(E + V(\lambda)\big)x_\lambda = 0$ admits non-trivial solution $x_\lambda \neq 0$. Then for each $\lambda \in \mathfrak{D} - \mathfrak{F}$ $E + V(\lambda)$ has a unique (continuous) inverse $E + K(\lambda)$: $\big(E + V(\lambda)\big)\big(E + K(\lambda)\big) = \big(E + K(\lambda)\big)\big(E + V(\lambda)\big) = E$. $K(\lambda)$ is regular in $\lambda \in \mathfrak{D} - \mathfrak{F}$ and is completely continuous for each $\lambda \in \mathfrak{D} - \mathfrak{F}$.

Proof. By Riesz's theorem $E + V(\lambda)$ has a unique (continuous) inverse $E + K(\lambda)$ for each $\lambda \in \mathfrak{D} - \mathfrak{F}$. By $K(\lambda) = -V(\lambda) - K(\lambda)V(\lambda)$, we see that $K(\lambda)$ is completely continuous.

Let $\lambda_0 \in \mathfrak{D} - \mathfrak{F}$, then the series $\Big[E + \sum_{n=1}^{\infty} \{E - \big(E + K(\lambda_0)\big)\big(E + V(\lambda)\big)\}^n\Big]$. $\big(E + K(\lambda_0)\big)$ are absolutely and uniformly convergent for sufficiently small $|\lambda - \lambda_0|$. It is easy to see that this series are the demanded inverse $E + K(\lambda)$.

Lemma 3.[1] Let T satisfy the condition (1). By the lemma 1, the proper values of T with modulus 1 are isolated proper values. Let these proper values be $\lambda_1, \lambda_2, \ldots, \lambda_k$. Then there exists a positive ε such that $E + \lambda T$ admits a unique (continuous) inverse $E + \lambda R_\lambda$ for each λ, $1 - 2\varepsilon < |\lambda| < 1 + 2\varepsilon$, except for $\lambda = -\lambda_1^{-1}, -\lambda_2^{-1}, \ldots, -\lambda_k^{-1}$. R_λ is regular in λ, $1 - 2\varepsilon < |\lambda| < 1 + 2\varepsilon$, except for poles $-\lambda_i^{-1}$ $(i = 1, 2, \ldots, k)$.

Proof. As $\lambda_1, \lambda_2, \ldots, \lambda_k$ are isolated proper values of T, there exists

---

1) If T is completely continuous this lemma reduces to the Satz 12 in Nagumo, loc. cit. p. 79: the resolvent R_λ of T defined by $(E + \lambda T)(E + \lambda R_\lambda) = (E + \lambda R_\lambda)(E + \lambda T) = E$ is meromorphic in $|\lambda| < \infty$. For, in this case, λT satisfies (1) for any λ.

a positive η such that $(E+\lambda T)x_\lambda=0$ does not admit non-trivial solution $x_\lambda \neq 0$ for any λ, $1-\eta<|\lambda|<1+\eta$, except for $\lambda=-\lambda_1^{-1}, -\lambda_2^{-1}, \ldots, -\lambda_k^{-1}$.

Put $T=V+U$. By $\|U\|=\delta<1$, $E+\lambda U$ admits a unique (continuous) inverse $E+\lambda U_\lambda=E+\sum_{n=1}^{\infty}(-\lambda U)^n$ which is regular in $|\lambda|<(1/\delta)$.

We have $(E+\lambda U_\lambda)(E+\lambda T)=E+\lambda V+\lambda^2 U_\lambda V$. Put $V(\lambda)=\lambda V+\lambda^2 U_\lambda V$. It is regular in $\lambda<(1/\delta)$ and is completely continuous with V for each λ, $\lambda<(1/\delta)$.

Let $2\varepsilon=\text{Min.}\big((1/\delta)-1, \eta\big)$. We denote by $\mathfrak{D}$ the domain $1-2\varepsilon<|\lambda|<1+2\varepsilon$, and let $\mathfrak{F}$ be the point set $(-\lambda_1^{-1}, -\lambda_2^{-1}, \ldots, -\lambda_k^{-1})$. Then the equation $\big(E+V(\lambda)\big)x_\lambda=0$ does not admit non-trivial solution $x_\lambda \neq 0$ for any $\lambda \in \mathfrak{D}-\mathfrak{F}$. Assume that there exists a $x_{\lambda_0}\neq 0$, $\lambda_0 \in \mathfrak{D}$, which satisfies $\big(E+V(\lambda_0)\big)x_{\lambda_0}=0$. Then we would have $(E+\lambda_0 T)x_{\lambda_0}=(E+\lambda_0 U)\big(E+V(\lambda_0)\big)x_{\lambda_0}=0$. This shows that $\lambda_0 \in \mathfrak{F}$.

Thus, by the lemma 2, $E+V(\lambda)$ admits a unique (continuous) inverse $E+K(\lambda)$ for each $\lambda \in \mathfrak{D}-\mathfrak{F}$, and $K(\lambda)$ is regular in $\lambda \in \mathfrak{D}-\mathfrak{F}$. We easily verify that $E+\lambda R_\lambda=\big(E+K(\lambda)\big)(E+\lambda U_\lambda)$ is the inverse of $E+\lambda T$ for each $\lambda \in \mathfrak{D}-\mathfrak{F}$.

Let the Laurent expansion of $R_\lambda=\big(K(\lambda)+\lambda U_\lambda+\lambda K(\lambda)U_\lambda\big)/\lambda$ at the isolated singular point $\lambda=-\lambda_j^{-1}$ be

$$(8) \qquad \sum_{n=-\infty}^{\infty}(\lambda+\lambda_j^{-1})^n C_n(j) .$$

By Cauchy's theorem $C_{-1}(j)=\dfrac{1}{2\pi i}\displaystyle\int R_\lambda d\lambda$. U_λ being regular at $\lambda=-\lambda_j^{-1}$, we have $C_{-1}(j)=\dfrac{1}{2\pi i}\displaystyle\int\big\{\big(K(\lambda)+\lambda K(\lambda)U_\lambda\big)/\lambda\big\}d\lambda$. As $K(\lambda)$ is completely continuous we see that $C_{-1}(j)$ is also completely continuous.

By substituting (8) in the resolvent equation $R_\lambda-R_\mu=(\lambda-\mu)R_\lambda R_\mu$ we obtain

$$(9) \qquad C_{-n}(j)C_m(j)=C_m(j)C_{-n}(j)=0 \qquad (n>0, m\geq 0) ,$$

$$(10) \qquad C_{-1}(j)^2=C_{-1}(j), \ C_{-n}(j)=C_{-n}(j)C_{-1}(j)=C_{-1}(j)C_{-n}(j),$$
$$C_{-(n+1)}(j)=C_{-2}^n(j) \qquad (n>0) .$$

$\mathfrak{B}$ is mapped on its linear subspace $\mathfrak{B}_j$ by $C_{-1}(j)$. By $C_{-1}^2(j)=C_{-1}(j)$ all the points of $\mathfrak{B}_j$ is invariant by $C_{-1}(j)$. The unit sphere in $\mathfrak{B}_j$ is compact since $C_{-1}(j)$ is completely continuous. Thus $\mathfrak{B}_j$ is of finite dimension by Riesz's theorem. By (10) $C_{-n}(j)$ maps $\mathfrak{B}_j$ in $\mathfrak{B}_j$ and hence $C_{-(n+1)}(j)$ is of the form $D_j^n C_{-1}(j)$, where D_j is a linear mapping of $\mathfrak{B}_j$ in $\mathfrak{B}_j$. Thus $\sum_{n=1}^{\infty}(\lambda+\lambda_j^{-1})^{-n}C_{-n}(j)=\sum_{n=0}^{\infty}(\lambda+\lambda_j^{-1})^{-(n+1)}D_j^n C_{-1}(j)$. As it converges for $|\lambda+\lambda_j^{-1}|>0$, the matrix D_j must be nilpotent: $D_i^n=0$ for large n.

Hence $\lambda=-\lambda_j^{-1}$ is a pole of R_λ.

§ 4. *The proof of the theorem.* By (1) and the lemma 3, the (continuous) inverse $E+\lambda R_\lambda$ of $E+\lambda T$ is regular in $1-2\varepsilon<|\lambda|<1+2\varepsilon$ except for poles $\lambda=-\lambda_1^{-1}, -\lambda_2^{-1}, \ldots, \lambda_k^{-1}$. By (2) we see that, for $|\lambda|<1$, R_λ is given by the absolutely and uniformly convergent series $\sum_{n=1}^\infty \lambda^{n-1}(-T)^n$: $\left\|\sum_{n=1}^\infty \lambda^{n-1}(-T)^n\right\| \leq a/(1-|\lambda|)$. Hence R_λ is regular in $|\lambda|<1+2\varepsilon$, except for simple poles $\lambda=-\lambda_1^{-1}, -\lambda_2^{-1}, \ldots, -\lambda_k^{-1}$. Let the Laurent expansion of R_λ at $\lambda=-\lambda_j^{-1}$ be

$$(11) \qquad (\lambda+\lambda_j^{-1})^{-1}T_j+\sum_{n=0}^\infty(\lambda+\lambda_j^{-1})^n T_{j,n}.$$

Then, by (8), (9), (10) and $R_0=-T$ we see that

$$T_j^2=T_j,\ T_iT_j=0 \quad (i\neq j), \quad (T-\sum_{i=1}^k \lambda_i T_i)T_j=T_j(T-\sum_{i=1}^k \lambda_i T_i)=0$$

$$(i,j=1, 2, \ldots, k).$$

Put $T=\sum_{i=1}^k \lambda_i T_i+S$. Then, by the above relations, we obtain for $|\lambda|<1$

$$R_\lambda=\sum_{n=1}^\infty \lambda^{n-1}(-T)^n=\sum_{j=1}^k \sum_{n=1}^\infty (-\lambda_j)^n \lambda^{n-1}T_j+\sum_{n=1}^\infty \lambda^{n-1}(-S)^n$$

$$=\sum_{j=1}^k(\lambda+\lambda_j^{-1})^{-1}T_j+\sum_{n=1}^\infty \lambda^{n-1}(-S)^n.$$

Therefore, by (11), we see that $\sum_{n=1}^\infty \lambda^{n-1}(-S)^n$ is regular in $|\lambda|<1+2\varepsilon$.

Hence, by Cauchy's theorem, $\|S^n\|\leq \beta/(1+\varepsilon)^n$, $\beta=\text{l. u. b.}_{|\lambda|\leq 1+\frac{3}{2}\varepsilon}\left\|\sum_{n=1}^\infty(-\lambda S)^n\right\|$

for $n=1, 2, \ldots$.

§5. *Smoluchousky's equation.*[1] Let a family $T(t)$ of continuous operators in $\mathfrak{B}$ satisfy the equation of Smoluchousky: $T(t+s)=T(t)T(s)$ $(0<t, s<\infty)$. We assume that $T(t)$ is continuous in t: $\lim_{t\to t_0}\| T(t)-T(t_0)\|=0$, and that there exists a positive t_1 such that $T=T(t_1)$ satisfies (1) and (2).

By the theorem we have the representation (3). We put

$$T(t)=\sum_{j=1}^k T_j(t)+S(t),\ T_j(t)=T_jT(t)T_j.$$

$T_j(t)$ and $S(t)$ is continuous in t. T_j is commutative with every $T(t)$ by (4), and hence we obtain, for $0<t, s<\infty$, $T_j(t+s)=T_j(t)T_j(s)$, $T_j(t)S(s)=S(s)T_j(t)=0$ and $S(t+s)=S(t)S(s)$.

As $S=S(t_1)$ satisfies (4) we obtain, by positive a and b, $\|S(t)\|\leq a\cdot \exp(-bt)$ for $t_1\leq t<\infty$.

By $T_j^2=T_j$, $T_j(t)=T_jT_j(t)=T_j(t)T_j$ and the complete continuity of T_j we see, as in the proof of the lemma 3, that $T_j(t)=M_j(t)T_j$, where the finite dimentional matrix $M_j(t)$ is continuous in t and satisfies the equation of Smoluchousky. As $M_j(t_1)=$ the unit matrix we see that

---

1) An analogus result is obtained by Kakutani also, by applying the theorem to the sequence $\{T(t/2^n)\}$.

$M_j(0) = \lim\limits_{t \to 0} M_j(t) = $ the unit matrix.   Hence,[1] if $\|M_j(t) - M_j(0)\| < 1$ for $t \leq t_0$, $t_0 > 0$, we have $M_j(t) = \exp(C_j t/t_0)$, where $C_j = \log\left(M_j(t_0)\right)$.   Thus, by $M_j(t_1) = $ the unit matrix, we see that $M_j(t)$ is similar to the matrix of the form

$$\begin{pmatrix} \exp(2\pi i a_{j_1} t/t_0) \cdots\cdots 0 \\ \vdots \quad \ddots \quad \vdots \\ \times \cdots\cdots \exp(2\pi i a_{jl_j} t/t_0) \end{pmatrix} \qquad \left(a\frac{t_1}{t_0} \text{ all real} \equiv 0 \ \text{mod. } 1 \right).$$

Therefore the theorem is extended to the continuous stochastic process.

---

1) K. Yosida: Jap. J. of Math. **13** (1936), 25.

Mean ergodic theorem in Banach spaces

Proc. Imp. Acad. Tokyo **14** (1938) 292–294

(Comm. by T. TAKAGI, M.I.A., Oct. 12, 1938.)

§1. *Introduction and the theorem.*

The mean ergodic theorem of J. von Neumann reads as follows:

Let T be a unitary operator in the Hilbert space $\mathfrak{H}$. Then, for any $x \in \mathfrak{H}$, the sequence

$$(1) \qquad x_n = \frac{T \cdot x + T^2 \cdot x + \cdots + T^n \cdot x}{n} \qquad (n = 1, 2, \ldots)$$

converges strongly to a point $\in \mathfrak{H}$.

Neumann's proof is based upon Stone's theorem concerning the one-parameter group of unitary operators in $\mathfrak{H}$. We find F. Riesz's elementary proof in E. Hopf's book.[1]

Recently C. Visser[2] gave the following theorem:

Let a linear operator T in $\mathfrak{H}$ satisfy the condition; $\| T^n \| \leq a$ constant for $n = 1, 2, \ldots$. Then, for any $x \in \mathfrak{H}$, the sequence (1) converges weakly to a point $\in \mathfrak{H}$.

He also showed that the mean ergodic theorem is easily obtained from this theorem. Thus we have another elementary proof of the mean ergodic theorem.

In the present note I intend to give a more general

Theorem. Let a linear operator T in the (real or complex) Banach space $\mathfrak{B}$ satisfy the two conditions:

(2) $\| T^n \| \leq a$ *constant C for $n = 1, 2, \ldots$,*

(3) $\begin{cases} T \text{ is weakly completely continuous, viz. } T \text{ maps the unit sphere} \\ \| x \| \leq 1 \text{ of } \mathfrak{B} \text{ on the point set which is weakly compact in } \mathfrak{B}.[3] \end{cases}$

Then, for any $x \in \mathfrak{B}$, the sequence (1) converges strongly to a point $\bar{x} \in \mathfrak{B}$. We have $T \cdot \bar{x} = \bar{x}$.

As the existence of the inverse T^{-1} of T is not assumed, this theorem may be applied in the problem of the temporally homogeneous stochastic process.[4] The applications will be published elsewhere.

I here express my hearty thanks to S. Kakutani who kindly communicated me that Visser's weak convergence theorem can be extended to $\mathfrak{B}$.[5]

---

1) Ergodentheorie, Berlin (1937), 23.

2) Proc. Amsterdam Acad. 16, 5 (1938), 487–495.

3) It is sufficient to assume that, for any $x \in \mathfrak{B}$, the sequence (1) is weakly compact in $\mathfrak{B}$. See the proof below. As the Hilbert space is weakly compact locally such conditions are not needed in Visser's theorem.

4) Cf. my preceding paper.

5) See the following paper of Kakutani, where we find his ingeneous arguments.

§ 2. *The proof of the theorem.*

Let x be any point of $\mathfrak{B}$. We have, by (2), $\left\|\dfrac{T \cdot x + T^2 \cdot x + \cdots + T^n \cdot x}{n}\right\|$
$\leq C \|x\|$ for $n = 1, 2, \ldots$. Hence the sequence (1) is weakly compact in $\mathfrak{B}$ by (3). Thus there exist a partial sequence $\{x_{n'}\}$ and an element $\bar{x} \in \mathfrak{B}$ such that

$$(4) \qquad\qquad \lim_{n' \to \infty} f(x_{n'}) = f(\bar{x})$$

for any *linear functional* f defined in $\mathfrak{B}$. We have

$$(5) \qquad\qquad T \cdot \bar{x} = \bar{x} ,$$

by
$$\| T \cdot x_{n'} - x_{n'} \| = \left\| \frac{T^{n'+1} \cdot x - T \cdot x}{n'} \right\| \leq \frac{2C \|x\|}{n'} .$$

We put $x = \bar{x} + (x - \bar{x})$. By (5) we have $T^n \cdot \bar{x} = \bar{x}$ and hence we have $x_n = \bar{x} + z_n$ where $z_n = \dfrac{T + T^2 + \cdots + T^n}{n} (x - \bar{x})$. Hence it is sufficient to prove that z_n converges *strongly* to zero.

Consider the linear *closed* subspace $\mathfrak{B}_0$ of $\mathfrak{B}$ spanned by the elements of the from $(y - T \cdot y)$, $y \in \mathfrak{B}$. Then, for any $w \in \mathfrak{B}_0$, $\dfrac{T + T^2 + \cdots + T^n}{n} w$ converges strongly to zero. The proof runs as follows. If w is of the form $(y - T \cdot y)$, we have

$$(6) \qquad \left\| \frac{T + T^2 + \cdots + T^n}{n} w \right\| = \left\| \frac{T \cdot y - T^{n+1} \cdot y}{n} \right\| \leq \frac{2C}{n} \| y \|$$

which converges strongly to zero. Let now w be not of the form $(y - T \cdot y)$, then for any $\varepsilon > 0$ there exists $y \in \mathfrak{B}$ such that $\| w - (y - T \cdot y) \| \leq \varepsilon$. Thus, by (1),

$$\left\| \frac{T + T^2 + \cdots + T^n}{n} w - \frac{T + T^2 + \cdots + T^n}{n} (y - T \cdot y) \right\| \leq C\varepsilon .$$

Therefore, at any $w \in \mathfrak{B}_0$, $\dfrac{T + T^2 + \cdots + T^n}{n} w$ converges strongly to zero.

Next assume that $(x - \bar{x})$ does not belong to $\mathfrak{B}_0$. Then by S. Banach's theorem,[1] there exists a linear functional f_0 such that

$$f_0(x - \bar{x}) = 1 , \qquad f_0(z) = 0 \text{ for any } z \in \mathfrak{B}_0 .$$

As $(T^m \cdot x - T^{m+1} \cdot x) \in \mathfrak{B}_0$ we have $f_0(T^m \cdot x) = f_0(T^{m+1} \cdot x)$. Hence $f_0\left(\dfrac{T + T^2 + \cdots + T^n}{n} x \right) = f_0(x)$ for $n = 1, 2, \ldots$. Therefore, by (4), $f_0(\bar{x}) = f_0(x)$ contrary to $f_0(x - \bar{x}) = 1$.

Thus $(x - \bar{x}) \in \mathfrak{B}_0$ and hence x_n converges strongly to $\bar{x}$. Q. E. D.

Remark 1. The above proof shows that

---

1) S. Banach: Théorie des opérations linéaires, Warszawa (1932), 57.

$$\left\| \frac{T+T^2+\cdots+T^n}{n}x-\bar{x} \right\| \leqq \frac{d(x)}{n} \qquad (n=1, 2, \ldots)$$

with a constant $d(x)$, if the image of $\mathfrak{B}$ by $(E-T)$[1] is closed.

Remark 2. The correspondence $x \to \bar{x}$ is given by a linear operator $T_1 : \bar{x}=T_1\cdot x$. By (5) we have $TT_1=T_1$. Thus $T^n T_1=T_1$ for any n and hence

$$(7) \qquad\qquad T_1^2=T_1 .$$

From the inequalities $\left\| \dfrac{T+T^2+\cdots+T^n}{n}x - \dfrac{T+T^2+\cdots+T^n}{n}T\cdot x \right\| \leqq \dfrac{2C}{n}\|x\|$

we see that $T_1 T=T_1$. Thus

$$(8) \qquad\qquad TT_1=T_1 T=T_1 .$$

Hence $T(T_1 \cdot x)=T_1 \cdot x$ for any $x \in \mathfrak{B}$. If $T \cdot y=y$, we have $T^n \cdot y=y$ for any n and thus $T_1 \cdot y=y$.

Therefore T_1 is the *projection operator* which maps $\mathfrak{B}$ on the proper subspace (Eigenraum) of T belonging to the proper value 1.

By applying the theorem to (T/λ), $|\lambda|=1$, we see that the strong limit T_λ of $T_{\lambda, n}=\dfrac{1}{n}\left\{ \dfrac{T}{\lambda}+\dfrac{T^2}{\lambda^2}+\cdots+\dfrac{T^n}{\lambda^n} \right\}$ satisfies

$$(9) \qquad T_\lambda^2=T_\lambda , \quad T_\lambda T=TT_\lambda=\lambda T_\lambda , \quad T_\lambda T_\mu=0$$

$$\text{for} \quad \lambda \neq \mu \quad \text{and} \quad |\lambda|=|\mu|=1 .$$

---

1) E is the identity operator in $\mathfrak{B}$.

Application of mean ergodic theorem to the problems of Markoff's process

(*with S. Kakutani*)

Proc. Imp. Acad. Tokyo **14** (1938) 333–339

(Comm. by T. Takagi, m.i.a., Nov. 12, 1938.)

§ **1.** *Two supplements to the Mean Ergodic Theorem.*

Mean Ergodic Theorem. Let $\mathfrak{B}$ be a (real or complex) Banach space, and denote by T a linear operator which maps $\mathfrak{B}$ in itself. If

(1) *there exists a constant C such that $\| T^n \| \leq C$ for $n = 1, 2, \cdots$,*

and

(2) $\begin{cases} \text{for any } x \in \mathfrak{B} \text{ the sequence } x_n = \dfrac{1}{n}(T + T^2 + \cdots + T^n)x \ (n = 1, \\ 2, \ldots) \text{ is weakly compact in } \mathfrak{B}, \end{cases}$

then

(3) $\begin{cases} \text{there exists a linear operator } T_1, \text{ which maps } \mathfrak{B} \text{ in itself, such} \\ \text{that } \lim\limits_{n \to \infty} \dfrac{1}{n}(T + T^2 + \cdots + T^n)x = T_1 x \text{ strongly for any } x \in \mathfrak{B}, \text{ and} \\ TT_1 = T_1 T = T_1^2 = T_1. \end{cases}$

T_1 is a projection operator which maps $\mathfrak{B}$ on the proper space $\mathfrak{B}_1$ of T belonging to the proper value 1. Because of (1), (2) is surely satisfied if T is *weakly completely continuous*, viz., if T maps the unit sphere $\| x \| \leq 1$ of $\mathfrak{B}$ on a point set weakly compact in $\mathfrak{B}$. These results were obtained in our previous notes.[1] We now prove the

Theorem 1. (2) *and hence* (3) *hold good if T satisfies* (1) *and if*

(2′) $\begin{cases} \text{there exist an integer } k \text{ and a weakly completely continuous} \\ \text{linear operator } V, \text{ which maps } \mathfrak{B} \text{ in itself, such that} \\ \| T^k - V \| < 1. \end{cases}$

Proof :[2] It is sufficient to prove the case $k = 1$. Put $\| T - V \| = a < 1$ and $x_{n, p} = \dfrac{1}{n}(T + T^2 + \cdots + T^p)x \ (n, p = 1, 2, \ldots)$. We have $T^p = V_p + D_p$, where $V_p = T^p - (T - V)^p$ is weakly completely continuous with $V_1 \equiv V$ and $\| D_p \| \leq a^p$. Hence $x_n = x_{n, p} + T^p x_{n, n-p} = x_{n, p} + V_p x_{n, n-p} + D_p x_{n, n-p}$. Since $\| x_{n, n-p} \| \leq C \cdot \| x \|$ for $n = 1, 2, \cdots$, there exists (for each p) a subsequence $\{ n' \}$ of $\{ n \}$ such that $\{ V_p x_{n', n'-p} \}$ converges weakly to a point $y_p \in \mathfrak{B}$. Consequently we have (since $\lim\limits_{n' \to \infty} | f(x_{n', p}) | = 0$)

(4) $$\varlimsup_{n' \to \infty} | f(x_{n'}) - f(y_p) | \leq \varlimsup_{n' \to \infty} | f(x_{n', p}) |$$

$$+ \varlimsup_{n' \to \infty} | f(D_p x_{n', n'-p}) | \leq a^p \cdot \| f \| \cdot C \cdot \| x \|$$

for any linear functional f on $\mathfrak{B}$.

---

1) K. Yosida: Mean Ergodic Theorem in Banach spaces, Proc. **14** (1938), 292.

S. Kakutani: Iteration of linear operations in complex Banach spaces, ibd., 295.

2) Cf. the arguments given by one of us. See the paper of S. Kakutani cited in (1).

Applying the diagonal method, we may assume that (4) holds for any linear functional f on $\mathfrak{B}$ and for $p=1, 2, \cdots$ (y_p may depend on p). Consider the sequence $\{y_p\}$ ($p=1, 2, \ldots$). From (4) we have $|f(y_p)-f(y_q)| \leq (a^p+a^q)\cdot\|f\|\cdot C\cdot\|x\|$, and, since f is an arbitrary functional on $\mathfrak{B}$, $\|y_p-y_q\| \leq (a^p+a^q)\cdot C\cdot\|x\|$ for any p and q, which shows that $\{y_p\}$ is a fundamental sequence in $\mathfrak{B}$. Put $y=\lim\limits_{p\to\infty} y_p$. Then it is easy to see that we have $\lim\limits_{n'\to\infty} f(x_{n'})=f(y)$ for any linear functional f on $\mathfrak{B}$. Hence the sequence $\{x_{n'}\}$ converges weakly to a point $y \in \mathfrak{B}$.

Next we shall prove a theorem which constitutes a generalisation of a theorem due to S. Mazur.[1]

Theorem 2. Let T satisfy (1) *and* (2). *Consider the proper value equation*

$$(5) \qquad\qquad Tx=x ,$$

and its conjugate equation[2]

$$(6) \qquad\qquad \overline{T}X=X .$$

Then, if p and q denote the numbers of the linearly independent solutions of (5) *and* (6) *respectively, we must have $p=q$.*

Proof: Put $\mathfrak{B}_1=T_1\mathfrak{B}$. Then $p=$dimension of $\mathfrak{B}_1$. Any linear functional $X(x)$ on $\mathfrak{B}_1$ defines a linear functional $X'(x)$ on $\mathfrak{B}$: $X'(x)=X(T_1x)$. By (3) we have, for any $x \in \mathfrak{B}$, $X'(Tx)-X'(x)=X(T_1Tx)-X(T_1x)=X(T_1x-T_1x)=0$. Hence the linear functional X' satisfies (6), from which follows $q \geq p$.

Conversely, let X be a linear functional on $\mathfrak{B}$ which satisfies (6). Then we have, for any $x \in \mathfrak{B}$, $X(x)=X(Tx)=X\left(\dfrac{1}{n}(T+T^2+\cdots+T^n)x\right)$ and hence $X(x)=X(T_1x)$ by (3). Thus X is, essentially, a linear functional on $\mathfrak{B}_1=T_1\mathfrak{B}$, and hence $q \leq p$.

§ **2.** *Applications to the problem of Markoff's process.*

Consider a Markoff's process by which each point x of the closed interval $\Omega=[0, 1]$ is transferred to a point $y \in \Omega$ after the elapse of a unit time. Denote by $P(x, E)$ its transition probability; that is, $P(x, E)$ is a probability that a point x comes into a Borel set E of Ω after the elapse of a unit time. We have $0 \leq P(x, E) \leq 1$ and $P(x, \Omega)=1$. Assume that $P(x, E)$ is measurable in x if E is fixed, and that, for any fixed x, $P(x, E)$ is a totally additive set function defined for all the Borel sets of Ω.

§ **2-1.** *Condition of J. L. Doob.*[3]

---

1) S. Mazur: Über die Nullstellen linearer Operatoren, Studia Math. **2** (1930), 11–20. The assumption in Theorem 2 is much weaker than that of Mazur's. He assumed that $\mathfrak{B}$ is locally weakly compact.

2) As to the notion of conjugate operators see the paper of S. Mazur cited in (1).

3) J. L. Doob: Stochastic processes with an integral valued parameter, Trans. Amer. Math. Soc. **44** (1938), 87–150.

$$(7) \begin{cases} \textit{There is a measurable function } p(x, y) \textit{ defined for } 0 \leq x, y \leq 1 \\ \textit{such that } P(x, E) = \int_E p(x, y)\,dy \textit{ for any } x \in \Omega \textit{ and for any Borel} \\ \textit{set } E \textit{ of } \Omega \textit{; and moreover, } p(x, y) \textit{ satisfies the uniform inte-} \\ \textit{grability condition: for any decreasing sequence } \{E_n\} \textit{ of mea-} \\ \textit{surable sets with } m(E_n) \to 0, \textit{ we have } \int_{E_n} p(x, y)\,dy \to 0 \textit{ uni-} \\ \textit{formly in } x. \end{cases}$$

It will be easily seen that this condition is equivalent to the following one:

$$(8) \begin{cases} \textit{for any positive number } \varepsilon > 0 \textit{ there exists a positive number} \\ \delta(\varepsilon) > 0 \textit{ such that } m(E) < \delta(\varepsilon) \textit{ implies } P(x, E) < \varepsilon \textit{ for any } x \in \Omega. \end{cases}$$

Theorem 3. Under the condition of Doob, the integral operator

$$f \to Tf = g: \quad g(y) = \int_0^1 f(x)\, p(x, y)\, dx$$

is a linear operator which maps the space (L)[1] in itself. This T is of norm 1 and is weakly completely continuous.

Proof: We have, by Fubini-Tonelli's theorem, $\|g\|_L = \int_0^1 |g(y)|\,dy$

$$\leq \int_0^1 \int_0^1 |f(x)\,p(x, y)|\,dx\,dy = \int_0^1 |f(x)| \left(\int_0^1 p(x, y)\,dy \right) dx = \int_0^1 |f(x)|\,dx = \|f\|_L.$$

Hence $\|T\|_L \leq 1$[2] By taking $f(x) \equiv 1$ we see that $\|T\|_L = 1$.

As the conjugate space of (L) is the space (M),[3] any linear functional k defined on the image $T(L)$ of (L) is given by

$$\int_0^1 g(y)\,k(y)\,dy = \int_0^1 \left(\int_0^1 f(x)\,p(x, y)\,dx \right) k(y)\,dy$$

$$= \int_0^1 f(x) \left(\int_0^1 p(x, y)\,k(y)\,dy \right) dx, \quad k(y) \in (M).$$

The subset $(M)'$ of (M) of all the functions of the form: $\int_0^1 p(x, y)\,k(y)\,dy$, $k(y) \in (M)$, $\|k\|_M \leq 1$, is separable in the topology of (M). This may be proved as follows:

Let S be the unit sphere $\|k\|_M \leq 1$ of (M). Since $S \subset (M) \subset (L)$ and since (L) is separable, there exists a countable subset $\{k_n(y)\}$ of S which is dense in S in the topology of (L); that is, for any $k(y) \in (M)$ with $\|k\|_M \leq 1$, there exists a subsequence $\{k_{n'}(y)\}$ of $\{k_n(y)\}$ such that $\lim_{n' \to \infty} \|k - k_{n'}\|_L = \lim_{n' \to \infty} \int_0^1 |k(y) - k_{n'}(y)|\,dy = 0$. Consequently, there exists a further subsequence $\{k_{n''}(y)\}$ of $\{k_{n'}(y)\}$, a decreasing sequence $\{E_{n''}\}$ of measurable sets and a sequence $\{\varepsilon_{n''}\}$ of positive numbers such that $\lim_{n'' \to \infty} m(E_{n''}) = 0$, $\lim_{n'' \to \infty} \varepsilon_{n''} = 0$ and $|k(y) - k_{n''}(y)| \leq \varepsilon_{n''}$ for any $y \bar{\in} E_{n''}$.

---

1) (L) is the linear space of all the measurable functions which are absolutely integrable in $[0, 1]$. For any $f(x) \in (L)$, we define its norm by $\|f\|_L = \int_0^1 |f(x)|\,dx$.

2) $\|T\|_L$ is a norm of T as an operator in (L). Analogous notations will be used for other Banach spaces.

3) (M) is the linear space of all the bounded measurable functions defined in $[0, 1]$. For any $k(x) \in (M)$, we define its norm by $\|k\|_M = \mathrm{ess.\ max.}_{0 \leq x \leq 1} |k(x)|$.

Hence $\left| \int_0^1 p(x, y) k(y)\, dy - \int_0^1 p(x, y) k_{n''}(y)\, dy \right| \leqq \int_0^1 p(x, y)\, |k(y) - k_{n''}(y)|\, dy$

$\leqq \int_{E_{n''}} + \int_{\Omega - E_{n''}} \leqq 2\int_{E_{n''}} p(x, y)\, dy + \varepsilon_{n''} \to 0$ uniformly in x. This proves the

separability of $(M)'$ in the topology of (M).

Since the space (L) is weakly complete, we see by the diagonal process that T is weakly completely continuous as an operator in (L).

Theorem 4. Under the condition of Doob, there exists a measurable function $p_\infty(x, y)$ defined for $0 \leqq x, y \leqq 1$ such that for any $f(x) \in (L)$ we have

$$\lim_{n \to \infty} \int_0^1 \left| \int_0^1 f(x) \left(\frac{p(x, y) + p^{(2)}(x, y) + \cdots + p^{(n)}(x, y)}{n} \right) dx - \int_0^1 f(x) p_\infty(x, y)\, dx \right| dy = 0$$

$$\left(p^{(n)}(x, y) = \int_0^1 p^{(n-1)}(x, z) p(z, y)\, dz, \ n = 2, 3, \ldots; \ p^{(1)}(x, y) = p(x, y) \right),$$

and

$$(9) \quad \int_0^1 p(x, z) p_\infty(z, y)\, dz = \int_0^1 p_\infty(x, z) p(z, y)\, dz = \int_0^1 p_\infty(x, z) p_\infty(z, y)\, dz = p_\infty(x, y),$$

$$(10) \qquad p_\infty(x, y) \geqq 0, \quad \int_0^1 p_\infty(x, y)\, dy = 1.$$

Theorem 5.[1] Under the condition of Doob, the proper value λ of modulus 1 of T is finite in number and satisfies the binomial equation: $\lambda^N = 1$, where N is a fixed positive integer.

Proof: The conjugate equation of $f(y) = \lambda \int_0^1 f(x) p(x, y)\, dx, \ f(x) \in (L)$, is given by

$$(11) \qquad g(x) = \lambda \int_0^1 p(x, y) g(y)\, dy, \quad g(y) \in (M).$$

Hence, by Theorem 2, it is sufficient to show that if (11) admits a solution $g(x)$, $\|g\|_M = 1$, we must have $\lambda^n = 1$, where n is an integer not greater than some constant determined only by the function $p(x, y)$. This may be done as follows:

For any δ with $0 < \delta < 1$ we have

$$|g(x)| \leqq \int_{|g(y)| \leqq 1-\delta} p(x, y)(1 - \delta)\, dy + \int_{|g(y)| > 1-\delta} p(x, y)\, dy$$

$$= \int_0^1 p(x, y)(1 - \delta)\, dy + \delta \int_{|g(y)| > 1-\delta} p(x, y)\, dy = 1 - \delta + \delta \int_{|g(y)| > 1-\delta} p(x, y)\, dy.$$

Since $\|g\|_M = 1$, there exists an x_0 with $|g(x_0)| > 1 - \dfrac{\delta}{2}$; and for this x_0

we have $1 \geqq \int_{|g(y)| > 1-\delta} p(x_0, y)\, dy \geqq \dfrac{1}{2}$. Therefore, by (8), there exists a con-

stant $\gamma > 0$ determined from the function $p(x, y)$ only, such that

---

1) This is a generalisation of a theorem of M. Fréchet. Fréchet assumed that $p^{(n)}(x, y)$ is uniformly bounded. See the paper of Fréchet: Sur l'allure asymptotique des densités itétées dans le problème des probabilités "en chaîne," Bull. de la Soc. math. de France, **62** (1934), 68-83.

$m\big(E[\,|\,g(y)\,|>1-\delta]\big)>\gamma$ for any $\delta>0$ and for any solution $g(y)$ of (11) with $\|g\|_M=1$.

The rest of the proof may be carried out as in the paper of Fréchet's.

Theorem 6. Under the condition of Doob, the proper value 1 of T is of finite multiplicity.

Proof: First we notice that there is a constant $\gamma>0$ such that $\int_E p(x,y)\,dy=1$ implies $m(E)\geq\gamma$ for any x and for any measurable set $E\subset\varOmega$.

Let now $f(y)=\int_0^1 f(x)\,p(x,y)\,dx$. Then $|f(y)|\leq\int_0^1|f(x)|\,p(x,y)\,dx$, and hence by integrating with respect to y and applying Fubini-Tonelli's theorem, we see that $|f(y)|=\int_0^1|f(x)|\,p(x,y)dx$ almost everywhere. Therefore, following the same arguments as were given by N. Kryloff and N. Bogolioùboff,[1] we see that, if the multiplicity of the proper value 1 is greater than $>\dfrac{1}{\gamma}$, there exists a system of $n\big(>\dfrac{1}{\gamma}\big)$ real non-negative measurable functions satisfying

$$\int_0^1 p_i(y)\,dy=1,\quad p_i(y)\cdot p_j(y)\equiv0 \text{ for } i\neq j,$$

and
$$p_i(y)=\int_0^1 p_i(x)\,p(x,y)\,dx \text{ almost everywhere.}$$

Let E_i be the set of y at which $p_i(y)>0$, then E_i are mutually disjoint. We have $1=\int_{E_i}p_i(y)\,dy=\int_{E_i}\Big(\int_{E_i}p_i(x)\,p(x,y)\,dx\Big)dy=\int_{E_i}p_i(x)\Big(\int_{E_i}p(x,y)\,dy\Big)dx$ by Fubini-Tonelli's theorem. Hence $\int_{E_i}p(x,y)\,dy=1$ almost everywhere in E_i, and thus $m(E_i)\geq\gamma$ for $i=1,2,\ldots,n$, which is a contradiction since $n>\dfrac{1}{\gamma}$. Consequently the multiplicity of the proper value 1 is not greater than $\dfrac{1}{\gamma}$.

§ 2-2. Condition of W. Doeblin.[2]

(12) $\begin{cases} \textit{There exist two positive numbers } b,\,\eta>0 \textit{ such that } m(E)<\eta \textit{ im-}\\ \textit{plies } P(x,E)<1-b \textit{ for any } x \textit{ and for any measurable set } E\subset\varOmega. \end{cases}$

Clearly the condition of Doeblin is much more general than that of Doob.

Theorem 7. Under the condition of Doeblin, the integral operator

$$\varphi\to T\varphi=\psi:\quad \psi(E)=\int_0^1\varphi(de_x)\,P(x,E)$$

---

1) N. Kryloff and N. Bogolioùboff: Sur les propriétés ergodiques de l'equation de Smoluchovski, Bull. de la Soc. math. de France, **64** (1936), 49-56.

2) W. Doeblin: Sur les propriétés asymptotiques de mouvements régis par certains types de chaines simples, Bull. math. de la Soc. Roumaine des Sciences, **39** (1937), (2), 3-61.

is a linear operator which maps the space $(\mathfrak{M})$[1] *in itself. This T is of norm 1 and there exists a weakly completely continuous linear operator V, which maps* $(\mathfrak{M})$ *in itself, such that* $\|T-V\|_{\mathfrak{M}} < 1$.

The same theorem may be stated for the space (BV).[2] For this purpose, denote by $I(y_0)$ the closed interval $0 \leq y \leq y_0$ and consider the function $F(x, y) \equiv P\big(x, I(y)\big)$. $F(x, y)$ is a measurable function defined for $0 \leq x, y \leq 1$, and is monotone in y if x is fixed.

Theorem 7′. Under the condition of Doeblin, the integral operator

$$\varphi \to T\varphi = \psi: \quad \psi(y) = \int_0^1 \varphi(dx)\, F(x, y)$$

is a linear operator which maps the space (BV) in itself. This T is of norm 1 and there exists a weakly completely continuous linear operator V, which maps (BV) in itself, such that $\|T-V\|_{BV} < 1$.

Proof: Since $F(x, y)$ is monotone in y if x is fixed, $p(x, y) = \dfrac{\partial F}{\partial y}$ exists almost everywhere (for each x). $p(x, y)$ is measurable in $0 \leq x, y \leq 1$. Put $q(x, y) = p(x, y)$ if $p(x, y) \leq \dfrac{1}{\eta}$ and $q(x, y) = 0$ if $p(x, y) > \dfrac{1}{\eta}$. Then $G(x, y) = \displaystyle\int_0^y q(x, t)\, dt$ and $H(x, y) = F(x, y) - G(x, y)$ are also measurable in $0 \leq x, y \leq 1$, and monotone in y if x is fixed. Now, consider the linear operators V and W which correspond to $G(x, y)$ and $H(x, y)$ respectively. Clearly $T = V + W$. We shall show that V is weakly completely continuous as an operator in (BV) and that $\|W\|_{BV} \leq 1 - b < 1$.

In order to prove that V is weakly completely continuous, let $\{\varphi_n(x)\}$ be a sequence of functions of bounded variation with $\|\varphi_n\|_{BV} \leq 1$, $n = 1, 2, \ldots$. We have to choose a subsequence $\{\varphi_{n'}(x)\}$ of $\{\varphi_n(x)\}$ and a function $\varphi_0(x) \in (BV)$ such that $V\varphi_{n'}$ converges weakly to $\varphi_0(x)$; that is, $f(\varphi_{n'})$ converges to $f(\varphi_0)$ for any linear functional f on (BV). It is disappointing that the general form of linear functionals on (BV) is not yet known, but we can evade this difficulty. Since $V\varphi$ is absolutely continuous for any $\varphi(x) \in (BV)$, and since the subspace (A) of (BV) of all the absolutely continuous functions of (BV) is isometric to (L), f may be considered as a functional on (L); and consequently, by a well-known result, f is represented by a function $k(x) \in (M)$. Moreover, since $V\varphi$ is absolutely continuous with uniformly bounded density $\left(\leq \dfrac{1}{\eta}\right)$ for any $\varphi(x) \in (BV)$ with $\|\varphi\|_{BV} \leq 1$, the range of V corresponding to a sphere $\|\varphi\|_{BV} \leq 1$ of (BV) is even isometric to a uniformly bounded $\big($in the topology of $(M)\big)$ part of (M), which is a linear subspace of (L).

Thus our problem is transformed into the following one: Given

---

1) $(\mathfrak{M})$ is the linear space of all the totally additive set functions defined for all the Borel sets of $\Omega = [0, 1]$. For any $\varphi(E) \in (\mathfrak{M})$, we define its norm by $\|\varphi\|_{\mathfrak{M}} = $ total variation of $\varphi(E) \equiv$ l. u. b. $\varphi(E) - $ g. l. b. $\varphi(E)$.

2) (BV) is the linear space of all the functions of bounded variation defined in $0 \leq x \leq 1$. For any $\varphi(x) \in (BV)$, we define its norm by $\|\varphi\|_{BV} = |\varphi(0)| + $ total variation of $\varphi(x)$ in $0 \leq x \leq 1$.

a sequence $\{g_n(x)\}$ of uniformly bounded measurable functions, $g_n(x) \in (M)$, we have to choose a subsequence $\{g_{n'}(x)\}$ of $\{g_n(x)\}$ and a function $g_0(x) \in (L)$,[1] such that we have $\lim_{n' \to \infty} \int_0^1 g_{n'}(x) k(x) dx = \int_0^1 g_0(x) k(x) dx$ for any function $k(x) \in (M)$.

This problem may be solved as follows: Since $(M) \subset (L)$ and since (L) is separable, there is a countable subset $\{k_m(x)\}$ of (M) which is dense in (M) in the topology of (L). Applying the diagonal method, we can choose a subsequence $\{g_{n'}(x)\}$ of $\{g_n(x)\}$ such that $\lim_{n' \to \infty} \int_0^1 g_{n'}(x) k_m(x) dx$ exists for $m = 1, 2, \cdots$. Since $\{g_{n'}(x)\}$ is uniformly bounded, $\lim_{n' \to \infty} \int_0^1 g_{n'}(x) k(x) dx$ exists for any $k(x) \in (M)$. The existence of a limiting function $g_0(x) \in (L)$ is now a direct consequence of the facts that $(M) \subset (L)$ and that (L) is weakly complete.

In order to prove that $\|W\|_{BV} \leqq 1 - b < 1$, it is sufficient to show that $H(x, 1) \leq 1 - b$ for any x. This may be easily seen from the condition of Doeblin, if we observe that, for any x, the set of y, where $H(x, y)$ actually increases, is of measure $< \eta$ by the construction of $H(x, y)$ $\big(\text{and } q(x, y)\big)$.

Remark. Theorems 4 and 6 are also true for the case when the condition of Doeblin is satisfied. This may be easily seen as in the preceding.

---

1) In general, it is impossible to take $g_0(x)$ in (M).

Operator-theoretical treatment of the Markoff's process

Proc. Imp. Acad. Tokyo **14** (1938) 363–367

(Comm. by T. TAKAGI, M.I.A., Dec. 12, 1938.)

§ 1. *Introduction.* Let $P(x, E)$ denote the transition probability that the point x of the interval $(0, 1)$ is transferred, by a simple Markoff's process, into the Borel set E on the interval $(0, 1)$ after the elapse of a unit time. We assume that $P(x, E)$ is completely additive for Borel sets E if x is fixed and that $P(x, E)$ is Borel measurable in x if E is fixed. Then the transition probability after the elapse of n units of time is given by $P^{(n)}(x, E) = \int_0^1 P^{(n-1)}(x, dy)P(y, E)$, $\left(P^{(1)}(x, E) = P(x, E)\right)$.

Under certain general condition given below, W. Doeblin[1] investigated the asymptotic behaviour of $P^{(n)}(x, E)$ for large n. His method of proof is based upon set-theoretical considerations. It may be termed as a direct method. In the present note I intend to give an operator-theoretical treatment of the problem, by virtue of the results of the preceding notes.[2] Our method of proof will make clear the spectral properties of the Markoff's process in question, and the results obtained is somewhat more precise than that of Doeblin. The author is indebted to S. Kakutani in the proof of the lemma 1 and 5 below. I want to express my hearty thanks to him.

§ 2. *Preliminary lemmas.* By definition we have

(1) $\quad P^{(n)}(x, E) \geq 0$ and $P^{(n)}(x, \Omega) \equiv 1$, where $\Omega =$ the interval $(0, 1)$.

We make on $P(x, E)$ the following assumptions due to Doeblin:

(2) $\quad \begin{cases} \text{there exist an integer } s \text{ and positive } b, \eta \ (<1) \text{ such that,} \\ \text{if mes } (E) < \eta, \quad P^{(s)}(x, E) < 1 - b \quad \text{uniformly in } x. \end{cases}$

Then it is easy to see that

(2)′ $\quad$ if mes $(E) < \eta$, $P^{(t)}(x, E) < 1 - b$ uniformly in x and $t \geq s$.

We may decompose $P^{(s)}(x, E)$ as follows[3]:

(3) $\quad \begin{cases} P^{(s)}(x, E) = \int_E q(x, y)dy + R(x, E), \\[2mm] 0 \leq q(x, y) \leq \dfrac{1}{\eta}, \quad 0 \leq R(x, E) < 1 - b. \end{cases}$

---

1) W. Doeblin: Sur les propriétés asymptotiques de mouvements régis par certains types de chaines simples, Bull. math. de la Soc. Roumaine des Sciences, **39** (1937), (2), 3–61.

2) K. Yosida: Abstract integral equations and the homogeneous stochastic process, Proc. **14** (1938), 286. K. Yosida and S. Kakutani: Application of mean ergodic theorem to the problem of Markoff's process, ibid. 333. K. Yosida, Y. Mimura and S. Kakutani: Integral operator with bounded kernels, ibid. 359. These notes will respectively be referred to as [I], [II] and [III].

3) [II], the proof of Theorem 7′.

Consider the integral operator P which transforms the Banach space $(\mathfrak{M})$ in $(\mathfrak{M})^{1)}$:

$$P \cdot f = g, \qquad g(E) = \int_0^1 P(x, E) f(dx).$$

By (1) and (3) we have

$$\begin{cases} P^s = Q + R, \quad \|P\| = 1, \quad \|Q\| \leqq 1, \quad \|R\| \leqq 1 - b, \quad \text{where} \\ Q \cdot f = \int_E dy \int_0^1 q(x, y) f(dx), \quad R \cdot f = \int_0^1 R(x, E) f(dx). \end{cases}$$

Lemma 1. There exist an integer n and a completely continuous operator V such that $\|P^n - V\| < 1$.

Proof. We have $P^{sk} = Q^k + Q^{k-1}R + Q^{k-2}RQ + \cdots + RQR^{k-2} + QR^{k-1} + R^k$. The term which contains Q at least two times as factor is *completely continuous*. Consider, for example, $RQRQ^{k-3}$. QR and Q^{k-3} are both integral operators with bounded kernels. Hence[2] QRQ^{k-3} is completely continuous in $(\mathfrak{M})$, and thus $RQRQ^{k-3}$ is also completely continuous in $(\mathfrak{M})$. The number of terms that contains Q at most once as factor is $k+1$, each of norm $\leqq (1-b)^{k-1}$. As $\lim_{k\to\infty} (k+1)(1-b)^{k-1} = 0$ we have the lemma.

Lemma 2. Let $\|P^n - V\| < 1$ as in lemma 1. Then, if $k \geqq n$, the proper value λ with modulus 1 of P^k satisfies $\lambda^{m_k} = 1$, where the integer m_k is bounded uniformly for k.

Proof. We have $\|P^k - P^{k-n}V\| \leqq \|P^{k-n}\| \|P^n - V\| \leqq \|P^n - V\| < 1$. $P^{k-n}V$ is completely continuous with V. Let $\lambda(|\lambda| = 1)$ be a proper value of P^k. Then[3] there exists a projection operator $P_{k,\lambda}(x, E)$ which maps $(\mathfrak{M})$ on the proper space belonging to the proper value λ of P^k:

$$\begin{cases} P_{k,\lambda}(x, E) = \lim_{t\to\infty} \frac{1}{t} \sum_{i=1}^{t} P^{(ki)}(x, E) \quad \text{(uniform limit)}, \\ \lambda P_{k,\lambda}(x, E) = \int_0^1 P_{k,\lambda}(y, E) P^{(k)}(x, dy) = \int_0^1 P^{(k)}(x, dy) P_{k,\lambda}(y, E), \\ P_{k,\lambda}(x, E) = \int_0^1 P_{k,\lambda}(x, dy) P_{k,\lambda}(y, E). \end{cases}$$

Hence, for any $g(y)$ of the Banach space $(\mathfrak{M})$, the element in $h(x) = \int_0^1 P_{k,\lambda}(x, dy) g(y)$ of $(\mathfrak{M})$ satisfies $h(x) = \lambda \int_0^1 P^{(k)}(x, dz) h(dz)$. Therefore, by $(1)'$, we obtain the lemma by applying M. Fréchet's arguments.[4]

Lemma 3. Let $\|P^n - V\| < 1$ as in lemma 1. Then, for any $k \geqq n$, the multiplicity of the proper value 1 of P^k is $< \dfrac{1}{\eta}$.

Proof. Let $f(E) = \int_0^1 P^{(k)}(x, E) f(dx)$. We have, by (1), $g(E) =$

---

1) $(\mathfrak{M})$ is the Banach space of all the totally additive set functions defined for all the Borel sets of $\Omega = (0, 1)$. For any $f(E) \in (\mathfrak{M})$, we define its norm by $\|f\|_{\mathfrak{M}} = $ total variation of f on $(0, 1)$.

2) [III], Theorem 2.

3) [I], Theorem. Cf. also Theorem 4 in S. Kakutani: Iteration of linear operations in complex Banach spaces, Proc. **14** (1938), 295.

4) [II], Theorem 5.

$\int_0^1 P^{(k)}(x, E)g(dx)$, where $g(E)=$ the total variation of f on E. Hence, by (1)′, we obtain the lemma by applying Kryloff-Bogoliouboff's arguments.[1]

Kryloff-Bogoliouboff's arguments for the proof of the above lemma also proves the following lemma simultaneously.

Lemma 4. Let $\| P^n - V \| < 1$ as in lemma 1. Then, for any $k \geq n$, there exists $f_{1_k}(E), f_{2_k}(E), \ldots, f_{l_k}(E)\left(l_k < \dfrac{1}{\eta}\right)$ with the properties:

$$P^k \cdot f_{i_k} = f_{i_k}, \quad f_{i_k}(E) \geq 0, \quad f_{i_k}(\Omega) = 1, \quad f_{i_k}(E)f_{j_k}(E) \equiv 0 \quad \text{for} \quad i \neq j,$$

such that any $f(E)$ satisfying $P^k \cdot f = f$, $f(E) \geq 0$, $f(\Omega) = 1$ is uniquely expressed as a linear combination $f(E) = \sum\limits_{i_k=1}^{l_k} c_{i_k} f_{i_k}(E)$, $c_{i_k} \geq 0$, $\sum\limits_{i_k=1}^{l_k} c_{i_k} = 1$.

Lemma 5. Let square matrix $C = \| c_{ij} \|$ $(i, j = 1, 2, \ldots, l_n)$ satisfy the conditions: $c_{ij} \geq 0$, $\sum\limits_{j=1}^{l_n} c_{ij} = 1$ $(i = 1, 2, \ldots, l_n)$, $C^m =$ unit matrix for a certain m. Then C represents a permutation of l_n indices $1, 2, \ldots, l_n$.

Proof. The matrix C^{m-1} also satisfies the same condition as C. Thus we obtain the lemma by performing the multiplication $C^{m-1} \cdot C =$ the unit matrix.

§3. *Asymptotic behaviour of the process P.* By lemma 1, 2 and 3 we see that there exist an integer n and *a completely continuous* operator V such that i) $\| P^n - V \| < 1$, ii) P^n admits of no proper values with modulus 1 other than 1. Hence we have[2]

$$(4) \quad \begin{cases} P^n = P_1 + S, \quad P^n P_1 = P_1 P^n = P_1^2 = P_1, \quad P_1 S = S P_1 = 0, \\[2mm] \| S^k \| \leq \dfrac{a}{(1+\varepsilon)^k} \qquad (k = 1, 2, \ldots) \text{ with positive } a \text{ and } k. \end{cases}$$

Here the integral operator P_1 is defined by the kernel

$$(5) \qquad P_1(x, E) = \lim_{t\to\infty} \frac{1}{t} \sum_{i=1}^t P^{(ni)}(x, E) \qquad \text{(uniform limit).}$$

Surely we have $P_1(x, E) \geq 0$, $P_1(x, \Omega) \equiv 1$ and thus, by lemma 4,

$$(6) \quad \begin{cases} P_1(x, E) = \sum\limits_{i=1}^{l_n} c_i(x) f_i(E) \qquad \left(l_n < \dfrac{1}{\eta}\right), \\[2mm] P_1 \cdot f_i = f_i, \quad f_i(E) \geq 0, \quad f_i(\Omega) = 1, \quad f_i(E)f_j(E) \equiv 0 \quad \text{for} \quad i \neq j, \\[2mm] c_i(x) \geq 0, \quad \sum\limits_{i=1}^{l_n} c_i(x) \equiv 1. \end{cases}$$

We put $P \cdot f_i = g_i$. Then, by $P^n \cdot P = P \cdot P^n$, we have $P^n \cdot g_i = g_i$. It is easy to see that $g_i(E) \geq 0$, $g_i(\Omega) = 1$. Hence by lemma 4

$$P \cdot f_i = \sum_{j=1}^{l_n} c_{ij} f_j, \quad c_{ij} \geq 0, \quad \sum_{j=1}^{l_n} c_{ij} = 1 \quad (i = 1, 2, \ldots, l_n).$$

The matrix $C = \| c_{ij} \|$ satisfies $C^n =$ the unit matrix by $P^n \cdot f_i = f_i$. Hence, by lemma 5, the l_n indices $1, 2, \ldots, l_n$ is divided into p classes $(p \leq l_n)$, each class being permutated *cyclically* in itself by C. Let K_a

---

1) [II], Theorem 6.
2) [I], Theorem. Cf. also Theorem 4 in S. Kakutani: loc. cit.

be the class which consists of the indices $1_a, 2_a, \cdots, d_a$: $\sum_{d=1}^{p} d_a = l_n$. As $C^n =$ the identical permutation we see that each d_a is a divisor of n.

$f_i(E)$ is completely additive and non-negative. Hence, by Radon-Nikodym's theorem, $f_i(E) = \int_E p_i(x)dx + f_i(E \cdot N_i)$ with $p_i(x) \geq 0$ and mes$(N_i) = 0$. Let $\bar{E}_i$ be the set at each point of which we have $p_i(x) > 0$. We put $E_i = \bar{E}_i + N_i$, $G_a = \sum_{i \in K_a} E_i$. By (6) E_i $(i = 1, 2, \cdots, l_n)$ are *mutually disjoint*, and hence G_a $(a = 1, 2, \cdots, p)$ are also mutually disjoint. G_a $(a = 1, 2, \cdots, p)$ are called the *final sets* of the Markoff's process P.

We now prove the following properties of the final sets.

i). l. u. b. $\underset{x \in \Omega}{} P^{(t)}(x, \Omega - \sum_{d=1}^{p} G_a) \leq \dfrac{a'}{(1+\varepsilon')^t}$ $(t = 1, 2, \cdots)$ with positive a', ε'.

ii). $P(x, G_a) = 1$ for almost all $x \in G_a$.

iii). There exists $E_i' \subset E_i$ with mes$(E_i - E_i') = 0$ such that, if $i \in K_a$,

$$\text{l. u. b.} \underset{x \in E_i', E \subset E_i}{} | P^{(td_a)}(x, E) - f_i(E) | \leq \dfrac{a''}{(1+\varepsilon'')^t} \qquad (t = 1, 2, \cdots),$$

where a'' and ε'' denote positive constants.

iv). Let $G_a' = \sum_{i \in K_a} E_i'$, then $G_a' \subset G_a$ and mes$(G_a - G_a') = 0$. For any $E \subset G_a$, mes$(E) > 0$ and for any $x \in G_a'$, there exists a positive integer $m = m(x, E)$ such that $P^{(m)}(x, E) > 0$.

v). By a suitable numerotation of the indices $1_a, 2_a, \cdots, d_a$, we have $P(x, E_{(i+1)_a}) = 1$ for almost all $x \in E_{i_a}$ $\left(i = 1, 2, \cdots, d, (d+1)_a = 1_a, a = 1, 2, \cdots, p \right)$.

vi). By ii) $P(x, E)$ defines a Markoff's process P_a in G_a $(a = 1, 2, \cdots, p)$. P_a admits of no proper values with modulus 1 other than 1, if and only if $d_a = 1$.

Proof of i). By (4) and (6) we have $P^{(ni)}(x, \Omega - \sum_{a=1}^{p} G_a) = P^{(ni)}(x, \Omega - \sum_{i=1}^{l_n} E_i) = S^{(i)}(x, \Omega - \sum_{i=1}^{l_n} E_i)$. Hence, by (4) and $\|P\| = 1$, we obtain i).

Proof of ii). By the definition of the class K_a we have $h_a(E) = \int_0^1 P(x, E)h_a(dx)$, where $h_a(E) = f_{1_a}(E) + f_{2_a}(E) + \cdots + f_{d_a}(E)$. By putting $E = G_a = \sum_{i \in K_a} E_i$ we obtain $1 = \int_0^1 P(x, G_a)h_a(dx)$. We have $1 \geq h_a(E) \geq 0$, $h_a(G_a) = 1$ and $1 \geq P(x. G_a) \geq 0$. Hence we must have $P(x, G_a) = 1$ for almost all $x \in G_a$.

Proof of iii). We obtain, by (5) and (6), $f_i(E) = \int_0^1 P_1(x, E)f_i(dx)$. Hence, by (6) $f_i(E) = \int_0^1 c_i(x)f_i(E)f_i(dx)$ if $E \subset E_i$. Hence, by putting $E = E_i$, $1 = \int_0^1 c_i(x)f_i(dx)$. Thus, by $1 \geq c_i(x) \geq 0, 1 \geq f_i(E) \geq 0$ and $f_i(E_i) = 1$ we obtain $c_i(x) = 1$ for almost all $x \in E_i$. Therefore there exists $E_i' \subset E_i$ with mes$(E_i - E_i') = 0$ such that $P_1(x, E) = c_i(x)f_i(E) = f_i(E)$ for $x \in E_i'$ if $E \subset E_i$. Then by (4) we have $P^{(nk)}(x, E) = f_i(E) + S^{(k)}(x, E)$ for $x \in E_i'$

and $E \subset E_i$. As d_a is a divisor of n we put $n = d_a s_a$. Then any integer of the form $t d_a$ can be expressed as $t d_a = m_t n + r_t d_a$, $s_a > r_t \geq 0$. Thus $P^{(t d_a)} = (P^{d_a})^{r_t} (P^n)^{m_t}$. As $P^{d_a} \cdot f_i = f_i$ if $i \in K_a$, we obtain iv) by (6) and $\| P \| = 1$.

Proof of iv):   From the above arguments we have $P_1(x, E) = \sum_{i \in K_a} f_i(E)$ for $x \in G'_a$, $E \subset G_a$. This proves iv) by (6).

Proof of v).   By the definition of the class K_a, we have $f_{(i+1)_a}(E) = \int_0^1 P(x, E) f_{i_a}(dx)$, by a suitable numerotation of the indices $1_a, 2_a, \ldots, d_a$ $((d+1)_a = 1_a)$. Hence we obtain $1 = \int_0^1 P(x, E_{(i+1)_a}) f_{i_a}(dx)$. As $1 \geq f_{i_a}(E) \geq 0$, $f_{i_a}(E_{i_a}) = 1$ and $1 \geq P(x, E_{(i+1)_a}) \geq 0$ we must have v).

Proof of vi).   Let $d_a = 1$. Then, by iii), the operator P_a does not admit of proper values with modulus 1 other than 1. Let $d_a > 1$. We put, as in the poof of v), $P \cdot f_{i_a} = f_{(i+1)_a}$ $(i = 1, 2, \ldots, d, (d+1)_a = 1_a)$. We have $P \cdot (f_{1_a} + \lambda f_{2_a} + \lambda^2 f_{3_a} + \cdots + \lambda^{d_a-1} f_{d_a}) = \lambda(f_{1_a} + \lambda f_{2_a} + \lambda^2 f_{3_a} + \cdots + \lambda^{d_a-1} f_{d_a})$ for any λ with $\lambda^{d_a} = 1$. As $f_{1_a}, f_{2_a}, \ldots, f_{d_a}$ are linearly independent we see that there exists at least one proper value $\lambda (\lambda \neq 1)$ with $\lambda^{d_a} = 1$. (E. Schmidt's arguments.)

Operator-theoretical treatment of Markoff's process. II

Proc. Imp. Acad. Tokyo **15** (1939) 127–130

(Comm. by T. TAKAGI, M.I.A., May 12, 1939.)

§ 1. Let $P(x, E)$ denote the transition probability that the point x of the interval $\Omega = (0, 1)$ is transferred, by a simple Markoff's process, into the Borel set E of Ω after the elapse of a unit time. It is naturally assumed that $P(x, E)$ is completely additive for Borel sets E if x is fixed and that $P(x, E)$ is Borel measurable in x if E is fixed. $P(x, E)$ defines a linear operator P on the complex Banach space $(\mathfrak{M})$ in $(\mathfrak{M})$[1]:

$$P \cdot f = g, \quad g(E) = \int_{\Omega} P(x, E) f(dx).$$

It is easy to see that the iterated operator P^n is defined by the kernel $P^{(n)}(x, E) = \int_{\Omega} P^{(n-1)}(x, dy) P(y, E) \left(P^{(1)}(x, E) = P(x, E) \right)$. In the preceding note,[2] it is proved that the following condition (D) implies the condition (K):

(D) $\quad \begin{cases} \text{there exist an integer } s \text{ and positive constants } b, \eta \ (<1) \text{ such} \\ \text{that, if mes } (E) < \eta, \ P^{(s)}(x, E) < 1 - b \text{ uniformly in } x, E. \end{cases}$

(K) $\quad \begin{cases} \text{there exist an integer } n \text{ and a completely continuous linear} \\ \text{operator } V \text{ such that } \| P^n - V \|_{\mathfrak{M}} < 1. \end{cases}$

The condition (K) is more general than (D), since there exists $P(x, E)$ which satisfies (K) but not (D). In [I] it is proved that, if $P(x, E)$ satisfies (D), then

(B) $\quad \begin{cases} \text{the proper values } \lambda \text{ with modulus 1 of } P \text{ are all roots of} \\ \text{unity.} \end{cases}$

Thus, combined with (K), we were able to give an operator-theoretical treatment of the Markoff's process $P(x, E)$ under the condition (D). (See [I].)

In the present note I intend to show that the condition (K) im-

---

1) $(\mathfrak{M})$ is the linear space of all the totally additive set functions defined for all the Borel sets of Ω. For any $f \in (\mathfrak{M})$ we define its norm $\|f\|_{\mathfrak{M}}$ by the total variation of f on Ω.

2) K. Yosida: Operator-theoretical Treatment of the Markoff's Process, Proc. **14** (1938), 363. This note will be referred to as [I] below. It contains many misprints. On page 364, line 27 and line 28 $(\mathfrak{M})$ is to be read (M^*). On page 364, line 28 $h(dz)$ is to be read $h(z)$. On page 365, line 7 "$f_{i_k}(E) \cdot f_{j_k}(E) \equiv 0$ for $i \neq j$" is to be read "$f_{i_k}(E_{i_k}) = 1$ where $E_{i_k} \cdot E_{j_k} = $ void for $i \neq j$." On page 367, line 4 and 5 "From… by (6)" is to be read "Evident from iii) and the equations $f_{(i+1)_a}(E) = \int_0^1 P(x, E) f_{i_a}(dx)$ below."

plies the property (B). Hence the results in [I] are, in essential, valid for the Markoff's process under the condition (K).

§ 2. Let $P(x, E)$ satisfy the condition (K). Then,[1] there exists completely continuous linear operators P_λ such that

$$(1) \quad \begin{cases} \lim\limits_{n\to\infty} \left\| \dfrac{1}{n} \sum\limits_{m=1}^{n} \left(\dfrac{P}{\lambda}\right)^m - P_\lambda \right\|_{\mathfrak{M}} = 0 , \\ P_\lambda^2 = P_\lambda , \quad PP_\lambda = P_\lambda P = \lambda P_\lambda \qquad (|\lambda|=1) . \end{cases}$$

P_λ is the projection operator which maps $(\mathfrak{M})$ on the proper space (Eigenraum) of P belonging to the proper value λ. Let P_1 be defined by the kernel $P_\lambda(x, E)$. We have, by $P^{(m)}(x, E) \geq 0$ and $P^{(m)}(x, \Omega) \equiv 1$,

$$(2) \qquad P_1(x, E) \geq 0 , \quad P_1(x, \Omega) \equiv 1 .$$

Thus $P_1 \neq 0$. Let $P \cdot f = f$, that is, $P_1 \cdot f = f$. Then, by (2), we obtain $P_1 \cdot \tilde{f} = \tilde{f}$, where $\tilde{f}(E) =$ the total variation of f on E. As P_1 is completely continuous, the number of the linearly independent solutions of $P_1 \cdot f = f$ is finite. Thus applying Kryloff-Bogoliouboff's arguments[2] we obtain the

Lemma. There exist $f_1, f_2, \ldots, f_k \in (\mathfrak{M})$ with the properties :

$$P \cdot f_i = f_i , \quad f_i(E) \geq 0 , \quad f_i(E_i) = 1 \quad (E_i \cdot E_j = \text{void for } i \neq j) ,$$

such that any f satisfying $P \cdot f = f$, $f(E) \geq 0$, $f(\Omega) = 1$ is uniquely expressed as a linear combination $f(E) = \sum\limits_{i=1}^{k} c_i f_i(E)$, $\sum\limits_{i=1}^{k} c_i = 1$, $c_i \geq 0$.

Hence, from $PP_1 = P_1$ and (2), we obtain

$$(3) \quad \begin{cases} P_1(x, E) = \sum\limits_{i=1}^{k} c_i(x) f_i(E) , \\ c_i(x) \text{ measurable with } c_i(x) \geq 0 , \quad \sum\limits_{i=1}^{k} c_i(x) \equiv 1 . \end{cases}$$

Let now $\lambda(|\lambda|=1)$ be a proper value of $P : P_\lambda \neq 0$. Let P_λ be defined by the kernel $P_\lambda(x, E)$. From $P_\lambda P^m = \lambda^m P_\lambda$ we see that the proper value equations

$$(4) \qquad \int_\Omega P^{(m)}(x, dy)\, g(y) = \lambda^m g(x) \qquad (m = 1, 2, \ldots)$$

admit bounded measurable solution $g(x) \neq 0$. We may assume that

$$(5) \qquad \left\| g \right\|_{M*} = \text{lowest upper bound}_{x \in \Omega} \left| g(x) \right| = 1 .$$

Then we may prove that

$$(6) \qquad \text{there exists } x_0 \in \Omega \text{ such that } |g(x_0)| = 1 .$$

---

1) K. Yosida: Abstract Integral Equations and the Homogeneous Stochastic Process, Proc. **14** (1938), 286. K. Yosida: Quasi-completely-continuous Linear Functional Operations, to appear soon in Jap. J. Math. Cf. also S. Kakutani: Iteration of Linear Operations in Complex Banach Spaces, Proc. **14** (1938), 292.

2) [I], Lemma 4.

Proof of (6). From (4) and $P^{(m)}(x, E) \geq 0$, $P^{(m)}(x, \Omega) \equiv 1$, we obtain

$$|g(x)| \leq \int_{|g(y)|\geq 1-\delta} P^{(m)}(x, dy) + (1-\delta) \int_{|g(y)|<1-\delta} P^{(m)}(x, dy) = 1 - \delta \int_{|g(y)|<1-\delta} P^{(m)}(x, dy),$$

$$(1 > \delta > 0).$$

Hence, by (1),

$$(7) \qquad |g(x)| \leq 1 - \delta \int_{|g(y)|<1-\delta} P_1(x, dy).$$

Let $x' \in \Omega$ be such that $|g(x')| \geq 1 - \delta\varepsilon$ $(1 > \varepsilon > 0)$, then, by (7) and (3),

$$\begin{cases} \varepsilon \geq P_1\big(x', E(\delta)\big) = \sum_{i=1}^{k} c_i(x') f_i\big(E_i - E_i \cdot E(\delta)\big), \\ E(\delta) = \underset{y}{E} \{|g(y)| \geq 1 - \delta\}. \end{cases}$$

As $c_i(x) \geq 0$, $\sum_{i=1}^{k} c_i(x) \equiv 1$ and ε, δ were arbitrary, we must have

$$f_i\big(E_i - E_i \cdot E(0)\big) = 0, \qquad E(0) = \underset{y}{E} \{|g(y)| = 1\}$$

for a certain i $(=1$ or 2 or $\cdots$ or $k)$. Thus, by $f_i(E_i) = 1$, $E(0)$ is not void. Q. E. D.

Next let $|g(x_0)| = 1$. Then, by (4) and $P^{(m)}(x_0, \Omega) = 1$, we have

$$(8) \qquad \int_{\Omega} P^{(m)}(x_0, dy) \left\{ 1 - \frac{g(y)}{\lambda^m g(x_0)} \right\} = 0. \qquad (m = 1, 2, \dots)$$

Put $g(y)/\lambda^m g(x_0) = h^{(m)}(y) = h_1^{(m)}(y) + \sqrt{-1}\, h_2^{(m)}(y)$, where $h_1^{(m)}(y)$ the real part of $h^{(m)}(y)$. From (5) we have $|h^{(m)}(y)| \leq 1$. Thus $h_1^{(m)}(y) \leq 1$, and if $h_1^{(m)}(y) = 1$ we must have $h^{(m)}(y) = 1$ viz. $g(y) = \lambda^m g(x_0)$.

From (8) we have

$$(9) \qquad \int_{\Omega} P^{(m)}(x_0, dy) \big(1 - h_1^{(m)}(y)\big) = 0. \qquad (m = 1, 2, \dots)$$

As $P^{(m)}(x_0, E) \geq 0$, $P^{(m)}(x_0, \Omega) = 1$, $1 \geq h_1^{(m)}(y)$ we must have

$$P^{(m)}\big(x_0, E(m)\big) = 1, \qquad E(m) = \underset{y}{E} \{g(y) = \lambda^m g(x_0)\}.$$

Hence, if

$$(10) \quad E(i) \cdot E(j) \neq \text{void for a certain couple of integers } i, j \text{ with } i \neq j,$$

then $\lambda^{i-j} = 1$, as was to be proved.

Now let (10) be not true. Then, by (1),

$$\lim_{n \to \infty} \frac{1}{n} \sum_{i=1}^{n} P^{(i)}\big(x_0, E(s)\big) = P_1\big(x_0, E(s)\big) = 0$$

for any s. Hence we would obtain $P_1\left(x_0, \sum_{s=1}^{\infty} E(s)\right) = \sum_{s=1}^{\infty} P_1\left(x_0, E(s)\right) = 0$. This is a contradiction, since, by (9) and (1),

$$P_1\left(x_0, \sum_{s=1}^{\infty} E(s)\right) = \lim_{n \to \infty} \frac{1}{n} \sum_{i=1}^{n} P^{(i)}\left(x_0; \sum_{s=1}^{\infty} E(s)\right)$$

$$= \lim_{n \to \infty} \frac{1}{n} \sum_{i=1}^{n} P^{(i)}\left(x_0, E(i)\right) = 1 .$$

Birkhoff's ergodic theorem and the maximal ergodic theorem

(*with S. Kakutani*)

Proc. Imp. Acad. Tokyo **15** (1939) 165–168

(Comm. by T. TAKAGI, M.I.A., June 12, 1939.)

1. *Statement of the theorem.* Let S be a space in which a measure of Lebesgue type is defined, and let T be a one-to-one measure-preserving transformation of S into itself. We do not assume ·that the total measure mes (S) is finite. For any real valued function $f(x)$ defined on S, we define the functions $\bar{f}(x), \underline{f}(x), f^*(x)$ and $f_*(x)$ as follows:

$$
\begin{cases}
\bar{f}(x)=\varlimsup_{n\to\infty}\frac{1}{n}\sum_{i=0}^{n-1}f(T^i x)\,, & \underline{f}(x)=\varliminf_{n\to\infty}\frac{1}{n}\sum_{i=0}^{n-1}f(T^i x)\,, \\[2em]
f^*(x)=\underset{0\leq n<\infty}{\text{l. u. b.}}\frac{1}{n}\sum_{i=0}^{n-1}f(T^i x)\,, & f_*(x)=\underset{0\leq n<\infty}{\text{g. l. b.}}\frac{1}{n}\sum_{i=0}^{n-1}f(T^i x)\,.
\end{cases}
$$

If $f(x)$ is measurable and absolutely integrable on S, then we can prove the following two theorems:

Theorem 1. For any pair of real numbers α and β, we have

$$
(1)\quad
\begin{cases}
\alpha \text{ mes}\,\big(E(\alpha, \beta)\big) \leqq \displaystyle\int_{E(\alpha,\beta)} f(x)\,dx \leqq \beta \text{ mes}\,\big(E(\alpha, \beta)\big)\,, \\[1.5em]
\text{where}\quad E(\alpha, \beta)=\underset{x}{E}\,[\bar{f}(x) > \alpha, \underline{f}(x) < \beta]\,.
\end{cases}
$$

Consequently, $\alpha > \beta$ implies mes $\big(E(\alpha, \beta)\big)=0$, and since this is true for any pair of real numbers α and β with $\alpha > \beta$, we have $\bar{f}(x)=\underline{f}(x)$ almost everywhere; that is,

$$
\lim_{n\to\infty}\frac{1}{n}\sum_{i=0}^{n-1}f(T^i x)=f_1(x)
$$

exists almost everywhere.

Theorem 2. For any real number α we have

$$
(2)\quad
\begin{cases}
\alpha \text{ mes}\,\big(E^*(\alpha)\big) \leqq \displaystyle\int_{E^*(\alpha)} f(x)\,dx\,, & \alpha \text{ mes}\,\big(E_*(\alpha)\big) \geqq \displaystyle\int_{E_*(\alpha)} f(x)\,dx\,, \\[1.5em]
\text{where}\quad E^*(\alpha)=\underset{x}{E}\,[f^*(x) > \alpha] \text{ and } E_*(\alpha)=\underset{x}{E}\,[f_*(x) < \alpha]\,.
\end{cases}
$$

Theorem 1 is the *Ergodic Theorem of Birkhoff* in its form given by A. Kolmogoroff.[1] Theorem 2 is new. We shall call Theorem 2

---

1) A. Kolmogoroff: Ein vereinfachter Beweis des Birkhoff-Khintchinschen Ergodensatzes, Recueil Math., **44** (1937), 366-368. See also E. Hopf: Ergodentheorie, Ergebnisse der Math., Heft 5·(1937).

the *Maximal Ergodic Theorem*. Recently N. Wiener[2] obtained the analogous result:

$$(2') \quad \begin{cases} \text{if } \operatorname{mes}(S) = \text{finite, and if } f(x) \geq 0 \text{ throughout on } S, \\ \text{then } \alpha \operatorname{mes}\left(E^*(\alpha)\right) \leq \int_S f(x)\,dx. \end{cases}$$

This result is clearly weaker than (2). Wiener's proof of (2') is based on the so-called Maximal Theorem of Hardy and Littlewood[3]; and using (2') he deduced from the Mean Ergodic Theorem of v. Neumann a new proof of the Ergodic Theorem of Birkhoff. Wiener has also obtained from (2') the so-called *Dominated Ergodic Theorem.*[4] It is to be noted that the latter is also possible even if we have no assumption that $\operatorname{mes}(S) = $ finite, while the former is not always possible without this assumption.

In the present note, we shall give a direct proof of Theorem 2. Our method of proof is a modification of that of Khintchine-Kolmogoroff,[1] which was used to prove Theorem 1; and it is to be noted that we can prove Theorem 1 (Birkhoff's Ergodic Theorem) and Theorem 2 (Maximal Ergodic Theorem) simultaneously by the same principle without appealing to Maximal Theorem nor to the Mean Ergodic Theorem.

2. *Proof of Theorem 2.* We define

$$(3) \qquad f_{ab}(x) = \frac{1}{b-a} \sum_{i=a}^{b-1} f(T^i x), \qquad a < b.$$

For any fixed $x \in S$, consider the pair of integers a and b such that $f_{ab}(x) > \alpha$ while $f_{ab'}(x) \leq \alpha$ for any b' with $a < b' < b$. Such an interval (a, b) is called a *maximal interval* (corresponding to α and x), and $b-a$ is called the *length* of this maximal interval. Of two maximal intervals (a, b) and (a', b') (both corresponding to α and x), the one may contain the other; but these cannot overlap each other. For, if $a < a' < b < b'$, we have

$$f_{ab}(x) = \frac{(a'-a)\cdot f_{aa'}(x) + (b-a')\cdot f_{a'b}(x)}{b-a}$$

and, since $f_{aa'}(x) \leq \alpha$ and $f_{a'b}(x) \leq \alpha$ by assumption, we have $f_{ab}(x) \leq \alpha$, contrary to the assumption that (a, b) is maximal. A maximal interval (a, b) (corresponding to α and x) of length $b-a \leq s$ is called *s-maximal* if it is contained in no other maximal interval (corresponding to α and to x) of length $\leq s$. Thus all *s*-maxinal intervals (corresponding to α and to x) lie outside each other.

---

2) N. Wiener: The Ergodic Theorem, Duke Math., Journ., 5 (1939), 1–18.

3) G. H. Hardy, J. E. Littlewood and G. Pólya: Inequalities. Cambridge (1935).

4) N. Wiener: The Homogeneous Chaos, Amer. Journ. of Math., 60 (1938), 897–936. In this paper Zygmund's class only was considered. The general case $L^p (p > 1)$ was obtained by N. Wiener and M. Fukamiya independently. N. Wiener: the paper cited in the footnote (2). M. Fukamiya: On Dominated Ergodic Theorem in $L^p (p \geq 1)$, to be published in Tôhoku Math. Journ. Fukamiya's proof also appeals to the Maximal Theorem of Hardy and Littlewood.

Now let $E_s^*(a)$ be the set of all the points $x \in S$ such that there exists an s-maximal interval (a, b) (corresponding to a and to x) with $a \leq 0 < b$. It is clear, by the argument above, that to any point $x \in E_s^*(a)$ there corresponds one and only one s-maximal interval of this sort. Since $f_{ab}(x) > a$ and since $f_{ab'}(x) \leq a$ for any b' with $a < b' < b$, we must have $f_{ob}(x) > a$ and consequently $E_s^*(a) \subset E^*(a)$ for any s. Moreover, by the definition of $E^*(a)$, we have

$$(4) \qquad \lim_{s \to \infty} E_s^*(a) = E^*(a) .$$

On the other hand, $E_s^*(a)$ may be divided into disjoint subsets $E_{pq}^*(a)$:

$$(5) \qquad E_s^*(a) = \sum_{q=1}^{s} \sum_{p=0}^{q-1} E_{pq}^*(a) ,$$

where $E_{pq}^*(a)$, $0 \leq p < q \leq s$, is the set of all the points $x \in E_s^*(a)$ whose corresponding s-maximal interval is $(-p, -p+q)$. From the identity:

$$\frac{1}{q} \sum_{i=-p}^{-p+q-1} f(T^i x) = \frac{1}{q} \sum_{i=0}^{q-1} f(T^i T^{-p} x)$$

we see that

$$T^{-p} E_{pq}^*(a) = E_{oq}^*(a) ,$$

and, since T is measure-preserving, we have

$$(6) \qquad \begin{cases} \mathrm{mes}\left(E_{pq}^*(a)\right) = \mathrm{mes}\left(E_{oq}^*(a)\right) , \\[2mm] \int_{E_{pq}^*(a)} f(x)\,dx = \int_{E_{oq}^*(a)} f(T^p x)\,dx . \end{cases}$$

Consequently, we have by (5) and (6)

$$(7) \quad \begin{cases} \displaystyle\int_{E_s^*(a)} f(x)\,dx = \sum_{q=1}^{s}\sum_{p=0}^{q-1} \int_{E_{pq}^*(a)} f(x)\,dx = \sum_{q=1}^{s}\sum_{p=0}^{q-1} \int_{E_{oq}^*(a)} f(T^p x)\,dx = \sum_{q=1}^{s} \int_{E_{oq}^*(a)} q \cdot f_{oq}(x)\,dx \\[4mm] \displaystyle \geq \sum_{q=1}^{s} \int_{E_{oq}^*(a)} q \cdot a\,dx = \sum_{q=1}^{s} q \cdot a\,\mathrm{mes}\left(E_{oq}^*(a)\right) = a \sum_{q=1}^{s}\sum_{p=0}^{q-1} \mathrm{mes}\left(E_{pq}^*(a)\right) \\[4mm] \hspace{7cm} = a\,\mathrm{mes}\left(E_s^*(a)\right) . \end{cases}$$

Hence, by (4), we obtain

$$\int_{E^*(a)} f(x)\,dx \geq a\,\mathrm{mes}\left(E^*(a)\right) .$$

Thus the first part of Theorem 2 is proved, and the second part may be proved analogously. We may also obtain the proof of Theorem 1, if we start from $E(a, \beta)$ instead of from $E^*(a)$, and if we consider $E_s(a, \beta) = E_s^*(a) \cdot E(a, \beta)$ and $E_{pq}(a, \beta) = E_{pq}^*(a) \cdot E(a, \beta)$ instead of $E_s^*(a)$ and $E_{pq}^*(a)$ respectively, remembering the invariance of $E(a, \beta)$: $E(a, \beta) = TE(a, \beta)$. This is indeed the proof of Theorem 1 due to A. Kolmogoroff.

3. *Integrability of the functions $f_1(x)$, $f^*(x)$ and $f_*(x)$.* We can see easily from Theorem 1 that, if $f(x)$ is absolutely integrable, the limit function $f_1(x)$ $(=\bar{f}(x)=\underline{f}(x)$ almost everywhere) is also absolutely integrable. In order to show this, it is sufficient to consider the case that $f(x) \geq 0$ throughout on S. Denoting again by $E(\alpha, \beta)$ the set of all the points $x \in S$ such that $\alpha < f_1(x) < \beta$, we have for any pair of real numbers α and β with $0 < \alpha < \beta$

$$\alpha \text{ mes } \big(E(\alpha, \beta)\big) \leq \int_{E(\alpha,\beta)} f(x)\,dx \leq \beta \text{ mes } \big(E(\alpha, \beta)\big) ,$$

and, since mes $\big(E(\alpha, \beta)\big) < \infty$, we have

$$\int_{E(\alpha,\beta)} f_1(x)\,dx = \int_{E(\alpha,\beta)} f(x)\,dx .$$

Since α and β $(0 < \alpha < \beta)$ are arbitrary, we have

$$(8) \qquad \int_S f_1(x)\,dx = \int_{E(0,\infty)} f_1(x)\,dx = \int_{E(0,\infty)} f(x)\,dx \leq \int_S f(x)\,dx .$$

Thus we have proved that $f_1(x)$ is absolutely integrable with the additional inequality (8).

If, moreover, $f(x)$ belongs to the Lebesgue's class L^p $(p > 1)$, then $f^*(x)$ and $f_*(x)$ belong also to the same class L^p; and if $f(x)$ belongs to the Zygmund's class:

$$\int_S |f(x)| \log^+ |f(x)|\,dx = \text{finite},$$

then $f^*(x)$ and $f_*(x)$ both belong to the class L^1. These results (Dominated Ergodic Theorem) were obtained from (2′) by N. Wiener, and directly from the Maximal Theorem of Hardy and Littlewood by M. Fukamiya, in case mes $(S) = $ finite. The same argument as that used by Wiener will lead us to the same conclusion for the class L^p $(p > 1)$ even in the general case mes $(S) = \infty$ from our (2); for, the assumption that mes $(S) = $ finite is not needed in this part of the proof of Wiener's. We therefore omit the proof.

Asymptotic almost periodicities and ergodic theorems

Proc. Imp. Acad. Tokyo **15** (1939) 255–259

(Comm. by T. TAKAGI, M.I.A., Oct. 12, 1939.)

1. *Introduction.* Two ergodic theorems, the mean ergodic theorem (M. E. T.) and a generalisation of Fréchet-Kryloff-Bogoliouboff's theorem (F-K-B E. T.), were obtained in the preceding notes [1] There are some strong difference or gap between these two ergodic theorems. The purpose of the present note is to fulfill this gap with new ergodic theorems and to show that these theorems (including the M. E. T. and the F-K-B E. T.) are intimately related to the properties of *asymptotic almost periodicity*, to be defined below.

Let T denote a continuous (bounded) linear operation defined on a Banach space B to B, and consider the sequence $\{T^n \cdot x\}$, $x \in B$, $n = 1, 2, \dots$. Corresponding to the various assumptions of *total boundedness* of $\{T^n \cdot x\}$, we may obtain various ergodic theorems together with the respective properties of asymptotic almost periodicity (in n) of $T^n \cdot x$. This simple idea was suggested by Bochner-Neumann's theory [2] (B-N theory) of almost periodic functions in groups. However, since we do not assume the existence of the inverse T^{-1} of T, we are here concerned with the *semi-group* of the addition of positive integers. We also remark that a new proof of the existence of the mean for Bohr's (Stepanoff's, Muckenhaupt's and other author's) almost periodic functions may be obtained by virtue of the ergodic theorems. Combined with the Fourier analysis in the B-N theory, the M. E. T. yields a Fourier expansion theorem and a theorem of existence of the proper values for *unitary* (isometric) *operator* T of B. Lastly it is to be noted that the B-N theory also suggests us not to confine ourselves to the Banach spaces; the (ergodic) theorems obtained may be extended to *linear topological spaces.*

2. *Ergodic theorems and asymptotic almost periodicities.*

Theorem 1. We assume that T satisfies the following total boundedness :

(1) $\qquad \| T^n \| \leq a \text{ constant } C \ (n = 1, 2, \dots),$

(2) $\quad \begin{cases} \text{for given } x \in B, \text{ the sequence } \{x_n\}, \ x_n = \dfrac{T + T^2 + \dots + T^n}{n} \cdot x \\ (n = 1, 2, \dots), \text{ contains a subsequence weakly convergent to a} \\ \text{point } x_0 \in B. \end{cases}$

Then x_n converges strongly to x_0 and we have $T \cdot x_0 = x_0$. If (2) is satisfied for all $x \in B$, then x_0 is defined by a continuous linear operation T_1 such that

---

1) Proc. **14** (1938), 286–294. Cf. also S. Kakutani: ibd., 295–298.

2) Trans. Amer. Math. Soc. **37** (1935), 21–50.

$$(3) \qquad TT_1 = T_1 T = T_1^2 = T_1 .$$

This is the M. E. T., expressed somewhat more precisely than in the preceding note. T_1 is the projection operator which maps B on the proper space of T belonging to the proper value 1. Since we have $T^n \cdot x = x_0 + (T^n \cdot x - x_0)$, $T^n \cdot x$ is periodic in n (with the period zero) except the error whose arithmetic mean $\dfrac{1}{m} \sum\limits_{n-k}^{k+m-1} (T^n \cdot x - x_0)$ tends strongly to zero *uniformly in k* when m tends to $+\infty$. If, in particular, T is a unitary operator in the hilbert space, then we have a more precise result concerning the almost periodicity in n of $T^n \cdot x$, i. e. E. Hopf's theorem.[1]

Theorem 2. We assume that T satisfies the condition (1) *and*

$$(4) \qquad \begin{cases} \{T^n \cdot x\} \text{ is totally bounded in the strong topology defined by} \\ \text{the norm in } B. \end{cases}$$

Then, since the convex closure of $\{T^n \cdot x\}$ is totally bounded, the M. E. T. of course applies to T. (It is to be noted that, in this case, the proof of the M. E. T. may be shortened; it can be obtained without appealing to the Hahn-Banach's extension theorem.[2]) Moreover, we have the asymptotic almost periodicity (in n) of $T^n \cdot x$:

$$(5) \qquad \begin{cases} \text{for any } \varepsilon > 0, \text{ there exists a positive integer } p_\varepsilon \text{ such that any} \\ \text{interval of length } p_\varepsilon \text{ with positive integers as its extremities con-} \\ \text{tains at least one integer } p \text{ satisfying } \overline{\lim_{n \to \infty}} \, \| T^{n+p} \cdot x - T^n \cdot x \| \leqq \varepsilon. \end{cases}$$

Proof of (5). There exists, by (4), a positive integer p_ε such that

$$\min_{1 \leq k \leq p_\varepsilon} \| T^m \cdot x - T^k \cdot x \| \leqq \frac{\varepsilon}{C} \quad \text{for any } m \ (=1, 2, \ldots).$$

Hence, by (1), there exists $k(m)$, $1 \leq k(m) \leq p_\varepsilon$, such that

$$\| T^{n+m} \cdot x - T^{n+k(m)} \cdot x \| \leqq \varepsilon \quad \text{for } n=1, 2, \ldots .$$

Thus we have

$$\min_{m-p_\varepsilon \leq p \leq m-1} \{ \overline{\lim_{n \to \infty}} \, \| T^{n+p} \cdot x - T^n \cdot x \| \} \leqq \varepsilon \quad \text{for } m = p_\varepsilon + 1,$$

$$p_\varepsilon + 2, \ldots . \qquad \text{Q. E. D.}[3]$$

---

1) E. Hopf: Ergodentheorie, Berlin (1937), 25.

2) For the sake of comprehension, we here sketch the proof for the (strong) convergence of the sequence $\{x_n\}$, $x_n = \dfrac{T + T^2 + \cdots + T^n}{n} \cdot x$ $(n=1, 2, \ldots)$. Since $\{x_n\}$ is totally bounded, $\{x_n\}$ contains a subsequence $\{x_{n'}\}$ such that $\lim\limits_{n' \to \infty} x_{n'} = x_0 \in B$. Clearly we have $T \cdot x_0 = x_0$, and hence we would obtain $\lim\limits_{n \to \infty} x_n = x_0$ if $\lim\limits_{n \to \infty} \dfrac{T + T^2 + \cdots + T_n}{n} \cdot (x - x_0) = 0$. Since T^n is of norm $\leqq C$ $(n=1, 2, \ldots)$, this last equation is clear when $(x-x_0)$ is of the form $(y - T \cdot y)$, $y \in B$; and it is also clear when $(x-x_0)$ is a limit element of the form $(y - T \cdot y)$, $y \in B$. Thus from $(x-x_0) = \lim\limits_{n' \to \infty} \left(x - \dfrac{T + T^2 + \cdots + T^{n'}}{n'} \cdot x \right) = \lim\limits_{n' \to \infty} (E - T) \left(\dfrac{n' E + (n'-1) T + \cdots + T^{n'}}{n'} \right) \cdot x$, $E =$ the identity operator, we see that $\lim\limits_{n \to \infty} x_n = x_0$.

3) The original proof is somewhat shortened by S. Kakutani's remark.

Example. Let a continuous point transformation P defined on a compactum R to R be a *contraction*:

$$\text{distance } (P \cdot t, P \cdot s) \leq \text{distance } (t, s) \quad \text{for any} \quad t, s \in R.^{1)}$$

Let B denote the Banach space of all the complex-valued continuous functions $f(t)$ on R with the norm $\|f\| = \max_{t \in R} |f(t)|$, then P defines a linear operation T on B to $B: T \cdot f = g$, $g(t) = f(P \cdot t)$. It is easy to see that the conditions (1) and (4) are satisfied for any $x \in B$. Another example is given from the theory of almost periodic functions. See the next paragraph.

The set of all the continuous linear operations T on B to B constitutes a Banach space $\widetilde{B}$ with the norm $\|T\|$ $(= \text{l. u. b.}_{|x| \leq 1} \|T \cdot x\|)$. Our next theorem reads as follows.

Theorem 3. We assume that T satisfies (1) and

(6) $\{T^n\}$ *is totally bounded in the topology defined by the norm in $\widetilde{B}$.*

Then the M. E. T. applies to T in the uniform sense :

$$\text{(7)} \qquad \lim_{n \to \infty} \left\| \frac{T + T^2 + \cdots + T^n}{n} - T_1 \right\| = 0 .$$

Moreover, we have the asymptotic almost periodicity in the uniform sense, viz. we have, in (5), $\overline{\lim}_{n \to \infty} \| T^{n+p} - T^n \| \leq \epsilon$ instead of $\overline{\lim}_{n \to \infty} \| T^{n+p} \cdot x - T^n \cdot x \| \leq \epsilon.$

We omit the proof.

Theorem 4. We assume, beside (1) and (6), that any limit element $(\in \widetilde{B})$ of the totally bounded set $\{T^n\}$ is a completely continuous linear operation on B to B. This case is precisely the case of the F-K-B E. T.

Proof. For the proof we have only to restate the F-K-B E. T. :

Let T satisfy (1), and let there exist an integer m and a completely continuous linear operation V on B to B such that $\| T^m - V \| < 1$, then the proper values with modulus 1 of T are finite in number. Let these proper values be $\lambda_1, \lambda_2, \ldots, \lambda_k$, then there exist completely continuous linear operations $T_{\lambda_1}, T_{\lambda_2}, \ldots, T_{\lambda_k}$ and a continuous linear operation S such that

$$\text{(8)} \quad \begin{cases} T = \sum_{i=1}^{k} \lambda_i T_{\lambda_i} + S, \quad T_{\lambda_i}^2 = T_{\lambda_i}, \quad T T_{\lambda_i} = T_{\lambda_i} T = \lambda_i T_{\lambda_i}, \\[2mm] T_{\lambda_i} T_{\lambda_j} = 0 \quad (i \neq j), \ ST_{\lambda_i} = T_{\lambda_i} S = 0 \quad (i, j = 1, 2, \ldots, k), \\[2mm] \| S^n \| \leq \dfrac{\partial}{(1+\epsilon)^n} \quad (n = 1, 2, \ldots) \ \text{with positive constants } \epsilon, \partial. \end{cases}$$

Since we obtain $T^n = \sum_{i=1}^{k} \lambda_i^n T_{\lambda_i} + S^n$ $(n = 1, 2, \cdots)$ from (8), the asymptotic almost periodicity in n of T^n is apparent.

3. *A proof of the existence of the mean for Bohr's almost periodic functions.*

---

1) It is sufficient to assume that distance $(P^n \cdot t, P^n \cdot s) \leq$ distance (t, s) multiplied by a constant independent of n.

Theorem 5. *Let a complex-valued continuous function $f(t)$ on the infinite interval $(-\infty, \infty)$ be almost periodic in Bohr's sense, then*

$$\lim_{u \to \infty} \frac{1}{u} \int_s^{s+u} f(t)\,dt \quad \text{exists uniformly in } s, \quad -\infty < s < \infty.$$

Proof. According to S. Bochner and J. Favard, the sequence $\{f^{(n)}(t)\}$, $f^{(n)}(t) = f(t+n)$ $(n=1, 2, \ldots)$, is totally bounded in the topology defined by the distance $\|f^{(n)} - f^{(m)}\| = \underset{-\infty < t < \infty}{\text{l. u. b.}} |f^{(n)}(t) - f^{(m)}(t)|$. Let B denote the Banach space spanned by $\{f^{(n)}\}$ with the above norm $\|\ \|$, and let T be the continuous linear operation on B to B defined by $T \cdot f = g$, $g(t) = f(t+1)$. T is surely of norm 1 and the Theorem 2 applies to T. Hence the. sequence $\{f_n\}$, $f_n = \dfrac{T + T^2 + \cdots + T^n}{n} \cdot f$ $(n=1, 2, \ldots)$, converges strongly to an element $\in B$.[1]

Hence, we have, uniformly in s $(-\infty < s < \infty)$,

$$\int_0^1 \left\{ \lim_{n \to \infty} \frac{1}{n} \sum_{m-1}^{n} f(s+t+m) \right\} dt = \lim_{n \to \infty} \frac{1}{n} \sum_{m-1}^{n} \int_0^1 f(s+t+m)\,dt$$

$$= \lim_{n \to \infty} \frac{1}{n} \int_1^{n+1} f(s+t)\,dt.$$

This proves the theorem, by the uniform boundedness of $f(t)$.

4. *Application of the Fourier analysis to the operator T.* Throughout this paragraph, we assume that the inverse T^{-1} of T exists.

Theorem 6. *We assume that*

(9) $\qquad \|T^n\| \leq a$ *constant C* $(n=0, \pm 1, \pm 2, \ldots)$,

(10) $\qquad \begin{cases} \textit{for any } x \in B, \textit{ the set } \{T^n \cdot x\} \ (n=0, \pm 1, \pm 2, \ldots) \textit{ is totally} \\ \textit{bounded in the strong topology defined by the norm in B.} \end{cases}$

Then there exists at least one proper value with modulus 1 of T.

Proof. From (9) and (10) we see that the set $\{F(n)\}$, $F(n) = T^n \cdot x$ $(n=0, \pm 1, \pm 2, \ldots)$, is totally bounded by the distance $\|F(n) - F(m)\| = \underset{-\infty < p < \infty}{\text{l. u. b.}} \|T^{n+p} \cdot x - T^{m+p} \cdot x\|$. Hence $F(n)$ is almost periodic in the group of addition of all the integers. Thus, by the Weierstrass approximation theorem in the B-N theory, we obtain the *Fourier expansion*:

(11) $\qquad T^n \cdot x = F(n) \sim \sum_{i-1}^{n} \lambda_i^n C_{\lambda_i}$ $(n=0, \pm 1, \pm 2, \ldots)$, $|\lambda_i| = 1$

$$(i = 1, 2, \ldots),$$

where the *Fourier coefficients* C_{λ_i} are given by

(12) $\qquad C_{\lambda_i} = \lim_{n \to \infty} \frac{1}{n} \sum_{m-1}^{n} \lambda_i^{-m} F(m) = \lim_{n \to \infty} \left(\frac{1}{n} \sum_{m-1}^{n} \lambda_i^{-m} T^m \right) \cdot x.$

By the M. E. T., $T_{\lambda_i} = \lim\limits_{n \to \infty} \dfrac{1}{n} \sum\limits_{m-1}^{n} \lambda_i^{-m} T^m$ is the projection operator which

---

2) Cf. the footnote 2).

maps B on the proper space of T belonging to the proper value λ_i. Hence, if T does not admit proper value with modulus 1, then the Fourier expansion (11) is nought for any $x \in B$. By the uniqueness theorem in the B-N theory, this means that $T^n \cdot x = 0$ for any $x \in B$ and n. Thus, in particular, $T^o \cdot x = x = 0$ for any $x \in B$, which is surely a contradiction.

Remark. If we put $\|\!|x|\!\| = \underset{-\infty < n < \infty}{\text{l. u. b.}} \|T^n \cdot x\|$, we have $\|\!|T^n \cdot x|\!\| = \|\!|x|\!\|$. This new norm $\|\!|x|\!\|$ gives an equivalent topology as $\|x\|$. Hence, the condition (9) means, in essential, that T is a unitary ($=$isometric) operator in B.

Corollary. As an application of the theorem we have the following result. Let a one-to-one isometric point transformation P of a compactum R on itself be $\neq$ the identical transformation, then there exist a complex-valued continuous function $f(t)$ on R and a complex number $\lambda \neq 1$, $|\lambda| = 1$, such that $f(P \cdot t) = \lambda f(t)$ for all $t \in R$. In other words, there exists at least one *angle variable* of the transformation P.

The proof of the Theorem 6 suggests the

Theorem 7. We assume (9), (10) *and*

$$(13) \quad \begin{cases} T \text{ admits at most enumerably infinite number of proper} \\ \text{values } \{\lambda_i\} \text{ with modulus 1, } (i=1, 2, \cdots), \end{cases}$$

then we have the Fourier expansion:

$$(14) \quad \begin{cases} T^n \sim \sum_{i=1}^{\infty} \lambda_j^n T_{\lambda_i} \ (n=0, \pm 1, \pm 2, \cdots), \ T_{\lambda_i} = \lim_{n \to \infty} \frac{1}{n} \sum_{m=1}^{n} \lambda_i^{-m} T^m, \\ TT_{\lambda_i} = T_{\lambda_i} T = \lambda_i T_{\lambda_i}, \ T_{\lambda_i}^2 = T_{\lambda_i}, \ T_{\lambda_i} T_{\lambda_j} = 0 \ (i \neq j), \ (i, j = 1, 2, \cdots). \end{cases}$$

If the expansion converges in weak or strong sense, then it represents T^n (in weak or strong sense). Moreover we have the Parseval theorem in the weak sense:

$$(15) \qquad \lim_{n \to \infty} \frac{1}{n} \sum_{m=1}^{n} |f(T^m \cdot x)|^2 = \sum_{i=1}^{\infty} |f(T_{\lambda_i} \cdot x)|^2 ,$$

for any $x \in B$ and for any linear functional f on B.

Thus B is decomposed into the proper spaces $T_{\lambda_i} \cdot B$ belonging to the proper values λ_i with modulus one of T.

Markoff process with an enumerable infinite number of possible states

(*with S. Kakutani*)

Japan. J. Math. **16** (1940) 47–55

(Received March 27, 1939.)

1. Introduction. Let P be a simple Markoff process with an enumerable infinite number of possible states $\Re = (S_1, S_2, \ldots)$. P is represented by an infinite matrix: $P = (p_{ij})$, $i, j = 1, 2, \ldots$, where p_{ij} is the transition probability that the state S_i is transferred to the state S_j after the elapse of a unit-time. Then the probability $p_{ij}^{(n)}$ that the state S_i is transferred to the state S_j after the elapse of n unit-times is given recurrently by:

$$(1.1) \qquad p_{ij}^{(n)} = \sum_{k=1}^{\infty} p_{ik} \cdot p_{kj}^{(n-1)}, \qquad n = 2, \quad \ldots \; ; \; p_{ij}^{(1)} = p_{ij} .$$

We have always

$$(1.2) \qquad 0 \leqq p_{ij}^{(n)} \leqq 1 , \qquad i, j, n = 1, 2, \ldots ,$$

$$(1.3) \qquad \sum_{j=1}^{\infty} p_{ij}^{(n)} = 1 , \qquad i, n = 1, 2, \ldots ,$$

and it will be easily seen that

$$(1.4) \qquad p_{ij}^{(m+n)} = \sum_{k=1}^{\infty} p_{ik}^{(m)} \cdot p_{kj}^{(n)} , \qquad i, j, m, n = 1, 2, \ldots .$$

Such a Markoff process was discussed in detail by A. Kolmogoroff[1]. He has proved among others that

$$(1.5) \qquad \lim_{n \to \infty} \frac{1}{n} \sum_{m=1}^{n} p_{ij}^{(m)} = P_{ij}$$

exists for any i and j. In the present paper we shall give two elementary proofs of this result, accompanied with some additional new results.

The first proof is given in §2. This is a direct proof, and the same method is also valid for the analogous problems concerning the iteration of positive linear transformations in Banach spaces. Using a new formula:

$$(1.6) \quad \sum_{k=1}^{\infty} P_{ik} \cdot p_{kj}^{(n)} = \sum_{k=1}^{\infty} p_{ik}^{(n)} \cdot P_{kj} = \sum_{k=1}^{\infty} P_{ik} \cdot P_{kj} = P_{ij} , \quad i, j, n = 1, 2, \ldots ,$$

---

[1] A. Kolmogoroff: Anfangsgründe der Theorie der Markoffschen Ketten mit unendlich vielen möglichen Zuständen, Recueil Math. **43** (1936), 606-609.

which we obtained in §2 as a by-product, we shall give in §3 the decomposition of the total state $\mathfrak{R}$ into a dissipative part $\mathfrak{D}$ and ergodic parts $\mathfrak{E}_\alpha$. It is to be noted that we have first proved the existence of the mean sojourn (1.5) and that then the decomposition of the total state is given, while these were in the reverse order by Kolmogoroff.

The second proof, which is given in §4, is entirely different from the first one. Our method is purely analytic and will probably become a useful tool in the further investigations of such Markoff processes. Moreover, our observations in §5 concerning the construction of a Markoff process with given $\{p_{11}^{(n)}\}$ or $\{k_{11}^{(n)}\}$ (the meaning of $k_{11}^{(n)}$ will be explained in §4) may be noted with some interests.

2. First proof of the existence of the mean sojourn.

Theorem 1. (1.5) *exists for any i and j. This limit P_{ij} satisfies* (1.6) *and*

$$(2.1) \qquad \sum_{j=1}^{\infty} P_{ij} \leq 1, \qquad i = 1, 2, \ldots.$$

Proof. Put $q_{ij}^{(n)} = \dfrac{1}{n} \sum_{m=1}^{n} p_{ij}^{(m)}$. Then we have from (1.2) and (1.3)

$$(2.2) \qquad 0 \leq q_{ij}^{(n)} \leq 1, \qquad i, j, n = 1, 2, \ldots,$$

$$(2.3) \qquad \sum_{j=1}^{\infty} q_{ij}^{(n)} = 1, \qquad i, n = 1, 2, \ldots.$$

Because of (2.2) there exists, by the diagonal method, an increasing sequence of positive integers $\{n_\nu\}$, $\nu = 1, 2, \ldots$, such that $\lim_{\nu \to \infty} q_{ij}^{(n_\nu)} = P_{ij}$ exists for any i and j. Before coming to the proof of the existence of (1.5), let us first prove that this P_{ij} satisfies the relations (1.6) and (2.1). (2.1) is a direct consequence of (2.3), and (1.6) may be proved as follows :

From $\quad \sum_{k=1}^{\infty} p_{ik} \cdot q_{kj}^{(n)} - q_{ij}^{(n)} = \dfrac{1}{n}(p_{ij}^{(n+1)} - p_{ij}) \leq \dfrac{2}{n}$, we have

$\lim_{\nu \to \infty} \sum_{k=1}^{\infty} p_{ik} \cdot q_{kj}^{(n_\nu)} = \lim_{\nu \to \infty} q_{ij}^{(n_\nu)} = P_{ij}$. Since $\sum_{k=1}^{\infty} p_{ij} = 1$ is absolutely convergent,

we have $\lim_{\nu \to \infty} \sum_{k=1}^{\infty} p_{ik} \cdot q_{kj}^{(n_\nu)} = \sum_{k=1}^{\infty} p_{ik} \cdot (\lim_{\nu \to \infty} q_{kj}^{(n_\nu)}) = \sum_{k=1}^{\infty} p_{ik} \cdot P_{kj}$.

Hence we have

$$\sum_{k=1}^{\infty} p_{ik} \cdot P_{kj} = P_{ij}, \qquad i, j = 1, 2, \ldots,$$

and consequently

$$(2.4) \qquad \sum_{k=1}^{\infty} p_{ik}^{(n)} \cdot P_{kj} = P_{ij}, \qquad i, j, n = 1, 2, \ldots,$$

(2.5)
$$\sum_{k=1}^{\infty} q_{ik}^{(n)} \cdot P_{kj} = P_{ij}, \qquad i,\ j,\ n = 1,\ 2,\ \ldots .$$

Analogously we have, from $\left| \sum_{k=1}^{\infty} q_{ik}^{(n)} \cdot p_{kj} - q_{ij}^{(n)} \right| = \left| \frac{1}{n}(p_{ij}^{(n-1)} - p_{ij}) \right| \leq \frac{2}{n}$,

$$\lim_{\nu \to \infty} \sum_{k=1}^{\infty} q_{ik}^{(n_\nu)} \cdot p_{kj} = \lim_{\nu \to \infty} q_{ij}^{(n_\nu)} = P_{ij}.$$

Since $q_{ik}^{(n)}$ and p_{kj} are all non-negative, we have, by making ν tend to ∞,

(2.6)
$$\sum_{k=1}^{\infty} P_{ik} \cdot p_{kj} \leq P_{ij}, \qquad i,\ j = 1,\ 2,\ \ldots .$$

It is to be noted that, from the arguments just given above, the equality in (2.6) does not follow directly; but we must have equal sign here for any i and j. For, if there exists a couple of integers i_0 and j_0 such that $\sum_{k=1}^{\infty} P_{i_0 k} \cdot p_{k j_0} < P_{i_0 j_0}$, then we shall have, by taking a sum with respect to j in (2.6) $(i = i_0)$,

$$\sum_{k=1}^{\infty} P_{i_0 k} = \sum_{k=1}^{\infty} \sum_{j=1}^{\infty} P_{i_0 k} p_{kj} < \sum_{j=1}^{\infty} P_{i_0 j},$$

which is clearly a contradiction. Hence we have

$$\sum_{k=1}^{\infty} P_{ik} \cdot p_{kj} = P_{ij}, \qquad i,\ j = 1,\ 2,\ \ldots ,$$

and consequently

(2.7)
$$\sum_{k=1}^{\infty} P_{ik} \cdot p_{kj}^{(n)} = P_{ij}, \qquad i,\ j,\ n = 1,\ 2,\ \ldots ,$$

(2.8)
$$\sum_{k=1}^{\infty} P_{ik} \cdot q_{kj}^{(n)} = P_{ij}, \qquad i,\ j,\ n = 1,\ 2,\ \ldots .$$

Putting $n = n_\nu$ in (2.8) and making ν tend to ∞, we have, since $\sum_{k=1}^{\infty} P_{ik}$ is absolutely convergent,

(2.9)
$$\sum_{k=1}^{\infty} P_{ik} \cdot P_{kj} = P_{ij}, \qquad i,\ j = 1,\ 2,\ \ldots .$$

Thus (1.6) is proved by (2.4), (2.7) and (2.9).

Now in order to prove (1.5), let us assume that $\lim_{n \to \infty} q_{ij}^{(n)}$ does not exist for a certain couple of integers $i = i_0$, $j = j_0$. Then there will be another increasing sequence of positive integers $\{m_\nu\}$, $\nu = 1, 2, \ldots$, such that $\lim_{\nu \to \infty} q_{ij}^{(m_\nu)} = Q_{ij}$ exists for any i and j, and such that $P_{i_0 j_0} \neq Q_{i_0 j_0}$. Q_{ij} clearly satisfies the same conditions as P_{ij}. If we now put $n = m_\nu$ in (2.5) and (2.8), and let ν tend to ∞, then we have

(2.10)
$$\sum_{k=1}^{\infty} Q_{ik} \cdot P_{kj} \leq P_{ij}, \qquad i,\ j = 1,\ 2,\ \ldots .$$

$$(2.11) \qquad \sum_{k=1}^{\infty} P_{ik} \cdot Q_{kj} = P_{ij} , \qquad i, j = 1, 2, \ldots .$$

Note that in (2.10) the equality does not follow directly, while we have always equal sign in (2.11). (The latter is a consequence of the absolute summability of $\sum_{k=1}^{\infty} P_{ik}$).

Since Q_{ij} has the same properties as P_{ij}, we have, by exchanging P_{ij} and Q_{ij} in (2.10) and (2.11),

$$(2.12) \qquad \sum_{k=1}^{\infty} P_{ik} \cdot Q_{kj} \leqq Q_{ij} , \qquad i, j = 1, 2, \ldots ,$$

$$(2.13) \qquad \sum_{k=1}^{\infty} Q_{ik} \cdot P_{kj} = Q_{ij} . \qquad i, j = 1, 2, \ldots ,$$

From (2.10) and (2.13) we have $Q_{ij} \leqq P_{ij}$ and from (2.11) and (2.12) we have $P_{ij} \leqq Q_{ij}$, which will lead us to a contradiction since we have, by construction, $P_{i_0 j_0} \neq Q_{i_0 j_0}$. Hence the limit (1.5) must exist for any i and j, and thus Theorem 1 is completely proved.

3. Decomposition of the total state $\mathfrak{R}$ into a dissipative part $\mathfrak{D}$ and ergodic parts $\mathfrak{E}_\alpha$.

Theorem 2. The total state $\mathfrak{R}$ is decomposed into a dissipative part $\mathfrak{D}$ and ergodic parts $\mathfrak{E}_\alpha (\alpha = 1, 2, \ldots ,$ finite or enumerably infinite) in such a way that

(3.1) *for any $S_j \in \mathfrak{D}$ we have $P_{ij} = 0$, $i = 1, 2 \ldots ,$*

(3.2) *for any $S_i \in \mathfrak{E}_\alpha$ we have $\sum\limits_{S_j \in \mathfrak{E}_\alpha} p_{ij}^{(n)} = \sum\limits_{S_j \in \mathfrak{E}_\alpha} P_{ij} = 1$, $n = 1, 2, \ldots ,$*

(3.3) *for any $S_i \in \mathfrak{E}_\alpha$, $S_j \bar{\in} \mathfrak{E}_\alpha$ we have $p_{ij}^{(n)} = P_{ij} = 0$, $n = 1, 2, \ldots ,$*

(3.4) *for any S_i, $S_j \in \mathfrak{E}_\alpha$ we have $P_{ij} > 0$,*

(3.5) *for any S_i, $S_j \in \mathfrak{E}_\alpha$ there exists a positive integer n such that $p_{ij}^{(n)} > 0$,*

(3.6) *for any S_i, $S_j \in \mathfrak{E}_\alpha$ P_{ij} is independent of i.*

Proof. Let $\mathfrak{D}$ be defined as the totality of all the states S_j such that $P_{ij} = 0$ for $i = 1, 2, \ldots .$ In order to prove that $\mathfrak{R} - \mathfrak{D}$ is decomposed into ergodic parts $\mathfrak{E}_\alpha$, which are mutually disjoint, we shall first show that

(3.7) *for any S_j, $S_k \in \mathfrak{R} - \mathfrak{D}$ $P_{jk} > 0$ implies $P_{kj} > 0$.*

For this purpose, assume that $P_{jk} > 0$ and $P_{kj} = 0$, and denote by $\mathfrak{A}_k$ the totality, inclusive S_k, of all the states S_α such that $P_{k\alpha} > 0$. We have $S_j \bar{\in} \mathfrak{A}_k$, and $\mathfrak{A}_k$ is closed in the sense that $P_{\alpha\beta} = 0$ for any $S_\alpha \in \mathfrak{A}_k$ and $S_\beta \bar{\in} \mathfrak{A}_k$. For, by (1.6), $P_{k\alpha} > 0$ and $P_{\alpha\beta} > 0$ imply $P_{k\beta} > 0$. Hence, again by (1.6), we have for any i

$$\sum_{S_\beta \bar{\in} \mathfrak{A}_k} P_{i\beta} = \sum_{S_\beta \bar{\in} \mathfrak{A}_k} \sum_{\alpha=1}^{\infty} P_{i\alpha} \cdot P_{\alpha\beta} = \sum_{S_\beta \bar{\in} \mathfrak{A}_k} \sum_{S_\alpha \bar{\in} \mathfrak{A}_k} P_{i\alpha} \cdot P_{\alpha\beta} \leq \sum_{\substack{S_\alpha \bar{\in} \mathfrak{A}_k \\ \alpha \neq j}} P_{i\alpha} + P_{ij} \cdot (1 - P_{jk})$$

$$= \sum_{S_\alpha \bar{\in} \mathfrak{A}_k} P_{i\alpha} - P_{ij} \cdot P_{jk} ,$$

which leads to a contradiction, since $P_{jk} > 0$ and since there exists, by assumption, a state S_i with $P_{ij} > 0$. Hence (3.7) is proved. Consequently, for any two states S_j, $S_k \in \mathfrak{R} - \mathfrak{D}$, we have either $P_{jk} > 0$, $P_{kj} > 0$ simultaneously or $P_{jk} = P_{kj} = 0$. In the first case, we call S_j and S_k to be mutually dependent, and denote this relation by $S_j \sim S_k$. The relation $\sim$ is symmetric and transitive (since, by (1.6), $P_{jk} > 0$ and $P_{kl} > 0$ imply $P_{jl} > 0$), and it will be easily seen that for any $S_j \in \mathfrak{R} - \mathfrak{D}$ we have $S_j \sim S_j$. This is clear from the argument above, if there exists a state S_k ($\in \mathfrak{R} - \mathfrak{D}$) with $P_{jk} > 0$; and if we have $P_{jk} = 0$ for any S_k, then we have for any i

$$\sum_{k=1}^{\infty} P_{ik} = \sum_{k=1}^{\infty} \sum_{\alpha=1}^{\infty} P_{i\alpha} \cdot P_{\alpha k} = \sum_{k=1}^{\infty} \sum_{\alpha \neq j} P_{i\alpha} \cdot P_{\alpha k} \leq \sum_{\alpha \neq j} P_{i\alpha} = \sum_{\alpha=1}^{\infty} P_{i\alpha} - P_{ij} ,$$

which is a contradiction, since there exists a state S_i with $P_{ij} > 0$. Thus $\mathfrak{R} - \mathfrak{D}$ is divided into the classes—the ergodic parts—$\mathfrak{E}_\alpha$ ($\alpha = 1, 2, \ldots,$ finite or enumerably infinite), which consist of mutually dependent states and each of which satisfies (3.4). Since (3.3) and (3.5) are the direct consequences of (3.2) and (3.4) respectively, we need only prove that these ergodic parts thus obtained have the properties (3.2) and (3.6).

From (1.6) and the definition of $\mathfrak{E}_\alpha$ we have for any $S_i \in \mathfrak{E}_\alpha$

$$(3.8) \qquad \sum_{S_j \in \mathfrak{E}_\alpha} P_{ij} = \sum_{S_j \in \mathfrak{E}_\alpha} \sum_{k=1}^{\infty} P_{ik} \cdot p_{kj}^{(n)} = \sum_{S_j \in \mathfrak{E}_\alpha} \sum_{S_k \in \mathfrak{E}_\alpha} P_{ik} \cdot p_{kj}^{(n)} = \sum_{S_k \in \mathfrak{E}_\alpha} P_{ik} \cdot \left(\sum_{S_j \in \mathfrak{E}_\alpha} p_{kj}^{(n)} \right).$$

Since $\sum_{S_j \in \mathfrak{E}_\alpha} p_{kj}^{(n)} \leq 1$ for any k and since $P_{ik} > 0$ for any S_i, $S_k \in \mathfrak{E}_\alpha$, the equality in (3.8) holds only when $\sum_{S_j \in \mathfrak{E}_\alpha} p_{kj}^{(n)} = 1$ for any k with $S_k \in \mathfrak{E}_\alpha$. Thus the first part of (3.2) is proved and the second part of it may be obtained analogously, if we start from the third relation of (1.6): $\sum_{k=1}^{\infty} P_{ik} \cdot P_{kj} = P_{ij}$.

Now, in order to prove (3.6), it is sufficient (by (1.6)) to show that if $\{\xi_i\}$ and $\{\eta_i\}$ are the systems of real numbers (defined for any i with $S_i \in \mathfrak{E}_\alpha$) such that

$$\sum_{S_i \in \mathfrak{E}_\alpha} |\xi_i| < \infty , \qquad \sum_{S_i \in \mathfrak{E}_\alpha} |\eta_i| < \infty ,$$

and

$$\sum_{S_i \in \mathfrak{C}_\alpha} \xi_i P_{ij} = \xi_j \,, \qquad \sum_{S_i \in \mathfrak{C}_\alpha} \eta_i P_{ij} = \eta_j \,,$$

then we have $\xi_i = \sigma \eta_i$ (σ : constant) for any i with $S_i \in \mathfrak{C}_\alpha$; or, it is sufficient to show that

$$\sum_{S_i \in \mathfrak{C}_\alpha} |\xi_i| < \infty \,, \qquad \sum_{S_i \in \mathfrak{C}_\alpha} \xi_i P_{ij} = \xi_j$$

imply $\xi_i \geq 0$ for any i, or $\xi_i \leq 0$ for any i.

For this purpose, let $\mathfrak{P}$ ($\mathfrak{N}$) be the totality of all the states $S_i \in \mathfrak{C}_\alpha$ such that $\xi_i > 0$ ($\xi_i < 0$), and assume that both $\mathfrak{P}$ and $\mathfrak{N}$ are not empty. Then, since $P_{ij} > 0$ for any S_i, $S_j \in \mathfrak{C}_\alpha$ by (3.4), we have

$$\sum_{S_j \in \mathfrak{P}} \xi_j = \sum_{S_i \in \mathfrak{P}} \sum_{S_i \in \mathfrak{C}_\alpha} \xi_i P_{ij} < \sum_{S_j \in \mathfrak{P}} \sum_{S_i \in \mathfrak{P}} \xi_i P_{ij} = \sum_{S_i \in \mathfrak{P}} \xi_i \left(\sum_{S_j \in \mathfrak{P}} P_{ij} \right)$$

which is a contradiction, since $\sum_{S_j \in \mathfrak{P}} P_{ij} \leq 1$ for any i.

Thus Theorem 2 is completely proved.

4. Second proof of the existence of the mean sojourn.

Following Kolmogoroff, let $k_{ij}^{(n)}$ be the probability that the state S_i is transferred to the state S_j after the elapse of n unit-times, without being previously transferred to the state S_j. We have clearly

(4.1) $$0 \leq k_{ij}^{(n)} \leq 1 \,, \qquad i, j, n = 1, 2, \ldots ,$$

(4.2) $$\sum_{n=1}^{\infty} k_{ij}^{(n)} \equiv L_{ij} \leq 1 \,, \qquad i, j = 1, 2, \ldots ,$$

(4.3) $$p_{ij}^{(n)} = k_{ij}^{(1)} \cdot p_{jj}^{(n-1)} + k_{ij}^{(2)} \cdot p_{jj}^{(n-2)} + \cdots + k_{ij}^{(n-1)} \cdot p_{jj}^{(1)} + k_{ij}^{(n)} \,.$$

From (4.3) we have easily

$$q_{ij}^{(n)} = k_{ij}^{(1)} \cdot \frac{n-1}{n} \cdot q_{jj}^{(n-1)} + k_{ij}^{(2)} \cdot \frac{n-2}{n} \cdot q_{jj}^{(n-2)} + \cdots + k_{ij}^{(n-1)} \cdot \frac{1}{n} \cdot q_{jj}^{(1)} + \frac{1}{n} \sum_{m=1}^{n} k_{ij}^{(m)} \,,$$

and consequently, by Toeplitz's summation, the existence of $\lim_{n \to \infty} q_{jj}^{(n)} = P_{jj}$ implies that of $\lim_{n \to \infty} q_{ij}^{(n)} = L_{ij} \cdot P_{jj}$, so that we need only prove the existence of the limit (1.5) in the case $i = j$.

For this purpose, consider two power series :

(4.4) $$P_{jj}(z) = p_{jj}^{(1)} z + p_{jj}^{(2)} z^2 + \cdots + p_{jj}^{(n)} z^n + \cdots ,$$

(4.5) $$K_{jj}(z) = k_{jj}^{(1)} z + k_{jj}^{(2)} z^2 + \cdots + k_{jj}^{(n)} z^n + \cdots .$$

These are convergent in $|z| < 1$ ((4.5) converges uniformly even in $|z| \leq 1$ by (4.2)), and by virtue of (4.3)($i = j$), we have the relation:

$$(4.6) \qquad 1 + P_{jj}(z) = \frac{1}{1 - K_{jj}(z)}, \qquad |z| < 1$$

The further arguments are divided into two cases:

1st case: $L_{jj} = 1$. Put $M_{jj} = \sum_{n=1}^{\infty} n \cdot k_{jj}^{(n)}$. M_{jj} might become infinite; in this case put $\frac{1}{M_{jj}} = 0$. Then we have from (4.6)

$$(4.7) \quad \lim_{z \to 1-0} (1-z)(1 + P_{jj}(z)) = \lim_{z \to 1-0} \frac{1-z}{1 - K_{jj}(z)}$$

$$= \lim_{z \to 1-0} \frac{1}{\sum_{n=1}^{\infty} k_{jj}^{(n)}(1 + z + \cdots + z^{n-1})} = \frac{1}{M_{jj}},$$

if we let z tend to 1 along the real axis. Since $\{p_{jj}^{(n)}\}$ is uniformly bounded, we deduce from (4.7), by a well-known argument, that([2])

$$(4.8) \qquad \lim_{n \to \infty} q_{jj}^{(n)} = \lim_{n \to \infty} \frac{1}{n}(p_{jj}^{(1)} + p_{jj}^{(2)} + \cdots + p_{jj}^{(n)}) = \frac{1}{M_{jj}}$$

Thus we have proved the existence of (1.5) and obtained the additional relation:

$$(4.9) \qquad P_{jj} = \frac{1}{M_{jj}}$$

2nd case: $L_{jj} < 1$. This case is easier to discuss. Making z tend to 1 along the real axis in (4.6), we have

$$(4.10) \qquad 1 + \sum_{n=1}^{\infty} p_{jj}^{(n)} = \frac{1}{1 - L_{jj}} < \infty .$$

Hence, a fortiori, $p_{jj}^{(n)}$ and $q_{jj}^{(n)}$ tend to 0 as $n \to \infty$

5. Construction of a Markoff process with given $\{k_{11}^{(n)}\}$.

If we consider the sequence of real numbers $\{p_{11}^{(n)}\}$, $n = 1, 2, \ldots$, where $p_{11}^{(n)}$ is defined by (1.1), $\{p_{11}^{(n)}\}$ clearly satisfies the relations:

$$(5.1) \qquad 0 \leq p_{11}^{(n)} \leq 1, \qquad n = 1, 2, \ldots,$$

$$(5.2) \qquad p_{11}^{(m+n)} \geq p_{11}^{(m)} \cdot p_{11}^{(n)}, \qquad m, n = 1, 2, \ldots,$$

([2]) See, for example, N. Wiener: The Fourier integral and certain of its applications, Cambridge, 1933, p. 105–106.

((5.2) follows from (1.4)). These are the necessary conditions that $\{p_{11}^{(n)}\}$ should satisfy, and it will be easily seen that these are not sufficient; that is, it is not always possible to construct a simple Markoff process $P = (p_{ij})$. $i, j = 1, 2, \ldots$, for which $\{p_{11}^{(n)}\}$ is the given one. On the contrary, however, if we consider the sequence of real numbers $\{k_{11}^{(n)}\}$, which satisfies the conditions (4.1) and (4.2) $(i = j = 1)$, then it is always possible to construct a simple Markoff process $P = (p_{ij})$, $i, j = 1, 2, \ldots$, for which $\{k_{11}^{(n)}\}$ (which is defined by the way given at the beginning of §4) is exactly the given one. This may be done as follows:

1st case: $L_{11} = 1$. Define the matrix $P = (p_{ij})$, $i, j = 1, 2, \ldots$, as follows:

$$P = \begin{pmatrix}
k_{11}^{(1)} & k_{11}^{(2)} & k_{11}^{(3)} & 0 & k_{11}^{(4)} & 0 & 0 & k_{11}^{(5)} & 0 & 0 & 0 & \ldots & \overbrace{k_{11}^{(n)}}^{\displaystyle n-1} & 0 & 0 & \ldots & 0 & 0 & \ldots \\
1 & 0 & 0 & 0 & 0 & 0 & 0 & 0 & 0 & 0 & 0 & \ldots & 0 & 0 & 0 & \ldots & 0 & 0 & \ldots \\
0 & 0 & 0 & 1 & 0 & 0 & 0 & 0 & 0 & 0 & 0 & \ldots & 0 & 0 & 0 & \ldots & 0 & 0 & \ldots \\
1 & 0 & 0 & 0 & 0 & 0 & 0 & 0 & 0 & 0 & 0 & \ldots & 0 & 0 & 0 & \ldots & 0 & 0 & \ldots \\
0 & 0 & 0 & 0 & 0 & 1 & 0 & 0 & 0 & 0 & 0 & \ldots & 0 & 0 & 0 & \ldots & 0 & 0 & \ldots \\
0 & 0 & 0 & 0 & 0 & 0 & 1 & 0 & 0 & 0 & 0 & \ldots & 0 & 0 & 0 & \ldots & 0 & 0 & \ldots \\
1 & 0 & 0 & 0 & 0 & 0 & 0 & 0 & 0 & 0 & 0 & \ldots & 0 & 0 & 0 & \ldots & 0 & 0 & \ldots \\
0 & 0 & 0 & 0 & 0 & 0 & 0 & 0 & 1 & 0 & 0 & \ldots & 0 & 0 & 0 & \ldots & 0 & 0 & \ldots \\
0 & 0 & 0 & 0 & 0 & 0 & 0 & 0 & 0 & 1 & 0 & \ldots & 0 & 0 & 0 & \ldots & 0 & 0 & \ldots \\
0 & 0 & 0 & 0 & 0 & 0 & 0 & 0 & 0 & 0 & 1 & \ldots & 0 & 0 & 0 & \ldots & 0 & 0 & \ldots \\
1 & 0 & 0 & 0 & 0 & 0 & 0 & 0 & 0 & 0 & 0 & \ldots & 0 & 0 & 0 & \ldots & 0 & 0 & \ldots \\
\cdots & \cdots & \cdots & \cdots & \cdots & \cdots & \cdots & \cdots & \cdots & \cdots & \cdots & \cdots & \cdots & \cdots & \cdots & \cdots & \cdots & \cdots & \cdots \\
0 & 0 & 0 & 0 & 0 & 0 & 0 & 0 & 0 & 0 & 0 & \ldots & 0 & 1 & 0 & \ldots & 0 & 0 & \ldots \\
0 & 0 & 0 & 0 & 0 & 0 & 0 & 0 & 0 & 0 & 0 & \ldots & 0 & 0 & 1 & \ldots & 0 & 0 & \ldots \\
0 & 0 & 0 & 0 & 0 & 0 & 0 & 0 & 0 & 0 & 0 & \ldots & 0 & 0 & 0 & \ldots & 0 & 0 & \ldots \\
\cdots & \cdots & \cdots & \cdots & \cdots & \cdots & \cdots & \cdots & \cdots & \cdots & \cdots & \cdots & \cdots & \cdots & \cdots & \cdots & \cdots & \cdots & \cdots \\
0 & 0 & 0 & 0 & 0 & 0 & 0 & 0 & 0 & 0 & 0 & \ldots & 0 & 0 & 0 & \ldots & 0 & 1 & \ldots \\
1 & 0 & 0 & 0 & 0 & 0 & 0 & 0 & 0 & 0 & 0 & \ldots & 0 & 0 & 0 & \ldots & 0 & 0 & \ldots \\
\cdots & \cdots & \cdots & \cdots & \cdots & \cdots & \cdots & \cdots & \cdots & \cdots & \cdots & \cdots & \cdots & \cdots & \cdots & \cdots & \cdots & \cdots & \cdots
\end{pmatrix}$$

2nd case: $L_{11} < 1$. Put $k'_{11} = 1 - L_{11}$ and define the matrix P, whose rows and columns are both of type $\omega + \omega$, as follows:

$$P = \begin{pmatrix}
 & & & & & k'_{11} & 0 & 0 & 0 & \ldots\ldots \\
 & & & & & 0 & 0 & 0 & 0 & \ldots\ldots \\
 & & K & & & 0 & 0 & 0 & 0 & \ldots\ldots \\
 & & & & & \ldots\ldots\ldots & & \ldots\ldots \\
 & & & & & \ldots\ldots\ldots & & \ldots\ldots \\
0 & 0 & 0 & \ldots\ldots & & 0 & 1 & 0 & 0 & \ldots\ldots \\
0 & 0 & 0 & \ldots\ldots & & 0 & 0 & 1 & 0 & \ldots\ldots \\
0 & 0 & 0 & \ldots\ldots & & 0 & 0 & 0 & 1 & \ldots\ldots \\
0 & 0 & 0 & \ldots\ldots & & 0 & 0 & 0 & 0 & \ldots\ldots \\
 & \ldots\ldots\ldots\ldots & & & & \ldots\ldots\ldots\ldots \\
 & \ldots\ldots\ldots\ldots & & & & \ldots\ldots\ldots\ldots
\end{pmatrix}$$

where K is a matrix whose rows and columns are of type ω and which is defined in just the same manner as in the 1st case.

It is to be remarked that our problem is in general impossible if the number of possible states is finite.

Mathematical Institute,
Osaka Imperial University.

Ergodic theorems of Birkhoff-Khintchine's type

Japan. J. Math. **17** (1940) 31–36

(Received February 27, 1940.)

§ 1. *Introduction and the theorems.* Let $p \geq 1$, then the class of all the real-valued measurable functions $f(t)$ on $(0,1)$ with $\int_0^1 |f(t)|^p \, dt < \infty$ constitutes a Banach space (L^p) with the norm $\|f\| = \left(\int_0^1 |f(t)|^p dt \right)^{\frac{1}{p}}$. The class of all the real-valued measurable functions $f(t)$ on $(0,1)$ which are essentially bounded also constitutes a Banach space with the norm $\|f\|_M = $ essential sup. $|f(t)|$. The purpose of the present note is to give the following three ergodic theorems.

Theorem 1. Let T be a linear operation on (L^p) to (L^p) and put $f^{(n)} = T^n \cdot f \ (n = 1, 2, \ldots)$ for any $f \in (L^p)$. We assume that

$$(1) \qquad \|T^n\| = \sup_{\|f\| = 1} \|T^n \cdot f\| \leq \ a \ constant \ C \quad (n = 1, 2, \ldots),$$

$$(2) \qquad \varlimsup_{n \to \infty} \left| \frac{1}{n} \sum_{m=1}^{n} f^{(m)}(t) \right| < \infty \quad almost \ everywhere \ for \ any \ f \in (L^p).$$

If an element $g \in (L^p)$ satisfies

$$(3) \qquad \lim_{n \to \infty} \frac{g^{(n)}(t)}{n} = 0 \quad almost \ everywhere,$$

$$(4) \qquad \begin{cases} the \ sequence \ \left\{ \dfrac{1}{n} \sum_{m=1}^{n} g^{(m)} \right\} \ contains \ a \ subsequence \ weakly \\ convergent \ to \ an \cdot element \ g^* \in (L^p) \, , \end{cases}$$

then we have

$$(5) \qquad \lim_{n \to \infty} \frac{1}{n} \sum_{m=1}^{n} g^{(m)}(t) = g^*(t) \quad almost \ everywhere,$$

$$(6) \qquad \lim_{n \to \infty} \left\| g^* - \frac{1}{n} \sum_{m=1}^{n} g^{(m)} \right\| = 0 \, , \quad T \cdot g^* = g^* \, .$$

$$\text{K\^osaku Yosida}$$

Remark. By (1) we have $\left\|\dfrac{1}{n}\sum_{m=1}^{n} f^{(m)}\right\| \leq C\|f\|$ $(n=1, 2, \ldots)$ for any $f \in (L^p)$. Hence, if $p=1$, (4) is equivalent to the equi-integrability of a certain subsequence of $\left\{\dfrac{1}{n}\sum_{m=1}^{n} g^{(m)}\right\}$. If $p>1$, (4) is superflous, since (L^p) with $p>1$ is locally weakly compact([1]).

Theorem 2. If we assume, besides (1) and (2)

$$(1)' \qquad \|T^n\|_M = \sup_{\|f\|_M=1} \|T^n \cdot f\|_M \leq a \ constant \ C' \quad (n = 1, 2, \ldots),$$

then (3), (4), (5) and (6) all hold good for any $g \in (L^p)$.

Theorem 3. Let T be a linear operation on (L^p) to (L^p) satisfying

$$(7) \qquad \begin{cases} \text{for any } f \in (L^p), \quad \text{there exists } F \in (L^p) \quad \text{such that} \\ |f^{(n)}(t)| \leq F(t) \quad (n = 1, 2, \ldots) \quad \text{almost everywhere,} \end{cases}$$

then (3), (4), (5) and (6) all hold good for any $g \in (L^p)$.

As will be shown in §3, Birkhoff-Khintchine's *individual ergodic theorem* ($B-K$ theorem) and J. von Neumann's *mean ergodic theorem*([2]) are both contained in theorem 1 and theorem 2. In two points, theorem 1 is more general than $B-K$ theorem. *Firstly*, the existence of the inverse T^{-1} of T is not necessary, while the linear operation T in (L^p) induced by *equi-measure point transformation* (Birkhoff-Khintchine's and J. von Neumann's case) necessarily admits the inverse T^{-1}, and moreover satisfies $(1)'$. Hence theorem 1 and 2 may be applied to the reversible as well as to the irreversible transition process, in particular to the simple Markoff process. *Secondly*, theorem 1 concerns with the individual ergodic theorem (5) for *individual point $g \in (L^p)$* under the condition (3) and (4); $B-K$ theorem stating the validity of (5) for all $g \in (L^p)$.

The proof of theorem 1 will be given in §2. It appeals to the mean ergodic theorem in Banach space([3]) and a theorem in the theory of the space of type (F)([4]). Since we want to obtain the almost everywhere convergence (5), it seems to me very natural to make use of the operation $\tilde{T}$ to be defined below.

([1]) If $p>1$, (3) is also superflous. See the remark 3 on page 36.

([2]) See, for example, E. Hopf's book: "Ergodentheorie", (1937).

([3]) K. Yosida: Mean ergodic theorem in Banach spaces, Proc. Imp. Acad. Japan, **14** (1938), 286-294; S. Kakutani: Iteration of linear operation in complex Banach spaces, ibid., 295-298; F. Riesz: Some mean ergodic theorems, Journ. London Math. Soc., **13** (1938), 274-278.

([4]) For the definition of the space of type (F) see S. Banach's book: "Théorie des opérations linéaires".

Theorem 3 is a precision to a result due to F. Riesz[5]. He proved the mean convergence (6) for any $g \in (L^p)$ under the condition (1) and (7) when $p = 1$. That Riesz's result includes the ergodic theorem of Garrett Birkhoff[6], formulated in abstract manner, was proved by S. Kakutani[7].

§ 2. *Proof of theorems.* Let (S) be the class of all the real-valued measurable functions $f(t)$ on $(0,1)$ which are finite almost everywhere. (S) is a space of type (F) by the norm $\|f\|_S = \int_0^1 \frac{|f(t)|}{1+|f(t)|} dt$.

Lemma. *Let T satisfy (2), and put*

$$\tilde{f}(t) = \overline{\lim_{n\to\infty}} \frac{1}{n} \sum_{m=1}^n f^{(m)}(t) - \underline{\lim_{n\to\infty}} \frac{1}{n} \sum_{m=1}^n f^{(m)}(t)$$

for any $f \in (L^p)$. Then the (not necessarily additive) operation $\tilde{T}$ on (L^p) to (S), $\tilde{f} = \tilde{T} \cdot f$, is continuous, viz. we obtain $\lim_{n\to\infty} \|\tilde{T} \cdot f_n - \tilde{T} \cdot f\|_S = 0$ from $\lim_{n\to\infty} \|f_n - f\| = 0$.

This lemma is a special case of a theorem due to S. Banach and S. Saks[8]. Since this lemma is the key to our method, I here sketch the proof of it following after S. Mazur and W. Orlicz[9].

Proof. Put $f_n = \frac{1}{n} \sum_{m=1}^n f^{(m)}$, $f_n'(t) = \sup_{m \le n} |f_m(t)|$, $f'(t) = \sup_m |f_m(t)|$ for any $f \in (L^p)$, and consider the operations $V_n \cdot f = f_n'$, $V \cdot f = f'$ on (L^p) to (S). Each V_n is continuous and $\lim_{n\to\infty} \|V_n \cdot f - V \cdot f\|_S = 0$ at every $f \in (L^p)$. Hence, from $\lim_{n\to\infty} \left\| \frac{1}{k} V_n \cdot f \right\|_S = \left\| \frac{1}{k} V \cdot f \right\|_S$ $(k = 1, 2, \ldots)$ and $\lim_{k\to\infty} \left\| \frac{1}{k} V \cdot f \right\|_S = 0$, we obtain the partition $(L^p) = \sum_{k=1}^\infty G_k$, $G_k = \underset{f}{E} \left(\left\| \frac{1}{k} V_n \cdot f \right\|_S \le \varepsilon ; n = 1, 2, \ldots \right)$, for any $\varepsilon > 0$. G_k is a closed set of (L^p) by the continuity of V_n. Since (L^p) is a complete metric space, there must exists a G_k which contains a sphere. Thus we have

---

(5) F. Riesz: loc. cit.

(6) G. Birkhoff: Dependent probabilities and spaces (L), Proc. Nat. Acad. Sci., **24** (1938), 155-159.

(7) S. Kakutani: Mean ergodic theorem in abstract (L)-spaces, Proc. Imp. Acad. Japan, **15** (1939), 121-123. F. Riesz also promises the same result in the paper cited above.

(8) S. Banach: Sur la convergence presque partout des fonctionnelles linéaires, Bult. Sc. Math. France, **50** (1926), 27-32 and 36-43. S. Saks: Sur le fonctionnelles de M. Banach et leurs application aux développements des fonctions, Fun . Math., **10** (1927), 186-196.

(9) S. Mazur und W. Orlicz: Über Folgen linearer Operationen, Stud. Math., **4** (1933), 152-157.

a point $f_0 \in (L^p)$ and a constant $\delta > 0$ such that $\sup\limits_{n} \left\| \dfrac{1}{k} V_n \cdot f \right\|_S \leq \varepsilon$

for $\|f - f_0\| \leq \delta$. Hence, by $\left\| \dfrac{1}{k} V_n \cdot (f - f_0) \right\|_S \leq \left\| \dfrac{1}{k} V_n \cdot f \right\|_S + \left\| \dfrac{1}{k} V_n \cdot f_0 \right\|_S$

and $\left\| V_n \cdot \left(\dfrac{1}{k} f \right) \right\|_S = \left\| \dfrac{1}{k} V_n \cdot f \right\|_S$, we see that $\lim\limits_{\|f\| \to 0} \| V_n \cdot f \|_S = 0$ uniformly

in n. $V \cdot f$ is thus continuous at $f = 0$ (the zero element of (L^p))
and $V \cdot f = 0$ for $f = 0$. Since we have $\| \tilde{T} \cdot f \|_S \leq 2 \| V \cdot f \|_S$,
$\| \tilde{T} \cdot f - \tilde{T} \cdot h \|_S \leq \| \tilde{T} \cdot (f - h) \|_S$, $\tilde{T}$ is continuous at every $f \in (L^p)$.

Proof of theorem 1. By (1) and (4) the mean ergodic theorem (6)
applies to g([10]). Put $g = g^* + (g - g^*)$, then $g^{(m)} = g^* + (g - g^*)^{(m)}$. Since
$\left\{ \dfrac{1}{n} \sum\limits_{m=1}^{n} (g - g^*)^{(m)} \right\}$ converges strongly to 0, it is sufficient to show that

$$(8) \qquad \lim\limits_{n \to \infty} \dfrac{1}{n} \sum\limits_{m=1}^{n} (g(t) - g^*(t))^{(m)} \quad \text{exists almost everywhere.}$$

Now, from (6), we obtain

$$(9) \qquad \lim\limits_{n \to \infty} \left\| (g - g^*) - (E - T) \cdot \dfrac{ng + (n-1)g^{(1)} + \cdots + g^{(n-1)}}{n} \right\| = 0,$$

where E denote the identity operator. We have, from (3),

$$(10) \qquad \tilde{T} \cdot (E - T) \cdot \left\{ \dfrac{ng + (n-1)g^{(1)} + \cdots + g^{(n-1)}}{n} \right\} = 0 \quad (n = 1, 2, \ldots).$$

Hence, by (9), (10) and lemma, we obtain $\tilde{T} \cdot (g - g^*) = 0$, which is equivalent to (8).

Proof of theorem 2. By (1)$'$ and the remark to theorem 1, (4) is
satisfied for all $g \in (L^p) \cdot (M)$. (3) is also satisfied for all $g \in (L^p) \cdot (M)$ by
(1)$'$. Thus, by theorem 1, (5) is satisfied for all $g \in (L^p) \cdot (M)$. Since, (M)
is strongly dense in (L^p) by the norm $\| \quad \|$ of (L^p), we have (5) for all
$g \in (L^p)$, by lemma. We have already the mean convergence (6) for all
$g \in (L^p) \cdot (M)$. Since (M) is strongly dense in (L^p), and since $\left\| \dfrac{1}{n} \sum\limits_{m=1}^{n} T^m \right\| \leq C$
$(n = 1, 2, \ldots)$ by (1), we have (6) for all $g \in (L^p)$.

Proof of theorem 3. From (7) and a well-known theorem of Banach,
we see that (1) is satisfied. (2), (3) and (4) for any $f \in (L^p)$ are also

---

([10]) That mean ergodic theorem applies to the individual point of the Banach space
was stressed by the author in the note: Asymptotic almost periodicities and ergodic
theorems, Proc. Imp. Acad. Japan, **15** (1939), 255-259.

satisfièd by (7); for (4), use the remark to theorem 1. Thus, by theorem 1, we obtain the theorem.

§ 3. *Deduction of Birkhoff-Khintchine's ergodic theorem.* Let P be an equi-measure point transformation of $(0,1)$ on $(0,1)$ which is one-to-one except the set of points of zero measure, and consider the linear operation T on (L^p) to (L^p) defined by $T \cdot f = f^{(1)}$, $f^{(1)}(t) = f(P \cdot t)$. T is surely of norm one and thus (1) is satisfied. According to N. Wiener([11]), we have

$$(11) \qquad \operatorname*{mes}\left(E_t(\bar{f}(t) \geq a)\right) \leq \frac{1}{a} \int_0^1 f(t) dt , \quad \bar{f}(t) = \sup. \frac{1}{n} \sum_{m=1}^n f^{(m)}(t) ,$$

for any constant $a > 0$ and for any non-negative function $f(t) \in (L^p)$. Since T is a positlve operation (i.e., we have $f^{(1)}(t) \geq 0$ almost everywhere if $f(t) \geq 0$ almost everywhere), (2) is satisfied by (11). As T is induced by the point transformation P, $(1)'$ is satisfied surely.

Therefore, by theorem 2, we obtain (5) and (6) for all $g \in (L^p)$, and thus $B-K$ theorem is obtained.

Remark 1. N. Wiener (loc. cit.) also deduced $B-K$ theorem from the *dominated ergodic theorem* (11) and the mean ergodic theorem, by an ingeneous argument. Our method shows that the *weaker* condition (2) of the dominated ergodic theorem's type also plays an essential rôle for general linear operation, not necessarily induced by equi-measure point transformation. It is to be noted, moreover, that $B-K$ theorem and the dominated ergodic theorem may simultaneously be obtained by a unified argument, if we deal with equi-measure transformation only([12]).

Remark 2. We may also deduce $B-K$ theorem from theorem 1 directly, without appealing to theorem 2. By the above arguments, it is sufficient to prove that (3) is satisfied for all the non-negative functions $g \in (L^p)$.

To this purpose, put $E_m^n(a) = E_t(ma \leq g^{(n)}(t) \leq (m+1)a)$, $(n, m = 1,$
2, ...), then $E_{k, l:m}(a) = E_t\left(a \leq \sup_{l \geq n \geq k} \frac{g^{(n)}(t)}{n} < (m+1)a\right) \leq \sum_{n=k}^l \sum_{i=n}^{nm} E_i^n(a)$.
Since P is an equi-measure transformation, mes $(E_i^n(a))$ is independent of $n\,(i = 1, 2, ...)$. Thus we have

---

([11]) N. Wiener: The ergodic theorem, Duke Math. Journ., **5** (1939), 1-18. Cf. K. Yosida and S. Kakutani: Birkhoff's ergodic theorem and the maximal ergodic theorem Proc. Imp. Acad. Japan, **15** (1939), 165-168. M. Fukamiya: On dominated ergodic theorems in $L^p\,(p \geq 1)$. Tôhoku Math. J., **46** (1939), 150-153.

([12]) K. Yosida and S. Kakutani: loc. cit.

$$\text{mes } (E_{k, l; m}(a)) \leq \sum_{n=k}^{l} \sum_{i=n}^{nm} \text{mes } (E_i^1(a)) \leq \sum_{j=k}^{\infty} j \text{ mes } (E_j^1(a)) .$$

Hence $\lim_{k \to \infty} \text{mes} \left(E \left(\left(\sup_{n \geq k} \frac{g^{(n)}(t)}{n} \right) \geq a \right) \right) = 0$, and thus (3) is proved([13]).

Remark 3 (added during the proof). As was kindly pointed out to me by M. Fukamiya, (3) is superflous if $p > 1$. For we have $\varepsilon^p \cdot \text{mes}\,(E(\varepsilon)) \leq (n^{-1} \cdot C \cdot \| g \|)^p$, $E(\varepsilon) = E_t \left(\left| \frac{g^{(n)}(t)}{n} \right| \geq \varepsilon \right)$, $n = 1, 2, \dots$, at all $g \in (L^p)$ and for any $\varepsilon > 0$; proving $\lim_{n \to \infty} \frac{g^{(n)}(t)}{n} = 0$ almost everywhere by $\infty > \sum_n \frac{1}{n^p}$.

At this juncture, I remark that the results of the present note may be extended so as to be applicable to functions taking values from semi-ordered linear spaces. We may also obtain extensions of n-parameter individual ergodic theorem of N. Wiener. The details will be published elsewhere.

---

(13) For the proof I owe to S. Kakutani.

The Markoff process with a stable distribution

Proc. Imp. Acad. Tokyo **16** (1940) 43–48

(Comm. by T. TAKAGI, M.I.A., March 12, 1940.)

§ 1. *Introduction and the theorems.* Let R be a space and let $B(R)$ be a completely additive family of "measurable" subsets of R. We assume that R itself belongs to $B(R)$. Let $P(x, E)$ denote the transition probability that the point $x \in R$ is transferred, by a simple Markoff process, into the set $E \in B(R)$ after the elapse of a unit-time. We naturally assume that $P(x, E)$ is completely additive for $E \in B(R)$ if x is fixed, and that $P(x, E)$ is "measurable" (with respect to $B(R)$) in x if E is fixed. Then the transition probability $P^{(n)}(x, E)$ that the point x is transferred into E after the elapse of n unit-times is given by $P^{(n)}(x, E) = \int P^{(n-1)}(x, dy) P(y, E)^{1)}$ $\left(n = 1, 2, \ldots; P^{(1)}(x, E) = P(x, E) \right)$. We have surely

$$(1) \qquad P^{(n)}(x, E) \geqq 0, \quad P^{(n)}(x, R) \equiv 1 \qquad (n = 1, 2, \ldots).$$

We now assume that there exists a non-negative set function $\varphi(E)$ which satisfies the conditions:

(2) $\varphi(E)$ is completely additive for $E \in B(R)$ and $\varphi(R) = 1$.

(3) $\begin{cases} \text{The space } B(R), \text{ if metrised by the distance } d(E_1, E_2) = \\ \varphi(E_1 + E_2 - E_1 \cdot E_2), \text{ is (complete and) separable.}^{2)} \end{cases}$

(4) $\int \varphi(dx) P(x, E) = \varphi(E)$ for any $E \in B(R)$.

We have, from (4), $\int \varphi(dx) P^{(n)}(x, E) = \varphi(E)$ for any n. Hence the mass distribution $\varphi(E)$ is *stable* with respect to the time. Such Markoff process (*with a stable distribution* $\varphi(E)$) is fairly general; it includes the *deterministic transition process* in the ergodic theory of the incomplessible stationary flow, originated by G. D. Birkhoff and J. von Neumann. In fact, let T be a one-to-one point transformation of R on R which maps any set $E \in B(R)$ on the set $T \cdot E \in B(R)$ in measure-preserving way: $\varphi(E) = \varphi(T \cdot E)$. Let $C_E(x)$ be the characteristic function of E and put $P(x, E) = C_E(T \cdot x)$, then it is easy to see that this $P(x, E)$ defines a Markoff process with stable distribution $\varphi(E)$. Another example is given by the Markoff process with *symmetric* φ-density: $P(x, E) = \int_E p(x, y) \varphi(dy)$, $p(x, y) \equiv p(y, x)$. Thus our general

---

1) The definite integral over R will be denoted by $\int \varphi(dx)$.

2) The completeness of the metrical space $B(R)$ follows from the complete additivity of $B(R)$. The separability hypothesis may be taken away, by suitably modifying the proof below. However, for the sake of brevity, I here assume it.

$P(x, E)$ applies to the deterministic transition process $x \to T \cdot x$ as well as to the *indeterministic (probabilistic) transition process.*

Let (L) denote the set of all the real-valued measurable functions $f(x)$ which are φ-integrable. (L) is surely a Banach space with the norm $\|f\| = \int |f(x)| \varphi(dx)$. We will prove the following theorems.

Theorem 1. For any $f \in (L)$,

$$f^{(m)}(x) = \int P^{(m)}(x, dy) f(y) \qquad (m = 1, 2, \ldots)$$

exists φ-almost everywhere and belongs to (L) in such a way that $\|f^{(m)}\| \leq \|f\|$. For any $f \in (L)$ and for any m $(= 1, 2, \ldots)$, there corresponds $f^{(-m)} \in (L)$, $\|f^{(-m)}\| \leq \|f\|$, such that

$$\int \varphi(dx) f(x) P^{(m)}(x, E) = \int_E f^{(-m)}(x) \varphi(dx) \quad \textit{for all} \quad E \in B(R).$$

Theorem 2. For any $f \in (L)$, there corresponds $f^{(*)}$, $f^{(-*)} \in (L)$ such that

$$(5) \quad \begin{cases} \lim_{n \to \infty} \int \left| f^{(*)}(x) - \frac{1}{n} \sum_{m=1}^{n} f^{(m)}(x) \right| \varphi(dx) = 0, \\[2mm] f^{(*)}(x) = \int P(x, dy) f^{(*)}(y) \quad \varphi\text{-almost everywhere,} \end{cases}$$

$$(6) \quad \begin{cases} \lim_{n \to \infty} \int \left| f^{(-*)}(x) - \frac{1}{n} \sum_{m=1}^{n} f^{(-m)}(x) \right| \varphi(dx) = 0, \\[2mm] \int \varphi(dx) f^{(-*)}(x) P(x, E) = \int_E f^{(-*)}(x) \varphi(dx) \quad \textit{for any} \quad E \in B(R). \end{cases}$$

In the deterministic case $P(x, E) = C_E(T \cdot x)$, $f^{(m)}(x)$ and $f^{(-m)}(x)$ corresponds to $f(T^m \cdot x)$ and $f(T^{-m} \cdot x)$ respectively. Theorem 2 is an extension of J. von Neumann's mean ergodic theorem for (L) to the indeterministic case. *Thus the problem of the Markoff process is reduced, in a certain sense, to the determination of all the possible stable distributions.* Under some topological assumptions, such determination is possible. This will be indicated in §3. As a physical application of the Markoff process with stable distribution, an interpretation may be given to the H-theorem of the statistical mechanics.

Remark. On reading the manuscript, S. Kakutani obtained that,[3] *if $f(x)$ is bounded and measurable,* the mean convergence in (5) may be replaced by the φ-almost-everywhere convergence, by modifying J. L. Doob's arguments.[4] This is more precise than (5), since from this result and the theorem 1 (5) may easily be deduced.[5] It is noted, however, that Doob did not obtain the mean convergence (5), nor its "conjugate" (6). The Doob-Kakutani's proof appeals to Birkhoff-Khintchine's ergodic theorem and to the measure theory in infinite product space. Our proof is operator-theoretical; it is a direct adaptation of the mean ergodic theorem in Banach spaces.

---

3) See the following paper of S. Kakutani.
4) Trans. Amer. Math. Soc., **44** (1938), 87–150.
5) See S. Kakutani: loc. cit.

§2. *Proof of the theorems.* The set (M') of all the bounded measurable functions $f(x)$ constitutes a Banach space with the norm $\|f\|_{M'} = \sup |f(x)|$. In the same way the set (M) of all the measurable functions $f(x)$ which are φ-essentially bounded also constitutes a Banach space with the norm $\|f\|_M = \text{ess. sup} |f(x)|$. Any element $f \in (M') \big((M)\big)$ may be considered as an element $\in (L)$, and $(M') \big((M)\big)$ is dense in (L) by the topology defined by the norm $\|\ \|$.

Proof of theorem 1. Let $f \in (M')$, then $f^{(m)}(x)$ is well defined and $\|f^{(m)}\|_{M'} \leqq \|f\|_{M'}$ by (1). We have $\|f^{(m)}\| = \int |f^{(m)}(x)| \varphi(dx) \leqq \int \varphi(dx) \left\{ \int P^{(m)}(x, dy) |f(y)| \right\} = \|f\|_{M'}$ by (4). Next let $f \in (L)$ be non-negative and put $f_n(x) = \min. \big(f(x), n\big)$. Then, since $f_n(x)$ is a bounded function, $f_n^{(m)}(x)$ is well defined and $f_n^{(m)}(x) \leqq f_{n+1}^{(m)}(x) \leqq \cdots$, $\|f_n^{(m)}\| \leqq \|f_n\| \leqq \|f\|$. Hence by Levi's theorem, the finite

$$\lim_{n \to \infty} f_n^{(m)}(x) = \lim_{n \to \infty} \int P^{(m)}(x, dy) f_n(y) = \int P^{(m)}(x, dy)$$

$$\{\lim_{n \to \infty} f_n(y)\} = \int P^{(m)}(x, dy) f(y)$$

exists φ-almost everywhere. Thus $f^{(m)}(x) = \lim_{n \to \infty} f_n^{(m)}(x) \in (L)$ and $\|f^{(m)}\| \leqq \|f\|$. This proves the first part of the theorem.

Since $P^{(m)}(x, E)$ is a bounded function, $\int \varphi(dx) f(x) P^{(m)}(x, E)$ exists for any $f \in (L)$ and we have from (4)

$$(7) \quad \sup_E \int \varphi(dx) f(x) P^{(m)}(x, E) \leqq \sup_E \int \varphi(dx) |f(x)| P^{(m)}(x, E) \leqq \|f\|.$$

Let $f \in (M)$, then by (4), $\int \varphi(dx) f(x) P^{(m)}(x, E)$ is φ-absolutely continuous: $\left| \int \varphi(dx) f(x) P^{(m)}(x, E) \right| \leqq \|f\|_M \cdot \varphi(E)$, and hence, in this case, $f^{(-m)}$ is defined and $\|f^{(-m)}\|_M \leqq \|f\|_M$. Next let $f \in (L)$ be non-negative and put $f_n(x) = \min. \big(f(x), n\big)$ as above. Since $\int \varphi(dx) f(x) P^{(m)}(x, E) = \lim_{n \to \infty} \int \varphi(dx) f_n(x) P^{(m)}(x, E)$ for any $E \in B(R)$, the left hand member is φ-absolutely continuous, by Vitali-Hahn-Saks' theorem. Thus, in this case also $f^{(-m)} \in (L)$ is defined. By (7), we have $\|f^{(-m)}\| \leqq \|f\|$. The last part of theorem is hereby proved.

Remark. The operation $P^m : f \to f^{(m)} (\bar{P}^m : f \to f^{(-m)})$ is thus a linear operation on (L) to (L) with the norm one. $P^m(P^{\bar{m}})$ also defines an operation on (M') to (M') $\big($on (M) to $(M)\big)$ with the norm one. This remark is the key to the proof of theorem 2.

Proof of theorem 2. Let $f \in (M')$, then by (1)

$$(8) \quad \left\| \frac{1}{n} \sum_{m=1}^{n} f^{(m)} \right\|_{M'} \leqq \|f\|_{M'} \qquad (n = 1, 2, \ldots).$$

Thus, by (3) and the diagonal method, we obtain a sequence of integers $\{n'\}$, $n' < (n+1)'$, such that

$$\lim_{n' \to \infty} \int_E \left\{ \frac{1}{n'} \sum_{m=1}^{n'} f^{(m)}(x) \right\} \varphi(dx) = F^{(*)}(E) \quad \text{exists for all } E \in R(R).$$

By (8) and Vitali-Hahn-Saks' theorem, $F^{(*)}(E)$ is φ-absolutely continuous. Let $F^{(*)}(E) = \int_E f^{(*)}(x)\,\varphi(dx)$, $f^{(*)} \in (L)$. Then it is easy to see that $f^{(*)}$ is the weak limit of the sequence $\left\{ \frac{1}{n'} \sum_{m=1}^{n'} f^{(m)} \right\}$: for any $g \in (M)$ we have $\lim_{n' \to \infty} \int \left\{ \frac{1}{n'} \sum_{m=1}^{n'} f^{(m)}(x) \right\} g(x)\,\varphi(dx) = \int f^{(*)}(x) g(x)\,\varphi(dx)$. Hence by the mean ergodic in Banach spaces,[6] we see that the sequence $\left\{ \frac{1}{n} \sum_{m=1}^{n} f^{(m)} \right\}$ itself converges strongly (in the sense of the norm $\| \ \|$) to $f^{(*)}$. Since the sequence of operators $\left\{ \frac{1}{n} \sum_{m=1}^{n} P^m \right\}$ are all of (L)—norm ≤ 1 and since (M') is strongly dense in (L), $\left\{ \frac{1}{n} \sum_{m=1}^{n} P^m \cdot f \right\}$ converges strongly for any $f \in (L)$. That $P \cdot f^{(*)} = f^{(*)}$ is proved at the same time. Thus (5) is proved. The proof of (6) may be obtained analogously.

§ 3. *The determination of the stable distributions.* Let R be a compactum and let $B(R)$ denote the family of all the Borel sets of R. We assume the Markoff process to satisfy the continuity hypothesis :

$$(9) \quad \left\{ \begin{array}{l} \int P(x, dy) f(y) \text{ is a continuous function of } x \text{ for continuous} \\ \text{function } f(y). \end{array} \right.$$

For such Markoff process $P(x, E)$ we may prove the following results.

Let (C) denote the Banach space of all the real-valued continuous functions $f(t)$ on R with the norm $\|f\|_C = \sup |f(x)|$. Then, i) the Markoff process $P(x, E)$ satisfying (9) may be characterised as the linear operation P on (C) to (C) which maps positive element $f(f(x) \geq 0$ for $x \in R)$ onto positive element $P \cdot f$ of (C) in such a way that $P \cdot f = f$ if $f(x) \equiv 1$. ii) Since (C) is separable and since we have (1), the sequence of distributions of the form

$$\left\{ \begin{array}{l} P_{n_i, m_i}(x_0, E) = \dfrac{1}{m_i - n_i} \sum_{k=n_i+1}^{m_i} P^{(k)}(x_0, E) \quad (i=1, 2, \ldots; \lim_{i \to \infty} (m_i - n_i) = \infty) \\ \text{for fixed } x_0, \end{array} \right.$$

is compact as a set of continuous linear functionals on (C). That is, there exists a partial sequence $\{P_{n_i', m_i'}(x_0, E)\}$ and a set function $\tilde{P}(x_0, E)$

---

6) K. Yosida: Proc., **14** (1938), 292–294, S. Kakutani: Proc., **14** (1938), 295–300 and F. Riesz: Journal of London Math. Soc., **13** (1938), 274–278. That mean ergodic theorem is valid for individual point of the Banach space was stressed on by the writer in the preceding note: Proc., **15** (1939), 255–259.

completely additive for $E \in B(R)$ such that $\int \widetilde{P}(x_0, dy) f(y) = \lim_{i \to \infty} \int P_{n_i', m_i'}$
$(x_0, dy) f(y)$ for any $f \in (C)$. Such limit distribution $\widetilde{P}(x_0, E)$ is stable, since by (9) we have

$$\int \left\{ \int \widetilde{P}(x_0, dy) P(y, dz) \right\} \; f(z) = \int \widetilde{P}(x_0, dy) \left\{ \int P(y, dz) f(z) \right\}$$

$$= \lim_{i \to \infty} \int P_{n_i', m_i'}(x_0, dy) \left\{ \int P(y, dz) f(z) \right\}$$

$$= \lim_{i \to \infty} \int P_{n_i', m_i'}(x_0, dz) f(z) = \int \widetilde{P}(x_0, dz) f(z)$$

for any $f \in (C)$. iii) Let $\varphi(E)$ be any stable distribution and let a sequence of elements $f_1, f_2, \ldots \in (C)$ be dense in (C). By (5) and the diagonal method, we may find a sequence of integers $\{n'\}$ and a set $E_0 \in B(R)$ of φ-measure zero such that $\lim_{n' \to \infty} \dfrac{1}{n'} \sum_{m=1}^{n'} f_i^{(m)}(x)$ exists for $x \in R - E_0$, $i = 1, 2, \ldots$ Hence $\lim_{n' \to \infty} \int P_{1, n'}(x, dy) f(y)$ exists for any $f \in (C)$ and for any $x \in R - E_0$. Put $\lim_{n' \to \infty} \int P_{1, n'}(x, dy) f(y) = \int \widetilde{P}(x, dy) f(y)$ for $x \in R - E_0$, $f \in (C)$. Then, since φ is stable,

$$\int \varphi(dy) f(y) = \lim_{n' \to \infty} \int \left\{ \int \varphi(dx) P_{1, n'}(x, dy) \right\}$$

$$f(y) = \int \varphi(dx) \left\{ \lim_{n' \to \infty} \int P_{1, n'}(x, dy) f(y) \right\} = \int \varphi(dx) \left\{ \int \widetilde{P}(x, dy) f(y) \right\}$$

$$= \int \left\{ \int \varphi(dx) \widetilde{P}(x, dy) \right\} f(y) \quad \text{for any} \quad f \in (C).$$

Thus it is proved that any stable distribution $\varphi(E)$ of $P(x, E)$ may be obtained as a "convex combination" of the limit distributions as $\widetilde{P}(x, E)$. In this way we may extend Kryloff-Bogoliouboff's results[7] to the indeterministic transition process. It is to be noted that similar results may be obtained for Markoff process with enumerably infinite number of possible states. However this idea was partly and unconsciously used in a joint work with S. Kakutani.[8]

§ 4. *An interpretation of the H-theorem.* Let R be the interval $(0, 1)$ and let $B(R)$ be the set of all the Borel sets of R. We assume that $P(x, E)$ is given by Borel-measurable density in the following manner :

(10).
$$P(x, E) = \int_E p(x, y) dy , \quad \int p(x, y) dx \equiv 1 .$$

The uniform distribution dx is thus stable for $P(x, E)$.

Let a non-negative $f(x)$ be such that $\int f(x) dx = 1$, $\int f(x) \log^{+} f(x) dx \neq$ infinite. Then we have

---

7) Ann. Math., **38** (1937), 65–113.

8) Jap. J. Math., **16** (1939), 47–55.

$$(11) \qquad \int H\big(f(y)\big)\,dy \geqq \int H\big(f^{(-1)}(y)\big)\,dy \geqq \int H\big(f^{(-2)}(y)\big)\,dy \geqq \cdots ,$$

$$H(z)=z\log z .$$

The proof follows from (10) and the convexity of the function $H(z)$. If, moreover, we assume the measurability of $p^{(s)}(y)=\inf\limits_{x} p^{(s)}(x, y)$

$\left(P^{(s)}(x, E)=\int p^{(s)}(x, y)\,dy\right)$ and $\int p^{(s)}(y)\,dy>0$ for a certain s, we obtain

$$(12) \qquad \sup_{f(x)\geqq 0,\ |f|=1} \int \big|\,f^{(-n)}(y)-1\,\big|\,dy \leqq \frac{\delta}{(1+\varepsilon)^n} \qquad (n=1, 2, \cdots) ,$$

for certain constants $\delta,\ \varepsilon>0$. The proof may be obtained easily by modifying the *ergodic principle* of A. Kolmogoroff.[9]

The above result may serve as an interpretation of the H-theorem in the statistical mechanics. For, by $f(x)\geqq 0$ and $\int f(x)\,dx=1$, $f(x)$ may be considered as the *weight* of the " state " x at the origin of the time. Then, *irrespective of the initial weight* $f(x)$, the weight $f^{(-n)}(x)$ of the state x after the elapse of n unit-times tends to the uniform weight $\big(f^{(-*)}(x)\equiv 1\big)$ as $n\to\infty$, in such a way that (6), (11) and (12) hold. The hypothesis (10) may be considered as a *generalised symmetric condition*, and the hypothesis concerning $p^{(s)}(y)$ amounts to a kind of *irreducibility* of the transition process $P(x, E)$. Thus the results may serve as a precision to the arguments of R. von Mises in his book.[10]

---

9) Math. Ann., **104** (1931), 415–458.

10) Wahrscheinlichkeitsrechnung (1931). For the discussion of this § I owe to K. Husimi and S. Kakutani.

An abstract treatment of the individual ergodic theorem

Proc. Imp. Acad. Tokyo **16** (1940) 280–284

(Comm. by T. Takagi, m.i.a., July 12, 1940.)

§ 1. *Introduction.* The analytical interests concerning the ergodic theorems may be observed from two viewpoints. Firstly, the ergodic theorems give us means for determining the fixed points und linear operations. Secondly, they concern with the deduction of the stronger convergences of linear operations from the weaker ones. Some entensive literatures on the mean ergodic theorem are more or less guided by these viewpoints. Concerning the individual ergodic theorem, however, we have only small number of literatures. The dominated ergodic theorem due to N. Wiener and M. Fukamiya,[1] and the writer's extensions[2] of Birkhoff-Khintchine's ergodic theorem both constitute examples on the individual ergodic theorem. The purpose of the present note is to extend the idea developed in [I]. The continuity theorems (theorem 1 and its corollary) have interests of their own. Theorem 2 is an abstract form of the individual ergodic theorem. We may deduce from this the individual ergodic theorem for m-parameter abelian group of equi-measure transformations, and more generally that for general semi-group of linear operations (theorem 3). The results in [I] in a somewhat extended form are also deducible from theorem 3.

§ 2. *A continuity theorem and an abstract form of the individual ergodic theorem.* Under abstract (S) space $(A-S)$ we mean a linear space of type-F, which satisfies the following axioms (we donote by $x, y, \cdots$ the elements of $(A-S)$, by $\|x\|_S$, $\|y\|_S$, $\cdots$ their quasi-norms and by λ a real scalar):

(1)   a semi-order relation $x > y$ is defined in $(A-S)$, relative to which $(A-S)$ is a linear lattice, viz.:

(1–1)   corresponding to any two elements x, y there exist the least upper bound sup (x, y) and the greatest lower bound inf (x, y),

(1–2)   translations $x \to x+y$ and homothetic expansions $x \to \lambda x$ with $\lambda > 0$ preserve the semi-ordering,

(1–3)   sup (x, y) and inf (x, y) are both continuous in x and y in the topology defined by the norm $\|\ \|_S$.

(2)   any sequence $\{x_n\}$ bounded from above (below) admits of the least upper bound $\sup_n x_n$ (the greatest lower bound $\inf_n x_n$).

(3)   if we write $\mathrm{Lim}_{n\to\infty} x_n = x$, in case $\overline{\mathrm{Lim}}_{n\to\infty} x_n = \inf_n \sup_{m \geq n} x_m = \underline{\mathrm{Lim}}_{n\to\infty} x_n = \sup_n \inf_{m \geq n} x_m$, then $\mathrm{Lim}_{n\to\infty} x_n = x$ implies $\lim_{n\to\infty} \|x_n - x\|_S = 0$.

(4) $x \geqq y \geqq -x$ implies $\|x\|_S \geqq \|y\|_S$.

---

1) N. Wiener: Duke Math. J., **5** (1939), 1–18. M. Fukamiya: Tôhoku Math. J., **46** (1939), 150–153. Cf. also K. Yosida and S. Kakutani: Proc. **15** (1939), 165–168.

2) K. Yosida: Jap. J. of Math., **15** (1940), 31–36, to be cited as [I] below.

Let (S) be the class of real-valued measurable functions $x(t)$ on finite or infinite interval (a, b), which are finite almost everywhere. By the quasi-norm $\|x\|_S = \int_b^a \frac{|\dot{x}(t)|}{1+|x(t)|} \frac{dt}{1+t^2}$ and the semi-order relation $x \geqq y$ $\big(x(t) \geqq y(t)$ almost everywhere$\big)$, (S) constitutes a concrete example of the space $(A-S)$.[1] In this case, $\underset{n \to \infty}{\mathrm{Lim}}\, x_n = x$ means $\underset{n \to \infty}{\lim}\, x_n(t) = x(t)$ almost everywhere. In the sequel, we will write $|x|$ for $\sup(x, 0) - \inf(x, 0)$.

After these preliminaries we may prove

Theorem 1.[2]   Let (F) be a space of type-F by the quasi-norm $\|x\|_F$, and consider a sequence $\{T_n\}$ of linear continuous operations $\cdot$ on (F) to $(A-S)$. Putting $x_n = T_n \cdot x$ for any $x \in (F)$, we assume that the set $(F)'$ of points $x \in (F)$ for which $\overline{\underset{n \to \infty}{\mathrm{Lim}}}\, x_n$ exist constitutes a set of second category in (F). Then $\overline{\underset{n \to \infty}{\mathrm{Lim}}}\, x_n$ and $\underline{\underset{n \to \infty}{\mathrm{Lim}}}\, x_n$ exist for all $x \in (F)$ and the operation $\widetilde{T}$, $\widetilde{T} \cdot x = \widetilde{x}$, $\widetilde{x} = \overline{\underset{n \to \infty}{\mathrm{Lim}}}\, x_n - \underline{\underset{n \to \infty}{\mathrm{Lim}}}\, x_n$ is a continuous operation on (F) to $(A-S)$.

Proof. Put $x'_n = \underset{n \geqq m}{\sup}\, |x_m|$, $x' = \underset{m}{\sup}\, |x_m|$ for any $x \in (F)'$, and consider the operations $V_n \cdot x = x'_n$, $V \cdot x = x'$ on $(F)'$ to $(A-S)$. Each V_n is continuous and $\underset{n \to \infty}{\lim} \|V_n \cdot x - V \cdot x\|_S = 0$. Hence, from $\underset{n \to \infty}{\lim} \left\|\frac{1}{k} V_n \cdot x\right\|_S = \left\|\frac{1}{k} V \cdot x\right\|_S$ $(k = 1, 2, \ldots)$ and $\underset{n \to \infty}{\lim} \left\|\frac{1}{k} V \cdot x\right\|_S = 0$, we obtain the inclusion $(F)' \leqq \sum_{k=1}^{\infty} G_k$, $G_k = \underset{x \in (F)}{E}\left(\underset{n}{\sup} \left\|\frac{1}{k} V_n \cdot x\right\|_S \leqq \varepsilon\right)$, for any $\varepsilon > 0$. G_k is a closed set of (F) by the continuity of V_n. Since $(F)'$ is of second category in the complete metric space (F), there must exist a G_k which contains a sphere of (F). Thus we have a point $x_0 \in (F)$ and a constant $\delta > 0$ such that $\underset{n}{\sup} \left\|\frac{1}{k} V_n \cdot x\right\|_S \leqq \varepsilon$ for $\|x - x_0\|_F \leqq \delta$. Hence, from $|V_n \cdot (x - x_0)| \leqq |V_n \cdot x| + |V_n \cdot x_0|$ and $V_n \cdot \left(\frac{1}{k} x\right) = \frac{1}{k} V_n \cdot x$, we see that $\underset{|x|_F \to 0}{\lim} \|V_n \cdot x\|_S = 0$ uniformly in n. $V \cdot x$ is thus defined for all $x \in (F)$ and is continuous at $x = 0$ with $V \cdot 0 = 0$. Thus, by $|\widetilde{T} \cdot x| \leqq 2V \cdot x$ and $\|\widetilde{T} \cdot x - \widetilde{T} \cdot y\|_S \leqq \|\widetilde{T} \cdot (x - y)\|_S$, $\widetilde{T}$ is defined and continuous at every $x \in (F)$.   Q. E. D.

Corollary.   The set $(F)''$ of points $x \in (F)$ for which $\underset{n \to \infty}{\mathrm{Lim}}\, x_n$ exists either coincides with the whole (F) or it constitutes a set of first category in (F).

Proof. Let $(F)''$ be of second category in (F), then $\widetilde{T} \cdot x$ is con-

---

1) In truth, (S) is a semi-ordered ring. The concrete representation of semi-ordered ring as subring of (S) will be published elsewhere. It is to be noted that the metrical completeness of $(A-S)$ is not necessary in the present note.

2) This is an extension of Banach-Saks' theorem as formulated by S. Mazur and W. Orlicz: Stud. Math., **4** (1933), 152–157.

tinuous on (F) to $(A-S)$. Thus $(F)''=E(\widetilde{T}\cdot x=0)$ is a closed set of (F). That $(F)''$ is a linear subspace of (F) is evident. A closed linear subspace which is of second category in a space of type-F must coincide with the whole space. Q. E. D.

Theorem 2. Let the above (F) be a linear subspace of $(A-S)$ such that $\lim_{n\to\infty}\|x_n-x\|_F=0$ implies $\lim_{n\to\infty}\|x_n-x\|_S=0$, and let $\{T_n\}$ be a sequence of linear continuous operations on (F) to (F). Assume, as above, that $\overline{\operatorname{Lim}}_{n\to\infty} x_n$ exists as point $\in(A-S)$ for those $x\in(F)$ which constitutes a set of second category in (F). If, for a point $y\in(F)$, there corresponds $\bar{y}\in(F)$, such that $\lim_{n\to\infty}\|y_n-\bar{y}\|_F=0$, $T_n\cdot\bar{y}=\bar{y}$ $(n=1, 2, \ldots)$, $\operatorname{Lim}_{n\to\infty}(T_n\cdot y-T_nT_m\cdot y)=0$ $(m=1, 2, \ldots)$, then we must have $\operatorname{Lim}_n y_n=\bar{y}$.

Proof. Put $y=\bar{y}+(y-\bar{y})$, then $0\leq\widetilde{T}\cdot y\leq\widetilde{T}\cdot(y-\bar{y})$ by $T_n\cdot\bar{y}=\bar{y}$ $(n=1, 2, \ldots)$. We have, by theorem 1 and $\lim_{m\to\infty}\|(y-\bar{y})-(y-y_m)\|_F=0$, $\widetilde{T}(y-\bar{y})=0$ from $\widetilde{T}\cdot(y-y_m)=0$ $(m=1, 2, \ldots)$. However we have $\widetilde{T}\cdot(y-y_m)=\overline{\operatorname{Lim}}_{n\to\infty}T_n(y-T_m\cdot y)-\underline{\operatorname{Lim}}_{n\to\infty}T_n(y-T_m\cdot y)=0$ from the assumption. Thus $\operatorname{Lim}_{n\to\infty} y_n=z$ exists, and by (3) $\lim_{n\to\infty}\|y_n-z\|_S=0$, which proves $\bar{y}=z$. Hence we have $\operatorname{Lim}_{n\to\infty} y_n=\bar{y}$.

§ 3. *The individual ergodic theorem for general semi-groups.* Let $p\geq 1$, then the class of real-valued measurable functions $x(t)$ on (a, b) with $\int_a^b|x(t)|^p dt<+\infty$ constitutes a Banach space (L^p) by the norm
$$\|x\|=\left(\int_a^b|x(t)|^p dt\right)^{\frac{1}{p}}.$$

Let G be a semi-group (multiplicative system) of linear continuous operations $T^{(g)}$ on (L^p) to (L^p) with uniformly bounded norms,

(5) $\quad\|T^{(g)}\|\leq C$ for all $T^{(g)}\in G$.

Following L. Alaoglu and Garrett Birkhoff,[1] we will call G "ergodic" when it possesses a sequence of measures $\varphi_n(V)$ satisfying i) $\varphi_n(G)=1$ for all n, and ii) given $T^{(g)}$ in G and $\epsilon>0$, N exists so large that $n\geq N$ implies $|\varphi_n(VT^{(g)})-\varphi_n(V)|+|\varphi_n(T^{(g)}V)-\varphi_n(V)|<\epsilon$. We assume that for any $x\in(L^p)$ the integral $T_n\cdot x=\int_G(T^{(g)}\cdot x)d\varphi_n$ exists in S. Bochner's or G. Birkhoff's sense $(n=1, 2, \ldots)$. We call the point $y\in(L^p)$ "strongly ergodic" if we have

(6) $\quad\lim_{n\to\infty}|y_{n,m}(t)-y_n(t)|=0$ almost everywhere for $m=1, 2, \ldots$,

where $\qquad y_n=T_n\cdot y$, $\quad y_{n,m}=T_nT_m\cdot y$.

Then we have the

Theorem 3. We assume that

(7) $\quad\overline{\lim}_{n\to\infty}|x_n(t)|<\infty$ *almost everywhere for those x which constitutes a set of second category in (L^p).*

---

1) L. Alaoglu and Garrett Birkhoff: Ann. of Math., **41** (1940), 293–309.

Let a strongly ergodic point $y \in (L^p)$ satisfy the condition.

(8) $\{y_n\}$ *contains a subsequence weakly convergent to a point* $\bar{y} \in (L^p)$.[1]

Then $\lim\limits_{n \to \infty} y_n(t) = \bar{y}(t)$ *almost everywhere,* $\lim\limits_{n \to \infty} \|y_n - \bar{y}\| = 0$ *and* $T^{(g)}\bar{y} = \bar{y}$ *for all* $T^{(g)} \in G$.

Proof. From (5) we have $\|T_n\| \leq C$ $(n = 1, 2, \ldots)$. Hence, by (8) and the ergodicity of G, we see[2] that $\lim\limits_{n \to \infty} \|y_n - \bar{y}\| = 0$ and $T^{(g)} \cdot \bar{y} = \bar{y}$ for all $T^{(g)} \in G$. Hence, by (6), (7) and theorem 2, we obtain the theorem.   Q. E. D.

§ 4. *Application of theorem 3.* Let $T^{(g)}$ be defined by equi-measure transformation $P^{(g)}$: $x^{(g)} = T^{(g)} \cdot x$, $x^{(g)}(t) = x(P^{(g)} \cdot t)$ for $x \in (L^p)$, and let $G = \{T^{(g)}\}$ be an m-parameter abelian group of such $T^{(g)}$. Then each $T^{(g)}$ satisfies (5) with $C = 1$ as linear operation on (L^1) and on (L^2). By putting $\varphi_n(V) =$ the proportion of the cube $0 \leq \lambda_1 \leq n$, $0 \leq \lambda_2 \leq n$, $\ldots$, $0 \leq \lambda_m \leq n$, to its portion occupied by V, we see that G is ergodic. We have, moreover, (7) for $p = 1$ and for $p = 2$, by the dominated ergodic theorem of N. Wiener.[3] That (6) is satisfied for all $y \in (L^2)$ may be proved in the following manner. For any $y \in (L^2)$ and k we have $\|y_{n,k} - y_n\| \leq C_k \dfrac{1}{n} \|y\|$ $(n = 1, 2, \ldots)$, by the definition of $\varphi_n(V)$. Hence, if we put

$$E(\varepsilon, n, k) = \underset{t}{E}(|y_{n,k}(t) - y_n(t)| \geq \varepsilon),$$ we have mes $\left(E(\varepsilon, n, k)\right) \leq \dfrac{C_k^2}{\varepsilon^2} \cdot \dfrac{1}{n^2} \|y\|^2$,

proving (6) by $\sum\limits_{n=1}^{\infty} \dfrac{1}{n^2} < \infty$.[4] Since (L^2) is locally weakly compact, we have (8) for any $y \in (L^2)$. Thus, by theorem 3, $\lim\limits_{n \to \infty} y_n(t)$ exists almost everywhere for all $y \in (L^2)$. Since (L^2) is strongly dense in (L^1) by the norm of (L^1), we have also the existence almost everywhere of $\lim\limits_{n \to \infty} y_n(t)$ for all $y \in (L^1)$, by theorem 1.

Thus we have proved N. Wiener's *m-parameter individual ergodic theorem.*[5]

Next let G be the semi-group generated by the iterations $\{T^n\}$ of a linear operation T on (L^p) to (L^p), and put $T_n = \dfrac{1}{n} \sum\limits_{m=1}^{n} T^m$ $(n = 1, 2, \ldots)$. Then, as above, we may prove by theorem 1 and 3 the following individual ergodic theorems. Since, their proofs are similar as those in [I], we omit them. However, it is to be noted that the results are somewhat more general than in [I], for we here deal with (L^p) on finite or infinite interval (a, b).

---

1) If $p > 1$, (8) is superfluous, since (L^p) with $p > 1$ is locally weakly compact.

2) L. Alaoglu and G. Birkhoff: loc. cit. It is to be noted that we may obtain the same result by the arguments employed in the proof of the mean ergodic theorem of F. Riesz, S. Kakutani and the present author.

3) N. Wiener: loc. cit. Though we here are concerned with (L^p) on finite or infinite interval (a, b), we may obtain (7) by Wiener's argument.

4) This device I owe to M. Fukamiya.

5) N. Wiener: loc. cit. Wiener's proof applies to (L^1) on finite interval (a, b) only; the mean ergodic theorem for (L^1) on infinite interval does not in general hold good. Our proof, however, applies to (L^1) on finite as well as on infinite interval (a, b).

Theorem 4. Let T be a linear continuous operation (L^1) to (L^1) and put $x^{(n)} = T^n \cdot x$ for any $x \in (L^1)$. We assume that

$$(9) \quad \| T^n \| \leq a \text{ constant } C \quad (n = 1, 2, \ldots),$$

$$(10) \quad \varlimsup_{n \to \infty} \left| \frac{1}{n} \sum_{m=1}^{n} x^{(m)}(t) \right| < \infty \quad almost \quad everywhere \quad for \quad those \quad x \quad which$$

constitute a set of second category in (L^1).

If for an element $y \in (L^1)$

$$\lim_{n \to \infty} \frac{y^{(n)}(t)}{n} = 0 \quad almost \quad everywhere,$$

the sequence $\left\{ \dfrac{1}{n} \sum_{m=1}^{n} y^{(m)} \right\}$ contains a subsequence weakly con-

vergent to an element $\bar{y} \in (L^1)$,

then we have

$$(11) \quad \lim_{n \to \infty} \left\| \frac{1}{n} \sum_{m=1}^{n} y^{(m)} - \bar{y} \right\| = 0, \quad \lim_{n \to \infty} \frac{1}{n} \sum_{m=1}^{n} y^{(m)}(t) = \bar{y}(t) \quad almgst \quad every-$$

where.

Theorem 5. Let $p > 1$, and let T be a linear continuous operation on (L^p) to (L^p) satisfying (9) and (10) in (L^p). Then we have (11) for all $y \in (L^p)$.

Theorem 6. Let T be a linear operation on (L^p) to (L^p), $p \geqq 1$. We assume that

for any $x \in (L^p)$ there corresponds a $X \in (L^p)$ such that

$$| x^{(n)}(t) | \leq X(t) \quad almost \quad everywhere \quad (n = 1, 2, \ldots).$$

Then (11) *hold good for any $y \in (L^p)$.*[1]

Remark. The application of theorem 1, 2 and 3 is not exhausted by the above theorems. We may obtain the individual ergodic theorem in the space of abstractly-valued functions. We may also prove and extend B. Jessen's theorem of approximation of Lebesgue integral by Riemann sums.[2] We will here not enter into the details.

---

1) This is more precise than the results of Garrett Birkhoff: Proc. Nat. Acad. Sc., **24** (1938), 154–159. F. Riesz: Proc. London Math. Soc., **13** (1938), 274–278 and S. Kakutani: Proc. **15** (1939), 121–123.

2) B. Jessen: Ann. of Math., **35** (1934), 248-251.

Operator-theoretical treatment of Markoff's process and mean ergodic theorem

(*with S. Kakutani*)

Ann. Math. **42** (1941) 188–228

(Received November 15, 1939)

Chapter I. Introduction

It is the purpose of the present paper to give a consistent treatment of the problems of Markoff's process and mean ergodic theorem from the standpoint of the theory of iteration of bounded linear operations in Banach spaces.

Let Ω be an abstract space where a measure of Lebesgue type is defined, and consider a simple Markoff's process, by which a moving point $t \, \epsilon \, \Omega$ is transferred stochastically inside Ω. If we denote by $P(t, E)$ the transition probability that a moving point $t \, \epsilon \, \Omega$ is transferred into a Borel set E of Ω after the elapse of a unit-time, then we have always $P(t, E) \geq 0$ and $P(t, \Omega) = 1$. We shall further assume that $P(t, E)$ is completely additive as a set function of Borel sets E if t is fixed, and that $P(t, E)$ is Borel measurable in t if E is fixed. Under these assumptions, the probability $P^{(n)}(t, E)$ that a moving point $t \, \epsilon \, \Omega$ is transferred into a Borel set E of Ω after the elapse of n unit-times is given recurrently by

$$P^{(n)}(t, E) = \int_{\Omega} P^{(n-1)}(t, ds)P(s, E), \qquad n = 2, 3, \cdots; \qquad P^{(1)}(t, E) = P(t, E),$$

where the integration is of Radon-Stieltjes type.

The problem of Markoff consists in the investigation of the behavior of the sequence of the iterations $P^{(n)}(t, E)$ and their arithmetic means $Q^{(n)}(t, E) = \frac{1}{n} \sum_{m=1}^{n} P^{(m)}(t, E)$ for large n.

In the special case, when Ω is composed of only a finite number $(= N)$ of points $\left(\text{with equal measure } \frac{1}{N}\right)$, our Markoff's process is reduced to the classical one, and our problem is nothing but the investigation of the behavior of the iterations P^n of a matrix $P = (p_{ij})$, $i, j = 1, 2, \cdots, N$, such that $p_{ij} \geq 0$, $i, j = 1, 2, \cdots, N$, and $\sum_{j=1}^{N} p_{ij} = 1$, $i = 1, 2, \cdots, N$. This case was first treated by A. Markoff in his classical papers, and in the last ten years many contributions to this problem were given by J. Hadamard, M. Fréchet, R. von Mises, A. Kolmogoroff, W. Doeblin, J. L. Doob and others. These results are all collected in a recent monograph of M. Fréchet [3], which is so complete that we find almost no need of giving further discussions on this case. We shall only remark that this case may be treated exactly in the same manner as in the general case.

In the general case, when Ω is a continuum,[1] this problem was discussed by

---

[1] We shall confine ourselves in this paper only to the case when Ω is the closed interval $0 \leq t \leq 1$. This is not an essential restriction.

B. Hostinsky, M. Fréchet, J. L. Doob, W. Doeblin and N. Kryloff-N. Bogolioù-boff. B. Hostinsky [1] treated the case when $P(t, E)$ is given by a continuous density $p(t, s)$: $P(t, E) = \int_E p(t, s)\, ds$, where $p(t, s)$ is a continuous function of two variables t and s in $\Omega \times \Omega$; M. Fréchet [1], [2] considered the case when $p(t, s)$ is bounded and measurable instead of being continuous; J. L. Doob [1] weakened this condition by requiring only that $p(t, s)$ is uniformly (in t) integrable in s, that is, that there exists for any $\epsilon > 0$ an $\eta > 0$ such that $\mathrm{mes}(E) < \eta$ implies $\int_E p(t, s)\, ds < \epsilon$ for any $t \, \epsilon \, \Omega$; and lastly, W. Doeblin [1] discussed the case when the existence of the density $p(t, s)$ is not necessarily assumed. He has only assumed that

(D) $\quad \begin{cases} \textit{there exist an integer } d \textit{ and positive constants } b,\ \eta \textit{ such that } \mathrm{mes}(E) < \eta \\ \textit{implies } P^{(d)}(t;\ E) \leqq 1 - b \textit{ for any } t \, \epsilon \, \Omega. \end{cases}$

It is clear that the condition (D) of W. Doeblin is more general than that of J. L. Doob, and under this condition, W. Doeblin has obtained, by a direct (point-set-theoretical) method, a considerably precise result.

In the present paper, we shall discuss these problems by another method, which is due to N. Kryloff-N. Bogolioùboff [1], [2], and which may be called an operator-theoretical method. Indeed, we shall treat these problems by considering $P^{(n)}(t, E)$ as a kernel of an integral operator in some special Banach spaces. This may be done in two ways: firstly,

$$x \to T(x) = y: \quad y(E) = \int_\Omega x(dt) P(t, E)$$

is a bounded positive[2] linear operation which maps the Banach space $(\mathbf{M})$[3] into itself; and secondly,

$$x \to \overline{T}(x) = y: \quad y(t) = \int_\Omega P(t, ds) y(s)$$

is a bounded positive linear operation which maps the Banach space (M^*)[4]

---

[2] A bounded linear operation T, which maps a semi-ordered Banach space into itself, is called to be positive if $x \geqq 0$ implies $T(x) \geqq 0$.

[3] $(\mathbf{M})$ is the space of all the (real- or complex-valued) completely additive set functions $x(E)$ defined for all Borel set E of Ω. For any $x(E) \, \epsilon \, (\mathbf{M})$, its norm is defined by: $\| x \| =$ total variation of $| x(E) |$ on Ω. In $(\mathbf{M})$, a real-valued completely additive set function $x(E)$ is called to be positive (and is denoted by $x \geqq 0$), if $x(E) \geqq 0$ for any Borel set $E \subset \Omega$.

[4] (M^*) is the space of all the (real- or complex-valued) bounded Borel measurable functions $x(t)$ defined on Ω. For any $x(t) \, \epsilon \, (M^*)$, its norm is defined by: $\| x \| = \text{l.u.b.} \, | x(t) |$. $\underset{t \, \epsilon \, \Omega}{}$ In (M^*), two functions, which differ from each other at least at one point, are considered to be different; it is to noted that the Banach space (M^*) is essentially different from (M), where the norm is defined by: $\| x \| = \text{ess. max.} \, | x(t) |$. $\underset{t \, \epsilon \, \Omega}{}$ In (M^*), a real-valued bounded Borel measurable function $x(t)$ is called to be positive if $x(t) \geqq 0$ for any $t \, \epsilon \, \Omega$.

into itself. If, in particular, $P(t, E)$ has the density $p(t, s)$, then these operations become

$$x \to T(x) = y: \quad y(E) = \int_E \left(\int_\Omega x(dt)p(t, s) \right) ds,$$

$$x \to \overline{T}(x) = y: \quad y(t) = \int_\Omega \cdot p(t, s)x(s)\, ds;$$

consequently, in the former case, T may also be considered as a bounded positive linear operation which maps the Banach space $(L)^5$ into itself:

$$x \to T(x) = y: \quad y(s) = \int_\Omega x(t)\, p(t, s)\, dt.$$

In all these cases, T and $\overline{T}$ are both of norm 1, and it will be easily seen that the iterates T^n and $\overline{T}^n$ of T and $\overline{T}$ respectively, which are defined by the iterated kernel $P^{(n)}(t, E)$:

$$x \to T^n(x) = y: \quad y(E) = \int_\Omega x(dt)\, P^{(n)}(t, E),$$

$$x \to \overline{T}^n(x) = y: \quad y(t) = \int_\Omega P^{(n)}(t, ds)\, x(s),$$

have also the same properties. Thus our problem is transformed into the investigation of the behavior of the iterates T^n of a bounded positive linear operation T, which maps the Banach space $(\mathbf{M})$, (L) or (M^*) into itself.

N. Kryloff-N. Bogolioùboff [1], [2] introduced the condition that

$$(\mathrm{K}) \quad \begin{cases} \textit{there exist an integer m and a strongly completely continuous}^6 \textit{ linear} \\ \textit{operation V, which maps } (\mathbf{M}) \textit{ into itself, such that } \| T^m - V \| < 1, \end{cases}$$

and under this condition they have obtained a remarkable result. In the present paper we shall develop this idea of N. Kryloff-N. Bogolioùboff and shall obtain a more precise result. This investigation will be carried out in §4, and our principal results may be summed up in the following two statements:

(1) *the condition* (D) *implies the condition* (K),
(2) *under the condition* (K), *all the results of W. Doeblin can be obtained even in a more precise form.*

Besides the *ergodic kernel* (= *ensemble final* due to W. Doeblin) we shall introduce a new notion of *ergodic part*, and it will be shown that this also plays a fundamental rôle in the investigation of the asymptotic behavior of the sequence $P^{(n)}(t, E)$ for large n.

In order to obtain these results, we shall develop in §§2 and 3 a general theory of iteration of bounded linear operations in Banach spaces. Although the

---

[5] (L) is, as usual, the space of all the (real- or complex-valued) measurable functions $x(t)$ which are absolutely integrable on Ω. For any $x(t) \,\epsilon\, (L)$, its norm is defined by: $\| x \| = \int_\Omega |x(t)|\, dt$, and in (L), an integrable function $x(t)$ is called to be positive if we have $x(t) \geqq 0$ almost everywhere in t.

iteration of bounded linear operations in concrete Banach space (for example, in Hilbert space) was discussed by many authors, such a theory has not hitherto been developed in general Banach spaces. We shall show now that this is possible under general conditions, and the results thus obtained will find their full applications in §4.

Thus §§2 and 3 may be considered as preparations to §4. Nevertheless, these are also the principal chapters of this paper. Indeed, Theorem 1 (mean ergodic theorem) and Theorem 4 (uniform ergodic theorem) are among the main results of our paper. In order to explain the meaning of these theorems, let us recall the *mean ergodic theorem* of J. v. Neumann:

Let S be a one-to-one measure-preserving transformation of a space Ω into itself. Then

$$x \rightarrow T(x) = y: \qquad y(t) = x(S(t))$$

is a positive unitary transformation which maps the Hilbert space (L^2) (defined on Ω) into itself, and the mean ergodic theorem of J. v. Neumann says that,

for any $x \, \epsilon \, (L^2)$, the sequence $\{x_n\}$ $(n = 1, 2, \cdots)$, where $x_n = \dfrac{1}{n} (T + T^2 \cdots +$

$T^n)x$, converges strongly to some element $\bar{x} \, \epsilon \, (L^2)$.

This theorem may be generalized in two ways: firstly, (L^2) may be substituted by an arbitrary Banach space (B), and secondly, T may be substituted by an arbitrary quasi-weakly completely continuous[6] linear operation which maps the Banach space (B) into itself. The positiveness of T and the existence of the inverse operation T^{-1} are both not assumed. This theorem is called the *mean ergodic theorem in Banach spaces* and will be proved in §2.[7]

Moreover, if T is quasi-strongly completely continuous,[6] i.e., if T satisfies the condition (K) of N. Kryloff-N. Bogolioùboff, then the strong convergence in the mean ergodic theorem may be substituted by the uniform one. This is a generalization of the results of M. Frechet [1], [2], C. Visser [1] and N. Kryloff-N. Bogolioùboff [1], [2], and will be called the *uniform ergodic theorem in Banach spaces*. We shall prove this in §3. This theorem is extremely powerful in the investigation of concrete Markoff's process in §4.

In concluding the introduction, we express our hearty thanks to Yukio Mimura and Shôkichi Iyanaga for their constant encouragements during the preparation of this paper.

Chapter 2. Mean Ergodic Theorem in Banach Spaces[8]

(Generalization of J. v. Neumann's mean ergodic theorem)

In this chapter we are concerned with weakly completely continuous and quasi-weakly completely continuous linear operations in general (real or com-

---

[6] This notion will be explained at the beginning of §2.

[7] Recently, the generalizations of the mean ergodic theorem were given also by N. Wiener [1], [2], G. Birkhoff [2] and N. Dunford [1]. We shall discuss these results in §2. See the remarks after Theorem 1 of §2.

[8] K. Yosida [2], S. Kakutani [1], K. Yosida and S. Kakutani [1].

plex) Banach spaces. By definition, a bounded linear operation T, which maps a Banach space (B) into itself, is called to be *weakly completely continuous* if it maps the unit sphere $\| x \| \leqq 1$ of (B) on a weakly compact set of (B). More generally, T is called to be *quasi-weakly completely continuous*, if there exist an integer m and a weakly completely continuous linear operation V, which maps (B) into itself, such that $\| T^m - V \| < 1$.

THEOREM 1. *Let T be a bounded linear operation which maps a Banach space (B) into itself. Let us further assume that*

(2.1)   *there exists a constant C such that $\| T^n \| \leqq C$ for $n = 1, 2, \cdots$, and that*

(2.2) $\begin{cases} \textit{for any } x \in (B), \textit{ the sequence } \{x_n\} \ (n = 1, 2, \cdots), \textit{ where } x_n = \dfrac{1}{n}(T + \\ T^2 + \cdots + T^n)x, \textit{ contains a subsequence which converges weakly to a} \\ \textit{point } \bar{x} \in (B). \end{cases}$

Under these assumptions,

(2.3)   *the sequence $\{x_n\}$ $(n = 1, 2, \cdots)$ converges strongly to a point $\bar{x}$,*

and if
we denote by T_1 the operation $x \rightarrow \bar{x}$, then T_1 is a bounded linear operation which maps (B) into itself and

$$(2.4) \qquad TT_1 = T_1 T = T_1^2 = T_1, \qquad \| T_1 \| \leqq C.$$

PROOF. Let x be an arbitrary point of (B). By (2.2), there exists a subsequence $\{x_{n_\nu}\}$ $(\nu = 1, 2, \cdots)$ of $\{x_n\}$ $(n = 1, 2, \cdots)$ which converges weakly to a point $\bar{x}$ of (B). We shall first prove that we have

$$(2.5) \qquad T(\bar{x}) = \bar{x}.$$

This is an easy consequence of the relation:

$$(2.6) \quad \begin{aligned} & \| T(x_n) - x_n \| \\ &= \left\| \frac{1}{n}(T^2 + T^3 + \cdots + T^{n+1})x - \frac{1}{n}(T + T^2 + \cdots + T^n)x \right\| \\ &= \left\| \frac{1}{n}(T^{n+1} - T)x \right\| \leqq \frac{2C}{n} \| x \|, \end{aligned}$$

which is valid for $n = 1, 2, \cdots$. Indeed, putting $n = n_\nu$ in (2.6) and making ν tend to ∞, we have (2.5) at once; for $\{x_{n_\nu}\}$ and $\{T(x_{n_\nu})\}$ converge weakly to $\bar{x}$ and $T(\bar{x})$ respectively.

We shall next prove (2.3). For this purpose, decompose x into two parts: $x = \bar{x} + (x - \bar{x})$. Then we have, by (2.5),

$$(2.7) \qquad x_n = \bar{x} + \frac{1}{n}(T + T^2 + \cdots + T^n)(x - \bar{x}),$$

and consequently, in order to prove (2.3), we have only to prove that the second term of the right hand side of (2.7) converges strongly to 0 as $n \rightarrow \infty$.

Let the range of the linear operation $I - T$ (I is the identical transformation) be R, and denote its closure (in the strong sense) by $\bar{R}$. Since R is a linear subspace of (B), $\bar{R}$ is also closed in the weak sense. Since the sequence

$$\left\{ I - \frac{1}{n}(T + T^2 + \cdots + T^n) \right\} x$$

$$= (I - T)\left(I + \frac{n-1}{n} T + \frac{n-2}{n} T^2 + \cdots + \frac{1}{n} T^{n-1} \right) x$$

converges weakly to $x - \bar{x}$, and since the right hand side clearly belongs to R, $x - \bar{x}$ belongs to $\bar{R}$. Consequently, in order to prove (2.3), we have only to prove that

$$(2.8) \qquad \frac{1}{n}(T + T^2 + \cdots + T^n)x \text{ converges strongly to } 0$$

for any $y \, \epsilon \, \bar{R}$. This may be performed in two steps:

1st case. $y \, \epsilon \, R$. Taking a point $z \, \epsilon \, (B)$ such that $y = z - T(z)$, we have

$$\left\| \frac{1}{n}(T + T^2 + \cdots + T^n)y \right\| = \left\| \frac{1}{n}(T + T^2 + \cdots + T^n)(z - T(z)) \right\|$$

$$= \left\| \frac{1}{n}(T - T^{n+1})z \right\| \leqq \frac{2C}{n} \|z\| \to 0.$$

2nd case. $y \, \epsilon \, \bar{R}$. For any $\epsilon > 0$ there exists a point $y' \, \epsilon \, R$ such that $\| y - y' \| < \epsilon$. For this y' we have

$$\left\| \frac{1}{n}(T + T^2 + \cdots + T^n)y \right\| \leqq \left\| \frac{1}{n}(T + T^2 + \cdots + T^n)y' \right\|$$

$$+ \left\| \frac{1}{n}(T + T^2 + \cdots + T^n)(y - y') \right\|$$

$$\leqq \left\| \frac{1}{n}(T + T^2 + \cdots + T^n)y' \right\| + C\epsilon.$$

Since the first term of the right hand side converges strongly to 0 as $n \to \infty$ (by the result obtained in the 1st case), we have

$$\overline{\lim_{n \to \infty}} \left\| \frac{1}{n}(T + T^2 + \cdots + T^n)y \right\| \leqq C\epsilon;$$

and, since $\epsilon > 0$ is arbitrary, we have (2.8). Thus (2.8) is proved for any $y \, \epsilon \, \bar{R}$ and hereby (2.3) is completely proved.

Let now T_1 be the transformation $x \to \bar{x}$. T_1 is clearly linear and bounded: $\| T_1 \| \leqq C$. Since (2.5) is always true, we have $TT_1 = T_1$. Thus $T^n T_1 = T_1$ and $\frac{1}{n}(T + T^2 + \cdots + T^n)T_1 = T_1$ for $n = 1, 2, \cdots$, and consequently $T_1^2 = T_1$. In order to prove the relation: $T_1 T = T_1$, we have only to start from the inequality:

$$\left\| \frac{1}{n}(T + T^2 + \cdots + T^n)T(x) - \frac{1}{n}(T + T^2 + \cdots + T^n)x \right\| \leqq \frac{2C}{n}\,\|x\| \to 0.$$

Since the terms on the left hand side converge strongly to $T_1 T(x)$ and $T_1(x)$ respectively, we have $T_1 T(x) = T_1(x)$ for any $x \,\epsilon\, (B)$.

Thus the proof of Theorem 1 is completed.

REMARK 1. T_1 is a projection operator which maps (B) on the proper space (B_1) of T belonging to the proper value 1.[9] That is, T_1 does not vanish identically if and only if 1 is a proper value of T, and $T_1(x) = x$ if and only if we have $T(x) = x$.

REMARK 2. In virtue of (2.1), (2.2) is surely satisfied if T is weakly completely continuous. Hence

COROLLARY 1. *Let T be a weakly completely continuous linear operation which maps a Banach space (B) into itself. If there exists a constant C such that $\| T^n \| \leqq C$ for $n = 1, 2, \cdots$, then the sequence of operations $\left\{ \frac{1}{n}(T + T^2 + \cdots + T^n) \right\}$ $(n = 1, 2, \cdots)$ converges strongly to a bounded linear operation T_1 which maps (B) into itself and which satisfies (2.4); i.e., mean ergodic theorem is valid in (B).*

The conditions of this Corollary are clearly satisfied if (B) is a Hilbert space and if T is a unitary operation in it. (Since Hilbert space itself is locally weakly compact.) Thus Theorem 1 is, in two respects, a generalization of the mean ergodic theorem of J. v. Neumann [1], [2]. (See also E. Hopf [1].) In the first place, our theorem is true in general complex Banach space, while the proof of J. v. Neumann is valid only in the case of a Hilbert space. In the second place, we do not assume that T is isometric, and the existence of the inverse operation T^{-1} is not required. The positiveness of the operation T is also not required. We have only to assume that the conditions (2.1) and (2.2) are satisfied. The first condition (2.1) is always satisfied with $C = 1$ in the case of ergodic theorems and Markoff's processes; and the second one (2.2) is satisfied in every weakly compact Banach space (for example, in regular Banach space and especially in (L^p) $(p > 1)$) for any bounded linear operation T. Consequently, mean ergodic theorem holds true in (L^p) $(p > 1)$ under the trivial condition (2.1). This result was also obtained independently by F. Riesz [1]. On the contrary, in the Banach space $(L) = (L^1)$, the condition (2.2) is not always satisfied for an arbitrary bounded linear operation T. In order that a bounded subset S of (L) be weakly compact, it is (necessary and) sufficient that the functions belonging to this subset S are uniformly integrable, that is, that there exists for any $\epsilon > 0$ a positive number $\delta > 0$ such that $\mathrm{mes}(E) < \delta$ implies

---

[9] It is to be noted that in general Banach space (B), there does not necessarily exist, for any closed linear subspace (B'), a projection operator (of norm 1 or of finite norm) which maps (B) in (B').

$\int_{E} |x(t)| \, dt < \epsilon$ for any $x(t) \, \epsilon \, S$. This condition is surely satisfied if there exists a function $x_0(t) \, \epsilon \, (L)$ such that $|x| \leqq x_0$ (that is, $|x(t)| \leqq x_0(t)$ almost everywhere in t) for any $x(t) \, \epsilon \, S$. Hence we have

COROLLARY 2. *Let T be a bounded linear operation which maps the Banach space (L) into itself. If there exists a constant C such that $\| T^n \| \leqq C$ for $n = 1, 2, \cdots$, and if there exists, for any $x(t) \, \epsilon \, (L)$, a function $x_0(t) \, \epsilon \, (L)$ such that*

$$T^n(x) \leqq x_0 \left(or, more generally, \frac{1}{n}(T + T^2 + \cdots + T^n)x \leqq x_0 \right) for \ n = 1, 2, \cdots,$$

then for any $x(t) \, \epsilon \, (L)$ the sequence $\left\{ \dfrac{1}{n}(T + T^2 + \cdots + T^n)x \right\}$ $(n = 1, 2, \cdots)$ *converges strongly to some function $\bar{x}(t) \, \epsilon \, (L)$; i.e., mean ergodic theorem is valid in (L).*

In order to prove this, we have only to notice that there exists also an x_0' such that $T^n(-x) \leqq x_0'$ i.e. $-x_0' \leqq T^n(x)$ for $n = 1, 2, \cdots$.

This theorem was proved by F. Riesz [2] and an analogous theorem was also stated by Garrett Birkhoff [1] for abstract (L)-spaces. The result of G. Birkhoff is more general in formulation, but it was pointed out that in essential the two theorems are equivalent.[10]

Moreover, if T is an integral operator with the probability density kernel $p(t, s)$:

$$x \to T(x) = y: \ y(s) = \int_{\Omega} x(t)p(t, s) \, dt$$

$$(p(t, s) \geqq 0 \quad \text{for any } t, s; \qquad \int_{\Omega} p(t, s) \, ds = 1 \quad \text{for any } t),$$

then a (necessary and) sufficient condition that T be weakly completely continuous is that $p(t, s)$ is uniformly (in t) integrable in s, that is, that there exists for any $\epsilon > 0$ a positive number $\delta > 0$ such that $\mathrm{mes}(E) < \delta$ implies $\int_{E} p(t, s) \, ds < \epsilon$ for any t. This condition was investigated by J. L. Doob [1] without being noticed that the linear operation T becomes weakly completely continuous under this condition. We have once[11] treated this case of the condition of J. L. Doob as an application of the mean ergodic theorem. But, on looking precisely into the detail of the fact, we have found that, in this case, the linear operation T satisfies even the (in some sense stronger) condition (K) of N. Kryloff-N. Bogolioùboff, and that the uniform ergodic theorem is true in this case. (Indeed, the condition (K) of N. Kryloff-N. Bogolioùboff follows from the condition (D) of W. Doeblin, which is weaker than that of J. L. Doob.)[12]

REMARK 3. Recently Garrett Birkhoff [2] proved that in every uniformly

---

[10] S. Kakutani [2].

[11] K. Yosida and S. Kakutani [1].

[12] K. Yosida [3].

convex Banach space mean ergodic theorem is valid for any bounded linear operation of norm not exceeding unity. This is also a generalization of the mean ergodic theorem of J. v. Neumann, since Hilbert space is uniformly convex. But this result of G Birkhoff is contained in our Theorem 1; for, every uniformly convex Banach space is regular and consequently locally weakly compact.[13] (It is, however, to be remarked that the proof of G. Birkhoff distinguishes itself by its extreme simplicity.)

REMARK 4. We can also prove the theorem of the following type:

COROLLARY 3. *Let S and T be two bounded linear operations which map a Banach space (B) into itself, and assume that these are commutative between themselves: $ST = TS$. If there exists a constant C such that $\| T^n \| \leq C$ and $\| S^n \| \leq C$ for $n = 1, 2, \cdots$, and if, for any $x \, \epsilon \, (B)$, the sequence $\{x_{mn}\}$ $(m, n = 1, 2, \cdots)$, where*

$$x_{mn} = \frac{1}{mn} (S + S^2 + \cdots + S^m)(T + T^2 + \cdots + T^n)x,$$

contains a subsequence $\{x_{m_\nu n_\nu}\}$ $(m_\nu \to \infty, n_\nu \to \infty)$ which converges weakly to a point $\bar{x}$ of (B), then we have $\lim\limits_{m, n \to \infty} x_{mn} = \bar{x}$ strongly for any $x \, \epsilon \, (B)$; and if we denote the transformation $x \to \bar{x}$ by U, then U is a bounded linear operation which maps (B) into itself and we have

$$US = SU = UT = TU = U^2 = U \quad and \quad \| U \| \leq C.$$

PROOF. It will be easily seen (exactly as in the proof of Theorem 1) that we have $S(\bar{x}) = T(\bar{x}) = \bar{x}$. Hence we have only to prove that the sequence $\left\{ \frac{1}{mn} (S + S^2 + \cdots + S^m)(T + T^2 + \cdots + T^n)(x - \bar{x}) \right\}$ converges strongly to 0 as $m \to \infty$ and $n \to \infty$ simultaneously. In order to prove this, denote by R the set of all the points $z \, \epsilon \, (B)$ of the form: $z = (x - S(x)) + (y - T(y))$ with $x, y \, \epsilon \, (B)$. R is clearly a linear subspace of (B) and $x - \bar{x}$ belongs to its strong (= weak) closure $\bar{R}$; for, we have

$$\left\{ I - \frac{1}{mn} (S + S^2 + \cdots + S^m)(T + T^2 + \cdots + T^n) \right\} x$$

$$= \left\{ I - \frac{1}{m} (S + S^2 + \cdots + S^m) \right\} x$$

$$+ \left\{ \frac{1}{m} (S + S^2 + \cdots + S^m) \right.$$

$$\left. - \frac{1}{mn} (S + S^2 + \cdots + S^m)(T + T^2 + \cdots + T^n) \right\} x$$

---

[13] S. Kakutani [3]. See also D. Milman [1]. B. J. Pettis [1].

$$= (I - S)\left(I + \frac{m-1}{m}S + \frac{m-2}{m}S^2 + \cdots + \frac{1}{m}S^{m-1}\right)x$$

$$+ \frac{1}{m}(S + S^2 + \cdots + S^m)(I - T)$$

$$\cdot\left(I + \frac{n-1}{n}T + \frac{n-2}{n}T^2 + \cdots + \frac{1}{n}T^{n-1}\right)x,$$

and the left hand side converges weakly to $x - \bar{x}$. Consequently, we have only to prove that $\left\{\frac{1}{mn}(S + S^2 + \cdots + S^m)(T + T^2 + \cdots + T^n)z\right\}$ converges strongly to 0 for any $z \,\epsilon\, \bar{R}$. This is clear if $z \,\epsilon\, R$, and the general case $z \,\epsilon\, \bar{R}$ may be treated in exactly the same manner as in the proof of Theorem 1.

Mean ergodic theorem of this type was first discussed by N. Wiener [1], [2]. N. Wiener treated the case when the inverse operators S^{-1} and T^{-1} both exist, and has also obtained the individual ergodic theorem of G. D. Birkhoff's type. Recently N. Dunford [1] has also announced the theorem of the same kind.

THEOREM 2. *Under the same assumptions as in Theorem 1, we have: for any complex number λ of absolute value 1, there exists a bounded linear operation T_λ, which maps (B) into itself, such that*

$$(2.9) \qquad \frac{1}{n}\left(\frac{T}{\lambda} + \frac{T^2}{\lambda^2} + \cdots + \frac{T^n}{\lambda^n}\right) \quad \textit{converges strongly to } T_\lambda,$$

$$(2.10) \qquad T_\lambda T = TT_\lambda = \lambda T_\lambda, \qquad T_\lambda^2 = T_\lambda, \qquad \|T_\lambda\| \leqq C,$$

$$(2.11) \qquad \lambda \neq \mu \quad \textit{implies} \quad T_\lambda T_\mu = 0,$$

$$(2.12) \qquad T_\lambda \neq 0 \textit{ if and only if } \lambda \textit{ is a proper value of } T.$$

In this case, T_λ is a projection operator which maps (B) on the proper space (B_λ) of T belonging to the proper value λ, and $T_\lambda(x) = x$ if and only if we have $T(x) = \lambda x$. If we further put $T' = T - \sum_{i=1}^{k} \lambda_i T_{\lambda_i}$ for any system $\{\lambda_1, \lambda_2, \cdots, \lambda_k\}$ of complex numbers of modulus 1 with $\lambda_i \neq \lambda_j \ (i \neq j)$ and $T_{\lambda_i} \neq 0$, then we have

$$(2.13) \qquad T_{\lambda_i}T' = T'T_{\lambda_i} = 0 \quad \textit{for} \quad i = 1, 2, \cdots, k,$$

$$(2.14) \qquad TT' = T'T = T'^2,$$

$$(2.15) \qquad T^n = \sum_{i=1}^{k} \lambda_i^n T_{\lambda_i} + T'^n \quad \textit{for } n = 1, 2, \cdots,$$

$$(2.16) \qquad \begin{cases} \lambda \textit{ is a proper value of } T' \textit{ if and only if it is a proper value of } T \textit{ and} \\ \lambda \neq \lambda_i \textit{ for } i = 1, 2, \cdots, k. \end{cases}$$

PROOF. (2.9) is clear from Theorem 1 if we consider T/λ instead of T. From this (2.10) follows at once. To prove (2.11), multiply (2.9) by T_μ from the right. Then we have, by (2.10),

$$\frac{1}{n}\left(\frac{\mu}{\lambda} + \frac{\mu^2}{\lambda^2} + \cdots + \frac{\mu^n}{\lambda^n}\right) T_\mu \quad \text{converges strongly to } T_\lambda T_\mu.$$

Since, in case $\lambda \neq \mu$, the left hand side converges strongly (even uniformly) to a zero operator, we have (2.11). (2.12) is clear from the Remark 1 to Theorem 1. (2.13), (2.14) and (2.15) may also be proved by easy calculations. To prove lastly (2.16), let λ, $|\lambda| \neq 0$, be a proper value of T and assume that $\lambda \neq \lambda_i$ for $i = 1, 2, \cdots, k$. Then there exists a point $x_0 \in (B)$ $(x_0 \neq 0)$ such that $T(x_0) = \lambda x_0$. Therefore, multiplying the both sides by T_{λ_i} from the left, we have, by (2.10), $\lambda_i T_{\lambda_i}(x_0) = \lambda T_{\lambda_i}(x_0)$ and consequently (since $\lambda \neq \lambda_i$ we have $T_{\lambda_i}(x_0) = 0$ for $i = 1, 2, \cdots, k$, so that we have $\lambda x_0 = T(x_0) = \sum_{i=1}^{k} \lambda_i T_{\lambda_i}(x_0) + T'(x_0) = T'(x_0)$. Hence λ is a proper value of T'. Conversely, let λ be a proper value of T'. Then there exists a point $x_0 \in (B)$ $(x_0 \neq 0)$ such that $T'(x_0) = \lambda x_0$. Since we have $T(x_0) = T\left(\frac{1}{\lambda} T'(x_0)\right) = \frac{1}{\lambda} TT'(x_0) = \frac{1}{\lambda} T'^2(x_0)$ (by (2.14)) $= \lambda x_0$, λ is also a proper value of T. In order to prove that $\lambda \neq \lambda_i$ for $i = 1, 2, \cdots, k$, assume that $\lambda = \lambda_i$ for some i. Then we have

$$\frac{1}{n}\left(\frac{T}{\lambda_i} + \frac{T^2}{\lambda_i^2} + \cdots + \frac{T^n}{\lambda_i^n}\right) x_0 = x_0 \quad \text{for } n = 1, 2, \cdots,$$

and consequently (by (2.9)) $T_{\lambda_i}(x_0) = x_0 \neq 0$, which leads to a contradiction, since $T_{\lambda_i}(x_0) = T_{\lambda_i}\left(\frac{1}{\lambda} T'(x_0)\right) = \frac{1}{\lambda} T_{\lambda_i} T'(x_0) = 0$ by (2.13).

The proof of Theorem 2 is hereby completed.

THEOREM 3. *In Theorems 1 and 2, the condition* (2.2) *may be substituted by the condition that T is quasi-weakly completely continuous.*

PROOF. We have only to prove that, under the condition (2.1) and the assumption that T is quasi-weakly completely continuous, there exists for any $x \in (B)$ a subsequence $\{x_{n_\nu}\}$ $(\nu = 1, 2, \cdots)$ of $\{x_n\}$ $(n = 1, 2, \cdots)$ which converges weakly to a point $\bar{x} \in (B)$. In order to prove this, put $T^m = V + D$ with $\|D\| = \alpha < 1$. Then we have

$$(2.17) \qquad\qquad T^{pm} = V_p + D^p,$$

where $V_p = T^{pm} - (T^m - V)^p$ is weakly completely continuous with $V_1 = V$ and $\|D^p\| \leq \alpha^p$.[14] Hence, putting

---

[14] If V and W are weakly completely continuous, then $V + W$, TV and VT are also weakly completely continuous for any bounded linear operation T. Thus the totality of all the weakly completely continuous linear operations constitutes an ideal in the ring of all bounded linear operations. Expanding $T^{pm} - (T^m - V)^p$, the term T^{pm} vanishes and there remain only those terms (finite in number) which contain at least one V-factor. Consequently V_p is weakly completely continuous.

It is to be remarked that the same is also true for strongly completely continuous linear operations.

$$x_{n,m} = \frac{1}{n}(T + T^2 + \cdots + T^m)x$$

for brevity's sake, we have for $n > pm$

$$x_n = \frac{1}{n}(T + T^2 + \cdots + T^n)x$$

$$= \frac{1}{n}(T + T^2 + \cdots + T^{pm})x + \frac{1}{n}T^{pm}(T + T^2 + \cdots + T^{n-pm})x$$

$$= x_{n,pm} + T^{pm}(x_{n,n-pm}) = x_{n,pm} + V_p(x_{n,n-pm}) + D^p(x_{n,n-pm}).$$

Since $\| x_{n,n-pm} \| \leq C \cdot \| x \|$ for any $n > pm$ (by (2.1)), there exists, for any p, a subsequence $\{n_\nu\}$ ($\nu = 1, 2, \cdots$) of $\{n\}$ ($n = 1, 2, \cdots$) such that $\{V_p(x_{n_\nu,n_\nu-pm})\}$ ($\nu = 1, 2, \cdots$) converges weakly to a point $\bar{x}_p \, \epsilon \, (B)$. Consequently we have

$$\overline{\lim_{\nu \to \infty}} \, |f(x_{n_\nu}) - f(\bar{x}_p)| \leq \overline{\lim_{\nu \to \infty}} \, |f(x_{n_\nu,pm})| + \overline{\lim_{\nu \to \infty}} \, |f(D^p(x_{n_\nu,n_\nu-pm}))|$$

for any bounded linear functional $f(x)$ defined on (B). Since

$$|f(x_{n_\nu,pm})| \leq \|f\| \cdot \| x_{n_\nu,pm} \| \leq \|f\| \cdot \frac{pm}{n_\nu} \cdot C \| x \| \to 0,$$

and

$$|f(D^p(x_{n_\nu,n_\nu-pm}))| \leq \|f\| \cdot \| D^p \| \cdot \| x_{n_\nu,n_\nu-pm} \| \leq \|f\| \cdot \alpha^p \cdot C \cdot \| x \|,$$

we have

$$(2.18) \qquad \overline{\lim_{\nu \to \infty}} \, |f(x_{n_\nu}) - f(\bar{x}_p)| \leq \|f\| \cdot \alpha^p \cdot C \| x \|$$

for any bounded linear functional $f(x)$ defined on (B).

Applying the diagonal method, we may assume that (2.18) holds for any bounded linear functional $f(x)$ defined on (B) and for $p = 1, 2, \cdots$. Of course, $\bar{x}_p$ may depend of p. Let us now consider the sequence $\{\bar{x}_p\}$ ($p = 1, 2, \cdots$). From (2.18) we have

$$|f(\bar{x}_p) - f(\bar{x}_q)| \leq (\alpha^p + \alpha^q) \cdot \|f\| \, C \, \| x \|,$$

and since $f(x)$ is an arbitrary bounded linear functional defined on (B), we have $\| \bar{x}_p - \bar{x}_q \| \leq (\alpha^p + \alpha^q) \cdot C \cdot \| x \|$ for any p and q. Consequently, (since $\alpha < 1$), $\{\bar{x}_p\}$ ($p = 1, 2, \cdots$) is a fundamental sequence in (B). If we put $\bar{x} = \lim_{p \to \infty} \bar{x}_p$, then we have, for any p, (from (2.18))

$$\overline{\lim_{\nu \to \infty}} \, |f(x_{n_\nu}) - f(\bar{x})| \leq \overline{\lim_{\nu \to \infty}} \, |f(x_{n_\nu}) - f(\bar{x}_p)| + |f(\bar{x}_p) - f(\bar{x})|$$

$$\leq \|f\| \cdot \alpha^p \cdot C \| x \| + \|f\| \cdot \| \bar{x}_p - \bar{x} \|.$$

Since p is arbitrary and since the right hand side tends to 0 as $p \to \infty$, we have $\lim_{\nu \to \infty} | f(x_{n_\nu}) - f(\bar{x}) | = 0$. Since $f(x)$ is an arbitrary bounded linear functional defined on (B), we have thus proved that the sequence $\{x_{n_\nu}\}$ $(\nu = 1, 2, \cdots)$ converges weakly to $\bar{x}$. The proof of Theorem 3 is hereby completed.

REMARK. All what we have obtained in Theorems 1 and 2 is also valid for quasi-weakly completely continuous linear operations. We can also prove several propositions concerning the conjugate operators of such bounded linear operations.[15] But we shall not go into the detail, which is not essential in the further discussions of this paper, and we shall proceed to the case of strongly completely continuous and quasi-strongly completely continuous linear operations.

CHAPTER 3. UNIFORM ERGODIC THEOREM IN BANACH SPACES[16]

(Generalization of the theorems of Fréchet-Kryloff-Bogoliouboff)

In this chapter we shall discuss strongly completely continuous and quasi-strongly completely continuous linear operations. By definition, a bounded linear operation T of a Banach space (B) into itself is called to be *strongly completely continuous* if it maps the unit sphere $\| x \| \leqq 1$ of (B) on a strongly compact set of (B). More generally, T is called to be *quasi-strongly completely continuous*, if there exist an integer m and a strongly completely continuous linear operation V, which maps (B) into itself, such that $\| T^m - V \| < 1$.

Since every strongly completely continuous linear operation is a fortiori weakly completely continuous, all what we have obtained in §2 is also valid for the case of strongly completely continuous or. quasi-strongly completely continuous linear operations. Moreover, we can show that in the present case, the *strong* convergence in Theorems 1, 2 and 3 may be substituted by the *uniform* one. These are, as will be seen from the proofs below, the consequences of a theorem of F. Riesz and its generalizations to the case of quasi-strongly completely continuous linear operations (Lemmas 3.1 and 3.2).

THEOREM 4. *Let T be a strongly completely continuous or quasi-strongly completely continuous linear operation which maps a Banach space (B) into itself. If there exists a constant C such that $\| T^n \| \leqq C$ for $n = 1, 2, \cdots$, then the proper values λ of T of modulus 1 (if such proper value ever exists) are finite in number and each of them is of finite multiplicity. Let us denote these by $\lambda_1, \lambda_2, \cdots, \lambda_k$. Then there exists a system of strongly completely continuous linear operations $T_{\lambda_1}, T_{\lambda_2}, \cdots, T_{\lambda_k}$ (each of them $\neq 0$), and a strongly completely continuous or quasi-strongly completely continuous linear operation S (which might vanish), such that T^n is decomposed into the form:*

$$(3.1) \qquad T^n = \sum_{i=1}^{k} \lambda_i^n T_{\lambda_i} + S^n, \qquad n = 1, 2, \cdots,$$

---

[15] K. Yosida and S. Kakutani [1], K. Yosida [4].

[16] K. Yosida [1], [4], S. Kakutani [1].

with

$$(3.2) \quad \begin{cases} TT_{\lambda_i} = T_{\lambda_i} T = \lambda_i T_{\lambda_i}, & T_{\lambda_i}^2 = T_{\lambda_i}, & T_{\lambda_i} T_{\lambda_j} = 0 \quad (i \ne j), \\ T_{\lambda_i} S = S T_{\lambda_i} = 0, & \| T_{\lambda_i} \| \le C, & i = 1, 2, \cdots, k, \end{cases}$$

and

$$(3.3) \quad \| S^n \| \le \frac{M}{(1 + \epsilon)^n}, \qquad n = 1, 2, \cdots,$$

where M and ϵ are positive constants which are independent of n.

REMARK. Theorem 4 is a generalization of the results of C. Visser, M. Fréchet and N. Kryloff-N. Bogolioùboff. C. Visser [1] discussed the iteration of strongly completely continuous linear operations in Hilbert space, and has obtained only strong convergence. M. Fréchet [1], [2] discussed the case of an integral operator T with bounded density kernel $p(t, s)$:

$$x \to T(x) = y: \quad y(s) = \int_\Omega x(t) p(t, s) \, dt,$$

and has obtained a considerably precise result after somewhat long calculations. It was, however, shown by Y. Mimura and the authors (Lemma 4.5 in §4) that, for such integral operation T with bounded density kernel, the second iterate T^2 already becomes strongly completely continuous as an operator which maps the Banach space (L) into itself. Hence the theorem of M. Fréchet may be considered as a special case of Theorem 4. N. Kryloff-N. Bogolioùboff [1], [2] treated the case of an integral operator with general probability kernel $P(t, E)$, which maps the Banach space $(\mathbf{M})$ into itself:

$$x \to T(x) = y: \quad y(E) = \int_\Omega x(dt) P(t, E),$$

and have obtained the same results (they announced these without proof) under the same conditions as in Theorem 4. It is, however, to be noted, that Theorem 4 holds true even for the general case when the Banach space (B) is arbitrary and when the positiveness of the operation T is not assumed. The more detailed discussions of such operations and their iterations, where the condition of positiveness plays its essential rôle, will be fully developed in §4.

In order to prove Theorem 4, we shall first prove two Lemmas which may be considered as generalizations of the results of F. Riesz [1].

LEMMA 3.1. *If T is quasi-strongly completely continuous, then the proper values λ of T do not accumulate to a point not interior to the unit circle $| \lambda | = 1$ of the complex plane.*

PROOF. Take an integer p so large that we have $\alpha^p < \frac{1}{4}$ where $T^m = V + D$, $\| D \| = \alpha < 1$. We have (exactly as we have obtained (2.17) in §2)

$$(3.4) \quad T^{pm} = V_p + D^p, \qquad \| D^p \| \le \alpha^p < \tfrac{1}{4},$$

where V_p is a strongly completely continuous linear operation which maps (B) into itself. From this it will be easily seen that we have only to consider the case when $m = 1$ and $\alpha < \frac{1}{4}$.

In order to prove Lemma 3.1 in this case, let $\{\lambda_n\}$ ($n = 1, 2, \cdots, \lambda_n \neq \lambda_m$ ($n \neq m$)) be a sequence of proper values of T which converges to a point λ_0 of modulus not smaller than 1. We shall deduce a contradiction from these assumptions. For this purpose, let $\{x_n\}$ ($n = 1, 2, \cdots$) be a proper element of T corresponding to the proper value λ_n : $T(x_n) = \lambda_n x_n$, $x_n \neq 0$, $n = 1, 2, \cdots$. We shall first prove that these x_n are mutually linearly independent. This may be done by mathematical induction. Let $x_1, x_2, \cdots, x_{n-1}$ be already linearly independent and assume that x_n depends linearly on $x_1, x_2, \cdots, x_{n-1}$: $x_n = \sum_{i=1}^{n-1} \alpha_i x_i$. Then we have

$$\sum_{i=1}^{n-1} \alpha_i(\lambda_n - \lambda_i)x_i = \lambda_n x_n - \sum_{i=1}^{n-1} \alpha_i \lambda_i x_i = T\left(x_n - \sum_{i=1}^{n-1} \alpha_i x_i\right) = 0.$$

Since $\{\alpha_i\}$ ($i = 1, 2, \cdots, n - 1$) do not vanish simultaneously, this is a contradiction to the assumption that $x_1, x_2, \cdots, x_{n-1}$ are mutually linearly independent.

Thus we have proved that x_n ($n = 1, 2, \cdots$) are mutually linearly independent. Consequently, the $(n - 1)$-dimensional subspace (X_{n-1}) of (B), which is spanned by $x_1, x_2, \cdots, x_{n-1}$, is a true subspace of the linear space (X_n) which is spanned by $x_1, x_2, \cdots, x_n$. Hence, by a well-known theorem of F. Riesz [1], there exists a sequence of points $\{y_n\}$ ($n = 1, 2, \cdots$) such that $y_n \in (X_n)$, $\| y_n \| = 1$ and $\| y_n - x \| > \frac{1}{2}$ for any $x \in (X_{n-1})$. Since each y_n is of the form: $y_n = \sum_{i=1}^{n} \alpha_i x_i$, we have

$$y_n - T\left(\frac{y_n}{\lambda_n}\right) = \sum_{i=1}^{n} \alpha_i x_i - \sum_{i=1}^{n} \frac{\alpha_i}{\lambda_n} T(x_i) = \sum_{i=1}^{n-1} \alpha_i\left(1 - \frac{\lambda_i}{\lambda_n}\right) x_i \in (X_{n-1})$$

and

$$T\left(\frac{y_m}{\lambda_m}\right) = \sum_{i=1}^{m} \frac{\alpha_i}{\lambda_m} T(x_i) = \sum_{i=1}^{m} \alpha_i \frac{\lambda_i}{\lambda_m} x_i \in (X_m)$$

for any n and m. Consequently, the trivial identity:

$$T\left(\frac{y_n}{\lambda_n}\right) - T\left(\frac{y_m}{\lambda_m}\right) = y_n - \left\{\left(y_n - T\left(\frac{y_n}{\lambda_n}\right)\right) + T\left(\frac{y_m}{\lambda_m}\right)\right\}$$

implies

$$(3.5) \qquad \left\| T\left(\frac{y_n}{\lambda_n}\right) - T\left(\frac{y_m}{\lambda_m}\right) \right\| > \tfrac{1}{2} \quad \text{for} \quad n > m.$$

On the other hand, since V is strongly completely continuous and since the sequence $\left\{\frac{y_n}{\lambda_n}\right\}$ ($n = 1, 2, \cdots$) is uniformly bounded, there exist two sequences of integers $\{n_\nu\}$ and $\{m_\nu\}$ ($n_\nu > m_\nu$, $\nu = 1, 2, \cdots$) such that $\lim_{\nu \to \infty} \left\| V\left(\frac{y_{n_\nu}}{\lambda_{n_\nu}}\right) - V\left(\frac{y_{m_\nu}}{\lambda_{m_\nu}}\right) \right\| = 0$. This is, however, a contradiction with (3.5), since we have by assumption

$$\left\| \left\{ T\left(\frac{y_{n_\nu}}{\lambda_{n_\nu}}\right) - T\left(\frac{y_{m_\nu}}{\lambda_{m_\nu}}\right) \right\} - \left\{ V\left(\frac{y_{n_\nu}}{\lambda_{n_\nu}}\right) - V\left(\frac{y_{m_\nu}}{\lambda_{m_\nu}}\right) \right\} \right\|$$

$$\leq \left\| T\left(\frac{y_{n_\nu}}{\lambda_{n_\nu}}\right) - V\left(\frac{y_{n_\nu}}{\lambda_{n_\nu}}\right) \right\| + \left\| T\left(\frac{y_{m_\nu}}{\lambda_{m_\nu}}\right) - V\left(\frac{y_{m_\nu}}{\lambda_{m_\nu}}\right) \right\|$$

$$= \left\| D\left(\frac{y_{n_\nu}}{\lambda_{n_\nu}}\right) \right\| + \left\| D\left(\frac{y_{m_\nu}}{\lambda_{m_\nu}}\right) \right\| \leq \alpha \left(\left\| \frac{y_{n_\nu}}{\lambda_{n_\nu}} \right\| + \left\| \frac{y_{m_\nu}}{\lambda_{m_\nu}} \right\| \right) \to \frac{2\alpha}{|\lambda_0|} < \tfrac{1}{2}.$$

Thus Lemma 3.1 is completely proved.

LEMMA 3.2. *If T is quasi-strongly completely continuous, then the proper space (B_λ) of T belonging to a proper value λ of T of modulus 1 is of finite dimension.*

PROOF. This may be done in exactly the same manner as in the proof of Lemma 3.1. We may again assume that $m = 1$ and $\alpha < \tfrac{1}{4}$.

If, for some proper value λ with $|\lambda| = 1$, the proper space (B_λ) is not of finite dimension, then there exists a sequence of points $x_n \in (B_\lambda)$ such that $\| x_n \| = 1$, $T(x_n) = \lambda x_n (n = 1, 2, \cdots)$ and $\| x_n - x_m \| > \tfrac{1}{2}$ for $n > m$.

On the other hand, since V is strongly completely continuous, there exist two sequences $\{n_\nu\}$ and $\{m_\nu\}$ $(n_\nu > m_\nu , \nu = 1, 2, \cdots)$ such that $\lim_{\nu\to\infty} \| V(x_{n_\nu}) - V(x_{m_\nu}) \| = 0$. This will, however, lead to a contradiction since we have

$$\tfrac{1}{2} < \| x_{n_\nu} - x_{m_\nu} \| = \| \lambda(x_{n_\nu} - x_{m_\nu}) \| = \| T(x_{n_\nu}) - T(x_{m_\nu}) \|$$

$$\leq \| V(x_{n_\nu}) - V(x_{m_\nu}) \| + \| D(x_{n_\nu}) - D(x_{m_\nu}) \|$$

$$\leq \| V(x_{n_\nu}) - V(x_{m_\nu}) \| + 2\alpha.$$

The proof of Lemma 3.2 is hereby completed.

PROOF OF THEOREM 4. Now, in order to prove Theorem 4, let us recall that Theorem 3 is valid in our case. From Lemma 3.1, there exists only a finite number of proper values λ of T on the unit circle $|\lambda| = 1$. Let us denote these proper values by $\lambda_1 , \lambda_2 , \cdots , \lambda_k$. Then there exists (by Theorem 2) for each λ_i a bounded linear operation T_{λ_i} , which maps (B) on the proper space (B_{λ_i}) of T belonging to the proper value λ_i . T_{λ_i} satisfies all the conditions of Theorem 2 and is strongly completely continuous since each (B_{λ_i}) is of finite dimension by Lemma 3.2. Let us put

$$(3.6) \qquad\qquad S = T - \sum_{i=1}^{k} \lambda_i T_{\lambda_i}$$

and consider the bounded linear operation S thus defined. We shall prove that this S is also quasi-strongly completely continuous and that its iterations S^n are uniformly bounded. This is an easy consequence of the relation $S^n = T^n - \sum_{i=1}^{k} \lambda_i^n T_{\lambda_i}$, which follows immediately from (3.6) and the result of Theorem 2. Indeed, there exists a constant $C' \equiv C + \sum_{i=1}^{k} \| T_{\lambda_i} \| \leq (k + 1)C$ such that $\| S^n \| \leq \| T^n \| + \sum_{i=1}^{k} \| T_{\lambda_i} \| \leq C'$ for $n = 1, 2, \cdots$, and also an integer m and a strongly completely continuous linear operation $V' \equiv V - \sum_{i=1}^{k} \lambda_i^m T_{\lambda_i}$ such that $\| S^m - V' \| = \| T^m - V \| < 1$.

Moreover, S has no proper value of modulus 1, so that in order to complete the proof of Theorem 4, we have only to prove the following

LEMMA 3.3. *If, in addition to the assumptions in Theorem 4, T has no proper value of modulus 1, then there exist two constants M and $\epsilon > 0$ such that*

$$(3.7) \qquad \| T^n \| \leq \frac{M}{(1 + \epsilon)^n} \quad \text{for} \quad n = 1, 2, \cdots.$$

PROOF. It is sufficient to prove the case $m = 1$. For, if Lemma 3.3 is true for the case $m = 1$, then there exist two constants M' and $\epsilon' > 0$ such that $\| T^{mn} \| \leq M'/(1 + \epsilon')^n$ for $n = 1, 2, \cdots$, and it is easy to deduce from this the existence of two constants M and $\epsilon > 0$ which satisfy (3.7).

Now, in order to prove Lemma 3.3 in the case $m = 1$, put $T = V + D$, where V is strongly completely continuous and $\| D \| = \alpha < 1$. By a well-known result, $I - \lambda D$ (I is the identical transformation) admits a unique inverse $I + \lambda D(\lambda) \equiv I + \sum_{n=1}^{\infty} \lambda^n D^n$ which is regular in $| \lambda | < \dfrac{1}{\alpha}$, and we have $(I + \lambda D(\lambda))$ $(I - \lambda T) = (I + \lambda D(\lambda))(I - \lambda V - \lambda D) = I - \lambda V - \lambda^2 D(\lambda) V \equiv I - V(\lambda)$, where $V(\lambda) \equiv \lambda V + \lambda^2 D(\lambda) V$ is regular in λ for $| \lambda | < \dfrac{1}{\alpha}$, and is strongly completely continuous for each λ with $| \lambda | < \dfrac{1}{\alpha}$. By Lemma 3.1, there exists a positive number $\eta > 0$ such that T has no proper value λ in $1 - \eta < | \lambda | < 1 + \eta$. Put $2\epsilon = \min \left(\dfrac{1}{\alpha} - 1, \eta \right)$ and consider the domain Δ: $1 - 2\epsilon < | \lambda | < 1 + 2\epsilon$. The equation $(I - V(\lambda))x = 0$ has no non-trivial solution $x \neq 0$ for each $\lambda \,\epsilon\, \Delta$. For, if there exists an $x_0 \neq 0$ with $(I - V(\lambda))x_0 = 0$, then we have $(I - \lambda T)x_0 = (I - \lambda D)(I - V(\lambda))x_0 = 0$, and this is a contradiction since T has no proper value in Δ. Consequently, since $V(\lambda)$ is strongly completely continuous for any $\lambda \,\epsilon\, \Delta$, there exists, by a theorem of F. Riesz, a unique inverse $I - K(\lambda)$ of $I - V(\lambda)$ for any $\lambda \,\epsilon\, \Delta$; and it will be easily seen that $I - \lambda R(\lambda) = (I - K(\lambda))(I + \lambda D(\lambda))$ is an inverse of $I - \lambda T$ for each $\lambda \,\epsilon\, \Delta$.

Thus we have proved that $I - \lambda T$ has an inverse for any $\lambda \,\epsilon\, \Delta$. Since it is clear from the uniform boundedness of $\{T^n\}$ ($n = 1, 2, \cdots$) that $I - \lambda T$ has an inverse $I + \sum_{n=1}^{\infty} \lambda^n T^n$ for any λ with $| \lambda | < 1$, we have thus proved the existence of an inverse $(I - \lambda T)^{-1}$ for any λ with $| \lambda | < 1 + 2\epsilon$. Consequently, by a theorem of M. Nagumo [1], $(I - \lambda T)^{-1}$ is regular in λ for $| \lambda | < 1 + 2\epsilon$ and the series of C. Neumann: $(I - \lambda T)^{-1} = I + \sum_{n=1}^{\infty} \lambda^n T^n$ converges in the uniform sense in $| \lambda | < 1 + 2\epsilon$. Hence there exists a constant M such that the inequality (3.7) is valid for $n = 1, 2, \cdots$.

The proof of Lemma 3.3 and herewith the proof of Theorem 4 are completed.

COROLLARY. *Under the same assumptions as in Theorem 4 we have:*

(i) *For any complex number λ with $| \lambda | = 1$, there exists a strongly completely continuous linear operation T_λ, which maps (B) into itself, such that*

$$\left\| \frac{1}{n}\left(\frac{T}{\lambda} + \frac{T^2}{\lambda^2} + \cdots + \frac{T^n}{\lambda^n} \right) - T_\lambda \right\| \leqq \frac{M}{n} \quad \text{for} \quad n = 1, 2, \cdots,$$

where M is a constant which is independent of n, and T_λ does not vanish identically if and only if λ is a proper value of T.

(ii) *In order that the sequence* $\{T^n\}$ $(n = 1, 2, \cdots)$ *converge in the uniform sense to a zero operation, it is necessary and sufficient that* T *has no proper value of absolute value* 1.

(iii) *In order that the sequence* $\{T^n\}$ $(n = 1, 2, \cdots)$ *converge in the uniform sense to a bounded linear operation* $T_1 \neq 0$, *it is necessary and sufficient that* 1 *is a proper value of* T *and that* T *has no other proper value of absolute value* 1.

In (ii) *and in* (iii), *if the sequence* $\{T^n\}$ $(n = 1, 2, \cdots)$ *converges in the uniform sense, then it is of the order of geometrical progression; that is, there exist a constant* M *and a positive number* ϵ *both independent of* n, *such that we have* $\| T^n \| \leqq M/(1 + \epsilon)^n$ *and* $\| T^n - T_1 \| \leqq M/(1 + \epsilon)^n$ *respectively for* $n = 1, 2, \cdots$

CHAPTER 4. MARKOFF'S PROCESS[17]

§4.1. Introduction. Let us denote by $P(t, E)$ the transition probability that a point t of the unit interval $\Omega = (0, 1)$ is transferred, by a simple Markoff's process, into a Borel set E of Ω after the elapse of a unit-time. We have always $P(t, E) \geqq 0$ and $P(t, \Omega) = 1$. We shall assume that $P(t, E)$ is completely additive for Borel sets E if t is fixed, and that $P(t, E)$ is Borel measurable in t if E is fixed. Then the transition probability $P^{(n)}(t, E)$ that a point $t \in \Omega$ is transferred into a Borel set E of Ω after the elapse of n unit-times is given recurrently by

$$P^{(n)}(t, E) = \int_\Omega P^{(n-1)}(t, ds)P(s, E) = \int_\Omega P(t, ds)P^{(n-1)}(s, E),$$

(4.1)
$$n = 2, 3, \cdots,$$

$$P^{(1)}(t, E) = P(t, E),$$

where the integration is of Radon-Stieltjes type.

Consider the complex Banach space **(M)** of the complex-valued completely additive set functions $x(E)$ defined for all Borel set E of Ω. For any $x(E) \in$ **(M)**, its norm is defined by: $\| x \| = $ total variation of $| x(E) |$ on Ω. Then we have

LEMMA 4.1. *The integral operator*

(4.2)
$$x \to T(x) = y: \quad y(E) = \int_\Omega x(dt)P(t, E)$$

is a bounded linear operation which maps the Banach space **(M)** *into itself and* $\| T \| = 1$.

---

[17] K. Yosida and S. Kakutani [1], K. Yosida [3], S. Kakutani [4].

On the other hand, if we consider the complex Banach space (M^*) of all the complex-valued bounded Borel measurable functions $x(t)$ defined on Ω, with $\| x \| = \underset{t \, \epsilon \, \Omega}{\text{l.u.b.}} | x(t) |$ as its norm, then we have

LEMMA 4.2. *The integral operator*

$$(4.3) \qquad x \to \overline{T}(x) = y: \quad y(t) = \int_\Omega P(t, ds)x(s)$$

is a bounded linear operation which maps the Banach space (M^*) into itself and $\| T \| = 1$.

LEMMA 4.3. *For any $x(E) \, \epsilon \, (\mathbf{M})$ and $y(t) \, \epsilon \, (M^*)$, we have*

$$(4.4) \qquad \int_\Omega x(dt) \left(\int_\Omega P(t, ds)y(s) \right) = \int_\Omega x_1(ds)y(s),$$

where $x_1(E) = \displaystyle\int_\Omega x(dt)P(t, E)$.

These three Lemmas are clear from the properties of $P(t, E)$.

REMARK. By virtue of Lemmas 4.1 and 4.2, $P^{(n)}(t, E)$ can be defined recurrently by (4.1). Hence $P^{(n)}(t, E)$ is completely additive for Borel sets E if t is fixed, and $P^{(n)}(t, E)$ is Borel measurable in t if E is fixed. Clearly we have

$$(4.5) \qquad P^{(n)}(t, E) \geqq 0, \qquad P^{(n)}(t, \Omega) = 1, \qquad n = 1, 2, \cdots.$$

Moreover, by the repeated use of Lemma 4.3, we have

$$(4.6) \qquad P^{(m+n)}(t, E) = \int_\Omega P^{(m)}(t, ds)P^{(n)}(s, E)$$

for any m and n, and it will be easily seen that the iterated operators T^n and $\overline{T}^n$ are defined by the kernel $P^{(n)}(t, E)$. We have clearly

$$(4.7) \qquad \| T^n \| = \| \overline{T}^n \| = 1, \qquad n = 1, 2, \cdots.$$

It is now the purpose of this chapter to investigate the asymptotic behavior of the sequence $\{P^{(n)}(t, E)\}$ for large n. We shall treat this problem by considering $P^{(n)}(t, E)$ as the kernel of the integral operators T^n and $\overline{T}^n$. Since the Banach spaces $(\mathbf{M})$ and (M^*) are not conjugate to each other, these two operators T^n and $\overline{T}^n$ are not the conjugate operators to each other in the strict sense which was given by S. Banach [1]. But in essential, these play the same rôle.

Our problem is not quite easy if we have no further assumptions on the kernel $P(t, E)$. Our fundamental assumptions are the conditions (D) and (K) which were stated in §1. The first condition (D), which is due to W. Doeblin [1] is more general than those given by B. Hostinsky, M. Fréchet and J. L. Doob. The second one (K) is due to N. Kryloff-N. Bogoliouboff [1], [2] and was introduced by them independently of W. Doeblin. We shall show (§4.7) that the condition (D) implies (K), and under the condition (K) all the results of W. Doeblin will be obtained in a more precise form (§§4.2–4.6). Our principal results are stated in Theorems 5–12. Theorem 5 is a restatement of the uniform ergodic theorem (Theorem 4 of §3), and this is a starting point of all the discussions of this chapter. Among other theorems, Theorem 6 is to be noticed. The formula (4.23) will show how the notions of the Banach spaces $(\mathbf{M})$ and (M^*) are essential in our problems. As a corollary to Theorem 6, we shall obtain the new notion of ergodic parts (Theorem 7); and the decomposition of Ω into ergodic kernels ($=$ "ensembles finals" of W. Doeblin) and the dissipative part (Theorem 8) is also a direct consequence of Theorem 6. Moreover, using the fact that under the condition (K) each proper value of T of modulus 1 is a root of unity (Theorem 9), the subdivision of the ergodic parts (and ergodic kernels) into cyclic parts will be easily obtained (Theorem 11 and its Corollary).

The classical results concerning the Markoff's process with a finite number of possible states may be easily obtained from our Theorems. In order to obtain these results, we have only to take a kernel $P(t, E)$ of the special type. This will be carried out in §4.8. In this way, the hitherto known results concerning the Markoff's process with a finite number or a continuum of possible states are obtained in a more precise form by a unified method.

§4.2. Spectral decomposition of $P^{(n)}(t, E)$ under the condition (K).

Theorem 5.   Under the condition (K), $P^{(n)}(t, E)$ *is decomposed into the form:*

$$(4.8) \qquad P^{(n)}(t, E) = \sum_{i=1}^{k} \lambda_i^n P_{\lambda_i}(t, E) + S^{(n)}(t, E), \qquad n = 1, 2, \cdots,$$

where $\{\lambda_i\}$ $(i = 1, 2, \cdots, k)$ *are the proper values of* T *of modulus 1, and*

$$(4.9) \qquad \underset{t \,\epsilon\, \Omega, E \subset \Omega}{\mathrm{l.u.b.}} \left| \frac{1}{n} \sum_{m=1}^{n} \frac{P^{(m)}(t, E)}{\lambda_i^m} - P_{\lambda_i}(t, E) \right| \leqq \frac{M}{n},$$

$$(4.10) \qquad \int_{\Omega} P^{(n)}(t, ds) P_{\lambda_i}(s, E) = \int_{\Omega} P_{\lambda_i}(t, ds) P^{(n)}(s, E) = \lambda_i^n P_{\lambda_i}(t, E),$$

$$(4.11) \qquad \int_{\Omega} P_{\lambda_i}(t, ds) P_{\lambda_j}(s, E) = P_{\lambda_i}(t, E) \quad \text{or 0 according as } i = j \text{ or } i \neq j,$$

$$(4.12) \qquad \int_{\Omega} P_{\lambda_i}(t, ds) S(s, E) = \int_{\Omega} S(t, ds) P_{\lambda_i}(s, E) = 0,$$

$$(4.13) \qquad \underset{t \,\epsilon\, \Omega, E \subset \Omega}{\mathrm{l.u.b.}} |S^{(n)}(t, E)| \leqq \frac{M}{(1 + \epsilon)^n},$$

$i = 1, 2, \cdots, k;\ n = 1, 2, \cdots,$ *where* M *and* ϵ *are positive constants which are independent of* n.

PROOF. This theorem follows directly from Theorem 4 (uniform ergodic theorem) of §3. We have only to notice that, by Theorem 4, only the decomposition of the operators T^n is given and that the decomposition of the kernels

$P^{(n)}(t, E)$ is not yet obtained. But, since the convergence $\lim\limits_{n \to \infty} \frac{1}{n} \sum_{m=1}^{n} T^m / \lambda_i^m =$ T_{λ_i} is uniform, the decomposition of the kernels is simultaneously obtained.

Remark. The uniform limit

$$\lim_{n \to \infty} \frac{1}{n} \sum_{m=1}^{n} \frac{P^{(m)}(t, E)}{\lambda^m} = P_\lambda(t, E)$$

exists for any λ with $|\lambda| = 1$. By putting $\lambda = 1$ and remembering (4.5), we see that $P_1(t, E)$ is not identically zero. Hence $\lambda = 1$ is a proper value of T. If we put $\lambda_1 = 1$, then the integral operator T_1 defined by the kernel

$$(4.14) \qquad P_1(t, E) = \lim_{n \to \infty} \frac{1}{n} \sum_{m=1}^{n} P^{(m)}(t, E) \qquad \text{(uniform limit)}$$

is a projection operator, which maps the Banach space (**M**) on the proper space of T belonging to the proper value 1. More precisely, we have

$$(4.15) \qquad \left\| \frac{T + T^2 + \cdots + T^n}{n} - T_1 \right\| \leq \frac{M}{n}, \qquad n = 1, 2, \cdots,$$

with a positive constant M. This is a result of N. Kryloff-N. Bogolioùboff. (The existence of the mean sojourn.) Moreover, the integral operators T_{λ_i} $(i = 1, 2, \cdots, k)$ defined by the kernels $P_{\lambda_i}(t, E)$ are all strongly completely continuous.

§4.3. Structure of the kernel $P_1(t, E)$. By (4.5) and (4.14). we have

$$(4.16) \qquad P_1(t, E) \geq 0, \qquad P_1(t, \Omega) \equiv 1.$$

and the proper value equation in (**M**):

$$(4.17) \qquad T(x) = x: \quad x(E) = \int_\Omega x(dt) P(t, E)$$

admits a non-trivial solution $x \neq 0$. In fact, by (4.10) of Theorem 5, for any $t_0 \epsilon \Omega$. $P_1(t_0, E)$ is a solution of (4.17):

$$(4.18) \qquad P_1(t_0, E) = \int_\Omega P_1(t_0, dt) P(t, E),$$

and $P_1(t, E)$ is not identically zero.

In this section, we shall study the general form of such a solution of (4.17). and using the results thus obtained, the structure of the kernel $P_1(t, E)$ will be determined.

In order to make our discussions clearer, we shall make use of some elementary notions from the theory of semi-ordered Banach spaces. A completely additive real-valued set function $x(E) \epsilon$ (**M**) is called to be positive and is denoted by $x \geq 0$, if we have $x(E) \geq 0$ for any Borel set E of Ω; and for any pair of real-valued set functions $x(E), y(E) \epsilon$ (**M**), denote by $x \geq y$ the relation $x - y \geq 0$.

Then the relation $x \geqq y$ determines a semi-ordering of the real Banach space $(\mathbf{M})$.

For any real-valued set function $x(E) \; \epsilon \; (\mathbf{M})$, its total variation $\tilde{x}(E)$ is a positive element of $(\mathbf{M})$, and if we put

$$x_+(E) = \tfrac{1}{2}(\tilde{x}(E) + x(E)), \qquad x_-(E) = \tfrac{1}{2}(\tilde{x}(E) - x(E)),$$

then these are also the positive elements of $(\mathbf{M})$. These are called the positive and the negative variations of x (on E) respectively. We have $x = x_+ - x_-$, $\tilde{x} = x_+ + x_-$ and, from the trivial relation: $(-x)_+ = x_-$, we have $x - (x - y)_+ = y - (y - x)_+$. This last common element is called the minimum of two elements x and y, and is denoted by $x \wedge y$.[18] The special case $x \wedge y = 0$ requires our special attention. It is to be remarked that for two positive elements x and y of $(\mathbf{M})$, $x \wedge y = 0$ is equivalent to the condition that there exist two disjoint Borel sets E_1 and E_2 of Ω such that $x(E_1) = x(\Omega)$ and $y(E_2) = y(\Omega)$. This fact is needed in the following discussions. It is also to be noted that we have always $x_+ \wedge x_- = 0$, $(x - y)_+ \wedge (y - x)_+ = 0$ and

$$(4.19) \qquad (x - (x \wedge y)) \wedge (y - (x \wedge y)) = 0.$$

Lemma 4.4. *If x and y are two real-valued solutions of* (4.17), *then $\tilde{x}$, x_+, x_- and $x \wedge y$ are also solutions of* (4.17).

Proof. It is sufficient to prove this for $\tilde{x}$. Since $P(t, E) \geqq 0$ for any Borel set $E \subset \Omega$, we have

$$\tilde{x}(E) \leqq \int_\Omega \tilde{x}(dt) P(t, E),$$

and, since $P(t, \Omega) \equiv 1$, here must stand the equal sign.

Lemma 4.5. *There exists a system of completely additive set functions $\{x_\alpha(E)\}$ $(\alpha = 1, 2, \cdots, l) \; \epsilon \; (\mathbf{M})$, with the properties:*

$$(4.20) \qquad T(x_\alpha) = x_\alpha, \qquad x_\alpha \geqq 0, \qquad x_\alpha(\Omega) = 1, \qquad x_\alpha \wedge x_\beta = 0 \qquad (\alpha \neq \beta),$$

such that any $x(E) \; \epsilon \; (\mathbf{M})$ which satisfies

$$(4.21) \qquad T(x) = x, \qquad x \geqq 0, \qquad x(\Omega) = 1,$$

is uniquely expressed as a linear combination:

$$(4.22) \qquad x(E) = \sum_{\alpha=1}^{l} c_\alpha x_\alpha(E), \qquad c_\alpha \geqq 0, \sum_{\alpha=1}^{l} c_\alpha = 1.$$

Proof. Let l be the maximum number of elements $x_1, x_2, \cdots, x_l \; \epsilon \; (\mathbf{M})$ which satisfy (4.20). The existence of such an l is clear from the quasi-strong complete continuity of the operation T. For, such $x_1, x_2, \cdots, x_l$ are clearly mutually linearly independent and the proper space of T belonging to the proper value 1 is of finite dimension by Lemma 3.2 of §3.

---

[18] Indeed, it will be easily seen that $x \wedge y$ is the minimum of x and y in the sense of lattice, that is, $x \wedge y \leqq x$, $x \wedge y \leqq y$, and for any $z \; \epsilon \; (M)$ with $z \leqq x$, $z \leqq y$, we have $z \leqq x \wedge y$.

We shall prove that this system $\{x_\alpha(E)\}$ ($\alpha = 1, 2, \cdots, l$) is just the required one. In order to show this, let $x(E)$ be an arbitrary element of (**M**) which satisfies (4.21). We shall show that $x(E)$ is a linear combination of $x_1, x_2, \cdots, x_l$. For this purpose, consider the minimum $x'_\alpha = x \wedge x_\alpha$ of x and x_α for $\alpha = 1, 2, \cdots, l$. By Lemma 4.4, x'_α is also a solution of (4.17). We shall first show that each x'_α is a constant multiple of $x_\alpha : x'_\alpha = c_\alpha x_\alpha$. Indeed, if this is not true for some α, then x_α and $x''_\alpha \equiv x'_\alpha / \| x'_\alpha \|$ are not equal, and consequently $(x_\alpha - x''_\alpha)_+$ and $(x''_\alpha - x_\alpha)_+$ are both $\neq 0$. Again by Lemma 4.4, these are also the solutions of (4.17). Hence, if we put $x_{\alpha 1} = (x_\alpha - x''_\alpha)_+ / \| (x_\alpha - x''_\alpha)_+ \|$, $x_{\alpha 2} = (x''_\alpha - x_\alpha)_+ / \| (x''_\alpha - x_\alpha)_+ \|$, then the system of $l + 1$ elements $x_1, x_2, \cdots, x_{\alpha-1}, x_{\alpha 1}, x_{\alpha 2}, x_{\alpha+1}, \cdots, x_l$ clearly satisfies (4.20), and this is a contradiction to the definition of l.

Thus we have proved that each $x'_\alpha = x \wedge x_\alpha$ is expressed in the form: $x'_\alpha = c_\alpha x_\alpha$, where c_α is a real number with $0 \leq c_\alpha \leq 1$. Next we shall prove that we have $x = \sum_{\alpha=1}^{l} x'_\alpha \equiv \sum_{\alpha=1}^{l} c_\alpha x_\alpha$. For this purpose, we shall show that $x' = x - \sum_{\alpha=1}^{l} x'_\alpha$ satisfies $x' \wedge x_\alpha = 0$ for $\alpha = 1, 2, \cdots, l$. In order to prove this, it is sufficient to show that we have $(x - x'_\alpha) \wedge x_\alpha = 0$ for $\alpha = 1, 2, \cdots, l$. This is, however, clear if $c_\alpha = 1$; for, $x'_\alpha \equiv x \wedge x_\alpha = x_\alpha$ implies $x = x_\alpha$ (since $x(\Omega) = x_\alpha(\Omega) = 1$), and $x' = x - x'_\alpha = 0$. And, in case $c_\alpha < 1$, this follows from $(1 - c_\alpha)((x - x'_\alpha) \wedge x_\alpha) \leq (x - x'_\alpha) \wedge (1 - c_\alpha)x_\alpha = (x - x'_\alpha) \wedge (x_\alpha - x'_\alpha) = 0$ (by (4.19)). Thus $x' \wedge x_\alpha = 0$ is proved for $\alpha = 1, 2, \cdots, l$. Consequently, if we have $x' \neq 0$, then the system of $l + 1$ elements $x'/\| x' \|, x_1, x_2, \cdots, x_l$ will again satisfy (4.20), and this is also a contradiction.

Thus we have proved that we have $x' = 0$, and consequently $x = \sum_{\alpha=1}^{l} x'_\alpha = \sum_{\alpha=1}^{l} c_\alpha x_\alpha$. Since the uniqueness of the expression and the condition $\sum_{\alpha=1}^{l} c_\alpha = 1$ are both clear, the proof of Lemma 4.5 is hereby completed.

COROLLARY. *$\{x_\alpha(E)\}$ ($\alpha = 1, 2, \cdots, l$) is a base of all the solutions of the proper value equation (4.17); i.e., any $x \in$ (**M**) which satisfies (4.17) is uniquely expressed as a linear combination of $x_1, x_2, \cdots, x_l$.*

PROOF. Clear from Lemmas 4.4 and 4.5.

THEOREM 6. *$P_1(t, E)$ is expressible in the form:*

$$(4.23) \qquad P_1(t, E) = \sum_{\alpha=1}^{l} y_\alpha(t) x_\alpha(E),$$

*where $\{x_\alpha(E)\}$ ($\alpha = 1, 2, \cdots, l$) is the system of completely additive set functions $\in$ (**M**), which was defined in Lemma 4.5, and $\{y_\alpha(t)\}$ ($\alpha = 1, 2, \cdots, l$) is a system of bounded Borel measurable functions $\in$ (M^*), which satisfy*

$$(4.24) \qquad T(y_\alpha) = y_\alpha, \qquad y_\alpha(t) \geq 0, \qquad \sum_{\alpha=1}^{l} y_\alpha(t) = 1,$$

and

$$(4.25) \qquad \int_\Omega x_\alpha(dt) y_\beta(t) = 1 \quad \text{or } 0 \text{ according as } \alpha = \beta \text{ or } \alpha \neq \beta.$$

Moreover, $\{y_\alpha(t)\}$ $(\alpha = 1, 2, \cdots, l)$ is a base of all the solutions of the proper value equation in (M^*):

$$(4.26) \qquad \overline{T}(y) = y: \quad y(t) = \int_\Omega P(t, ds)y(s),$$

and any solution of (4.26) with $y(t) \geq 0$ can be expressed uniquely in the form:

$$(4.27) \qquad y(t) = \sum_{\alpha=1}^{l} c_\alpha y_\alpha(t)$$

with non-negative constants c_α $(\alpha = 1, 2, \cdots, l)$.

PROOF. Since $P_1(t, E)$ is a solution of (4.17) for any $t \, \epsilon \, \Omega$, the expression (4.23) follows directly from Lemma 4.5. If we take a Borel set E'_α such that $x_\alpha(E'_\alpha) = 1$ and $x_\beta(E'_\alpha) = 0$ for any $\beta \neq \alpha$ (the existence of such E'_α follows from the fact that $x_\alpha(\Omega) = 1$ and $x_\beta \wedge x_\alpha = 0$ for any $\beta \neq \alpha$), then (4.23) becomes $P_1(t, E'_\alpha) = y_\alpha(t)$. Hence each $y_\alpha(t)$ is a bounded Borel measurable function of t. We shall next prove (4.24). Since the second and the third relation of (4.24) are clear from (4.16), we have only to prove the first one. From (4.10) of Theorem 5 we have

$$\int_\Omega P(t, ds)P_1(s, E) = P_1(t, E),$$

or, by (4.23),

$$\sum_{\alpha=1}^{l} \left(\int_\Omega P(t, ds)y_\alpha(s) \right) x_\alpha(E) = \sum_{\alpha=1}^{l} y_\alpha(t)x_\alpha(E),$$

and, if we put $E = E'_\alpha$, then we have

$$\int_\Omega P(t, ds)y_\alpha(s) = y_\alpha(t).$$

Thus (4.24) is proved. In order to prove (4.25), start from the relation $T(x_\alpha) = x_\alpha$. Since $T(x) = x$ is equivalent to $T_1(x) = x$, we have $T_1(x_\alpha) = x_\alpha$, or

$$\int_\Omega x_\alpha(dt)P_1(t, E) = x_\alpha(E),$$

and, putting $E = E'_\beta$, we have the required relation (4.25).

Thus the first part of the theorem is proved. The second part may be proved as follows: Let $y(t)$ be a solution of (4.26). From (4.26) we have (exactly as in the preceding case)

$$y(t) = \int_\Omega P_1(t, ds)y(s),$$

and, by (4.23),

$$y(t) = \sum_{\alpha=1}^{l} y_\alpha(t) \left(\int_\Omega x_\alpha(ds)y(s) \right) = \sum_{\alpha=1}^{l} c_\alpha y_\alpha(t)$$

with $c_\alpha = \int_\Omega x_\alpha(ds)y(s)$. Since $y(s) \geqq 0$ implies $c_\alpha \geqq 0$, the proof of Theorem 6 is hereby completed.

§4.4. Ergodic decomposition of Ω under the condition (K). Let us consider the sets $\bar{E}_\alpha = E_t[y_\alpha(t) = 1]$, $\alpha \doteq 1, 2, \cdots, l$. By (4.24), these are mutually disjoint Borel sets.

THEOREM 7. *There exists a system of mutually disjoint Borel sets* $\{\bar{E}_\alpha\}$ $(\alpha = 1, 2, \cdots, l)$, *such that*

$$(4.28) \qquad x_\alpha(\bar{E}_\beta) = 1 \text{ or } 0 \text{ according as } \alpha = \beta \text{ or } \alpha \neq \beta,$$

$$(4.29) \qquad P(t, \bar{E}_\alpha) = 1, \qquad t \in \bar{E}_\alpha,$$

$$(4.30) \qquad \operatorname*{l.u.b.}_{t \in \bar{E}_\alpha, E \subset \Omega} \left| \frac{1}{n} \sum_{m=1}^{n} P^{(m)}(t, E) - x_\alpha(E) \right| \leqq \frac{M}{n}, \qquad n = 1, 2, \cdots,$$

where M is a constant which is independent of n.

REMARK. (4.29) means that, for any α, each point $t \in \bar{E}_\alpha$ is transferred by the Markoff's process $P(t, E)$ inside $\bar{E}_\alpha$, and (4.30) means that the uniform limit

$$\lim_{n \to \infty} \frac{1}{n} \sum_{m=1}^{n} P^{(m)}(t, E) = P_1(t, E)$$

is independent of the initial point $t \in \bar{E}_\alpha$. Because of these properties, $\bar{E}_\alpha$ $(\alpha = 1, 2, \cdots, l)$ will be called the *ergodic parts* of Ω.

PROOF OF THEOREM 7. (4.28) and (4.29) are the consequences of (4.25) and (4.26) respectively. In order to show this, we have only to prove

LEMMA 4.6. *If $x(E)$ is a completely additive real-valued set function ϵ (M) such that $x(\Omega) = 1$ and $x(E) \geqq 0$ for any Borel set $E \subset \Omega$, and if $y(t)$ is a bounded Borel measurable real-valued function ϵ (M*) such that $0 \leqq y(t) \leqq 1$ for any $t \in \Omega$, then*

$$\int_\Omega x(dt)y(t) = 1$$

implies ·

$$x(E_0) = 1, \qquad \text{where } E_0 = \operatorname*{E}_t[y(t) = 1].$$

PROOF. If we put

$$E_n = \operatorname*{E}_t\left[1 - \frac{1}{n} \leqq y(t) < 1 - \frac{1}{n+1}\right],$$

then we have $\Omega = E_0 + \sum_{n=1}^{\infty} E_n$ and

$$1 = \int_\Omega x(dt)y(t) = \int_{E_0} x(dt)y(t) + \sum_{n=1}^\infty \int_{E_n} x(dt)y(t)$$

$$\leq x(E_0) + \sum_{n=1}^\infty \left(1 - \frac{1}{n+1}\right) x(E_n)$$

$$\leq x(E_0) + \sum_{n=1}^\infty x(E_n) = x(\Omega) = 1.$$

Since the equality holds only if we have $x(E_n) = 0$ for $n = 1, 2, \cdots$, we have $x(E_0) = 1$. Thus Lemma 4.6 and hereby (4.28) and (4.29) are proved.

(4.30) is a restatement of the relation (4.9) of Theorem 5 $(i = 1, \lambda_1 = 1)$:

$$\operatorname*{l.u.b.}_{t\,\epsilon\,\Omega, E \subset \Omega} \left| \frac{1}{n} \sum_{m=1}^n P^{(m)}(t, E) - P_1(t, E) \right| \leq \frac{M}{n}, \qquad n = 1, 2, \cdots,$$

if we only observe that we have

$$P_1(t, E) = x_\alpha(E) \quad \text{for} \quad t \,\epsilon\, \bar{E}_\alpha .$$

As is easily seen, $\bar{E}_\alpha$ is not necessarily the set of the smallest measure with the property (4.28). Indeed, there might exist a Borel set $E \subset \bar{E}_\alpha$ such that mes $(E) <$ mes $(\bar{E}_\alpha)$ and $x_\alpha(E) = 1$. If we denote by $E_\alpha(\subset \bar{E}_\alpha)$ the Borel set of the smallest measure among those which satisfy (4.28), then E_α is determined up to a set of measure zero and $E \subset E_\alpha$, mes $(E) > 0$ imply $x_\alpha(E) > 0$. We shall show that, if we take suitably the sets $E_\alpha \subset \bar{E}_\alpha$ $(\alpha = 1, 2, \cdots, l)$, then the following theorem is true:

THEOREM 8. *The Borel sets E_α $(\alpha = 1, 2, \cdots, l)$ and $\Delta = \Omega - \sum_{\alpha=1}^l E_\alpha$ satisfy*

(4.31)
$$P_1(t, E_\alpha) = 1, \qquad t \,\epsilon\, \bar{E}_\alpha ,$$

(4.32)
$$P(t, E_\alpha) = 1, \qquad t \,\epsilon\, E_\alpha ,$$

(4.33)
$$\begin{cases} \textit{for any } t \,\epsilon\, \bar{E}_\alpha \textit{ and } E \subset E_\alpha, \textit{ mes } (E) > 0 \textit{ implies} \\ P_1(t, E) > 0 \textit{ and consequently there exists a posi-} \\ \textit{tive integer } n = n(t, E) \textit{ such that } P^{(n)}(t, E) > 0, \end{cases}$$

(4.34)
$$\operatorname*{l.u.b.}_{t\,\epsilon\,\Omega} \frac{1}{n} \sum_{m=1}^n P^{(m)}(t, \Delta) \leq \frac{M}{n}, \qquad n = 1, 2, \cdots,$$

where M is a constant which is independent of n.

REMARK. (4.31) means that, for any α, each point $t \,\epsilon\, \bar{E}_\alpha$ is transferred finally into E_α (in the sense of arithmetic mean), and (4.32) means that each point $t \,\epsilon\, E_\alpha$ is transferred inside E_α. Moreover, by (4.33), E_α is indecomposable into two sets with the property (4.32). In Theorem 12, we shall obtain a more precise result than (4.34):

$$\operatorname*{l.u.b.}_{t\,\epsilon\,\Omega} P^{(n)}(t, \Delta) \leq \frac{M}{(1 + \epsilon)^n}, \qquad n = 1, 2, \cdots,$$

with positive constants M and ϵ.

Because of these properties, E_α $(\alpha = 1, 2, \cdots, l)$ and Ω will be called the *ergodic kernels* and the *dissipative part* of Ω respectively. It is to be noted that the ergodic kernels are only determined up to a set of measure zero (even under the condition (4.32)), while the ergodic parts $\bar{E}_\alpha$ are strictly determined by $\bar{E}_\alpha = E[y(t) = 1]$.

PROOF OF THEOREM 8. Let $E_\alpha^0 (\subset \bar{E}_\alpha)$ be the Borel set of the smallest measure among those which satisfy (4.28). E_α^0 clearly satisfies (4.31) and (4.33) (since $P_1(t, E) = x_\alpha(E)$ for $t \epsilon \bar{E}_\alpha$), but (4.32) is not always automatically satisfied for any $t \epsilon E_\alpha^0$. In order to obtain the required Borel set E_α, we shall construct a sequence of Borel sets $E_\alpha^0 \supset E_\alpha^1 \supset E_\alpha^2 \cdots \supset E_\alpha^n \supset \cdots$ by mathematical induction. Let E_α^n be already defined and assume that we have $x_\alpha(E_\alpha^n) = 1$. Then we define E_α^{n+1} as the set of all $t \epsilon E_\alpha^n$ which satisfies $P(t, E_\alpha^n) = 1$. Since

$$\int_{E_\alpha^n} x_\alpha(dt) P(t, E_\alpha^n) = \int_\Omega x_\alpha(dt) P(t, E_\alpha^n) = x_\alpha(E_\alpha^n) = 1,$$

we have (by Lemma 4.6) $x_\alpha(E_\alpha^{n+1}) = 1$. If we now consider the set $E_\alpha^\infty = \prod_{n=1}^\infty E_\alpha^n$, then $E_\alpha = E_\alpha^\infty$ is the required set. For, we have

$$x_\alpha(E_\alpha^\infty) = \lim_{n \to \infty} x_\alpha(E_\alpha^n) = 1$$

and

$$P(t, E_\alpha^\infty) = \lim_{n \to \infty} P(t, E_0^n) = 1 \quad \text{for } t \epsilon E_\alpha^\infty.$$

Thus we have proved the existence of the Borel set $E_\alpha \subset \bar{E}_\alpha$ which satisfies (4.31), (4.32) and (4.33). Since (4.34) is clear from

$$P(t, \Delta) = \sum_{\alpha=1}^l y_\alpha(t) x_\alpha(\Delta) = 0, \qquad t \epsilon \Omega,$$

the proof of Theorem 8 is completed.

§4.5. Proper values of modulus 1 of the operator T.

THEOREM 9. *Under the condition* (K), *each proper value of modulus 1 of the operation T is a root of unity.*

PROOF. Let λ, $|\lambda| = 1$, be a proper value of the bounded linear operation T, which maps (**M**) into itself. By Theorem 5, λ may also be considered as a proper value of the bounded linear operation $\bar{T}$, which maps (M*) into itself. Indeed, (since $P_\lambda(t, E)$ is not identically zero) if we put $x(t) \equiv P_\lambda(t, E)$ for a suitable Borel set E, then $x(t)$ is a non-trivial solution of the proper value equation:

$$(4.35) \qquad \bar{T}(x) = \lambda x: \lambda x(t) = \int_\Omega P(t, ds) x(s).$$

$x(t)$ is clearly a complex-valued bounded Borel measurable function $\epsilon\,(M^*)$. We shall prove that λ is a root of unity.

In order to prove this, we shall first show that the real-valued bounded measurable function $\tilde{x}(t) \equiv |\,x(t)\,|\,\epsilon\,(M^*)$ attains its maximum. From (4.35), we have

$$\tilde{x}(t) \leqq \int_\Omega P(t,\,ds)\tilde{x}(s)$$

and consequently

$$\tilde{x}(t) \leqq \int_\Omega P^{(n)}(t,\,ds)\tilde{x}(s), \qquad\qquad n = 1,\,2,\,\cdots.$$

Taking the arithmetic mean and considering its limit, we have

$$\tilde{x}(t) \leqq \int_\Omega P_1(t,\,ds)\tilde{x}(s)$$

or, by (4.23),

$$\tilde{x}(t) \leqq \sum_{\alpha=1}^{l} y_\alpha(t)\left(\int_\Omega x_\alpha(ds)x(s)\right) \equiv \sum_{\alpha=1}^{l} \xi_\alpha y_\alpha(t),$$

where $\xi_\alpha = \displaystyle\int_\Omega x_\alpha(ds)x(s)$ is a real non-negative number. Let ξ be the maximum of ξ_α ($\alpha = 1,\,2,\,\cdots,\,l$). Since $\sum_{\alpha=1}^{l} y_\alpha(t) \equiv 1$ by (4.24), we have $\tilde{x}(t) \leqq \xi$ for any $t\,\epsilon\,\Omega$. We shall prove that this ξ is attained by $\tilde{x}(t)$ at some point $t_0\,\epsilon\,\Omega$. Indeed, by the definition of ξ, there exists at least one integer α such that $\xi = \displaystyle\int_\Omega x_\alpha(ds)x(s)$, and since $x_\alpha(E) \geqq 0$, $x_\alpha(\Omega) = 1$ and $\tilde{x}(s) \leqq \xi$ for any $s\,\epsilon\,\Omega$, there must exist, by Lemma 4.6, a point $t_0\,\epsilon\,\Omega$ such that $\tilde{x}(t_0) = \xi$ (or more precisely, the set $E_0 = E[\tilde{x}(s) = \xi]$ satisfies $x_\alpha(E_0) = 1$).

We have thus proved that there exists a point $t_0\,\epsilon\,\Omega$, such that $\tilde{x}(t_0) = |\,x(t_0)\,| = \max_{t\,\epsilon\,\Omega} |\,x(t)\,| = \xi$. We shall next prove that the set

$$(4.36) \qquad\qquad E(n) = E_t[x(t) = \lambda^n x(t_0)]$$

satisfies

$$(4.37) \qquad\qquad P^{(n)}(t,\,E(n)) = 1$$

for $n = 1,\,2,\,\cdots$. From (4.35), we have

$$\lambda^n x(t_0) = \int_\Omega P^{(n)}(t_0,\,ds)x(s)$$

and, dividing by $\lambda^n x(t_0)$ and taking the real part,

$$1 = \int_\Omega P^{(n)}(t_0,\,ds)R\left(\frac{x(s)}{\lambda^n x(t_0)}\right), \qquad\qquad n = 1,\,2,\,\cdots.$$

Since $P^{(n)}(t_0, E) \geqq 0$, $P^{(n)}(t_0, \Omega) \equiv 1$ and $R\left(\dfrac{x(s)}{\lambda^n x(t_0)}\right) \leqq 1$ for any $s \in \Omega$, we have (4.37) by Lemma 4.6. For, we have

$$\underset{s}{E}\left[R\left(\frac{x(s)}{\lambda^n x(t_0)}\right) = 1\right] = \underset{t}{E}\left[x(t) = \lambda^n x(t_0)\right].$$

Thus we have proved (4.37) for $n = 1, 2, \cdots$. From this follows the existence of two positive integers m and n, such that $E(m) \cdot E(n) \neq 0$. For, if we have $E(m) \cdot E(n) = 0$ for any couple of integers m, n ($m \neq n$), then we have

$$(4.38) \qquad P^{(m)}(t_0, E(l)) = 0 \quad \text{for} \quad m \neq l,$$

and consequently, by (4.37),

$$P_1(t_0, E(l)) = \lim_{n \to \infty} \frac{1}{n} \sum_{m=1}^{n} P^{(m)}(t_0, E(l)) = 0.$$

Since $P_1(t_0, E)$ is completely additive for Borel sets E, this implies

$$P_1\left(t_0, \sum_{l=1}^{\infty} E(l)\right) = 0.$$

This is, however, a contradiction, since we have, by (4.37) and (4.38),

$$P^{(m)}\left(t_0, \sum_{l=1}^{\infty} E(l)\right) = 1, \qquad\qquad m = 1, 2, \cdots,$$

and consequently

$$P_1\left(t_0, \sum_{l=1}^{\infty} E(l)\right) = \lim_{m \to \infty} \frac{1}{n} \sum_{m=1}^{n} P^{(m)}\left(t_0, \sum_{l=1}^{\infty} E(l)\right) = 1.$$

Thus we have $E(m) \cdot E(n) \neq 0$ for a certain couple of integers m, n ($m \neq n$). Consequently we have (by (4.36)) $\lambda^m x(t_0) = \lambda^n x(t_0)(\neq 0)$ or $\lambda^{m-n} = 1$, and hereby the proof of Theorem 9 is completed.

§4.6. Decomposition of each ergodic part (and ergodic kernel) into subergodic parts (and subergodic kernels).

THEOREM 10. *Under the condition* (K), *there exists a positive integer* N *such that* $P^{(nN)}(t, E)$ *is decomposed into the form:*

$$(4.39) \qquad P^{(nN)}(t, E) = P_1^*(t, E) + S^{*(n)}(t, E), \qquad n = 1, 2, \cdots,$$

in such a way that we have

$$\int_\Omega P^{(N)}(t, ds)P_1^*(s, E) = \int_\Omega P_1^*(t, ds)P^{(N)}(s, E)$$

$$(4.40)$$

$$= \int_\Omega P_1^*(t, ds)P_1^*(s, E) = P_1^*(t, E),$$

$$(4.41) \qquad \int_\Omega P_1^*(t, ds)S^*(s, E) = \int_\Omega S^*(t, ds)P_1^*(s, E) = 0$$

and

$$(4.42) \qquad \operatorname*{l.u.b.}_{t \in \Omega, E \subset \Omega} | S^{*(n)}(t, E) | \leqq \frac{M}{(1 + \epsilon)^n}, \qquad n = 1, 2, \cdots,$$

where M and ϵ are positive constants which are independent of n.

PROOF. In Theorem 9, we have seen that each proper value λ of modulus 1 of T is a root of unity. Since T has only a finite number of such proper values, there exists a (sufficient large) positive integer N such that $\lambda^N = 1$ for any proper value λ of modulus 1 of T. Hence the bounded linear operation T^N corresponding to the kernel $P^{(N)}(t, E)$ has no proper values of modulus 1 other than 1, and Theorem 10 is a direct consequence of Theorem 5.

REMARK. It is to be noted that we have, by (4.39) and (4.42),

$$(4.43) \qquad \operatorname*{l.u.b.}_{t \in \Omega, E \subset \Omega} | P^{(nN)}(t, E) - P_1^*(t, E) | \leqq \frac{M}{(1 + \epsilon)^n}, \qquad n = 1, 2, \cdots$$

with positive constants M and ϵ.

Moreover, just as in Lemma 4.5, there exists a system of real-valued completely additive set functions $\{x_i^*(E)\}$ $(i = 1, 2, \cdots, L) \epsilon$ (**M**) with the properties:

$$(4.44) \qquad T^N(x_i^*) = x_i^*, \quad x_i^* \geqq 0, \quad x_i^*(\Omega) = 1, \quad x_i^* \wedge x_j^* = 0 \qquad (i \neq j),$$

such that each $x^*(E) \epsilon$ (**M**) which satisfies

$$(4.45) \qquad \quad T^N(x^*) = x^*, \quad x^* \geqq 0, \quad x^*(\Omega) = 1$$

is uniquely expressed in the form:

$$(4.46) \qquad x^*(E) = \sum_{i=1}^{L} c_i^* x_i^*(E), \qquad c_i^* \geqq 0, \qquad \sum_{i=1}^{L} c_i^* = 1.$$

In particular, just as in Theorem 6, $P_1^*(t, E)$ is expressed in the form:

$$(4.47) \qquad P_1^*(t, E) = \sum_{i=1}^{L} y_i^*(t) x_i^*(E),$$

where $\{y_i^*(t)\}$ $(i = 1, 2, \cdots, L)$ is a system of real-valued bounded Borel measurable functions ϵ (M^*) which satisfy

$$(4.48) \qquad \bar{T}^N(y_i^*) = y_i^*, \qquad y_i^*(t) \geqq 0, \qquad \sum_{i=1}^{L} y_i^*(t) \equiv 1.$$

Let us denote by $\bar{E}_i^*$ the set $\underset{t}{E}[y^*(t) = 1]$, $i = 1, 2, \cdots, L$. These will be called the *subergodic parts* of Ω. Exactly as in Theorem 7, these are mutually disjoint Borel sets, and we have

$$(4.49) \quad \begin{cases} x_i^*(\bar{E}_j^*) = 1 \text{ or } 0 \text{ according as } i = j \text{ or } i \neq j. \\[2mm] P^{(N)}(t, \bar{E}_i^*) = 1, \qquad t \in \bar{E}_i^*, \\[2mm] \underset{t \in \bar{E}_i^*, E \subset \Omega}{\text{l.u.b.}} \; | P^{(nN)}(t, E) - x_i^*(E) | \leq \dfrac{M}{(1 + \epsilon)^n}, \qquad n = 1, 2, \cdots \end{cases}$$

with positive constants M and ϵ. More precisely we can prove

THEOREM 11. *The totality of all the subergodic parts $\bar{E}_i^*$ ($i = 1, 2, \cdots, L$) is divided into l classes $(\bar{E}_{\alpha_1}^*, \bar{E}_{\alpha_2}^*, \cdots, \bar{E}_{\alpha_{d_\alpha}}^*)$ ($\alpha = 1, 2, \cdots, l$), where d_α is a divisor of N with $\sum_{\alpha=1}^l d_\alpha = L$, in such a way that*

$$(4.50) \qquad P(t, \bar{E}_{\alpha_{i+1}}^*) = 1, \qquad t \in \bar{E}_{\alpha_i}^*, \qquad i = 1, 2, \cdots, d_\alpha(\alpha_{d_\alpha+1} = \alpha_1),$$

$$(4.51) \qquad \underset{t \in \bar{E}_{\alpha_i}^*, E \subset \Omega}{\text{l.u.b.}} \; | P^{(nd_\alpha)}(t, E) - x_{\alpha_i}^*(E) | \leq \dfrac{M}{(1 + \epsilon)^n}, \qquad n = 1, 2, \cdots,$$

where M and ϵ are positive constants which are independent of n.

Moreover, all $\bar{E}_{\alpha_i}^*$ ($i = 1, 2, \cdots, d_\alpha$) belonging to the same class are contained in the same ergodic part $\bar{E}_\alpha$, and if we denote by $x_{\alpha_i}^*(E)$ and $y_{\alpha_i}^*(t)$ the corresponding elements of $(\mathbf{M})$ and (M^*) respectively (which are obtained by the arguments given above), then we have

$$(4.52) \qquad T(x_{\alpha_i}^*) = x_{\alpha_{i+1}}^*, \qquad i = 1, 2, \cdots, d_\alpha(\alpha_{d_\alpha+1} = \alpha_1),$$

$$(4.53) \qquad \overline{T}(y_{\alpha_{i+1}}^*) = y_{\alpha_i}^*, \qquad i = 1, 2, \cdots, d_\alpha(\alpha_{d_\alpha+1} = \alpha_1),$$

$$(4.54) \qquad x_\alpha(E) = \frac{1}{d_\alpha} \sum_{i=1}^{d_\alpha} x_{\alpha_i}^*(E),$$

$$(4.55) \qquad y_\alpha(t) = \sum_{i=1}^{d_\alpha} y_{\alpha_i}^*(t).$$

REMARK 1. (4.50) means that, for any α, each point $t \in \sum_{i=1}^{d_\alpha} \bar{E}_{\alpha_i}^*$ is transferred, by the Markoff's process $P(t, E)$, cyclically in $\bar{E}_{\alpha_1}^*, \bar{E}_{\alpha_2}^*, \cdots, \bar{E}_{\alpha_{d_\alpha}}^*$; and (4.51) means that, in each $\bar{E}_{\alpha_i}^*$, $P^{(d_\alpha)}(t, E)$ defines a Markoff's process, whose n^{th} iterate $P^{(nd_\alpha)}(t, E)$ is uniformly convergent to a limit which is independent of the initial point $t \in \bar{E}_{\alpha_i}^*$.

REMARK 2. The equality $\bar{E}_\alpha = \sum_{i=1}^{d_\alpha} \bar{E}_{\alpha_i}^*$ is not necessarily true. Indeed, $D_\alpha = \bar{E}_\alpha - \sum_{i=1}^{d_\alpha} \bar{E}_{\alpha_i}^*$ is the set of all $t \in \Omega$ such that $y_\alpha(t) \equiv \sum_{i=1}^{d_\alpha} y_{\alpha_i}^*(t) = 1$ and $y_{\alpha_i}^*(t) < 1$ for $i = 1, 2, \cdots, d_\alpha$. From the proof of (4.51), we see that

$$(4.51') \qquad \underset{t \in \bar{E}_\alpha, E \subset \Omega}{\text{l.u.b.}} \; \Big| P^{(nd_\alpha)}(t, E) - \sum_{i=1}^{d_\alpha} y_{\alpha_i}^*(t) x_{\alpha_i}^*(E) \Big| \leq \frac{M}{(1 + \epsilon)^n}, \qquad n = 1, 2, \cdots$$

with positive constants M and ϵ.

PROOF OF THEOREM 11. We begin with some preliminary considerations. Since $T^N T(x_i^*) = T T^N(x_i^*) = T(x_i^*)$, $T(x_i^*)$ satisfies (4.43) for $i = 1, 2, \cdots, L$. Hence there exists a system of real constants c_{ij}, $i, j = 1, 2, \cdots, L$, such that

$$(4.56) \qquad T(x_i^*) = \sum_{j=1}^{L} c_{ij} x_j^*, \qquad c_{ij} \geqq 0, \qquad \sum_{j=1}^{L} c_{ij} = 1.$$

Thus T may be considered as a linear transformation on the system $\{x_i^*(E)\}$ $(i = 1, 2, \cdots, L)$. Consider the matrix $C = (c_{ij})$, $i, j = 1, 2, \cdots, L$. This matrix clearly satisfies

$$(4.57) \qquad C^N = \text{unit matrix.}$$

We shall show that C is a matrix composed of only 0 and 1, which defines a permutation of the indices $1, 2, \cdots, L$. Indeed, denoting by $c_{ij}^{(N-1)}$ the ij-elements of the matrix $C^{(N-1)}$, we have $\sum_{k=1}^{L} c_{ik}^{(N-1)} = 1$, $\sum_{k=1}^{L} c_{ik}^{(N-1)} c_{ki} = 1$ for $i = 1, 2, \cdots, L$. Since $0 \leqq c_{ki} \leqq 1$ for any k and i, there must exist, for each i, an index k_i such that $c_{k_i i} = 1$. In other words, each column of the matrix C must contain 1 at least once. Since $\sum_{i=1}^{L} c_{ki} = 1$ for each k, we must have $k_i \neq k_j$ for $i \neq j$. Hence, $(k_1, k_2, \cdots, k_L)$ is a permutation of the indices $(1, 2, \cdots, L)$, and consequently we have $c_{ki} = 0$ for $k \neq k_i$.

Thus we have proved that C defines a permutation of the indices $1, 2, \cdots, L$. Hence the indices $1, 2, \cdots, L$ are divided into $l'(\leqq L)$ classes, and each class is permuted cyclically inside itself by the matrix C. Let these classes be K_α $(\alpha = 1, 2, \cdots, l')$ and the number of the indices belonging to K_α be d_α. By (4.57), each d_α is a divisor of N.

For each α, consider the set of all the indices which belong to K_α. By a suitable numbering $\alpha_1, \alpha_2, \cdots, \alpha_{d_\alpha}$ of these indices, we must have $c_{\alpha_i \alpha_{i+1}} = 1$ and $c_{\alpha_i \alpha_j} = 0$ for $j \neq i + 1$ $(i = 1, 2, \cdots, d_\alpha$; once for all, we put $\alpha_{d_\alpha+1} = \alpha_1)$. Consequently, we have (4.52).

After these preliminaries, we shall proceed to the proof of Theorem 11. We shall first prove that we have $l = l'$ and that there is a one-to-one correspondence between the ergodic parts $\bar{E}_\alpha$ and the classes K_α in such a way that (4.52), (4.53), (4.54) and (4.55) are true. ((4.52) is already proved.)

For this purpose, recall that each $x_\alpha(E)$ satisfies $T(x_\alpha) = x_\alpha$. Hence $x_\alpha(E)$ satisfies (4.45), and consequently $x_\alpha(E)$ is uniquely expressed in the form:

$$(4.58) \qquad x_\alpha(E) = \sum_{i=1}^{L} c_i^* x_i^*(E) \equiv \sum_{\alpha=1}^{l'} \sum_{i=1}^{d_\alpha} c_{\alpha_i}^* x_{\alpha_i}^*(E).$$

We shall first prove that there exists a class K_α such that $c_i^* = 1/d_\alpha$ for $i \,\epsilon\, K_\alpha$ and $c_i^* = 0$ for $i \,\bar\epsilon\, K_\alpha$. In the first place, it is clear that c_i^* is independent of i in each class K_α. For, since $T(x_\alpha) = x_\alpha$ and $T(x_{\alpha_i}) = x_{\alpha_{i+1}}$, we have, from (4.58),

$$x_\alpha(E) = \sum_{\alpha=1}^{l} \sum_{i=1}^{d_\alpha} c_{\alpha_i}^* x_{\alpha_{i+1}}^*(E),$$

and this implies $c_{\alpha_i}^* = c_{\alpha_{i+1}}^*$ for $i = 1, 2, \cdots, d_\alpha$. In the second place, all indices i with $c_i^* > 0$ belong to the same class K_α. For, if this is not the case, then $x_\alpha(E)$ will be decomposed into two non-trivial parts x_α' and x_α'', which are both invariant under T: $x_\alpha = x'_\alpha + x''_\alpha$, $x'_\alpha > 0$, $x''_\alpha > 0$, $x'_\alpha \wedge x''_\alpha = 0$, $T(x'_\alpha) = x'_\alpha$, $T(x''_\alpha) = x''_\alpha$. If we put $x_{\alpha 1} = x'_\alpha / \| x'_\alpha \|$ and $x_{\alpha 2} = x''_\alpha / \| x''_\alpha \|$, then the system of $l + 1$ elements $x_1, x_2, \cdots, x_{\alpha-1}, x_{\alpha 1}, x_{\alpha 2}, x_{\alpha+1}, \cdots, x_l$ will satisfy the condition (4.20), and this is a contradiction to the definition of l. Thus there must exist a class K_α such that $c_i^* = c > 0$ for $i \, \epsilon \, K_\alpha$ and $c_i^* = 0$ for $i \, \bar\epsilon \, K_\alpha$. Since it is clear that we have $c = 1/d_\alpha$, the relation (4.54) is hereby proved.

Thus to each $x_\alpha(E)$ ($\alpha = 1, 2, \cdots, l$) there corresponds a class of indices K_α. Conversely, as is easily seen, to each class K_α there corresponds a completely additive set function $x_\alpha(E) = \dfrac{1}{d_\alpha} \sum_{i=1}^{d_\alpha} x_{\alpha_i}^*(E)$ (which clearly belongs to the system determined in Lemma 4.5) in such a way that the correspondence $x_\alpha(E) \leftrightarrow K_\alpha$ is one-to-one. Hence we must have $l = l'$, and the one-to-one correspondence between the ergodic part $\bar{\bar E}_\alpha$ and the class of indices K_α is also established.

We shall next prove (4.55). From the relation

$$\frac{1}{N} \sum_{m=1}^{N} \int_\Omega P_1^*(t, ds) P^{(m)}(s, E) = P_1(t, E)$$

we have, by (4.23) and (4.47),

$$\frac{1}{N} \sum_{m=1}^{N} \sum_{\alpha=1}^{l} \sum_{i=1}^{d_\alpha} y_{\alpha_i}^*(t) \left(\int_\Omega x_{\alpha_i}^*(ds) P^{(m)}(s, E) \right) = \sum_{\alpha=1}^{l} y_\alpha(t) x_\alpha(E),$$

or, by (4.52),

$$\frac{1}{N} \sum_{m=1}^{N} \sum_{\alpha=1}^{l} \sum_{i=1}^{d_\alpha} y_{\alpha_i}^*(t) x_{\alpha_{i+m}}^*(E) = \sum_{\alpha=1}^{l} y_\alpha(t) x_\alpha(E).$$

Using the fact that d_α is a divisor of N, we have

$$\sum_{\alpha=1}^{l} \left(\sum_{i=1}^{d_\alpha} y_{\alpha_i}^*(t) \right) \left(\frac{1}{d_\alpha} \sum_{i=1}^{d_\alpha} x_{\alpha_i}^*(E) \right) = \sum_{\alpha=1}^{l} y_\alpha(t) x_\alpha(E),$$

or, by (4.54),

$$\sum_{\alpha=1}^{l} \left(\sum_{i=1}^{d_\alpha} y_{\alpha_i}^*(t) \right) x_\alpha(E) = \sum_{\alpha=1}^{l} y_\alpha(t) x_\alpha(E);$$

and this implies (4.55) (put $E = \bar{\bar E}_\alpha$).

In order to prove (4.53), we start from the trivial relation:

$$\int_\Omega P(t, ds) P_1^*(s, E) = \int_\Omega P_1^*(t, ds) P(s, E).$$

By (4.45) and (4.52), this becomes

$$\sum_{\alpha=1}^{l} \sum_{i=1}^{d_\alpha} \left(\int_\Omega P(t, ds) y_{\alpha_i}^*(s) \right) x_{\alpha_i}^*(E) = \sum_{\alpha=1}^{l} \sum_{i=1}^{d_\alpha} y_{\alpha_i}^*(t) \left(\int_\Omega x_{\alpha_i}^*(ds) P(s, E) \right)$$

$$= \sum_{\alpha=1}^{l} \sum_{i=1}^{d_\alpha} y_{\alpha_i}^*(t) x_{\alpha_{i+1}}^*(E),$$

and, putting $E = \bar{E}^*_{\alpha_{i+1}}$, we have

$$\int_\Omega P(t, ds)y^*_{\alpha_{i+1}}(s) = y^*_{\alpha_i}(t);$$

i.e., (4.53) is proved.

Thus we have proved (4.52), (4.53), (4.54) and (4.55). From (4.55), it is clear that each $\bar{E}^*_{\alpha_i}$ is contained in $\bar{E}_\alpha$, and (4.50) is a direct consequence of (4.53) (use Lemma 4.6).

Lastly, we shall prove (5.51). For any positive integers n and k we have, by (4.6) and (4.37),

$$P^{(nN+kd_\alpha)}(t, E) = \int_\Omega P^{(nN)}(t, ds)P^{(kd_\alpha)}(s, E) = \int_\Omega P^*_1(t, ds)P^{(kd_\alpha)}(s, E)$$

$$+ \int_\Omega S^{*(n)}(t, ds)P^{(kd_\alpha)}(s, E).$$

Since we have

$$\int_\Omega P^*_1(t, ds)P^{(kd_\alpha)}(s, E) = \int_\Omega x^*_{\alpha_i}(ds)P^{(kd_\alpha)}(s, E) = x^*_{\alpha_i}(E)$$

for any $t \in \bar{E}^*_{\alpha_i}$, and since, by (4.40),

$$\left| \int_\Omega S^{*(n)}(t, ds)P^{(kd_\alpha)}(s, E) \right| \leqq \operatorname*{l.u.b.}_{t \in \Omega, E \subset \Omega} | S^{*(n)}(t, E) | \leqq \frac{M}{(1 + \epsilon)^n}$$

for $n = 1, 2, \cdots$, we have

$$| P^{(nN+kd_\alpha)}(t, E) - x^*_{\alpha_i}(E) | \leqq \frac{M}{(1 + \epsilon)^n}$$

for $t \in \bar{E}^*_{\alpha_i}$, $k = 1, 2, \cdots, N/d_\alpha$; $n = 1, 2, \cdots$, where M and ϵ are positive constants which are independent of n and k. Hence we have (4.51), by a suitable change of M and ϵ.

Thus Theorem 11 is completely proved.

REMARK 3. We can also define *subergodic kernel* $E^*_{\alpha_i}$ in each subergodic part $\bar{E}^*_{\alpha_i}$; namely, $E^*_{\alpha_i}$ is the set of the smallest measure which satisfies $x^*_{\alpha_i}(E^*_{\alpha_i}) = 1$. If we suitably take E_α and $E^*_{\alpha_i}$ (which are all determined only up to a set of measure zero), then we have

COROLLARY.

$$(4.59) \qquad E_\alpha = \sum_{i=1}^{d_\alpha} E^*_{\alpha_i}, \qquad E^*_{\alpha_i} = E_\alpha \bar{E}^*_{\alpha_i},$$

$$(4.60) \qquad P(t, E^*_{\alpha_i+1}) = 1. \qquad t \in E^*_{\alpha_i}.$$

PROOF. (4.59) is clear from (4.54), and (4.60) follows from (4.29), (4.50) and the second relation of (4.59).

It is to be noted that these E_α and $E^*_{\alpha_i}$ are exactly the final sets (ensembles finals) and their cyclic subsets which were discussed by W. Doeblin [1].

THEOREM 12.

$$(4.61) \qquad \underset{t \epsilon \Omega}{\text{l.u.b.}} \; P^{(n)}(t, \Delta) \leqq \frac{M}{(1 + \epsilon)^n}, \qquad n = 1, 2, \cdots,$$

where M and ϵ are positive constants which are independent of n.

PROOF. Since each $x^*_{\alpha_i}(E)$ satisfies

$$x^*_{\alpha_i}(\Delta) \leqq \sum_{i=1}^{d_\alpha} x^*_{\alpha_i}(\Delta) = d_\alpha \cdot x_\alpha(\Delta) = 0,$$

we have

$$P^*_1(t, \Delta) = \sum_{i=1}^{L} y^*_i(t) x^*_i(\Delta) \equiv 0$$

for any $t \epsilon \Omega$, and consequently we have, by Theorem 10,

$$\underset{t \epsilon \Omega}{\text{l.u.b.}} \; P^{(nN)}(t, \Delta) \leqq \frac{M}{(1 + \epsilon)^n}, \qquad n = 1, 2, \cdots.$$

Since $P^{(n)}(t, \Delta)$ is monotone decreasing in n, we can deduce from this easily the relation (4.61) (by a suitable change of M and ϵ).

§4.7. Deduction of the condition (K) from the condition (D).

LEMMA 4.7. Let us denote by $I(s_0)$ the closed interval $0 \leqq s \leqq s_0$. Then $Q(t, s) \equiv P(t, I(s))$ is Borel measurable as a function of two variables t and s in $0 \leqq t, s \leqq 1$.

PROOF. By assumption, $Q(t, s)$ is Borel measurable in t if s is fixed, and if t is fixed $Q(t, s)$ is monotone increasing in s and is continuous on the right: $\lim_{s \to s_0 + 0} Q(t, s) = Q(t, s_0)$.

We shall prove that, for any α, the set $E(\alpha) = \underset{(t, s)}{E} [Q(t, s) < \alpha]$ is Borel measurable as a two-dimensional point set. For this purpose, put $E_s(\alpha) = E [Q(t, s) < \alpha]$. Since $Q(t, s)$ is monotone increasing in s, we have $E_{s_1}(\alpha) \supset E_{s_2}(\alpha)$ for $s_1 < s_2$, and consequently

$$E(\alpha) = \sum_{s \epsilon \Omega} E_s(\alpha) \times I(s),$$

where $\times$ denotes the Cartesian product. Since $Q(t, s)$ is continuous in s on the right, the section of $E(\alpha)$ by the straight line $t = t_0$ is, if not empty, a semi-open interval of the form: $0 \leqq s < s_0$. Hence we have

$$E(\alpha) = \sum_{\substack{s-\text{rational} \\ s \epsilon \Omega}} E_s(\alpha) \times I(s),$$

and this shows that $E(\alpha)$ is Borel measurable as a two-dimensional point set.

LEMMA 4.8.[19] *Let $K(t, E)$ and $n(t, E)$ be two kernels with bounded density*:

$$K(t, E) = \int_E k(t, s)\, ds, \qquad |\, k(t, s)\,| \leqq ||\, K\,||,$$

$$N(t, E) = \int_E n(t, s)\, ds, \qquad |\, n(t, s)\,| \leqq ||\, N\,||,$$

where $k(t, s)$ and $n(t, s)$ are both bounded measurable functions defined on $0 \leqq t$, $s \leqq 1$. If we consider the corresponding integral operators K and N, which map the Banach space $(\mathbf{M})$ *into itself, then the integral operator NK defined by the kernel* $\int_E \left(\int_0^1 k(t, u) n(u, s)\, du \right) ds$ *is strongly completely continuous as an operator which maps* $(\mathbf{M})$ *into itself.*

PROOF. For any $x(E) \in (\mathbf{M})$, put $y = K(x)$ and $z = N(y) = NK(x)$. The set function $z(E)$ is absolutely continuous:

$$z(E) = \int_E z'(s)\, ds,$$

and its density $z'(s)$ is **given by**

$$z'(s) = \int_0^1 y(du) n(u, s) = \int_0^1 \left(\int_0^1 x(dt) k(t, u) \right) n(u, s)\, du.$$

Hence $||\, x\,|| \leqq 1$ implies

$$\operatorname*{l.u.b.}_{0 \leqq s \leqq 1} |\, z'(s)\,| \leqq ||\, K\,|| \cdot ||\, N\,||$$

and

$$\int_{-\infty}^{+\infty} |\, z'(s + \delta) - z'(s)\,|\, ds \leqq ||\, K\,|| \int_{-\infty}^{+\infty} \left(\int_0^1 |\, n(u, s + \delta) - n(u, s)\,|\, du \right) ds$$

if we put $n(u, s) = 0$ for $s < 0$ and $s > 1$.

Consequently, we have

$$\lim_{\delta \to 0} \int_{-\infty}^{+\infty} |\, z'(s + \delta) - z'(s)\,|\, ds = 0$$

uniformly for all $x(E) \in (\mathbf{M})$ with $||\, x\,|| \leqq 1$. Hence, by a theorem of A. Kolmogoroff [1] and M. Riesz [1], (if we consider $z'(s)$ as an element of the Banach space (L)) the totality of all $z'(s)$ corresponding to the unit sphere $||\, x\,|| \leqq 1$ of $(\mathbf{M})$ is strongly compact in (L). In other words, NK is strongly completely continuous as an operator which maps $(\mathbf{M})$ into (L). Since (L) is isometric to a closed linear subspace of $(\mathbf{M})$, NK is also strongly completely continuous as an operator which maps $(\mathbf{M})$ into itself. Thus the proof of Lemma 4.8 is completed.

---

[19] K. Yosida, Y. Mimura and S. Kakutani [1].

LEMMA 4.9. *If $P(t, E)$ satisfies the condition (D), then we have*

$$(4.62) \qquad P^{(d)}(t, E) = \int_E q(t, s)\, ds + R(t, E),$$

where $q(t, s)$ is a bounded $\left(\leqq \dfrac{1}{\eta}\right)$ Borel measurable function defined for $0 \leqq t,$ $s \leqq 1,$ and

$$(4.63) \qquad 0 \leqq R(t, E) \leqq 1 - b$$

for any $t \,\epsilon\, \Omega$ and $E \subset \Omega.$

PROOF. Define $Q(t, s)$ newly by

$$Q(t, s) = P^{(d)}(t, I(s)).$$

By Lemma 4.7, $Q(t, s)$ is Borel measurable as a function of two variables t and $s.$ Since $Q(t, s)$ is monotone in s for any $t,$ $Q(t, s)$ is almost everywhere differentiable in s for any fixed $t.$ If we put

$$p(t, s) = \lim_{n \to \infty} n\left(Q\left(t, s + \frac{1}{n}\right) - Q(t, s)\right)$$

and define $q(t, s)$ by

$$q(t, s) = \min\left(p(t, s), \frac{1}{\eta}\right),$$

then $q(t, s)$ is bounded $\left(\leqq \dfrac{1}{\eta}\right)$ and Borel measurable as a function of two variables t and $s.$ We shall prove that the kernel $R(t, E)$ defined by

$$R(t, E) = P^{(d)}(t, E) - \int_E q(t, s)\, ds$$

satisfies (4.63) for any $t \,\epsilon\, \Omega$ and $E \subset \Omega.$

For this purpose, we have only to notice that there exists for any $t \,\epsilon\, \Omega$ a Borel set N_t of measure zero such that

$$P^{(d)}(t, E) = \int_E p(t, s)\, ds + P^{(d)}(t, N_t \cdot E)$$

for any Borel set $E \subset \Omega.$ Then we have

$$0 \leqq R(t, E) = \int_E (p(t, s) - q(t, s))\, ds + P^{(d)}(t, N_t \cdot E)$$

$$= \int_{E \cdot E_t(\eta)} p(t, s)\, ds + P^{(d)}(t, N_t \cdot E)$$

$$\leqq P^{(d)}(t, E_t(\eta) + N_t),$$

where $E_t(\eta) = E_s\left[p(t, s) > \dfrac{1}{\eta}\right]$. Since clearly mes $(E_t(\eta)) < \eta$ for any $t \,\epsilon\, \Omega$, we have mes $(E_t(\eta) + N_t) < \eta$ and consequently, by (D), $0 \leqq R(t, E) \leqq 1 - b$ for any $t \,\epsilon\, \Omega$ and $E \subset \Omega$.

THEOREM 13. *The condition* (D) *implies the condition* (K).

PROOF. By Lemma 4.9, we have

$$(4.64) \qquad\qquad T^d = Q + R, \qquad \| R \| \leqq 1 - b,$$

where Q and R are the two integral operators which are defined by the kernels $Q(t, E) = \displaystyle\int_E q(t, s)\, ds$ and $R(t, E)$ respectively. If we expand $T^{dm} = (Q + R)^m$ in 2^m terms:

$$T^{dm} = Q^m + Q^{m-1}R + Q^{m-2}RQ + \cdots + QR^{m-1} + R^m,$$

then the terms which contain Q at least twice as factor are all strongly completely continuous. In order to see this, consider for example the term $RQRQ^{m-3}$. By (4.22), QR and Q^{m-3} are both integral operators with bounded density kernels. Hence, by Lemma 4.8, QRQ^{m-3} and consequently $RQRQ^{m-3}$ are strongly completely continuous. Since the number of terms which contain Q at most once as factor is $m + 1$, and since the norm of each such term is $\leqq (1 - b)^{m-1}$ by (4.64), we see that there exists, for each m, a strongly completely continuous operator V_m, which maps (**M**) into itself, such that

$$\| T^{dm} - V_m \| \leqq (m + 1)(1 - b)^{m-1}.$$

Since the right hand side converges to zero as $m \rightarrow \infty$, the proof of Theorem 13 is hereby completed.

REMARK. The converse of Theorem 13 is not true. To see this, take an arbitrary point $s_0 \,\epsilon\, \Omega$, and define $P(t, E)$ by

$$P(t, E) = 1 \quad \text{if} \quad s_0 \,\epsilon\, E,$$

$$= 0 \quad \text{if} \quad s_0 \,\bar{\epsilon}\, E,$$

for any $t \,\epsilon\, \Omega$. This kernel $P(t, E)$ defines a strongly completely continuous integral operator, but the condition (D) is clearly not satisfied.

§4.8. Markoff's process with a finite number of possible states. Consider a Markoff's process with a finite number $(= N)$ of possible states. Let p_{ij} $(i, j = 1, 2, \cdots, N)$ be the transition probability that the ith state is transferred to the jth state after the elapse of a unit-time. Then the transition probability $p_{ij}^{(n)}$ that the ith state is transferred to the jth state after the elapse of n unit-times is given recurrently by

$$(4.65) \qquad\qquad p_{ij}^{(n)} = \sum_{k=1}^{N} p_{ik}\, p_{kj}^{(n-1)}, \qquad p_{ij}^{(1)} = p_{ij},$$

and we have always

$$(4.66) \qquad p_{ij}^{(n)} \geqq 0, \qquad \sum_{j=1}^{N} p_{ij}^{(n)} = 1,$$

for $i, j = 1, 2, \cdots, N; n = 1, 2, \cdots$.

We shall investigate the asymptotic behavior of $p_{ij}^{(n)}$ for large n. This problem was discussed by many authors (see the introduction at the beginning of the paper), and sometimes direct methods were successful in this case. We shall, however, treat this problem as a special case of our general Markoff's process.

Consider the finitely valued function $p(t, s)$ defined in the square $0 \leqq t$, $s \leqq 1$ by

$$(4.67) \qquad p(t, s) = N \cdot p_{ij} \quad \text{for} \quad \frac{i-1}{N} \leqq t < \frac{i}{N}, \quad \frac{j-1}{N} \leqq s < \frac{j}{N},$$

$i, j = 1, 2, \cdots, N$ (in case $i = N$ or $j = N$, $<$ is to be replaced by $\leqq$). Then $P(t, E) = \int_{E} p(t, s) \, ds$ defines a simple Markoff's process on the interval $\Omega = (0, 1)$, and we have

$$(4.68) \qquad P^{(n)}(t, E) = \int_{E} p^{(n)}(t, s) \, ds,$$

where

$$(4.69) \qquad p^{(n)}(t, s) = N \cdot p_{ij}^{(n)} \quad \text{for} \quad \frac{i-1}{N} \leqq t < \frac{i}{N}, \quad \frac{j-1}{N} \leqq s < \frac{j}{N},$$

$i, j = 1, 2, \cdots, N$ (in case $i = N$ or $j = N$, $<$ is again to be replaced by $\leqq$).

Thus the Markoff's process $P = (p_{ij})$ $(i, j = 1, 2, \cdots, N)$ is reduced to the continuous case $P = P(t, E)$. Since it is clear that the corresponding integral operator T is strongly completely continuous in this case, we have, by the results obtained above,

THEOREM 14. (i) *The limit*

$$(4.70) \qquad \lim_{n \to \infty} \frac{1}{n} \sum_{m=1}^{n} p_{ij}^{(m)} = p_{ij}^{(\infty)}$$

exists for any i and j, and there exists a constant M such that

$$(4.71) \qquad \left| \frac{1}{n} \sum_{m=1}^{n} p_{ij}^{(m)} - p_{ij}^{(\infty)} \right| \leqq \frac{M}{n}, \qquad i, j = 1, 2, \cdots, N; n = 1, 2, \cdots.$$

(ii) *There exists a system of mutually disjoint ergodic parts $\bar{E}_\alpha$ ($\alpha = 1, 2, \cdots, l$) such that*

$$(4.72) \qquad \sum_{i \, \epsilon \, \bar{E}_\alpha} p_{ij} = 1 \quad \text{if} \quad i \, \epsilon \, \bar{E}_\alpha,$$

(4.73) $$p_{ij}^{(\infty)} \text{ is independent of } i \text{ in each } \bar{E}_\alpha.$$

(iii) *Each ergodic part $\bar{E}_\alpha$ contains an ergodic kernel E_α such that*

(4.74) $$\sum_{j \,\epsilon\, E_\alpha} p_{ij}^{(\infty)} = 1 \quad if \quad i \,\epsilon\, \bar{E}_\alpha,$$

(4.75) $$\sum_{j \,\epsilon\, E_\alpha} p_{ij} = 1 \quad if \quad i \,\epsilon\, E_\alpha.$$

(iv) $\Delta = \Omega - \sum_{\alpha=1}^{l} E_\alpha$ *is called the dissipative part of Ω and we have*

(4.76) $$p_{ij}^{(n)} \leqq \frac{M}{(1+\epsilon)^n} \qquad j \,\epsilon\, \Delta,\, i = 1, 2, \cdots, N;\, n = 1, 2, \cdots$$

with positive constants M and ϵ.

(v) *To each ergodic part $\bar{E}_\alpha$ there corresponds a positive integer d_α such that*

(4.77) $$|\, p_{ij}^{(nd_\alpha)} - d_\alpha p_{ij}^{(\infty)} \,| \leqq \frac{M}{(1+\epsilon)^n}, \qquad i \,\epsilon\, \bar{E}_\alpha,\, j = 1, 2, \cdots, N;\, n = 1, 2, \cdots$$

with positive constants M and ϵ. Moreover, each $\bar{E}_\alpha$ contains d_α (mutually disjoint) subergodic parts $\bar{E}_{\alpha_1}^*,\, \bar{E}_{\alpha_2}^*,\, \cdots,\, \bar{E}_{\alpha_{d_\alpha}}^*$ such that

(4.78) $$\sum_{j \,\epsilon\, \bar{E}_{\alpha_{k+1}}^*} p_{ij} = 1 \quad if \quad i \,\epsilon\, \bar{E}_{\alpha_k}^*, \qquad k = 1, 2, \cdots, d_\alpha \quad (\alpha_{d_\alpha+1} = \alpha_1).$$

(vi) *If we further put $E_{\alpha_k}^* = E_\alpha \cdot \bar{E}_{\alpha_k}^*$ for any α and k, then we have*

(4.79) $$\sum_{j \,\epsilon\, E_{\alpha_{k+1}}^*} p_{ij} = 1 \quad if \quad i \,\epsilon\, E_{\alpha_k}^*, \qquad k = 1, 2, \cdots, d_\alpha \quad (\alpha_{d_\alpha+1} = \alpha_1).$$

Added in proof. Recently, N. Dunford and B. J. Pettis [1] obtained some new results concerning weakly completely continuous operators defined on the space (L). Among others, they proved that if K and N are weakly completely continous operators which map (L) into itself, then NK is strongly completely continuous. This result is more precise than Lemma 4.8.

MATHEMATICAL INSTITUTE
OSAKA IMPERIAL UNIVERSITY.

BIBLIOGRAPHY

S. Banach [1]: *Théorie des opérations linéaires*, Warsaw, 1933.

G. Birkhoff [1]: *Dependent probabilities and the space (L)*, Proc. Nat. Acad. Sci., U. S. A., 24(1938), 154–159.

[2]: *The mean ergodic theorem*, Duke Math. Journ., 5(1939), 19–20.

W. Doeblin [1]: *Sur les propriétés asymptotiques de mouvements régis par certains types de chaînes simples*, Bull. Soc. Math. Roumaine, 39–1(1937), 57–115; 39–2(1938), 3–61.

J. L. Doob [1]: *Stochastic processes with an integral valued parameter*, Trans. Amer. Math. Soc., 44(1938), 87–150.

N. Dunford [1]: *An ergodic theorem for n-parameter groups*, Proc. Nat. Acad. Sci., U. S. A., 25(1939), 195–196.

N. Dunford and B. J. Pettis [1]: *Linear operations on summable functions*, Trans. Amer. Math. Soc. 47(1940), 323–392.

M. Fréchet [1]: *Sur l'allure asymptotoque des densités itérées dans le problème des probabilités "en chaîne"*, Bull. Soc. Math. France, 62(1934), 68–83.

[2]: *Sur l'allure asymptotique de la suite des itérés d'un noyau de Frédholm*. Quart Journ. Math., 5(1934), 106–144.

[3]: *Théorie des événements en chaîne dans le cas d'un nombre fini d'états possibles*. Recherches théoriques modernes sur le calcul des probabilités, 2nd livre, Paris, 1938.

E. Hopf [1]: *Ergodentheorie, Ergebnisse der Math. und ihrer Grenzgebiete*, 5, Berlin, 1937.

B. Hostinsky [1]: *Méthodes générales du calcul des probabilités*, Mémorial des Sci. Math., 52, Paris, 1931.

S. Kakutani [1]: *Iteration of linear operations in complex Banach spaces*, Proc. Imp. Acad. Japan, 14(1938), 295–300.

[2]: *Mean ergodic theorem in abstract (L)-spaces*, Proc. Imp. Acad. Japan, 15(1939), 121–123.

[3]: *Weak topology and regularity of Banach spaces*, Proc. Imp. Acad. Japan, 15(1939), 167–173.

[4]: *Some results in the operator-theoretical treatment of the Markoff's process*, Proc. Imp. Acad. Japan, 15(1939), 260-264.

A. Kolmogoroff [1]: *Über die Kompaktheit der Funktionenmengen bei der Konvergenz im Mittel, Göttinger Nachrichten*, 1931, 60–63.

N. Kryloff and N. Bogolioùboff [1]: *Sur les propriétés en chaîne*, C.R. Paris, 204(1937), 1386-1388.

[2]: *Les propriétés ergodiques des suites des probabilités en chaîne*, C.R. Paris, 204(1937), 1454-1456.

D. Milman [1]: *On some criteria for the regularity of spaces of type (B)*, C.R. URSS, 20(1938), 243-246.

M. Nagumo [1]: *Einige analytische Untersuchungen in linearen normierten metrischen ringen*, Jap. Journ. of Math., 12(1936), 61–80.

J. V. Neumann [1]: *Proof of the quasi-ergodic hypothesis*, Proc. Nat. Acad. Sc., U. S. A., 18(1932), 587–642.

[2]: *Zur Operationsmethode in der klassischen Mechanik*, Annals of Math., 33(1932), 587-642.

B. J. Pettis [1]: *A proof that every uniformly convex space is reflexive*, Duke Math. Journ., 5(1939), 249-253.

F. Riesz [1]: *Über lineare Funktionalgleichungen*, Acta Math., 41(1918), 71–98.

[2]: *Some mean ergodic theorems*, Journ. of the London Math. Soc., 13(1938), 274–278.

M. Riesz [1]: *Sur les ensembles compacts de fonctions sommables*, Acta Szeged, 6(1932–34), 130–142.

C. Visser [1]: *On the iteration of linear operations in a Hilbert space*, Proc. Acad. Amsterdam, 41(1938), 487–495.

N. Wiener [1]: *Homogeneous chaos*, Amer. Journ. of Math., 61(1938), 71–98.

[2]: *The ergodic theorem*, Duke Math. Journ., 5(1939), 1–18.

K. Yosida [1]: *Abstract integral equation and the homogeneous stochastic process*, Proc. Imp. Acad. Japan, 14(1938), 286–291.

[2]: *Mean ergodic theorem in Banach spaces*, Proc. Imp. Acad. Japan, 14(1938), 292–294.

[3]: *Operator theoretical treatment of the Markoff's process*, Proc. Imp. Acad. Japan, 14(1938), 363–367. The second report under the same title, Proc. Imp. Acad. Japan, 15(1939), 127–130.

[4]: *Quasi-completely continuous linear functional equations*, Jap. Journ. of Math., 15(1939), 297–301.

K. Yosida and S. Kakutani [1]: *Application of mean ergodic theorem to the problem of Markoff's process*, Proc. Imp. Acad. Japan, 14(1938), 333–339.

K. Yosida, Y. Mimura and S. Kakutani [1]: *Integral operator with bounded kernel*, Proc. Imp. Acad. Japan, 14(1938), 359–362.

Simple Markoff process with a locally compact phase space

Math. Japonica 1 (1948) 99–103

1. *Introduction.* Let X be a metrical space whose bounded closed subset is compact. Consider a simple Markoff process in X and let $P(x, A)$ denote the transition probability that the point x is, by this process, transferred into the Borel set A of X after the elapse of unit time. We assume, as usual, that i) $P(x, A)$ is a completely additive function of Borel set A for fixed x, and that ii) $P(x, A)$ is Borel measurable in x for fixed Borel set A. Then the transition probability $P^{(n)}(x, A)$ that x is transferred into A after the elapse of n unit time is given recurrently by

$$(1) \quad P^{(n)}(x, A) = \int_x P^{(n-1)}(x, dy) P(y, A), \quad (n = 2, 3 \cdots), \quad P^{(1)}(x, A) = P(x, A).$$

We further assume the continuity condition:

(2) if $C(X)$ denotes the totality of real-valued continuous functions $f(x)$ such that the closure of the set $\{x \,;\, f(x) \neq 0\}$ is compact, then $f(x) \,\epsilon\, C(X)$ implies $f_1(x) = \int_x P(x, dy) f(y) \,\epsilon\, C(X)$.

The purpose of the present note is to give the docomposition of X into ergodic parts and dissipative part as an extension of Kryloff-Bogoliouboff's case[2] of the deterministic, reversible transition process in a compactum. The possibility of such extension was, in the case when X is a compactum together with the condition (2), was remarked by the present author[3] and was carried out by M. Beboutoff[4]. Our extension to the case of locally compact X will be of some interest in view of the application. By virtue of the systematic use of a generalisation of Riesz-Markoff-Kakutani's theorem[5] on linear functionals of the linear normed spaca $C(X)$, our treatment is somewhat simplified than that of Beboutoff.

2. *Three lemmas for the existence of invariant measures.*

Lemma 1. $C(X)$ is a separable linear normed space according to the norm

$$(3) \qquad \qquad \|f\| = \sup_x |f(x)| \,.$$

Proof. Easy.

---

1) Read May 12 (1947) at the annual meeting of the Mathematical Society of Japan.
2) Annals of Math., **38** (1937).
3) Proc. Imp. Acad. Tokyo, **16** (1940).
4) Rec. Math., **10** (1042).
5) A. Markoff: Rec. Math., **46** (1938). S. Kakutani: Proc. Imp. Acad. Tokyo, **19** (1943).

Lemma 2. Let a linear functional $L(f)$ on $C(X)$ be non-nagative, viz. $L(f) \geqq 0$ if $f(x) \geqq 0$ on X. The $L(f)$ is represented by a uniquely determined regular measure $\varphi(A)$ completely additive for Borel sets A with $\varphi(X) < \infty$ in the following integral form:

$$(4) \qquad L(f) = \int_x f(x)\, \varphi(dx).$$

Proof. The proof may be obtained by suitably modifying Riesz-Markoff-Kakutani's result for the case when X is a compactum.

A regular measure $\varphi(A)$ completely additive for Borel sets A with $\varphi(x) \leqq 1$ is callad an invariant measure (or a stable distribution), if, for any A,

$$(5) \qquad \varphi(A) = \int_x \varphi(dx)\, P(x, A).$$

Lemma 3. For any $f \in C(X)$ and for any invariant measure $\varphi(A)$

$$(6) \qquad \lim_{n \to \infty} n^{-1} \sum_{m=1}^{n} f_m(x) = f^*(x) \quad \left(f_m(x) = \int_x P^{(m)}(x, dy)\, f(y) \right) \quad \text{exists}$$

φ-almost everywhere, and

$$(7) \qquad \int_x f^*(x)\, \varphi(dx) = \int_x f(x)\, \varphi(dx).$$

This is Doob-Kakutani's individual ergodic theorem[6].

Let $_1f, _2f, \cdots$ be dense in the linear normed space $C(X)$. The existence of such sequence is guaranteed by the lemma 1. Applying the lemma 3 to $_1f, _2f, \cdots$ and summing up the exceptional sets of φ-measure zero, we see that there exists a set N of φ-measure zero with the property:

$$(8) \quad \text{for any } x \bar\in N \text{ and for any } f \in C(X),\ \lim_{n \to \infty} n^{-1} \sum_{m=1}^{n} f_m(x) = f^*(x) \text{ exists}$$

A Borel set $X' \leqq X$ is called to be of *the maximal probability* if $\varphi(X - X') = 0$ for every invariant measure φ. Thus the set X' for each point x of which $f^*(x)$ exists for all $f \in C(X)$ is of the maximal probability. Hence[7] if there exists an invariant measure φ with $\varphi(X) > 0$, then there exists a $g \in C(X)$ and a point x_0 such that

$$(9) \qquad g^{**}(x_0) = \varlimsup_{n \to \infty} n^{-1} \sum_{m=1}^{n} g_m(x_0) > 0.$$

Proof. If otherwise we would have $f^*(x) = 0$ on X for all $f \in C(X)$, and hence, by (7), $\int_x f(x)\, \varphi(dx) = 0$.

---

6) J. L. Doob: Trans. Amer. Math. Soc., **44** (1938). S. Kakutani: Proc. Imp. Acad. Tokyy, **16** (1940).

7) Cf. J. C. Oxtoby and S. Ulam: Annals of Math., **40** (1939).

Conversely let (9) be satisfied for a certain $g \varepsilon C'(X)$ and for a certain x_0.
Let a sequence $\{n'\}$ of natural number be such that $\lim\limits_{n'\to\infty} (n')^{-1} \sum\limits_{m=1}^{n'} g_m(x_0) = g^{**}(x_0)$.
By a diagonal method, we may choose a subsequence $\{n''\}$ of $\{n'\}$ such that
$\lim (n'')^{-1} \sum\limits_{m=1}^{n''} ({}_kf)_m(x)$ exists for $k=1, 2, \cdots$. By the denseness of $\{{}_kf\}$ in
$C(X)$, we see that $\lim (n'')^{-1} \sum\limits_{m=1}^{n''} f_m(x_0) = f^{***}(x_0)$ exists for every $f \varepsilon C(X)$.
We put $Lx_0(f) = f^{***}(x_0)$, then

$$(10) \qquad Lx_0(f) = Lx_0(f_1).$$

(Proof. $f_1 \varepsilon C(X)$ with f by the condition (2). Hence

$$Lx_0(f) = \lim_{n''\to\infty} (n'')^{-1} \sum_{m=1}^{n''} f_{m+1}(x_0) = f_1^{***}(x).)$$

By the lemma 2, there exists a measure $\varphi x_0(A)$ such that

$$Lx_0(f) = f^{***}(x_0) = \int_x f(x)\ \varphi x_0(dx).$$

Surely

$$(11) \qquad 0 \leqq \varphi x_0(X) \leqq 1.$$

We have, by (10),

$$\int_x f(x)\ \varphi x_0(dx) = \int_x\Big(\int_x P(x, dy)f(y)\Big)\ \varphi x_0(dx).$$

By letting $f(x)$ tend to the characteristic function of the Borel set A we see that
$\varphi x_0(A)$ is an invariant measure. $\varphi x_0(X) > 0$, since by the Lemma 3, $Lx_0(g) = g^{***}(x_0) = g^{**}(x_0) = \int_x g(x)\ \varphi x_0(dx) > 0$. Hence *a necessary and sufficient
condition for the non-existence of non-trival invariant measure is*

$$(12) \qquad \lim_{n\to m} n^{-1} \sum_{m=1}^{n} f_m(x) = f^*(x) = 0 \text{ on } X \text{ for any } f \varepsilon C(X).$$

We will call the process $P(x, A)$ in X *dissipative* if (12) is satisfied.

3. *Properties of the invariant measure* $\varphi_x(A)$. We assume that $P(x, A)$ is
not dissipative in X. Let D denote the set

$$\Big\{x;\, f^*(x) = \lim_{n\to\infty} n^{-1} \sum_{m=1}^{n} f(x) = 0 \text{ for all } f \varepsilon C(X)\Big\}$$

Since it is equal to

$$\{x;\, ({}_kf)^*(x) = 0 \text{ for } k=1, 2, \cdots\cdots\},$$

D is a Borel set. We call D the *dissipative part of X*. We have, as will be proved
below, $\varphi(D) = 0$ for any invariant measure φ. Since the process is assumed to
be not dissipative in X, $X_0 = XD$ is not void.

By the argument in 2, *there exists a Borel set* $X_1 \leqq X_0$ *with the property*:

(13)　for any $x \varepsilon X_1$, there corresponds a non-trival invariant measure

$$\varphi_x(A) \text{ and } f^*(x) = \lim_{n \to \infty} n^{-1} \sum_{m=1}^{n} f_m(x) = \int_{X_1} f(y)\, \varphi_x(dy) = \int_{X_1} f^*(y)\, \varphi_x(dy).$$

We have, for any invariant measure φ, (7) and hence, by (13),

$$\int_{X_1} f(y)\, \varphi(dy) = \int_{X_1} \left(\int_X f(z)\, \varphi_y(dz) \right) \varphi(dy).$$

Thus

(14)
$$\varphi(A) = \int_{X_1} \varphi_y(A)\, \varphi(dy),$$

which shows that *any invariant measure* φ *may be obtained as a convex combination of the invariant measure* $\varphi_y(A)$ *with y as parameter.*

We see from (11) and (14), that *the set*

$$X_2 = \{x;\ x \varepsilon X_1,\ \varphi_x(X_1) = 1\}$$

is of the maximal probability.

For any $f \varepsilon C(X)$ and for any invariant measure φ we have

(15)
$$\int_{X_2} \varphi(dx) \left(\int_{X_2} (f^*(y) - f^*(x))^2\, \varphi_x(dy) \right) = 0$$

since the left hand member is equal to

$$\int_{X_2} \varphi(dx) \left(\int_{X_2} f^*(y)^2\, \varphi_x(dy) \right) - 2 \left(\int_{X_2} f^*(x)\, \varphi(dx) \right) \left(\int_{X_2} f^*(y)\, \varphi_x(dy) \right)$$

$$+ \int_{X_2} f^*(x)^2\, \varphi(dx) \cdot \int_{X_2} \varphi_x(dy)$$

$$= \int_{X_2} f^*(y)^2\, \varphi(dy) - 2 \int_{X_2} f^*(x)^2\, \varphi(dx) + \int_{X_2} f^*(x)^2\, \varphi(dx)$$

by (13), (14), (7) and the definition of X_2.　By applying (15) to $_1f,\ _2f, \cdots\cdots$ *we see that the set*

$$X_3 = \{x;\ x \varepsilon X_2,\ \int_{X_2} (f^*(y) - f^*(x))^2\, \varphi_x(dy) = 0 \text{ for all } \varepsilon C(X)\}$$

is of the maximal probability.

4.　*Ergodic decomposition of* X_3. *For any* $x \varepsilon X_3$ *put*

$$E_x = \{y;\ y \varepsilon X_3,\ f^*(y) = f^*(x) \text{ for all } f \varepsilon C(X)\}$$

Then each E_x *contains a set* $\hat{E}_x$ *with the properties*:

(16)
$$\varphi_x(E_x) = \varphi_x(\hat{E}_x) \text{ and } P(y,\ \hat{E}_x) = 1 \text{ for any } y \varepsilon \hat{E}_x.$$

Proof. By the definition of X_3, $f^*(y) = f^*(x)$ if the measure $\varphi_x(A)$ has variation at y. Thus $\varphi_x(E_x) = \varphi_x(X_3) = 1$. Hence, by the invariance of φ_x,

$$1 = \varphi_x(E_x) = \int_{X_3} P(z,\ E_x)\, \varphi_x(dz) = \int_{E_x} P(z,\ E_x)\, \varphi_x(dz).$$

Since $0 \leqq P(z, E_x) \leqq 1$, there exists a Borel set $E' \leqq E_x$ such that

$$\varphi_x(E') = \varphi_x(E'_x) \text{ and } z \varepsilon E' \text{ implies } P(z, E_x) = 1.$$

Put $E^2 = \{z; z \varepsilon E', P(z, E') = 1\}$. Since

$$\int_{E'} P(z, E') \varphi_x(dz) = \varphi_x(E') = \varphi_x(E_x) = \int_{E'} P(z, E_x) \varphi_x(dz),$$

we have

$$\int_{E'} (P(z, E_x) - P(z, E')) \varphi_x(dz) = 0.$$

As $z \varepsilon E^2$ implies $P(z, E') = 1$ by the definition of E^2, we have $\varphi_x(E' - E^2) = 0$.

Next put $E^3 = \{z; z \varepsilon E^2, P(z, E^2) = 1\}$, then $\varphi_x(E^3) = \varphi_x(E_x)$. In this way, we obtain a sequence $\{E^n\}$ such that

$$E_x \geqq E' \geqq E^2 \cdots\cdots, \quad \varphi_x(E^n) = \varphi_x(E_x) \text{ and}$$

$$P(z, E^n) = 1 \text{ if } z \varepsilon E^{n+k} \ (n \geqq 0, \ k \geqq 1, \ E^0 = E_x).$$

We have thus obtained $\hat{E}_x = \overset{\infty}{\underset{n=1}{\frown}} E^n$ with the demanded property.   Q.E.D.

Therefore $P(y, A)$ defines a Markoff process in $\hat{E}_x$. We will show that this process is ergodic. Here the ergodicity is defined by the irreducibility of the process :

(17) $\hat{E}_x$ is not decomposable in two parts A, B such that $\varphi(A) \varphi(B) > 0$ for a certain invariant measure φ and $P(a, B) = 0$ when $a \varepsilon A$ and $P(b, A) = 0$ when $b \varepsilon B$.

Proof. Let $\varphi(A)$ be any invariant measure in $\hat{E}_x$ with $\varphi(\hat{E}_x) = 1$. Then, by (14), $\varphi(A) = \int_{\hat{E}_x} \varphi(dz) \varphi_x(A)$, $A \leqq \hat{E}_x$ Since, by the definition of $\hat{E}_x$, $\varphi_z(A) = \varphi_x(A)$ for $z \varepsilon \hat{E}_x$, we have

$$\varphi(A) = \varphi_x(A) \int_{\hat{E}_x} \varphi(dz) = \varphi_x(A).$$

Therefore *there is essentially a unique invariant measure $\varphi_x(A)$ in $\hat{E}_x$.* This proves the ergodicity. For let $\hat{E}_x$ be decomposed as in (17), then by the invariance of φ.

$$\varphi(C) = \int_{\hat{E}_x} P(z, C) \varphi(dz), \quad C \leqq \hat{E}_x.$$

Hence the measure ψ defnied by

$$\psi(C) = \varphi(C)/\varphi(A), \ C \leqq A$$
$$= 0, \ C \leqq B$$

is invariant in $\hat{E}_x$ and differs from the unique invariant measure φ_x.

Ergodic theorems for pseudo-resolvents

Proc. Japan Acad. **37** (1961) 422–425

(Comm. by Z. Suetuna, m.j.a., Oct. 12, 1961)

1. The theorem. Let X be a complete locally convex linear topological space, and $L(X, X)$ the algebra of all continuous linear operators on X into X. A pseudo-resolvent J_λ is a function on a subset $D(J)$ of the complex plane with values in $L(X, X)$ satisfying the resolvent equation

$$(1) \qquad J_\lambda - J_\mu = (\mu - \lambda) J_\lambda J_\mu.$$

We have, denoting by I the identity operator,

$$(2) \qquad (I - \lambda J_\lambda) = (I - (\lambda - \mu) J_\lambda)(I - \mu J_\mu)$$

and

$$(3) \qquad \lambda J_\lambda (I - \mu J_\mu) = (1 - \mu(\mu - \lambda)^{-1}) \lambda J_\lambda - \lambda(\lambda - \mu)^{-1} \mu J_\mu.$$

We see, by (1), that all J_λ, $\lambda \in D(J)$, have a common null space $N(J)$ and a common range $R(J)$. We also see, by (2), that all $(I - \lambda J_\lambda)$, $\lambda \in D(J)$, have a common null space $N(I - J)$ and a common range $R(I - J)$. $N(J)$ and $N(I - J)$ are closed linear subspace of X, but $R(J)$ and $R(I - J)$ need not be closed; we shall denote by $R(J)^a$ and $R(I - J)^a$ their closures respectively.

To formulate our ergodic theorems we prepare two lemmas.

Lemma 1. Let there exist a sequence $\{\lambda_n\}$ of numbers $\in D(J)$ such that

$$(4) \qquad \lim_{n \to \infty} \lambda_n = 0 \text{ and the family of operators } \{\lambda_n J_{\lambda_n}\} \text{ is equi-continuous.}$$

Then we have

$$(5) \qquad R(I - J)^a = P(J) = \{x \in X; \lim_{n \to \infty} \lambda_n J_{\lambda_n} x = 0\},$$

and hence

$$(6) \qquad N(I - J) \cap R(I - J)^a = \{0\}.$$

Lemma 1′. Let there exist a sequence $\{\lambda_n\}$ of numbers $\in D(J)$ such that

$$(4)' \qquad \lim_{n \to \infty} |\lambda_n| = \infty \text{ and the family of operators } \{\lambda_n J_{\lambda_n}\} \text{ is equi-continuous.}$$

Then we have

$$(5)' \qquad R(J)^a = I(J) = \{x \in X; \lim_{n \to \infty} \lambda_n J_{\lambda_n} x = x\}$$

and hence

$$(6)' \qquad N(J) \cap R(J)^a = \{0\}.$$

Our ergodic theorems read as follows.

Theorem 1. Let (4) be satisfied. Let, for a given $x \in X$, there exist a subsequence $\{\lambda_{n'}\}$ of $\{\lambda_n\}$ such that

$$(7) \qquad \text{weak-}\lim_{n'\to\infty} \lambda_{n'} J_{\lambda_{n'}} x = x_h \quad \text{exists.}$$

Then $x_h = \lim_{n\to\infty} \lambda_n J_{\lambda_n} x$ and $x_h \in N(I-J)$, $x_p = x - x_h \in P(J)$.

Corollary 1. Let (4) be satisfied, and let X be locally sequentially weakly compact. Then

$$(8)' \qquad X = N(I-J) \oplus R(I-J)^a = N(I-J) \oplus P(J) \quad \text{(direct sum).}$$

Theorem 1'. Let (4)' be satisfied. Let, for a given $x \in X$, there exist a subsequence $\{\lambda_{n'}\}$ of $\{\lambda_n\}$ such that

$$(7)' \qquad \text{weak-}\lim_{n'\to\infty} \lambda_{n'} J_{\lambda_{n'}} x = x_{h'} \quad \text{sxists.}$$

Then $x_{h'} = \lim_{n\to\infty} \lambda_n J_{\lambda_n} x$ and $x_{h'} \in I(J)$, $x_{p'} = x - x_{h'} \in N(J)$.

Corollary 1'. Let (4)' be satisfied, and let X be locally sequentially weakly compact. Then

$$(8)' \qquad X = N(J) \oplus R(J)^a = N(J) \oplus I(J) \quad \text{(direct sum).}$$

Remark. The pseudo-resolvent J_λ is a resolvent of a closed linear operator A if and only if $N(J) = 0$; in this case $R(J)$ coincides with the domain $D(A)$ of A. When J_λ is the resolvent of a closed linear operator A with domain $D(A)$ and range $R(A)$ both in a Banach space X, the Theorem 1 and Theorm 1' respectively correspond to the "abelian ergodic theorem at ∞ for semi-groups" and the "abelian ergodic theorem at 0 for semi-groups". Our formulation is more general than those due to E. Hille and R.S. Phillips.[1] The Theorem 1' for the case of a Banach space X is due to T. Kato,[2] and our formulation of two theorems is modelled after him.

2. Proofs of the theorems.

Proof of Lemma 1. We see, by (3) and (4), that $x \in R(I-J)$ implies $x \in P(J)$. Let $y \in R(I-J)^a$. Then, for any semi-norm q on X and $\varepsilon > 0$, there exists $x \in R(I-J)$ such that $q(y-x) < \varepsilon$. By (4), we have, for any semi-norm q' on X, $q'(\lambda_n J_{\lambda_n}(y-x)) \leqq Mq(y-x)$ where the positive constant M depends on q and q'. Thus we see that y must belong to $P(J)$.

Let conversely $x \in P(J)$. Then, for any semi-norm q on X and $\varepsilon > 0$, there exists λ_n such that $q(x - (x - \lambda_n J_{\lambda_n} x)) < \varepsilon$. Hence x must belong to $R(I-J)^a$.

The proof of Lemma 1' may be obtained similarly by making use of (3) and (4)'.

Proof of Theorem 1. Setting $\mu = \lambda_{n'}$ in (2) and letting $n' \to \infty$, we see, by (4), that $x_h \in N(I-J)$. We have thus

$$(9) \qquad \lambda_n J_{\lambda_n} x = x_h + \lambda_n J_{\lambda_n}(x - x_h),$$

and hence we have to prove that $(x - x_h) \in P(J)$. We prove it from

---

1) Functional analysis and semi-groups, Providence, 502 (1957).

2) Remarks on pseudo-resolvents and infinitesimal generators of semi-groups, Proc. Japan Acad. **35**, 467-468 (1959).

Lemma 1 and $(x-\lambda_{n'}J_{\lambda_{n'}}x)\in R(I-J)$, observing that, in a locally convex linear topological space X, any closed linear subspace is weakly closed.

Proof of Theorem 1'. Setting $\mu=\lambda_{n'}$ in (3) and letting $n'\to\infty$, we see, by (4)', that $\lambda J_\lambda(x-x_{h'})=0$, that is, $x_{p'}=x-x_{h'}\in N(J)$. On the other hand, we have $x_{h'}\in R(J)^a$. For, $\lambda_{n'}J_{\lambda_{n'}}x\in R(J)$ and the closed linear subspace $R(J)^a$ is weakly closed. Thus, by Lemma 1', $x_{h'}\in I(J)$.

3. Some applications.

i) *An application to fractional powers of closed operators.* Let J_λ be the resolvent $(\lambda I-A)^{-1}$ of a closed linear operator A and $D(J)=\{\lambda;\lambda>0\}$. If we assume that

(4)'' the family of operators $\{\lambda(\lambda I-A)^{-1};\lambda>0\}$ is equi-continuous, then the fractional power A^α of A, $0<\alpha<1$, may be defined as the smallest closed extension of the operator

$$(10)\qquad A^\alpha x=\frac{\sin\alpha\pi}{\pi}\int_0^\infty \lambda^{\alpha-1}(\lambda I-A)^{-1}(-Ax)d\lambda \text{ for } x\in D(A).^{3)}$$

Then the equation

$$(11)\qquad \lim_{\alpha\downarrow 0} A^\alpha x=x$$

is satisfied for $x\in D(A)\cap P(J)$. This we see from the fact that such an x satisfies the "Poisson equation"

$$(12)\qquad \lim_{\alpha\downarrow 0}(\lambda I-A)^{-1}(-Ax)=x.$$

ii) *An application to potential theory.* Suggested by a special case when J_λ is the resolvent of the Laplacian, we may call the elements of $N(I-J)$ "harmonic" and the elements of $P(J)$ "potentials." Then the Theorem 1 may be considered as an analogue of F. Riesz decomposition for subharmonic functions. The analogy is more close if we introduce the notion of "subharmonicity". To this end, we assume that a notion of "positivity", denoted by $x\geq 0$, is defined in X in such a way that X is a semi-ordered linear space satisfying the condition:

(13) a monotone increasing bounded sequence of elements $\in X$ converges weakly to an element of X which is greater than the elements of the sequence.

We further assume that $D(J)$ contains an open interval $(0,\lambda_0)$ and the operator J_λ is positive for $\lambda\in(0,\lambda_0)$ in the sense that $x\geq 0$ implies $J_\lambda x\geq 0$. We shall call an element $x\in X$ "subharmonic" if it satisfies

(14) $\lambda J_\lambda x\geq x$ for some and hence for all $\lambda\in(0,\lambda_0)$.

Then we can prove the following corollary of Theorem 1.

---

3) A. M. Balakrishnan: Fractional powers of closed operators and semi-groups generated by them, Pacific J. of Math.; K. Yosida: Fractional powers of infinitesimal generators and the analyticity of the semi-groups generated by them, Proc. Japan Acad. **36**, 86–89 (1960); T. Kato: Note on fractional powers of linear operators, Ibid., 94–96.

Corollary 2.　Let

(15)　　the family of operators $\{\lambda J_\lambda; 0<\lambda<\lambda_0\}$ is equi-continuous.

Then a "subharmonic" element x is uniquely decomposed as the sum of a "harmonic" element x_h and a "potential" x_p. The "harmonic" part x_h of x is given by $x_h=\lim_{\lambda\downarrow 0}\lambda J_\lambda x$ and x_h is the "least harmonic majorant" of x.

Proof.　By (2) and the positivity of J_λ, we see that

(16)　　　　　　　　　$\lambda<\mu$ implies $\lambda J_\lambda x\geq\mu J_\mu x\geq x$.

Therefore, by (13) and (12), weak-$\lim_{\lambda\downarrow 0}\lambda J_\lambda x=x_h$ exists, and the first part of the Corollary is proved.

Let a "harmonic" element x_H satisfy $x_H\geq x$. Then, by the positivity of λJ_λ and the "harmonicity" of x_H, we have

$$x_H=\lambda J_\lambda x_H\geq\lambda J_\lambda x \text{ and hence } x_H\geq x_h.$$

iii)　*An application to semi-group theory.*　The following corollaries of Theorem 1′ are obtained after T. Kato, loc. cit. in the footnote 2.

Corollary 2′.　If J_λ is a pseudo-resolvent satisfying (4)′ and if $R(J)$ is dense in X, then $N(J)=\{0\}$ and hence J_λ is a resolvent.

Corollary 3′.　If X is locally sequentially weakly compact and J_λ is the resolvent of a closed linear operator A satisfying (4)′, then the domain $D(A)$ of A is dense in X.

IV. Spectral Theorems, Vector Lattices and Miscellanea

Comments (Hikosaburo Komatsu)

Professor Yosida's earliest contribution to Functional Analysis is the paper [5], in which he extended Hilbert's theory of integral operators to operators K satisfying the conditions $K^3 = KK^*K$ and/or $K^*K^2 = (K^*)^2K$.

The most important contribution in his earliest career is his dissertation [14]. In the same issue of the journal he and M. Nagumo introduced Banach algebras under the name of metric complete rings five years earlier than I. M. Gelfand (Mat. Sb., **9** (1941)). Yosida's main purpose in [14] was to prove that a locally compact group embedded in a Banach algebra is a Lie group as a partial answer to Hilbert's fifth problem. To do so he defined exponentials exp A, logarithms ln A and inverses $(1 - A)^{-1}$ by power series. On the other hand, Nagumo employed the complex integration to obtain the Jordan decomposition of an element at its isolated spectrum. They did not reach, however, the Gelfand-Mazur theorem asserting that a division Banach algebra is isomorphic to C, nor the Gelfand representation.

In the meantime, he extended in [29] the Fredholm-Riesz-Schauder theory to operators K satisfying

$$(1) \qquad \qquad \|K^m - V\| < 1$$

for a natural number m and a completely continuous operator V.

In his joint paper [27] with Mimura and Kakutani it is proved that an integral operator $K: L^1([0, 1]) \to L^1([0, 1])$ with bounded measurable kernel $K(x, y)$ is not necessarily completely continuous but its product KN with another integral operator N with bounded measurable kernel is completely continuous and hence K satisfies condition (1). The same is true if K is regarded as an operator in the space of bounded measures $\mathfrak{M}^1([0, 1])$ on Borel sets. They also prove that K is weakly completely continuous.

Later N. Dunford and B. J. Pettis (Trans. Amer. Math. Soc., **40** (1940)) proved that a weakly completely continuous operator $K: L^1([0, 1]) \to L^1([0, 1])$ is an integral operator and its product with another weakly completely continuous operator is completely continuous. This became a fundamental result on the Radon-Nikodym property of Banach spaces.

In the same year as the paper [27] was written, Yosida formulated in [25] the mean ergodic theorem in terms of weakly completely continuous operators. Probably weakly completely continuous operators were made use of for the first time in these papers.

In 1940–1944 Yosida concentrated his efforts on abstract formulation of spectral theorems and new proofs. At first his approach was lattice-theoretical. The paper [37] defines Pythagorian rings as σ-complete vector lattices with the commutative ring structure over $\mathbf{R}$ such that $X > 0$ if and only if $X = Y^2$ for a $Y \neq 0$. Then every X has the spectral decomposition

$$(2) \qquad X = \int_{-\infty}^{\infty} \lambda dE_\lambda$$

with a monotone class E_λ of idempotent elements.

In his papers [40], [42], [43], [44] and [45] he sought for the analogy of the Gelfand theory in the setting of vector lattices. For example, he shows in [42] that if E is a vector lattice with an Archemedean unit I, then the quotient lattice E/R modulo the "radical" R is isomorphic to a dense subspace of the space $C(\mathfrak{N})$ of all continuous functions on the compact space $\mathfrak{N}$ consisting of all "maximal ideals" N of E. If E is a Banach space under the norm $\|X\| = \inf\{\alpha \geq 0; \ -\alpha I \leq X \leq \alpha I\}$, then E is isomorphic to $C(\mathfrak{N})$ as a Banach lattice. This is the famous characterization as an abstract M-space of the space of all continuous functions on a compact set by S. Kakutani (Ann. of Math., **41** (1941)) and by M. and S. Krein. If E is, moreover, a ring and σ-complete, then it is proved that every $X \in E$ has the spectral decomposition (2) in his joint works [44] and [45] with Nakayama.

The paper [41] gives a new proof of the Random-Nikodym theorem and its abstract version for σ-complete vector lattices, which is closely connected with Kakutani's concrete representation of abstract L-spaces (ibid.).

The series [47], [48], [49], [50], [51] and [53] are applications of the Gelfand theory to spectral theorems and others. The paper [47] establishes the simultaneous spectral decompositions of a commutative family A of normal operators closed under adjoint by the Gelfand representation of the W^*-algebra A'' of double commutant. His textbook "Functional Analysis" adopts this method for the proof of spectral theorems. The Fredholm-Riesz-Schauder theory is discussed in the next paper [48].

The papers [49] and [50] concern inequalities of spectral functions which are useful to estimate the greatest and other eigenvalues.

In [51] and [53] he tries to simplify M. Krein's proof of the Plancherel theorem and D. A. Raikov's proof of the Bochner theorem for locally compact abelian groups G by representing the group ring $L^1(G)$ by the C^*-algebra R generated by the convolutions by $x \in L^1(G)$ on $L^2(G)$. Unfortunately he misses proving the surjectivity of the correspondence of the maximal ideals in R to those in $L^1(G)$ (cf. Kadison-Ringrose's book "Fundamental of the Theory of Operator Algebras", Vol. I., pp. 195–197.). Moreover, the function space $L^1(G) \cap C(G)$ has to be replaced by the linear combinations of the convolutions $x * y$ for $x, y \in L^1(G) \cap L^2(G)$.

In [55] he proves that two self-adjoint operators T_1 and T_2 are unitarily equivalent if and only if the commutants T_1' and T_2' are $*$-isomorphic under a correspondence which maps T_1 to T_2.

His joint paper [68] with Hewitt is a fundamental study of finitely additive measures on σ-algebras of sets. It is proved, in particular, that the dual of L^∞ is the space of all bounded finitely additive measures which vanish on negligible sets.

The paper [39] with Fukamiya is by now a standard proof of the Krein-Milman theorem.

In [54] he gives necessary and sufficient conditions in order that a bounded continuous function $f(t)$ on $\mathbf{R}$ be the inverse Fourier transform of a bounded measure, a bounded absolutely continuous measure, or a bounded singular measure, respectively, by the behavior of the Fourier transform of $f(t)(\sin n/t)^2/(n/t)^2$ as $n \to \infty$.

For physicists and engineers he gives in [81] an elementary proof of the fact that a sequentially continuous linear functional on the distributions $\mathscr{D}'(\mathbf{R}^n)$ is a function in $\mathscr{D}(\mathbf{R}^n)$ using only the Riesz representation theorem in Hilbert spaces and Fourier series. The paper [104] is an elementary proof of the strict monotonicity of functions $f(x)$ on $[a, b]$ with derivatives $f'(x) > 0$ on (a, b).

Integral operator with bounded kernel

(*with Y. Mimura and S. Kakutani*)

Proc. Imp. Acad. Tokyo **14** (1938) 359–362

(Comm. by T. TAKAGI, M.I.A., Dec. 12, 1938.)

§ 1. Let $K(x, y)$ be bounded and measurable in the square $0 \leq x \leq 1,\ 0 \leq y \leq 1$. Consider the integral operator K which transforms the Banach space $(L)^{1)}$ in (L).

$$(1) \qquad f \to Kf = g: \quad g(y) = \int_0^1 f(x)\, K(x, y)\, dx.$$

It is to be noted that such an operator is not always completely continuous$^{2)}$ in (L). This may be shown by an example (§ 3). We can, however, prove the following

Theorem 1. Let $N(x, y)$ and $K(x, y)$ be bounded and measurable in $0 \leq x \leq 1,\ 0 \leq y \leq 1$. Then the integral operator P defined by the bounded Kernel $P(x, y) = \int_0^1 N(x, z)\, K(z, y)\, dz$ is completely continuous as an operator which maps (L) in (L).

Remark. The integral operator (1) may also be considered as a linear operator which maps (L) in $(M),^{3)}$ (M) in (M) or (M) in (L).

Proof of Theorem 1 : Denote by N and K the integral operators which correspond to the kernels $N(x, y)$ and $K(x, y)$ respectively. P may be considered as a combination of two operators N and K performed successively in this order, where N is an operator which maps (L) in (M) and K is the one which maps (M) in (L): $f \in (L) \to Nf = g \in (M) \to Kg(=Pf) = h \in (L)$.

The unit sphere $\|f\|_L \leq 1$ of (L) is mapped by N on a set contained in the sphere $\|g\|_M \leq n$ of (M), where $n = \underset{0 \leq x,\, y \leq 1}{\text{l. u. b.}} |N(x, y)|$. Hence it is sufficient to prove the

Theorem 2. The integral operator K with bounded kernel $K(x, y)$ is completely continuous as an operator which maps (M) in (L).

Proof : We extend the definition domain of $K(x, y)$ to the infinite square $-\infty < x < +\infty,\ -\infty < y < +\infty$, by putting $K(x, y) = 0$ if the point (x, y) is outside the square $0 \leq x \leq 1,\ 0 \leq y \leq 1$. Let $Kg = h$, where $g \in (M),\ \|g\|_M \leq 1$. By Fubuni-Tonelli's theorem, we have

---

1) (L) is the space of all the measurable functions $f(x)$ which are absolutely integrable in $0 \leq x \leq 1$. For any $f \in (L)$, we define its norm by $\|f\|_L = \int_0^1 |f(x)|\, dx$.

2) A linear operator which maps the Banach space E_1 in another Banach space E_2 is called to be completely continuous if it maps the unit sphere $\|x\| \leq 1$ of E_1 on a compact (in E_2) set of E_2.

3) (M) is the space of all the bounded measurable functions defined in $0 \leq x \leq 1$. For any $f \in (M)$ we define its norm by $\|f\|_M = \underset{0 \leq x \leq 1}{\text{ess. max.}} |f(x)|$.

$$\int_{-\infty}^{\infty} |h(y+\delta)-h(y)|\,dy = \int_{-\infty}^{\infty} dy \left| \int_{-\infty}^{\infty} g(x)\left(K(x,y+\delta)-K(x,y)\right)dx \right|$$

$$\leqq \int_{-\infty}^{\infty}\int_{-\infty}^{\infty} |K(x,y+\delta)-K(x,y)|\,dxdy .$$

The last integral (which is independent of the particular choice of $g \in (M)$, with $\|g\|_M \leqq 1$) tends to zero as δ tends to zero, by Lebesgue's theorem. Further we have $|h(y)| \leqq k$ for any $y(0 \leqq y \leqq 1)$ and for any $g \in (M)$ with $\|g\|_M \leqq 1$, where $k = \text{l. u. b.}_{0 \leqq x,\, y \leqq 1} |K(x,y)|$. Thus by Kolmogoroff-Riesz's theorem[1] the image by K of the unit sphere $\|g\|_M \leqq 1$ of (M) is compact $\left(\text{in } (L)\right)$ in the topology of (L).

Remark. Another proof given below does not appeal to Kolmogoroff-Riesz's theorem.

Consider P as a succession of two operators N and K, each transforming (L) in (L). N is weakly completely continuous as an operator which maps (L) in (L); viz. N maps the unit sphere $\|f\|_L \leqq 1$ of (L) on a set of (L) which is weakly compact in (L).[2] Thus any sequence $\{f_n\}$ with $\|f_n\|_L \leqq 1$ contains a subsequence $\{f_{n_\nu}\}$ such that the sequence $\{g_{n_\nu}\}$ $(g_{n_\nu}=Nf_{n_\nu})$ converges weakly to some element (function) g_0 of (L). As the conjugate space of (L) is (M), we have, by the boundedness of $K(x,y)$

$$\lim_{\nu \to \infty} \int_0^1 g_{n_\nu}(x)\,K(x,y)\,dx = \int_0^1 g_0(x)\,K(x,y)\,dx \qquad \text{for any } y(0 \leqq y \leqq 1).$$

We have $\text{l. u. b.}_{0 \leqq y \leqq 1} |h_{n_\nu}(y)| \leqq n \cdot k$, $h_{n_\nu}(y)=\int_0^1 g_{n_\nu}(x)\,K(x,y)\,dx$. Hence by Lebesgue's theorem, the sequence $\{h_{n_\nu}(y)\}$ must converge to $h_0(y)= \int_0^1 g_0(x)\,K(x,y)\,dx$ in the strong topology of (L).

§ **2.** Let $(\mathfrak{M})$ denote the Banach space consisting of all the set functions $\varphi(E)$, which are completely additive for Borel set of the interval $0 \leqq x \leqq 1$. For any $\varphi \in (\mathfrak{M})$ the norm of φ is given by $\|\varphi\|=$ total variation of $\varphi(E)$ in $0 \leqq x \leqq 1$. A bounded measurable kernel $K(x,y)$ defines an integral operator which maps $(\mathfrak{M})$ in $(\mathfrak{M})$:

$$\varphi \to K\varphi = \psi: \quad \psi(E)=\int_E dy \int_0^1 \varphi(dx)\,K(x,y) .$$

Such an operator is not always completely continuous in $(\mathfrak{M})$. The example for the space (L) shows this fact. Corresponding to Theorem 1 we may give the

---

1)　A. Kolmogoroff: Über Kompaktheit der Funktionenmengen bei der Konvergenz im Mittel, Nachr. Ges. Wiss. Göttg., Math.-phys. Kl. 1931, 60–63.

　　M. Riesz: Sur les ensembles compacts de fonctions sommables, Acta Litt. Sci. Szeged, **6** (1933), 136–142.

2)　K. Yosida and S. Kakutani: Applications of Mean Ergodic Theorem to the problem of Markoff's process, Proc. **14** (1938), 333.

Theorem 3. Let $N(x, y)$ and $K(x, y)$ be bounded and measurable in $0 \leq x \leq 1$, $0 \leq y \leq 1$. Then the integral operator P defined by the bounded kernel $P(x, y) = \int_0^1 N(x, z)\, K(z, y)dz$ is completely continuous as an operator which maps $(\mathfrak{M})$ in $(\mathfrak{M})$.

Proof: This may be carried out as in the case of Theorem 1. Firstly, N may be considered as a bounded linear operator which maps $(\mathfrak{M})$ in (M):

$$\varphi \to K\varphi = g: \quad g(y) = \int_0^1 \varphi(dx)\, K(x, y) .$$

Secondly, K is a completely continuous linear operator which maps (M) in $(\mathfrak{M})$:

$$g \to Kg = \psi: \quad \psi(E) = \int_E dy \int_0^1 g(x)\, K(x, y)\, dx .$$

Remark. Another proof analogous to the remark above is also possible. Firstly, N is weakly completely continuous as an operator which maps $(\mathfrak{M})$ in (L). Secondly K is a linear operator which maps the weakly convergent sequence of (L) into the strongly convergent sequence of $(\mathfrak{M})$.

§ **3.** *Example.* We shall construct in this chapter a bounded measurable kernel $K(x, y)$ defined in $0 \leq x \leq 1$, $0 \leq y \leq 1$, such that the corresponding integral operator is not completely continuous as an operator which maps (L) in (L).

Put

$$K(x, y) = 2 \quad \text{if} \quad \frac{1}{2^n} < x \leqq \frac{1}{2^{n-1}}, \quad \frac{2k}{2^n} < y \leqq \frac{2k+1}{2^n} ,$$

$$k = 0, 1, \ldots, 2^{n-1}-1,$$

$$K(x, y) = 0 \quad \text{if} \quad \frac{1}{2^n} < x \leqq \frac{1}{2^{n-1}}, \quad \frac{2k+1}{2^n} < y \leqq \frac{2k+2}{2^n} ,$$

$$k = 0, 1, \ldots, 2^{n-1}-1 ,$$

$$n = 1, 2, \ldots .$$

$K(x, y)$ is defined in $0 < x \leqq 1$, $0 < y \leqq 1$. At the points where $x = 0$ or $y = 0$ put $K(x, y) = 1$. This $K(x, y)$ is a required one. It is clear that K is bounded and measurable. In order to prove that the corresponding integral operator is not completely continuous, put

$$f_n(x) = 2^n \quad \text{if} \quad \frac{1}{2^n} < x \leqq \frac{1}{2^{n-1}}$$

$$= 0 \quad \text{elsewhere in } 0 \leqq x \leqq 1 .$$

Then we have $\|f_n\|_L = 1$ $(n = 1, 2, \ldots)$ and $\{g_n\}$ $\left(g_n(y) = \int_0^1 f_n(x)\, K(x, y)\, dx\right)$ is not compact in (L) in the topology of (L). For we have, by easy calculation, $\|g_m - g_n\|_L = 1$ for any $m \neq n$.

The importance of this example consists in the fact that $K(x, y)$

may be considered as a density of transition probability of a simple Markoff's process. (See the following paper of K. Yosida: Operator-theoretical Treatment of the Markoff's Process.)[1]

---

1) After the present paper was completed, we found that Theorem 1 was already obtained by J. Sirvint in another way. *Cf.* J. Sirvint: Sur les transformations inté-grales de l'espace *L*, C. R. URSS. **18** (1938), 255–257. He also obtained an example of a bounded and measurable kernel of the said property. His example coincides with ours up to an additive constant 1.

Quasi-completely-continuous linear functional operations

Japan. J. Math. **15** (1939) 297–301

(Received January 1, 1939.)

§ 1. We owe to F. Riesz and J. Schauder the "determinantenfrei" treatment of the Fredholm's theory of integral equations[1]. They showed that the so-called Fredholm's alternative is also valid for completely continuous (c.c.) linear functional equations in complex Banach spaces. It is to be remarked that there exists an important class of integral equations to which the Riesz-Schauder's theory cannot be applied directly. Consider the integral equation

$$(1) \qquad f = \lambda K \cdot f \qquad \left(f(x) = \lambda \int_0^1 K(x, y) f(y) dy \right),$$

where $K(x, y)$ is bounded and measurable in the square $O \leq x, y \leq 1$ and $f(x)$ belongs to the Banach space (L). Such linear operator K is not, in general, c.c. as an operator in (L)[2]. Nevertheless the classical analysis of Fredholm also applies to (1), as was shown by M. Fréchet[3]. In a preceding note it is shown that the operator K^2 is c.c. in (L)[4]. Hence we are lead to introduce the following definition:

A linear operator K in complex Banach space $\mathfrak{B}$ is called quasi-completely-continuous (q. c. c.) if there exist a positive interger m and a c. c. linear operator V in $\mathfrak{B}$ such that

$$(2) \qquad \| K^m - V \| < 1 .$$

In particular, K is called *m-quasi-completely-continuous* (*m*-q. c. c.) if K^m is c. c. The 1-q. c. c. is equivalent to the usual c. c.

In the present note I intend to show that the validity of the Riesz-Schauder's theory can be extended to the q. c. c. linear functional equa-

---

[1] F. Riesz: Acta Math. **41** (1918), 71-98. J. Schauder: Studia Math. **2** (1930), 183-196.

[2] K. Yosida, Y. Mimura and S. Kakutani: Proc. Imp. Acad. **14** (1938), 359-362.

[3] M. Fréchet: Quart. J. of Math. **5** (1934), 106-144.

[4] See the note cited in the footnote 2). Cf. also J. Sirvint: C.R. URSS **18** (1938), 255-257.

tions([5]). Fréchet's results quoted above are of course contained in it. The extension of Fréchet - Kryloff - Bogoliouboff's theorem given in a preceding note will also be generalised([6]).

§ 2. In $[A]$ I obtained some results concerning the q. c. c. linear operation K with $m = 1$. The Lemma 1, 2 below generalise these results to the case $m \geq 1$. Throughout this paper we assume that K satisfies (2) unless otherwise is stated.

Lemma 1. The proper values([7]) of K does not accumulate to the point not exterior of the unit cercle of the complex λ plane.

Proof. The proof for the case $m = 1$ ($[A]$, Lemma 1) also proves the Lemma for the general case $m \geq 1$.

Remark. The only accumulation point of the proper values of a m-q. c. c. linear operator K is the point ∞. For λK is q. c. c. with any complex number λ. The well known theorem of F. Riesz corresponds to the case $m = 1$.

Lemma 2. There exists a positive constant ε such that the resolvent R_λ([8]) of K is meromorphic for $1 - \varepsilon < |\lambda| < 1 + \varepsilon$, the poles of R_λ being the proper values of K with modulus 1. Let the Laurent expansion of R_λ at the pole $\lambda_0(|\lambda_0| = 1)$ be $R_\lambda = \sum\limits_{n = -\infty}^{+\infty} (\lambda - \lambda_0)^n K_n$. Then $K_n (n < 0)$ are all c. c.

Proof. This Lemma is proved for the case $m = 1$ in $[A]$. Hence there exists a positive constant ε such that the resolvent S_λ of K^m is meromorphic for $1 - \varepsilon < |\lambda| < 1 + \varepsilon$. By the identity

$$(E - \lambda^m K^m) = (E - \lambda K)(E + \lambda K + \cdots + \lambda^{m-1} K^{m-1})$$

$$= (E + \lambda K + \cdots + \lambda^{m-1} K^{m-1})(E - \lambda K) ,$$

we see that R_λ defined by

$$E + \lambda R_\lambda = (E + \lambda K + \cdots + \lambda^{m-1} K^{m-1})(E + \lambda^m S_{\lambda^m})$$

$$= (E + \lambda^m S_{\lambda^m})(E + \lambda K + \ldots + \lambda^{m-1} K^{m-1})$$

---

(5) After the present work is completed, I knew the paper of S. Nikolskij: C. R. URSS **16** (1926), 315-319. He proves that the Riesz-Schauder's results can be extended to $m - $q.c.c. linear functional equations. His method is somewhat different from ours.

(6) K. Yosida: Proc. Imp. Acad. **14** (1938), 286-291. This note will be cited as $[A]$. Cf. S. Kakutani: Ibid.

(7) λ is called a proper value of K if there exists $x \neq 0$ of $\mathfrak{B}$ such that $x = \lambda K \cdot x$.

(8) A linear continuous operator R_λ is called the resolvent of K if $(E + \lambda R_\lambda)$ constitutes the inverse $(E - \lambda K)^{-1}$ of $E - \lambda K : E = (E - \lambda K)(E + \lambda R_\lambda) = (E + \lambda R_\lambda)(E - \lambda K)$. Here E denotes the identity operator in $\mathfrak{B}$.

is the resolvent of K. The rest of the Lemma is easily be proved as it holds for S_λ.

Remark. If K is m-q. c. c., the resolvent R_λ of K is meromorphic for $|\lambda| < +\infty$. For λK is q. c. c. with any complex number λ. This constitutes a generalisation of Nagumo-Nikolskij's theorem([9]).

Lemma 3. Let $\lambda_0(|\lambda_0| = 1)$ be a proper value of K. Then we have the decomposition

$$(3) \qquad K = A + B, \qquad AB = BA = 0$$

such that the resolvents of A, B respectively coincides with the principal part, regular part of R_λ in the vicinity of $\lambda = \lambda_0$. A is completely continuous and B is continuous.

Proof. By substituting the Laurent expansion $R_\lambda = \sum\limits_{n=-\infty}^{+\infty} (\lambda-\lambda_0)^n K_n$ in the resolvent equation $R_\lambda - R_\mu = (\lambda-\mu)R_\lambda R_\mu$ we obtain the Lemma, as R_λ is given by $\sum\limits_{n=1}^{+\infty} \lambda^{n-1} K^n$ for sufficiently small λ: $A = \sum\limits_{n=-\infty}^{-1} (-\lambda_0)^n K_n$, $B = K - A$([10]).

Remark. The above Lemma holds at any proper value of K if K is m-q. c. c. This corresponds to the classical results in the theory of the integral equations.

An extension of Fréchet-Kryloff-Bogoliouboff's theorem. Let a q. c. c. linear operator K in $\mathfrak{B}$ satisfy condition

$$(4) \qquad \| K^n \| \leqq \text{ a constant } \alpha \qquad (n = 1, 2, \ldots).$$

Then the proper values of K with modulus 1 are isolated proper values of finite multiplicities. Let these proper values be denoted by $\lambda_1, \lambda_2, \ldots, \lambda_k$. There exist c. c. linear operators $T_1, T_2, \ldots, T_k$ and a continuous linear operator S such that

$$(5) \quad \begin{cases} K = \sum\limits_{i=1}^{k} \lambda_i^{-1} T_i + S, \quad T_i^2 = T_i, \quad T_i T_j = 0 \ (i \neq j), \quad T_i S = S T_i = 0, \\[2mm] \| S^n \| \leqq \dfrac{\varepsilon}{(1+\varepsilon)^n} \ (n = 1, 2, \ldots) \text{ with positive constants } \varepsilon, \delta. \end{cases}$$

Proof. The proof given in [A] for the case $m = 1$ also applies to the general case $m \geqq 1$, by virtue of the above Lemmas.

([9]) M. Nagumo: Jap. J. of Math. **13** (1936), 75-80. S. Nikolskij: loc. cit.

([10]) For the detail see M. Nagumo: loc. cit.

§ 3. Let T be a continuous linear operator in $\mathfrak{B}$. The linear functionals $f(x)$ on $\mathfrak{B}$ constitutes a Banach space $\bar{\mathfrak{B}}$ with the norm $\|f\| = \text{l. u. b.}_{\|x\|=1} |f(x)|$. $\bar{\mathfrak{B}}$ is called the conjugate space of $\mathfrak{B}$. T defines a continuous linear operator $\bar{T}$ in $\bar{\mathfrak{B}}$, the conjugate operator of T, by the equation

$$g = \bar{T} \cdot f, \qquad g(x) = f(T \cdot x).$$

Let K be a q. c. c. linear operator in $\mathfrak{B}$. Consider the linear homogeneous functional equation

$$(6) \qquad x - K \cdot x = 0 \qquad (x \in \mathfrak{B})$$

and its conjugate equation

$$(7) \qquad f - \bar{K} \cdot f = 0 \qquad (f \in \bar{\mathfrak{B}})$$

Then

i). *The equations (6), (7) admit the same numbers of linearly independent solutions.*

Next consider the inhomogeneous equation

$$(8) \qquad x - K \cdot x = y \qquad (x,\, y \in \mathfrak{B})$$

and its conjugate equation

$$(9) \qquad f - \bar{K} \cdot f = g \qquad (f,\, g \in \mathfrak{B})$$

Then

ii). *For a given y, (8) admits solution x if and only if y is orthogonal to all the solutions f of (7): $f(y) = 0$. For a given g, (9) admits solution f if and only if g is orthogonal to all the solutions x of (6): $g(x) = 0$.*

Proof. We have, by the Lemma 3, $K = A + B$, $AB = BA = 0$, where A is c. c. and $(E - B)$ admits continuous inverse $(E - B)^{-1}$. Hence we obtain

$$E - K = (E - A)(E - B) = (E - B)(E - A).$$

i). From $x - K \cdot x = (E - B)(E - A)x$, $f(x) - f(K \cdot x) = f(x - K \cdot x)$

$= f\big((E - A)(E - B)x\big)$ we see that (6) and (7) are respectively equivalent to

$$x - A \cdot x = 0, \qquad f - \bar{A} \cdot f = 0,$$

as $(E-B)^{-1}$ exists. Since A is c. c., these last two equations admit the same number of linearly independent solutions (Schauder's result).

Q. E. D.

ii). The necessities of the propositions are evident. We will show the sufficiencies.

Let y be orthogonal to all the solutious f of (7). Put $(E-B)^{-1}\cdot y=z$, then $f\big((E-B)z\big)=0$. Hence, as $(E-B)^{-1}$ exists, z is orthogonal to all the solutions f' of $f'-\bar{A}\cdot f'=0$. A being c. c. , there exists a solution x of $x-A\cdot x=z$ (Schauder's result). Therefore $x-K\cdot x=(E-B)(x-A\cdot x)=(E-B)z=y$.

Secondly let g be orthogonal to all the solutions x of (6). $x-K\cdot x=0$ is equivalent to $x-A\cdot x=0$ as shown above. Thus $g'\big(g'(x)=g((E-B)^{-1}\cdot x)\big)$ is orthogonal to all the solutions x of $x-A\cdot x=0$. A being c. c., there exists a solution f of $f-\bar{A}\cdot f=g'$ (Schauder's result). Therefore $f(y-A\cdot y)=g\big((E-B)^{-1}\cdot y\big)$ for all $y\in\mathfrak{B}$. Hence

$$f(x-K\cdot x)=f\big((E-A)(E-B)x\big)=g(x) \text{ for all } x\in\mathfrak{B}, \text{ viz. } f-\bar{K}\cdot f=g.$$

Q. E. D.

Remark. Let K be a m-q. c. c. linear operator. Then λK is q. c. c. for any complex number λ. Therefore, in this case, i) and ii) hold for the equations

$$x-\lambda K\cdot x=0, \qquad f-\lambda\bar{K}\cdot f=0,$$

and

$$x-\lambda K\cdot x=y, \qquad f-\lambda\bar{K}\cdot f=g$$

respectively, λ being any complex numbers. This is Nikolskij's results. When K is c. c. (1-c. c.), we obtain thus the Riesz-Schauder's result.

———

On the theory of spectra

Proc. Imp. Acad. Tokyo **16** (1940) 378–383

(Comm. by T. TAKAGI, M.I.A., Oct. 12, 1940.)

The " algebraization " of the spectral theory, inaugurated by J. von Neumann, H. Freudenthal and S. Steen, was taken up recently by S. Kakutani, F. Riesz, M. H. Stone and B. Vulich,[1] and was treated with their respective methods and results. The purpose of the present note is to give a ring-lattice-theoretic treatment of the problem, stressing the analogy to the field of real numbers. Without assuming metrical (even topological) nor divisibility axiom, a *characterisation of the function ring of the Borel-measurable functions*[2] is obtained. Thus the results may be applied to the operator theory as well as to the theory of probability.

§ 1. *Axioms of Pythagorean ring.* A system $\Re$ of elements A, $B, \ldots, X, Y, Z$ is called a " Pythagorean ring " if it satisfies the following axioms.

(A–1) $\Re$ is a commutative, associative ring with unit I, admitting the field of real numbers as coefficients (Operatoren).—The real numbers will be denoted by small greak letters.

(A–2) $X^2 = 0$ implies $X = 0$.

(A–3) If non-zero element X of the form $X = Y^2$ is called " positive " (in symbol $X > 0$), then the sum of positive element and " non-negative " element is positive, viz. $X^2 \neq 0$ or $Y^2 \neq 0$ implies the existence of $Z^2 \neq 0$ such that $X^2 + Y^2 = Z^2$.

(A–4) By the semi-order relation $X > Y (X - Y > 0)$, there exists, for all X, the lowest upper bound (l. u. b.) $\sup (X, 0)$ of X and 0.— We will write $\sup (X, 0) = X^+$, $\sup (-X, 0) = X^-$ and $|X| = X^+ + X^-$.

(A–5) $X^+ \cdot X^- = 0$ for all X.

(A–6) Monotone increasing sequence $\{X_n\}$ bounded from above admits the l. u. b. $\sup_{n \geq 1} X_n$.

(A–7) If $A > 0$, $X_i \geqq 0$, $X_{i+1} \geqq X_i$ and $\sup_{i \geq 1} X_i$ exists, then $A \cdot \sup_i X_i = \sup_i (A \cdot X_i)$.

Remark 1. The " real " character of $\Re$ is expressed by the " Pythagorean axiom " (A–3) together with (A–2). (A–4) and (A–6) are lattice-theoretic axioms.[3] (A–5) is equivalent to $|X^2| = |X|^2$, and (A–7) means a generalised distributive law.

---

1) J. von Neumann: Rec. Math., **43** (1936), 415–484. H. Freudenthal: Proc. Akad. Amsterdam, **39** (1936), 641–651. S. Steen: Proc. London Math. Soc., **41** (1936), 361–392. S. Kakutani: Proc. **15** (1939), 121–123. F. Riesz: Ann. Math., **41** (1940), 174–206. M. H. Stone: Proc. Nat. Acad. Sci., **26** (1940), 280–283. B. Vulich: C. R. URSS, **26** (1940), 850–859. The last two papers appeared during the preparation of the present note. In the redaction, the writer is much suggested by Steen's paper.

2) Not necessarily bounded!

3) Concerning lattice see G. Birkhoff: Lattice Theory, New York (1940).

Remark 2. If we assume the "boundedness axiom," i. e. for any X, there exist $\alpha, \beta \geq 0$ with $-\alpha I \leq X \leq \beta I$, then it is easy to see that the axioms (A-5) and (A-7) are redundant. In this case we may replace (A-3) and (A-4) by the postulate of the existence of semi-ordering satisfying (2-1) below. For we are able to make use of the power series expansion of "square root." Thus, in this bounded case, our results coincides with that of M. H. Stone. The details and applications will be published elsewhere.

§ 2. *Some preliminary consequences from the axioms.*

(2-1) $I > 0$. $X > Z$, $Y \geq W$ and $\alpha \gtrless 0$ imply $X + Y > Z + W$ and $\alpha X \gtrless \alpha Z$. If $X > Z > 0$, $Y \geq W \geq 0$, then $XY \geq ZW \gtrless 0$. $X^2 > 0$ if $X \neq 0$.

(2-2) By the semi-order relation $X > Y$, $\Re$ is a lattice, viz. to any pair X, Y there corresponds the l. u. b. $\sup(X, Y) = (X - Y)^+ + Y = (Y - X)^+ + X$ and the greatest lower bound (g. l. b.) $\inf(X, Y) = -\sup(-X, -Y)$.

Proof. By (2-1), the translations $A \to A + B$ and the expansions $A \to \alpha A$ $(\alpha > 0)$ induce one-one transformations of $\Re$ which preserve the semi-ordering; the one-one transformation $A \to -A$ of $\Re$ inverts the semi-ordering.

(2-3) $\sup(X + Z, Y + Z) = \sup(X, Y) + Z$.

(2-4) $\sup(X, Y) + \inf(X, Y) = X + Y$, in particular $X = X^+ - X^-$.

Proof. Add X, Y to $\sup(X, Y) + \inf(-X, -Y) = 0$ and apply (2-3).

(2-5) The pair (X^+, X^-) is characterised by $X = X^+ - X^-$, $X^+ \cdot X^- = 0$, $X^+ \geq 0$ and $X^- \geq 0$.

Proof. Let $X = X' - X''$, $X' \cdot X'' = 0$, $X' \geq 0$, $X'' \geq 0$, then $(X^+ - X')^2 = (X^+ - X') \cdot (X^- - X'') = -(X^+ \cdot X'' + X' \cdot X^-)$. Thus $(X^+ - X')^2 \leq 0$, and so $X^+ = X'$ by (A-2).

(2-6) $A \cdot \sup(X, Y) = \sup(A \cdot X, A \cdot Y)$ $\big(A \cdot \inf(X, Y) = \inf(A \cdot X, A \cdot Y)\big)$, if $A \geq 0$.

Proof. By (2-3), it is sufficient to prove the case $Y = 0$, i. e. $A \cdot X^+ = (A \cdot X)^+$. Now $AX = AX^+ - AX^-$, $AX^+ \geq 0$, $AX^- \geq 0$, $AX^+ \cdot AX^- = A^2 \cdot (X^+ \cdot X^-) = 0$, and hence $A \cdot X^+ = (A \cdot X)^+$ by (2-5).

(2-7) Any sequence $\{X_n\}$ bounded from above (below) admits the l. u. b. $\sup_{n \geq 1} X_n$ (g. l. b. $\inf_{n \geq 1} X_n$).

Proof. Put $X'_n = \sup_{n \geq m} X_m$ and apply (A-6) to $\{X'_n\}$.

(2-8) If $\sup_{n \geq 1} X_n$, $\sup_{n \geq 1} Y_n$ ($\inf_{n \geq 1} X_n$, $\inf_{n \geq 1} Y_n$) exist, then

$$\sup_{n, m}(X_n + Y_m) = \sup_n X_n + \sup_n Y_n \ \big(\inf_{n, m}(X_n + Y_m) = \inf_n X_n + \inf_n Y_n\big).$$

Proof. Surely $\sup(X_n + Y_m) \leq \sup X_n + \sup Y_n$. Let the equality does not hold good, then $\sup X_n \nleq \sup(X_n + Y_m) - \sup Y_n$. Thus X_n exists such that $X_n \nleq \sup(X_n + Y_m) - \sup Y_n$, and so Y_m exists such that $X_n + Y_m \nleq \sup(X_n + Y_m)$, which is a contradiction.

(2-9) If $X_n \geq 0$, $Y_n \geq 0$ and $\sup_{n \geq 1} X_n$ and $\sup_{n \geq 1} Y_n$ ($\inf_{n \geq 1} X_n$, $\inf_{n \geq 1} Y_n$)

exist, then $(\sup X_n) \cdot (\sup Y_n) = \sup (X_n \cdot Y_m)$ $\big((\inf X_n) \cdot (\inf Y_n) = \inf (X_n \cdot Y_m)\big)$.

Proof. By (A-7) and (2-6), we have $X_i \cdot \sup_n \sup_{n \geq n} Y_m = \sup_n \sup_{n \geq m} (X_i \cdot Y_m) = \sup_n (X_i \cdot Y_n)$. Thus we have $\sup_{n, m} (X_n \cdot Y_m) = \sup_n (X_n \cdot \sup_m Y_m) = (\sup X_n) \cdot (\sup Y_n)$

(2-10) Let $X > 0$, then $\inf_{n \geq 1} (X/n) = 0$.

Proof. Let $\inf (X/n) = Y > 0$, then $0 < \sup_n (nY) \leq X$. By $Y > 0$ we have $\sup (nY) \lneqq \sup (nY) - Y$ and hence $mY \lneqq \sup (nY) - Y$ for an m. Thus $(m+1)Y \lneqq \sup (nY)$, which is a contradiction.

(2-11) Let $0 < X$, $0 < Y$, then $X^2 < Y^2$ implies $X < Y$.

Proof. $(Y-X)(Y+X) > 0$ and hence $(Y-X)^- \cdot (Y-X)(Y+X) = -\big((Y-X)^-\big)^2 \cdot (Y+X) \geq 0$ by (A-5). Thus $\big((Y-X)^-\big)^2 \cdot (Y+X) = 0$, and hence by (A-3) and (2-1), we obtain $\big((Y-X)^-\big)^2 \cdot Y = 0$, $\big((Y-X)^-\big)^2 \cdot X = 0$. Thus $\big((Y-X)^-\big)^2 \cdot (Y-X) = -\big((Y-X)_-\big)^3 = 0$ and so $(Y-X)^- = 0$ by (A-2). Therefore we must have $Y > X$.

§ 3. *Spectral theorem for positive elements.*

(3-1) Let $X > 0$, then $X = Y^2$. By (A-5) we have $(Y^+ + Y^-)^2 = (Y^+)^2 + (Y^-)^2 = (Y^+ - Y^-)^2 = Y^2 = X$. We call $(Y^+ + Y^-)$ the " positive square root" $X^{\frac{1}{2}}$ of X. Because of (2-11), $X^{\frac{1}{2}}$ is characterised by $X^{\frac{1}{2}} \geq 0$, $(X^{\frac{1}{2}})^2 = X$.

(3-2) Let $X > 0$ and put $X_1 = \inf_{n \geq 1} X^n$, $X_2 = X - X_1$. Then $X_1^2 \geq X_1$, $X_1 \cdot X_2 = 0$ and $X_2^2 \leq X_2$.

Proof. By (2-9) $X_1^2 = \inf_{n, m} X^{n+m} \geq X_1$, $X_1 \cdot X_2 = X_1 \cdot X - X_1^2 = \inf_n X^{n+1} - \inf_{n, m} X^{n+m} = 0$. From $X^{n+1} - X^2 - X^n + X = X(X-I)^2 \cdot (X^{n-2} + X^{n-3} + \cdots + I) \geq 0$ we obtain $\inf_n X^{n+1} - X^2 \geq \inf_n X^n - X$, i. e. $XX_1 - X^2 = X_1^2 - X^2 \geq X_1 - X$. This proves $X_2^2 \leq X_2$ by $X_2^2 = X^2 - 2XX_1 + X_1^2 = X^2 - X_1^2$.

(3-3) By (3-2) and (2-11) we have $X_1 \geq X_1^{\frac{1}{2}} \geq X_1^{\frac{1}{4}} = (X_1^{\frac{1}{2}})^{\frac{1}{2}} \geq \cdots \geq X_1^{\frac{1}{2^n}} \geq \cdots \geq 0$. Thus, by (2-7) and (2-9), $\inf_{n \geq 1} X_1^{\frac{1}{2^n}} = E$ exists and $E^2 = E \leq X_1$.

(3-4) $XE = X_1 = X_1 E \geq E$, $X(I-E) = X_2 = X_2(I-E) \leq (I-E)$.

Proof. $X_2 E = X_2 \cdot \inf X_1^{\frac{1}{2^n}} = \inf (X_2 X_1^{\frac{1}{2^n}}) = \inf \big((X_2^{2^n} \cdot X_1)^{\frac{1}{2^n}}\big) = \inf \big((X_2^{2^n-1} \cdot X_2 X_1)^{\frac{1}{2^n}}\big) = 0$. Thus $XE = X_1 E + X_2 E = X_1 \cdot E$. Next from $X_1 \leq X_1^2$ we obtain $X_1^{2^n} \leq X_1^{2^n+1}$ and hence $X_1 \leq \inf (X_1 X_1^{\frac{1}{2^n}}) = X_1 \cdot \inf X_1^{\frac{1}{2^n}} = X_1 E$. On the other hand $E \leq I$ by $I - E = I - 2E + E^2 = (I-E)^2 \geq 0$ and thus $X_1 \geq X_1 E$. Therefore $X_1 = X_1 E$. We have already $X(I-E) = X - X_1 = X_2 = X_2(I-E)$. To show that $X_2(I-E) \leq I-E$, it is sufficient to prove $X_2 \leq I$. Now we have $0 \leq (I - X_2)^2 = I - 2X_2 +$

$X_2^2 \leq I - 2X_2 + X_2 = I - X_2$ by $X_2^2 \leq X_2$, i. e. $X_2 \leq I$.

(3-5) For any $X > 0$ and $\lambda > 0$, put

$$E_\lambda = I - \inf_{m \geq 1} \left(\inf_{n \geq 1} (X/\lambda)^n \right)^{\frac{1}{2m}}$$

Then, by (3-4), $E_\lambda^2 = E_\lambda$, $E_\lambda \cdot X = E_\lambda \cdot (X/\lambda)\lambda \leq \lambda E_\lambda$, $X(I - E_\lambda) = \lambda(X/\lambda) \cdot (I - E_\lambda) \geq \lambda(I - E_\lambda)$.

(3-6) If $\lambda > \mu > 0$, we have $E_\mu \leq E_\lambda$ and $E_\mu \cdot E_\lambda = E_\mu$, i. e. $(E_\lambda - E_\mu)^2 = (E_\lambda - E_\mu)$.

Proof. That $E_\lambda \geq E_\mu$ is evident, and hence $E_\mu = E_\mu^2 \leq E_\mu \cdot E_\lambda$. We have, by $E_\lambda \leq I$, $E_\mu \cdot E_\lambda \leq E_\mu$ and thus $E_\mu \cdot E_\lambda = E_\mu$.

(3-7) Let $\lambda > \mu > 0$, then $\mu(E_\lambda - E_\mu) \leq X(E_\lambda - E_\mu) \leq \lambda(E_\lambda - E_\mu)$.

Proof. $X(E_\lambda - E_\mu) = XE_\lambda(E_\lambda - E_\mu) = XE_\lambda(I - E_\mu) \geq E_\lambda\big(\mu(I - E_\mu)\big) = \mu(E_\lambda - E_\mu)$ by (3-6). The last inequality may be proved in the same manner.

(3-8) If we put $\inf_{\lambda > 0} E_\lambda = E_{+0}$ and $\sup_{\lambda > 0} E_\lambda = E_\infty$, then $XE_{+0} = 0$ and $E_\infty = I$.

Proof. $XE_{+0} = 0$ follows from $0 \leq XE_\lambda \leq \lambda E_\lambda$. We have $X/\lambda \geq (I - E_\lambda) \geq 0$. Thus, by (2-10), $E_\infty = I$.

(3-9) $E_\lambda = E_{\lambda-0} = \sup_{\lambda > \mu} E_\mu$.

Proof. We have $\lambda(E_\lambda - E_{\lambda-0}) = X(E_\lambda - E_{\lambda-0})$. Hence, by the definition of E_λ and the idempotent character of $(E_\lambda - E_{\lambda-0})$, $(E_\lambda - E_{\lambda-0}) = (X/\lambda)^n \cdot (E_\lambda - E_{\lambda-0}) = (I - E_\lambda) \cdot (E_\lambda - E_{\lambda-0}) = 0$.

(3-10)—*Spectral theorem.* Let $X > 0$, $\mu > 0$, $0 = \lambda_0 < \lambda_1 < \lambda_2 < \cdots < \lambda_k = \mu$ and $\lambda_i - \lambda_{i-1} < \varepsilon$, $\lambda_{i-1} \leq \lambda_{i-1}' \leq \lambda_i$ $(i = 1, 2, \ldots, k)$. Then, if we put $E_0 = 0$, we have $-\varepsilon I \leq XE_\mu - \sum_{i=1}^{k} \lambda_{i-1}'(E_{\lambda_i} - E_{\lambda_{i-1}}) < \varepsilon I$, and thus we may write $X \cdot E_\mu = \int_0^\mu \lambda dE_\lambda$. Therefore, by $\sup_{\mu > 0} E_\mu = I$, we obtain the spectral representation $X = \int_0^\infty \lambda dE_\lambda$.

Proof. $XE_\mu - \sum_i \lambda_{i-1}'(E_{\lambda_i} - E_{\lambda_{i-1}}) = \sum_i (X - \lambda_{i-1}'I)(E_{\lambda_i} - E_{\lambda_{i-1}})$, and moreover, by (3-7), we obtain $(\lambda_{i-1} - \lambda_{i-1}')(E_{\lambda_i} - E_{\lambda_{i-1}}) \leq (X - \lambda_{i-1}'I)(E_{\lambda_i} - E_{\lambda_{i-1}}) \leq (\lambda_i - \lambda_{i-1}')(E_{\lambda_i} - E_{\lambda_{i-1}})$.

§4. *Spectral theorem for general elements.* Let X be not necessarily positive. By (3-10) we have the spectral representation of $(X + nI)^+$, $n = 1, 2, \ldots$: $(X + nI)^+ = \int_0^\infty \lambda dE_\lambda(n)$, $E_\lambda(n) = I - \inf_{k \geq 1} \left(\inf_{m \geq 1} \left((X + nI)^+/\lambda\right)^m \right)^{\frac{1}{2k}}$. Hence, for any $\lambda > \mu > 0$, we have $\mu\big(E_\lambda(n) - E_\mu(n)\big) \leq \big(E_\lambda(n) - E_\mu(n)\big)(X + nI)^+ \leq \lambda\big(E_\lambda(n) - E_\mu(n)\big)$. Since $(X + nI)^+ \cdot (X + nI)^- = 0$ by (A-5), we have $(X + nI)^- \cdot \big(E_\lambda(n) - E_\mu(n)\big) = 0$. Hence $(\mu - n)\big(E_\lambda(n) - E_\mu(n)\big) \leq X\big(E_\lambda(n) - E_\mu(n)\big) \leq (\lambda - n)\big(E_\lambda(n) - E_\mu(n)\big)$. Putting $E_\lambda(n) = E_{\lambda-n, n}$, we thus have $\mu(E_{\lambda, n} - E_{\mu, n}) \leq X(E_{\lambda, n} - E_{\mu, n}) \leq \lambda(E_{\lambda, n} - E_{\mu, n})$, $\lambda > \mu > -n$. Hence, if we put

$$E_\lambda = \sup_{n \geq 1} E_{\lambda, n} = \sup_{n > 1} E_{\lambda+n}(n),$$

we obtain $E_\lambda^2 = E_\lambda$, $E_\lambda \geq E_\mu (\lambda > \mu > -\infty)$, $\mu(I - E_\mu) \leq X(I - E_\mu)$ and $XE_\lambda \leq \lambda E_\lambda$. Therefore, as in (3-10), we have the spectral representation

$$\begin{cases} X = \int_{-\infty}^{\infty} \lambda dE_\lambda, \ E_\lambda = E_\lambda^2 = E_{\lambda-0}, \ E_{-\infty} = \mathrm{g.\,l.\,b.} \ E_\lambda = 0, \ E_\infty = \mathrm{l.\,u.\,b.} \ E_\lambda = I, \\ E_\lambda \geq E_\mu \ (\lambda \geq \mu). \end{cases}$$

It is easy to see that the spectral system $\{E_\lambda\}$ is uniquely determined by the above properties.

§ 5. *Concrete representation of the idempotents as point sets.* The set $\mathfrak{J}$ of the "idempotents" $E(E^2 = E)$ of $\mathfrak{R}$ satisfies the following three conditions.

(5-1) $A, B \in \mathfrak{J}$ implies $A \vee B = A + B - AB \in \mathfrak{J}$, $A \wedge B = AB \in \mathfrak{J}$.

(5-2) $0, I \in \mathfrak{J}$, and $A \in \mathfrak{J}$ implies $I - A \in \mathfrak{J}$.

(5-3) If $A, B \in \mathfrak{J}$ are different, then either $B \leq A$ or $A \leq B$. Let $B \leq A$, then there exists $C \in \mathfrak{J}$ such that $B \wedge C \neq 0$, $A \wedge C = 0$. (Take, for example, $C = B \wedge (I - A)$.)

$\mathfrak{J}$ is thus a "complemented, distributive lattice" by the "join" $\vee$ and the "meet" $\wedge$, satisfying the "disjunction property" (5-3). Hence, by G. Birkhoff-Stone-Wallman's theory of Boolean ring, there exists a totally disconnected, bicompact T_1-space $\widetilde{\mathfrak{J}}$ and a closed, open base $\{\widetilde{A}\}$ of $\widetilde{\mathfrak{J}}$ with the properties:

By a suitable correspondence $\mathfrak{J} \ni A \leftrightarrow \widetilde{A} \in \widetilde{\mathfrak{J}}$, $\mathfrak{J}$ is lattice-isomorphic to $\{\widetilde{A}\}$, i. e. $\widetilde{(A \vee B)} = \widetilde{A} + \widetilde{B}$, $\widetilde{(A \wedge B)} = \widetilde{A} \cdot \widetilde{B}$.—Here $+$ and $\cdot$ denote the set-theoretic sum and product.

For the proof of this fact see, for example, H. Wallman's paper.[1]

Moreover, by (A-6), $\mathfrak{J}$ is countably additive and countably multiplicative as lattice. This fact implies the following property:

Let $A_1 < A_2 < \cdots$ and $\bigvee_{i=1}^{\infty} A_i = A$. Then surely $\widetilde{A} \geq \sum_{i=1}^{\infty} \widetilde{A}_i$. The closed set $(\widetilde{A} - \sum_{i=1}^{\infty} \widetilde{A}_i)$ does not contain any open set, i. e. $(\widetilde{A} - \sum_{i=1}^{\infty} \widetilde{A}_i)$ is "non-dense."

Proof. If otherwise, $(\widetilde{A} - \sum_{i=1}^{\infty} \widetilde{A}_i)$ would contain a certain $\widetilde{B}$ with $B \neq 0$. Then, by the isomorphism $C \leftrightarrow \widetilde{C}$, we must have $B \wedge A = B$ and $B \wedge A_i = 0$ $(i = 1, 2, \ldots)$. This contradicts to $A = \bigvee_{i=1}^{\infty} A_i$.

Therefore the closed and open base $\{\widetilde{A}\}$ of the topological space $\widetilde{\mathfrak{J}}$ is countably additive and countably multiplicative, if we neglect the set of "first category," i. e. enumerable sum of non-dense sets. Hence, $\{\widetilde{A}\}$ constitutes a "Borel field" except the set of first category. Let a real-valued function $f(t)$ on $\widetilde{\mathfrak{J}}$ satisfies i) for any $\lambda > \mu$, the set $\underset{t}{E}\left(\lambda > f(t) > \mu\right)$ coincides with a certain $\widetilde{A}$ except the set of first cate-

---

1) Ann. Math., **39** (1938), 112–126.

gory, and ii) the set $E_t(|f(t)|=\infty)$ is of first category, then we may call such function $f(t)$ as " Borel-measurable."

§ 6. *Concrete representation of $\mathfrak{R}$ as function ring.* By the spectral theorem we have, for all $x \in \mathfrak{R}$, $x=\int_{-\infty}^{\infty} \lambda dE_\lambda$. Let $0<\lambda_i-\lambda_{i-1}<1/n$ $(i=0 \pm 1, \pm 2, \ldots)$, $\lim_{i\to-\infty} \lambda_i= -\infty$, $\lim_{i\to\infty} \lambda_i= \infty$ and consider the step functions $f_X^{(n)}(t)$ on $\widetilde{\mathfrak{F}}$: $f_X^{(n)}(t)=\lambda_i$ for $t \in (E_{\lambda_i}-E_{\lambda_{i-1}})$. Surely $f_X(t)=\lim_{n\to\infty} f_X^{(n)}(t)$ is Borel-measurable. It is easy to see that, by the correspondence $X \to f_X(t)$, $\mathfrak{R}$ is ring-isomorphically and lattice-isomorphically represented upon $\{f_X(t)\}$.—Here sums, products in $\{f_X(t)\}$ are ordinary functional sums and products, and $f_X(t)$ is called positive if the function $f_X(t)$ is positive except the t-set of first category.

However, to assure that any Borel-measurable function $f(t)$ is the image of some $X \in \mathfrak{R}$, it is necessary and sufficient to assume the following axiom.

(A-8) For any sequence $\{X_i\}$ with $\inf(|X_i|,|X_j|)=0 \ (i \neq j)$, we have

$$\sup_{n\geq 1} \inf_{m\leq n} \sum_{i=1}^{m} X_i = \inf_{n\geq 1} \sup_{m\leq n} \sum_{i=1}^{m} X_i \in R.$$

Proof. We have only to approximate $f(t)$ by finitely-valued Borel-measurable step functions and to put $X=$ the limit (which surely exists by (A-8)) of the inverse images in $\mathfrak{R}$ of these step functions.

Remark 1. The axiom (A-8) is redundant, if we assume the " boundedness axiom " as stated in the Remark 2 after (A-7). For, in this case, we deal only with " bounded " Borel-measurable functions. Thus, in the " bounded case," $\mathfrak{R}$ constitutes a characterisation of the function ring of ' bounded " Borel-measurable functions. Such characterisation is also announced by M. H. Stone, loc. cit. However, stone's method is different from ours.

Remark 2. In (3-3) we may take $E=\inf(X_1, I)=\inf(I, X, X^2, X^3, \ldots)$. Thus, in (3-5), we may put

$$(3-5)' \qquad E_\lambda=I-\inf\left(I, X/\lambda, (X/\lambda)^2, (X/\lambda)^3, \ldots\right).$$

The proof is easy. Accordingly, the axioms (A-2) and (A-3) may be replaced by (2-1). (3-5) together with (3-10) give a lattice-theoretic interpretation of Stone-Lengyel's proof of the spectral theorem (Ann. Math. **37** (1936), 853–864). To this point I hope to discuss in another occasion.

278

On regularly convex sets

(*with M. Fukamiya*)

Proc. Imp. Acad. Tokyo **17** (1941) 49–52

(Comm. by T. Takagi, M.I.A., March 12, 1941.)

§ 1. *Introduction.* We denote by E a Banach space and by $\bar{E}$ its conjugate space. A set $X \subseteq E$ is called *convex* if $x, y \in E$ implies $ax + (1-a)y \in E$ for all a $(1 \geq a \geq 0)$. According to M. Krein and V. Smulian[1], a set $F \subseteq \bar{E}$ is called *regularly convex* if for every $g \bar{\in} F$ $(g \in \bar{E})$ there exists $x_0 \in E$ such that $\sup_{f \in F} f(x_0) < g(x_0)$. It is easy to see that only convex sets in $\bar{E}$ may be regularly convex. Moreover we may prove

Theorem 1. A convex set $F \subseteq \bar{E}$ is regularly convex if and only if F is closed in the weak topology of $\bar{E}$ as functionals.

Hereby, for any $f_0 \in \bar{E}$, its weak neighbourhood $U(f_0, x_1, x_2, \cdots, x_n, \varepsilon)$ is defined as the totality of $f \in \bar{E}$ such that $\sup_{1 \leq i \leq n} |f(x_i) - f_0(x_i)| < \varepsilon$, where $\{x_i\}_{i=1, 2, \dots, n}$ is an arbitrary system of points $\in E$ and ε is an arbitrary positive number[2].

The purpose of the present note is to show that there exists a kind of duality between (strongly) closed convex sets $\subseteq E$ and regularly convex sets $\subseteq \bar{E}$. By this duality we may give, to almost all the theorems in Chapter 1 of $[K-S]$, geometrical interpretations and new proofs. We then give a proof to Krein-Milman's[3]

Theorem 2. If a (strongly) bounded set $F \subseteq \bar{E}$ is regularly convex, then F has extreme points f_0, viz. points f_0 such that $f_0 \neq \frac{1}{2}(g+h)$ for any two $g, h \in F$, $g \neq f_0$, $h \neq f_0$.

If E is separable, the above duality shows that Theorem 2 is an immediate corollary of a theorem due to S. Mazur[4]. The proof for non-separable E is also given by tranfinite induction. This was obtained by one of us (Fukamiya): Zenkoku Shijyô-Sugaku Danwakai, 207 (Japanese). After the present note is completed, we received a letter from S. Kakutani, now in Princeton, and we knew that F. Bohnenblust

---

1) Ann. of Math., **41** (1940), 556–583, to be referred as $[K-S]$.

2) The importance of weak topologies in the theory of Banach spaces was especially stressed on by S. Kakutani with much success: Proc. **15** (1939), 169–173 and **16** (1940), 63–67. We omit the easy proof of Theorem 1, since it is similar to his proof of the equivalence between the transfinite closure and the regular closure of linear subspaces $\subseteq \bar{E}$. The equivalence of the two notions: transfinitely closed convex sets $\subseteq \bar{E}$ and regularly convex sets $\subseteq \bar{E}$, was proved in $[K-S]$, 569.

3) The vol. 9 of the Stud. Math. is not yet arrived at our institute. We knew their result from M. and S. Krein's paper: C. R. URSS, **27**, 5 (1940), 427–430.

4) Stud. Math., **4** (1933), 70–84.

also obtained the same proof. Kakutani also says in the letter that Chap. 1 of $[K-S]$ may be rewritten by his method (Theorem 1).

§2. *The duality.* For any $X \subseteq E$, let X^* denote the totality of $f \in \bar{E}$ such that $\sup_{x \in X} f(x) \leqq 1$. In the same way, for any set $F \subseteq \bar{E}$, let F' denote the totality of $x \in E$ such that $\sup_{f \in F} f(x) \leqq 1$. Then we have[1]

Lemma 1. $X^{*/*} = X^*$, $F'^{*/} = F'$.

Proof. Surely $X \leqq X^{*/}$ and $F \leqq F'^*$. Moreover we have $X_1^* \leqq X_2^*$ $(F_1' \leqq F_2')$ from $X_1 \geqq X_2$ $(F_1 \geqq F_2)$.

Theorem 3. For any $X \leqq E$, $X^{*/}$ is the (strong) closure $\overline{conv}(X, 0)$ of the convex hull $conv(X, 0)$ of X and the zero vector 0 of E.

Proof. Surely we have $X^{*/} \geqq \overline{conv}(X, 0)$. If $x_0 \in X^{*/} - \overline{conv}(X, 0)$, there exists, by Ascoli-Mazur's theorem, an $f_0 \in \bar{E}$ such that $f_0(x_0) > 1$ and $f_0(x) \leqq 1$ at every $x \in \overline{conv}(X, 0)$. Then $f_0 \in X^*$ and hence x_0 must be $\bar{\in} X^{*/}$, contrary to the hypothesis. Thus we must have $X^{*/} = \overline{conv}(X, 0)$.

Theorem 4. For any $F \leqq \bar{E}$, F'^* is the closure $\overline{conv}(F, 0)$ (in the sense of weak topology of $\bar{E}$ as functionals) of the convex hull $\overline{conv}(F, 0)$ of F and the zero vector 0 of $\bar{E}$.

Proof. Surely we have $F'^* \geqq \overline{conv}(F, 0)$. If $f_0 \in F'^* - \overline{conv}(F, 0)$, then there exists, by the definition of weak topology, $\varepsilon > 0$ and $x_1, x_2, \ldots, x_n, \in E$ such that $\sup_{1 \leqq i \leqq n} |f(x_i) - f_0(x_i)| > \varepsilon$ at every $f \in \overline{conv}(F, 0)$. Thus the point $\xi_{f_0} = \big(f_0(x_1), f_0(x_2), \ldots, f_0(x_n)\big)$ in n-dimensional euclidian space E_n is of euclidian distance $> \varepsilon$ from the convex point set $\{\xi_f\}$ in E_n, where $\xi_f = \big(f(x_1), f(x_2), \ldots, f(x_n)\big)$, $f \in \overline{conv}(F, 0)$. Hence, by Ascoli-Mazur's theorem, there exists real numbers $a_1, a_2, \ldots, a_n$ such that $\sum_{i=1}^{n} a_i f_0(x_i) > 1$ and $\sum_{i=1}^{n} a_i f(x_i) \leqq 1$ at every $f \in \overline{conv}(F, 0)$. Put $x_0 = \sum_{i=1}^{n} a_i x_i$, then $x_0 \in F'$ by $f(x_0) \leqq 1$. This contradicts to $1 < f_0(x_0)$, $f_0 \in F'^*$. Therefore we must have $F'^* = \overline{conv}(F, 0)$.

By Lemma 1, Theorem 1, 3 and 4 we obtain the following duality

Theorem 5. *Regularly convex set* $F \leqq \bar{E}$ *containing* 0 *is characterised by the property that it is expressible as* $F = X^*$, $X \leqq E$. *(Strongly) closed convex set* $X \leqq E$ *containing* 0 *is characterised by the property that it is expressible as* $X = F'$, $F \leqq \bar{E}$.

§3. *Boundedness of convex sets and regularly convex sets.* Regularly convex set F need not be (strongly) bounded. However we may prove

Theorem 6. *In order that a regularly convex set* F *be (strongly) bounded, it is necessary and sufficient that* 0 *is the inner point of the (strongly) closed convex set* $F' \subseteq E$.

Proof. Let $\sup_{f \in F} \|f\| = a < \infty$, and assume that 0 is not an inner point of F'. Then we have a sequence $\{x_i\} \leqq E$ with the properties

---

1) Cf. Garrett Birkhoff: Lattice Theory, New York (1940), 24.

$x_i \bar{\in} F'(i=1, 2, \ldots)$, $\lim\limits_{i\to\infty} \|x_i\|=0$. This contradicts to $|f(x_i)| \leqq \|f\| \cdot \|x_i\| \leqq a\|x_i\|$, $(i=1, 2, \ldots)$, $f \in F$. For we have $|f(x_i)| \leqq 1$, viz. $x_i \in F'$ if $\|x_i\| \leqq \dfrac{1}{a}$. Next let the sphere $\|x\| \leqq a$, $a>0$, of E be $\subseteq F'$, and assume that F is not (strongly) bounded. Then there exists $f_0 \in F$ such that $f_0(x_0)>1$ or $f_0(x_0)<-1$ at a point x_0 with $\|x_0\|=a$. Since $\|-x_0\|=\|x_0\|$, x_0 and $-x_0$ both $\in F'$, and thus we must have a point $x_1 \in F'$ such that $f_0(x_1)>1$. This contradicts to the fact that $f_0 \in F$.

Dually we may prove

Theorem 7. In order that a regularly convex set F contains 0 as an inner point, it is necessary and sufficient that the (strongly) closed convex set F' be (strongly) bounded.

Proof. Let $\sup\limits_{x \in F'} \|x\|=a$, $0<a<\infty$, then $f \in \bar{E}$ with $\|f\| \leqq \dfrac{1}{a}$ surely belongs to $F'^{*}=F$. Next let the sphere $\|f\| \leqq a$, $0<a<\infty$, be contained in F. By Hahn-Banach's theorem, there exists, for any $x_0 \in E$, an $f_0 \in \bar{E}$ such that $\|f_0\|=1$, $f_0(x_0)=\|x_0\|$. Thus we must have $\|x\| \leqq \dfrac{1}{a}$ for any $x \in F'$.

§ 4. *Existence of extreme points.* Let E be separable and let a (strongly) closed convex set $X \subseteq E$ contain 0 as an inner point. Then, by S. Mazur's theorem cited in § 1, there exists boundary point x_0 of X at which we have one and only one tangential hyperplane. This means that there exists one and only one $f_0 \in \bar{E}$ with the properties $f_0(x_0)=1$, $\sup\limits_{x \in X} f_0(x) \leqq 1$. Such f_0 is surely an extreme point of the (strongly) bounded regularly convex set X^{*}. Hence, combined with theorem 5 and 6, we obtain a new proof of Theorem 2 in the special case when E is separable.

In order to prove Theorem 2 for non-separable E we first prove the

Lemma 2. A (strongly) bounded regularly convex set $F \subseteq \bar{E}$ is bicompact in the weak topology of $\bar{E}$ as functionals.

Proof. Let $\sup\limits_{f \in F} \|f\|=a$, $a<\infty$. The sphere $G: \|g\| \leqq a$ of $\bar{E}$ is bicompact in the weak topology of $\bar{E}$ as functionals[1]. Since, by Theorem 1, $F \subseteq G$ is closed in the weak topology of $\bar{E}$ as functionals, F is bicompact with G.

Proof to Theorem 2. Let, as above, $\sup\limits_{f \in F} \|f\|=a$, $0<a<\infty$, where F is regularly convex. Well-order the points of the unit-sphere $\|x\| \leqq 1$ of E as follows:

(1) $x_0, x_1, x_2, \ldots, x_a, \ldots$ $(a<\eta)$.

F is bicompact in the weak topology of $\bar{E}$ as functionals, by Lemma 2. Hence, for any $x \in E$, the continuous function $f(x)$ on $F(f \in F)$ attains its supremum and infimum on F. Let $\sup\limits_{f \in F} f(x_0)=f_1(x_0)$, $f_1 \in F$,

---

1) See S. Kakutani: loc. cit.

and let F_1 be the totality of such $f_1 \in F$. F_1 is convex and closed (in the weak topology of $\bar{E}$ as functionals), and is thus bicompact. If F_1 contains only one point f_1^0, f_1^0 is surely an extreme point of F and the Theorem 2 is proved. Assume that F_1 consists of more than one point, and let x_a be the first point in the well-ordered sequence (1) which satisfies $f(x_a) \not\equiv$ constant on F_1. Put $a = a_1$ and let $\sup_{f_1 \in F_1} f_1(x_{a_1}) = f_2(x_{a_1})$, $f_2 \in F_1$, and denote by F_2 the totality of such $f_2 \in F_1$. In this way let possibly transfinite sequence of convex, bicompact sets $F \supseteqq F_1 \supseteqq F_2 \supseteqq \cdots \supseteqq F_\xi \supseteqq \cdots$ be defined for all $\xi < \xi_1(< \eta)$. If ξ_1 is not a limit ordinal, let $a_{\xi_1 - 1}$ be the least a such that $f(x_a) \not\equiv$ constant on $F_{\xi_1 - 1}$. Let $\sup_{f \in F_{\xi_1 - 1}} f(x_{a_{\xi_1 - 1}}) = f_{\xi_1}(x_{a_{\xi_1 - 1}})$, $f_{\xi_1} \in F_{\xi_1 - 1}$, and define F_{ξ_1} as the totality of such $f_{\xi_1} \in F_{\xi_1 - 1}$. If ξ_1 is a limit ordinal, put $F'_{\xi_1} =$ the intersection $\bigwedge_{\xi < \xi_1} F_\xi$. As F_ξ are all bicompact, F'_{ξ_1} is not void. We then define F_{ξ_1}, taking F'_{ξ_1} in place of $F_{\xi_1 - 1}$ in the above argument.

Thus, assuming the process not to give any extreme point of F, we define bicompact sets F_ξ for all $\xi < \eta$ such that: $F \supseteqq F_1 \supseteqq \cdots \supseteqq F_\xi \supseteqq \cdots$. Then $\bigwedge_{\xi < \eta} F_\xi$ is not void. $\bigwedge_{\xi < \eta} F_\xi$ consists of only one point. For, if $f, g \in \bigwedge_{\xi < \eta} F_\xi$, $f \not\equiv g$, then $f(x) \not\equiv g(x)$ for a certain $x = x_a$, $a < a_\xi$. This contradicts $f(x_a) = g(x_a)$ (by $f, g \in F_\xi$). Thus let $f_0 = \bigwedge_{\xi < \eta} F_\xi$, f_0 is an extreme point of F. For the proof, take any two $g, h \in F$, $g \not\equiv f_0$, $h \not\equiv f_0$. Then there exists the least ξ such that g, h both $\in F_\xi$. Thus $g(x_a), h(x_a)$ both $< f_0(x_a)$ for all $a > a_\xi$. This proves that $f_0 \not\equiv \dfrac{1}{2}(g+h)$.

On vector lattice with a unit

Proc. Imp. Acad. Tokyo **17** (1941) 121–124

(Comm. by T. TAKAGI, M.I.A., May 12, 1941.)

§ 1. *Introduction and the theorems.* The purpose of the present note is to give a new proof to Kakutani-Krein's lattice-theoretic characterisation of the space of continuous functions on a bicompact space[1]. We first represent algebraically the vector lattice as point functions and then introduce the topology to the represented function lattice, while Kakutani-Krein's (independent and different) proofs both make use of the assumed norm and hence the conjugate space of the vector lattice. Our treatment may be compared with Birkhoff-Stone-Wallman's representation of Boolean algebra as field of sets or with Gelfand's[2] representation of normed ring as function ring[3].

A *vector lattice E* is a partially ordered real linear space, some of whose elements f are "non-negative" (written $f \geq 0$) and in which

(V 1): If $f \geq 0$ and $a \geq 0$, then $af \geq 0$.

(V 2): If $f \geq 0$ and $-f \geq 0$, then $f = 0$.

(V 3): If $f \geq 0$ and $g \geq 0$, then $f + g \geq 0$.

(V 4): E is a lattice by the semi-order relation $f \geq g$.

In this note we further assume the *Archimedean axiom*

(V 5): If $f > 0$ and $a_n \downarrow 0$, then $a_n f \downarrow 0$ (in order-topology),

and the existence of a *unit $I > 0$* satisfying

(V 6): For any $f \in E$ there exists $a > 0$ such that $-aI \leq f \leq aI$.

A linear subspace N of E is called an *ideal*[4] if N contains with f all x satisfying $|x| \leq |f|$. Here we put, as usual, $|f| = f^+ - f^-$, $f^+ = f \vee 0 = \sup(f, 0)$, $f^- = f \wedge 0 = \inf(f, 0)$. An ideal $N \neq E$ is called *maximal* if it is contained in no other ideal $\neq E$. Denote by $\mathfrak{N}$ the set of all the maximal ideals N of E. It is proved below that the residual class E/N of E mod any maximal ideal N is linear-lattice-isomorphic to the vector lattice of real numbers, the non-negative elements $\in E$ and I respectively being represented by non-negative numbers and the number 1. We denote by $f(N)$ the real number which

---

1) S. Kakutani: Proc. **16** (1940), 63–67. M. and S. Krein: C. R. URSS, **27** (1940), 427–430.

2) I. Gelfand: C. R. URSS, **23** (1939), 430–432.

3) After the present paper was completed, I knew, by Y. Kawada's remark, the paper of M. H. Stone: Proc. Nat. Acad. Sci. **27** (1941), 83–87, which arrived at our institute only recently. In this paper, Stone sketches a proof of Kakutani's theorem which is essentially the same as ours. (He seems to have not read Krein's paper.) Stone proves firstly the case of Banach vector lattice and then reduces the general case to it, while we first prove the algebraic case and then deduce from it the case of Banach vector lattice. It is to be noted that we do not, in the proof of the representation theorem, make use of the metrical nor order completeness other than the Archimedean axiom (V 5).

(4) Normal subspace in the terminology of Garrett Birkhoff: Lattice Theory, New York (1940).

corresponds to $f \in E$ by the homomorphism $E \to E/N$, $N \in \mathfrak{N}$. Then we have the

Theorem 1. By the correspondence $f \to f(N)$, E is linear-lattice-isomorphically mapped on the vector lattice $F(\mathfrak{N})$ of real-valued bounded functions on $\mathfrak{N}$ such that i) $|f| \to |f(N)|$, ii) $I(N) \equiv 1$ on $\mathfrak{N}$ and iii) $F(\mathfrak{N})$ is *separable* on $\mathfrak{N}$, viz.

(V 7): For any two different points N_1, $N_2 \in \mathfrak{N}$ and for any two real numbers a_1, a_2 there exists $f \in E$ satisfying $f(N_1) = a_1$, $f(N_2) = a_2$.

Next introduce a topology in $\mathfrak{N}$ by calling open the set of all points $N \in \mathfrak{N}$ which satisfy $|f_i(N) - f_i(N_0)| < \varepsilon$, $i = 1, 2, \ldots, n$, where $N_0 \in \mathfrak{N}$, $\varepsilon > 0$, n and $f_i(-I \leqq f_i \leqq I)$ are arbitrary. Then $\mathfrak{N}$ is bicompact since it may be identified with a closed subset of a topological product (of the same potency as the cardinal number of elements $\in E$ between $-I$ and I) of the real intervals $(-1, +1)$. Each function $f(N) \in F(\mathfrak{N})$ is continuous on the bicompact space $\mathfrak{N}$ by the above topology. The set $C(\mathfrak{N})$ of all the real-valued continuous functions $C(N)$ on $\mathfrak{N}$ is a Banach vector lattice with the norm $\|c\| = \sup |c(N)|$, by calling "non-negative" the non-negative functions. As a precision to theorem 1 we have

Theorem 2. $F(\mathfrak{N})$ is dense in $C(\mathfrak{N})$ by the norm $\|c\|$.

Theorem 3. If we assume that E is a Banach space by the norm $\|f\| = \inf a$, $-aI \leqq f \leqq aI$, then the representation $f \to f(N)$ is not only linear-lattice-isomorphism but also an isometry and $F(\mathfrak{N})$ coincides with the whole $C(\mathfrak{N})$.

The theorem 3 is the Kakutani-Krein's characterisation mentioned above.

§ 2. *The proofs.* The theorem 1 may be proved by the following series of lemmas.

Lemma 1. A linear subspace N of E determines a lattice homomorphism $E \to E/N$ if and only if N is an ideal.

Proof. See G. Birkhoff: loc. cit., p. 109.

Lemma 2. Unless $E \neq \{aI\}$, $-\infty < a < \infty$, E contains a non-trivial $(\neq 0, E)$ ideal.

Proof. Let $f \neq \gamma I$ for any γ. Let $a = \inf a'$, $a'I \geqq f$, $\beta = \sup \beta'$, $\beta'I \leqq f$, then $aI \geqq f \geqq \beta I$ and $a > \beta$. Hence $(f - \delta I)^+ \neq 0$, $(f - \delta I)^- \neq 0$ for $a > \delta > \beta$. The set N of all the elements $g \in E$ satisfying $|g| \leqq \eta(f - \delta I)^+$ is a non-trival ideal since $N \bar{\ni} (f - \delta I)^-$.

Lemma 3. For any non-trivial ideal N_0, there exists a maximal ideal N containing N_0.

Proof. Construct a sequence of ideals

$$N_0, N_1, \ldots, N_\eta, \ldots \quad (N_\eta \subset N_{\eta+1} \neq E), \quad \eta < \omega.$$

If ω is a limit ordinal, define $f \equiv g \bmod N_\omega$ to mean $f \equiv g \bmod N_\eta$ for some $\eta < \omega$. That N_ω is an ideal follows from lemma 1, moreover $N_\omega \neq E$ since $I \not\equiv 0 \bmod N_\eta$, $\eta < \omega$. This process defines a transfinite sequence of linear-lattice-congruence relations on E, each more inclusive than the last. Hence it cannot continue indefinitely, for the number of different congruence relations are bounded. Thus, by the lemma 2, we would obtain the demanded maximal ideal $N \supset N_0$.

Lemma 4. If $f \neq 0$, then there exists a maximal ideal N not containing f.

Proof. If $|f| \geq aI$ for a certain $a > 0$, then the lemma is evident from lemma 2 and 3. Let such a do not exist, and assume $f^+ > 0$ and $f^+ \geq aI$ for no $a > 0$, without losing the generality. Let $\beta = \inf \gamma$, $\gamma I \geq f^+$, then $\beta > 0$ by $f^+ > 0$. We have $\beta I - f^+ > 0$ and $\beta I - f^+ \geq \delta f^+$ for no $\delta > 0$, by the definition of β. Thus the ideal N_0 of all the elements $g \in E$ satisfying $|g| \leq \eta(\beta I - f^+)$ does not contain f^+. Hence the maximal ideal N containing N_0 does not contain f^+, since otherwise N would contain I and hence would coincide with E. A fortiori N does not contain f.

We have incidentally proved that for any two different points N_1. $N_2 \in \mathfrak{N}$, there exists $f \in E$ such that $f \equiv 0 \bmod N_1$, $f \equiv 1 \bmod N_2$. Hence the function lattice $F(\mathfrak{N})$ on the bicompact space $\mathfrak{N}$ satisfies the separability axiom (V 7). Thus, by M. and S. Krein's lemma in the cited paper[1], $F(\mathfrak{N})$ is dense in $C(\mathfrak{N})$ by the norm $\|c\|$. This proves theorem 2, and the proof of theorem 3 is now evident.

§ 3. *Determination of the maximal ideals.* Let T be a set of points t and let $E(T)$ be a real linear set of real-valued bounded functions $f(t)$ on T such that: i) $E(T)$ contains the constant function 1, ii) $E(T)$ is a lattice such that we have $|f| = |f(t)|$, iii) for any two points $t_1, t_2 \in T(t_1 \neq t_2)$ there exists a function $f(t) \in E(T)$ satisfying $f(t_1) = 0$, $f(t_2) = 1$ and iv) $E(T)$ is a Banach space by the norm $\|f\| = \sup |f(t)|$. Then any point $t \in T$ induces a linear-lattice-homomorphism $f \to f(t)$ of $E(T)$ on the vector lattice of real numbers. Hence each $t \in T$ defines a maximal ideal $N_t(f \to f(t)$ by $E(T) \to E(T)/N_t)$ of $E(T)$. The set $\{N_t\}$ is dense in the bicompact space $\mathfrak{N}$ of all the maximal ideals N of $E(T)$. For if an element $f \in E(T)$ satisfies $f(t) \equiv 0$ on T, then f is the zero-element of $E(T)$ and hence the continuous function on $\mathfrak{N}$, which represent f by the theorem 3, must be $\equiv 0$ on $\mathfrak{N}$. Thus we obtain, by making use of iii)

Theorem 4. T is mapped by $t \to N_t$ one-to-one on a dense subset of $\mathfrak{N}$.

If T is a *completely regular* topological space and if $E(T)$ is the set of all the real-valued bounded *continuous* functions on T, then $E(T)$ surely satisfies i)-iv). In this case we easily verify, by the definition of the topology of $\mathfrak{N}$ (§ 2), that the correspondence $t \leftrightarrow N_t$ is a homeomorphism. Thus, by theorem 4, we obtain Tychonoff's theorem[2] concerning the imbedding of the completely regular space in bicompact space[3]. In particular, when T is bicompact, T may be identified with $\mathfrak{N}$ by the homeomorphism $t \leftrightarrow N_t$. Hence two bicompact spaces T_1 and

---

1) They stated this lemma without proof. I express my thanks to T. Nakayama and M. Fukamiya for discussing the proof of this lemma.

2) A. Tychonoff: Math. Annalen, **102** (1930), 544–561.

3) Similar treatment, which makes use of the conjugate space of $E(T)$ (without regarding $E(T)$ as a lattice), was given by S. Kakutani: Isôsûgaku (Japanese), **2** (1940), 14–21.

T_2 are homeomorphic if and only if the vector lattices $E(T_1)$ and $E(T_2)$ are linear-lattice-isomorphic in such a way that the constant function 1 corresponds to the constant function 1. This may be considered as lattice-theoretic counterpart of Gelfand-Kolmogoroff's theorem[1].

1) I. Gelfand and A. Kolmogoroff: C. R. URSS, **22** (1939), 11–15.

Vector lattices and additive set functions

Proc. Imp. Acad. Tokyo **17** (1941) 228–232

(Comm. by T. Takagi, m.i.a., July 12, 1941.)

§ 1. *Introduction.* One of the fundamental theorems in the theory of the integral is that of Radon-Nikodym concerning the countably additive set functions. The proof given below (§ 2) of this theorem will be shorter than the standard one in S. Saks' book[1] or the one recently published by C. Carathéodory[2]. Our proof is carried out by a *maximal method*, by making use of a lemma which is a simple modification of H. Hahn's decomposition theorem[3]. The same method is also applied (§ 3 and § 4) to give lattice-theoretic formulations of Radon-Nikodym's theorem. Of these ten years, such formulations were given more or less explicitly by many authors, F. Riesz, H. Freudenthal, Garrett Birkhoff, S. Kakutani, F. Maeda and S. Bochner-S. R. Phillips[4]. Our maximal method also makes use of the ideas of Riesz and Freudenthal, but it seems to be more direct than those of the cited authors. Thus, without appealing to Freudenthal's spectral theorem, we may obtain Kakutani's lattice-theoretic characterisation of the Banach space (L) from our result in § 3. Moreover the result in § 4 will give a simplified proof and extension of Freudenthal's spectral theorem.

§ 2. *The concrete case.* A class $\mathfrak{X}$ of sets in a space X is called *countably additive* if (i) the empty set belongs to $\mathfrak{X}$, (ii) with E its complement CE also belongs to $\mathfrak{X}$ and (iii) the sum $\bigvee_{n=1}^{\infty} E_n$ of sequence $\{E_n\}$ of sets $\epsilon\, \mathfrak{X}$ belongs to $\mathfrak{X}$. A real-valued finite function $F(E)$ defined on $\mathfrak{X}$ is called *countably additive* if $F(\bigvee_{n=1}^{\infty} E_n) = \sum_{n=1}^{\infty} F(E_n)$ for any sequence $\{E_n\}$ of mutually disjoint sets $\epsilon\, \mathfrak{X}$. Let $\varphi(E)$ be a *measure* on X, that is, $\varphi(E)$ be countably additive and non-negative on X. Without losing generality we assume that any subset of a set $\epsilon\, \mathfrak{X}$ of φ-measure zero also belongs to $\mathfrak{X}$. A countably additive set function

---

1) Theory of the integral, Warsaw (1937), 36.

2) Ueber die Differentiation von Maszfunktionen, Math. Zeits., **42** (1940), 181–189. J. von Neumann also gave an interessant proof by making use of the Banach space (L_2). See his Rings of operators, III, Ann. of Math., **41** (1940), 126.

3) S. Saks: loc. cit., 32.

4) F. Riesz: Sur la décomposition des opérations linéaires, Bologna Congress, III (1928), 143–148. Sur quelques notions fondamentales dans la théorie générale des opérations linéaires, Ann. of Math., **41** (1940), 174–206. Sur la théorie ergodique des espaces abstraits, Acta Szeged, **10** (1941), 1–20, to be cited as [R—3]. H. Freudenthal: Teilweise geordnete Modulen, Proc. Acad. Amsterdam, **39** (1936), 641–651, to be cited as [F—1]. G. Birkhoff: Dependent probabilities and spaces (L), Proc. Nat. Acad., **24** (1938), 154–158. S. Kakutani: Mean ergodic theorem in abstract L-spaces, Proc., **15** (1939), 121–123. Concrete representations of abstract L-spaces and the mean ergodic theorem, Ann. of Math., **42** (1941), 523–537, to be cited as [K—2]. F. Maeda: Partially ordered linear spaces, J. Sci. Hirosima Univ., **10** (1940), 137–150. S. Bochner-S. R. Phillips: Additive set functions and vector lattices, Ann. of Math., **42** (1941), 316–324.

$F(E)$ is called *absolutely continuous* if $\varphi(E)=0$ implies $F(E)=0$. $F(E)$ is called *singular* if there exists a set E_0 of φ-measure zero such that $F(E)=0$ for all $E \subset CE_0$. A real-valued function $f(x)$ on X is called *measurable* if the set $\underset{x}{E}\big(f(x) > a\big) \in \mathfrak{X}$ for all real number a. We may define the *integral* $\int_X f(x)\varphi(dx)$ of Lebesgue's type of *integrable*, measurable function $f(x)$. It is well-known that the set function $F(E)=\int_E f(x)\varphi(dx)$ on $\mathfrak{X}$, the *indefinite integral*, is countably additive and absolutely continuous.

Radon-Nikodym's theorem states that any countably additive set function may uniquely be expressed as the sum of an indefinite integral and a singular set function. In particular, countably additive and absolutely continuous set functions may be identified with the indefinite integrals.

For the proof we need a

Lemma 1. Let $G(E)$ be a non-negative, countably additive set function and suppose that $G(E)$ is not identically zero on $\mathfrak{X}$. Then either $G(E)$ is singular or there exist a rational number $a>0$ and a set E_a of φ-measure >0 such that $G(E) \geq a\varphi(E)$ for all $E \subset E_a$.

Proof. By Hahn's decomposition theorem, there exists, for any (rational) number $a>0$, a set $E_a \subset \mathfrak{X}$ such that $G(E) \geq a\varphi(E)$ for $E \subset E_a$ and $G(E) \leqq a\varphi(E)$ for $E \subset CE_a$. Assume that $\varphi(E_a)=0$ for all rational number $a>0$, then the sum $\underset{a>0}{\bigvee} E_a = E_0$ is of φ-measure zero and $G(E)=0$ for all $E \subset CE_0$.

Proof of Radon-Nikodym's theorem. It will be sufficient to prove the case of non-negative, countably additive set function $F(E)$. Let μ be the supremum of the numbers $\int_X f(x)\varphi(dx)$, when $f(x)$ runs through the set $[F]$ of all the non-negative, integrable functions such that $\int_E f(x)\varphi(dx) \leqq F(E)$ all over $\mathfrak{X}$. Then there exists a sequence $\{f_n(x)\}$ of functions $\in [F]$ such that $\lim_{n \to \infty} \int_X f_n(x)\varphi(dx)=\mu$. We have, for any $E \in \mathfrak{X}$,

$$\int_E \sup_{m \leq n}\big(f_m(x)\big)\varphi(dx)= \sup_{E=\underset{m=1}{\overset{n}{\bigvee}} E_m}\Big(\sum_{m=1}^{n}\int_{E_m} f_m(x)\,\varphi(dx)\Big)$$

$$\leqq \sup_{E=\underset{m=1}{\overset{n}{\bigvee}} E_m}\Big(\sum_{m=1}^{n} F(E_m)\Big)=F(E)\,,$$

where the sup on the second and the third terms are to be taken for all the disjoint decomposition of E. Hence, by putting $f(x)=\sup_{1 \leq n} f_n(x)$, we have $f(x) \in [F]$ and $\int_X f(x)\varphi(dx)=\mu$. Next put $G(E)=F(E)-\int_E f(x)\varphi(dx)$, then our task is to show that $G(E)$ is either singular or $G(E)=0$ on $\mathfrak{X}$. Let $G(E)$ be not singular and $G(E) \not\equiv 0$ on $\mathfrak{X}$, then there exist, by the lemma 1, a rational number $a>0$ and a set E_a of φ-measure >0 such that $G(E) \geqq a\varphi(E)$ for all $E \subset E_a$. Let $c_a(x)/a$ be the characteristic function of E_a, then we would obtain $\big(f(x)+c_a(x)\big) \in [F]$ and $\int_{\mathfrak{X}} \big((f(x)+c_a(x)\big)\varphi(dx) > \mu$, contrary to the definition of μ. The uniqueness of the decomposition may easily be proved. For a countably additive set function is identically zero if it is absolutely continuous and singular simultaneously.

§3. *The metrical case.* The space (A) of all the countably additive set functions $F(E)$ on $\mathfrak{X}$ is a semi-ordered linear space if we call $F \in (A)$ *non-negative* (written $F \geq 0$) when $F(E) \geqq 0$ for all $E \in \mathfrak{X}$:

(1) If $F \geq 0$ and $a \geq 0$, then $aF \geq 0$,
(2) If $F \geq 0$ and $-F \geq 0$, then $F=0$,
(3) If $F \geq 0$ and $G \geq 0$, then $F+G \geq 0$,
(4) (A) is a lattice by the semi-order relation $\geq$.

In fact $F^+ = F \vee 0 = \sup(F, 0)$, $F^- = F \wedge 0 = \inf(F, 0)$ are respectively defined by $F^+(E) = \sup_{E' \subset E} F(E')$, $F^-(E) = \inf_{E' \subset E} F(E')$. As is well-known, $F = F^+ + F^-$ and $|F| = F^+ - F^- = \sup(F, -F)$. By putting $\|F\| = |F|(X) =$ the total variation of F on $\mathfrak{X}$, we have

(5) (A) is a Banach space by the norm $\|F\| = \|(|F|)\|$ such that $F \geq 0$, $G \geq 0$ imply $\|F+G\| = \|F\| + \|G\|$.

Let (A) be any abstract space satisfying (1)–(5). Choose any *positive* element $\varphi \in (A)$ (written $\varphi > 0$), that is, an element φ satisfying $\varphi \geq 0$, $\varphi \not\equiv 0$. Call φ a *unit* of (A) and write 1 for φ; we also write a for $a \cdot 1$ when there occur no ambiguities. An element G is called *singular* if $|G| \wedge 1 = 0$. A non-negative element E is called a *quasi-unit* if $E \wedge (1-E)=0$. A finite linear combination $\sum_{i=1}^{n} a_i E_i$ of quasi-units E_i is called a *step-element* and we call *absolutely continuous* the element which can be expressed as the strong limit of step-elements.

We will show that any element $\in (A)$ may uniquely be expressed as the sum of an absolutely continuous element and singular element.

For the proof we need four lemmas.

Lemma 2. A *vector lattice* satisfying (1)–(4) is a *distributive lattice*, viz. we have $(A \vee B) \wedge C = (A \wedge C) \vee (B \wedge C)$, $(A \wedge B) \vee C = (A \vee C) \wedge (B \vee C)$ for any three elements A, B and C.

Proof. See [F–1], 642.

Lemma 3. Let $0 \leq F_1 \leq F_2 \leq \cdots$ and let the sequence of numbers $\{\|F_n\|\}$ be bounded. Then the $\sup(F_1, F_2, \ldots)=F$ exists and we have $\lim_{n \to \infty} \|F-F_n\|=0$.

Proof[1]. The set of all the non-negative elements is strongly closed. For we have $\|F\|-\|G\| \leq \|(|F|-|G|)\| \leq \|F-G\|$ from the well-known inequality $|F|-|G| \leq |F-G|$. Since $\|F_n-F_m\|=\|F_n\|-$

---

1) due to [R–3], 7. Cf. also [K–2], 526.

$\|F_m\|$, $n \geq m$, tends to 0 as m tends to ∞, F_n converges in norm to an element F. From $F_n - F_m \geq 0$, $n \geq m$, and $\lim_{n \to \infty} \|(F_n - F_m) - (F - F_m)\| = 0$ we obtain $F \geq F_m$ ($m = 1, 2, \dots$) by the strong closure of the non-negative part of (A). Let $G \geq F_m$ ($m = 1, 2, \dots$), then, in the same way, we obtain $G \geq F$. Thus we have $F = \sup (F_1, F_2, \dots)$.

Lemma 4. The set $\mathfrak{T}$ of all the step-elements constitutes a sub-lattice of (A).

Proof. Since the symmetric relation $E \wedge (1 - E) = 0$ is equivalent to $2E \wedge 1 = E$, the set $\mathfrak{E}$ of all the quasi-units forms a Boolean algebra. For $E_1, E_2 \in \mathfrak{E}$ implies $2(E_1 \vee E_2) \wedge 1 = (2E_1 \wedge 1) \vee (2E_2 \wedge 1) = E_1 \vee E_2$, $2(E_1 \wedge E_2) \wedge 1 = (2E_1 \wedge 1) \wedge (2E_2 \wedge 1) = E_1 \wedge E_2$. Therefore any two elements $F, G \in \mathfrak{T}$ may be expressed as $F = \sum_{i=1}^{n} a_i E_i$, $G = \sum_{i=1}^{n} \beta_i E_i$, $E_i \wedge E_j = 0$ ($i \neq j$), $E_i \in \mathfrak{E}$ ($i = 1, 2, \dots, n$), and hence $F \vee G = \sum_{i=1}^{n} \max (a_i, \beta_i) E_i$ $F \wedge G = \sum_{i=1}^{n} \min (a_i, \beta_i) E_i$ both $\in \mathfrak{T}$. Cf. $[K-2]$, 531.

Lemma 1'. Let $G > 0$ and let $G \wedge 1 \neq 0$. Then there exists a rational number $a > 0$ and a quasi-unit $E_a > 0$ such that $G \geq a E_a$.

Proof. There exists a rational number $a > 0$ such that $(G - a)^+ \wedge 1 \neq 0$. Assume the contrary, then $0 \leq (G - a)^+ \wedge 1 = \big((G - a) \wedge 1\big)^+ = \big((G \wedge (1 + a) - a\big)^+ = 0$ and hence $0 \leq G \wedge (1 + a) \leq a$ for all rational number $a > 0$. Thus we would have $G \wedge 1 = 0$, contrary to the hypothesis. Let now $(G - a)^+ \wedge 1 \neq 0$ and hence $(G/a - 1)^+ \wedge 1 \neq 0$. Then, by the lemma 3, $E_a = \sup_{1 \leq n} \big(n(G/a - 1)^+ \wedge 1\big)$ exists and > 0. Since $2E_a \wedge 1 = \big(\sup_{1 \leq n} \big(2n(G/a - 1)^+ \wedge 2\big)\big) \wedge 1 = E_a$, E_a is a quasi-unit. Next from $n(G/a - 1)^+ \wedge 1 = \big(\big(n(G/a - 1)\big) \wedge 1\big)^+ = \big(\big(nG/a \wedge (n+1)\big) - n\big)^+ \leq \big((n+1)(G/a \wedge 1) - n\big)^+$ we obtain $E_a \leq G/a \wedge 1$. Thus $a E_a \leq G$.

The decomposition of (A). Let $F \in (A)$ be positive, and denote by $[F]$ the set of all the non-negative step-elements T such that $T \leq F$. Put $\mu = \sup_{T \in [F]} \|T\|$, then there exists a sequence $\{T_n\}$ of elements $\in [F]$ such that $\lim_{n \to \infty} \|T_n\| = \mu$. By the lemma 4, $F_n = \sup_{m \leq n} T_m \in [F]$. We have, by the lemma 3, $\|\bar{F}\| = \mu$ and $\lim_{n \to \infty} \|\bar{F} - F_n\| = 0$, where $\bar{F} = \sup_{1 \leq n} F_n$. We next prove that $G = F - \bar{F}$ is either singular or $= 0$. Assume the contrary, then, by the lemma 1', there exists a quasi-unit $E_a > 0$ such that $G \geq a E_a$, $a > 0$. Thus $\bar{F} + a E_a$ is a strong limit of elements $\in [F]$ and hence we must have $\|\bar{F} + a E_a\| = \|\bar{F}\| + a \| E_a\| \leq \mu$, contrary to $\|\bar{F}\| = \mu$, $\| E_a\| > 0$.

The uniqueness of the decomposition. It will be sufficient to show than an absolutely continuous, singular element G is $= 0$. Since G is absolutely continuous, we have $\lim_{n \to \infty} \|G - G_n\| = 0$, $G_n \in \mathfrak{T}$. As $0 \leq E \leq 1$ for any $E \in \mathfrak{E}$ we have $|G_n| \leq a_n$ ($n = 1, 2, \dots$) and hence, by the singularity of G, $|G| \wedge |G_n| = 0$ ($n = 1, 2, \dots$). As is well-known, we have

$|A \wedge B - A' \wedge B| + |A \vee B - A' \vee B| = |A - A'|$ in any vector lattice. Thus we have $\|G\| = \|(|G|)\| = \|(|G| \wedge |G|)\| \leq \|(|G| - |G_n|)\| \leq \|G - G_n\|$ $(n = 1, 2, \ldots)$, proving $G = 0$.

§4. *The general case.* The abstract space (A) satisfying (1)–(5) is, by lemma 3, *σ-complete*[1], viz.

(5)′ Any sequence $\{F_n\}$, $n = 1, 2, \ldots$, bounded from above (below) admit the supremum (infimum) in (A).

Let now (A) be any abstract space satisfying (1)–(5)′, then we can extend the maximal method to obtain the decomposition of (A). In this case we call absolutely continuous the element which can be expressed as the order-limit of an enumerable number of step-elements. We first prove a

Lemma 1″. Let $G \geq 0$, $G \geq aE$, $E \in \mathfrak{E}$ and $a > 0$. Then, for any β, $0 < \beta < a$, $E_\beta = \sup\limits_{1 \leq n} \left(n(G/\beta - 1)^+ \wedge 1 \right) \geq E$ and $G \geq \beta E_\beta$.

Proof. That $G \geq \beta E_\beta$ was already proved in lemma 1′. We will prove $E_\beta \geq E_a$ in the special case $a = 1$. The proof reads as follows. Let $0 < \delta < 1$, then we have $G - (1-\delta) = G - (1-\delta)E - (1-\delta)(1-E) \geq \delta E - (1-\delta)(1-E)$ and $\left(\delta E - (1-\delta)(1-E) \right)^+ = \left(\delta E \vee (1-\delta)(1-E) - (1-\delta)(1-E) \right) = \delta E + (1-\delta)(1-E) - (1-\delta)(1-E) = \delta E$ by $E \in \mathfrak{E}$. Thus $n\left(G/(1-\delta) - 1 \right)^+ \wedge 1 \geq n\delta E \wedge 1 \geq n\delta E \wedge E$ for $n = 1, 2, \ldots$ and hence $E_{1-\delta} \geq E$.

The decomposition of (A). Let $F > 0$ and put

$$\bar{F} = \sup_{a > 0} (aE_a), \qquad E_a = \sup_{1 \leq n} \left(n(F/a - 1)^+ \wedge 1 \right),$$

where the supremum of the first term is to be taken over all the rational numbers $a > 0$. If $G = F - \bar{F}$ is not singular, then there exist, by the lemma 1′, a rational number $\beta > 0$ and a quasi-unit $E > 0$ such that $G \geq \beta E$. We have, by the lemma 1″, $E_{\beta^{(1)}} \geq E$ if $0 < \beta^{(1)} < \beta$. By $F \geq \bar{F}$, $\bar{F} \geq \beta^{(1)} E_{\beta^{(1)}}$ and $F - \bar{F} \geq \beta E$ we obtain $F \geq 2\beta^{(1)} E$. Again by the lemma 1″ we have $E_{2\beta^{(2)}} \geq E$ if $0 < \beta^{(2)} < \beta^{(1)}$. Thus by $F \geq \bar{F}$, $\bar{F} \geq 2\beta^{(2)} E_{2\beta^{(2)}}$ and $\bar{F} - F \geq \beta E$ we obtain $F \geq 3\beta^{(2)} E$. In this way we would have $F \geq (n+1)\beta^{(n)} E$ $(n = 1, 2, \ldots)$ if $0 < \beta^{(n)} < \beta$, which is a contradiction. Thus G must be singular.

The uniqueness of the decomposition. Let G be absolutely continuous and singular. As in §3 we have $|G| \wedge |G_n| = 0$ $(n = 1, 2, \ldots)$, where $G = \text{order-lim}\limits_{n \to \infty} G_n$, $G_n \in \mathfrak{T}$. Thus $|G| = |G| \wedge |G| = \text{order-lim}\limits_{n \to \infty} (|G| \wedge |G_n|) = 0$. This prove the uniqueness of decomposition.

Remark. The above result is fairly general and it may be applied to the theory of probability or to the operator theory in Hilbert or Banach spaces.

---

1) In truth, it may be proved ([R-3], 7) that the vector lattice (A) satisfying (1)–(5) is *complete*, viz.

(5)″ any set bounded from above (below) admits the supremum (infimum) in (A). The second paper of Riesz and those of Maeda, Bochner-Phillips treat the vector lattice satisfying (1)–(5)″. However, the decomposition theorem would be more general if we assume the weaker condition (5)′ instead of (5)″.

On vector lattice with a unit, II

(with M. Fukamiya)

Proc. Imp. Acad. Tokyo 17 (1941) 479–482

(Comm. by T. Takagi, m.i.a., Dec. 12, 1941.)

§ 1. *Introduction and the theorems.* In a preceding note[1] one of the authors gave a representation of the vector lattice with a unit to obtain an algebraic proof of Kakutani-Krein's lattice-theoretic characterisation[2] of the space of continuous functions on a bicompact Hausdorff space. The purpose of the present note is to extend the result and to show that there exists a close analogy between the structures of the vector lattice and the algebras as in the case of the normed ring and the algebras[3].

A vector lattice E is a partially ordered real linear space, some of whose elements f are "non-negative" (written $f \geq 0$) and in which[4]

(V 1): If $f \geq 0$ and $a \geq 0$, then $af \geq 0$.

(V 2): If $f \geq 0$ and $-f \geq 0$, then $f = 0$.

(V 3): If $f \geq 0$ and $g \geq 0$, then $f + g \geq 0$.

(V 4): E is a lattice by the semi-order relation $f \geq g$ ($f - g \geq 0$).

In this note we further assume the existence of a "unit" $I > 0$ satisfying

(V 5): For any $f \in E$ there exists $a > 0$ such that $-aI \leq f \leq aI$.

An element $f \in E$ is called "nilpotent" if $n|f| < I (n = 1, 2, \ldots)$. The set R of all the nilpotent elements f is called the "radical" of E. Surely R constitutes a linear subspace of E. Moreover it is easy to see that R is an "ideal" of E, viz. $f \in R$ and $|g| \leq |f|$ imply $g \in R$. Here we put as usual $|f| = f^+ - f^-$, $f^+ = f \vee 0 = \sup(f, 0)$, $f^- = f \wedge 0 = \inf(f, 0)$.

Let N be a linear subspace of E. Then the linear congruence $a \equiv b \pmod{N}$ is also a lattice-congruence:

$$a \equiv b, \quad a' \equiv b' \pmod{N} \quad \text{implies} \quad ab \equiv a'b' \pmod{N},$$

if and only if N is an ideal of E[5]. An ideal N is called "non-trivial" if $N \neq 0, E$. A non-trivial ideal N is called "maximal" if it is contained in no other ideal $\neq E$. Denote by $\mathfrak{N}$ the set of all the maximal ideals N of E. The residual class E/N of E mod. any ideal $N \in \mathfrak{N}$ is "simple", viz. E/N does not contain non-trivial ideals. It is proved

---

1) K. Yosida: Proc. **17** (1941), 121–124. Cf. also M. H. Stone: Proc. Nat. Acad. Sci. **27** (1941), 83–87, and H. Nakano: Proc. **17** (1941), 311–317.

2) S. Kakutani: Proc. **16** (1940), 63–67. M. and S. Krein: C. R. URSS, **27** (1940), 427–430.

3) I. Gelfand: Rec. Math. **9** (1941), 1–24. We here express our hearty thanks to Tadasi Nakayama for his discussions during the preparation of the present note. He also obtained another proof of the theorem 1 below by considering the embedding of "lattice-groups" in a direct product of linearly ordered lattice-groups. See his paper shortly to appear in these Proceedings.

4) Small roman letters and small greek letters respectively denote elements $\in E$ and real numbers. We write $f > 0$ if $f \geq 0$ and $f \neq 0$.

5) Garrett Birkhoff: Lattice Theory, New York (1940), 109.

below that simple vector lattice with a unit is linear-lattice-isomorphic to the vector lattice of real numbers, the non-negative elements and the unit being represented by non-negative numbers and the number 1. We denote by $f(N)$ the real number which corresponds to $f \in E$ by the linear-lattice-homomorphism $E \to E/N$, $N \in \mathfrak{N}$.

After these preliminaries we may state our

Theorem 1. The radical R coincides with the intersection ideal $\bigwedge\limits_{N \in \mathfrak{N}} N$.

The vector lattice $\bar{E} = E/R$ is again a vector lattice with a unit $\bar{I}$. By the theorem 1 the intersection ideal $\bigwedge\limits_{\bar{N} \in \bar{\mathfrak{N}}} \bar{N}$ of all the maximal ideals $\bar{N}$ of E is the zero ideal and hence $\bar{E}$ does not contain nilpotent element $\neq 0$. Thus $\bar{E}$ satisfies the "Archimedean axiom":

$$(V\,6): \quad \text{order-limit}_{n \to \infty} \frac{1}{n} |\bar{f}| = 0 \quad \text{for all} \quad \bar{f} \in \bar{E}.$$

Therefore, by the result of the preceding note, we may add a precision to the theorem 1:

Theorem 2. By the correspondence $f \to \bar{f}(\bar{N})$, $\bar{E}$ is linear-lattice-isomorphically mapped on the vector lattice $F(\bar{\mathfrak{N}})$ of real-valued bounded functions on $\bar{N}$ such that i) $\bar{f} \to \bar{f}(\bar{N})$, ii) $\bar{I}(\bar{N}) \pm 1$ *on* $\bar{\mathfrak{N}}$ *and* iii) $F(\bar{\mathfrak{N}})$ *is dense in the set of all the real-valued continuous functions $c(\bar{N})$ on $\bar{\mathfrak{N}}$ by the "norm" $\|c\| = \sup\limits_{\bar{N}} |c(\bar{N})|$. Here the topology in $\bar{\mathfrak{N}}$ is defined by calling open the set of all the points $\bar{N} \in \bar{\mathfrak{N}}$ which satisfy $|\bar{f}_i(\bar{N}) - \bar{f}_i(\bar{N}_0)| < \varepsilon_i$, $i = 1, 2, \ldots, n$, where $\bar{N}_0, \bar{N} \in \bar{\mathfrak{N}}$, $\varepsilon_i > 0$, n and $\bar{f}_i (-\bar{I} \leq \bar{f}_i \leq \bar{I})$ are arbitrary.*

The theorems 1 and 2 show the analogy to a fundamental theorem in the theory of algebras, viz. the theorem stating that the residual class of an algebra mod. its maximal nilpotent ideal is a direct sum of total matric algebras.

§ 2. *The proof of the theorem 1* may be obtained by the following four lemmas.

Lemma 1. Let E be a simple vector lattice with a unit I, then we must have $E = \{aI\}$, $-\infty < a < \infty$.

Proof. E does not contain a nilpotent element $f \doteq 0$, for otherwise E would contain the non-trivial ideal $N_0 = \mathop{\mathcal{E}}\limits_{g}(|g| \leq \eta |f|, \eta < \infty)$. Hence E satisfies the Archimedean axiom (V 6). Let $E \ni f \neq \gamma I$ for any γ. Let $\alpha = \inf \alpha'$, $\alpha' I \geq f$, $\beta = \sup \beta'$, $\beta' I \leq f$, then $\beta I \leq f \leq \alpha I$ and $\alpha > \beta$. Hence $(f - \delta I)^+ \neq 0$, $(f - \delta I)^- \neq 0$ for $\beta < \delta < \alpha$. Then the set $N_0 = \mathop{\mathcal{E}}\limits_{g}(|g| \leq \eta(f - \delta I)^+, \eta < \infty)$ is a non-trivial ideal, contrary to the hypothesis.

Lemma 2. For any non-trivial ideal N_0 there exists a maximal ideal $N \supset N_0$.

Proof. Let $N_0 \subset N_1 \subset N_2 \subset \cdots \subset N_\eta \subset \cdots$, $\eta < \omega$, be a transfinite sequence of non-trivial ideals. If ω is a limit ordinal, define $f \equiv g$

(mod. N_ω) to mean $f \equiv g$ (mod. N_η) for some $\eta < \omega$. That N_ω is a non-trivial ideal follows from the fact that $I \not\equiv 0$ (mod. N_η), $\eta < \omega$. This process defines a transfinite sequence of linear-lattice-congruence relations on E, each more inclusive than the last. Hence it cannot continue indefinitely. Therefore we would obtain the demanded maximal ideal $N \supset N_0$.

Lemma 3. We have $R \underline{\underline{\leq}} \bigwedge_{N \in \mathfrak{N}} N$.

Proof. Let $f > 0$ and $nf < I$ $(n = 1, 2, \ldots)$, then for any $N \in \mathfrak{N}$ we have $n \cdot f(N) \leq I(N) = 1$ $(n = 1, 2, \ldots)$ and hence $f(N) = 0$, that is, $f \in N$.

Lemma 4. We have $R \underline{\underline{\geq}} \bigwedge_{N \in \mathfrak{N}} N$.

Proof. Let $f > 0$ be not nilpotent, then we have to show that there exists an ideal $N \in \mathfrak{N}$ such that $f \bar{\in} N$. This may be proved as follows.

Let $n \cdot f \underline{\not\leq} I$. Such an integer $n \geq 1$ surely exists, since f is not nilpotent. We may assume that $n \cdot f \underline{\not\geq} I$, since otherwise $f \bar{\in} N$ for any $N \in \mathfrak{N}$. Thus $p = I - (n \cdot f) \wedge I > 0$. For any positive integer m we do not have $m \cdot p \geq I$. If otherwise we would have $\dfrac{1}{m} I \underline{\leq} I - (n \cdot f) \wedge I$ and hence

$$(1) \qquad n \cdot f \wedge I = n \cdot f \wedge \left(1 - \frac{1}{m}\right) I,$$

which implies

$$(2) \qquad n \cdot f \underline{\leq} \left(1 - \frac{1}{m}\right) I,$$

contrary to $n \cdot f \underline{\not\leq} I$. Thus the set $N_0 = \mathop{\mathscr{E}}\limits_{g} (|g| \underline{\leq} \eta p, \eta < \infty)$ is a non-trivial ideal and hence there exists a maximal ideal $N \ni N_0$, by the lemma 2. Then $O = p(N) = 1 - (n \cdot f(N)) \wedge 1$, and thus $f(N) > 0$, that is, $f \bar{\in} N$.

The deduction of (2) from (1). From (1) we have

$$\left(n \cdot f - \left(1 - \frac{1}{m}\right) I\right) \wedge \frac{1}{m} I = \left(n \cdot f - \left(1 - \frac{1}{m}\right) I\right) \wedge 0 \underline{\leq} 0,$$

and hence, by the distributivity of the vector lattice,

$$0 = \left\{\left(n \cdot f - \left(1 - \frac{1}{m}\right) I\right) \wedge \frac{1}{m} I\right\} \vee 0 = \left(n \cdot f - \left(1 - \frac{1}{m}\right) I\right)^{+} \wedge \frac{1}{m} I.$$

Thus $\left(n \cdot f - \left(1 - \frac{1}{m}\right) I\right)^{+} \wedge I = 0$. Put $b = \left(n \cdot f - \left(1 - \frac{1}{m}\right) I\right)^{+}$ and assume that $b > 0$. By (V 5) we have $b < aI$ with $a > 1$. Then $0 < b = b \wedge aI$, and hence $0 < \dfrac{b}{a} \wedge I \underline{\leq} b \wedge I$, contrary to $b \wedge I = 0$. Thus $b = 0$, which is equivalent to (2).

§ 3. *An example due to T. Nakayama.* The following example shows that the existence of the unit is important for the theorem 1. Consider linear functions $\alpha x + \beta$ with an indeterminate symbol x. We put $\alpha x + \beta \geq \gamma x + \delta$ if, and only if, 1) $\alpha > \delta$ or 2) $\alpha = \gamma$ and $\beta \geq \delta$. Then the totality of the vectors $f = (\alpha_1 x + \beta_1,\ \alpha_2 x + \beta_2,\ \ldots)$ forms a vector lattice by componentwise addition and componentwise order relation. Now, consider the sublattice E consisting of those f such that almost all α_i are zero. This vector lattice possesses no unit. Further, if we call an element g nilpotent when $n\,|\,g\,| < f$ $(n = 1, 2, \ldots)$ for a certain $f > 0$, then g is nilpotent in E if and only if all its α_i vanish and almost all its β_i vanish. On the other hand, the intersection of all the maximal ideals in E contains the totality of those f such that all its α_i are zero. This last property may be proved by the fact that a simple vector lattice is linear-lattice-isomorphic to real numbers (proof similar as in the lemma 1). In fact, let $c = (\gamma_1, \gamma_2, \ldots)$ with all $\gamma_i \geq 0$ be $\bar{\in}$ a maximal ideal N, then, since E/N is isomorphic to real numbers, we have $(2x, 0, 0, \ldots) \equiv \partial c$ (mod. N). Hence $(x, 0, 0 \ldots) \in N$ and similarly $(0, x, 0, 0, \ldots)$, $(0, 0, x, 0, 0, \ldots)$, $\ldots \in N$. Thus if only a finite number of $\alpha_i x + \beta_i \neq 0$, then $(\alpha_1 x + \beta_1,\ \alpha_2 x + \beta_2,\ \ldots) \in N$. Let M denote the totality of such elements. N/M is a maximal ideal of E/M. Since $n\gamma_i < i\gamma_i$ for almost all i, we have $nc < c_1 = (\gamma_1, 2\gamma_2, 3\gamma_3, \ldots)$ (mod. M). Thus c (mod. M) is contained in any maximal ideal of E/M and hence c (mod. $M) \in N/M$. Therefore $c \in N$, contrary to the assumption.

On the representation of the vector lattice

Proc. Imp. Acad. Tokyo **18** (1942) 339–342

(Comm. by T. Takagi, m.i.a., July 13, 1942.)

§1. *Introduction.* In a preceding note the author gave, jointly with M. Fukamiya[1], a representation of the vector lattice with an *Archimedean-unit* to obtain an algebraic proof of Kakutani-Krein's lattice-theoretic characterisation[2] of the space of continuous functions on a bicompact Hausdorff space. Recently, by different approaches, H. Nakano[3] and F. Maeda-T. Ogasawara[4] treated a more general case when the existence of an Archimedean-unit is not assumed. The purpose of the present note is to show that our method is also applicable to this case as a short-cut to the representation theory. Their representation space is totally disconnected and so their results will not be a direct extension of our preceding note. T. Nakayama, who stressed[5] the applicability of the Lorenzen-Clifford's procedure to the representation of the vector lattice, kindly read the manuscript and discussed with me the difference of their method and that of ours. The conclusion may, in short, be stated as follows. The point of their representation space is perhaps, so to speak, a *minimal prime ideal*, while our point is a *maximal prime ideal*. In concluding the introduction I express may hearty thanks to T. Nakayama.

§2. *Preliminaries.* A vector lattice E is a real linear space, some of whose elements f are *non-negative* (written $f \geq 0$) and in which

(V 1): If $f \geq 0$ and $a \geq 0$, then $af \geq 0$.

(V 2): If $f \geq 0$ and $-f \geq 0$, then $f = 0$.

(V 3): If $f \geq 0$ and $g \geq 0$, then $f + g \geq 0$.

(V 4): E is a lattice by the semi-order relation $f \geq g$ $(f - g \geq 0)$.

We put, as usual, $|f| = f^{+} - f^{-}$, $f^{+} = f \vee 0$, $f^{-} = f \wedge 0$. Two elements f and g is called *disjoint* (or *orthogonal*) if $|f| \wedge |g| = 0$. Let $\{u_a\}$ be a maximal set of mutually disjoint positive $(u_a \geq 0$ but $\neq 0)$ elements of E. The maximality means that if $x > 0$ then $x \wedge u_a > 0$ for at least one u_a. An element f is called *nilpotent* (with respect to $\{u_a\}$) if $n(|f| \wedge u_a) < u_a$ $(n = 1, 2, \dots)$ for all u_a. The totality R of the nilpotent elements is called the *radical* of E. R constitutes a linear subspace of E. *Proof:* Let f and g be nilpotent, then $n(|f+g| \wedge u_a) \leq$
$$n\big(2(|f| \vee |g|)\big) \wedge u_a \leq 2n\big((|f| \wedge u_a) \vee (|g| \wedge u_a)\big) < u_a \vee u_a = u_a \ (n = 1,$$

---

1) Proc. **17** (1941), 479–482.

2) S. Kakutani: Proc. **16** (1940), 63–67. M. and S. Krein: C. R. URSS, **27** (1940), 427–430.

3) Proc. Physico-Math. Soc. Japan, **23** (1941), 485–511.

4) 全國紙上數學談話會, 第 231 號.

5) 全國紙上數學談話會, 第 233 號.

$2, \ldots)$. Moreover R is an *ideal* of E, viz. $f \in R$ and $|g| \leq |f|$ implies $g \in R$.

Lemma 1. Let N be a linear subspace of E. Then the *linear-congruence* $a \equiv b \pmod{N}$ is also a *lattice-congruence*:

$$c \equiv c', \quad d \equiv d' \pmod{N} \quad \text{implies} \quad c \times d \equiv c' \times d' \pmod{N}$$

if and only if N is an ideal of E.

Proof. See, for example, Garrett Birkhoff: Lattice Theory (1940), 109.

An ideal $N \neq E$ is called *prime*, if the *residual vector lattice* E/N of E mod N is *simply ordered*, viz. $f \geq g$ or $f < g \pmod{N}$ for any two elements f, g. Since $x^+ \wedge (-x)^+ = 0$ for any x, we see that E is simply ordered if and only if $|f| \wedge |g| = 0$ implies $f = 0$ or $g = 0$.

Lemma 2[1]. For any $f \neq 0$, there exists a prime ideal $N \ni f$.

Proof. Let E be not simply ordered and suppose $g > 0$, $h > 0$, $g \wedge h = 0$. Then at least one of the ideals

$$N^{(1)} = \mathop{\mathcal{E}}_{g'}(|g'| \leq ag, \, a < \infty) \neq 0, \quad N^{(2)} = \mathop{\mathcal{E}}_{h'}(|h'| \leq ah, \, a < \infty) \neq 0$$

does not contain f. Let $N_1 = N^{(1)} \ni f$ and let $0 \subset N_1 \subset N_2 \subset \cdots \subset N_\eta \subset \cdots$ $(\eta < \omega)$ be a properly increasing (transfinite) sequence of ideals not containing f. If ω is a limit ordinal, define $x \equiv y \pmod{N_\omega}$ to mean $x \equiv y \pmod{N_\eta}$ for some $\eta < \omega$. N_ω does not contain f. Thus we may obtain, at a certain step, an ideal $N \ni f$ which is not contained in no other ideal $\ni f$. By this maximality N is a prime ideal.

§3. *The representation theorem.* By the lemma 2, there exists an ideal $N \ni u_a$ which is not contained in no other ideal $\ni u_a$. Let $\mathfrak{N}(u_a)$ be the totality of such ideals and let $\mathfrak{N}$ be the totality of the ideals $\in$ some $\mathfrak{N}(u_a)$. Since each $N \in \mathfrak{N}$ is a prime ideal, there exists, for any $N \in \mathfrak{N}$, exactly one $u_a = u_{a(N)}$ which satisfies $u_{a(N)} \bar{\in} N$. Thus we may write u_N for $u_{a(N)}$. For any $x \in E$ and for any $N \in \mathfrak{N}$ we put

$$
(1) \quad
\begin{cases}
x(N) = \text{l. u. b. } \lambda, \text{ where } x \geq \lambda u_N \pmod{N} \\
\quad\;\; = \text{g. l. b. } \mu, \text{ where } x \leq \mu u_N \pmod{N}.
\end{cases}
$$

The equivalence of the two definitions of $x(N)$ follows from the fact that E/N is simply ordered. Of course, we put $x(N) = +\infty$ if there exists no μ such that $x \leq \mu u_N \pmod{N}$; similarly for $x(N) = -\infty$. By the lemma 1, we have

$$(2) \quad (x \vee y)(N) = \max\big(x(N), y(N)\big), \quad (x \wedge y)(N) = \min\big(x(N), y(N)\big),$$

$$(3) \qquad\qquad (\alpha x + \beta y)(N) = \alpha x(N) + \beta y(N).$$

It is to be noted that (3) is ambiguous in case $x(N) = \pm\infty$, $y(N) = \pm\infty$. If E satisfies the *Archimedean axiom* (V 6) below, this ambiguity will be removed by introducing a topology in $\mathfrak{N}$ (§ 4).

Remark. Let there exist an *Archimedean-unit* u:

---

1) 全國紙上數學談話會, 第 227 號.

$$(V\,5): \quad \left\{ \begin{array}{l} \text{For any } x \in E, \text{ there exists a positive number } a = a(x) \text{ such} \\ \quad \text{that } -a(x)u \leqq x \leqq a(x)u. \end{array} \right.$$

If we take the one-element-set $\{u\}$ for $\{u_a\}$, then every function $x(N)$ is bounded on $\mathfrak{N}\big(|x(N)| \leqq a(x)\big)$. In this (Archimedean-unit) case, $N \in \mathfrak{N}$ means that N is a *maximal non-trivial* ideal.

Returning to our representation (1), we have

$$(4) \qquad x(N) = 0 \text{ identically on } \mathfrak{N} \text{ if and only if } x \in R.$$

Proof. Let $x \geqq 0$ be nilpotent, then, by (1) and (2), we have $n\big(\min\big(x(N), 1\big)\big) \leqq 1$ $(n = 1, 2, \ldots)$ and hence $x(N) = 0$ on $\mathfrak{N}$. Conversely let $0 \leqq x \leqq u_a$ and $nx \leqslant u_a$, $n \geqq 1$. By the lemma 2, there exists a prime ideal $N(y) \ni y = (nx - u_a)^+$, viz. $(nx - u_a) > 0$ $\big(\text{mod } N(y)\big)$. $N(y)$ does not contain u_a, for otherwise, we would obtain $0 = (0 - 0) > 0$ $\big(\text{mod } N(y)\big)$. Let N be an ideal $\supseteq N(y)$, $\ni u_a$, which is not contained in no other ideal $\supseteq N(y)$, $\ni u_a$. Surely we have $N \in \mathfrak{N}$ and hence $u_a = u_N$. Since $N \supseteq N(y)$, we have $(nx - u_N) \geqq 0$ $(\text{mod } N)$ and thus $nx(N) \geqq u_N(N)$ or $x(N) \geqq 1/n$.

§ 4. *Introduction of a topology and the Archimeden axiom.* For any $x \geqq 0$, we call x-set the totality of $N \in \mathfrak{N}$ such that $N \ni x$, Then we have

$$(5) \quad \left\{ \begin{array}{l} (x \vee y)\text{-set} = \text{the sum } (x\text{-set}) \vee (y\text{-set}), \\ (x \wedge y)\text{-set} = \text{the intersection } (x\text{-set}) \wedge (y\text{-set}). \end{array} \right.$$

Proof. That $(x\text{-set}) \vee (y\text{-set}) \subseteq (x \vee y)\text{-set}$ is evident from the definition of the ideal. Let $x \vee y \bar\in N$ and let $x \in N$, $y \in N$. Then, since N is prime, $x \vee y \equiv x$ or y $(\text{mod } N)$, that is, $x \vee y \in N$, contrary to the hypothesis. Next we have $(x\text{-set}) \wedge (y\text{-set}) \geqq (x \wedge y)\text{-set}$ from the definition of the ideal. Let $x \bar\in N$, $y \bar\in N$, then, since N is prime, $x \wedge y \equiv x$ or y $(\text{mod } N)$ and thus $x \wedge y \bar\in N$. Q. E. D.

Hence, if we call *open* the x-set's, $\mathfrak{N}$ is a topological space. In the truth, $\mathfrak{N}$ is a *Hausdorff* space. Proof: If $N_1 \neq N_2$, then there exist $x_1 > 0$ and $x_2 > 0$ such that $x_1 \bar\in N_1$, $x_1 \in N_2$, $x_2 \bar\in N_2$, $x_2 \in N_1$. Since N_1 is prime, $(x_1 - x_2) \geqq 0$ $(\text{mod } N_1)$ or $(x_1 - x_2) < 0$ $(\text{mod } N_1)$. The latter inequality is excluded by $x_1 > 0$ $(\text{mod } N_1)$, $x_2 = 0$ $(\text{mod } N_1)$. Thus $(x_1 - x_2)^+ \bar\in N_1$ and $(x_2 - x_1)^+ \bar\in N_2$ similarly. By the idantity $x^+ \wedge (-x)^+ = 0$ and (5), the intersection of $(x_1 - x_2)^+$-set and $(x_2 - x_1)^+$-set is void.

The continuity of the function $x(N)$ on $\mathfrak{N}$ may be proved as follows. Let $x(N_0) = \lambda \neq \pm \infty$ and let ε be any positive number. Then we have $(\lambda - \varepsilon) \leqq x(N) \leqq (\lambda + \varepsilon)$ if N belongs to

$$u_{N_0} \wedge \Big(\big(x - (\lambda - \varepsilon)u_{N_0}\big)^+\Big) \wedge \big((\lambda + \varepsilon)u_{N_0} - x\big)^+\big)\text{-set} \ni N_0.$$

Similarly for the case $x(N_0) = \pm \infty$.

Next we assume that the *Archimedean axiom*:

(V 6): $\quad \bigwedge_{n \geq 1}\left(\dfrac{1}{n}x\right)=0$ for any $x \geq 0$

is satisfied in E. We have, in this case, $R=0$. Moreover we have the result:

(6) The set of N at which $x(N)=\pm\infty$ is non-dense on $\mathfrak{N}$.

Proof. We assume $x>0$ and will prove that, for any $y>0$, there exists a point $N_0 \in y$-set such that $x(N_0)<+\infty$. Assume the contrary and let $x>nu_N \pmod{N}$ $(n=1, 2, \ldots)$ for every $N \in y$-set. Then

(*) $\quad x>n(u_N \wedge y) \pmod{N}$ $(n=1, 2, \ldots)$ for every $N \in y$-set.

By the maximality of $\{u_a\}$, there exists u_a such that $u_a \wedge y>0$. By (V 6), we have $x \ngeq z=n(u_a \wedge y)$ for some $n \geq 1$. Thus $(z-x)^+>0$ and hence, by $R=0$, $\max\left\{\left(z(N_0)-x(N_0)\right),\, 0\right\}>0$ for some N_0. Therefore

$$n \cdot \min\left(u_a(N_0),\, y(N_0)\right)>x(N_0) \geq 0\,.$$

This contradicts to (*), for from $u_a(N_0)>0$, $y(N_0)>0$ we must have $u_a=u_{N_0}$, $N_0 \in y$-set.

Remark 1. In general, our vicinity, the x-set, does not disconnect the space $\mathfrak{N}$. While, in the treatments of H. Nakano and F. Maeda-T. Ogasawara cited above, the representation space is totally disconnected.

Remark 2. In the Archimedean-unit case, our topology is equivalent to the *weak topology* obtained by calling open the set of the form

$$\mathop{\mathscr{E}}_{N}\left(\,|\,x_i(N)-x_i(N_0)\,|<\varepsilon_i \quad (i=1, 2, \ldots, n)\right)$$

where $-u \leq x_i \leq u$, $\varepsilon_i>0$ $(i=1, 2, \ldots, n)$ and n are arbitrary. In this case, $\mathfrak{N}$ is bicompact and any continuous function on $\mathfrak{N}$ may be approximated uniformly on $\mathfrak{N}$ by the functions $x(N)$, $x \in E$. For the proof, see the preceding note.

On the semi-ordered ring and its application to the spectral theorem

(with T. Nakayama)

Proc. Imp. Acad. Tokyo **18** (1942) 555–560

(Comm. by T. Takagi, m.i.a., Nov. 12, 1942.)

This note deals with some remarks about semi-ordered rings and their application to the spectral theorem. Semi-ordered rings have been treated jointly by Messrs. I. Vernikoff, S. Krein and A. Tovbin[1]. We first observe that for their result the assumption of the associative law of multiplication is unnecessary ; it follows, as the commutativity, from the other axioms, and this fact will be of use in applications. Further, their theorems may be obtained rather easily also from Clifford-Lorenzen's theorem concerning semi-ordered abelian groups[2] by considering operator-domains. As for application, we prove a spectral theorem in the semi-ordered rings without appealing to the spectral theorem in vector lattice but relying upon Baire's category theorem. We thus obtain a new approach to the spectral theorem for bounded self-adjoint operators in a Hilbert space.

1. *Elementary observations about semi-ordered abelian groups.*
Let G be a semi-ordered abelian group, that is, an abelian group which possesses a semi-order $x \geq y$ (equivalent to $x - y \geq 0$) such that

(i) if $x \geq 0$ and $y \geq 0$ then $x + y \geq 0$,

(ii) if $x \geq 0$ and $-x \geq 0$ then $x = 0$.

We assume a further condition :

(iii) if $nx \geq 0$ for a certain natural number n, then $x \geq 0$.

Let moreover G possess an Archimedean unit e :

$$\text{(iv)} \quad \begin{cases} \text{for any } x \text{ there exists a natural number } n = n(x) \text{ such that} \\ -ne \leq x \leq ne. \end{cases}$$

And we call the totality N of those elements x in G satisfying $-e \leq tx \leq e$ (for every $t = 1, 2, \ldots$) the radical of G. N is a normal subgroup[3] of G, and the factor group G/N is also a semi-ordered group.

In virtue of the condition (iii) G is, according to Clifford-Lorenzen's

1) Sur les anneaux semi-ordonnés, C. R. URSS, **30** (1941). Cf. also H. Nakano Teilweise geordnete Algebra, Jap. J. Math., **17** (1941).

2) A. H. Clifford : Partially ordered abelian groups, Ann. of Math., **41** (1940). P. Lorenzen : Abstracte Begründung der multiplikativen Idealtheorie, Math. Zeitschr. **45** (1939).

3) Here we call a subgroup of G normal when it is a kernel of an order-homomorphism of G. Thus a subgroup H is normal if and only if $x \epsilon H$, $0 \leq y \leq x$ implies $y \epsilon H$.

theorem[4], order-isomorphically embedded in a direct sum of linearly ordered groups :

$$G \leqq \cdots G_\sigma + \cdots, \quad G \ni x \leftrightarrow (\ldots, x_\sigma, \ldots) \quad (x_\sigma \in G_\sigma);$$

here the order relation in the direct sum is explained, as usual, component-wise, and without losing generality we may suppose that when x runs over G its component x_σ exhausts G_σ. The image (component) e_σ of e is, for each σ, an Archimedean unit of G_σ, and we denote the radical of G_σ by N_σ. Evidently an element x belongs to N when and only when x_σ belongs to N_σ for every σ. Thus the group G/N is embedded isomorphically in the direct sum $\bar{G}_\sigma = G_\sigma/N_\sigma$:

$$x \bmod. N \leftrightarrow (\ldots, x_\sigma \bmod. N_\sigma, \ldots).$$

The order relation is preserved in the direction " $\rightarrow$ ". Further, each $\bar{G}_\sigma = G_\sigma/N_\sigma$ is, as an Archimedean linearly ordered group, a subgroup of the ordered group of real numbers, and the kernel M_σ of the homomorphism $G \rightarrow G_\sigma$ is a maximal normal subgroup of G. If, in particular, N consists of 0 only, that is, if the condition :

(v) $\qquad$ if $\quad -e \leqq tx \leqq e \quad (t=1,2,\ldots)$ then $\quad x=0,$

is satisfied, *then G itself is mapped isomorphically into the direct sum of the Archimedean linearly ordered groups $\bar{G}_\sigma$, order being preserved in the direct direction " $\rightarrow$ ".* Furthermore, if the stronger condition :

(vi) $\qquad$ if $\quad tx \leqq e \quad (t=1,2,\ldots)$ then $\quad x \leqq 0,$

is fulfilled, the order relation is preserved in the both directions " $\rightarrow$ " and " $\leftarrow$ ", that is, *G is order-isomorphically embedded in the direct sum of $\bar{G}_\sigma$.* For, if $x_\sigma \leqq 0_\sigma \bmod. N_\sigma$ then $x_\sigma \leqq z_\sigma$ for a certain element z_σ in N_σ, whence $tx_\sigma \leqq tz_\sigma \leqq e_\sigma$ for every natural number t. When this is the case for every σ, the assumption in (vi) is fulfilled and we have $x \leqq 0$. Now, suppose that G possesses a domain of operators $\Omega = \{A\}$, which is by itself a semi-ordered abelian group, satisfying the axioms (i), (ii) and such that

(vii) if $\quad x \geqq 0$ (in G), $\quad A \geqq 0$ (in Ω) then $\quad Ax \geqq 0$ (in G),

---

4) See 2). As the proof suggested by Clifford shows, we may take as G_σ linearly ordered groups which are (not order-but) group-isomorphic with G. Namely : consider subsets P in G which satisfy the conditions : i) if $x > 0$ then $x \in P$, ii) if $x \in P$, $y \in P$ then $x+y \in P$, iii) $0 \bar{\in} P$, iv) if $mx \in P$ $(m > 0)$ then $x \in P$. When P is such a subset (for instance, the set of all the positive elements of G), we can re-order G by calling the elements of P positive; denote the semi-ordered abelian group thus obtained by $G(P)$. Further, if there is an element x in G such that neither x nor $-x$ is contained in P, then there exists a second P which contains P and x both. We may, for example, take the set $\underset{z}{\mathfrak{S}}(z; \ kz = mp + nx, \ k > 0, \ m \geqq 0, \ n \geqq 0, \ m+n > 0)$ for a new P. From this follows that if P is a maximal such subset then $G(P)$ is linearly ordered. Moreover any non-zero and non-negative element is contained in at least one maximal such subset P, whence positive in $G(P)$. Thus G is order-isomorphically embedded in the direct sum of $G(P)'s$, P running over all the maximal such subsets in G, by letting x correspond for each P to x itself in $G(P)$.

(viii) $(A+B)x = Ax + Bx , \qquad A(x+y) = Ax + Ay .$

Let moreover

(ix) $\begin{cases} \Omega \text{ possess an Archimedean unit } I \text{ which satisfies } Ix = x \\ \text{for every } x \in G. \end{cases}$

Then we have the *Lemma: Every normal subgroup of G is allowable with respect to Ω. Proof:* Let H be a normal subgroup in G, and let $x \in H$, $A \in \Omega$. There is a natural number n such that $-nI \leq A \leq nI$ whence $-nIx \leq Ax \leq nIx$. Thus $-nx \leq Ax \leq nx$ and Ax belongs to H too. In particular the maximal normal subgroups M_σ are Ω-allowable, and the above isomorphisms are operator-isomorphisms with respect to Ω.

2. *Semi-ordered rings.* Let R be a ring with a unit element e and real multipliers. Neither the commutativity nor the associativity of the multiplication is assumed. Let in R be defined a semi-order such that

(I)   if $x \geq 0$ and $y \geq 0$ than $x+y \geq 0$ and $xy \geq 0$,

(II)   if $x \geq 0$ and $-x \geq 0$ then $x = 0$,

(III)   if $x \geq 0$ and a (real number) ≥ 0 then $ax \geq 0$.

Further we assume that the ring unit e is an Archimedean unit:

(IV) $\begin{cases} \text{for any } x \text{ there is a positive number } a = a(x) \text{ such} \\ -ae \leq x \leq ae. \end{cases}$

Then R satisfies, considered as an abelian group possessing R itself as an (either left or right) operator domain, the conditions (i)–(iv), (vii)–(ix) of the preceding section. And, every residue class mod. an $M = M_\sigma$ is represented by a multiple ae. Hence, we have, expressing the condition (v), (vi) in a modified fashion,

Theorem 1. Let R satisfy, besides the above conditions,

(V) $if \quad -\dfrac{1}{t} e \leq x \leq \dfrac{1}{t} e \quad (t = 1, 2, \dots) \quad then \quad x = 0.$

Then R is ring-isomorphic to a certain ring $R(\mathfrak{M})$ of (real-valued bounded functions over a certain space $\mathfrak{M} = \{M\}$: $x \leftrightarrow x(M)$, such that e is represented by 1: $e(M) \equiv 1$. In particular, R is both associative and commutative. The order is preserved in the direction $x \rightarrow x(M)$, when we order $R(\mathfrak{M})$ in the usual manner.

Theorem 2. If R satisfies the stronger condition:

(VI)[5] $\inf\limits_{t>0} \dfrac{1}{t} e \quad$ *exists and is equal to* 0,

then the order is preserved in the both directions, so that R is ring-order-isomorphic to $R(\mathfrak{M})$.

---

5)   Whence for every x in R an order-limit $\dfrac{1}{t} x$ exists and $=0$.

Here, according to our construction, $\mathfrak{M}$ is a certain set of maximal normal[6] ideals of R, but not necessarily all of them. However, the theorems are still the more true if $\mathfrak{M}$ represents the totality of the maximal normal ideals of R. So, assume this be the case. Then $\mathfrak{M}$ is a bicompact Hausdorff space by the so-called *weak topology*, under which the functions in R are continuous. Furthermore, since 1 is contained in $R(\mathfrak{M})$ and since there exists for any two distinct points M, M' in $\mathfrak{M}$ an element x in R such that $x(M) \neq x(M')$, the ring $R(M)$ is dense in the ring of all the continuous functions on $\mathfrak{M}$ with respect to the metric defined by the greatest absolute value taken by a function as its norm[7]. Hence

Theorem 3. Let R satisfy besides (I)–(VI) *the condition :*

(VII) *R is a Banach space by the norm* $\|x\| = \inf\,(-ae \leq x \leq ae)$.

Then R is ring-order-isomorphic to the ring $R(\mathfrak{M})$ of all the continuous functions over a bicompact Huasdorff space $\mathfrak{M}$. In particular, R is a vector lattice, viz. lattice-ordered abelian group with real multipliers.

Remark. Let, conversely, R be a vector lattice which satisfies (I)′–(VII) :

(I)′ if $x \geq 0$ and $y \geq 0$ then $x + y \geq 0$.

Such a vector lattice R is called, by S. Kakutani[8], an abstract (M) space. We may, following after F. Riesz and Y. Kawada[9], define a multiplication xy in R by

$$4xy = (x+y)^2 - (x-y)^2, \quad x^2 = \sup_{\lambda > 0}\,(2\lambda\,|x| - \lambda^2 e), \quad x = \sup\,(x, 0) - \inf\,(x, 0).$$

It is easy to see that R now satisfies the axioms (I)–(VIII). In this way, *the equivalence of the semi-ordered ring and the abstract (M) space may be proved appealing neither to the spectral theorem of H. Freudenthal[10] nor to the representation theorem[11] of the abstract (M) space.*

Another method of reducing the above theorems of our semi-ordered rings to the known results is, in case the condition (vi) is satisfied, to complete by cuts and apply the representation theory of vector lattices and lattice-ordered rings[12]. When we have only the

---

6) Defined similarly as in 3). "Fundamental" in the sense of Vernikoff-Krein-Tovbin, loc. cit.

7) See H. Nakano: 連續函數ノ ring 及ビ vector lattice, 全國紙上數學談話會, **218** (1941).

8) Weak topology, bicompact set and the principle of duality, Proc. **16** (1940). See also the literatures referred to in K. Yosida: On the representation of the vector lattice, Proc., **18** (1942).

9) F. Riesz: Sur la théorie ergodique des espaces abstraits, Acta Szeged, **10** (1941), 1. Y. Kawada: 抽象 M-空間ノ表現ニ就テ, 全國紙上數學談話會, **227** (1941).

10) Teilweise geordnete Moduln, Proc. Amsterdam Acad., **39** (1936).

11) See 8).

12) B. Vulich: Une définition du produit dans les espaces semiordonnés linéaires, C.R. URSS., **26** (1940). H. Nakano: loc. cit. in 1). T. Ogasawara: Ring lattice 公理系及ビ表現論, 全國紙上數學談話會, **230** (1942).

condition (v), then we have first to re-order the ring by Vernikoff-Krein-Tovbin's procedure so as to have (vi).

3. *An abstract spectral theorem.* Let R be a semi-ordered ring, again neither associativity nor commutativity being assumed, satisfying the conditions (I)–(IV), (VII) and, furthermore,

(VIII) $\begin{cases} \text{for any increasing sequence } \{x_n\} \text{ bounded from above} \\ (x_1 \leq x_2 \leq \cdots \leq y), \quad \sup x_n = \text{order-limit } x_n \text{ exists in } R. \end{cases}$

Then we have the

Theorem 4.   There exists, for any $x \in R$, a resolution of the identity $\{e_\lambda\}$ with the properties :

(1) $e_\lambda^2 = e_\lambda \leq e_\mu = e_\mu^2$ *if* $\lambda \leq \mu$,

(2) *if* $\lambda_n \downarrow \lambda$, *then order-limit* $e_{\lambda_n} = e_\lambda$,

(3) $e_\lambda = e$ *for* $\lambda \geq \|x\|$ *and* $e_\lambda = 0$ *for* $\lambda < -\|x\|$,

(4) $\begin{cases} \text{for ang } \varepsilon > 0, \quad x = \displaystyle\int_{-\|x\|-\varepsilon}^{\|x\|} \lambda \, de_\lambda \quad (\text{Riemann-Stieltjes integral} \\ \text{in, semi-order sense}), \end{cases}$

(5) $\{e_\lambda\}$ *is determined uniquely by the properties* (1)–(4).

Proof. By the theorem 3, there exists a bicompact Hausdorff space $\mathfrak{M}$ such that R is ring-order isomorphic to the ring $R(\mathfrak{M})$ of all the continuous functions on $\mathfrak{M}$. Let the isomorphism be given by $x \leftrightarrow x(M)$. We will prove the following property of the representation $R \to R(\mathfrak{M})$. Let $x_1 \leq x_2 \leq \cdots \leq y$ and let order-limit $x_n = x$. By Baire's theorem, the discontinuities of the function $\bar{x}(M) = \lim_{n \to \infty} x_n(M)$ constitute a set of first category, viz. enumerable sum of non-dense sets. We have surely $x(M) \geq \bar{x}(M)$. *In the truth, the set* $\underset{M}{\mathscr{E}}\left(x(M) - \bar{x}(M) > 0\right)$ *is of first category.* *Proof:* If otherwise, we would have a point M_0 such that $x(M)$ is continuous at M_0 and $x(M_0) > \bar{x}(M_0)$, and thus we would obtain a continuous function $x^*(M)$ such that $x(M_0) > x^*(M_0)$ and $x(M) \geq x^*(M) \geq \bar{x}(M)$ on $\mathfrak{M}$. This contradicts to the isomorphism $R \leftrightarrow R(\mathfrak{M})$ and the definition of x as order-limit x_n. Here use is made of the fact that a bicompact Hausdorff space is not of first category.

Next consider the set $R'(\mathfrak{M})$ of all the bounded functions $x'(\mathfrak{M})$ on $\mathfrak{M}$ such that $x'(M)$ is different from a continuous function $x(M)$ only on a set of first category. We then identify two functions from $R'(\mathfrak{M})$ if they differ on a set of first category. Thus $R'(\mathfrak{M})$ is divided into classes, each class x' containing exactly one continuous function $x(M)$ which corresponds to an element $x \in R$ by the isomorphism $R \leftrightarrow R(\mathfrak{M})$. This results from the fact that the complementary to a set of first category is dense on the bicompact Hausdorff space $\mathfrak{M}$.

The proof of the theorem is now immediate. We have only to put $e_\lambda = $ the element $\in R$ which corresponds to the class containing the characteristic function $e_\lambda'(M)$ of the set $\underset{M}{\mathscr{E}}\left(x(M) \geq \lambda\right)$. For then we would have $|x(M) - \sum_{i=1}^{n} \lambda_i e'_{\lambda_i}(M)| \leq \varepsilon$ and thus $|x(M) - \sum_{i=1}^{n} \lambda_i e_{\lambda_i}(M)| \leq \varepsilon$,

viz. $-\varepsilon e \leq x - \sum_{i=1}^{n} \lambda_i e_{\lambda_i} \leq \varepsilon e$ for $\lambda_1 = -\|x\| - \varepsilon < \lambda_2 < \lambda_3 < \cdots < \lambda_n = \|x\|$,

$\max_i (\lambda_{i+1} - \lambda_i) \leq \varepsilon$. Perhaps the fact that $e'_\lambda(M) \in R'(\mathfrak{M})$ will demand proof.

However the function $e'_\lambda(M)$ defined by $1 - \sup_{n \geq 1} \left(\inf \left(1, n(x(M) - \lambda)^+ \right) \right)$

belongs to $R'(\mathfrak{M})$ by the above property of the representation $R \to R(\mathfrak{M})$.

4. *Application to the Hilbert space.* Let (T) be a set of mutually commutative, bounded self-adjoint operators in Hilbert space $\mathfrak{H}$, and denote by $(T)'$ the totality of the bounded self-adjoint operators commutative with every operator $\in (T)$. Similarly we define $(T)'' = \left((T)' \right)'$, $(T)''' = \left((T)'' \right)'$ etc. $R = (T)''$ is a ring with unit operator ($=$ the identiy operator) I and is commutative, since from $(T) \leq (T)'$ we obtain $(T)' \geq (T)''$ and hence $(T)''' \geq (T)''$. We define a semi-order in R by writing $T \geq 0$ if and only if $(T \cdot f, f) \geq 0$ for all $f \in \mathfrak{H}$. Then R satisfies (I)–(IV), (VIII). Hence the theorem 4 is directly applicable to R. Only the proof of the axioms (I) and (VIII) would be non-trivial. However these may be proved following after F. Riesz's idea[13].

Remark. The above procedure also gives a simultaneous resolutions $T = \int \lambda dE_\lambda(T), \quad S = \int \lambda dE_\lambda(S)$ such that $E_\lambda(T)E_\mu(S) = E_\mu(S)E_\lambda(T)$, if T and S and mutually commutative bounded self-adjoint operators. Hence our method also gives the spectral theorem of the bounded *normal* (and of course *unitary*) operators, for such operators are of the form $T = \sqrt{-1}\,S$, where T and S are mutually commutative, bounded self-adjoint operators.

---

13) Über die linearen Transformationen des komplexen Hilbertschen Raumes, Acta Szeged, **5** (1930). Namely: *Ad.* (*I*). It will be sufficient to show that $TS \geq 0$, if $I \geq T, S \geq 0$. Put $T_1 = T, T_{n+1} = T_n - T_n^2$ ($n \geq 1$), Then we obtain $I \geq T_n \geq 0$ ($n \geq 1$) by induction, because of the identities $T_{n+1} = T_n^2 (I - T_n) + T_n (I - T_n)^2$, $I - T_{n+1} = (I - T_n) + T_n^2$. Hence $T \geq \sum_{m=1}^{n} T_m^2$ ($n \geq 1$) and thus $\lim \|T_n \cdot f\|^2 = \lim (T_n^2 \cdot f, f) = 0$, proving $T = \sum_{m=1}^{\infty} T_m^2$. Similarly we have $S = \sum_{m=1}^{\infty} S_m^2$ and thus $TS = \sum_{i,j} T_i^2 S_j^2 = \sum_{i,j} (T_i S_j)^2 \geq 0$. *Ad.* (*VIII*). We will prove the existence of the order-limit $T_n = T$ from $0 \leq T_1 \leq T_2 \leq \cdots \leq S$. By (I), $\{(T_n^2 \cdot f, f)\}$ is a bounded increasing sequence for any f, and hence $\lim (T_n^2 \cdot f, f)$ exists. We have, again by (I), $T_{n+k}^2 \geq T_{n+k}T_n \geq T_n^2$. Thus $\lim (T_{n+k}^2 \cdot f, f) = \lim (T_n^2 \cdot f, f) = \lim (T_{n+k}T_n \cdot f, f)$ and hence $\lim \left((T_n - T_m)^2 \cdot f, f \right) = \lim \|T_n \cdot f - T_m \cdot f\|^2 = 0$. Therefore the *strong limit* $T_n \cdot f = T \cdot f$ exists. T is surely the order-limit T_n.

On the semi-ordered ring and its application to the spectral theorem, II

(*with T. Nakayama*)

Proc. Imp. Acad. Tokyo **19** (1943) 144–147

(Comm. by T. Takagi, m.i.a., March 12, 1943.)

In the first place, our note[1] "On the semi-ordered ring and its application to the spectral theorem" contained, in its proof of algebraic part, a falsy argument, which we shall correct here. Namely, its lemma in §1 (p. 557) was incorrect; it ought to have referred only to a normal subgroup generated by positive elements. The following revised proof runs more or less in the same line as Vernikoff-Krein-Tovbin's,[2] but we may put emphasis on that neither associativity (nor commutativity) nor ring property is used; we simply deal with abelian groups with operators. Indeed, as an application of such mode of our approach, we can determine the structure of the additive group of *bounded automorphisms* of a semi-ordered abelian group (satisfying certain conditions); this forming the second purpose of the present supplementary note.

Let G be a semi-ordered abelian group with real multipliers[3], such that[4]

(i) if $x \geq 0$ and $y \geq 0$ then $x+y \geq 0$,

(ii) if $x \geq 0$ and $-x \geq 0$ then $x=0$,

(iii) if $x \geq 0$ and a (real number) ≥ 0 then $ax=0$.

Let G possess further an operator domain $\varOmega=\{A\}$ which is by itself a semi-ordered abelian group (in the same sense as G) such as

(vii) if $x \geq 0$ (in G), $A \geq 0$ (in $\varOmega$) then $Ax \geq 0$ (in G),

(viii) $(A+B)x=Ax+Bx$, $A(x+y)=Ax+Ay$, $A(ax)=aAx$,

and let moreover

(ix) $\varOmega$ possess an Archimedean unit I which satisfies $Ix=x$, $x \in G$.

Then we have

Lemma. Every normal subgroup of G generated by a certain system of positive elements is always allowable with respect to $\varOmega$.

For, our former proof remains valid certainly in this case.

Suppose now

(iv) G itself possess an Archimedean unit e,

---

1) Proc. **18** (1942), 555.
2) Sur les anneaux semi-ordonnés, C. R. URSS, **30** (1941), p. 758.
3) Cf. a remark below.
4) The numbers for the conditions are in accordance with our former note.

and furthermore[5]

   (v) if $-\varepsilon e < x < \varepsilon e$ for every $\varepsilon > 0$ then $x = 0$.

   Consider the totality P of elements x in G such that

$$x > -\varepsilon e \quad \text{for every} \quad \varepsilon > 0.$$

Then we immediately verify the followings: every element $x \geq 0$ lies in P; if $x, y \in P$ then $x + y \in P$; if both $x, -x \in P$ then $x = 0$ $\big(\text{cf. (v)}\big)$; if $x \in P$ and a (real number) ≥ 0 then $ax \in P$; if $x \in P$ and $A(\in \Omega) \geq 0$ then $Ax \in P$ $\big(\text{cf. (ix)}\big)$.

   But these mean that we may introduce in G a new semi-order, under which P is the totality of positive elements (and 0), every positive element in the original sense is also positive in the new sense, and moreover, (vii) remains true with respect to the new semi-order. Furthermore, from the above construction we readily see that under this new sense of semi-order G satisfies

   (vi) if $x < \varepsilon e$ for every $\varepsilon > 0$ then $x \leq 0$,

or

   (vi)′ $\inf\limits_{\varepsilon > 0} \varepsilon e$ exists and $= 0$.

   Put now, for an element x in G,

$$a(x) = \inf \, (a \, ; \, ae > x)^{6)}.$$

Then always $a(x)e - x \geq 0$ (in the new sense), according to (vi). Hence the normal subgroup generated by an element of the form $a(x)e - x$ is, by the above lemma, allowable (for Ω); we have $a(x)e \equiv x$ modulo this subgroup.

   On returning to the original sense of semi-order we deduce from these consideration the following: Let G satisfy (i)-(v), (vii)-(ix). Then, firstly, every element x of G is congruent to the real multiple $a(x)e$ of e modulo a suitable allowable normal subgroup not coinciding with G, whence modulo a certain maximal allowable normal subgroup of G; thus, secondly, the intersection of all the maximal allowable normal subgroups consists of 0 only $\big(\text{cf. (v)}\big)$; thirdly, the factor group G/M of G modulo a maximal allowable normal subgroup M is isomorphic with the additive group of real numbers. Hence, *G is operator-isomorphic to an additive group of real-valued bounded functions over the space of maximal allowable normal subgroups; order being preserved in the direct sense. If moreover the condition* (vi) *is fulfilled, then this isomorphism is also an order-isomorphism. For,* $x \not\leq 0$ *implies then* $a(x) > 0$.

---

5) When (v) is not satisfied, the elements x such as $-\varepsilon e < x < \varepsilon e$ for every $\varepsilon > 0$ form a normal subgroup N, the *radical*, which is allowable for Ω, as one sees readily, and the semi-ordered factor group of G modulo N fulfills (v).

6) Either in the new or in the old sense of semi-order; the results are the same.

From these follow now immediately the assertions in Theorems 1 and 2 of our former note which we restate here:

Let R be a semi-ordered ring with real multipliers and possessing a ring-unit e which is at the same time an Archimedean unit. *Suppose R satisfy* (v). *Then R is ring-isomorphic to a certain ring $R(\mathfrak{M})$ of real-valued bounded functions over the space $\mathfrak{M}$ of maximal normal ideals: $x \leftrightarrow x(M)$, such that e is represented by 1: $e(M) \equiv 1$. In particular, R is both associative and commutative. The order is preservec in the direction $x \to x(M)$. If R fulfills the stronger condition* (vi), *then R is ring-order-isomorphic with $R(\mathfrak{M})$.*

Remark. The above deduction depends on the existence of real multipliers. But we may replace it, as in our former note[1], by

(iii)′ if $nx \geq 0$ for a certain natural number n, then $x \geq 0$,

$\Big($and, of course, (v) $\big($or (vi)$\big)$ by the corresponding condition in our former note$\Big)$. Namely using (iii)′ we extend G so as, firstly, to possess rational multipliers, and then complete it by the norm $\|x\| = \inf(a; -ae < x < ae)$ according to (v). The extension thus obtained has real multipliers (and satisfies the above prescribed conditions), and we may apply our deduction to it.

Now, as an application of our proof to consider groups with operators rather than rings, let us study bounded automorphisms of a semi-ordered group (without operators). Namely, let G be a semi-ordered abelian group $\Big($such as (i), (ii), (iii) $\big($or (iii)′$\big)\Big)$ (without operators), and A be an automorphism of G (simply as a group)[7]. Then, if there exists a natural number n such as

$$nx \geqq x^A \geqq -nx \quad \text{for every } x(\in G) \geqq 0,$$

we call A bounded. The totality of the bounded automorphisms of G forms a semi-ordered ring R_G by the usual operations

$$x^{AB} = (x^A)^B, \quad x^{A+B} = x^A + x^B$$

and the semi-order:

$$A \geqq 0 \quad \text{meaning: } x^A \geqq 0 \quad \text{for every } x \geqq 0.$$

Now we have: *if G satisfies $\big($besides (i)-(iii)$\big)$ (iv) and (v), then the ring R_G of bounded automorphisms is, merely as an additive group, isomorphic to a svbgroup of G; order being preserved in the direct sense of correspondence.* To see this, again we first assume the existence of real multipliers. And, we consider R_G as an operator domain Ω for G. Then the above conditions (vii)-(ix) are satisfied. Therefore, the intersection of all the maximal R_G-allowable normal subgroups is 0. Denote their totality by $\mathfrak{M} = \{M\}$. G/M, with a certain $M \in \mathfrak{M}$, is the additive group of real numbers, and every element in G is con-

---

7) Thus A needs not preserve order relation in G.

gruent modulo M to a real multiple of e. Hence, if an $A \in R_G$ maps e into M, then A maps the whole G into M. Further, if $e^A = 0$ then $e^A \in \bigcap M$, whence $G^A \leq \bigcap M = 0$, $G^A = 0$. Therefore, since $e^{A+B} = e^A + e^B$, $A(\in R_G)$ is determined uniquely by the image e^A of e, and $A \to e^A$ gives an isomorphism of the additive group R_G with the subgroup $\{e^A\}$ of G. (Indeed, if we put

$$e^A \equiv \omega_A(M)e \quad \text{mod. } M,$$

then $\omega_{A+B}(M) = \omega_A(M) + \omega_B(M)$, $\omega_{AB}(M) = \omega_A(M)\omega_B(M)$ and the ring R_G is represented faithfully by the function ring $\{\omega_A(M)\}$ over $\mathfrak{M}$). As to the case without real multipliers, we have only to apply the above remark and to observe that A is a difference of two positive operators, say nI and $nI - A$, which secures $\big($in combination with (v), of course,$\big)$ the possibility of extending A from G to its complete extension with real multipliers. We see in this way that again A is determined uniquely by $e^A(\in G)$, and therefore, R_G is isomorphic with $\{e^A\}$. Since $A \geq 0$ implies $e^A \geq 0$, the last assertion of our theorem is evident.

Finally, we take this opportunity to correct some misprints in our note[1]: $\underset{M}{\mathcal{E}}\big(x(M) \geq \lambda\big)$ at the last line of p. 559 reads $\underset{M}{\mathcal{E}}\big(x(M) \leq \lambda\big)$; $\sum \lambda_i e'_{\lambda_i}(M)$ and $\sum \lambda_i e_{\lambda_i}(M)$ at the first line of p. 560 read respectively $\sum \lambda_i \big\{e'_{\lambda_i}(M) - e'_{\lambda_{i-1}}(M)\big\}$ and $\sum \lambda_i \big\{e_{\lambda_i}(M) - e_{\lambda_{i-1}}(M)\big\}$.

Normed rings and spectral theorems

Proc. Imp. Acad. Tokyo **19** (1943) 356–359

(Comm. by T. TAKAGI, M.I.A., July 12, 1943.)

1. *Introduction.* The purpose of the present note is to show that the simultaneous resolution of the identity for commutative ring of normal operators may easily be deduced from the theory of normed ring[1] by making use of two elementary lemmas in operator theory and of Baire's category theorem. As in the preceding notes[2], our treatment is rather algebraical and integration-free.

2. *Preliminaries.* Let B denote the totality of bounded linear operators in a general euclid space $\mathfrak{H}$. Let a subset A of B satisfy

(1)   the commutativity: $TS = ST$ for $T, S \in A$

and

(2)   the conjugated condition: if $T \in A$, then its adjoint T^* also $\in A$.

Let A' denote the totality of operators $\in B$ that commute with every operator of A, then it is easy to see that $R = A'\,' = (A')'$ is a commutative, conjugated ring with unit and with complex multipliers. R is a normed ring[3] by the norm $\|T\| = \sup\limits_{|f| \leqq 1} \|T \cdot f\|$. Moreover it is easy to see that R is closed in the sense of the *strong convergence*:

(3) $\begin{cases} \text{let a sequence } \{T_n\} \subseteqq R \text{ be such that } \operatorname*{strong\,lim}\limits_{n \to \infty} T_n \text{ exist[4]},\\ \text{then the operator } T = \operatorname*{strong\,lim}\limits_{n \to \infty} T_n \text{ also belongs to } R. \end{cases}$

Lemma 1[5]. Let $H \in B$ be hermitian, viz. $H = H^*$, then

(4) $\|H\| = \sup\limits_{|f| \leqq 1} \|H \cdot f\| = \sup\limits_{|f| \leqq 1} |(H \cdot f, f)|$.

Lemma 2[6]. Let $T \in B$ and let I denote the identity operator, then $(I + TT^*)$ admits the inverse $(I + TT^*)^{-1} \in B$. It is easy to see that $(I + TT^*)^{-1} \in R$ in case $T \in R$.

From lemma 1 we obtain

(5) $\|T\| = \|T^*\|$, $\|T^2\| = \|T\|^2$ for every $T \in R$.

Proof. We have, by (4), since $H = TT^*$ is hermitian $\big(= (TT^*)^*\big)$,
$\|T\|^2 = \sup\limits_{|f| \leqq 1} (T \cdot f, T \cdot f) = \sup\limits_{|f| \leqq 1} |(T^*T \cdot f, f)| = \|T^*T\| = \|TT^*\| = \|H\|$. Thus
$\|T\|^2 = \|T^*\|^2$ and $\|T^2\|^2 = \|T^{*2}T^2\| = \|(T^*T)^2\|$ by the commutativity of R.

---

1) I. Gelfand: Rec. Math., **9** (1941).
2) K. Yosida and T. Nakayama: Proc., **18** (1942) and **19** (1943).
3) R is a Banach space by the norm $\|T\|$ and satisfies $\|TS\| \leqq \|T\| \cdot \|S\|$.
4) strong $\lim\limits_{n \to \infty} T_n = T$ means that $\lim\limits_{n \to \infty} T_n \cdot f = T \cdot f$ strongly for every $f \in \mathfrak{H}$.
5) See, for example, F. J. Murray's book: Princeton (1941), 41.
6) Murray's book, 42.

Again, by (4), we have $\|H\|^2 = \sup_{|f|\leq 1}(H\cdot f, H\cdot f) = \sup_{|f|\leq 1}|(H^2\cdot f, f)| = \|H^2\|$.
Therefore $\|T^2\|^2 = \|(T^*T)^2\| = (\|T\|)^4$ and hence $\|T^2\| = \|T\|^2$.   Q. E. D.
From (5) we have

(6) $\|T\| = \|T^*\| = \lim_{n\to\infty}\sqrt[n]{\|T^n\|}$ for $T \in R$.

Proof. At first $\|T\| = \lim_{n\to\infty}\sqrt[2n]{\|T^{2n}\|}$ by (5). And since $\|T^{2n+1}\| \leq \|T^{2n}\|\cdot\|T\|, \|T^{2n+2}\| \leq \|T^{2n+1}\|\cdot\|T\|$, we must have (6).   Q. E. D.

The ring R is, by (6), *semi-simple* viz. R does not contain a *generalised nilpotent element*: $\lim_{n\to\infty}\sqrt[n]{\|T^n\|} \neq 0$ if $T \neq 0$. Hence[1], R is ring-isomorphic (with complex multipliers) to the function ring $R(\mathfrak{M})$ of complex-valued continuous functions $T(M)$ on the bicompact Hausdorff space $\mathfrak{M}$ which consists of the totality of the maximal ideals of R:

(7) $\begin{cases} T \leftrightarrow T(M),\quad I(M)\equiv 1 \quad \text{on} \quad \mathfrak{M}, \quad \sup_M|T(M)| = \lim_{n\to\infty}\sqrt[n]{\|T^n\|} = \|T\| \\ \text{(by (6)).} \end{cases}$

We next show that the above correspondence $T \leftrightarrow T(M)$ satisfies the following two conditions.

(8) $\begin{cases} T(M) \text{ is real-valued on } \mathfrak{M} \text{ if and only if } T \text{ is hermitian.} \\ \quad \text{Thus, in particular, } T^*(M)=\overline{T(M)} \text{ (bar indicates complex} \\ \quad \text{conjugate).} \end{cases}$

(9) $\begin{cases} T(M) \text{ is non-negative on } \mathfrak{M} \text{ if and only if } T \text{ is non-negative} \\ \quad \text{definite}^{2)}. \end{cases}$

Proof of (8). Let $T \in R$ be hermitian and let $T(M_0)=a+ib$, $i=\sqrt{-1}$, with $b \neq 0$. Then the hermitian operator $S=(T-aI)/b \in R$ satisfy $(I+S^2)(M_0)=1+i^2=0$. Thus the function $(I+S^2)(M)$ and hence, by the isomophism $R \leftrightarrow R(\mathfrak{M})$, the operator $(I+S^2)$ does not have inverse, contrary to lemma 2. Next let $T \in R$ be not hermitian and let it be decomposed into hermitian components: $T = \dfrac{T+T^*}{2} + i\dfrac{T-T^*}{2i}, \dfrac{T-T^*}{2i} \neq 0$. Then, by the isomorphism $R \leftrightarrow R(\mathfrak{M})$, $\dfrac{T-T^*}{2i}(M)$ is not identically zero on $\mathfrak{M}$.   Q. E. D.

For the proof of (9) we need the

Lemma 3. $R(\mathfrak{M})$ is the totality of the complex-valued continuous functions on $\mathfrak{M}$.

Proof. The function ring $R(\mathfrak{M})$ satisfies i) for any pair M_1, $M_2 \in \mathfrak{M}$ there exists $T(\mathfrak{M}) \in R(\mathfrak{M})$ such that $T(M_1) \neq T(M_2)$, ii) for any $T(M) \in R(\mathfrak{M})$ there exists complex-conjugated function $T^*(M)=\overline{T(M)} \in R(\mathfrak{M})$ (by (8)). Thus, by Gelfand-Silov's abstraction of Weierstrass' polynomial approximation theorem[3], every continuous function on $\mathfrak{M}$

---

1) by Gelfand's theorem, loc. cit.
2) viz. $(T\cdot f, f) \geq 0$ for every $f \in \mathfrak{H}$.
3) Rec. Math., **9** (1941). Cf. also H. Nakano: 全國紙上數學談話會, **218** (1941).

may be uniformly approximated by functions $\in R(M)$. Since R is complete by the norm $\| T \|$ which satisfies (7) we have the lemma.

Q. E. D.

Proof of (9). Let $T(M) \geq 0$ on $\mathfrak{M}$, then the function $S(M) = \sqrt{T(M)}$ belongs to $R(\mathfrak{M})$ by the lemma 3. Hence, by the isomorphism $R \leftrightarrow R(\mathfrak{M})$ and (8), we have $T = S^2$, $S = S^* \in R$. Thus $(T \cdot f, f) = (S^2 \cdot f, f) = (S \cdot f, S \cdot f) \geq 0$. Conversely let T be hermitian. We have, by (8), lemma 3, the isomorphism $R \leftrightarrow B(\mathfrak{M})$ and the result just above obtained, $T = T_1 - T_2$, $T_1 \cdot T_2 = 0$, $T_i = $ non-negative definite, $i = 1, 2$. For the proof we have only to put $T_1(M) = \sup \big(T(M), 0 \big)$, $T_2(M) = T_1(M) - T(M)$. Let, moreover, T be non-negative definite, then we have, by the above, $0 \leq (TT_2 \cdot f, T_2 \cdot f) = (-T_2^2 \cdot f, T_2 \cdot f) = (-T_2^3 \cdot f, f)$, $0 \leq (T_2 T_2 \cdot f, T_2 \cdot f) = (T_2^3 \cdot f, f)$ for $f \in \mathfrak{H}$. Thus $(T_2^3 \cdot f, f) = 0$ for $f \in \mathfrak{H}$ and hence $T_2^3 = 0$ by (4). Since R is semi-simple, we have $T_2 = 0$. Therefore $T(M) = T_1(M) \geq 0$ on $\mathfrak{M}$, as was to be proved. Q. E. D.

Summing up we have the

Theorem. R is ring-isomorphic (with complex multipliers) to the function ring $R(\mathfrak{M})$ of all the complex-valued continuous functions on a bicompact Hausdorff space $\mathfrak{M}$ such that (7), (8) and (9) hold good.

3. *The spectral resolution.* The simultaneous resolution of the identity of R may now be obtained by essentially the same idea as in the preceding notes referred to above. For the sake of completeness we will repeat it.

Lemma 4[1]. Let a sequence $\{ T_n \}$ of hermitian operators $\in R$ be such that $T_1 \leq T_2 \leq \cdots \leq T_n \leq \cdots \leq S \in B$, then the strong $\lim_{n \to \infty} T_n = T$ exists. We have $T \in R$ by (3). Here $T_n \leq S$ means that $(S - T_n)$ is non-negative definite.

From lemma 4, theorem and Baire's category theorem we obtain the

Lemma 5[2]. Let a sequence $\{ T_n(M) \}$ of real-valued functions $\in R(\mathfrak{M})$ be such that $T_1(M) \leq T_2(M) \leq \cdots \leq T_n(M) \leq \cdots \leq$ a constants all over $\mathfrak{M}$. Then, if we put $T = $ strong $\lim_{n \to \infty} T_n$ (its existence is assured by lemma 4 and theorem), we have $T(M) = \lim_{n \to \infty} T_n(M)$ on $\mathfrak{M}$ except possibly on a set of first category.

Now consider the set $R'(\mathfrak{M})$ of all the complex-valued bounded functions $X'(M)$ on $\mathfrak{M}$ such that $X'(M)$ is different from a continuous function $X(M)$ only on a set of first category. We identify two functions from $R'(\mathfrak{M})$ if they differ on a set of first category, then $R'(\mathfrak{M})$ is divided into classes. Since the complementary to a set of first category is dense on a bicompact Hausdorff space, each class X' contains exactly one continuous function $X(M)$ which corresponds, by the isomorphism $R \leftrightarrow R(\mathfrak{M})$, to an element $X \in R$.

For any $T \in R$ and for any complex number $z = \lambda + i\mu$, we put $E_z = $ the element $\in R$ which corresponds to the class E_z' containing the

<hr>

1) Yosida and Nakayama: Proc., **18**, loc. cit., 560.
2) Yosida and Nakayama: Proc., **18**, loc. cit., 559.

character isticfunction $E'_z(M)$ of the set $\underset{M}{\mathfrak{E}}\{\Re T(M)\leqq\lambda,\ \Im T(M)\leqq\mu\}^{1)}$ We have then

$$\left| T(M)-\sum_{j=1}^{n}(\lambda_j+i\mu_j)\Big(E'_{\lambda_j+i\mu_j}(M)+E'_{\lambda_{j-1}+i\mu_{j-1}}(M)-\right.$$
$$\left. E'_{\lambda_{j-1}+i\mu_j}(M)-E'_{\lambda_j+i\mu_{j-1}}\Big)(M)\right|\leqq\varepsilon$$

on $\mathfrak{M}$, if $\lambda_2=-a-\dfrac{\varepsilon}{\sqrt{2}}\leqq\lambda_2\leqq\cdots\leqq\lambda_n=a=\sup_M|\Re T(M)|,\ \mu_1=-\beta-$

$\dfrac{\varepsilon}{\sqrt{2}}\leqq\mu_2\leqq\cdots\leqq\mu_n=\beta=\sup_M|\Im T(M)|,\ \Big(\sup_j(\lambda_j-\lambda_{j-1})^2+\sup_j(\mu_j-\mu_{j-1})^2\Big)^{\frac{1}{2}}$

$\leqq\varepsilon.$ Thus, by the definition of $X'(M)$ and lemma 5, we have

$$\left| T(M)-\sum_{j=1}^{n}(\lambda_j+i\mu_j)\Big(E_{\lambda_j+i\mu_j}(M)+E_{\lambda_{j-1}+i\mu_{j-1}}(M)-\right.$$
$$\left. E_{\lambda_{j-1}+i\mu_j}(M)-E_{\lambda_j+i\mu_{j-1}}(M)\Big)\right|\leqq\varepsilon$$

on $\mathfrak{M}$, since the complementary to a set of first category is dense on a bicompact Hausforff space. Therefore, by theorem, we obtain

$$\left\| T-\sum_{j=1}^{n}(\lambda_j+i\mu_j)\Big(E_{\lambda_j+i\mu_j}+E_{\lambda_{j-1}+i\mu_{j-1}}-E_{\lambda_{j-1}+i\mu_j}-E_{\lambda_j+i\mu_{j-1}}\Big)\right\|\leqq\varepsilon.$$

We have thus arrived at the desired resolution of the identity $T=\int zdE_z.$

---

1) $\Re a$ and $\Im a$ denote respectively the real and imaginary part of a.

Normed rings and spectral theorems, II

Proc. Imp. Acad. Tokyo **19** (1943) 466–470

(Comm. by T. Takagi, m.i.a., Oct. 12, 1943.)

§ 1. *Introduction.* The purpose of this note is to give an algebraic treatment and version of Fredholm-Riesz-Schauder's theory[1] of completely continuous (c.c.) functional equations with the aide of the theory of normed ring[2]. Our method seems to be suited to obtain the results concerning conjugate equations. We also give a proof to S. Nikolski's extension[3], which is of importance in view of applications, of F-R-S's theory. Lastly we extend the existence theorem of proper values $\neq 0$ for c.c. hermitian operator $\neq 0$ as a corollary of our arguments.

§ 2. *Preliminaries and lemmas.* Let V be a linear operator from a complex Banach space E into E. V is called c.c. if it transforms any bounded set into a compact set. Let R be the commutative ring generated by the c.c. V and the identity operator I, completed by the uniform limit defined by the norm $\| T \| = \sup_{|x| \leq 1} \| T \cdot x \| \cdot R$ is a normed ring with unit I and the norm $\| T \|$, such that any element $T \in R$ may be represented as $T = \lambda I - U$, U being c.c. Let E^* denote the conjugate space $\big(=$ the space of all the linear functionals f on E with the norm $\| f \| = \sup_{|x| \leq 1} |f(x)| \big)$, then the operator T^* conjugate to T is defined by $T^* \cdot f = g$, $f(T \cdot x) = g(x)$, $f \in E^*$.

Lemma 1. Since[4] $\| T \| = \| T^* \|$, the set R^* of all the operators T^*, $T \in R$, is also a normed ring linear-isomorphic and linear-isometric with R by the correspondence $T \leftrightarrow T^*$.

Lemma 2[5]. R is, as a normed ring, linear homomorphically mapped upon a complex-valued function ring defined on the space $\mathfrak{M}$ of all the maximal ideals M of R: $R \ni T \to T(M)$, $I \to I(M) \equiv 1$ such that $T \equiv T(M)I \pmod{M}$, $\sup_{M} | T(M) | = \lim_{n \to \infty} \sqrt[n]{\| T^n \|}$. Moreover T admits inverse T^{-1} in R if and only if $T(M) \neq 0$ on $\mathfrak{M}$.

Lemma 3[6]. Let $I_1 \in R$ be an idempotent viz. $I_1^2 = I_1$, then the

---

1) See, for example, S. Banach's book: Théorie des opérations linéaires, Warsaw (1932), 151. Cf. M. Nagumo: Jap. J. of Math., **13** (1936), 6.
2) I. Gelfand: Rec. Math., **9** (1941), 3.
3) C. R. URSS, **16** (1926), 315. Cf. also K. Yosida: Jap. J. Math., **15** (1939), 297.
4) S. Banach: loc. cit., 100.
5) I. Gelfand: loc. cit.
6) Cf. I. Gelfand: loc. cit., 18. For the sake of completeness we will give the proof below. For any $T \in R$ we have $(T-T(M)I)I_1 \in M \frown R_1$, since $(T-T(M)I) I_1(M) = (T(M)-T(M)1)(I_1(M)) = 0$, $(T-T(M)I)I_1 \cdot I_1 = (T-T(M)I)I_1$. Thus $TI_1 = T(M)I_1 + (T-T(M)I)I_1 \equiv T(M)I_1 \pmod{M \frown R_1}$, proving that $M' = M \frown R_1$ is a maximal ideal of R_1. Next let M' be a maximal ideal of R_1. We will show that there exists a maximal ideal M of R such that $M \geq M'$, $M \bar{\ni} I_1$. To this purpose consider $M = M' + R(I-I_1)$. That $M \bar{\ni} I_1$ is trivial. Since $T = TI_1 + T(I-I_1)$, $TI_1 \equiv \lambda I_1 \pmod{M'}$ by the lemma 2), we obtain $T \equiv \lambda I_1 \pmod{M}$, proving that M is a maximal ideal of R,

subring $R_1 = RI_1$ of R constitutes normed ring with unit I_1. For any maximal ideal $M \ni I_1$ of R, $M' = M \cap R_1$ is a maximal ideal of R_1. Conversely for any maximal ideal M' of R_1, there exists a maximal ideal $M \ni I_1$ of R such that $M' = M \cap R_1$.

Lemma 4[1]. Let E_1 be a closed linear proper subspace of E, then there exists $x_0 \in E - E_1$ such that $\|x_0\| = 1$, $\mathrm{dis}\,(x_0, E_1) = \inf_{x \in E_1} \|x_0 - x\| = \dfrac{1}{2}$.

§3. *A deduction of F-R-S's theory.* By the lemma 2

(1) *the range $R(V)$ of the function $V(M)$ coincides with the spectra of V,*

viz. with the set of complex numbers λ such that $T = \lambda I - V$ does not have inverse T^{-1} in R. Thus, by the isomorphism $R \leftrightarrow R^*$,

(2) *the spectra are the same for V and for V^*.*

We first show that

(3) *any complex number $\neq 0$ from the spectra of V is a proper value of V.*

Proof. Let $\lambda = 1 \in R(V)$ and let $\lambda = 1$ be not a proper value of V. Then $T = I - V$ maps E on $T \cdot E$ in one-to-one manner. The inverse mapping T^{-1} is continuous from $T \cdot E$ on E viz. there exists a positive number $a > 0$ such that $\|T \cdot x\| \geq a \|x\|$ on E. Assume the contrary and let $\lim_{n \to \infty} \|T \cdot x_n\| = 0$, $\|x_n\| = 1$ $(n = 1, 2, \ldots)$. By the c.c. of V we may suppose that $\lim_{n \to \infty} V \cdot x_n = y$ exists, whence $\lim_{n \to \infty} (x_n - V \cdot x_n) = 0$ or $\lim_{n \to \infty} x_n = y$, $\|y\| = 1$, $y = V \cdot y$, contrary to the hypothesis. Thus T^{-1} must be continuous from $T \cdot E$ on E and hence $T \cdot E$ is a closed set of E. $T \cdot E \neq E$, since otherwise T^{-1} exists in R and thus $\lambda = 1 \bar{\in} R(V)$. Hence, if we put $E_1 = T \cdot E$, $E_2 = T \cdot E_1$, $\ldots$, E_{n+1} is a closed linear proper subspace of E_n $(n = 1, 2, \ldots)$. By the lemma 4, there exists a sequence $\{y_n\}$ such that $y_n \in E_n$, $\|y_n\| = 1$, $\mathrm{dis}\,(y_n, E_{n+1}) = \dfrac{1}{2}$. Hence, for $n > m$,
$$V \cdot y_m - V \cdot y_n = y_m - (y_n + T \cdot y_m - T \cdot y_n) = y_m - y, \quad y \in E_{m+1} \text{ and therefore}$$
$\|V \cdot y_m - V \cdot y_n\| \geq \dfrac{1}{2}$, contrary to the c.c. of V.   Q. E. D

Remark 1. Let E be a general euclid space and let V be a c.c. *normal* operator in E. Since[2], by the normality, $\|T\| = \lim_{n \to \infty} \sqrt[n]{\|T^n\|}$, $T = \lambda I - U$, the mapping $T \to T(M)$ is isomorphic. Thus, if $V \neq 0$, $R(V)$ contains complex numbers $\neq 0$. Hence V admits proper values $\neq 0$, in accordance with the well-known theorem due to Hilbert-Schmidt[3].

Since V is c.c. we have, by (3),

---

1) S. Banach: loc. oit., 83.

2) See K. Yosida: Proc. **19** (1943), 338.

3) A short cut to H-S's theorem is given by Y. Mimura: 全國紙上數學談話會, 第 170 號 (昭和 14 年).

    (4) *the spectra of V constitute an enumerable set which may accumulate only at 0[1].*

Next let $\lambda=1$ be a proper value of V. *We will show that $\lambda=1$ is also a proper value of V^*, without making use of the fact that V^* is c.c. with V.*

Proof. By (1) and (4) there exists $\varepsilon>0$ such that $(\lambda I-V)^{-1}$ exists in R if $0<|\lambda-1|<3\varepsilon$. Consider the resolvent integral

$$(5) \qquad I_1=\frac{1}{2\pi i}\int_{|\lambda-1|=\delta}(\lambda I-V)^{-1}d\lambda, \qquad 0<\delta<3\varepsilon.$$

By Cauehy's theorem, I_1 is independent of δ. Thus

$$(6) \qquad I_1^2=\frac{1}{2\pi i}\int_{|\lambda-1|=\varepsilon}(\lambda I-V)^{-1}d\lambda\cdot\frac{1}{2\pi i}\int_{|\mu-1|=2\varepsilon}(\mu I-V)^{-1}d\mu$$

$$=\frac{1}{2\pi i}\int_{|\lambda-1|=\varepsilon}\left(\frac{1}{2\pi i}\int_{|\mu-1|=2\varepsilon}\frac{d\mu}{\mu-\lambda}\right)(\lambda I-V)^{-1}d\lambda$$

$$-\frac{1}{2\pi i}\int_{|\mu-1|=2\varepsilon}\left(\frac{1}{2\pi i}\int_{|\lambda-1|=\varepsilon}\frac{d\lambda}{\mu-\lambda}\right)(\mu I-V)^{-1}d\mu$$

$$=\frac{1}{2\pi i}\int_{|\lambda-1|=\varepsilon}(\lambda I-V)^{-1}d\lambda=I_1.$$

Hence we obtain the direct decomposition

$$(7) \qquad\qquad R=RI_1+R(I-I_1).$$

By Cauchy's theorem and the lemma 2 we have

$$\begin{cases} I_1(M_1)=\dfrac{1}{2\pi i}\displaystyle\int_{|\lambda-1|=\varepsilon}(\lambda\cdot1-1)^{-1}d\lambda=1 & \text{if } V(M_1)=1, \\[2ex] I_1(M_\mu)=\dfrac{1}{2\pi i}\displaystyle\int_{|\lambda-1|=\varepsilon}(\lambda\cdot1-\mu)^{-1}d\lambda=0 & \text{if } V(M_\mu)=\mu\neq1. \end{cases}$$

Hence, by the lemma 3, we obtain the results:

$$\begin{cases} M_1'=M_1\cap RI_1 \text{ is the only maximal ideal of } RI_1, \\[1ex] M_\mu'=M_\mu\cap R(I-I_1) \text{ with } \mu\neq1 \text{ exaust the maximal ideals of } R(I-I_1). \end{cases}$$

Therefore, siece $(I-V)(I-I_1)(M_\mu)_{\mu\neq1}=(1-\mu)(1-0)\neq0$,

    (8) $(I-V)(I-I_1)$ *admits inverse* $\big($*in* $R(I-I_1)\big)$ *as an element of the ring* $R(I-I_1)$.

Moreover, since $(I-V)I_1(M_1)=(1-1)1=0$,

    (9) $(I-V)I_1$ *is, as an element of* RI_1, *contained in all the (in the truth, only one) maximal ideals of* RI_1.

---

1) S. Banach: loc. cit., 160.

On the other hand, if we put $(\lambda I - V)^{-1} = \dfrac{I}{\lambda} - R_\lambda$ then $R_\lambda = \dfrac{R_\lambda}{\lambda} \cdot V - \dfrac{V}{\lambda^2}$ is c.c. with V. Thus, by Cauchy's theorem,

$$(10) \quad I_1 = \frac{1}{2\pi i} \int_{|\lambda-1|=\varepsilon} \left(\frac{I}{\lambda} - R_\lambda \right) d\lambda = \frac{-1}{2\pi i} \int_{|\lambda-1|=\varepsilon} R_\lambda d\lambda \qquad is\ c.c.^{1)}$$

A Banach space is, by the lemma 4, of finite dimension if and only if its unit sphere is compact[2]. Hence, by the c.c. and the idempotent character of I_1, $I_1 \cdot E$ is of finite dimension. Therefore

(11) *the ring $RI_1 = I_1 R I_1$ may be considered as a commutative matrix ring with unit I_1 operating upon the vector space $I_1 \cdot E$ of finite dimension.*

Thus, by (9), the matrix $(I-V)I_1$ must be nilpotent. Hence

(12) *$(I-V)I_1$ is nilpotent.*

$I_1 \neq 0$, since otherwise $T = I - V$ admits inverse in R by (8), contrary to the hypothesis $\lambda = 1 \in R(V)$. By the isomorphism $R \leftrightarrow R^*$, (5)–(12) hold good when we assign *-notation to every operator or set in these equations (M^*, for example, means the set $\underset{T^*}{\mathscr{E}}\{T \in M\}$). Thus, by the formula corresponding to (12), we see that $\lambda = 1$ is also a proper value of V^*. $\left(\text{We will denote the formulae corresponding to (5)-(12) by } (5)^*-(12)^*\right).$ Q. E. D.

Now let $\lambda = 1$ be a proper value of V and of V^*. Then

(13) *the dimension of the proper space belonging to the proper value 1 is the same for V and for V^*.*

Proof. By (7) and (8) the proper value equation

$$(14) \qquad\qquad x = V \cdot x, \qquad x \in E$$

is equivalent to $(I-V)I_1 \cdot x = 0$, $(I-I_1) \cdot x = 0$. Thus (14) is equivalent to

$$(14)' \qquad\qquad x = V \cdot x, \qquad x \in I_1 \cdot E.$$

Similarly, by $(7)^*$ and $(8)^*$, the proper value equation

$$(15) \qquad\qquad f = V^* \cdot f, \qquad f \in E^*$$

is equivalent to

$$(15)' \qquad\qquad f = V^* \cdot f, \qquad f \in I_1^* \cdot E^*.$$

Since $I_1^* \cdot f = f$ means $f\big((I-I_1) \cdot x\big) = 0$ on E, $I_1^* \cdot E$ must be of the same dimension as $I_1 \cdot E$. Therefore the mutual conjugate matrix equation

---

1) A uniform limit of a sequence of c.c. linear operators is c.c. also. See S. Banach: loc. cit., 96.
2) S. Banach: loc. cit., 84.

(14)' and (15)' admit respectively the same number of linearly independent solutions.   Q. E. D.

Lastly we will prove that, if $\lambda=1$ is a proper value of V and of V^*, then

(16) *the equation $y=(I-V)\cdot x$ admits solution x if and only if $f(y)=0$ when $f=V^*\cdot f$,*

and similarly

(16)' *the equation $g=(I^*-V^*)\cdot f$ admits solution f if and only if $g(x)=0$ when $x=V\cdot x$.*

Proof of (16). The necessity is trivial. Because of (8) and (11), $(I-V)\cdot E$ is a closed set of E. Thus, if $y\,\bar{\epsilon}\,(I-V)\cdot E$, there exists by Hahn-Banach's theorem $f\epsilon E^*$ such that $f(y)\neq 0$, $f\big((I-V)\cdot E\big)=0$, contrary to the hypothesis.   Q. E. D.

§ 4. *An extension.* F-R-S's theory may be extended to linear operator V which satisfies

(17) V^n *is c.c. for a certain $n\geq 1$.*

We will show that this extension, first pointed out by S. Nikolski, is obtained as a corollary to our arguments in § 3.

Since, by the lemma 2, the spectra of V are contained in $\{\lambda^{\frac{1}{n}}\}$, $\lambda\in R(V^n)$, (4) holds goods for our V. Moreover if $\lambda=1\in R(V)$, then I_1 is, as will be proved below, c.c. These two facts would be sufficient for the validity of our extension, as will be verified reflecting upon the arguments in § 3.

Proof of the c.c. of I_1. By the assumption $\lambda=1\in R(V^n)$. Let $(\lambda^n I-V^n)=\dfrac{I}{\lambda^n}-R_\lambda(n)$, then $R_\lambda(n)$ is c.c. with V. From

$$I=(\lambda I-V)\Big(\frac{I}{\lambda}-R_\lambda\Big),$$

$$I=(\lambda^n I-V^n)\Big(\frac{I}{\lambda^n}-R_\lambda(n)\Big)=(\lambda I-V)(\lambda^{n-1}I+\lambda^{n-2}V+\cdots+V^{n-1})\Big(\frac{I}{\lambda^n}-R_\lambda(n)\Big)$$

we obtain

$$\frac{I}{\lambda}-R_\lambda=\frac{I}{\lambda}+\frac{V}{\lambda^2}+\cdots+\frac{V^{n-1}}{\lambda^n}-R_\lambda(n)(\lambda^{n-1}I+\lambda^{n-2}V+\cdots+V^{n-1}).$$

Thus, by Cauchy's theorem and the c.c. of $R_\lambda(n)$,

$$I_1=\frac{1}{2\pi i}\int_{|\lambda-1|=\varepsilon}\Big(\frac{I}{\lambda}-R_\lambda\Big)d=c.c.\qquad \text{Q. E. D.}$$

Remark 2. The proposition stated in the remark 1 is valid for normal operators satisfying (17).

Normed rings and spectral theorems, III

Proc. Imp. Acad. Tokyo **20** (1944) 71–73

(Comm. by T. TAKAGI, M.I.A., Feb. 12, 1944.)

§1. *A Spectral theorem.* Let R be a function ring of real-valued continuous functions $S(M)$ on a bicompact Hausdorff space $\mathfrak{M}$. We assume that R satisfies

(1) for any pair $M_1, M_2 \in \mathfrak{M}$ there exists $S(M) \in R$ such that $S(M_1) \neq S(M_2)$.

Then, by Gelfand-Silov's abstraction of Weierstrass' polynomial approximation theorem[1],

(2) every continuous function on $\mathfrak{M}$ may be uniformly approximated by functions $\in R$.

Next let $F(S)$ be a linear functional on R:

$$(3) \quad \begin{cases} F(aS+\beta U)=aF(S)+\beta F(U) \quad (a, \beta = \text{scalars}), \\ F(S_n) \to F(S) \ (n \to \infty) \ \text{ if } \ \sup_M |S_n(M)-S(M)| \to 0 \ (n \to \infty). \end{cases}$$

We further assume that

$$(4) \qquad F(S) \geq 0 \quad \text{if} \quad S(M) \geq 0 \quad \text{on } \mathfrak{M},$$

$$(5) \qquad F(I)=1, \quad \text{where} \quad I(M) \equiv 1 \quad \text{on } \mathfrak{M}.$$

Then, by (2)–(5) and Riesz-Markoff-Kakutani's theorem[2], we have the representation:

$$(6) \qquad F(S)=\int_{\mathfrak{M}} S(M)\varphi(dM) \qquad S \in R,$$

where φ is a non-negative, continuous from above set function countably additive on Borel sets $\subseteq \mathfrak{M}$ and $\varphi(\mathfrak{M})=1$. Here the continuity from above means that the value of the function on a set is equal to the infimum of its values on open sets covering this set.

We have, from (6),

$$(7) \quad \begin{cases} F(T)=\int_{\mathfrak{M}} T(M)\varphi(dM)=\int_{\lambda_0}^{\lambda_1} \lambda\, d\tau(\lambda), \quad \lambda_0=\inf_M T(M), \quad \lambda_1=\sup_M T(M), \\ \tau(\lambda)=\varphi(M; T(M)<\lambda). \end{cases}$$

Put

$$(8) \qquad \mu=\sup_\lambda \left(\lambda ; \tau(\lambda+\varepsilon)-\tau(\lambda-\varepsilon)>0 \ \text{ for all } \ \varepsilon>0\right).$$

μ may be called as *the maximal spectrum of F referring to T.*

We will prove the

---

1) Rec. Math., **9** (1941), Cf. also H. Nakano: 全國紙上數學談話會, **218** (1941).
2) A. Markoff: Rec. Math., **4** (1938). S. Kakutani: Proc. **19** (1943).

Theorem. Let λ_0 be > 0, then we have

(i) $$\sqrt{\frac{F(T^{2n})}{F(T^{2(n+1)})}} \geq \sqrt{\frac{F(T^{2(n+1)})}{F(T^{2(n+2)})}} \geq \frac{1}{\mu} \qquad (n \geq 0)$$

(ii) $$\sqrt{\frac{F(T^{2n})}{F(T^{2(n+1)})}} \geq \frac{F(T^{2n+1})}{F(T^{2(n+1)})} \geq \frac{1}{\mu} \qquad (n \geq 0)$$

(iii) $$\lim_{n\to\infty} \sqrt{\frac{F(T^{2n})}{F(T^{2(n+1)})}} = \lim_{n\to\infty} \frac{F(T^{2n+1})}{F(T^{2(n+1)})} = \frac{1}{\mu}$$

(iv) $$0 \leq \frac{F(T^{2n+1})}{F(T^{2(n+1)})} - \frac{1}{\mu} \leq \sqrt{\frac{F(T^{2n})}{F(T^{2(n+1)})} - \frac{F(T^{2n+1})^2}{F(T^{2(n+1)})^2}} \quad (n\to\infty).$$

Proof. By Schwarz's inequality, we have

$$F(T^m) = \int \lambda^m d\tau(\lambda) \leq \sqrt{\int \lambda^{2k} d\tau(\lambda)} \sqrt{\int \lambda^{2(m-k)} d\tau(\lambda)} = \sqrt{F(T^{2k})}\,\sqrt{F(T^{2(m-k)})}.$$

Thus, by taking $m = 2k+2$ and $m = 2k+1$, we obtain the first parts of (i) and of (ii). We have, by (8),

$$\sqrt{\frac{F(T^{2n})}{F(T^{2(n+1)})}} \geq \frac{F(T^{2n+1})}{F(T^{2(n+1)})} = \frac{\int \lambda^{2n+1} d\tau(\lambda)}{\int \lambda^{2(n+1)} d\tau(\lambda)} \geq \frac{\int \lambda^{2n+1} d\tau(\lambda)}{\mu \int \lambda^{2n+1} d\tau(\lambda)} = \frac{1}{\mu},$$

and, by (i) and

$$\lim_{n\to\infty} \sqrt[2n]{F(T^{2n})} = \lim_{n\to\infty} \sqrt[2n]{\int \lambda^{2n} d\tau(\lambda)} = \mu,$$

$$\sqrt{\frac{F(T^{2n})}{F(T^{2(n+1)})}} = \sqrt{\frac{\int \lambda^{2n} d\tau(\lambda)}{\int \lambda^{2(n+1)} d\tau(\lambda)}} \to \frac{1}{\mu} \qquad (n\to\infty).$$

Next put

$$\frac{F(T^{2n+1})}{F(T^{2(n+1)})} = a_{n+1}, \qquad \sqrt{\frac{F(T^{2n})}{F(T^{2(n+1)})}} = \beta_{n+1}, \qquad F\Big((T^{-1} - a_{n+1}I)^2 \cdot \frac{T^{2(n+1)}}{F(T^{2(n+1)})}\Big)$$
$$= \gamma_{n+1}.$$

We have

(9) $$\gamma_{n+1} = \frac{F(T^{2n} - 2a_{n+1}T^{2n+1} + a_{n+1}^2 T^{2n+2})}{F(T^{2(n+1)})} = \beta_{n+1}^2 - 2a_{n+1}\cdot a_{n+1} + a_{n+1}^2 \cdot 1$$
$$= \beta_{n+1}^2 - a_{n+1}^2 .$$

Since

(10) $$F_{n+1}(S) = F\Big(S \cdot \frac{T^{2(n+1)}}{F(T^{2(n+1)})}\Big), \qquad S \in R,$$

satisfies (3), (4) and (5), we have

(11) $$F_{n+1}(S) = \int_{\mathfrak{M}} S(M)\varphi_{n+1}(dM), \qquad F_{n+1}(T) = \int \lambda d\tau_{n+1}(\lambda)$$

as in (6) and (7). We have

$$(12) \quad \mu_{n+1}=\sup\left(\lambda\,;\,\tau_{n+1}(\lambda+\varepsilon)-\tau_{n+1}(\lambda-\varepsilon)>0 \quad \text{for all} \quad \varepsilon>0\right)=\mu,$$

for, by applying (iii) to $F_{n+1}(S)$,

$$\frac{1}{\mu_{n+1}}=\lim_{k\to\infty}\sqrt{\frac{F_{n+1}(T^{2k})}{F_{n+1}(T^{2(k+1)})}}=\lim_{k\to\infty}\sqrt{\frac{F(T^{2k+2(n+1)})}{F(T^{2(k+1)+2(n+1)})}}=\frac{1}{\mu}.$$

Therefore, by (iii), (9)-(12), we obtain, for any $\varepsilon>0$ $(\mu>\varepsilon)$,

$$\beta_{n+1}^2-\alpha_{n+1}^2=F_{n+1}\left((T^{-1}-a_{n+1}I)^2\right)=\int\left(\frac{1}{\lambda}-a_{n+1}\right)^2 d\tau_{n+1}(\lambda)$$

$$\geq\left(\frac{1}{\mu-\varepsilon}-a_{n+1}\right)^2\tau_{n+1}(\mu-\varepsilon),$$

as $n\to\infty$. Since $\tau_{n+1}(\nu)=\tau\left(\sqrt[2n+3]{\nu\int\lambda^{2(n+1)}d\tau(\lambda)}\right)$ we have $\lim\limits_{n\to\infty}\tau_{n+1}(\mu-\varepsilon)=1$.

§2. *Application to Hilbert space.* Let B denote the totality of bounded, self-adjoint operators in Hilbert space $\mathfrak{H}$ and $T\in B$ be positive-definite :

$$\lambda_1\|f\|^2\geq(Tf,f)\geq\lambda_0\|f\|^2,\quad f\in H,\quad \lambda_0>0.$$

Let $(T)'$ denote the totality of operators $\in B$ that commute with T, and $(T)''$ be the totality of operators $\in B$ that commute with every operators $\in(T)'$. $(T)''$ is a (real) normed ring by the norm $\|S\|=\sup\limits_{|f|=1}\|Sf\|$ and the unit I (=the identity operator). In the first note[1], it is proved that $(T)''$ is ring-isomorphic (with real multipliers) and linear-isometric with the function ring R of all the real-valued continuous functions $S(M)$ on the bicompact Hausdorff space $\mathfrak{M}$ of all the maximal ideals M of $(T)''$:

$$(T)''\ni S\leftrightarrow S(M)\in R,\quad I\leftrightarrow I(M)\equiv1,\quad \|S\|=\sup_M|S(M)|,$$

and moreover

$S(M)$ is non-negative if and only if $(Sf,f)\geq0$, $f\in\mathfrak{H}$.

Thus, by putting

$F(S)=(Sf_0,f_0)$, (f_0 is any element $\in\mathfrak{H}$ subject to the condition $\|f_0\|=1$),

$$\mu=\inf_\lambda(\lambda\,;\,E_\lambda\cdot f_0=f_0),\qquad T=\int\lambda dE_\lambda,$$

we may apply the theorem to $(T)''$ $(=R)$. In this way the well-known procedure of E. Schmidt concerning completely continuous self-adjoint operators is extended in a more precise form[3].

---

1)　Proc. **19** (1943).

3)　Math. Ann. **63** (1907).

3)　Concerning (iv), cf. N. Kryloff-N. Bogoliouboff: Bult. de l'acad. Sci. URSS. (1929). D. H. Weinstein: Proc. Nat. Acad. Sci., **20** (1934). S. Huruya: Mem. Fac. Sci. Kyûsyû Imp. Univ. **1** (1941).

Normed rings and spectral theorems, IV

Proc. Imp. Acad. Tokyo **20** (1944) 183 185

(Comm. by T. Takagi, m.i.a., April 12, 1944.)

1. *Introduction.* The spectral theorem given in the third note[1] constitutes, in the special case when $T(M)$ is an enumerably-valued step function, a refinement with an error estimation of E. Schmidt's procedure of the approximate calculation of the greatest proper value of the integral equation. The result is then similar to that due to Temple and Collatz[2]. The purpose of the present note is to show that our treatment may be extended to obtain an approximate calculation of the lower proper values.

At this juncture, I intend to correct the misprints in III: 1) the right hand side of (iv) on page 72, line 5 must be read as

$$\frac{1}{\sqrt{\tau(\mu)}}\sqrt{\frac{F(T^{2n})}{F(T^{2(n+1)})}-\frac{F(T^{2n+1})^2}{F(T^{2(n+1)})^2}} \qquad (\text{as} \quad n \longrightarrow \infty)$$

2) $\lim\limits_{n\to\infty}\tau_{n+1}(\mu-\varepsilon)=1$ on page 73, line 6 must be read as $\lim\limits_{n\to\infty}\tau_{n+1}(\mu-\varepsilon)=\tau(\mu-0)=\tau(\mu)$.

2. *The Theorem.* As in III, let $\boldsymbol{R}$ be the totality of the real-valued continuous functions $S(M)$ on a bicompact Hausdorff space $\mathfrak{M}$, and let $F(S)$ be a positive linear functional on $\boldsymbol{R}$ such that $F(I)=1$ where $I(M)\equiv1$. Then

$$F(T)=\int_{\mathfrak{M}}T(M)\varphi(dM)=\int_{\lambda_0}^{\lambda_1}\lambda d\tau(\lambda),\ \lambda_0=\inf_M T(M),\ \lambda_1=\sup_M T(M),$$

$$\tau(\lambda)=\varphi\big(M;\ T(M)<\lambda\big).$$

We assume that $\lambda_0>0$ and that $\tau(\lambda)$ be of the form

$$\text{(A)}\quad\begin{cases}\tau(\mu^{(2)}+0)>\tau(\mu^{(2)}-0),&\tau(\lambda)=\text{constant for }\mu^{(2)}<\lambda<\mu^{(1)}\\[4pt]\tau(\mu^{(1)}+0)>\tau(\mu^{(1)}-0),&\tau(\lambda)=\text{constant for }\mu^{(1)}<\lambda<\mu^{(0)}\\[4pt]\tau(\mu^{(0)}+0)>\tau(\mu^{(0)}-0),&\tau(\lambda)=\tau(\mu^{(0)}+0)\text{ for }\lambda>\mu^{(0)}\end{cases}$$

$\mu^{(0)}, \mu^{(1)}$ may respectively be called as *the maximal, the next maximal spectrum of F referring to T.*

We put

$$\frac{F(T^{2n+1})}{F(T^{2(n+1)})}=\frac{1}{\mu_n^{(0)}},\qquad \sqrt{\frac{F(T^{2n})}{F(T^{2(n+1)})}}=\frac{1}{\nu_n^{(0)}},$$

---

* The cost of this research has been defrayed from the Scientific Research Expenditure of the Department of Education.

1) Proc. **20** (1944), 71. This note will be referred to as III.

2) G. Temple: Proc. London Math. Soc., **29** (1929), 257. L. Collatz: Math. Zeitschr., **46** (1940), 692. Our formula (iv) ((iv)′) does not contain the unknown value $\mu^{(1)}(\mu^{(2)})$ explicitly. Moreover in (iv)″ the values $\mu^{(0)}$, $\mu^{(1)}$, $\mu^{(2)}$ are only implicitly needed. These are the main difference of our results from Temple-Collatz's.

then from the proof of the spectral theorem in III we obtain

(i) $\qquad 1/\nu_n^{(0)} \geqq 1/\nu_{n+1}^{(0)} \geqq 1/\mu^{(0)} \qquad (n \geqq 0)$,

(ii) $\qquad 1/\nu_n^{(0)} \geqq 1/\mu_n^{(0)} \geqq 1/\mu^{(0)} \qquad (n \geqq 0)$,

(iii) $\qquad \lim_{n \to \infty} 1/\nu_n^{(0)} = \lim_{n \to \infty} 1/\mu_n^{(0)} = 1/\mu^{(0)}$,

(iv) $\qquad 0 \leqq (1/\mu_n^{(0)}) - (1/\mu^{(0)}) \leqq \sqrt{(\nu_n^{(0)})^{-2} - (\mu_n^{(0)})^{-2}}$

if $\qquad (1/\mu^{(1)}) - (1/\mu_n^{(0)}) > (1/\mu_n^{(0)}) - (1/\mu^{(0)})$,

because, by (A),

$$\int \left(\frac{1}{\lambda} - \frac{1}{\mu_n^{(0)}} \right)^2 d\tau_{n+1}(\lambda) = \int \left(\frac{1}{\lambda} - \frac{1}{\mu_n^{(0)}} \right)^2 \frac{\lambda^{2(n+1)}}{F(T^{2(n+1)})} d\tau(\lambda) \geqq \left(\frac{1}{\mu^{(0)}} - \frac{1}{\mu_n^{(0)}} \right)^2$$

if $\qquad (1/\mu^{(1)}) - (1/\mu_n^{(0)}) > (1/\mu_n^{(0)}) - (1/\mu^{(0)})$.

We will prove the

Theorem. Put

$$F^{(1)}(S) = F \left(\frac{(\mu^{(0)}I - T)}{F(\mu^{(0)}I - T)} \cdot S \right),$$

and define

$$\frac{F^{(1)}(T^{2n+1})}{F^{(1)}(T^{2(n+1)})} = \frac{1}{\mu_n^{(1)}}, \qquad \sqrt{\frac{F^{(1)}(T^{2n})}{F^{(1)}(T^{2(n+1)})}} = \frac{1}{\nu_n^{(1)}},$$

then

(i)′ $\qquad 1/\nu_n^{(1)} \geqq 1/\nu_{n+1}^{(1)} \geqq 1/\mu^{(1)} \qquad (n \geqq 0)$,

(ii)′ $\qquad 1/\nu_n^{(1)} \geqq 1/\mu_n^{(1)} \geqq 1/\mu^{(1)} \qquad (n \geqq 0)$,

(iii)′ $\qquad \lim_{n \to \infty} 1/\nu_n^{(1)} = \lim_{n \to \infty} 1/\mu_n^{(1)} = 1/\mu^{(1)}$,

(iv)′ $\qquad 0 \leqq (1/\mu_n^{(1)}) - (1/\mu^{(1)}) \leqq \sqrt{(\nu_n^{(1)})^{-2} - (\mu_n^{(1)})^{-2}}$

if $\qquad (1/\mu^{(1)}) - (1/\mu_n^{(1)}) > (1/\mu_n^{(1)}) - (1/\mu^{(1)})$.

Proof. $F^{(1)}(S)$ is a positive linear functional on R such that $F^{(1)}(I) = 1$. Since, by the hypothesis concerning $\tau(\lambda)$,

$$F \left((\mu^{(0)}I - T)T^n \right) = \int_{\lambda_0}^{\lambda_1} (\mu^{(0)} - \lambda)\lambda^n d\tau(\lambda) = \int_{\lambda_0}^{\mu^{(1)}+0} (\mu^{(0)} - \lambda)\lambda^n d\tau(\lambda),$$

we have $F^{(1)}(T^n) = \int_{\lambda_0}^{\mu^{(1)}+0} \lambda^n d\tau^{(1)}(\lambda)$. Thus (i)′–(iv)′ may be proved as (i)–(iv).

3. *A practical formula.* (i)′–(iv)′ is applicable only when the exact value $\mu^{(0)}$ is obtained. We will give a practical approximate formula by making use of the approximations $\nu_k^{(0)}$, $\mu_k^{(0)}$, of $\mu^{(0)}$.

Lemma.

$$G(\nu) = \frac{F\left((\nu I - T)T^{2n+1} \right)}{F\left((\nu I - T)T^{2(n+1)} \right)} \quad \text{and} \quad H(\nu) = \frac{F\left((\nu I - T)T^{2n} \right)}{F\left((\nu I - T)T^{2(n+1)} \right)}$$

are decreasing in ν.

Proof. We have

$$G'(\nu) = \frac{F(T^{2n+2})^2 - F(T^{2n+1})F(^{2n+3})}{F\big((\nu 1 - T)T^{2(n+1)}\big)} \leq 0 ,$$

since, by Schwartz's inequality,

$$F(T^{2n+2}) = \int \lambda^{2n+2} d\tau(\lambda) \leq \sqrt{\int \lambda^{2n+1} d\tau(\lambda) \int \lambda^{2n+3} d\tau(\lambda)}$$

$$= \sqrt{F(T^{2n+1})F(T^{2n+3})} .$$

That $H'(\nu) \leq 0$ will be proved similarly.   Q. E. D.

By (i)–(iv) we have

$$\mu_k^{(0)} \leq \mu^{(0)} \qquad (k \geq 0)$$

$$\bar{\mu}_k^{(0)} \geq \mu^{(0)} \qquad \left(\text{if } (1/\mu^{(1)}) - (1/\mu_k^{(0)}) > (1/\mu_k^{(0)}) - (1/\mu^{(0)}) \right) ,$$

where
$$1/\bar{\mu}_k^{(0)} = (1/\mu_k^{(0)}) - \sqrt{(\nu_k^{(0)})^{-2} - (\mu_k^{(0)})^{-2}} .$$

Hence, by the lemma and (i)′–(iv)′, we have a *Practical formula* :

$$0 \leq \frac{F\big((\mu_k^{(0)}I - T)T^{2n+1}\big)}{F\big((\mu_k^{(0)}I - T)T^{2(n+1)}\big)} - \frac{1}{\mu^{(1)}} \qquad \text{which is}$$

$$\leq \sqrt{ \frac{F\big((\mu_k^{(0)}I - T)T^{2n}\big)}{F\big((\mu_k^{(0)}I - T)T^{2(n+1)}\big)} - \frac{F\big((\bar{\mu}_k^{(0)}I - T)T^{2n+1}\big)^2}{F\big((\bar{\mu}_k^{(0)}I - T)T^{2(n+1)}\big)^2} }$$

$$+ \frac{F\big((\mu_k^{(0)}I - T)T^{2n+1}\big)}{F\big((\mu_k^{(0)}I - T)T^{2(n+1)}\big)} - \frac{F\big((\bar{\mu}_k^{(0)}I - T)T^{2n+1}\big)}{F\big((\bar{\mu}_k^{(0)}I - T)T^{2(n+1)}\big)}$$

for sufficiently large k, n, viz. if k, n are large such that

$$(C) \quad \begin{cases} (1/\mu^{(1)}) - (1/\mu_k^{(0)}) > (1/\mu_k^{(0)}) - (1/\mu^{(0)}) , \\ (1/\mu^{(2)}) - (1/\mu_n^{(1)}) > (1/\mu_n^{(1)}) - (1/\mu^{(1)}) . \end{cases}$$

Remark. It is to be noted that, in practice, the condition (C) is often satisfied only when $k, n \geq 3$ or 4. Our procedure may be repeated to obtain the approximations to the next smaller proper values $\mu^{(2)}$ etc.

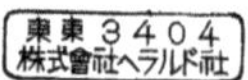

Normed rings and spectral theorems, V

Proc. Imp. Acad. Tokyo **20** (1944) 269–273

(Comm. by T. Takagi, m.i.a., May 12, 1944.)

  1. *Introduction.* Recently, M. Krein[1] published a generalisation of the Plancherel's theorem to the case of locally compact (=bicompact) abelian group. The result is much important, since it reveals the hitherto hidden algebraic character of the classical Fourier analysis. However, Krein's proof of the positivity of the functional J is somewhat complicated and moreover it seems that his paper lacks the proof of $3°$ which is the key to the proof of the positivity of J. The purpose of the present note is to show that a complete proof may be obtained by making use of the preceding note[2]. It is also to be remarked that the theorem below constitutes an extension of $3°$ and that, by virtue of this extension, Krein's arguments may be much simplified.

  2. *A theorem of positivity.* Let G be a locally compact, separable abelian group and let X be the group (without topology for the moment) of continuous characters $\chi(g)$ of G. Then, by Haar's invariant measure dg, we may define the linear space $L_p(G)\,(\infty > p \geq 1)$ of complex-valued measurable functions $x(g)$ such that $|x(g)|^p$ is summable over G :

$$(1) \qquad \|x\|_p = \sqrt[p]{\int |x(g)|^p \, dg} < \infty .$$

A multiplication $x * y$ is introduced in $L_1(G)$ by the convolution :

$$(2) \qquad x * y(g) = \int x(g\text{-}h)y(h)dh .$$

By adjoining formally[3] a unit e to $L_1(G)$ we obtain a normed ring $R(G)$ by the norm $\|z\|$ and the multiplication $*$:

$$(3) \quad \begin{cases} z = \lambda e + x(g) , & \|z\| = |\lambda| + \|x\| \qquad (\lambda = \text{complex number}), \\ z_1 = \lambda_1 e + x_1(g) , & z_2 = \lambda_2 e + x_2(g) , \\ z_1 * z_2 = \lambda_1 \lambda_2 e + \lambda_1 x_2(g) + \lambda_2 x_1(g) + x_1 * x_2(g) . \end{cases}$$

Such ring is considered by *I.* Gelfand and *D.* Raikov[4].

  We next introduce a new normed ring to be denoted as $\bar{R}_{op}(G)$. For any $x \in L_1(G)$ and for any $y \in L_2(G)$ we have

$$(4) \qquad x * y(g) \in L_2(G) , \qquad \|x * y\|_2 \leq \|x\|_1 \cdot \|y\|_2 ,$$

---

  * The cost of this research has been defrayed from the Scientific Research Expenditure of the Department of Education.

 1) C. R. URSS, **30** (1941), No. 6.
 2) Proc. **19** (1943), p. 356. This note will be referred to as (I).
 3) The trivial case of the discrete group G is excluded in the following lines.
 4) C. R. URSS, **28** (1940), No. 3.

since, by the invariance of Haar's measure,

$$\int |x*y(g)|^2\,dg \leqq \int \left|\int |x(g\text{-}h)|^{\frac{1}{2}}\,|x(g\text{-}h)|^{\frac{1}{2}}\,|y(h)|\,dh\right|^2 dg$$

$$\leqq \int\left\{\int |x(g\text{-}h)|\,dh.\ \int |x(g\text{-}h)|\,|y(h)|^2 dh\right\}dg \leqq \|x\|_1^2\cdot\|y\|_2^2\,.$$

Thus $x\in L_1(G)$ induces a linear operator T_x of the Hilbert space $L_2(G)$:

$$(5)\qquad T_x\cdot y=x*y\,,\qquad \|T_x\|=\sup_{|y|_2\leqq 1}\|T_x\cdot y\|_2\leqq\|x\|_1\,.$$

The set $\{T_x\,;\,x\in L_1(G)\}$ together with the identity operator E constitute a ring $R_{op}(G)$ isomorphic to the ring $R(G)$ by the correspondence $x\leftrightarrow T_x$:

$$(6)\qquad \begin{cases} e\leftrightarrow E,\quad \alpha x\leftrightarrow \alpha T_x,\quad x+y\leftrightarrow T_x+T_y,\quad x*y\leftrightarrow T_xT_y,\\[2mm] x^*\leftrightarrow T_x^*,\quad \text{where } x^*(g)=\overline{x(-g)} \text{ and } T^*=\text{adjoint of } T. \end{cases}$$

The closure $\bar{R}_{op}(G)$ by the topology defined by the norm $\|T\|$ of the ring $R_{op}(G)$ surely constitutes a normed ring by the norm $\|T\|$.

It is proved in (I) that $\bar{R}_{op}(G)$ is representable isomorphically as a function ring $R(\bar{\mathfrak{M}})$ of complex-valued continuous functions $T(\bar{M})$ on the compact Hausdorff space $\bar{\mathfrak{M}}$ of all the maximal ideals $\bar{M}$ of $\bar{R}_{op}(G)$:

$$(7)\qquad T\leftrightarrow T(\bar{M})\,,\qquad E\leftrightarrow E(\bar{M})\equiv 1\,,$$

such that

$$(8)\qquad \begin{cases} \|T\|=\sup_{\bar{M}}|T(\bar{M})|\,,\\[2mm] T(\bar{M}) \text{ is real-valued if and only if } T \text{ is symmetric } (T=T^*),\\[2mm] T(\bar{M}) \text{ is non-negative-valued if and only if } T \text{ is symmetric}\\[1mm] \quad \text{and positive definite (as operator).} \end{cases}$$

Hence

$$(9)\qquad \begin{cases} T_z(\bar{M}) \text{ is real-valued if and only if } z=z^* \text{ viz.}\\[2mm] \lambda=\bar{\lambda} \quad\text{and}\quad x(g)=x^*(g)=\overline{x(-g)}\,,\\[2mm] T_z(\bar{M}) \text{ is non-negative-valued if and only if } z=z^* \text{ and}\\[2mm] (T_z\cdot y,y)=z*y*y^*(0)\geqq 0 \text{ for all } y\in L_2(G). \end{cases}$$

Lemma 1. Let $\bar{M}$ be a maximal ideal of $\bar{R}_{op}(G)$, then the image $M\leqq R(G)$ of $\bar{M}\cap R_{op}(G)$ by the correspondence $z\leftrightarrow T_z=\lambda E+T_x$ is a maximal ideal of $R(G)$.

Proof. Since $T_z\equiv\big(\lambda+T_x(\bar{M})\big)E$ (mod $\bar{M}$), we have $z\equiv\big(\lambda+T_x(\bar{M})\big)e$ (mod M). Q. E. D.

Now, by Gelfand-Raikov's results[1], we have

$$(10) \quad \begin{cases} \lambda + T_x(\overline{M}_\infty) = \lambda & \text{if} \quad M_\infty = L_1(G), \\ \lambda + T_x(\overline{M}) = \lambda + \int x(g)\chi_M(g)dg, & \chi_M \in X, \quad \text{if} \quad M \neq L_1(G). \end{cases}$$

Therefore, by (9), (10) and the proof of the Lemma 1, we obtain the

Theorem. Let the Fourier transform $(P \cdot z)(\chi) = \varphi_z(\chi)$ of $z = \lambda e + x \in R(G)$ be defined by

$$(11) \quad (P \cdot z)(\chi) = \varphi_z(\chi) = \lambda + \int x(g)\chi(g)dg \qquad (\chi \in X).$$

If $\varphi_z(\chi)$ is real-valued on X, then $z = z^*$. If $\varphi_z(\chi)$ is non-negative-valued on X, then $z = z^*$ and $z * y * y^*(0) \geq 0$ for all $y \in L_2(G)$.

Corollary. Let $x(g) \in L_1(G)$ be continuous on G, viz. let $x \in L_1(G) \cap C(G)$ and let $\varphi_x(\chi) = \int x(g)\chi(g)dg \geq 0$ on X, then $x(g)$ is positive definite in Bochner's sense:

$$x * y * y^*(0) \geq 0 \quad \text{for all} \quad y \in L_1(G).$$

Thus a complete proof of 3° is obtained.

2. *Plancherel's theorem.* The Fourier transform P is defined by Krein only for $x \in L_1(G)$, $L_2(G)$. Our extension of P for all $z \in R(G)$ together with the theorem (instead of the corollary) will much simplifies Krein's proof of the Plancherel's theorem.

Following Krein, we introduce the additive homogeneous functional $J(\varphi_z(\chi))$:

$$(12) \quad J(\varphi_z(\chi)) = \lambda + x(0) \quad \text{for} \quad z = \lambda e + x, \quad x(g) \in L_1(G) \cap C(G).$$

Lemma 2. J is positive viz. $J(\varphi_z(\chi)) \geq 0$ if $\varphi_z(\chi) \geq 0$ on X.

Proof. Since, by the theorem,

$$z * y * y^*(0) = \lambda \iint y(s)\overline{y(t)}dsdt + \iint x(s-t)y(s)\overline{y(t)}dsdt \geq 0$$

$$\text{for all} \quad y \in L_2(G),$$

we have $\lambda + x(0) \geq 0$.   Q. E. D.

Therefore J is a linear (= continuous additive) functional on the space of the continuous functions $\varphi_z(\chi)$, where $z = \lambda e + x$, $x \in L_1(G) \cap C(G)$. Here the topology on X is defined by the weak topology induced by the neighbourhood:

$$(13) \quad U(\chi_0) = \{\chi ; |\varphi_{z_i}(\chi) - \varphi_{z_i}(\chi_0)| < \varepsilon, \quad i = 1, 2, \ldots, n\}.$$

Moreover we have

---

1) See 4).

$$(14) \qquad J\big(\varphi_z(\chi+\chi_1)\big)=J\big(\varphi_z(\chi)\big) \quad \text{for any} \quad \chi_1 \in X,$$

$$(15) \qquad x(g)=J\big(\varphi_z(\chi)\cdot\chi(-g)\big), \qquad x \in L_1(G)\cap C(G),$$

since

$$\varphi_z(\chi+\chi_1)=\lambda+\int x(g)(\chi\chi_1)(g)dg=\lambda+\int\{x(g)\chi_1(g)\}\chi(g)dg,$$

$$\int x(g+h)\chi(h)dh=\int x(g+h)\chi(g+h)\chi(-g)dh=\chi(-g)\int x(k)\chi(k)dk$$

(by the invariance of the Haar's measure).

Now define a formal character $\chi_\infty(g)$ by

$$(16) \qquad \varphi_z(\chi_\infty)=\lambda, \qquad z=\lambda e+x,$$

then the set $X\cup\chi_\infty$ is compact separable by the weak topology defined by the neighbourhood (13), since $X\cup\chi_\infty$ corresponds to the totality of the maximal ideals M of $R(G)$ in one to one manner[1]:

$$\chi_M \leftrightarrow M, \qquad z\equiv\varphi_z(\chi_M)e \qquad (\text{mod } M).$$

Thus X is a locally compact separable abelian group.

Since $x \in L_1(G)\cap C(G)$ is dense in $L_1(G)$, $\{\varphi_z(\chi)\,; \ z=\lambda e+x,$ $x \in L_1(G)\cap C(G)\}$ is dense[2] (by the topology of the uniform convergence on $X\cup\chi_\infty$) in the space of complex-valued continuous functions on $X\cup\chi_\infty$.

Thus the positive linear functional J is of the form

$$(17) \qquad J\big(\varphi_z(\chi)\big)=\int_{X\cup\chi_\infty}\varphi_z(\chi)d\chi,$$

with a measure $d\chi$ countably additive on Borel sets $\subseteqq X\cup\chi_\infty$. By (16), we have $\varphi_x(\chi_\infty)=0$ and hence

$$(18) \qquad x(0)=J\big(\varphi_x(\chi)\big)=\int_X\varphi_x(\chi)d\chi \quad \text{for} \quad x \in L_1(G)\cap C(G).$$

Because of the invariance (14), the measure $d\chi$ in (18) must be the Haar measure on X. Thus we obtain, by (11) and (15), the duality relation for $x \in L_1(G)\cap C(G)$:

$$(19) \qquad \begin{cases} \varphi_x(\chi)=\displaystyle\int_G x(g)\chi(g)dg, \qquad x(g)=\displaystyle\int_X\varphi_x(\chi)\chi(-g)d\chi, \\ \text{where } dg,\ d\chi \text{ being respectively Haar measures on } G,\ X. \end{cases}$$

Let $y \in L_1(G)\cap L_2(G)$, then, since $x=y*y^*\in L_1(G)\cap C(G)$, we have by (18)

$$x(0)=\int|y(g)|^2 dg=\int\varphi_x(\chi)d\chi=\int\varphi_y(\chi)\varphi_{y^*}(\chi)d\chi=\int|\varphi_y(\chi)|^2 d\chi$$

---

1) See 4). The separability follows from the separability of the normed ring $R(G)$: Cf. I. Gelfand's paper in Rec. Math., **9** (1941). No. 1.
2) See 4).

and thus $(P \cdot y)(\chi) \in L_2(X)$ and

$$(20) \quad \begin{cases} \| P \cdot y \|_2 = \int_X | \varphi_y(\chi) |^2 \, d\chi = \| y \|_2 = \int_G | y(g) |^2 \, dg, \\ Q(P \cdot y) = y, \quad \text{where} \quad (Q \cdot \varphi)(g) = \int_X \varphi(\chi)\chi(-g)d\chi, \end{cases}$$

by (19). Since $L_1(G) \cap L_2(G)$ is dense in $L_2(G)$, (20) is valid also for all $y \in L_2(G)$.

Thus we arrived at the generalised Plancherel's theorem.

（東京 八五四）

Normed rings and spectral theorems, VI

Proc. Imp. Acad. Tokyo **20** (1944) 580–583

(Comm. by T. TAKAGI, M.I.A., Oct. 12, 1944.)

1. *Introduction.* The arguments in 3 of the fifth note[1] was insufficient since the lemma 2 is valid for $z \in L_1(G)$ only. The purpose of the present note is i) to give a complete proof of (19)—the Plancherel's theorem—in 3 of the fifth note and ii) to show that the Bochner-Raikov's representation theorem[2] may be obtained easily from the Plancherel's theorem. In this way, the Fourier analysis may be subsumed under the operator theory in Hilbert space formulated in terms of the normed ring.

We will make use of, in this note, the results and the notations in the fifth note.

2. *Proof of (19).* The set $\left\{ \varphi_z(\chi) = \lambda + \int_G x(g)\chi(g)dg \; ; \; z = \lambda e + x, \right.$ $\left. x \in L_1(G) \right\}$ is dense in the space $C(X \cup \chi_\infty)$ of continuous functions $T(\chi)$ on the character group X of G compactified by adjoining the formal character χ_∞. This results from the Gelfand-Silov's abstraction[3] of Weierstrass' polynomial approximation theorem. We have

$$\| T_z \| = \lim_{n \to \infty} \sqrt[n]{\| T_z^n \|} \leq \lim_{n \to \infty} \sqrt[n]{\| z^n \|_1} = \sup_x | \varphi_z(\chi) |$$

by $\| T_z \| \leq \| z \|_1$. Hence,

$$(*) \quad \begin{cases} \text{for any continuous function } T(\chi) \text{ on } X \cup \chi_\infty, \text{ there exists one} \\ \text{and only one operator } T \text{ such that } \lim_{n \to \infty} \sup_\chi | \varphi_{z_n}(\chi) - T(\chi) | = 0 \\ \text{implies } \lim_{n \to \infty} \| T_{z_n} - T \| = 0. \end{cases}$$

Let $x \in L_1(G) \cap C(G)$ and put

$$J\big(\varphi_x(\chi) \big) = x(0) .$$

J is additive, homogeneous and positive on $\{ \varphi_x(\chi) \}$:

$$\varphi_x(\chi) \geq 0 \quad \text{implies} \quad J\big(\varphi_x(\chi) \big) \geq 0 .$$

The proof was given by the lemma 3. The Plancherel's theorem may be proved if we show that

$$(**) \quad J\big(\varphi_x(\chi) \big) = \int_X \varphi_x(\chi)d\chi , \quad d\chi = \text{the Haar's measure on } X.$$

This formula together with the definition

---

1) Proc. **20** (1944), 269.

2) C. R. URSS: **28**, 4 (1940).

3) Rec. Math., **9** (51), (1941).

$$\varphi_x(\chi)=\int_G x(g)\chi(g)dg\,, \qquad dg=\text{the Haar's measure on G.}$$

constitutes (19).

Let $\mathfrak{N}=\{\varphi_x(\chi)\,;\ x\in L_1(G)\cap C(G)\}$ and let $\Gamma=\{\varphi(\chi)\,;\ \varphi(\chi)$ is continuous and $=0$ in a certain vicinity U_φ of $\chi_\infty\}$ and denote by $\mathfrak{N}_1$ the linear envelope of $\mathfrak{N}$ and Γ. We first show that J may be extended additive homogeneous and positive on $\mathfrak{N}_1$. This extension is surely possible if the following conditions are satisfied:

$$(***)\quad \begin{cases} \text{Let } \widetilde{\varphi}(\chi)\in\Gamma, \text{ then} \\[2mm] \inf_{\substack{\varphi(\chi)\in\mathfrak{N} \\ \varphi(\chi)\geq\widetilde{\varphi}(\chi)}} J\big(\varphi(\chi)\big)= \sup_{\substack{\varphi(\chi)\in\mathfrak{N} \\ \varphi(\chi)\leq\widetilde{\varphi}(\chi)}} J\big(\varphi(\chi)\big). \end{cases}$$

For the proof of $(***)$ let $X_0=\{\chi\,;\ \widetilde{\varphi}(\chi)\neq 0\}$ and take $y\in L_1(G)\cap C(G)$ such that the closure $\overline{X}_0\subseteq\{\chi\,;\ \varphi_y(\chi)\neq 0\}$. Put $T(\chi)=\widetilde{\varphi}(\chi)/|\varphi_y(\chi)|^2$ and let

$$\begin{cases} \lim_{n\to\infty}\sup_x |\varphi_{z_n}(\chi)-T(\chi)|=0\,, \qquad \varphi_{z_n}(\chi)\geqq T(\chi)\,, \\[2mm] \lim_{n\to\infty}\sup_\chi |\varphi_{w_n}(\chi)-T(\chi)|=0\,, \qquad \varphi_{w_n}(\chi)\leqq T(\chi)\,, \\[2mm] z_n=\lambda_n e+x_n,\ x_n\in L_1(G)\cap C(G),\ w_n=\mu_n e+y_n,\ y_n\in L_1(G)\cap C(G). \end{cases}$$

Then, by $(*)$ there exists one and only one operator T such that $\lim_{n\to\infty}\|T_{z_n}-T\|=0,\ \lim_{n\to\infty}\|T_{w_n}-T\|=0$.
We have thus

$$\lim_{n\to\infty} J\big(\varphi_{z_n}(\chi)\,|\,\varphi_y(\chi)\,|^2\big)=\lim_{n\to\infty} z_n*y*y^*(0)$$

$$=\lim_{n\to\infty}(T_{z_n}\cdot y,y)=\lim_{n\to\infty}(T_{w_n}\cdot y,y)=\lim_{n\to\infty} w_n*y*y^*(0)$$

$$=\lim_{n\to\infty} J\big(\varphi_{w_n}(\chi)\,|\,\varphi_y(\chi)\,|^2\big).$$

Since z_n*y*y^* and $w_n*y*y^*\in L_1(G)\cap C(G)$, $(***)$ is completely proved.

Therefore[1], remembering $J\big(\varphi_x(\chi+\chi')\big)=\big(\chi'(g)x(g)\big)(0)=x(0)=J\big(\varphi_x(\chi)\big)$, we have for $\widetilde{\varphi}(x)\in\Gamma$,

$$(**)'\qquad J\big(\widetilde{\varphi}(\chi)\big)=\int_X\widetilde{\varphi}(\chi)d\chi, \quad d\chi=\text{the Haar measure on } X.$$

Hence, by the proof of $(***)$, we have

$$\begin{cases} J\big(T(\chi)\,\big|\,\varphi_y(\chi)\,\big|^2\big)=(T\cdot y,y)=\int_X T(\chi)\,|\,\varphi_y(\chi)|^2 d\chi\,, \\[2mm] T(\chi)\in\Gamma\,, \qquad y\in L_1(G)\cap C(G). \end{cases}$$

Thus, by letting $T(\chi)$ tend to 1, we see that $\varphi_y(\chi)\in L_2(X)$ and hence

$$(**)''\quad \begin{cases} J\big(\varphi_x(\chi)\,\big|\,\varphi_y(\chi)\,\big|^2\big)=(T_x\cdot y,y)=\int_X\varphi_x(\chi)\,|\,\varphi_y(\chi)|^2 d\chi\,, \\[2mm] x,y\in L_1(G)\cap C(G). \end{cases}$$

---

1) M. Krein: C. R. URSS, **30**, 6 (1941). The closure of the set $\{x\,;\ \widetilde{\varphi}(x)\neq 0\}$ is compact in X.

Since $L_1(G)$ is dense in $L_1(G) \cap L_2(G)$, $(**)''$ also holds good for $y \in L_1(G) \cap L_2(G)$. Let V_n be a symmetric and compact neighbourhood in G of 0 such that $0 < \int_{V_n} dg < \infty$, $\lim_{n \to \infty} V_n = 0$ and put $y_n(g) =$ characteristic function of V_n divided by $\int_{V_n} dg$. Then, we have, from $(**)''$,

$$x(0) = \lim_{n \to \infty} (T_x y_n, y_n) = \lim_{n \to \infty} \int_X \varphi_x(\chi) |\varphi_{y_n}(\chi)|^2 d\chi = \int_X \varphi_x(\chi) d\chi,$$

since[1] $\chi(g)$ is uniformly continuous in the aggregate of variables χ, g such that $\chi(0) = 1$.   Q. E. D.

3. *Bochner-Raikov's theorem.*   A measurable function $f(g)$ is called pesitive definite if $f(-g) = \overline{f(g)}$ and

$$\iint f(g-h) x(g) \overline{x(h)} dg dh \geqq 0$$

for every $x \in L_1(G)$.   Such $f(g)$ is essentially bounded and we may assume

$$f(0) = \operatorname*{ess.\ sup}_{g} |f(g)|.$$

We will prove Bochner-Raikov's theorem to the effect that

$$f(g) = \int_X \chi(g) F(d\chi) \qquad \text{a.e. on } G$$

with a uniquely determined continuous from above measure countably additive on Borel sets $\leqq X$.

Proof.   Let $X_i =$ symmetric and compact neighbourhood in X of the zero 0 such that $\int_{X_i} d\chi \neq 0$, $\lim_{i \to \infty} X_i = 0$. Then, since $\chi(g)f(g)$ is positive definite with $f(g)$,

$$f_i(g) = f(g) \left(\int_{X_i} \chi(g) d\chi \right)^2 \Big/ \left(\int_{X_i} d\chi \right)^2$$

is also positive definite, viz.

$$(f_i * y, y) \geqq 0 \qquad \text{for every} \qquad y \in L_1(G).$$

By the Plancherel's theorem, the continuous function $\left(\int_{x_i} \chi(g) d\chi \right)^2 \in L_1(G)$ and hence $\in L_2(G)$.   Thus $f_i(g) \in L_1(G) \cap L_2(G)$.   Therefore, again by Plancherel's theorem,

$(****)$ $\qquad f_i(g) = \int_X \chi(g) \varphi_{f_i}(\chi) d\chi \qquad$ a.e. on G in the mean.

Moreover we have

$$(f_i * y, y) = \int_X \varphi_{f_i}(\chi) |\varphi_y(\chi)|^2 d\chi \geqq 0$$

---

1) See 2).

by the positive-definiteness of $f_i(g)$ and the Plancherel's theorem. Since $\{\varphi_y(\chi)\,;\ y \in L_1(G)\}$ is dense in the space of continuous functions on $X \cup \chi_\infty$, vanishing at χ_∞, we have $\varphi_{f_i}(\chi) \geq 0$ on X. By taking average of (****) over a vicinity of 0 in G, we see that $\int_X \varphi_{f_i}(\chi)d\chi \leq f_i(0) = f(0)$. Hence the set $\left\{\int \varphi_{f_i}(\chi)d\chi\right\}$ is compact as a set of measures. Therefore, there exists a subsequence $\{i'\}$ of $\{i\}$ such that

$$F(d\chi) = \lim_{i' \to \infty} \varphi_{f_{i'}}(\chi)d\chi \qquad \text{(limit as measures)},$$

$$\lim_{i' \to \infty} f_{i'}(g) = f(g) = \int_X \chi(g)F(d\chi) \qquad \text{a.e. on } G$$

Q. E. D.

(東京 八五四)

On the representation of functions by Fourier integrals

Proc. Imp. Acad. Tokyo **20** (1944) 655–660

(Comm. by T. TAKAGI, M.I.A., Nov. 13, 1944.)

1. *Introduction and the theorems.* The purpose of the present note is to give the following representation theorems of complex-valued bounded continuous functions $f(t)$ on $(-\infty, \infty)$. The theorems may be applied in Fourier analysis as well as in probability theory. Since the proofs are carried through by virtue of the Plancherel's duality theorem, our results may be extended to the case of separable, locally compact abelian groups instead of the infinite line $(-\infty, \infty)$[1].

Theorem 1. $f(t)$ is positive definite[2] if and only if

$$(1) \qquad \varphi_n(\lambda) = \frac{1}{\sqrt{2\pi}} \int_{-\infty}^{\infty} f(t) \left(\frac{\sin \dfrac{t}{n}}{\dfrac{t}{n}} \right)^2 e^{-it\lambda} dt \geq 0$$

$$(n = 1, 2, \cdots),$$

and if (1) is satisfied, we have the representation[3]:

$$(2) \quad \begin{cases} f(t) = \displaystyle\int_{-\infty}^{\infty} e^{it\lambda} dv(\lambda) \quad \text{with a monotone increasing, right-con-} \\ \text{tinuous bounded function } v(\lambda). \end{cases}$$

Theorem 2.[4] $f(t)$ is positive definite if and only if $f(t)$ is expressible as

$$(3) \quad \begin{cases} f(t) = \displaystyle\lim_{n\to\infty} \int_{-\infty}^{\infty} g_n(t+s)g_n(s)ds \quad \text{uniformly in every finite inter-} \\ \text{val of } t, \text{ where} \end{cases}$$

$$(4) \qquad \sup_{n\geq 1} \int_{-\infty}^{\infty} |g_n(t)|^2 dt \leq f(0).$$

Theorem 3. $f(t)$ is representable in the form:

$$(5) \quad \begin{cases} f(t) = \displaystyle\int_{-\infty}^{\infty} e^{it\lambda} dv(\lambda) \quad \text{with a complex-valued right-continuous} \\ \text{function } v(\lambda) \text{ of bounded variation,} \end{cases}$$

if and only if

---

1) Cf. Proc. **20** (1944), 560–563.

2) $\overline{f(-t)}=f(t)$ and $\overset{n}{\sum} f(t_j-t_k)\xi_j\overline{\xi_k} \geq 0$ for any integer n and for arbitrary complex numbers ξ.

3) S. Bochner: Vorlesungen über Fouriersche Integrale, Leipzig (1932), 76.

4) A. Khintchine: Bullt. de l'université d'état à Moscou. Sect. A, vol. **1**, fasc. 5, 1–3.

$$(6) \qquad \sup_{n \geq 1} \int_{-\infty}^{\infty} \left\{ \left| \int_{-\infty}^{\infty} f(t) \left(\frac{\sin \frac{t}{n}}{\frac{t}{n}} \right)^2 e^{-it\lambda} dt \right| \right\} d\lambda < \infty .$$

Theorem 4. $f(t)$ is representable in the form (5) if and only if $f(t)$ is expressible as

$$(7) \qquad \left\{ \begin{array}{l} f(t) = \lim_{n \to \infty} \int_{-\infty}^{\infty} g_n(t+s) \overline{k_n(s)} ds \qquad \text{uniformly in every finite inter-} \\ \text{val of } t, \text{ where} \end{array} \right.$$

$$(8) \qquad \left\{ \begin{array}{l} \int_{-\infty}^{\infty} |g_n(t)|^2 dt = \int_{-\infty}^{\infty} |k_n(t)|^2 dt \leq a \text{ constant} < \infty \text{ independent of} \\ n.^{5)} \end{array} \right.$$

Theorem 5. Let (6) be satisfied, then, if we put,

$$(9) \qquad \psi_{n,m}(\lambda) = \frac{1}{\sqrt{2\pi}} \int_{-m}^{m} f(t) \frac{\sin \frac{t}{n}}{\frac{t}{n}} e^{-it\lambda} dt ,$$

$\lim_{m \to \infty} \psi_{n,m}(\lambda) = \psi_n(\lambda)$ exists. Concerning the representation (5) we have the results : i) $v(\lambda)$ is absolutely continuous, viz. $v(\lambda) = \int_{-\infty}^{\lambda} v'(\lambda) d\lambda$ if and only if

$$(10) \qquad \lim_{n, n' \to \infty} \int_{-\infty}^{\infty} |\psi_n(\lambda) - \psi_{n'}(\lambda)| \, d\lambda = 0 .$$

ii) $v(\lambda)$ is singular, viz. $v'(\lambda) = 0$ almost everywhere if and only if

$$(11) \qquad \lim_{n \to \infty} \psi_n(\lambda) = 0 \qquad \text{almost everywhere.}$$

Theorem 6. i) Concerning the representation (5) we have the result : $v(\lambda)$ is absolutely continuous if and only if

$$(12) \qquad \lim_{n, n' \to \infty} \int |\varphi_n(\lambda) - \varphi_{n'}(\lambda)| \, d\lambda = 0 .$$

ii) Concerning the representation (2) we have the result : $v(\lambda)$ is singular if

$$(13) \qquad \lim_{n \to \infty} \varphi_n(\lambda) = 0 \qquad \text{almost everywhere .}$$

2. *Proofs of the the theorems.*
Theorem 1. $f(t)e^{ist}$ is positive definite with $f(t)$ for any real s

---

5) The constant may be taken as the left hand side of (6).

and hence $f(t)\dfrac{\sin\dfrac{t}{n}}{\dfrac{t}{n}}=\dfrac{n}{2}\displaystyle\int_{-\frac{1}{n}}^{\frac{1}{n}}f(t)e^{ist}ds$ is positive definite. In the same manner we see that

$$(14) \qquad f_n(t)=f(t)\left(\frac{\sin\dfrac{t}{n}}{\dfrac{t}{n}}\right)^2$$

is also positive definite and hence

$$(15) \qquad |f_n(t)|\leqq f_n(0)=f(0)\ .$$

Thus $\quad f_n(t)\in L_1(-\infty,\ \infty),\in L_2(-\infty,\ \infty).\quad$ The continuous function

$$(16) \qquad \varphi_n(\lambda)=\frac{1}{\sqrt{2\pi}}\int_{-\infty}^{\infty}f_n(t)e^{-it\lambda}dt$$

is non-negative, since for any $\alpha<\beta$

$$(17) \qquad \int_{a}^{\beta}\varphi_n(\lambda)d\lambda=\int_{-\infty}^{\infty}f_n(t)dt\left\{\frac{1}{\sqrt{2\pi}}\int_{a}^{\beta}e^{-it\lambda}d\lambda\right\}$$

$$=\int_{-\infty}^{\infty}f_n(t)h(t)dt=\int_{-\infty}^{\infty}\int_{-\infty}^{\infty}f_n(t)h(t+s)\overline{h(s)}dtds\geqq 0$$

by the positive definite character of $f_n(t)$. That $h(t)=\dfrac{1}{\sqrt{2\pi}}\displaystyle\int_{a}^{\beta}e^{-it\lambda}d\lambda$ $=\displaystyle\int_{-\infty}^{\infty}h(t+s)h(s)ds$ follows from the Plancherel's theorem. Next we will prove that the non-negative function $\varphi_n(\lambda)$ is $\in L_1(-\infty,\ \infty)$. If we put

$$(18) \qquad v_n(\lambda)=\frac{1}{\sqrt{2\pi}}\int_{0}^{\lambda}\varphi_n(\lambda)d\lambda\ ,$$

then we have from (16)

$$\int_{-\delta}^{\delta}e^{is\lambda}dv_n(\lambda)=\frac{1}{\pi}\int_{-\infty}^{\infty}\frac{\sin\delta(s-t)}{s-t}f_n(t)dt$$

and hence

$$(18)' \qquad \frac{1}{\gamma}\int_{0}^{\gamma}d\delta\left\{\int_{-\delta}^{\delta}e^{is\lambda}dv_n(\lambda)\right\}=\frac{1}{\pi}\int_{-\infty}^{\infty}\frac{1-\cos\gamma(s-t)}{\gamma(s-t)^2}f_n(t)dt$$

$$=\frac{1}{\pi}\int_{-\infty}^{\infty}f_n\left(s+\frac{\tau}{\gamma}\right)\frac{1-\cos\tau}{\tau^2}d\tau\ ,$$

in particular (upon putting $s=0$), by (15).

$$\frac{1}{\gamma}\int_{0}^{\gamma}\{v_n(\delta)-v_n(-\delta)\}d\delta\leqq\sup_{t}|f_n(t)|=f(0)\ .$$

Therefore the monotone increasing function $v_n(\lambda)$ satisfies

$$(19) \qquad v_n(\infty)-v_n(-\infty)\leqq f(0),$$

and hence $\varphi_n(\lambda)\in L_1(-\infty,\infty)$.

Thus, by (16) and the Plancherel's theorem, we have

$$(2)' \qquad f_n(t)=\int_{-\infty}^{\infty}e^{it\lambda}dv_n(\lambda).$$

By (19) and the Helly's selection theorem, there exists a monotone increasing right-continuous function $v(\lambda)$ with $v(\infty)-v(-\infty)\leqq f(0)$ and a subsequence $\{v_{n_p}(\lambda)\}$ such that

$$\lim_{p\to\infty} v_{n_p}(\lambda)=v(\lambda) \text{ at the continuity points } \lambda \text{ of } v(\lambda).$$

Therefore we have, by taking $\lim_{r\to\infty}\lim_{n\to\infty}$ of (18)',

$$f(t)=\int_{-\infty}^{\infty}e^{it\lambda}dv(\lambda).$$

Theorem 2. Put

$$g_n(t)=\mathrm{l.\,i.\,m.}_{m\to\infty}\frac{1}{\sqrt{2\pi}}\int_{-m}^{m}\sqrt{u_n(\lambda)}\,e^{it\lambda}d\lambda$$

$$(\mathrm{l.\,i.\,m.}=\text{limit in the mean})$$

and apply the Parseval form of the Plancherel's theorem to (2)'. That $f(t)$ of the form (3) is positive definite may easily be verified.

Theorem 3 and 4 will be clear from the above proofs of theorem 1 and 2.

Theorem 5. We have, from (5) and (9),

$$\psi_{n,m}(\lambda)=\int_{-\infty}^{\infty}dv(\lambda')\left\{\frac{1}{\sqrt{2\pi}}\int_{-m}^{m}e^{it(\lambda'-\lambda)}\frac{\sin\dfrac{t}{n}}{\dfrac{t}{n}}dt\right\},$$

and hence, if $v(\lambda')$ is continuous at $\lambda+\dfrac{1}{n}$, $\lambda-\dfrac{1}{n}$,

$$(20) \qquad \psi_n(\lambda)=\lim_{m\to\infty}\psi_{n,m}(\lambda)=\int_{\lambda-\frac{1}{n}}^{\lambda+\frac{1}{n}}\frac{n}{2}\,dv(\lambda')$$

$$=\frac{n}{2}\left\{v\left(\lambda+\frac{1}{n}\right)-v\left(\lambda-\frac{1}{n}\right)\right\}=v(\lambda,n)\text{ say}.$$

ii) The condition (11) thus becomes

$$(11)' \qquad \lim_{n\to\infty}\frac{n}{2}\left\{v\left(\lambda+\frac{1}{n}\right)-v\left(\lambda-\frac{1}{n}\right)\right\}=0 \qquad \text{almost everywhere, and}$$

hence ii) is proved

i) The condition (10) becomes

$$(10)' \qquad \lim_{n,n'\to\infty}\int_{-\infty}^{\infty}|v(\lambda,n)-v(\lambda,n')|\,d\lambda=0.$$

Since $v(\lambda)$ is of bounded variation, we have

$$\lim_{n\to\infty} v(\lambda, n) = v'(\lambda) = \frac{dv(\lambda)}{d\lambda} \qquad \text{almost everywhere,}$$

and hence $(10)'$ is equivalent to

$$(10)'' \qquad\qquad \lim_{n\to\infty}\int_{-\infty}^{\infty} |v(\lambda, n) - v'(\lambda)|\, d\lambda = 0 .$$

Let $(10)''$ be satisfied, then for any $\alpha < \beta$

$$(21) \qquad\qquad \lim_{n\to\infty}\int_{\alpha}^{\beta} v(\lambda, n) d\lambda = \int_{\alpha}^{\beta} v'(\lambda) d\lambda .$$

On the other hand,

$$\int_{\alpha}^{\beta} v(\lambda, n) d\lambda = \frac{n}{2}\int_{\alpha}^{\beta}\left(v\left(\lambda + \frac{1}{n}\right) - v\left(\lambda - \frac{1}{n}\right)\right) d\lambda$$

$$= \frac{n}{2}\int_{-\frac{1}{n}}^{\frac{1}{n}}\left(v(\beta + \lambda) - v(\alpha + \lambda)\right) d\lambda .$$

Therefore, if α and β are continuity points of $v(\lambda)$, we have from (21)

$$v(\beta) - v(\alpha) = \int_{\alpha}^{\beta} v'(\lambda) d\lambda .$$

Thus the condition (10) is sufficient.

The necessity of $(10)''$ may be proved as follows. From $v(\lambda) = \int_{-\infty}^{\lambda} v'(\lambda) d\lambda$ we have

$$v(\lambda, n) = \frac{n}{2}\int_{\lambda - \frac{1}{n}}^{\lambda + \frac{1}{n}} v'(\lambda') d\lambda' .$$

Since $v'(\lambda) \in L_1(-\infty, \infty)$, we must have

$$\lim_{n\to\infty}\int_{-\infty}^{\infty}\left|\frac{n}{2}\int_{\lambda - \frac{1}{n}}^{\lambda + \frac{1}{n}} v'(\lambda') d\lambda' - v'(\lambda)\right| d\lambda = 0 ,$$

which is $(10)''$.

Theorem 6. We have, from (1) and (5),

$$(22) \qquad \frac{1}{\sqrt{2\pi}}\varphi_n(\lambda) = \int_{-\infty}^{\infty} dv(\lambda')\left\{\frac{1}{2\pi}\int_{-\infty}^{\infty} e^{it(\lambda'-\lambda)}\left(\frac{\sin\dfrac{t}{n}}{\dfrac{t}{n}}\right)^{2} dt\right.$$

$$= \int_{|\lambda'-\lambda|\leq\frac{2}{n}} dv(\lambda')\left\{\frac{n}{2}\left(1 - \frac{n}{2}\left|\lambda' - \lambda\right|\right)\right\}$$

$$= \frac{1}{2}\int_{-1}^{1}\left(v\left(\lambda + \frac{2}{n}\sigma + \frac{2}{n}\right) - v\left(\lambda + \frac{2}{n}\sigma - \frac{2}{n}\right)\right)\bigg/\frac{4}{n}\, d\sigma .$$

ii) Since $v(\lambda)$ is monoton increasing we have, from (22) and Fatou's theorem

$$\lim_{n\to\infty} \frac{1}{\sqrt{2\pi}}\, \varphi_n(\lambda) \geqq \frac{1}{2}\int_{-1}^{1} v'(\lambda)d\sigma = v'(\lambda)\,.$$

This proves ii).

i) Let $v(\lambda)=\int_{-\infty}^{\lambda} v'(\lambda)d\lambda$, then we have, from (22),

$$\frac{1}{\sqrt{2\pi}}\,\varphi_n(\lambda) = \frac{1}{2}\int_{-1}^{1} d\sigma \cdot \frac{n}{4}\int_{\lambda+\frac{2}{n}\sigma-\frac{2}{n}}^{\lambda+\frac{2}{n}\sigma-\frac{2}{n}} v'(\lambda')d\lambda'\,.$$

Hence, by $v'(\lambda')\in L_1(-\infty, \infty)$, we must have (12).

Next let (12) be satisfied, then the indefinite integrals

$$\frac{1}{\sqrt{2\pi}}\int_{M} \varphi_n(\lambda)d\lambda \qquad (n=1, 2\ \ldots)$$

converges at every measurable set M on $(-\infty, \infty)$. Hence, by Vitali-Hahn-Saks' theorem, the limit $v(\lambda)=\lim_{n\to\infty}\dfrac{1}{\sqrt{2\pi}}\int_{0}^{\lambda} \varphi_n(\lambda)d\lambda$ must be absolutely continuous.

$$\left(\ \text{東京 八五四}\ \right)$$

On the unitary equivalence in general Euclid space

Proc. Japan Acad. **22** (1946) 242–245

(Comm. by T. TAKAGI, M. I. A., Sept. 12, 1946.)

I. *Introduction and the theorem.* The problem of the unitary equivalence of two bounded self-adjoint (s. a.) operators in Hilbert space was solved by E. Hellinger[1] and H. Hahn;[2] the result was extended by M. H. Stone[3] to the case of not necessarily bounded s. a. operators. Later, K. Friedrichs[4] and H. Nakano[5] obtained respectively new forms of the condition for the unitary equivalence; and their results were respectively extended by F. Wecken[6] and H. Nakano[7] to the case of general euclid space R (the space in which all the axioms of the Hilbert space are satisfied except the axiom of separability). The purpose of the present note is to give a condition of the unitary equivalence in a form somewhat more simple and more algebraical than those of the above cited authors. It is easy to see[8] that we may reduce the problem to the case of bounded s. a. operators T_1 and T_2. For any bonnded s. a. operator T let (T)' be he totality of the bounded linear operators commutative with T, and let (T)'' be the totality of the bounded linear operators commutative with every operator ε (T)'. Then (T)' and (T)'' are operator rings (with complex multipliers) and satisfy the condition (1) if $S \varepsilon$ (T)' ((T)'') the conjugate operator S* also ε (T)' ((T)'').
Moreover the ring (T)'' is commutative. In terms of the operator-ring theory our result reads as follows.

Theorem. For the unitary equivalence of T_1 and T_2 it is necessary and sufficient that the ring $(T_1)'$ is isomorphic (with complex multipliers) to the ring $(T_2)'$ by a correspondence C which maps T_1 onto T_2 and which maps conjugate operators onto conjugate operators.

---

(1) Dissertation, Göttingen 1907.
(2) Monatsheft Math. u. Phys. *23* (1912), 169–224.
(3) Linear transformations in Hilbert space, New York 1932.
(4) Jahresber. d. D. Math. Ver. *45* (1935) II, 79–82.
(5) Ann. of Math. *42* (1941), 657–664.
(6) Math. Ann. *116* (1939), 422–455.
(7) Math. Ann. 118 (1941), 112-133.
(8) Consider $\mathrm{Tan}^{-1} T_1$ and $\mathrm{Tan}^{-1} T_2$ if T_1 and T_2 are unbounded.

2. *Proof of the theorem.* The necessity is evident. We will prove the sufficiency. The isomorphism C maps s. a. operators onto s. a. operators and positive definite operators onto positive definite operators. The latter fact may be proved by taking the square root of the positive definite operator. We will wri e $A \geqq B$ if the operator (A–B) is positive definite. Let $\{T_{1n}\}$ be a sequence of s. a. operators ϵ $(T_1)''$ such that $T_{11} \leqq T_{12} \leqq \ldots \leqq T_{1n} \leqq \ldots$ a s. a. operators ϵ $(T_1)''$, and let $T_{1n} \longleftrightarrow T_{2n}$ by the isomorphism C, then we have

(2)   strong $\lim_{n \to \infty} T_{1n} =$ strong $\lim_{n \to \infty} T_{2n}$ by C.

This results from the fact that the strong $\lim_{n \to \infty} T_{1n} = \sup_{n \geqq 1} T_{1n}$ in $(T_1)''$ (in the sense of the semi-order $\geqq$), and hence the strong $\lim_{n \to \infty} T_{2n} = \sup_{n \geqq 1} T_{2n}$ in $(T_2)''$. Thus we have the

Lemma. Let $T_1 = \int \lambda dE_1(\lambda)$ and $T_2 = \int \lambda dE_2(\lambda)$ be the spectral resolution of T_1 and T_2, then if $G(\lambda)$ denotes the characteristic function of a Borel measurable set $\mathfrak{A}$ on $(-\infty, \infty)$

(3) $G(T_1) = \int_{\mathfrak{A}} G(\lambda) dE_1(\lambda) \longleftrightarrow G(T_2) = \int_{\mathfrak{A}} G(\lambda) dE_2(\lambda)$ by C.

It is easy to see, by the isomorphism C, that the dimensions of the closed linear manifolds $N(T_1) = \{x ; T_1 x = 0\}$, $N(T_2) = \{y ; T_2 y = 0\}$ are the same. We put, for any $x \epsilon R \ominus N(T_1)$

$M_{T_1}(x) = \{F(T_1)x = \int F(\lambda) dE_1(\lambda)x ; \int |F(\lambda)|^2 d \| E_1(\lambda)x \|^2 < \infty$,where

F (λ) denote complex-valued Borel measurable functions$\}$

As is well-known, $M_{T_1}(x)$ is a separable closed linear manifold determined by the set of elements $\{E_1(\lambda)x\}, -\infty < \lambda < \infty$; it reduces both $E_1(\lambda)$ and T_1 viz. the projection $P(M_{T_1}(x))$ upon the manifold $M_{T_1}(x)$ is commutative with $E_1(\lambda)$ and with T_1. Let $P_2 \epsilon (T_2)'$ be the operator which corresponds to $P_1 = P(M_{T_1}(x))$ by the isomorphism C, then P_2 is also a projection and $P_2 R \leqq R \ominus N(T_2)$. As $M_{T_1}(x')$ is orthogonal to $M_{T_1}(x)$ if x' is in $R \ominus N(T_1)$ and orthogonal to $M_{T_1}(x)$, our theorem will be proved if we show that there exists an isometric mapping V from $P_1 R$ onto $P_2 R$ such that

(4) $P_1 T_1 P_1 = V^{-1} P_2 T_2 P_2 V.$

First we will show that the closed linear manifold $M = P_2 R$ is separable. —Proof. Lett $\{y_a\}$ be a complete orthonormal system in $P_2 R$, and we classi-

---

(9) The existence of the strong $\lim_{n \to \infty} T_{1n}$ may be proved following F. Riesz's idea. See the footnote in K. Yosida and T. Nakayama: Proc. Imp. Acad Tokyo, **18** (1942), 555–560.

(10) Acta Sci. Math. Szeged, **7** (1935), 147–159.

fy the set $\{M_{T_2}(y_a)\}$ as follows; $M_{T_2}(y_a)$ and $M_{T_2}(y_\beta)$ belong to the same class if and only if there exists a finite number of elements $y_{a\,1} = y_a$, $y_{a\,2}\ldots$, $y_{an} = y_\beta$ such that $M_{T_2}(y_{a\,+1})$ is not orthogonal to $M_{T_2}(y_a)$. Let the set of these classes k be K, then the closed linear manifold $M^{(k)}$ spanned by M_{T_2} (y_a) ε k is a separable closed linear manifold $\leq P_2R$ which reduces T_2 and P_2. Clearly $P_2R = \sum_{k \,\epsilon\, K} M^{(k)}$; here the cardinal number of K must be at most $\aleph_0$. This results from the fact that since $M_{T_1}(x)$ is separable there exists at most countable number of mutually orthogonal projections $P(1)$ ε $(T_1)'$ which satisfy $P(1)P_1 = P_1 P(1)$ and hence, because of the isomorphism C, there exists at most countable number of mutually orthogonal projections $P(2)$ ε $(T_2)'$ which satisfy $P(2)P_2 = P_2P(2) = P(2)$.

As P_2R is separable, there exists an element y ε P_2R such that, for any z ε P_2R, the monotone increasing function $\sigma(\lambda) = \| E_2(\lambda)z \|^2$ is absolutely continuous with respect to the monotone increasing function $\kappa(\lambda) = \| E_2(\lambda)y \|^2$. We will show that $M^T{}_2(y) = P_2R$. Proof. If otherwise, the projection $P(M_{T_2}(y))$ satisfies

$$(5) \qquad P_2P(M_{T_2}(y)) = P(M_{T_2}(y))P_2 = P(M_{T_2}(y)) \neq P_2.$$

Let Q be the projection ε $(T_1)'$ which corresponds to $P(M_{T_2}(y))$ by the isomorphism C, then we have

$$(6) \qquad 0 \neq Q = QP_1 = P_1Q \neq P_1.$$

Since QR is separable, there exists $x^{(1)}$ε QR such that, for any $z^{(1)}$ εQR, $\sigma^{(1)}(\lambda) = \| E_1(\lambda)z^{(1)} \|^2$ is absolutely continuous with respect to $\kappa^{(1)}(\lambda) = \| E_1(\lambda)x^{(1)} \|^2$. Then there exists Borel measurable set $\mathfrak{A}$ such that

$$(7) \qquad \int_{\mathfrak{A}} d\| E_1(\lambda)x \|^2 \neq 0, \quad \int_{\mathfrak{A}} d\| E_1(\lambda)x^{(1)} \|^2 = 0.$$

For, if otherwise, $\rho_1(\lambda) = \| E_1(\lambda)x \|^2$ is absolutely continuous with respect to $\kappa^{(1)}(\lambda)$. And since $\kappa^{(1)}(\lambda)$ is absolutely continuous with respect to $\rho_1(\lambda)$ by $Q = P_1\,Q = QP_1$, we would have $M_{T_1}(x) = M_{T_1}(x^{(1)})$ viz. $Q = P_1$, contrary to (6). Let $G(\lambda)$ be the characteristic function of $\mathfrak{A}$ then we have from (7)

$$G(T_1)x \neq 0, \qquad G(T_1)x^{(1)} = 0.$$

Hence we have $G(T_1)P_1 \neq 0$ and, for any $z^{(1)}$ε QR, $G(T_1)z^{(1)} = 0$ or $G(T_1)Q = 0$, because $\sigma^{(1)}(\lambda)$ is of the form $\int_{-\infty}^{\lambda} F(\lambda)d\,\kappa^{(1)}(\lambda)$ and thus $\| G(T_1)z^{(1)} \|^2 = \int F(\lambda)d\| E_1(\lambda)x^{(1)} \|^2 = 0$. Therefore, by (3), $G(T_2)P_2 \neq 0$ and $G(T_2)P(M_{T_2}(y)) = 0$. This contradicts to the choice of y. Hence we must have $M_{T_2}(y) = P_2R$.

By a similar argument we may prove that the two monotone increasing functions $\rho_1(\lambda) = \| E_1(\lambda)x \|^2$ and $\rho_2(\lambda) = \| E_2(\lambda)y \|^2$ are mutually absolutely continuous with respect to each other. Hence, by Radon-Nikodym's theorem, there exists a Borel measurable non-negative function $F(\lambda)$ such that

$$\rho_1(\lambda) = \int_{-\infty}^{\lambda} F(\lambda)\,d\rho_2(\lambda), \quad \rho_2(\lambda) = \int F(\lambda)^{-1}\,d\rho_1(\lambda).$$

hence, if we put $y(x) = \int \sqrt{F(\lambda)}\,dE_2(\lambda)\,y$,   we have

$$M_{T_2}(y(x)) = M_{T_2}(y) = P_2 R, \quad \rho(\lambda) = \| E_1(\lambda)x \|^2 = \| E_2(\lambda)y(x) \|^2.$$

Thus it is easy to see that the isometric operator V demanded in (4) is given by

$$VF(T_1)x = F(T_2)y(x).$$

Remork. Our heorem may easily be extended to the case where T_1 and T_2 are normal operators.

In concluding this note I express my hearty thanks to Dr. Kiyosi Itô for the discussion of the result.

Finitely additive measures

(*with E. Hewitt*)

Trans. Amer. Math. Soc. **72** (1952) 46–66

0. **Introduction.** The present paper is concerned with real-valued measures which enjoy the property of finite additivity but not necessarily the property of countable additivity. Our interest in such measures arose from two sources. First, the junior author has been concerned with the space of all finitely additive complex measures on a certain family of sets, which under certain conditions can be made into an algebra over the complex numbers. Second, both of us were informed by S. Kakutani of a result similar to our Theorem 1.23, the proof given by Kakutani being different from ours. The problem of characterizing finitely additive measures in some specific way occurred to us as being quite natural, and this problem we have succeeded in solving under fairly general conditions (Theorem 1.22).

The body of the paper is divided into four sections. In §1, we consider finitely additive measures in a reasonably general context, obtaining first a characterization of such measures in terms of a countably additive part and a purely finitely additive part. Purely finitely additive measures are then characterized explicitly. In §2, we extend the theorem of Fichtenholz and Kantorovič [4]([1]) and thereby characterize the general bounded linear functional on the Banach space of bounded measurable functions on a general measurable space. In §3, we consider a number of phenomena which appear in the special case of the real number system. Here we exhibit a number of finitely additive measures which have undeniably curious properties. In §4, we describe connections between our finitely additive measures and certain countably additive Borel measures defined on a special class of compact Hausdorff spaces. We are indebted to Professor S. Kakutani for comments on and improvements in the results obtained.

Throughout the present paper, the symbol R designates the real number system, and points are denoted by lower-case Latin letters, sets by capital Latin letters, families of sets by capital script letters. Sets of functions are denoted by capital German letters. For any set X and any $A \subset X$, the characteristic function of A is denoted by χ_A.

1. **General finitely additive measures.**

1.1 DEFINITION. Let X be an abstract set, and let $\mathcal{M}$ be a family of subsets of X closed under the formation of finite unions and of complements. Let $\mathcal{M}$ be the smallest family of sets containing $\mathcal{M}$ and closed under the formation of countable unions and of complements.

Presented to the Society, April 28, 1951; received by the editors April 9, 1951.

([1]) Numbers in brackets refer to the references at the end of the paper.

1.2 DEFINITION. Let ϕ be a single-valued function defined on $\mathfrak{M}$ such that

1.2.1 $\qquad -\infty < \phi(A) < +\infty \qquad\qquad\qquad\qquad\qquad$ for all $A \in \mathfrak{M}$,

1.2.2 $\qquad \phi(0) = 0$,

1.2.3 $\qquad \sup_{A \in \mathfrak{M}} |\phi(A)| < +\infty$,

1.2.4 $\qquad \phi(A \cup B) = \phi(A) + \phi(B) \qquad$ for all $A, B \in \mathfrak{M}$ such that $A \cap B = 0$.

Then ϕ is said to be a finitely additive measure on $\mathfrak{M}$. The set of all such measures for fixed X and $\mathfrak{M}$ is denoted by the symbol $\Phi(X, \mathfrak{M})$.

1.3 NOTE. We restrict all measures considered to be only finite, partly for simplicity of proofs and partly because the applications envisaged deal exclusively with finite measures. A number of the results of §1 can be proved for measures assuming infinite values, but the proofs involve a number of complications.

1.4 NOTE. Throughout the present section, we shall be concerned only with a single $\Phi(X, \mathfrak{M})$ and hence shall use the symbol Φ to denote $\Phi(X, \mathfrak{M})$.

1.5 NOTE. One cannot prove 1.2.3 from 1.2.1, 1.2.2, and 1.2.4. Let $X = \{1, 2, 3, \cdots, n, \cdots\}$, let $\sum_{n=1}^{\infty} a_n$ be a convergent but not absolutely convergent series, and let $\mathfrak{M}$ consist of finite sets and their complements. Then if $\phi(A) = \sum_{n \in A} a_n$ for all $A \in \mathfrak{M}$, one sees immediately that 1.2.3 is violated while 1.2.1, 1.2.2, and 1.2.4 are satisfied.

1.6 DEFINITION. Let ψ be an element of Φ such that for every sequence $\{A_n\}_{n=1}^{\infty} \subset \mathfrak{M}$ such that $A_1 \supset A_2 \supset \cdots \supset A_n \supset \cdots$ and $\bigcap_{n=1}^{\infty} A_n = 0$, we have $\lim_{n \to \infty} \psi(A_n) = 0$. Then ψ is said to be countably additive.

1.7 THEOREM. *A measure γ of Φ is countably additive if and only if, for every sequence $\{E_n\}_{n=1}^{\infty} \subset \mathfrak{M}$ of pairwise disjoint sets such that $\bigcup_{n=1}^{\infty} E_n \in \mathfrak{M}$, the equality $\gamma(\bigcup_{n=1}^{\infty} E_n) = \sum_{n=1}^{\infty} \gamma(E_n)$ obtains.*

The proof is simple and is omitted.

1.8 THEOREM. *Let ψ be any countably additive measure. Then ψ admits a unique extension $\bar{\psi}$ over the family $\overline{\mathfrak{M}}$ which is countably additive on $\overline{\mathfrak{M}}$.*

This result is proved in [6, p. 54, Theorem A].

We now consider general properties of the set of measures Φ.

1.8a NOTE. As G. Birkhoff has shown (see [2, p. 185]), any measure in Φ can be extended over $\overline{\mathfrak{M}}$ so as to remain finitely additive. However, uniqueness is often not obtainable. For example, let N_0 be a countably infinite set, and let $\mathcal{J}$ be all subsets of N_0 which are finite or have finite complements. $\mathcal{J}$ is then the algebra of all subsets of N_0. The measure π such that $\pi(A)$ is 0 or 1 as A or A' is finite is obviously finitely additive. Its extensions over $\overline{\mathcal{J}}$ can be identified with the finite Borel measures on the space $\beta N_0 \cap N_0'$; and there are $2^{2^{\aleph_0}}$ such measures. For any completely regular space X, the space βX is a compact Hausdorff space such that X is a dense subspace of βX and every bounded continuous real-valued function on X has a continuous extension over βX. (See [8, pp. 64–67] and [16, p. 59].)

1.9 THEOREM. *For arbitrary $\phi \in \Phi$ and $\alpha \in R$, let the set-function $\alpha\phi$ be defined by the relation $(\alpha\phi)(E) = \alpha \cdot \phi(E)$ for all $E \in \mathfrak{M}$. For arbitrary ϕ, $\gamma \in \Phi$, let the set-function $\phi + \gamma$ be defined by the relation $(\phi + \gamma)(E) = \phi(E) + \gamma(E)$ for all $E \in \mathfrak{M}$. Under these definitions, Φ is a linear space.*

This result being very simple, we omit the proof.

1.10 DEFINITION. For an arbitrary $\phi \in \Phi$, we write $\phi \geqq 0$ if the inequality $\phi(E) \geqq 0$ is valid for all $E \in \mathfrak{M}$. For arbitrary ϕ and $\gamma \in \Phi$, we write $\phi \leqq \gamma$ if $\gamma - \phi \geqq 0$.

1.11 THEOREM. *Under the partial ordering 1.10, the space Φ is a lattice. For arbitrary ϕ and $\gamma \in \Phi$, the measure $\phi \wedge \gamma$ is defined by the relation*

$$1.11.1 \qquad (\phi \wedge \gamma)(E) = \inf_{T \subset E, T \in \mathfrak{M}} (\phi(T) + \gamma(E \cap T')),$$

for all $E \in \mathfrak{M}$. The measure $\phi \vee \gamma$ is defined by the relation

$$1.11.2 \qquad \phi \vee \gamma = - ((-\phi) \wedge (-\gamma)).$$

For a proof of this result, see [3, p. 319].

1.12 THEOREM. *Let ϕ be an arbitrary element of Φ. Writing $\phi \vee 0$ as ϕ_+ and $(-\phi) \vee 0$ as ϕ_-, we have the relations $\phi = \phi_+ - \phi_-$ and $\phi_+ \wedge \phi_- = 0$.*

This result is obvious from 1.11.

Theorem 1.12 permits us in many cases to confine our attention to nonnegative measures $\phi \in \Phi$. We now define a class of finitely additive measures which are, so to say, as unlike countably additive measures as possible.

1.13 DEFINITION. Let ϕ be a measure in Φ such that $0 \leqq \phi$. If every countably additive measure ψ such that $0 \leqq \psi \leqq \phi$ is identically zero, then ϕ is said to be purely finitely additive. If $\phi \in \Phi$ and both ϕ_+ and ϕ_- are purely finitely additive, then ϕ is said to be purely finitely additive.

We now describe a number of properties possessed by countably additive and purely finitely additive measures.

1.14 THEOREM. *Let ψ_1 and ψ_2 be countably additive. Then $\psi_1 + \psi_2$, $\psi_1 \wedge \psi_2$, and $\psi_1 \vee \psi_2$ are also countably additive. If ψ is countably additive and $\alpha \in R$, then $\alpha\psi$ is countably additive.*

The assertions made about $\psi_1 + \psi_2$ and $\alpha\psi$ are immediate consequences of 1.9 and 1.6. Consider now $\psi_1 \wedge \psi_2$. Let $\{A_n\}_{n=1}^{\infty}$ be any decreasing sequence of elements of $\mathfrak{M}$ such that $\bigcap_{n=1}^{\infty} A_n = 0$. Since $(\psi_1 \wedge \psi_2)(A_n) \leqq \min (\psi_1(A_n), \psi_2(A_n))$, it follows from 1.6 that $\liminf_{n \to \infty} (\psi_1 \wedge \psi_2)(A_n) \leqq 0$. Assume that lim $\inf_{n\to\infty} (\psi_1\wedge\psi_2)(A_n)<0$. Then there exists a real number $t<0$ and a sequence of positive integers $n_1<n_2<\cdots<n_k<\cdots$ such that $(\psi_1\wedge\psi_2)(A_{n_k})$ $<t$ $(k=1, 2, 3, \cdots)$. By 1.11.1, there exist sets T_{n_k} such that $T_{n_k}\in\mathfrak{M}$, $T_{n_k}\subset A_{n_k}$, and $\psi_1(T_{n_k})+\psi_2(A_{n_k}\cap T'_{n_k})<t$ $(k=1, 2, 3, \cdots)$. For all k, then, $\psi_1(T_{n_k})<t/2$ or $\psi_2(A_{n_k}\cap T'_{n_k})<t/2$. This obviously implies that an infinite subsequence of at least one of the sequences $\{\psi_1(T_{n_k})\}_{k=1}^{\infty}$ and $\{\psi_2(A_{n_k}\cap T'_{n_k})\}_{n=1}^{\infty}$ consists entirely of numbers less than $t/2$. Since $\bigcap_{k=1}^{\infty}T_{n_k}$ $=\bigcap_{k=1}^{\infty}(A_{n_k}\cap T'_{n_k})=0$, this contradicts the hypothesis that ψ_1 and ψ_2 are countably additive, and establishes the fact that lim $\inf_{n\to\infty} (\psi_1\wedge\psi_2)(A_n)\geqq 0$. We may now infer that $\lim_{n\to\infty}(\psi_1\wedge\psi_2)(A_n)=0$ and that $\psi_1\wedge\psi_2$ is countably additive. The fact that $\psi_1\vee\psi_2$ is countably additive now follows from 1.11.2 and the fact that $-\psi$ is countably additive if ψ is.

1.15 THEOREM. *Let $\phi\in\Phi$ have the property that for certain countably additive measures ψ_1, ψ_2, the inequality $\psi_1\leqq\phi\leqq\psi_2$ obtains. Then ϕ is countably additive.*

Considering $\phi-\psi_1$ and noting that if $\phi-\psi_1$ is countably additive, then so is ϕ, we reduce the present theorem to the case $\psi_1=0$. Then, if $\{A_n\}_{n=1}^{\infty}$ is any decreasing sequence of sets in $\mathfrak{M}$ such that $\bigcap_{n=1}^{\infty}A_n=0$, the inequalities $0\leqq\phi(A_n)\leqq\psi_2(A_n)$ are evident; by 1.6, $\lim_{n\to\infty}\psi_2(A_n)=0$, and hence $\lim_{n\to\infty}\phi(A_n)=0$. Again by 1.6, it follows that ϕ is countably additive.

1.16 THEOREM. *Let ϕ be a non-negative measure in Φ. Then ϕ is purely finitely additive if and only if $\phi\wedge\psi=0$ for all non-negative countably additive measures ψ.*

Suppose that ϕ is purely finitely additive. Then, if $\gamma\in\Phi$ and $0\leqq\gamma\leqq\psi$, γ is countably additive, by 1.15, and if also $0\leqq\gamma\leqq\phi$, 1.13 shows that $\gamma=0$. Hence $\phi\wedge\psi=0$. Conversely, if $\phi\wedge\psi=0$ for all countably additive $\psi\geqq 0$, it follows at once that ϕ is purely finitely additive.

1.17 THEOREM. *Let π_1 and π_2 be purely finitely additive. Then $\pi_1+\pi_2$, $\pi_1\vee\pi_2$, and $\pi_1\wedge\pi_2$ are also purely finitely additive. If π is purely finitely additive and $\alpha\in R$, then $\alpha\pi$ is purely finitely additive.*

Consider first the case π_1, $\pi_2\geqq 0$. To show that $\pi_1+\pi_2$ is purely finitely additive, it suffices, in view of 1.16, to prove $(\pi_1+\pi_2)\wedge\psi=0$ for all countably additive $\psi\geqq 0$. Thus for all $E\in\mathfrak{M}$, we must compute the number

$$(\pi_1+\pi_2)\wedge\psi(E) = \inf_{A\subset E,\ A\in\mathfrak{M}} [\pi_1(A)+\pi_2(A)+\psi(E\cap A')].$$

Let ϵ_1, ϵ_2, ϵ_1', and ϵ_2' be arbitrary positive real numbers. Then, since $\pi_1\wedge\psi=\pi_2\wedge\psi=0$, there exist sets A_1, $A_2\subset E$ $(A_1, A_2\in\mathfrak{M})$ such that $\pi_1(A_1)$ $<\epsilon_1$, $\pi_2(A_2)<\epsilon_2$, $\psi(A_1')<\epsilon_1'$, $\psi(A_2')<\epsilon_2'$. Then $\pi_1(A_1\cap A_2)+\pi_2(A_1\cap A_2)$ $+\psi((A_1'\cup A_2')\cap E)\leqq\epsilon_1+\epsilon_2+\epsilon_1'+\epsilon_2'$; as this sum may be made arbitrarily small, it follows that $(\pi_1+\pi_2)\wedge\psi=0$, and that $\pi_1+\pi_2$ is purely finitely additive. The case π_1, π_2 arbitrary follows immediately. If π_1, π_2 are purely finitely additive, then $(\pi_1 \vee 0)$ and $(\pi_2 \vee 0)$ are also. Since $(\pi_1 + \pi_2) \vee 0 \leq \pi_1 \vee 0 + \pi_2 \vee 0$, and since we have just proved that $\pi_1 \vee 0 + \pi_2 \vee 0$ is purely finitely additive, we infer that $(\pi_1 + \pi_2) \vee 0$ is also. By an identical argument, we show that $-((\pi_1 + \pi_2) \wedge 0) = (-\pi_1 - \pi_2) \vee 0$ is purely finitely additive; hence $\pi_1 + \pi_2$ is also. If π_1, $\pi_2 \geq 0$, then $0 \leq \pi_1 \wedge \pi_2 \leq \pi_1 \vee \pi_2 \leq \pi_1 + \pi_2$. These inequalities make it obvious that $\pi_1 \wedge \pi_2$ and $\pi_1 \vee \pi_2$ are purely finitely additive. For arbitrary π_1, π_2, we note that $(\pi_1 \vee \pi_2) \vee 0 = (\pi_1 \vee 0) \vee (\pi_2 \vee 0)$ and $(\pi_1 \vee \pi_2) \wedge 0 = (\pi_1 \wedge 0) \vee (\pi_2 \wedge 0)$. This reduces the case of arbitrary π_1, π_2 to non-negative π_1, π_2. The final statement of the theorem is established by computing

$$\inf_{A \subset E,\, A \in \mathfrak{M}} [\alpha \pi(A) + \psi(E \cap A')],$$

which is obviously 0 for $\pi \geq 0$, if ψ is countably additive, and in general by writing $\alpha \pi$ as $\alpha \pi_+ - \alpha \pi_-$.

1.18 THEOREM. *A non-negative measure $\phi \in \Phi$ is purely finitely additive if and only if, for every non-negative countably additive ψ, every $A \in \mathfrak{M}$, and every $\alpha, \beta > 0$, there exists a set T such that $T \subset A$, $T \in \mathfrak{M}$, $\psi(T) < \alpha$, and $\phi(A \cap T') < \beta$.*

This observation follows immediately from 1.16 and 1.11.1.

We now obtain another characterization of purely finitely additive measures in a restricted case.

1.19 THEOREM. *Suppose that $\mathfrak{M} = \mathfrak{N}$. Let π be any non-negative purely finitely additive measure and let ψ be any non-negative countably additive measure. Then for every positive real number ϵ, there exists a set $A \in \mathfrak{M}$ such that $\pi(A') = 0$ and $\psi(A) < \epsilon$.*

For A, α, β, and T as in 1.18, we have, taking ϕ of 1.18 to be the measure π now under consideration, $\pi(T) + \pi(T' \cap A) = \pi(A)$; hence $\pi(T) = \pi(A) - \pi(T' \cap A) > \pi(A) - \beta$. Now let ϵ be any positive real number and let $\{\epsilon_n\}_{n=1}^{\infty}$ be any sequence of positive real numbers such that $\sum_{n=1}^{\infty} \epsilon_n < \epsilon$. Next, consider the set X itself and let D_1 be a set in $\mathfrak{M}$ such that $\pi(D_1) < \pi(X)/2$ and $\psi(D_1') < \epsilon_1$. Let D_1' be denoted by the symbol A_1. Then $\pi(A_1) > \pi(X)/2$ and $\psi(A_1) < \epsilon_1$. Now consider the set A_1'. By 1.18, there exists a subset D_2 of A_1' such that $D_2 \in \mathfrak{M}$, $\pi(D_2) < \pi(A_1')/2$, and $\psi(D_2' \cap A_1') < \epsilon_2$. Plainly, $\pi(D_2' \cap A_1') > \pi(A_1')/2$. Write $D_2' \cap A_1'$ as A_2. We then have $\pi(A_2) \geq \pi(A_1')/2$ and $\psi(A_2) < \epsilon_2$. It is also clear that $\pi(A_1 \cup A_2) = \pi(A_1) + \pi(A_2) > \pi(A_1) + \pi(A_1')/2 = \pi(A_1) + [\pi(X) - \pi(A_1)]/2 = \pi(X)/2 + \pi(A_1)/2 > \pi(X)/2 + \pi(X)/4$. Suppose now that disjoint sets $A_1, \cdots, A_n$ in $\mathfrak{M}$ have been found such that $\pi(A_1 \cup \cdots \cup A_n) = \sum_{i=1}^{n} \pi(A_i) \geq \sum_{i=1}^{n} 2^{-i} \pi(X) = (1 - 2^{-n}) \pi(X)$ and $\psi(A_1 \cup \cdots \cup A_n) < \epsilon_1 + \cdots + \epsilon_n$. There exists a set $D_{n+1} \in \mathfrak{M}$ such that $D_{n+1} \subset (A_1 \cup \cdots \cup A_n)'$, such that $\pi(D_{n+1}) < \pi((A_1 \cup \cdots \cup A_n)')/2$ and $\psi(D_{n+1}' \cap (A_1 \cup \cdots \cup A_n)') < \epsilon_{n+1}$. Writing $D_{n+1}' \cap (A_1 \cup \cdots \cup A_n)'$ as A_{n+1}, we immediately see that $\pi(A_{n+1}) > \pi((A_1 \cup \cdots \cup A_n)')/2$ and $\psi(A_{n+1})$ $< \epsilon_{n+1}$. We next observe that $\pi(A_1 \cup \cdots \cup A_{n+1}) = \pi(A_1 \cup \cdots \cup A_n)$ $+ \pi(A_{n+1}) > \pi(A_1 \cup \cdots \cup A_n) + \pi(A_1 \cup \cdots \cup A_n)'/2 \geqq (1 - 2^{-n})\pi(X)$ $+ (1 - 2^{-n})\pi(X)/2 = (1 - 2^{-(n+1)})\pi(X)$. It is also obvious that $\psi(A_1 \cup \cdots$ $\cup A_{n+1}) < \sum_{i=1}^{n+1} \epsilon_i$. By finite induction, therefore, we may produce a countable sequence $\{A_n\}_{n=1}^\infty$ of pairwise disjoint sets in $\mathfrak{M}$ enjoying, for all n, the properties just described. Now write $\cup_{n=1}^\infty A_n$ as A. We have $\pi(A) > (1 - 2^{-n})\pi(X)$, for $n = 1, 2, 3, \cdots$, and hence $\pi(A) = \pi(X)$, $\pi(A') = 0$. Since ψ is countably additive, we have $\psi(A) = \sum_{n=1}^\infty \psi(A_n) < \sum_{n=1}^\infty \epsilon_n < \epsilon$.

1.20 NOTE. Theorem 1.19 may fail if $\mathfrak{M} \neq \overline{\mathfrak{M}}$. To see this, consider the following example. Let X be the half-open interval $[0, 1)$ and let $\mathfrak{M}$ be the smallest algebra of sets which contains all intervals $[\alpha, \beta) = E[t; t \in R, \alpha \leqq t < \beta]$. Let ω be Lebesgue's singular measure constructed on the Cantor set, which we denote by the symbol C (see [14, pp. 100–101]), and let λ be ordinary Lebesgue measure. One may of course define the measure ω to be countably additive on a class of sets much larger than $\mathfrak{M}$; certainly ω, with its domain restricted to $\mathfrak{M}$, is a countably additive measure assuming only values between 0 and 1.

Now, let t be any number such that $0 \leqq t \leqq 1$. We shall prove later (Theorem 4.1) that there exists a purely finitely additive measure ζ_t on the σ-algebra of all λ-measurable subsets of $[0, 1]$ such that ζ_t assumes only the values 0 and 1, $\zeta_t(N) = 0$ if $\lambda(N) = 0$, and $\zeta_t(I) = 1$ if I is any open interval containing t. Let $\{t_1, \cdots, t_n, \cdots\}$ be any dense subset of C'. (The complement is relative to $[0, 1)$.) Let $\xi = \sum_{n=1}^\infty 2^{-n}\zeta_{t_n}$. It is easy to see that ξ is a purely finitely additive measure on the σ-algebra of all λ-measurable sets and that ξ remains purely finitely additive when its domain is restricted to $\mathfrak{M}$. Now consider any set A in $\mathfrak{M}$ such that $\xi(A) = 1$. A clearly must contain every point t_n. Thus $A^- \supset \{t_1, \cdots, t_n, \cdots\}^- = C'^- = [0, 1)$. Since $A = \cup_{i=1}^k [\alpha_i, \beta_i)$, $A^- = \cup_{i=1}^k [\alpha_i, \beta_i]$, and this implies that A' is a finite set. However, the only finite set in $\mathfrak{M}$ is void. Therefore $A = [0, 1)$ and $\omega(A) = 1$. This example shows that 1.19 may fail if $\mathfrak{M} \neq \overline{\mathfrak{M}}$. We note also a generalized form of 1.19.

1.21 THEOREM. *Suppose that $\mathfrak{M} = \overline{\mathfrak{M}}$. Let ϕ and γ be elements of Φ such that $\phi \wedge \gamma = 0$. Let α and β be arbitrary positive numbers. Then there exists a set $A \in \mathfrak{M}$ such that $\phi(A) < \alpha$ and $\gamma(A') < \beta$. If ϕ is countably additive, then $\gamma(A')$ can be made zero. If ϕ and γ are both countably additive, then $\phi(A)$ and $\gamma(A')$ can both be made zero.*

The first two assertions are essentially proved in 1.11 and 1.19. The third can be obtained in much the same way as 1.19, and we omit the proof.

The hypothesis that $\mathfrak{M} = \overline{\mathfrak{M}}$ is essential in the third assertion of 1.21. Using the same X, $\mathfrak{M}$, and ω as in 1.20, consider the measure λ. Both λ and ω are countably additive on $\mathfrak{M}$, and plainly $\lambda \wedge \omega = 0$. However, one can clearly find no set $A \in \mathfrak{M}$ such that $\omega(A) = \lambda(A') = 0$.

1.22 Theorem. *Let $\mathfrak{M}=\bar{\mathfrak{M}}$. Then if π is purely finitely additive and not less than 0 and ψ is countably additive and not less than 0, there exists a decreasing sequence $B_1, B_2, \cdots, B_n, \cdots$ of elements of $\mathfrak{M}$ such that $\lim_{n\to\infty} \psi(B_n)=0$ and $\pi(B_n)=\pi(X)$ $(n=1, 2, 3, \cdots)$. Conversely, if $\phi\in\Phi$ and the above conditions hold for all countably additive ψ, then ϕ is purely finitely additive.*

According to 1.18, there exists a set $C_n\in\mathfrak{M}$ such that $\pi(C_n)=\pi(X)$ and $\psi(C_n)<1/n$ $(n=1, 2, 3, \cdots)$. Putting $B_n=\bigcap_{i=1}^{n} C_i$, we find $\psi(B_n)\leqq\psi(C_n)$ $<1/n$ and $\pi(B_n')\leqq\pi(C_1')+\cdots+\pi(C_n')=0$. Hence $\pi(B_n)=\pi(X)$. The converse is virtually obvious; if ϕ satisfies the conditions enunciated, and if ψ is countably additive, then $\phi\wedge\psi=0$; and by 1.16, ϕ is purely finitely additive.

Every $\phi\in\Phi$ can be written uniquely as the sum of a countably additive part and a purely finitely additive part. We consider several cases.

1.23 Theorem. *Let ϕ be a measure in Φ such that $0\leqq\phi$. Then there exist measures ϕ_c and ϕ_p in Φ such that $0\leqq\phi_c$, $0\leqq\phi_p$, ϕ_c is countably additive, ϕ_p is purely finitely additive, and $\phi=\phi_c+\phi_p$.*

Let Γ be the set of all countably additive measures γ in Φ such that

$$1.24.1 \qquad\qquad 0 \leqq \gamma \leqq \phi.$$

Let the real number α be defined by the relation

$$1.24.2 \qquad\qquad \alpha = \sup_{\gamma\in\Gamma} \gamma(X).$$

It is obvious that $0\leqq\alpha\leqq\phi(X)<+\infty$. Select any sequence $\{\gamma_n\}_{n=1}^{\infty}$ of elements of Γ such that $\lim_{n\to\infty} \gamma_n(X)=\alpha$. Let $\tilde{\gamma}_n$ be defined as $\gamma_1\vee\gamma_2\vee\cdots\vee\gamma_n$, for $n=1, 2, 3, \cdots$. As proved in 1.14, $\tilde{\gamma}_n$ is countably additive. Using 1.8, we extend all of the measures $\tilde{\gamma}_n$ over the family of sets $\bar{\mathfrak{M}}$, on which they are countably additive; we retain the symbol $\tilde{\gamma}_n$ to denote this extended measure. It is clear that $\tilde{\gamma}_1\leqq\tilde{\gamma}_2\leqq\cdots\leqq\tilde{\gamma}_n\leqq\cdots$ and that accordingly, for all $E\in\bar{\mathfrak{M}}$, the number $\lim_{n\to\infty} \tilde{\gamma}_n(E)$ exists and is finite. Let $\phi_c(E)=\lim_{n\to\infty} \tilde{\gamma}_n(E)$, for all $E\in\bar{\mathfrak{M}}$. It follows from a theorem of Nikodým [13], often referred to as the Vitali-Hahn-Saks theorem, that ϕ_c, with its domain restricted to $\mathfrak{M}$, is a countably additive measure in Φ. We now define ϕ_p as $\phi-\phi_c$. It is obvious that ϕ_p is a non-negative measure in Φ, and we need only show that ϕ_p is purely finitely additive. Suppose that $\psi\in\Phi$ satisfies the inequalities $0\leqq\psi\leqq\phi-\phi_c$ and is countably additive. Then $\phi_c\leqq\psi+\phi_c\leqq\phi$. If $\psi\neq0$, then $\psi(X)>0$, and accordingly $(\psi+\phi_c)(X)>\phi_c(X)=\alpha$. This is an obvious contradiction, and ϕ_p is therefore purely finitely additive.

1.24 Theorem. *Let ϕ be any measure in Φ. Then ϕ can be uniquely written as the sum of a countably additive measure ϕ_c and a purely finitely additive measure ϕ_p.*

Write ϕ as $\phi_+-\phi_-$. By 1.23, $\phi_+=\phi_c^++\phi_p^+$ and $\phi_-=\phi_c^-+\phi_p^-$. Then ϕ

$=(\phi_c^+ - \phi_c^-) + (\phi_p^+ - \phi_p^-)$. By 1.14, $\phi_c^- - \phi_c^-$ is countably additive, and by 1.17, $\phi_p^+ - \phi_p^-$ is purely finitely additive. To prove uniqueness, suppose that π_i are purely finitely additive, ψ_i are countably additive ($i = 1, 2$), and that $\pi_1 + \psi_1 = \pi_2 + \psi_2$. Then $\psi_1 - \psi_2 = \pi_2 - \pi_1$, and by 1.17, $(\pi_2 - \pi_1) \vee 0$ is purely finitely additive. Thus $(\pi_2 - \pi_1) \vee 0 = (\psi_1 - \psi_2) \vee 0$, and as $(\psi_1 - \psi_2) \vee 0$ is countably additive, we have $(\psi_1 - \psi_2) \vee 0 = 0$. Similarly, $(\psi_2 - \psi_1) \vee 0 = 0$, and it follows that $\psi_1 = \psi_2$. This implies naturally that $\pi_1 = \pi_2$ also.

1.25 NOTE. A result related to 1.24 has been announced by Woodbury [18].

2. The space $\mathfrak{L}_\infty$ and its conjugate space.

2.1 DEFINITION. Let T be any set, and let $\mathfrak{M}$ be a family of subsets of X closed under the formation of countable unions and of complements. Let $\mathcal{N}$ be a proper subfamily of $\mathfrak{M}$ closed under the formation of countable unions and having the additional property that $N \in \mathcal{N}$, $A \subset T$, and $A \subset N$ imply $A \in \mathcal{N}$. For an $\mathfrak{M}$-measurable real-valued function $x(t)$ on T, let $\|x\|_\infty$, the ess sup of $|x|$, be the infimum of the set of numbers α such that $E[t; |x(t)| < \alpha] \in \mathcal{N}$, if this set is nonvoid. Otherwise let $\|x\|_\infty = +\infty$.

2.2 THEOREM. *Let $\mathfrak{L}_\infty(T, \mathfrak{M}, \mathcal{N})$ be the space obtained from the space of all measurable real-valued functions on T with finite norm $\|\cdot\|_\infty$ by identifying two functions x and x' if $\|x - x'\|_\infty = 0$. With the usual definitions of sum, multiplication by real numbers, and norm, $\mathfrak{L}_\infty(T, \mathfrak{M}, \mathcal{N})$ is a real Banach space.*

This result is well known and is stated merely for completeness. We write $\mathfrak{L}_\infty$ for $\mathfrak{L}_\infty(T, \mathfrak{M}, \mathcal{N})$ where no ambiguity is possible.

We now generalize a theorem of Fichtenholz and Kantorovič and Hildebrandt [4] and [9], obtaining the general bounded linear functional [1, p. 27], on $\mathfrak{L}_\infty$. The set of all such functionals is denoted by $\mathfrak{L}_\infty^*$.

2.3 THEOREM. *Let F be a real-valued functional defined on $\mathfrak{L}_\infty$ such that*

2.3.1 $F(x + y) = F(x) + F(y)$ *for all $x, y \in \mathfrak{L}_\infty$,*

2.3.2 $F(\alpha x) = \alpha F(x)$ *for all $\alpha \in R$ and $x \in \mathfrak{L}_\infty$,*

2.3.3 $|F(x)| \leq A\|x\|_\infty$,   *for some $A \geq 0$ and all $x \in \mathfrak{L}_\infty$.*

Then there exists a finitely additive measure ϕ on $\mathfrak{M}$, in the sense of 1.2, such that

$$2.3.4 \qquad\qquad F(x) = \int_T x(t) d\phi(t)$$

for all $x \in \mathfrak{L}_\infty$, and such that $\phi(N) = 0$ if $N \in \mathcal{N}$. Conversely, if ϕ is a finitely additive measure on $\mathfrak{M}$, in the sense of 1.2, such that $\phi(N) = 0$ for all $N \in \mathcal{N}$, then 2.3.4 defines a bounded linear functional on $\mathfrak{L}_\infty$. For F and ϕ as in 2.3.4, we have $\|F\| = \phi_+(T) + \phi_-(T) = |\phi|(T)$.

We first note that the integral in 2.3.4 is to be interpreted just as if ϕ

were a countably additive measure and 2.3.4 were a Lebesgue integral. The standard proofs for existence, additivity, homogeneity, and positivity (for $\phi \geq 0$) can be carried out almost exactly as they are, for example, in [17, pp. 332–337]. The only change required is to replace the lower and upper bounds of $x(t)$ respectively by ess inf $x(t)$ = supremum of real numbers α such that $E[t; t \in T, x(t) < \alpha] \in \mathcal{N}$, and ess sup $x(t)$ = infimum of real numbers β such that $E[t; t \in T, x(t) > \beta] \in \mathcal{N}$.

Starting, then, with a bounded linear functional on $\mathfrak{L}_\infty$, we consider a general set $E \in \mathfrak{M}$ and the characteristic function χ_E. Plainly, $\chi_E \in \mathfrak{L}_\infty$, and $\|\chi_E\|_\infty = 0$ or 1 as $E \in \mathcal{N}$ or $E \notin \mathcal{N}$. In either case, we apply the functional F to χ_E, obtaining

$$2.3.5 \qquad\qquad F(\chi_E) = \phi(E).$$

It is obvious that $\phi(E)$ is a real number for all $E \in \mathfrak{M}$. From 2.3.1, it is clear that ϕ is finitely additive on $\mathfrak{M}$, and from 2.3.3 that $|\phi(E)| \leq A$ for all $E \in \mathfrak{M}$. Now let $x(t)$ be an arbitrary element of $\mathfrak{L}_\infty$, let a = ess inf $x(t)$, and let b = ess sup $x(t)$. Let $A_i = E[t; t \in T, n^{-1}[(n-i)a+ib] \leq x(t) < n^{-1}[(n-i-1)a +(i+1)b]]$ $(i = 0, 1, 2, \cdots, n-2)$ and let $A_{n-1} = E[t; t \in T, n^{-1}[(n-1)b+a] \leq x(t) \leq b]$. Let $P_n = \sum_{i=0}^{n-1} n^{-1}[(n-i)a+ib]\chi_{A_i}$. It is obvious that $\|P_n - x\|_\infty \leq n^{-1}(b-a)$. Hence $F(x) = \lim_{n \to \infty} F(P_n) = \lim_{n \to \infty} \sum_{i=0}^{n-1} n^{-1}[(n-i)a+ib]\phi(A_i) = \int_T x(t)d\phi(t)$. Thus F has the integral representation 2.3.4 as asserted. The converse, of course, is simple: any finitely additive ϕ of the kind described produces a functional satisfying 2.3.1–2.3.3. We now consider the norm $\|F\| = \sup_{\|x\|_\infty \leq 1} |F(x)|$. Writing the finitely additive ϕ which corresponds to F as $\phi_+ - \phi_-$ and noting that $\phi_+ \wedge \phi_- = 0$, we find for every $\epsilon > 0$ a subset A of T such that $\phi_+(A') < \epsilon$, $\phi_-(A) < \epsilon$ (see Theorem 1.21). Let $g = \chi_A - \chi_{A'}$. Plainly $g \in \mathfrak{L}_\infty$, and $\|g\|_\infty \leq 1$. (If $\phi \neq 0$, $\|g\|_\infty = 1$.) Consider

$$
\begin{aligned}
F(g) &= \int_T g(t)d\phi_+(t) - \int_T g(t)d\phi_-(t) \\
&= \int_A g(t)d\phi_+(t) + \int_{A'} g(t)d\phi_+(t) - \int_A g(t)d\phi_-(t) - \int_{A'} g(t)d\phi_-(t) \\
&= \phi_+(A) - \phi_+(A') - \phi_-(A) + \phi_-(A') > \phi_+(A) + \phi_-(A') - 2\epsilon \\
&> \phi_+(A) + \phi_-(A) + \phi_+(A') + \phi_-(A') - 4\epsilon = |\phi|(A) + |\phi|(A') - 4\epsilon \\
&= |\phi(T)| - 4\epsilon.
\end{aligned}
$$

From this reckoning, it follows that $\|F\| \geq |\phi|(T)$. On the other hand, let $x(t) \in \mathfrak{L}_\infty$ and let $\|x\|_\infty$ be ≤ 1. Then $E[t; t \in T, |x(t)| > 1] \in \mathcal{N}$, and $|\int_T x(t)d\phi(t)| = |\int_T \max[\min(x(t), 1), -1]d\phi(t)|$. It is thus no restriction to suppose that $|x(t)| \leq 1$ for all $t \in T$. Then

$$
\left| \int_T x(t)d\phi(t) \right| = \left| \int_T x(t)d\phi_+(t) - \int_T x(t)d\phi_-(t) \right|
$$

$$\leq \left| \int_T x(t) d\phi_+(t) \right| + \left| \int_T x(t) d\phi_-(t) \right|$$

$$\leq \phi_+(T) + \phi_-(T) = |\phi|(T).$$

We have therefore proved that $\|F\| = |\phi|(T)$.

Provided with the information that finitely additive measures on $\mathfrak{M}$ are identifiable with linear functions on $\mathfrak{L}_\infty$, in accordance with 2.2, we are able to prove the existence of large numbers of purely finitely additive measures on $\mathfrak{M}$. It may very well be the case, of course, that no countably additive measures other than 0 exist on $\mathfrak{M}$. (For a discussion of this matter, and further references, see [2, pp. 185–187].) In this case, every finitely additive measure on $\mathfrak{M}$ is purely finitely additive.

The Hahn-Banach theorem of course shows that there exist finitely additive measures not equal to 0 on $\mathfrak{M}$. Thus:

2.4 THEOREM. *Let $x(t)$ be any element of $\mathfrak{L}_\infty$ such that $\|x\|_\infty > 0$ and let α be any real number. Then there exists a finitely additive measure ϕ on $\mathfrak{M}$ such that $\int_T x(t) d\phi(t) = \alpha$.*

Using a result somewhat more refined than the Hahn-Banach theorem, we can prove somewhat more.

2.5 THEOREM. *Let x be an element of $\mathfrak{L}_\infty$ such that $\|x\|_\infty > 0$ and ess inf $x(t) \geq 0$. Let α be any non-negative real number. Then there exists a finitely additive measure ϕ on $\mathfrak{M}$ such that $\phi \geq 0$ and $\int_T x(t) d\phi(t) = \alpha$.*

A result of M. G. Kreĭn (see [12, Theorem 1.1]) is applied here. The conditions imposed on x imply that there exists a positive real number, say δ, such that $E_0 = E[t, t \in T, x(t) > \delta] \notin \mathcal{N}$. The space $\mathfrak{S}$ of all $\mathfrak{M}$-measurable real-valued functions of the form $y \cdot \chi_{E_0}$, where $y \in \mathfrak{L}_\infty$, obviously forms a Banach space, under the algebraic operations and norm of $\mathfrak{L}_\infty$. Let $\mathfrak{P}$ denote the set of all $z \in \mathfrak{S}$ such that ess inf $z(t) \geq 0$. It is clear that $x\chi_{E_0}$ is an interior element of the set $\mathfrak{P}$, and that $\mathfrak{P}$ is a linear semigroup in the sense of Kreĭn. The linear subspace $\{ax\chi_{E_0}\}_{a \in R}$ of $\mathfrak{S}$ has the obvious linear functional

$$f(a \cdot x \cdot \chi_{E_0}) = a \cdot \alpha.$$

According to Kreĭn's theorem, f admits an extension F over all of $\mathfrak{S}$ such that $F(p) \geq 0$ for all $p \in \mathfrak{P}$. We apply 2.3 to the space $\mathfrak{S}$ and the functional F in an obvious way, and find that $F(y) = \int_{E_0} y(t) d\phi(t)$ for a finitely additive ϕ defined on all $\mathfrak{M}$-measurable subsets of E_0. For $A \in \mathfrak{M}$ and $A \subset E_0$, we have $0 \leq F(\chi_A) = \phi(A)$, and hence ϕ is non-negative. For arbitrary $A \in \mathfrak{M}$, let $\phi(A) = \phi(A \cap E_0)$. This ϕ clearly satisfies all conditions of the present theorem.

In our further investigations (§3), we require the following result.

2.6 THEOREM. *Let ϕ be any measure in $\mathfrak{L}_\infty^*(T, \mathfrak{M}, \mathcal{N})$. Then ϕ is purely* *finitely additive as a measure on $\mathfrak{M}$ if and only if $\phi_+ \wedge \psi = \phi_- \wedge \psi = 0$ for all countably additive measures ψ in $\mathfrak{L}_\infty^*(T, \mathfrak{M}, \mathcal{N})$ which are not less than 0.*

The sufficiency being obvious, consider a measure ϕ satisfying the condition stated. Then, plainly, any finitely additive measure γ such that $0 \leq \gamma \leq \phi_+$ must vanish for all sets in $\mathcal{N}$; and the conditions stated then imply that ϕ_+ is purely finitely additive in the sense of 1.13. ϕ_- is treated similarly.

3. **The real line.** We here consider finitely additive measures on R, applying results of §§1 and 2 to this special case. All of the results of the present section can be extended by obvious devices to n-dimensional Euclidean space R^n; others admit extensions to more or less arbitrary topological spaces. To avoid technical complications, we confine ourselves to R. Throughout the present section, we denote the family of Lebesgue measurable sets in R by $\mathcal{L}$, the family of sets of Lebesgue measure zero by Z, and by λ a fixed countably additive measure on $\mathcal{L}$ such that $0 \leq \lambda(E) \leq 1$ for all $E \in \mathcal{L}$, $\lambda(R) = 1$, and such that $\lambda(E) = 0$ if and only if $E \in Z$. (For example, $\lambda(E)$ could be taken as $\pi^{-1/2} \int_E e^{-t^2} dt$.) In the present section, we denote the space $\mathfrak{L}_\infty(R, \mathcal{L}, Z)$ simply by $\mathfrak{L}_\infty(R)$. We refer to $\mathfrak{L}_\infty^*(R)$ either as a space of functionals or of finitely additive measures.

We first characterize the measures in $\mathfrak{L}_\infty^*(R)$ which are purely finitely additive.

3.1 THEOREM. *Let ϕ be any measure in $\mathfrak{L}_\infty^*(R)$. Then ϕ is purely finitely additive if and only if there exists a monotone decreasing sequence $A_1, A_2, \cdots, A_n, \cdots$ of sets in $\mathcal{L}$ such that $\lim_{n \to \infty} \lambda(A_n) = 0$ and such that $|\phi|(A_n') = 0$ $(n = 1, 2, 3, \cdots)$.*

The necessity follows from 1.22, since $\mathcal{L}$ is certainly a σ-algebra, and since λ is countably additive. Sufficiency is proved as follows. Let ψ be any finite-valued countably additive measure on $\mathcal{L}$ (not necessarily absolutely continuous with respect to λ) such that $|\phi| \geq \psi \geq 0$. Then $\psi(N) = 0$ necessarily for $N \in Z$, and ψ is absolutely continuous with respect to λ. For $P \in \mathcal{L}$, $\psi(P) = \psi(A_1' \cap P) + \sum_{n=1}^\infty \psi(P \cap A_n \cap A_{n+1}') + \psi(P \cap \bigcap_{n=1}^\infty A_n) = 0$, since $\psi \leq |\phi|$. Therefore $|\phi|$ is purely finitely additive, and as $\phi_+, \phi_- \leq |\phi|$, ϕ is purely finitely additive.

We now explore the various peculiar properties that finitely additive measures in $\mathfrak{L}_\infty^*(R)$ may exhibit.

3.2 THEOREM. *There exist nonzero measures ϕ in $\mathfrak{L}_\infty^*(R)$ such that $\int_{-\infty}^\infty x(t) d\phi(t) = 0$ for every $x \in \mathfrak{L}_\infty$ which vanishes outside of some bounded set. Such measures are necessarily purely finitely additive. The relations $\phi_+ = 0$, $\phi_- = 0$, and $\phi_+ \neq 0 \neq \phi_-$ are all possible.*

Let $\mathfrak{B}$ be the set of all $x \in \mathfrak{L}_\infty(R)$ such that for some $a > 0$ (dependent on x), $\|x(1 - \chi_{[-a,a]})\|_\infty = 0$. It is clear that $\mathfrak{B}$ is a linear subspace of $\mathfrak{L}_\infty$. Let $\tilde{\mathfrak{B}}$ be the subspace of all $x + \alpha \cdot 1$, for $x \in \mathfrak{B}$ and $\alpha \in R$. Let $f(x + \alpha \cdot 1) = \alpha$. By Kreĭn's extension theorem (see 2.4), we know that f admits a positive extension F over all of $\mathfrak{L}_\infty$. F is bounded because positive. The measure corresponding to F by 2.3, which we write as ϕ, plainly has the property that $\phi(E) = F(\chi_E) = f(\chi_E) = 0$ for all bounded sets E.

We may replace $\mathfrak{B}$ in the construction above by the subspace $\mathfrak{B}_{-\infty}$ of all functions $x \in \mathfrak{L}_\infty$ such that for some $t \in R$ (dependent upon x), $\|x(1 - \chi_{(-\infty, t]})\|_\infty = 0$ or by the subspace $\mathfrak{B}_{+\infty}$ of all $x \in \mathfrak{L}_\infty$ such that for some $t \in R$, $\|x(1 - \chi_{[t, +\infty)})\|_\infty = 0$. Doing so, we obtain positive measures $\phi_{+\infty}$ and $\phi_{-\infty}$ which vanish for all sets bounded above and below, respectively. Such measures are said to be confined to $+\infty$ or to $-\infty$, respectively. By taking appropriate linear combinations of these measures, we can produce all measures specified in the present theorem.

Finally, it is easy to see that any measure $\phi \in \mathfrak{L}_\infty^*(R)$ which vanishes for all bounded sets must be purely finitely additive. We note that ϕ_+ and ϕ_- must both vanish for all bounded sets if ϕ does so. Using A_n as $(-\infty, -n) \cup (n, +\infty)$ $(n = 1, 2, 3, \cdots)$, we apply 3.1 and see that ϕ must be purely finitely additive.

Fubini's theorem completely loses its validity for finitely additive measures.

3.3 THEOREM. *Let α and β be any two real numbers. There exists $y \in \mathfrak{L}_\infty(R)$ and positive measures ϕ_1 and $\phi_2 \in \mathfrak{L}_\infty^*$ such that*

$$3.3.1 \qquad \int_{-\infty}^{+\infty} \left[\int_{-\infty}^{+\infty} y(t + u) d\phi_1(t) \right] d\phi_2(u) = \alpha$$

and

$$3.3.2 \qquad \int_{-\infty}^{+\infty} \left[\int_{-\infty}^{+\infty} y(t + u) d\phi_2(u) \right] d\phi_1(t) = \beta.$$

From 3.2, it is plain that there are positive measures $\phi_1, \phi_2 \in \mathfrak{L}_\infty^*$ such that $\phi_1(R) = \phi_2(R) = 1$, $\phi_1((a, +\infty)) = 1$, and $\phi_2((-\infty, a)) = 1$ for all $a \in R$. Let $y(t)$ be defined as $\beta \chi_{(-\infty, 0)} + \alpha \chi_{[0, +\infty)}$. For all $u \in R$, we have $y(t + u) = \alpha$ for $t \geq -u$. Hence $\int_{-\infty}^{+\infty} y(t+u) d\phi_1(t) = \alpha$ for all $u \in R$. The equality 3.3.1 follows immediately, and 3.3.2 is established by an identical argument.

3.4 THEOREM. *There exists a nonzero finitely additive measure ζ such that*

$$3.4.1 \qquad \int_{-\infty}^{+\infty} c(t) d\zeta(t) = 0$$

for all bounded continuous functions c. Any measure satisfying 3.4.1 *cannot be positive, however, and in fact $\zeta_+(R) = \zeta_-(R)$. Furthermore, any measure ζ satisfying* 3.4.1 *must be purely finitely additive.*

We prove this result by using the Hahn-Banach theorem in its general form (see [1, pp. 27–28]). For an arbitrary but fixed $a \in R$, and $x \in \mathfrak{L}_\infty(R)$, write ess lim $\sup_{t \downarrow a} x(t)$ for the number $\inf_{\epsilon > 0}$ ess $\sup_{a < t < a + \epsilon} x(t)$, and write ess lim $\inf_{t \downarrow a} x(t)$ for $\sup_{\epsilon > 0}$ ess $\inf_{a < t < a + \epsilon} x(t)$. The numbers ess lim $\sup_{t \uparrow a} x(t)$ and ess lim $\inf_{t \uparrow a} x(t)$ are defined dually. Let $p(x) =$ ess lim $\sup_{t \downarrow a} x(t)$ $-$ess lim $\inf_{t \uparrow a} x(t)$, for all $x \in \mathfrak{L}_\infty(R)$. It is easy to verify that $p(x+y) \leq p(x)$ $+ p(y)$ for all x, $y \in \mathfrak{L}_\infty(R)$ and that $p(\alpha x) = \alpha p(x)$ for $\alpha \geq 0$. Consider now the subspace $\mathfrak{C}(R)$ of $\mathfrak{L}_\infty(R)$ consisting of all bounded continuous functions. It is clear that $p(c) = 0$ for all $c \in \mathfrak{C}$; thus the linear functional $Z(c) = 0$ has the property that $Z(c) \leq p(c)$ for $c \in \mathfrak{C}$. By the Hahn-Banach theorem, Z admits a linear extension Z over $\mathfrak{L}_\infty$ such that $-p(-x) \leq Z(x) \leq p(x)$ for all $x \in \mathfrak{L}_\infty$. Since $|p(x)| \leq 2 \cdot \|x\|_\infty$ for all $x \in \mathfrak{L}_\infty$, it follows that $Z \in \mathfrak{L}_\infty^*(R)$, and that $Z(x) = p(x)$ whenever $p(x) = -p(-x)$. For example $p(\chi_{[a, +\infty)}) = 1$, $p(-\chi_{[a, +\infty)})$ $= -1$, so that $Z(\chi_{[a, +\infty)}) = 1$. If $c(t) \in \mathfrak{C}(R)$, then it is obvious that $p(c) = 0$. Therefore $Z(c) = 0$. If we write ζ for the finitely additive measure associated with Z by 2.3, we obviously have $\zeta((a, b)) = \zeta((a, +\infty)) = 1$ for all $b > a$, $\zeta((b, a)) = ((-\infty, a)) = -1$ for all $b < a$, and $\zeta((b, c)) = 0$ for $b < a < c$.

It is obvious that $\zeta(R) = \int_{-\infty}^\infty 1 d\zeta(t) = 0$ for any ζ satisfying 3.4.1, and hence $\zeta_+(R) = \zeta_-(R)$. The fact that any ζ satisfying 3.4.1 is of necessity purely finitely additive may be established by a straightforward argument; we omit the proof.

We obtain even more bizarre measures by selecting different $p(x)$. For example:

3.5 THEOREM. *Let $\alpha \in R$. There exists a finitely additive measure ϕ such that*

$$3.5.1 \qquad \int_{-\infty}^{+\infty} x(t) d\phi(t) = 0$$

for all $x \in \mathfrak{L}_\infty$ such that ess $\lim_{t \downarrow a} x(t)$ exists. Such a measure ϕ cannot be positive and must be purely finitely additive.

This is obtained by taking $p(x) =$ ess lim $\sup_{t \downarrow a} x(t) -$ ess lim $\inf_{t \downarrow a} x(t)$. The proof of 3.4 is then repeated *verbatim*.

For purposes of later application, we prove the existence of a measure in $\mathfrak{L}_\infty^*(R)$ having a certain unit property.

3.6 THEOREM. *Let a be any real number. There exists a measure $\epsilon_a \in \mathfrak{L}_\infty^*(R)$ such that*

$$3.6.1 \qquad \int_{-\infty}^{+\infty} x(t + u) d\epsilon_a(u) = x(t + a)$$

for all $x \in \mathfrak{L}_\infty(R)$.

We again use a transfinite device, which appears, indeed, to be an essential adjunct of the present topic. Here we consider the nonlinear functional

$$3.6.2 \qquad d(x) = \text{ess} \lim_{t \downarrow 0} \sup \frac{1}{t} \int_a^{a+t} x(u)du,$$

where the integral is the ordinary Lebesgue integral and x is a generic element of $\mathfrak{L}_\infty(R)$. It is easy to see that $d(tx) = td(x)$ for all $t \geq 0$ and that $p(x+y) \leq p(x) + p(y)$. Furthermore,

$$3.6.3 \qquad -d(-x) = \text{ess} \lim_{t \downarrow 0} \inf \frac{1}{t} \int_a^{a+t} x(u)du.$$

Finally, $|d(x)| \leq \|x\|_\infty$. Let $\mathfrak{F}$ be the linear subspace of $\mathfrak{L}_\infty(R)$ consisting of all x such that $d(x) = -d(-x)$. On $\mathfrak{F}$, d is a bounded linear functional. By the Hahn-Banach theorem, d admits an extension D over $\mathfrak{L}_\infty(R)$ such that $-d(-x) \leq D(x) \leq d(x)$ for all $x \in \mathfrak{L}_\infty(R)$. Consider any function $x \in \mathfrak{L}_\infty(R)$, and let v be any fixed real number. Then $D(x(v+u)) =$

$$3.6.4 \qquad \text{ess} \lim_{t \downarrow 0} \frac{1}{t} \int_a^{a+t} x(v+u)du$$

if this limit exists. 3.6.4 may be re-written as

$$3.6.5 \qquad \text{ess} \lim_{t \downarrow 0.} \frac{1}{t} \int_{a+v}^{a+v+t} x(u)du;$$

a celebrated theorem of Lebesgue (see for example [17, p. 362]) asserts that 3.6.5 exists for almost all $a+v$ and equals $x(a+v)$. Thus, a being fixed, the limit 3.6.4 exists for almost all v and equals $x(a+v)$. Therefore, in the sense of $\mathfrak{L}_\infty$, $D(x(v+u)) = x(a+v)$. If we write the measure associated with D as ϵ_a, the relation 3.6.1 follows at once.

4. Representation theory for $\mathfrak{L}_\infty$. We now return to the case of a general $\mathfrak{L}_\infty(X, \mathfrak{M}, \mathcal{N})$ as treated in §2, finding that the finitely additive measures considered in 2.3 can be regarded as countably additive Borel measures defined on a certain compact Hausdorff space. The countably additive and purely finitely additive parts of each of the former measures are thereupon characterized in terms of their countably additive counterparts.

4.1 THEOREM. *Let $\mathcal{E}_0$ be any subfamily of $\mathfrak{M}$ such that for every finite family $E_1, E_2, \cdots, E_m$ of sets in $\mathcal{E}_0$, the set $\bigcap_{i=1}^n E_i \notin \mathcal{N}$. Then there exists a measure ω in $\mathfrak{L}_\infty^*$ such that:*

    4.1.1 *ω assumes only the values 0 and 1;*

    4.1.2 *$\omega(E) = 1$ for all $E \in \mathcal{E}_0$.*

Let P be the collection of all subfamilies $\mathcal{A}$ of $\mathfrak{M}$ such that:

    4.1.3 $\mathcal{A} \cap \mathcal{N} = 0$;

    4.1.4 $\mathcal{E}_0 \subset \mathcal{A}$;

    4.1.5 $A, B \in \mathcal{A}$ imply that $A \cap B \in \mathcal{A}$;

4.1.6 $A \in \mathcal{A}$ and $C \in \mathcal{M}$ and $C \supset A$ imply $C \in \mathcal{A}$.

There is obviously some family in P: namely, the family of all $A \in \mathcal{M}$ such that $A \supset E$ for some $E \in \mathcal{E}_0$. It is obvious that if Q is a subcollection of P completely ordered by set-inclusion, then $\cup_{\mathcal{A} \in Q} \mathcal{A} \in P$. Hence, by Zorn's Lemma, P contains a maximal element $\mathcal{A}_0$. We note that if B_1 and B_2 are elements of $\mathcal{M}$ such that $B_1 \cap B_2 \in \mathcal{N}$, then at most one of B_1 and B_2 is in $\mathcal{A}_0$. This follows at once from 4.1.5 and 4.1.3. Next, let B be any set in $\mathcal{M}$. If $B \notin \mathcal{A}_0$, a simple argument based on the maximality of $\mathcal{A}_0$ shows that $B \cap A_0 \in \mathcal{N}$ for some $A_0 \in \mathcal{A}_0$. One sees similarly that $N \in \mathcal{N}$ implies $N' \in \mathcal{A}_0$. Hence $B' \cup A_0' \in \mathcal{A}_0$; this implies that $B' \cap A_0 \in \mathcal{A}_0$ and hence that $B' \in \mathcal{A}_0$. Therefore, exactly one of the sets B and B' is in $\mathcal{A}_0$. Now, we write

$$4.1.7 \qquad\qquad \omega(B) = \begin{cases} 1 & \text{if } B \in \mathcal{A}_0, \\ 0 & \text{if } B \notin \mathcal{A}_0, \end{cases}$$

and obtain in this fashion a measure satisfying 4.1.1 and 4.1.2.

4.2. THEOREM. *Let Ω be the space of all measures in $\mathfrak{L}_\infty^*$ which assume only the values 0 and 1. For $E \in \mathcal{M}$, let Δ_E consist of all $\omega \in \Omega$ such that $\omega(E) = 1$. Let Δ_E be assigned as a neighborhood of every $\omega \in \Delta_E$. Under this definition of neighborhoods, Ω is a compact Hausdorff space.*

The usual neighborhood axioms are obviously satisfied; we note that $\Delta_{E_1} \cap \Delta_{E_2} = \Delta_{E_1 \cap E_2}$. If $\omega_1 \neq \omega_2$, then there exists an $E \in \mathcal{M}$ such that $\omega_1(E) = 1$ and $\omega_2(E) = 0$. It is plain that $\omega_1 \in \Delta_E$, $\omega_2 \in \Delta_{E'}$, and that $\Delta_E \cap \Delta_{E'} = 0$. Hence Ω is a Hausdorff space. Compactness is easily established. Let $\{\Delta_E\}_{E \in \mathcal{A}}$ be any covering of Ω by sets Δ_E. We note the equality $\Delta_{E'} = (\Delta_E)'$. Thus $\{\Delta_{E'}\}$ is a family of closed subsets of Ω with total intersection void. If $\{\Delta_E\}$ admits no finite subcovering, then $\{\Delta_{E'}\}$ admits no finite subfamily with total intersection void. That is, for every $E_1, \cdots, E_n \in \mathcal{A}$, there exists a measure $\omega \in \Omega$ such that $\omega(E_i') = 1$ $(i = 1, \cdots, m)$. Hence $\cap_{i=1}^n E_i' \notin \mathcal{N}$, and it follows that there exists some $\omega_0 \in \Omega$ such that $\omega_0(E') = 1$ for all $E \in \mathcal{A}$ (see 4.1). This evident contradiction proves the present theorem.

4.3 THEOREM. *Let $x \in \mathfrak{L}_\infty$. Then, writing*

$$4.3.1 \qquad\qquad \bar{x}(\omega) = \int_T x(t) d\omega(t),$$

we represent x as a real-valued function on Ω. Every such function $\bar{x}$ is continuous on Ω; and conversely, every continuous real function on Ω is obtainable as a function $\bar{x}$ for some $x \in \mathfrak{L}_\infty$. The correspondence $x \rightarrow \bar{x}$ has the properties that

$$4.3.2 \qquad\qquad \overline{(\alpha x + \beta y)} = \alpha \bar{x} + \beta \bar{y},$$

$$4.3.3 \qquad\qquad \overline{xy} = \bar{x}\bar{y},$$

4.3.4
$$\|x\|_\infty = \sup_{\omega \in \Omega} |\bar{x}(\omega)|$$

Hence $\mathfrak{L}_\infty$ can be identified with the algebra $\mathfrak{C}(\Omega, R)$ of all real-valued continuous functions on Ω.

We first show that $\bar{x}$, defined by 4.3.1, is continuous on Ω. Let ω_0 be a fixed element of Ω. In the topology induced by the norm, the map $x \rightarrow \int_T x(t) d\omega_0(t)$ is obviously continuous. Let $\alpha = \text{ess inf}_{t \in T} \, x(t)$, $\beta = \text{ess sup}_{t \in T} \, x(t)$, and let A_i be defined as $E[t; t \in T, 2^{-n}(i\beta + (2^n - i)\alpha) \leq x(t) < 2^{-n}((i+1)\beta + (2^n - i - 1)\alpha)]$ for $i = 0, 1, 2, \cdots, 2^n - 2$, and let $A_{2^n - 1}$ be $E[t; t \in T, \beta - 2^{-n}(\beta - \alpha) \leq x(t) \leq \beta]$. Let $s_n(t) = \sum_{i=0}^{2^n - 1} 2^{-n}(i\beta + (2^n - i)\alpha)\chi_{A_i}$. Plainly $\|x - s_n\|_\infty \leq (\beta - \alpha)2^{-n}$, and since $\|\omega_0\| = 1$, considered as a linear functional, it follows that

$$\left| \int_T x(t) d\omega_0(t) - \int_T s_n(t) d\omega_0(t) \right| \leq (\beta - \alpha)2^{-n}.$$

However, $\omega_0(A_i) = 1$ for exactly one value of i. For every positive integer n, let B_n be the set A_i such that $\omega_0(A_i) = 1$, and let C_n be the number $2^{-n}(i\beta + (2^n - i)\alpha)$. It is clear that $B_1 \supset B_2 \supset \cdots \supset B_n \supset \cdots$, that $C_n \rightarrow$ some limit C, and that $C = \int_T x(t) d\omega_0(t)$. Now, if ω is any element of Ω which is in Δ_{B_n}, we have $\omega(B_n) = 1$, and $\int_T s_n(t) d\omega(t) = C_n$, and hence $|\bar{x}(\omega) - \bar{x}(\omega_0)| \leq (\beta - \alpha)2^{-n}$. This shows that $\bar{x}$ is continuous.

Property 4.3.2 is obvious. To prove 4.3.3, we must prove

4.3.5
$$\int_T x(t) y(t) d\omega(t) = \int_T x(t) d\omega(t) \int_T y(t) d\omega(t).$$

For any sets $E, F \in \mathfrak{M}$, we have

$$\int_T \chi_E(t)\chi_F(t) d\omega(t) = \int_T \chi_{E \cap F}(t) d\omega(t) = \omega(E \cap F) = \omega(E) \cdot \omega(F)$$

$$= \int_T \chi_E(t) d\omega(t) \cdot \int_T \chi_F(t) d\omega(t).$$

For any function

4.3.6
$$x = \sum_{i=1}^n \alpha_i \chi_{E_i} \qquad (\alpha_i \in R, \, E_i \in \mathfrak{M}),$$

we have

$$\int_T x^2(t) d\omega(t) = \sum_{i,j=1}^n \alpha_i \alpha_j \omega(E_i)\omega(E_j) = \left(\int_T x(t) d\omega(t) \right)^2.$$

Since functions of the form 4.3.6 are dense in $\mathfrak{L}_\infty$, we verify 4.3.5 by continuity for the case $x=y$. Therefore, for arbitrary x and y,

$$4.3.7 \qquad \int_T [x(t) + y(t)]^2 d\omega(t) = \left[\int_T x(t) d\omega(t)\right]^2 + 2\int_T x(t)y(t)d\omega(t)$$
$$+ \left[\int_T y(t)d\omega(t)\right]^2;$$

also

$$4.3.8 \qquad \int_T [x(t) + y(t)]^2 d\omega(t) = \left[\int_T [x(t) + y(t)]d\omega(t)\right]^2$$
$$= \left[\int_T x(t) d\omega(t)\right]^2 + 2\int_T x(t)d\omega(t)\int_T y(t)d\omega(t)$$
$$+ \left[\int_T y(t)d\omega(t)\right]^2.$$

Comparing 4.3.7 and 4.3.8, we have 4.3.5.

We now prove 4.3.4. Let x be any nonzero element of $\mathfrak{L}_\alpha$. Write α $= \text{ess inf}_{t\in T}\, x(t)$, $\beta = \text{ess sup}_{t\in T}\, x(t)$, and define a as β if $|\beta| \geq |\alpha|$ and as α if $|\beta| < |\alpha|$. Then for every $\epsilon > 0$, the set $D_\epsilon = E[t, t\in T, |x(t) - a| < \epsilon]$ is not in $\mathcal{N}$. By 4.1, there exists a measure $\omega \in \Omega$ such that $\omega(D_\epsilon) = 1$, and for such an ω, it is clear that $|\int_T x(t)d\omega(t) - a| < \epsilon$. 4.3.4 follows at once, since $\|x\|_\infty$ $= \max\,(|\alpha|, |\beta|)$.

To show that every $f \in \mathfrak{C}(\Omega, R)$ is representable as some $\bar{x}$, we use the Stone-Weierstrass approximation theorem [7]. For $\omega_1 \neq \omega_2$, consider any set $E \in \mathcal{M}$ such that $\omega_1(E) = 0$, $\omega_2(E) = 1$. Then, setting $x(t) = \chi_E(t)$, we have $\bar{x}(\omega_1) = 0$, $\bar{x}(\omega_2) = 1$. The collection of functions $\bar{x}$ on Ω is therefore a dense subring of $\mathfrak{C}(\Omega, R)$, and being uniformly closed, in view of 4.3.4, it coincides with $\mathfrak{C}(\Omega, R)$. This observation completes the present proof.

4.4 THEOREM. *Let F be any functional in $\mathfrak{L}_\infty^*$ such that $F(xy) = F(x)F(y)$ for all x, $y \in \mathfrak{L}_\infty$. Then the measure corresponding to F under 2.3 is a measure ω in Ω.*

Any multiplicative linear functional on $\mathfrak{C}(\Omega, R)$ has the form $f\rightarrow f(\omega_0)$ for some $\omega_0 \in \Omega$, by a well known theorem. Since $\mathfrak{L}_\infty$ is identifiable with $\mathfrak{C}(\Omega, R)$, as 4.3 shows, we infer that $F(x)$ has the form $x\rightarrow \int_T x(t)d\omega_0(t)$ for the same ω_0. (This fact is also apparent from examining $F(\chi_E)$ for $E\in\mathcal{M}$.)

4.5 Having established the representation theorem 4.3, we now apply a theorem of Kakutani [10] to characterize all bounded linear functionals f on $\mathfrak{C}(\Omega, R) = \mathfrak{L}_\infty$. The theorem in question states that for every such f, there exists a countably additive, finite-valued, regular, Borel measure ϕ on Ω such that

4.5.1
$$f(\tilde{x}) = \int_\Omega \tilde{x}(\omega) d\overline{\phi}(\omega).$$

The set of all such measures is denoted by $\mathfrak{C}^*(\Omega)$. The converse assertion, that every such ϕ produces a bounded linear functional on $\mathfrak{C}(\Omega, R)$, is obvious. Thus the space $\mathfrak{L}_\infty^*$ has two distinct representations: first, as the space of all finitely additive measures on $\mathfrak{M}$; second, as the space of all countably additive, regular, Borel measures on Ω. Therefore we are in possession of a one-to-one mapping carrying all of our finitely additive $\mathfrak{M}$-measures on T into all countably additive regular Borel measures on Ω. We denote the image of $\gamma \in \mathfrak{L}_\infty^*$ by $\overline{\gamma}$. It is obvious that something must happen to change mere finite additivity on T into countable additivity on Ω. We proceed to study this phenomenon. First, we note a simple consequence of the representation 4.5.1.

4.6 THEOREM. *Every measure γ in $\mathfrak{L}_\infty^*$ is the weak limit of linear combinations of 0–1 measures, in the sense that for every $\epsilon > 0$ and $x_1, \cdots, x_m \in \mathfrak{L}_\infty$, there exists a measure $\delta = \sum_{i=1}^k \alpha_i \omega_i$ such that*

4.6.1
$$\left| \int_T x_j(t) d\gamma(t) - \int_T x_j(t) d\delta(t) \right| < \epsilon, \qquad j = 1, 2, \cdots, m.$$

This is proved by transferring the corresponding theorem for $\mathfrak{C}^*(\Omega)$, proved by Kakutani [10], back to the space $\mathfrak{L}_\infty$.

4.7 Let A be any set in $\mathfrak{M}$. Let us denote by $\overline{A}$ the set Δ_A contained in Ω. Plainly $\overline{A}$ is an open-and-closed set, and it is easy to show that the correspondence $A \to \overline{A}$ maps $\mathfrak{M}$ in a one-to-one fashion into the family $\overline{\mathfrak{M}}$ of all open-and-closed subsets of Ω, with preservation of finite unions, intersections, and complements. Furthermore, it is obvious that $\overline{\chi_A} = \chi_{\overline{A}}$. From these facts we infer, for an arbitrary $\gamma \in \mathfrak{L}_\infty^*$, the equalities

$$\gamma(A) = \int_T \chi_A(t) d\gamma(t) = \int_\Omega \overline{\chi_A}(\omega) d\overline{\gamma}(\omega) = \int_\Omega \chi_{\overline{A}}(\omega) d\overline{\gamma}(\omega) = \overline{\gamma}(\overline{A}).$$

Hence the measures γ and $\overline{\gamma}$ are duplicates of each other on the families of sets $\mathfrak{M}$ and $\overline{\mathfrak{M}}$ respectively. The difference in behavior arises, naturally, from the facts that the correspondence between $\mathfrak{M}$ and $\overline{\mathfrak{M}}$ need not preserve countably infinite set operations, and that $\overline{\gamma}$ is defined on a family of sets which may properly contain $\overline{\mathfrak{M}}$.

4.8 THEOREM. *The measure $\overline{\omega}$ on Ω is the measure such that $\overline{\omega}(\Gamma) = 1$ or 0 as $\omega \in \Gamma$ or $\omega \notin \Gamma$, for all Borel sets Γ. Conversely, any Borel measure μ on Ω assuming the values 0 and 1 and no other values is some $\overline{\omega}$.*

For Γ an open-and-closed set in Ω, that is, $\Gamma = \overline{A}$ for some $A \in \mathfrak{M}$, it is obvious that $\overline{\omega}(\overline{A}) = \omega(A)$ and $\overline{\omega}(\overline{A}) = 1$ or 0 as $\omega \in \Delta_A$ or $\omega \in \Delta_{A'}$. Suppose that Γ is any closed set in Ω, and that $\omega \in \Gamma$. Since $\overline{\omega}$ is regular, there exists an open set P such that $\Gamma \subset P$ and $\bar{\omega}(P) - \bar{\omega}(\Gamma)$ is arbitrarily small. It is clear that P may be taken as open-and-closed, and therefore $\bar{\omega}(\Gamma) = 1$. If Γ is open and $\omega \in \Gamma$, then there exists an open-and-closed set Σ such that $\omega \in \Sigma \subset \Gamma$ and $\bar{\omega}(\Sigma)$ is arbitrarily close to $\bar{\omega}(\Gamma)$. This implies that $\bar{\omega}(\Gamma) = 1$. Now suppose that Γ is an arbitrary Borel set. If $\omega \in \Gamma$, then, using again the regularity of $\bar{\omega}$, we have $\bar{\omega}(\Gamma) = \sup \bar{\omega}(P)$ taken over all closed $P \subset \Gamma$. This number is clearly 1. If $\omega \in \Gamma'$, then $\bar{\omega}(\Gamma') = 1$ and $\bar{\omega}(\Gamma) = 0$.

To prove the converse, we note that a measure μ of the type described must be regular. Hence $\mu = \bar{\gamma}$ for some $\gamma \in \mathfrak{L}_{\infty}^{*}$; γ therefore assumes only the values 0 and 1 and is thus a measure ω. This brings us back to the first assertion of the theorem.

4.9 DEFINITION. A set $A \in \mathfrak{M} \cap \mathcal{N}'$ is said to be atomic if $B \in \mathfrak{M}$, $B \subset A$, and $B' \cap A \notin \mathcal{N}$ imply $B \in \mathcal{N}$.

4.10 THEOREM. *A point $\omega \in \Omega$ is isolated if and only if the family $E[M; M \in \mathfrak{M}, \omega(M) = 1]$ contains an atomic element. Any such ω is countably additive.*

This theorem is easy to prove from 4.9 and 4.2; accordingly, we omit the proof. A curious result concerning G_δ's in Ω is the following.

4.11 THEOREM. *Let Γ be a non-open closed G_δ in Ω. Then there exists a sequence $A_1 \supset A_2 \supset \cdots \supset A_n \supset \cdots$ of sets in $\mathfrak{M}$ where $A_n \cap A'_{n+1} \notin \mathcal{N}$ $(n = 1, 2, 3, \cdots)$ such that Γ consists of all $\omega \in \Omega$ with $\omega(A_n) = 1$ for all n. The interior of Γ consists of all ω such that $\omega(\cap_{n=1}^{\infty} A_n) = 1$. The boundary of Γ contains at least $2^{2^{\aleph_0}}$ elements. Conversely, for any sequence $\{A_n\}_{n=1}^{\infty}$ of sets as described above, the set of all $\omega \in \Omega$ such that $\omega(A_n) = 1$ for all n is a non-open closed G_δ in Ω.*

Since Ω is a normal space, there exists a continuous function $\bar{x}$ such that $\bar{x}(\omega) = 0$ for all $\omega \in \Gamma$ and $\bar{x}(\omega) > 0$ for all $\omega \in \Gamma'$ (this is easily proved for any closed G_δ in any normal space). Let $A_n = E[t; t \in T, x(t) \leq n^{-1}]$ $(n = 1, 2, 3, \cdots)$. On account of the properties of the ω-integral set forth in 4.3, we see that $\omega(A_n) = 1$ if $\omega \in \Gamma$ $(n = 1, 2, 3, \cdots)$ and $\omega(A_n) = 0$ for some positive integer n if $\omega \in \Gamma'$. Thus Γ is characterized as the set of all measures in Ω such that $\omega(A_n) = 1$ for all n. If all but a finite number of the A_n are equal, up to sets in $\mathcal{N}$, then Γ is plainly the open-and-closed subset of Ω consisting of all ω such that $\omega(\cap_{n=1}^{\infty} A_n) = 1$. Therefore, we incur no curtailment of generality by supposing that $B_n = A_n \cap A'_{n+1} \notin \mathcal{N}$ $(n = 1, 2, 3, \cdots)$. Next, writing A_0 for $\cap_{n=1}^{\infty} A_n$, we clearly have $\Delta_{A_0} \subset \Gamma'^{-1}$, since Δ_{A_0} is an open-and-closed set contained in Γ. The other elements of Γ are the measures ω such that $\omega(A_0) = 0$ and $\omega(A_n) = 1$ $(n = 1, 2, 3, \cdots)$. An argument based on the Čech-Stone β [8] for the integers shows that there are at least $2^{2^{\aleph_0}}$ distinct measures ω such that $\omega(A_0) = 0$ and $\omega(\cup_{k=1}^{\infty} B_{n_k}) = 1$ for some sequence $n_1 < n_2 < n_3 < \cdots < n_k < \cdots$ of positive integers. Let ω be any of these measures and let C be a set in $\mathfrak{M}$ such that $\omega(C) = 1$. Then for some positive integer n_0, $C \cap A'_{n_0} \notin \mathcal{N}$. For, if $C \cap A'_n \in \mathcal{N}$ for all n, then $\bigcup_{n=1}^{\infty}(C \cap A'_n) = C \cap \bigcup_{n=1}^{\infty} A'_n = C \cap (\bigcap_{n=1}^{\infty} A_n)' = C \cap A'_0 \in \mathcal{N}$, and $\omega(C) = \omega(C \cap A'_0) = 1$ would be impossible. There is by 4.1 a measure δ such that $\delta(C \cap A'_{n_0}) = 1$. It is plain that $\delta \in \Delta_C$, that $\delta(A'_{n_0}) = 1$, and that $\delta \notin \Gamma$. Hence $\omega \notin \Gamma'^{-1}$. This proves the first part of the present theorem. The converse is now obvious.

4.12 COROLLARY. *The interior of a closed G_δ is closed.*

4.13 COROLLARY. *A point ω of Ω is a G_δ if and only if it is isolated.*

Using 4.11, we obtain a characterization of countably additive measures in $\mathfrak{L}_\infty^*$ as follows. (Of course, there may be no such measures $\neq 0$.)

4.14 THEOREM. *A measure ϕ in $\mathfrak{L}_\infty^*$ is countably additive if and only if its counterpart $\bar{\phi}$ in $\mathfrak{C}^*(\Omega)$ vanishes for every nowhere-dense closed G_δ in Ω.*

Suppose that ϕ is countably additive. Then, if $A_1 \supset A_2 \supset \cdots \supset A_n \supset \cdots$ and $\bigcap_{n=1}^{\infty} A_n \in \mathcal{N}$, we have $\lim_{n \to \infty} \phi(A_n) = 0$. For Γ any nowhere-dense G_δ in Ω, there exist $A_1, A_2, \cdots$ as above such that $\Gamma = \bigcap_{n=1}^{\infty} \Delta_{A_n}$, as shown in 4.11. Clearly $\bar{\phi}(\Gamma) = \lim_{n \to \infty} \bar{\phi}(\Delta_{A_n}) = \lim_{n \to \infty} \phi(A_n) = 0$. The converse is proved in the same way.

A description of purely finitely additive measures analogous to **4.14** can be obtained in a certain special case.

4.15 THEOREM. *Suppose that $\mathfrak{L}_\infty^*$ contains a countably additive measure λ such that $\lambda(A) = 0$ if and only if $A \in \mathcal{N}$. Then, if $\phi \in \mathfrak{L}_\infty^*$ and ϕ is purely finitely additive, there exists a nowhere-dense closed G_δ, Π, in Ω such that $\bar{\phi}(\mathbf{B}) = 0$ for every Borel set $\mathbf{B} \subset \Pi'$.*

To obtain this result, we use 1.22. Considering $|\phi|$ and $|\lambda|$ instead of ϕ and λ, we may suppose $\phi \geq 0$, $\lambda \geq 0$. Let ϕ be purely finitely additive. Let $\{B_n\}_{n=1}^{\infty}$ be as described in 1.22. Then $\lambda(\bigcap_{n=1}^{\infty} B_n) = \lim_{n \to \infty} \lambda(B_n) = 0$, and hence $C = \bigcap_{n=1}^{\infty} B_n \in \mathcal{N}$. This shows that $\phi(B_n \cap C') = \phi(B_n) = \phi(T)$ $(n = 1, 2, 3, \cdots)$ and hence $\bar{\phi}(\Delta_{B_n}) = \bar{\phi}(\Omega)$ $(n = 1, 2, 3, \cdots)$. Since $\bar{\phi}$ is countably additive, $\bar{\phi}(\bigcap_{n=1}^{\infty} \Delta_{B_n}) = \bar{\phi}(\Omega)$. Since $C \in \mathcal{N}$, $\bigcap_{n=1}^{\infty} \Delta_{B_n}$ is a nowhere-dense closed G_δ (see 4.11) which can be used as Π in the present theorem.

4.16 THEOREM. *For a general $\mathfrak{M}$, $\mathcal{N}$, and T, any measure ϕ such that $\bar{\phi}$ is confined to a closed nowhere dense G_δ is purely finitely additive.*

We omit the proof of this result.

The question arises as to extending 4.15 to the case where $\mathfrak{M}$ does not admit a countably additive measure vanishing exactly on $\mathcal{N}$. It may be conjectured, for example, that if $\mathfrak{L}_\infty^*$ contains no countably additive measures at all, then every counterpart measure $\bar{\phi}$ is confined to a closed nowhere dense G_δ in Ω. We have not succeeded in deciding this.

BIBLIOGRAPHY

1. Stefan Banach, *Théorie des opérations linéaires*, Monografje Matematyczne, Warsaw, 1932.

2. Garrett Birkhoff, *Lattice theory*, Rev. ed., Amer. Math. Soc. Colloquium Publications, vol. 25, New York, 1948.

3. S. Bochner, and R. S. Phillips, *Additive set functions and vector lattices*, Ann. of Math. (2) vol. 42 (1941) pp. 316–324.

4. G. Fichtenholz and L. Kantorovič, *Sur les opérations linéaires dans l'espace des fonctions bornées*, Studia Math. vol. 5 (1934) pp. 69–98.

5. Hans Hahn and Arthur Rosenthal, *Set functions*, Albuquerque, University of New Mexico Press, 1948.

6. Paul R. Halmos, *Measure theory*, New York, Van Nostrand, 1950.

7. Edwin Hewitt, *Certain generalizations of the Weierstrass approximation theorem*, Duke Math. J. vol. 14 (1947) pp. 419–427.

8. ———, *Rings of real-valued continuous functions*. I, Trans. Amer. Math. Soc. vol. 64 (1948) pp. 45–99.

9. T. H. Hildebrandt, *On bounded functional operations*, Trans. Amer. Math. Soc. vol. 36 (1934) pp. 868–875.

10. Shizuo Kakutani, *Concrete representations of abstract M-spaces*, Ann. of Math. vol. 42 (1941) pp. 994–1024.

11. L. V. Kantorovič, B. Z. Vulih, and A. G. Pinsker, *Funktsional'niĭ analiz v poluuporyadočennih prostranstvah*, Moscow, Gosudarstvennoe Izdatel'stvo Tehniko-Teoretičeskoi Literatury, 1950.

12. M. G. Kreĭn and M. A. Rutman, *Lineĭnye operatory, ostavlayuŝčie invariantnym konus v prostranstve Banaha*, Uspehi Matematičeskih Nauk (3) vol. 1 (1948) pp. 3–95.

13. O. M. Nikodým, *Sur les suites convergentes de fonctions parfaitement additives d'ensemble abstrait*, Monatshefte für Mathematik und Physik vol 40 (1933) pp. 427–432.

14. Stanisław Saks, *Theory of the integral*, Monografje Matematyczne, Warsaw, 1937.

15. Alfred Tarski, *Ideale in vollständigen Mengenkörpern*, I, Fund. Math. vol. 32 (1938) pp. 45–63.

16. ———, *Ideale in vollständigen Mengenkörpern*, II, Fund. Math. vol. 33 (1945) pp. 51–65.

17. E. C. Titchmarsh, *The theory of functions*, Oxford University Press, 1939.

18. M. A. Woodbury, Bull. Amer. Math. Soc. Abstract 56-2-167.

19. Kôsaku Yosida, *Vector lattices and additive set functions*, Proc. Imp. Acad. Tokyo vol. 17 (1941) pp. 228–232.

NAGOYA UNIVERSITY,
   NAGOYA, JAPAN.
THE UNIVERSITY OF WASHINGTON
   SEATTLE, WASH.

On the reflexivity of the space of distribution

Sci. Papers Coll. Gen. Ed. Univ. Tokyo 7 (1957) 151–155

(Received November 2, 1957)

§ 0. It is about ten years since L. Schwartz generalized the notion of functions by that of distributions. The theory of distributions gives not only a clear-cut explanation of the δ-function technique of the physicist Dirac, but it is also very powerful in treating and extending the theory of partial differential equations and Fourier analysis. It is desirable that the theory should be appreciated by physicists and engineers. Schwartz himself has published two volumes, "Théorie des distributions, I et II, Paris (1950)", and these seem, at present, the only treatise of the theory in book form. His exposition is extremely fine. However, some physicists, friends of mine, complain that they feel certain difficulties in reading the beginning chapters of Schwartz book. For they encounter, in connection with the proof of the reflexivity of the space of distributions, with the notion of linear topological spaces and Mackey-Arens theorem without proof.

The physicists naturally dislike topological spaces, they prefer the notion of convergence. They are familiar with the theory of Hilbert space and Fourier series. I have devised another proof of the reflexivity which makes use of the convergence and of Hilbert space technique. I hope the proof be accepted by the physicists.

§ 1. To begin with the proof of the reflexivity, I shall recall some of the basic notions and results from the theory of distributions.

Let $\mathfrak{E}_{R^n}$ be the linear space of the totality of infinitely differentiable functions defined on an n-dimentional euclidean space R^n, and let $\mathfrak{D}_{R^n}$ be its subspace consisting of the totality of functions with *compact supports*, i.e., those functions vanishing outside of some bounded subsets of R^n. The notion of the the convergence in $\mathfrak{E}_{R^n}$, denoted by

$$\varphi_h \Rightarrow 0(\mathfrak{E}_{R^n}) \quad \text{as} \quad h \to \infty ,$$

---

1) The invited address delivered at the 1956 December Meeting of the Indian Mathematical Society, held at Baroda.

means that $\varphi_h(x)$, $x=(x_1, \ldots\ldots, x_n) \in R^n$, and all of its derivatives converge to 0 uniformly on every bounded subset of R^n. The covergence in $\mathfrak{D}_{R^n}$, denoted by

$$\varphi_h \Rightarrow 0 \ (\mathfrak{D}_{R^n}) \quad \text{as} \quad h \to \infty ,$$

is defined by two conditions:

i) there exists a bounded subset K of R^n such that every $\varphi_h(x)$ vanishes outside of K,

ii) $\varphi_h \Rightarrow 0 \ (\mathfrak{E}_{R^n})$.

A *functional* T defined on $\mathfrak{D}_{R^n}$ with values in the field of complex numbers is called a *distribution* if it is

linear: $T(\alpha_1\varphi_1+\alpha_2\varphi_2)=\alpha_1 T(\varphi_1)+\alpha_2 T(\varphi_2)$,

and

continuous: $\displaystyle\lim_{\varphi_h \Rightarrow 0 \, (\mathfrak{D}_{R^n})} T(\varphi_h)=0$.

The totality of distributions will be denoted by $\mathfrak{D}_{R^n}'$. It is a linear space by

$$(\alpha_1 T_1+\alpha_2 T_2)(\varphi)=\alpha_1 T_1(\varphi)+\alpha_2 T_2(\varphi) .$$

The mapping

$$\varphi \to (-1) T\!\left(\frac{\partial \varphi}{\partial x_k}\right)$$

defines also a distribution. It is, by definition, the *derivative of the distribution* T:

$$\frac{\partial T}{\partial x_k}(\varphi)=(-1) T\!\left(\frac{\partial \varphi}{\partial x_k}\right) .$$

Hence *a distribution is infinitely differentiable.*

Any function $f(x)$ which is locally integrable on R^n, i.e., a function $f(x)$ which is integrable on every bounded closed subset of R^n, defines a distribution T_f by

$$T_f(\varphi)=\int_{R^n} f(x)\varphi(x)dx \quad (dx=dx_1\ldots\ldots dx_n) .$$

Moreover, if $f(x)$ is continuously differentiable with respect to x_k, then we have, by partial integration,

$$\frac{\partial T_f}{\partial x_k}(\varphi) = (-1) T_f\!\left(\frac{\partial \varphi}{\partial x_k}\right)=T_{\partial f/\partial x_R}(\varphi).$$

A functional T on $\mathfrak{E}_{R^n}$ which is linear and continuous:

$$\lim_{\varphi_h \Rightarrow 0 \, (\mathfrak{E}_{R^n})} T(\varphi)=0$$

is identical with a distribution $\hat{T}$ with *compact support*, i.e., with such distribution $\hat{T}$ for which there exists a bounded subset K of R satisfying

$$\hat{T}(\varphi)=0 \quad \text{if} \quad \varphi(x)\equiv 0 \quad \text{on} \quad K.$$

By virtue of the so-called *Baire category argument*, it is easily proved that, if a sequence $\{T_h\}\subseteq\mathfrak{D}_{R^n}{}'$ satisfies

$$\lim_{h\to\infty} T_h(\varphi)=T(\varphi) \quad \text{for all} \quad \varphi\in\mathfrak{D}_{R^n},$$

then $T(\varphi)$ is also a distribution. The notion of the convergence:

$$T_h \Rightarrow T \ (\mathfrak{D}_{R^n}{}') \quad \text{as} \quad h\to\infty,$$

is defined by the above limit condition. It is a simple consequence that the term-by-term differentiation rule holds in $\mathfrak{D}_{R^n}{}'$, i.e.,

$$T_h \Rightarrow T(\mathfrak{D}_{R^n}{}') \quad \text{implies} \quad D^{(s)}T_h \Rightarrow D^{(s)}T \ (\mathfrak{D}_{R^n}{}')$$

for any differentiation

$$D^{(s)}=\partial^{s_1+s_2+\cdots+s_n}/\partial x_1{}^{s_1}\partial x_2{}^{s_2}\ldots\ldots\partial x_n{}^{s_n}.$$

§2. We are now able to explain and prove the reflexivity of the space of distributions.

For any $\varphi\in\mathfrak{D}_{R^n}$,

$$\Phi(T)=T(\varphi)$$

is a linear, continuous functional on $\mathfrak{D}_{R^n}{}'$, because

$$\Phi(\alpha_1 T_1+\alpha_2 T_2)=(\alpha T_1+\alpha_2 T_2)(\varphi)=\alpha_1 T_1(\varphi)+\alpha_2 T_2(\varphi)$$
$$=\alpha_1\Phi(T_1)+\alpha_2\Phi(T_2),$$

and

$$T_h \Rightarrow 0 \ (\mathfrak{D}_{R^n}{}') \quad \text{means that} \quad T_h(\varphi)\to 0 \quad \text{for all} \quad \varphi\in\mathfrak{D}_{R^n}.$$

The reflexivity theorem say, conversely, that any linear continuous functional $F(T)$ on $\mathfrak{D}_{R^n}{}'$ is given as

$$F(T)=T(f) \quad \text{with a certain fixed} \quad f\in\mathfrak{D}_{R^n}.$$

It is known, by a *convolution argument*, that the distributions of the special form T_ψ, $\psi\in\mathfrak{E}_{R^n}$, are dense in $\mathfrak{D}_{R^n}{}'$, i.e., for any $T\in\mathfrak{D}_{R^n}{}'$, we can find a sequence $\{\psi_h\}\subseteq\mathfrak{E}_{R^n}$ such that $T_{\psi_h}\Rightarrow(\mathfrak{D}_{R^n}{}')$. Thus, for the proof of the reflexivity theorem, we have only to show that

$$F(T_\psi)=T_\psi(f) \quad \text{for every} \quad \psi\in\mathfrak{E}_{R^n} \quad \text{simultaneously.}$$

Now

$$L^{(0)}(\psi) = F(T_\psi)$$

is a distribution with compact support, because

$\psi_h \Rightarrow 0(\mathfrak{E}_{R^n})$ implies $T_{\psi_h}(\varphi) \to 0$ for all $\varphi \in \mathfrak{D}_{R^n}$ and hence $T_{\psi_h} \Rightarrow 0(\mathfrak{D}_{R^n}{}')$, asserting that $L^{(0)}(\psi_h) = F(T_{\psi_h}) \to 0$.

Let the support of $L^{(0)}$ be K_0, i.e., let $L^{(0)}(\psi) = 0$ if $\psi(x) \equiv 0$ on the fixed bounded subset $K_0 \subseteq R^n$. Thus the value of $\psi(x)$ outside a bounded set $K \supseteq K_0$ does not affect the value of $F(T_\psi)$.

Moreover, the convergence

$$\int_{R^n} |\psi_h(x)|^2 dx \to 0$$

of such functions $\psi_h \in \mathfrak{E}_{R^n}$ implies

$$|T_{\psi_h}(\varphi)|^2 = |\int_{R^n} \psi_h(x)\varphi(x)dx|^2 \leq \int_{R^n} |\psi_h(x)|^2 dx \cdot \int_{R^n} |\varphi(x)|^2 dx \to 0,$$

and hence

$$T_{\psi_h} \Rightarrow 0(\mathfrak{D}_{R^n}{}'), \text{ implying } L^{(0)}(\psi_h) = F(T_{\psi_h}) \to 0.$$

Thus, by Riesz theorem on the representation of continuous linear functionals on Hilbert space, as applied to the space of square integrable functions on R^n, there exists a function $f^{(0)}(x)$ with the properties:

$$L^{(0)}(\psi) = F(T_\psi) = \int_K \psi(x)\overline{f^{(0)}}(x)dx,$$

$$\int_{R^n} |f^{(0)}(x)|^2 dx < \infty \text{ and } f^{(0)}(x) \equiv 0 \text{ outside of } K.$$

Similarly we have the representation:

$$F(D^{(s)}T_\psi) = \int_K \psi(x)\overline{f^{(s)}}(x)dx,$$

$$\int_{R^n} |f^{(s)}(x)|dx < \infty \text{ and } f^{(s)}(x) \equiv 0 \text{ outside of } K,$$

because

$$\int_{R^n} |\psi_h(x)|^2 dx \to 0$$

implies, by the term-by-term differentiation theorem,

$$D^{(s)}T_{\psi_h}=T_{D^{(s)}\psi_h}\Rightarrow 0(\mathfrak{D}_{R^n}{}') \quad \text{and hence} \quad F(D^{(s)}T_{\psi_h})\to 0.$$

Therefore we have

$$\int_K \psi(x)f^{(s)}(x)dx = F(D'^s T_\psi) = F(T_{D^{(s)}\psi}) = \int_K f^{(0)}(x)D^{(s)}\psi(x)dx,$$

which means

$$D^{(s)}T_{f^{(0)}}=(-1)^{s_1+s_2\cdots+s_n}T_{f^{(s)}} \quad \text{for every differentiation } D^{(s)}.$$

By this formula, we deduce that $f^{(s)}(x)$ can be considered to be differentiable any number of times. To this purpose, we imagine the functions $f^{(s)}(x)$ to be square integrable on a periodic parallelogram

$$0\leq x_i \leq 2\pi \quad (j=1, 2, \ldots\ldots, n)$$

which contains the bounded subset K. By the Parseval relation of Fourier series, we obtain, denoting by Δ the Laplacian,

$$\sum_k \overline{f_k^{(0)}}(1+|k|^2)^{|s|}\psi_k = \int_K \overline{f}^{(0)}(x)(1-\Delta)^{|s|}\psi(x)dx$$

$$= \int_K \overline{f}^{(s)}(x)\psi(x)\,dx = \sum \overline{f_k^{(s)}}\psi_k, \quad |s|=\sum s_j.$$

for

$$f^{(s)}(x)\sim\sum_k f_k^{(s)} exp(ik\cdot x) \quad \text{with} \quad \sum_k |f_k^{(s)}|^2 < \infty,$$

where

$$k=(k_1, \ldots\ldots, k_n), \ x=(x_1, \ldots\ldots, x_n), \ |k|^2=\sum_j k_j^2, \ k\cdot x=\sum_j k_j\cdot x_j.$$

Thus

$$f_k^{(0)}=f_k^{(s)}(1+|k|^2)^{-|s|},$$

and hence, for sufficiently large $|s|$,

$$f^{(0)}(x)=\sum_k f_k^{(s)}(1+|k|^2)^{-|s|}\,exp\,(ik\cdot x)$$

in the periodic parallelogram and there it is continuously differentiable any number of times.

Therefore $f^{(0)}(x)$ is equal to an infinitely differentiable function $f(x)$ vanishing outside of K, i.e., $f(x)$ belongs to $\mathfrak{D}_{R^n}$ and $F(T_\psi)=T_\psi(f)$.

A note on the fundamental theorem of calculus

Proc. Japan Acad. **57** A no. 5 (1981) 241

(Communicated May 12, 1981)

The entitled theorem is sometimes called as

The criteria for monotonicity. It reads: *Let $y = f(x)$ be a real-valued continuous function defined on a closed interval $[a, b]$. If the derivative $f'(x)$ exists and > 0 for all x of the open interval (a, b), then, for all c, d with $a < c < d < b$, we must have $f(c) < f(d)$.*

To this theorem, the present author should like to propose a proof which does not appeal to the *Mean Value Theorem* (cf. L. Bers [1], 223–224 and P. Lax-S. Burstein-A. Lax [2], 103) and which also gives the proof of the *Intermediate Value Theorem*.

Proof. Assume the contrary and let $f(c) \geq f(d)$. Since $f'(c) > 0$, there exists e with $c < e < d$ and $f(e) > f(d)$. Let m be any number satisfying $f(e) > m > f(d)$. Consider the graph $G(f; e, d)$ of f starting from the point $\{e, f(e)\}$ and ranging towards the point $\{d, f(d)\}$. Take *the first encounter point* $\{g, f(g)\}$ of the graph $G(f; e, d)$ with the line $y = m$. The existence of such point $\{g, f(g)\}$ is proved as follows.

Let S be the set of all points $x_1 \in [e, d]$ satisfying the condition that $f(x) > m$ for all $x \in [e, x_1]$. Let x_∞ be the least upper bound of the set S. Then $x_\infty \bar{\in} S$. If otherwise, $f(x_\infty) > m$ so that, by the continuity of f, $f(x) > m$ for all x sufficiently close to x_∞. Hence there should exist a point $x_1 \in S$ which is to the right of x_∞. This is absurd. Hence $x_\infty \bar{\in} S$ is a limit point of the set S and so $f(x_\infty) = m$. Therefore, $e < x_\infty < d$ and we can take x_∞ for g.

Since $\{g, f(g)\}$ is the first encounter point of the graph $G(f; e, d)$ with the line $y = m$, we must have $f(x) > m$ for all x with $e \leq x < g$. This implies, by $f(g) = m$, that $f'(g) \leq 0$, contrary to the hypothesis $f'(x) > 0$.

Remark. The "first encounter argument" is also applicable to the proof of the fact that, for a convex function $y = f(x)$ with $f''(x) > 0$, the graph $G(f; a, b)$ has no point which lies on the upper side of the line segment (the secant) connecting two points $\{a, f(a)\}$ and $\{b, f(b)\}$. We omit the details.

References

[1] L. Bers: Calculus. Holt, Rinehart and Winston Inc. (1969).
[2] P. Lax, S. Burstein, and A. Lax: Calculus with Applications and Computing, Volume I. Springer-Verlag (1976).

V. Semigroups and Evolution Equations

Comments (Tosio Kato)

Paper [57]

This is a monumental paper in which the celebrated Hille-Yosida theorem on strongly continuous, contraction semi-groups of linear operators in a Banach space is proved. In view of the special circumstances of its genesis, some historical remarks will be in order.

The Japanese mathematical environment had been cut off from international communication due to the war and the ensuing chaos. The results of western mathematics were only beginning to trickle into Japan. Yosida's work was done without knowledge of the western mathematics after 1940 or so. His result was presented in 1947 at the fall meeting of the new Mathematical Society of Japan, and was published in 1948 in Vol. 1, No. 1 of its Journal. Though a short paper, as is characteristic of Yosida's style, it is complete with all the details of proof. Hille's result, on the other hand, was given in his book *Functional Analysis and Semi-Groups*, which is a comprehensive treatise on the subject published in 1948. Hille's proof of the generation theory—an essential ingredient of the Hille-Yosida theorem—was given in this book for the first time, though it had been presented, without proof, at the Science Academy of Paris in September 1947. This sketch of the events will be enough to show how the two mathematicians arrived at the same results independently and simultaneously, an event not rare in the history of mathematics.

Although the theorems by the two authors are equivalent, their proofs are different. In particular, there is a marked difference in the generation theory, in which a semigroup T_t is to be constructed with a given operator A as its (infinitesimal) generator. Yosida introduces a sequence of approximating bounded operators $A_n = A(1 - n^{-1}A)^{-1}$ (nowadays called *Yosida approximation*), and constructs $T_t = \lim_n \exp(tA_n)$, where $\exp(B)$ is defined by the Taylor series. On the other hand, Hille uses the exponential formula $T_t = \lim_n (1 - tA/n)^{-n}$. It looks as if Hille's proof is simpler inasmuch as it uses a single limiting process, whereas Yosida's involves double limits. But the convergence proof in Hille was more complicated, depending partially on the inverse Laplace transformation. (Incidentally, this required a complex Banach space, while in Yosida's proof complex structure was not needed, although it was formally assumed.)

A few years later, Hille published a simplified version of his proof (Kungl. Fysiografiska Sällskapets i Lund Förhandlingar, Bd 21, Nr 14 (1952), 13pp.), which did not use complex structure. Even in this paper, he states that "While simpler than my original proof, it is still not as simple as Yosida's." In the second edition of his book published in 1957 (jointly with R. S. Phillips), however, the proof was much simplified.

It may be noted that an analog of Hille-Yosida theorem for continuous semi-groups (or groups, equivalently) in a Banach algebra had been proved in 1936 by Nagumo and by Yosida ([14] of II). The Hille-Yosida theorem for unitary groups in a Hilbert space is known as Stone's theorem; it had been partially generalized by Galfand (1939) to weakly measurable groups of operators in a Banach space (without the generation theory).

Paper [82]

The main theorem of this paper is concerned with the Hille-Yosida theorem for a special class of holomorphic semi-groups (a semi-group T_t which has analytic continuation on a sector containing the real axis, with $\|T_t\|$ bounded for bounded t). Such semi-groups are considered in Hille's books (1948, 1957), including a characterization of the generator A. Yosida's contribution lies in the discovery of another necessary and sufficient condition in the form $\lim\sup_{t \searrow 0} t\|dT_t/dt\| = C < \infty$.

This condition is not found explicitly in Hille's books, although its necessity is a simple consequence of the assumptions made. Actually Hille had considered such an inequality in 1950, in connection with the study of differentiable semi-groups. He showed that if $C < 1/e$ then T_t can be extended to an entire function (so that the generator is bounded), but he did not consider the case $C > 1/e$. Apparently that paper was the source of inspiration for Yosida. The importance of Yosida's condition is demonstrated in paper [84]. Moreover, it has become an indispensable tool in the theory of abstract evolution equations of "parabolic" type, which was originated by Tanabe (1959) and has been developed by numerous authors to this day. The second theorem of this paper deals with another kind of infinitely differentiable semi-groups.

Paper [84]

In this paper Yosida constructs fractional powers $(-A)^\alpha$ for the generator A of a C_0-semi-group T_t. The construction makes use of an averaging process applied to T_t (sometimes called subordination), which had been used by Bochner and others. The novelty of this paper is that $-(-A)^\alpha$ is shown to be the generator of a holomorphic semi-group. The proof uses effectively the condition deduced in paper [82]. More details of the proof are given in *Functional Analysis*, Chapter IX.

Paper [92]

A perturbation theorem for the generator A of a contraction C_0-semi-group is proved. If B is a similar operator with $D(B) \supset D((-A)^\alpha)$ for some $\alpha \in (0, 1)$, then $A + B$ (with domain $D(A)$) also generates a contraction C_0-semi-group. If, in addition, A generates a holomorphic semi-group, so does $A + B$.

Paper [93]

This paper deals with the equation of evolution (see [88]) in a locally convex linear topological space, the main assumption being that $D(A(t))$ is independent of t. Unlike [88], it uses the method of Kato (1953), with greatly simplified proof. The contents of this paper are (essentially) reproduced in the second and later editions of *Functional Analysis*, where the underlying space is taken to be a Banach space.

On the differentiability and the representation of one-parameter semi-group of linear operators

By Kôsaku YOSIDA

(Received Oct. 25, 1947.)

1. *The theorems.* Let $\{U_t\}$, $0 \leqq t < \infty$, be a one-parameter semi-group of linear ($=$additive, continuous) operators from a complex Banach space E to E:

$$U_t U_s = U_{t+s}, \qquad U_0 = I \ (=\text{the identity operator}), \tag{1.1}$$

such that

$$\sup_t \|U_t\| \leqq 1, \tag{1.2}$$

$$\lim_{t \to t_0} U_t x = U_{t_0} x \ (\lim = \text{strong limit}), 0 \leqq t_0 < \infty, x \in E. \tag{1.3}$$

The purpose of the present note is to prove the following two theorems[1].

Theorem 1. If we denote by D the totality of x for which

$$\underset{h \downarrow 0}{weak \ lim} \ h^{-1}(U_h - I)x = Ax \tag{1.4}$$

exists, then D is dense in E. Moreover A is a closed additive operator from D to E with the properties:

$$for \ any \ x \in D, \ \lim_{h \to 0} h^{-1}(U_{t+h} - U_t)x = AU_t x = U_t Ax, \tag{1.5}$$

there exists a sequence $\{I_n\}$ of linear operators each commutative with every U_t and A such that i) the range $R(I_n) = \{I_n x; x \in E\} \subseteqq D$, $AI_n = n(I_n - I)$, ii) $\|I_n\| \leqq 1$, $\lim_{n \to \infty} I_n x = x$, iii) $U_t' x = \lim_{n \to \infty} exp(tAI_n)x = \lim_{n \to \infty} \sum_{m=0}^{\infty} (m!)^{-1}(tAI_n)^m x$ uniformly for t in any finite interval[2], (1.6)

$\|(A - nI)x\| \geqq n\|x\| \ (n = 1, 2, \ldots)$ *for* $x \in D$ *and the range* $R(A - nI)$ *coinsides with* $E \ (n = 1, 2, \ldots)$, (1.7)

let, by (1.7), y_n be the unique solution of $(A - nI)y_n = y \ (n = 1, 2, \ldots)$, then

$$\lim_{n \to \infty} A(-ny_n) = \lim_{n \to \infty} (-n(y + ny_n)) = Ay \ for \ y \in D. \tag{1.8}$$

Theorem 2. Let, conversely, A be an additive operator from a dense linear subset D of E to E such that (1.7) and (1.8) are satisfied. Then there exists a one-parameter semi-group U_t which satisfies (1.1)–(1.3) and (1.5).

We have, thus, a characterisation of the differential quotient A of the one-parameter semi-group of linear operators. This may be applied to an operator-theoretical treatment of temporally homogeneous stochastic process[3]. As an application of the theorem 1, we give a new proof of Stone's theorem[4] (see 4 below).

2. *Proof of the theorem 1. The differentiability.* We may define the integral

$$C_\varphi \cdot x = \int_0^\infty \varphi(s) U_s x ds \qquad (2.1)$$

for complex-valued continuous function $\varphi(s)$ following after S. Bochner, G. Birkoff, I. Gelfand, B. J. Pettis and other authors[5]. Let $\varphi(s)$ satisfy

$$\lim_{h \downarrow 0} \int_h^\infty \left| \frac{\varphi(s-h) - \varphi(s)}{h} + \varphi'(s) \right| ds = 0. \qquad (2.2)$$

We have, by (1.1)

$$\frac{1}{h}(U_h - I)C_\varphi x = \frac{1}{h} \int_0^\infty \varphi(s) U_h U_s x ds - \frac{1}{h} \int_0^\infty \varphi(s) U_s x ds$$

$$= \int_h^\infty \frac{\varphi(s-h) - \varphi(s)}{h} U_s x ds - \frac{1}{h} \int_0^h \varphi(s) U_s x ds.$$

Thus, by (2.2) and $U_0 = I$,

$$\lim_{h \downarrow 0} \frac{1}{h}(U_h - I)C_\varphi x \qquad \text{exists and} \qquad = C_{-\varphi'} \cdot x - \varphi(0)x. \qquad (2.3)$$

Such $C_\varphi x$ is dense in E, since for any $\delta > 0$

$$\varphi_\delta(s) = \delta \exp(-\delta s) \qquad (2.4)$$

satisfies (2.2) and moreover, for any $x \in E$, $\lim_{\delta \to \infty} C_{\varphi_\delta} \cdot x = x$.

We have, by (1.4), for $x \in D$,

$$\text{weak} \lim_{h \downarrow 0} \frac{1}{h}(U_h - I)U_t x = \text{weak} \lim_{h \downarrow 0} \frac{1}{h}(U_{t+h} - U_t)x$$

$$= U_t \cdot \text{weak} \lim_{h \downarrow 0} \frac{1}{h}(U_h - I)x.$$

Hence $U_t \cdot D \subseteqq D$ and $AU_t x = U_t Ax$ for any $x \in D$, viz. A is commutative with every U_t, and

$$\begin{cases} \text{the right weak derivative } D^+ U_t x \text{ exists and} \\ = AU_t x = U_t Ax \text{ for any } x \in D. \end{cases} \qquad (2.5)$$

Hence, by the continuity (1.3) we have, for any $f \in E^*$ ($=$ the conjugate space of E)

$$f(U_t x) - f(x) = \int_0^t D^+ f(U_s x)\, ds = \int_0^t f(U_s Ax)\, ds = f\left(\int_0^t U_s Ax\, ds \right)$$

and therefore

$$U_t x - x = \int_0^t U_s A x \, ds \quad \text{for} \quad x \in D. \tag{2.6}$$

Thus we have the strong differentiability (1.5).

The closedness of A. Let $x_n \in D$ $(n = 1, 2, \ldots)$ and let $\lim_{n \to \infty} x_n = x$, $\lim_{n \to \infty} A x_n = z$. Then, by (2.6),

$$U_t x - x = \int_0^t U_s z \, ds.$$

The representation (1.6). Put

$$I_n = C_{\varphi_n} \quad (n = 1, 2, \ldots), \tag{2.7}$$

then, by (2.3) and (2.4), i) and ii) are surely satisfied. In fact, we have by (1.2) and (2.4)

$$\|I_n\| \leq \int_0^\infty n \exp(-ns) \, ds = 1. \tag{2.8}$$

Thus, since

$$A I_n = n(I_n - I) \tag{2.9}$$

by (2.3), we have

$$\exp(t A I_n) = \sum_{m=0}^\infty \frac{(t A I_n)^m}{m!} = \exp(tn(I_n - I)), \quad 0 \leq t < \infty. \tag{2.10}$$

Thus

$$\|\exp(t A I_n)\| = \|\exp(tn I_n) \exp(-tn I)\| \leq \exp(tn) \exp(-tn) = 1. \tag{2.11}$$

Since $A I_n$ is commutative with each U_t, we have, for $x \in D$,

$$\begin{aligned}
\|U_t x - \exp(t A I_n) x\| &= \left\| \int_0^t \frac{d}{ds} ((\exp(t - s) A I_n) U_s x) \, ds \right\| \\
&= \left\| \int_0^t (\exp(t - s) A I_n) U_s (A - A I_n) x \, ds \right\| \\
&\leq \int_0^t \|(A - A I_n) x\| \, ds \quad \text{by (1.2) and (2.11)} \\
&= t \|(A - A I_n) x\|.
\end{aligned} \tag{2.12}$$

We have

$$A I_n x = I_n A x \quad \text{for} \quad x \in D, \tag{2.13}$$

for A is a closed operator commutative with each U_t. Thus, by (2.12),

$$\|U_t x - \exp(t A I_n) x\| \leq t \|(I - I_n) A x\| \quad \text{for } x \in D. \tag{2.14}$$

Since U_t and $\exp(tAI_n)$ are both of norm ≤ 1, we have iii) by the fact that D is dense in E.

The proof of (1.7). We first show that $R(A - nI)$ is dense in E. If otherwise, there exists $f \in E^*$, $f \neq 0$, such that $f(Ax - nx) = 0$ on D. Thus, by $U_t D \subseteq D$ we have $f(AU_t x) = nf(U_t x)$ and hence

$$\frac{d}{dt} f(U_t x) = nf(U_t x).$$

Therefore, by $f(U_0 x) = f(x)$, we obtain $f(U_t x) = f(x) \exp(nt)$. This is a contradiction. Proof. By $f \neq 0$ and by the fact D is dense in E there exists $x \in D$ such that $f(x) \neq 0$. Then $f(x) \exp(nt)$ is unbounded in t when $t \to \infty$, contrary to $|f(U_t x)| \leq \|f\| \cdot \|U_t\| \cdot \|x\| \leq \|f\| \cdot \|x\|$. Next we show that

$$\|(A - nI)x\| \geq n\|x\| \quad \text{for} \quad x \in D.$$

Assume the contrary and let $\|(A - nI)x\| = a < n$ for a certain $x \in D$ with $\|x\| = 1$. Let $f \in E^*$ be such that $f(x) = 1$, $\|f\| = 1$. Then, by

$$\frac{d}{dt} U_t x = U_t Ax = nU_t x + U_t(A - nI)x,$$

we obtain

$$\begin{cases} \dfrac{d}{dt}\varphi(t) = n\varphi(t) + \psi(t), \quad \text{where} \\[2mm] \varphi(t) = f(U_t x), \quad \psi(t) = f(U_t(A - nI)x). \end{cases}$$

Since $\varphi(0) = 1$ we have

$$\varphi(t) = \exp(nt)\left(\int_0^t \exp(-nt)\psi(t)\, dt + 1 \right)$$

and hence, by $|\psi(t)| \leq \|f\| \cdot \|U_t\| \cdot \|(A - nI)x\| \leq a$,

$$|\varphi(t)| \geq \exp(nt)(1 - an^{-1}(1 - \exp(-nt))).$$

Thus $\varphi(t)$ is unbounded in t when $t \to \infty$, contrary to $|\varphi(t)| \leq \|f\| \cdot \|U_t\| \cdot \|x\| \leq 1$. Therefore, for any $y \in E$, there exists a sequence $\{x_h\} \subseteq D$ such that $\lim_{h \to \infty} (A - nI)x_h = y$. Because of $\|(A - nI)(x_h - x_k)\| \geq n\|x_h - x_k\|$, $\{x_h\}$ is a Cauchy sequence. Let $\lim_{h \to \infty} x_h = x$, then $\lim_{h \to \infty} (A - nI)x_h = y$ and by the closedness of A we have $y = (A - nI)x$.

The proof of (1.8). We have, by (2.9), $AI_n y - nI_n y = -ny$ and hence

$$-ny_n = I_n y. \tag{2.15}$$

Thus (1.6) and (2.13) imply (1.8).

3. *Proof of the theorem* 2. By (1.7), the operator J_n defined by

$$J_n y = -ny_n \quad (n = 1, 2, \ldots) \tag{3.1}$$

satisfies

$$\|J_n\| \leq 1. \tag{3.2}$$

Since

$$AJ_n y = A(-ny_n) = -nAy_n = -n(y + ny_n) = n(J_n - I)y, \tag{3.3}$$

we have, by (3.2),

$$\|\exp(tAJ_n)y\| = \|\exp(ntJ_n)\exp(-ntI)y\| \leq \exp(nt)\exp(-nt)\cdot\|y\|,$$

hence the linear operator defined by

$$U_t^{(n)} = \exp(tAJ_n) \tag{3.4}$$

satisfies

$$\|U_t^{(n)}\| \leq 1, \tag{3.5}$$

$$U_t^{(n)}x - x = \int_0^t U_s^{(n)} AJ_n x\, ds, \tag{3.6}$$

$$\lim_{h\to 0} h^{-1}(U_{t+h}^{(n)} - U_t^{(n)})x = U_t^{(n)} AJ_n x. \tag{3.7}$$

We have

$$J_n J_m = J_m J_n, \tag{3.8}$$

for, by (3.1),

$$J_n = -n^{-1}(A - nI)^{-1}. \tag{3.9}$$

Thus AJ_n is commutative with $U_t^{(m)}$ and hence

$$\begin{aligned}
\|(U_t^{(m)} - U_t^{(n)})x\| &= \left\| \int_0^t \frac{d}{ds}\left((\exp(t - s)AJ_n)U_s^{(m)}x\right) ds \right\| \\
&= \left\| \int_0^t (\exp(t - s)AJ_n)U_s^{(m)}(AJ_m - AJ_n)x\, ds \right\| \\
&\leq \int_0^t \|(AJ_m - AJ_n)x\|\, ds \quad \text{by (3.5) and (3.7)} \\
&= t\|(AJ_m - AJ_n)x\|.
\end{aligned}$$

Therefore, by (1.8),

$$U_t y = \lim_{n\to\infty} U_t^{(n)} y \quad (y \in D) \tag{3.10}$$

exists uniformly for t in any finite interval. Since D is dense in E and since we have (3.5), we see that the limit $U_t^{(n)}$ exists for all $y \in E$ and that U_t satisfies (1.1)–(1.3). Hence, by letting $n \to \infty$ in (3.6), we obtain

$$U_t y - y = \int_0^t U_s Ay\, ds, \quad y \in D. \tag{3.11}$$

K. Yosida

4. *Stone's theorem.* If E is a Hilbert space and if U_t is, for any $t \geq 0$, a unitary operator, then $H = -iA$ $(i = \sqrt{-1})$ is self-adjoint and

$$U_t = \int_{-\infty}^{\infty} \exp(i\lambda t)\, dE(\lambda), \quad \text{where } H = \int_{-\infty}^{\infty} \lambda\, dE(\lambda). \tag{4.1}$$

This theorem due to M. H. Stone may be obtained as follows[6]. Put $U_{-t} = U_t^{-1} = U_t^*$ for $t \geq 0$, then $\{U_t\}$, $-\infty < t < \infty$, is a one-parameter group of unitary operators strongly continuous in t. Thus it is easy to see, by (1.5),

$$\frac{dU_t x}{dt} = iHU_t x = iU_t Hx \qquad \text{for } x \in D,\ -\infty < t < \infty. \tag{1.5}'$$

Hence, if $x, y \in D$,

$$(Hx, y) = \frac{1}{i}\left(\frac{d}{dt}(U_t x, y)\right)_{t=0} = \frac{1}{i}\left(\frac{d}{dt}(x, U_{-t}y)\right)_{t=0} = (x, Hy), \tag{4.2}$$

which shows that H is a symmetric operator. We proved in 2 that $R(H + iI) = E$ ((1.7)) by letting $t \to \infty$. Letting $t \to -\infty$, similar argument shows that $R(H - iI) = E$ also. Therefore the Cayley transform of H is unitary and hence H is self-adjoint. Let $H = \int_{-\infty}^{\infty} \lambda\, dE(\lambda)$ be its spectral resolution, then, as in (3.9), $I_n = (I - n^{-1}A)^{-t}$ and hence

$$AI_n = n(I_n - I) = n((I - in^{-1}H)^{-1} - I) = \int_{-\infty}^{\infty} n((1 - i\lambda n^{-1})^{-1} - 1)\, dE(\lambda)$$

$$= \int_{-\infty}^{\infty} i\lambda(1 - i\lambda n^{-1})^{-1}\, dE(\lambda).$$

Hence

$$U_t x = \lim_{n \to \infty} \exp(tAI_n)x = \lim_{n \to \infty} \int_{-\infty}^{\infty} \exp(ti\lambda(1 - i\lambda n^{-1})^{-1})\, dE(\lambda)x$$

$$= \int_{-\infty}^{\infty} \exp(i\lambda t)\, dE(\lambda)x.$$

Mathematical Institute,
Nagoya University.

References

1. Cf. I. Gelfand: C. R. URSS, 25 (1939), 713–718 and N. Dunford-I. E. Segal: Bullet. Amer. Math. Soc., **52** (1946), 911–914.
2. According to N. Dunford and I. E. Segal's paper referred to in 1, E. Hille (Proc. Nat. Acad. Sci., **28** (1942), 175–178, 421–424) obtained the representation $U_t x = \lim_{n \to \infty} \exp(tn(U_{\frac{1}{n}} - I))x$. Hille's paper is not available to the author.

3. The question of the characterization of A together with differentiability of U_t is proposed to the author by Dr. K. Itô in connection with his theory of stochastic differential equations. See his forthcoming paper in Jap. J. of Math.
4. Proc. Nat. Acad. Sci., **16** (1930), 172–175. See also J. von Neumann: Ann. of Math., **33** (1932), 567–573. Proofs of Stone's theorem are given by many authors: F. Riesz, B. von Sz. Nagy and H. Nakano.
5. S. Bochner: Fund. Math., **20** (1933), 262–276. G. Birkhoff: Trans. Amer. Math. Soc. **38** (1935), 357–378. I. Gelfand: Commun. Inst. Sci. Math. et Mech. Univ. Kharkoff, **13** (1936), 35–40. B. J. Pettis: Trans. Amer. Math. Soc., **44** (1938), 277–304.
6. In the case of Stone's theorem, we may replace (1.3) by the separability of $\{U_t x; -\infty < t < \infty\}$ and the weak measurability of U_t. This may be carried out by virtue of N. Dunford's theorem in Ann. of Math., **39** (1938), 567–573. This fact is, however, already proved by J. von Neumann in another way. See his paper referred to in 4.

Added during the proof. Meanwhile, E. Hille kindly communicated me that he also obtained essentially the same results as above by a different method. See his paper in C. R., 8 September (1947) and PNAS notes referred to in 2.

On the differentiability of semi-groups of linear operators

Proc. Japan Acad. **34** (1958) 337–340

(Comm. by Z. SUETUNA, M.J.A., June 12, 1958)

§ 1. The purpose of the present note is to prove the two theorems given below which may be applied to the "abstract Cauchy problem" connected with the semi-group theory.[1] We recall, for the sake of exposition, some of the basic results from the theory. Let a one-parameter family T_t, $0 \leq t < \infty$, of bounded linear operators on a complex Banach space X into X satisfy the semi-group conditions:

(1.1) $\qquad\qquad T_t T_s = T_{t+s}, \ T_0 = I$ (the identity operator),

(1.2) $\qquad\qquad s - \lim_{t \to t_0} T_t x = T_{t_0} x, \ t_0 \geq 0, \ x \in X,$

(1.3)[2] $\qquad\qquad\qquad \| T_t \| \leq 1, \quad t \geq 0.$

Then the infinitesimal generator A of T_t defined by

(1.4) $\qquad\qquad Ax = s - \lim_{h \downarrow 0} h^{-1}(T_h - I)x$

is a closed linear operator with its domain $D(A)$ dense in X such that

(1.5) $\qquad R(\lambda, A)x = (\lambda I - A)^{-1}x = \int_0^\infty e^{-\lambda t} T_t x\, dt, \ Re(\lambda) > 0$

exists as a bounded linear operator on X into X satisfying

(1.6) $\qquad\qquad \| R(\lambda, A) \| \leq (Re(\lambda))^{-1}, \quad Re(\lambda) > 0.$

Conversely, let a linear operator A with its domain $D(A)$ dense in X and taking the values in X admit the resolvent $(\lambda I - A)^{-1} = R(\lambda, A)$ which satisfies (1.6). Then A is the infinitesimal generator of a semi-group T_t which enjoys (1.1)–(1.5).

We have the representations

(1.7) $\qquad\qquad T_t x = s - \lim_{n \to \infty} (I - n^{-1}tA)^{-n}x, \ x \in X,$

(1.7)′ $\qquad T_t x = s - \lim_{n \to \infty} \exp(tA(I - n^{-1}A)^{-1})x, \ x \in X,$

uniformly in t in any compact interval of t. We have moreover,

(1.8) $\quad T_t x = s - \lim_{|\tau| \uparrow \infty} (2\pi)^{-1} \int_{\sigma - i\tau}^{\sigma + i\tau} e^{\lambda t} R(\lambda, A)x\, d\lambda, \ x \in D(A), \ t > 0, \ \sigma = Re(\lambda) > 0.$

In these senses, we may write $T_t = \exp(tA)$.

(1.1), (1.2) and (1.4) imply that $T_t = \exp(tA)$ solves, for given A,

---

1) E. Hille and R. S. Phillips: Functional Analysis and Semi-groups, New York (1957). K. Yosida: On the differentiability and the representation of one-parameter semi-group of linear operators, J. Math. Soc. Japan, **1**, 15–21 (1948).

2) (1.1) and (1.2) imply that $\lim_{t \to \infty} t \log \| T_t \| = \beta < \infty$. Thus, by taking $e^{-\beta t}T_t$ in place of the original T_t, we may assume that the condition (1.3) is satisfied.

"the abstract Cauchy problem" (ACP):

(1.9)   for $x \in D(A)$ and $t \geqq 0$, we have
$$T_t'x = s - \lim_{h \to 0} h^{-1}(T_{t+h} - T_t)x = AT_t x = T_t Ax, \ s - \lim_{t \downarrow 0} T_t x = x.$$

However, there are semi-groups satisfying the condition:

(1.10)   for every $x \in X$ and every $t > 0$, we have $T_t x \in D(A)$.

In such case, T_t solves the $(ACP)_0$:

(1.9)′   for every $x \in X$ and every $t > 0$, we have
$$T_t'x = AT_t x, \ s - \lim_{t \downarrow 0} T_t x = x.$$

There are also semi-groups T_t which can be extended to T_λ which are analytic in a sector of the complex λ-plane of the form

(1.11) $$|\operatorname{Arg} \lambda| < \alpha < \pi/2.$$

(1.10) implies that $T_t' = AT_t$ is, as a closed linear operator defined on the Banach space X, a bounded linear operator and hence we have, by (1.1) and (1.10),
$$T_t'' = (T_t')' = (T_{t-s}AT_s)' = (AT_{t/2})^2 = (T_{t/2}')^2$$

or more generally

(1.12) $$T_t^{(n)} = (AT_{t/n})^n = (T_{t/n}')^n.$$

E. Hille and R. S. Phillips have discussed such differentiability properties of semi-groups and obtained conditions upon the behaviour of the resolvent $R(\lambda, A)$ which imply the differentiability for $t > 0$ of $T_t = \exp(tA)$. The following two theorems are suggested by their researches and may be applied to discuss the behaviour near $t = 0$ of solutions $T_t x = \exp(tA)x$ of $(ACP)_0$.

Theorem 1. *T_t' exists for $t > 0$ and*

(1.13) $$\overline{\lim_{t \downarrow 0}} \, t \, \| T_t' \| < \infty$$

if and only if

(1.14) $$\overline{\lim_{|\tau| \uparrow \infty}} \, |\tau| \cdot \| R(1 + i\tau, A) \| < \infty.$$

Moreover, the semi-group T_t of this class can be extended to T_λ which is analytic in a sector of the form

(1.11)′ $$|\lambda - t| < t/eC, \ \text{where } C = \sup_{t>0} \| (e^{-t/2}T_t)' \|.$$

Theorem 2. *T_t' exists for $t > 0$ and*

(1.13)′ $$\overline{\lim_{t \downarrow 0}} \, t \cdot \log \| T_t' \| = 0$$

if and only if

(1.14)′ $$\overline{\lim_{|\tau| \uparrow \infty}} \, \log |\tau| \cdot \| R(1 + i\tau, A) \| = 0.$$

Remark. The semi-groups T_t are assumed to satisfy (1.1)–(1.3) so that the theorems are exhibited under the assumption (1.6).

§ 2.   *Proof of Theorem* 1. Let $\overline{\lim_{t \downarrow 0}} \, t \, \| T_t' \| < \infty$. Then, since $\| T_t \| \leqq 1$, we have $\sup_{t>0} t \, \| (e^{-t/2}T_t)' \| = C < \infty$. Hence, by applying (1.12) to the semi-group $S_t = e^{-t/2}T_t$, we see that

$$(t/n)^n \, \|S_t^{(n)}\| = (t/n)^n \, \|(S'_{t/n})^n\| \leq C^n \text{ for } t>0, \ n\geq 1.$$

Thus

(2.1) $$(n!)^{-1} |\lambda - t|^n \cdot \|S_t^{(n)}\| \leq (n!)^{-1}(nCt^{-1}|\lambda - t|)^n,$$

and the analytical continuation

(2.2) $$S_\lambda = e^{-\lambda/2} T_\lambda = e^{-1/2} T_t + \sum_{n=1}^{\infty} (n!)^{-1}(\lambda - t)^n (e^{-t/2} T_t)^{(n)}$$

is possible for λ satisfying (1.11)'. Moreover, we have, from (2.1)–(2.2),

(2.3) $$\|S_\lambda\| = \|e^{-\lambda/2} T_\lambda\| \leq (1 - eCt^{-1}|\lambda - t|)^{-1} \text{ when } eCt^{-1}|\lambda - t| < 1.$$

Let $\tau > 0$. Then, because of the analyticity of S_λ in λ, we can deform the path of integration $0 \leq t < \infty$ of the integral

$$R(1 + i\tau, A) = \int_0^{\infty} e^{-(1+i\tau)t} T_t dt = \int_0^{\infty} e^{-i\tau t} e^{-t/2} S_t dt$$

to the path $te^{-i\varphi}$, $0 \leq t < \infty$, with a fixed φ such that $0 < \tan \varphi < 1/eC$:

$$R(1 + i\tau, A) = \int_0^{\infty} \exp(-i\tau te^{-i\varphi}) \exp(-te^{-i\varphi}/2) S_{te^{-i\varphi}} e^{-i\varphi} dt.$$

Such deformation is possible thanks to the convergence factor $\exp(-te^{-i\varphi}/2)$ and the estimate (2.3). Thus it is easy to obtain (1.14).

Let conversely $\|R(1 + i\tau, A)\| \leq C/|\tau|$ for large $|\tau|$. Then, by virtue of the resolvent equation

(2.4) $$R(\lambda, A) = \sum_{n=0}^{\infty} R(1 + i\tau, A)^{n+1}(1 + i\tau - \lambda)^n$$

valid for $|1 + i\tau - \lambda| \cdot \|R(1 + i\tau, A)\| < 1$, we see that $R(\lambda, A)$ exists and is analytic in λ for large value of $|\tau| = |Im(\lambda)|$ outside a sector of the left half λ-plane defined by the boundary curve $\lambda(s) = \sigma(s) + i\tau(s)$ which satisfies

$$\lim_{\tau(s) \uparrow \infty} (-\sigma(s)/\tau(s)) = \tan \varepsilon = \lim_{\tau(s) \downarrow -\infty} \sigma(s)/\tau(s), \ \varepsilon > 0.$$

Moreover, (2.4) shows that $R(\lambda, A)$ is of order $|\tau^{-1}|$ when $|\tau| = |Im(\lambda)|$ tends to ∞ outside the above sector and lying in the left half λ-plane. Therefore, by deforming the path of integration $-\infty < \tau < \infty$ of the integral

$$T_t x = (2\pi)^{-1} i \int_{-\infty}^{\infty} e^{(1+i\tau)t} R(1 + i\tau, A) x \, d\tau, \ x \in D(A), t > 0,$$

to the path $\tilde{\lambda}(s) = 2^{-1}\sigma(s) + i\tau(s)$, we see that $T'_t x$ exists for every $x \in X$ and every $t > 0$. Hence we easily see that T'_t satisfies (1.13).

Remark. The above proof shows that T_t is differentiable even at $t = 0$ if T'_t exists for $t > 0$ and $\overline{\lim}_{t \downarrow 0} t \|T'_t\| < e^{-1}$. This remarkable result was proved by E. Hille[3] in another way.

---

3) E. Hille: On the differentiability of semi-group operators, Acta Sci. Math. Szeged, **12**, 19–24 (1950).

Proof of Theorem 2.[4] We have, by partially integrating (1.5) and making use of the existence of T_t' for $t>0$,

$$R(1+i\tau, A)=\int_0^{\delta} e^{-(1+i\tau)t}T_t dt+(1+i\tau)^{-1}e^{-(1+i\tau)\delta}T_\delta$$

$$+(1+i\tau)^{-1}\int_\delta^\infty e^{-(1+i\tau)t}T_t' dt.$$

$\|T_t'\|$ is, by $T_t'=AT_t$, (1.1) and (1.2), monotone decreasing in t. Thus

$$(2.5) \qquad \|R(1+i\tau, A)\|\leq\delta+(1+\tau^2)^{-1/2}(1+\|T_\delta'\|).$$

Now $t^{-1}\|T_t'\|$ is monotone decreasing in t with $\|T_t'\|$ and so there are two cases: $\lim_{t\downarrow 0} t^{-1}\|T_t'\|=\infty$ and $\lim_{t\uparrow 0} t^{-1}\|T_t'\|<\infty$. In the latter case, we have, by Theorem 1, $\lim_{|\tau|\uparrow\infty} \tau\cdot\|R(1+i\tau, A)\|<\infty$ and hence (1.14)′. In the former case, we take $|\tau|$ as equal to $\delta^{-1}\|T_\delta'\|$. Then, since $\lim_{t\downarrow 0} t\cdot\log\|T_t'\|=0$ by the assumption, we have $\lim_{\delta\downarrow 0}\delta(\log\delta+\log\tau)=0$, viz. $\lim_{\delta\downarrow 0}\delta\log\tau=0$. This proves (1.14)′ by (2.5).

Let, conversely, (1.14)′ hold good. By virtue of the resolvent equation (2.4) and the assumption (1.14)′, we see that $R(\lambda, A)$ is analytic in λ for large value of $|\tau|=Im(\lambda)$ outside a curved sector of the left half λ-plane with boundary curve $\lambda(s)=\sigma(s)+i\tau(s)$ which satisfies the condition

$$|\sigma(s)|=\varepsilon(s)^{-1}\log|\tau(s)|, \; 0<\varepsilon(s) \quad \text{and} \quad \lim_{|s|\uparrow\infty}\varepsilon(s)=0,$$

$$\lim_{s\uparrow\infty}\tau(s)=\infty, \quad \lim_{s\downarrow-\infty}\tau(s)=-\infty.$$

Moreover, by (2.4), $\|R(\lambda, A)\|$ is of order $o(1/\log|\tau|)$ when $|\tau|=|Im(\lambda)|$ tends to ∞ outside the above sector and lying in the left half λ-plane. Thus, by deforming the path of integration $-\infty<\tau<\infty$ of the integral

$$T_t x=(2\pi)^{-1}i\int_{-\infty}^\infty e^{-(1+i\tau)t}R(1+i\tau, A)x d\tau, \; x\in D(A), t>0,$$

to the path $\tilde\lambda(s)=2^{-1}\sigma(s)+i\tau(s)$, we see that $T_t'x$ exists for every $x\in X$ and every $t>0$. Hence we easily see that (1.13)′ holds good.

---

4) The proof of the "only if part" is suggested by E. Hille, loc. cit. in 3). That the condition (1.14)′ implies the existence of T_t' for $t>0$ has already been proved in Hille and Phillips, loc. cit. in 1).

Fractional powers of infinitesimal generators and the analyticity of the semi-groups generated by them

Proc. Japan Acad. **36** (1960) 86–89

(Comm. by Z. SUETUNA, M.J.A., March 12, 1960)

1. Consider a one-parameter semi-group of bounded linear operators $T_t(t \geq 0)$ on a Banach space X into X:

(1) $\qquad\qquad T_t T_s = T_{t+s}, \quad T_0 = I$ (the identity operator),

(2) $\qquad\qquad \text{strong-lim}_{t \to t_0} T_t x = T_{t_0} x, \; x \in X,$

(3) $\qquad\qquad \sup_t \| T_t \| < \infty.$

The infinitesimal generator A of the semi-group T_t is defined by

(4) $\qquad\qquad Ax = \text{strong-lim}_{h \downarrow 0} h^{-1}(T_h - I)x.$

It is known that A is a closed linear operator whose domain $D(A)$ is strongly dense in X. A fractional power

(5) $\qquad\qquad -(-A)^\alpha, \; (0 < \alpha < 1),$

of A was defined by S. Bochner[2] and R. S. Phillips[3] as the infinitesimal generator of the semi-group

(6) $\qquad\qquad \widehat{T}_t x = \widehat{T}_{t,\alpha} x = \int_0^\infty T_\lambda x \, d\gamma_{t,\alpha}(\lambda),$

where the measure $d\gamma_{t,\alpha}(\lambda) \geq 0$ is defined through the Laplace integral

(7) $\qquad \exp(-ta^\alpha) = \int_0^\infty \exp(-\lambda a) d\gamma_{t,\alpha}(\lambda), \; (t, a > 0 \text{ and } 0 < \alpha < 1).$

The purpose of the present note is to prove that this semi-group $\widehat{T}_t = \widehat{T}_{t,\alpha}$ is analytic in t,[4] or more precisely, that $\widehat{T}_t$ belongs to the class of semi-groups introduced in a previous note.[5]

For any $x \in X$ and for any $t > 0$, $\widehat{T}_t x = \widehat{T}_{t,\alpha} x$ is strongly differentiable in t, and $\widehat{T}_t' x = \text{strong-lim}_{h \downarrow 0} h^{-1}(\widehat{T}_{t+h} - \widehat{T}_t)x$ satisfies

---

1) Dedicated to Prof. Zyoiti Suetuna on his 60th Birthday.

2) Diffusion equations and stochastic processes, Proc. Nat. Acad. Sci., **35**, 369–370 (1949).

3) On the generation of semi-groups of linear operators, Pacific J. Math., **2**, 343–369 (1952).

4) Originally the author proved the analyticity for the case $0 < \alpha \leq 1/2$. It was communicated to Prof. Tosio Kato, and he has proved the analyticity for the case $0 < \alpha < 1$ by a more general approach. See the following paper by Prof. Kato. The author wishes to express his hearty thanks to Prof. Kato for the friendly discussion.

5) K. Yosida: On the differentiability of semi-groups of linear operators, Proc. Japan Acad., **34**, 337–340 (1958). Cf. E. Hille's class $H(\Phi_1, \Phi_2)$ of semi-groups in his book: Functional Analysis and Semi-groups, New York (1948).

(8) $$\overline{\lim_{t\downarrow 0}}\, t\,\|\,\widehat{T}'_t\,\| < \infty,$$

so that[6] the semi-group $\widehat{T}_t$ can, as an abstract function of t, be extended analytically into a sector of the complex λ-plane defined by

(9) $|\lambda - t| < Ct$, where C is a positive constant.

Remark 1. The proof given below in 2 is based upon an explicit representation of the semi-group $\widehat{T}_t$: For any θ with $\pi/2 \leqq \theta \leqq \pi$, we have

(10) $$\widehat{T}_t x = \widehat{T}_{t,\alpha} x = \int_0^\infty f_{t,\alpha}(\lambda) T_\lambda x \, d\lambda, \ x \in X,$$

where

(11) $$f_{t,\alpha}(\lambda) = \pi^{-1} \int_0^\infty \exp(\lambda r \cdot \cos\theta - tr^\alpha \cos\alpha\theta)[\sin(\lambda r \cdot \sin\theta - tr^\alpha \sin\alpha\theta + \theta)]dr.$$

From this representation we easily derive the following formulae for $-(A)^\alpha$, announced recently by A. V. Balakrishnan:[7]

(12) $$-(-A)^\alpha x = (-\Gamma(-\alpha))^{-1} \int_0^\infty \lambda^{-\alpha-1}(T_\lambda - I)x \, d\lambda, \ x \in D(A),$$

(13) $$-(-A)^\alpha x = \pi^{-1} \sin\alpha\pi \int_0^\infty \lambda^{\alpha-1}(\lambda I - A)^{-1}Ax \, d\lambda, \ x \in D(A).$$

Remark 2. Let A, $-A$ and A^2 be infinitesimal generators of semi-groups. Then a "Hilbert transform C_A associated with A" shall be defined through

$$C_A \cdot Ax = -(-A^2)^{1/2}x = \pi^{-1} \int_0^\infty \lambda^{-1/2}(\lambda I - A^2)^{-1}A^2 \cdot x \, d\lambda$$

(14) $$= \pi^{-1} \int_0^\infty 2^{-1}\lambda^{-1/2}\{(\lambda^{1/2}I - A)^{-1} - (\lambda^{1/2}I + A)^{-1}\}A \cdot x \, d\lambda$$

$$= \pi^{-1} \int_0^\infty \{(\lambda I - A)^{-1} - (\lambda I + A)^{-1}\}Ax \, d\lambda, \ x \in D(A^2).$$

This definition is suggested by the following situation: Let $(Ax)(s) = dx(s)/ds$ for $x(s) \in C[-\infty, \infty]$. Then

$$((\lambda I - A)^{-1}x)(s) = \int_0^\infty \exp(-\lambda t)x(s+t) \, dt,$$

$$((\lambda I + A)^{-1}x)(s) = \int_0^\infty \exp(-\lambda t)x(s-t) \, dt,$$

so that

---

6) See the note referred to in 5).

7) Representation of abstract Riesz potentials of the elliptic type, Bull. Amer. Math. Soc., **64**, no. 5, 288–289 (1958). Fractional powers of closed operators and the semi-groups generated by them, ibid., abstract, no. 558-23 (1959). Cf. M. A. Krasno-selski and P. E. Sobolevski: Fractional power operators defined on Banach spaces (in Russian), Doklàdy Academy Nauk, **129**, no. 3, 499–502 (1959).

$$[-(-A^2)^{1/2}x](s)=\pi^{-1}\lim_{\epsilon\downarrow 0}\int_0^\infty d\lambda\left\{\int_\epsilon^\infty \exp(-\lambda t)(x'(s+t)-x'(s-t))dt\right\}$$

$$(15) \qquad =\pi^{-1}\lim_{\epsilon\downarrow 0}\int_\epsilon^\infty t^{-1}(x'(s+t)-x'(s-t))\,dt$$

$$=\text{the Hilbert transform of } (Ax)(s).^{8)}$$

2. We shall give the proof of the result in 1. Inverting the Laplace integral (7), we see that the measure $d\gamma_{t,\alpha}(\lambda)$ has the density $f_{t,\alpha}(\lambda)$ given by

$$(16) \qquad f_{t,\alpha}(\lambda)=(2\pi i)^{-1}\int_{\sigma-i\infty}^{\sigma+i\infty}\exp(z\lambda-z^\alpha t)dz \quad \text{(for any } \sigma>0).$$

Hence we obtain (10)–(11) by deforming the path of integration in (16) to the union of two paths: $r\cdot\exp(-i\theta)$ $(\infty>r>0)$ and $r\cdot\exp(i\theta)$ $(0<r<\infty)$. Taking $\theta=\theta_\alpha=\pi/(1+\alpha)$ in (10)–(11) and differentiating with respect to t, we obtain

$$(17) \quad \widehat{T'_t}x=\pi^{-1}\int_0^\infty T_\lambda x\,d\lambda\left\{\int_0^\infty \exp((\lambda r+tr^\alpha)\cos\theta_\alpha)\cdot[\sin((\lambda r-tr^\alpha)\sin\theta_\alpha)]r^\alpha dr\right\}.$$

This formal differentiation is justified, since the right hand side reduces, upon changing the variables of integration, to

$$(18) \quad (t\pi)^{-1}\int_0^\infty T_{\nu t^{1/\alpha}}\cdot x\,d\nu\left\{\int_0^\infty \exp((s\nu+s^\alpha)\cos\theta_\alpha)[\sin((s\nu-s^\alpha)\sin\theta_\alpha)]\,s^\alpha ds\right\},$$

which is, by $\cos\theta_\alpha<0$ and (3), uniformly convergent in $t\geq t_0$ for any fixed $t_0>0$. At the same time we have proved (8).

By a similar argument as above, we see, by (7), that

$$(19) \qquad \int_0^\infty (\partial f_{t,\alpha}(\lambda)/\partial t)\,d\lambda=0.$$

Hence we obtain, from (17),

$$\widehat{T'_t}x=\pi^{-1}\int_0^\infty (T_\lambda-I)x\,d\lambda\left\{\int_0^\infty \exp((\lambda r+tr^\alpha)\cos\theta_\alpha)\right.$$

$$(20) \qquad \left.\cdot[\sin((\lambda r-tr^\alpha)\sin\theta_\alpha)]r^\alpha\,dr\right\}.$$

If $x\in D(A)$, then $\lim_{\lambda\downarrow 0}\|(T_\lambda-I)\lambda^{-1}x\|=\|Ax\|$ and $\overline{\lim}_{\lambda\to\infty}\|(T_\lambda-I)x\|<\infty$. Thus we obtain, by letting $t\downarrow 0$ in (20),

$$\text{strong-}\lim_{t\downarrow 0}T'_tx=\pi^{-1}\int_0^\infty (T_\lambda-I)x\,d\lambda\left\{\int_0^\infty \exp(\lambda r\cdot\cos\theta_\alpha)\right.$$

$$(21) \qquad \left.\cdot[\sin(\lambda r\cdot\sin\theta_\alpha)]r^\alpha dr\right\}$$

$$=(-\Gamma(-\alpha))^{-1}\int_0^\infty \lambda^{-\alpha-1}(T_\lambda-I)x\,d\lambda,$$

because

---

8) Cf. p. 605 in E. Hille and R. S. Phillips: Functional Analysis and Semi-groups, Providence (1957).

$$\pi^{-1}\int_0^\infty \exp(\lambda r\cdot\cos\theta_\alpha)\cdot\sin(\lambda r\cdot\sin\theta_\alpha)r^\alpha dr$$

$$(22)\quad =(2\pi)^{-1}\cdot i\cdot\Gamma(1+\alpha)[(-\lambda\cos\theta_\alpha+i\lambda\sin\theta_\alpha)^{-1-\alpha}$$
$$-(-\lambda\cos\theta_\alpha-i\lambda\sin\theta_\alpha)^{-1-\alpha}]$$
$$=(-\Gamma(-\alpha))^{-1}\lambda^{-\alpha-1}.$$

Therefore (12) is proved by $\widehat{T}'_t x=(-(-A)^\alpha)\widehat{T}_t x$ $(t>0)$, the strong continuity in t of $\widehat{T}_t$ and the closure property of the infinitesimal generator $-(-A)^\alpha$.

Lastly, by making use of

$$\Gamma(1+\alpha)\lambda^{-\alpha-1}=\int_0^\infty \exp(-\lambda t)t^\alpha\, dt$$

and the resolvent formula

$$(23)\qquad (\lambda I-A)^{-1}x=\int_0^\infty \exp(-\lambda t)T_t x dt,\ x\in X,$$

in semi-group theory, we obtain (13) from (12) because of
$$(\lambda I-A)^{-1}\cdot A\cdot x=\{\lambda(\lambda I-A)^{-1}-I\}x,\ \ x\in D(A).$$

Remark 3. If we take $\theta=\pi$ in (10)–(11) and make use of (23) to the semi-group $\widehat{T}_t$, we obtain another proof of the formula due to T. Kato:[9]

$$(\mu+(-A)^\alpha)^{-1}=\int_0^\infty \exp(-\mu t)\,\widehat{T}'_{t,\alpha}dt$$

$$(24)\ =\pi^{-1}\int_0^\infty dr\int_0^\infty \exp(-\lambda r)T_\lambda d\lambda\int_0^\infty \exp(-\mu t-tr^\alpha\cos\alpha\pi)\sin(tr^\alpha\sin\alpha\pi)dt$$

$$=\frac{\sin\alpha\pi}{\pi}\int_0^\infty (r-A)^{-1}\frac{r^\alpha}{\mu^2-2\mu r^\alpha\cos\alpha\pi+r^{2\alpha}}\,dr.$$

---

9) See Kato's paper referred to in 4).

On a class of infinitesimal generators and the integration problem of evolution equations

Proc. Fourth Berkeley Sympos. on Math. Stat. and Prob. (Univ. Calif. Press)
II (1961) 623–633

1. Introduction

The theory of semigroups of bounded linear operators deals with exponential functions in infinite dimensional function spaces. It has been used, as an operator-theoretical substitute for the Laplace transform method, in the integration problem of temporally homogeneous evolution equations, especially of diffusion equations and wave equations (see Hille and Phillips [5] and Yosida [20], [21]).

The purpose of my paper is to call attention to a class of semigroups which is characterized by either one of the three mutually equivalent conditions to be explained below; one of them reads that the semigroup T_t satisfies

$$(1) \qquad \lim_{t \downarrow 0} t \left\| \frac{d}{dt} T_t \right\| < \infty.$$

The semigroups arising from the integration in L_2 of temporally homogeneous diffusion equations belong to this class. And the unique continuation theorem of diffusion equations, inaugurated by Yamabe and Itô [6] may be explained by the time-like analyticity of the corresponding semigroups. The situation has an intimate connection with the theory of analytical vectors published recently by Nelson [14]. There is a procedure to obtain semigroups of our class. Let A be the infinitesimal generator of a contraction semigroup. We can define, following Bochner [3], Feller [4], Phillips [15], and Balakrishnan [1], the fractional powers $-(-A)^\alpha$ of A and the semigroups generated by them belong to our class. Balakrishnan gave an interesting application of the operator $-(-A)^{1/2}$ to Hille's reduced Cauchy problem for equations $d^2u/dt^2 + Au = 0$.

Tanabe [19] has recently devised an ingenious method of integration of temporally inhomogeneous evolution equations in Banach spaces: $du/dt = A(t)u$. He assumes, for fixed t, that $A(t)$ is the infinitesimal generator of a semigroup of our class. He further assumes a certain regularity condition with respect to t of $A(t)$ which is the same as that introduced by Kato [8] for the integration of such equations. Under these conditions, Tanabe proved that the solution may be obtained by successive approximation starting with the first approximation $\exp[(t - s)A(s)]$. In this way, he has shown that Levi's classical construction

623

of the fundamental solution $U(t, s)$, with $t > s$, of a diffusion equation may be adapted to the evolution equations in Banach spaces. Komatsu [10] gave an important remark that, if $A(t)$ is analytic in t, then the fundamental solution $U(t, s)$ of Tanabe is also analytic in t and s. In this way, Komatsu proved a unique continuation theorem for temporally inhomogeneous diffusion equations, which was proved in a direct way by Shirota [17].

2. A class of semigroups

Let X be a complex Banach space. A one-parameter family T_t, with $t \geqq 0$, of bounded linear operators in X is said to be a semigroup of type $S(M, \beta)$ if it satisfies the conditions

$$(2) \qquad T_t T_s = T_{t+s}, \qquad T_0 = I \text{ (the identity)},$$

$$(3) \qquad s - \lim_{t \to t_0} T_t x = T_{t_0} x, \qquad x \in X; t_0 \geqq 0.$$

Such a semigroup satisfies, as was proved by Hille [5]

$$(4) \qquad ||T_t|| \leqq M e^{\beta t}, \qquad t \geqq 0,$$

with positive constants M and β. It is well known that the infinitesimal generator A of T_t defined by $A \cdot x = s - \lim_{t \downarrow 0} t^{-1} (T_t - I)x$ generates T_t by various equivalent procedures; one of them states that

$$(5) \qquad T_t x = s - \lim_{n \to \infty} \left(I - \frac{t}{n} A \right)^{-n} x.$$

Hence we shall write $T_t = \exp (tA)$.

Let θ be a positive constant $\leqq \pi/2$, and Σ_θ be the sector $|\arg z| < \theta$ in the complex z-plane. If a semigroup $T_t \in S(M, \beta)$ is analytically continuable into Σ_θ in such a way that $T_{t \exp(i\varphi)}$ is, for all φ with $|\varphi| < \theta$, of type $S(M', \beta')$ with a fixed pair (M', β'), then we say that T_t is of type $H(\theta, M', \beta')$.

Our class of semigroups is characterized by any one of the three conditions in the following

THEOREM 1. *Let A be a closed linear operator with domain $D(A)$ dense in X and range in X. Then the following three conditions are mutually equivalent*

$$(6) \qquad \exp (tA) \in H(\theta, M, \beta) \qquad \textit{for some } \theta, M, \textit{ and } \beta,$$

$(7) \qquad \exp (tA) \in S(M', \beta')$ *for some M', β', and $\exp (tA)$ is strongly*

differentiable in t in such a way that $\overline{\lim}_{t \downarrow 0} t \left\| \dfrac{d}{dt} \exp (tA) \right\| < \infty,$

$(8) \qquad$ *there exist positive constants M'' and β'' such that*

$$||(\lambda I - A)^{-n}|| \leqq M''(\lambda - \beta'')^{-n} \qquad for \qquad \lambda > \beta'', \qquad n = 1, 2, \cdots,$$

$$\overline{\lim_{|\tau| \uparrow \infty}} |\tau| \cdot ||[(\sigma + \sqrt{-1}\, \tau)I - A]^{-1}|| < \infty, \qquad for \qquad \sigma > \beta''.$$

The constants $M, \beta, M', \beta', M''$, and β'' are dependent on each other. For the proof of the equivalence of (6) and (7) see Yosida [22], and for that of (6) and (8) see Yosida [22] and Hille and Phillips [5]. We remark that the following facts are used in these proofs. First, (7) implies

$$(7') \qquad \frac{d^n}{dt^n} \exp(tA) = A^n \exp(tA) = \left[A \exp\left(\frac{t}{n} A\right)\right]^n.$$

Secondly, the last condition in (8) implies that $(\lambda I - A)^{-1}$ exists and is analytic in λ for large value of $|\tau| = |\mathrm{Im}(\lambda)|$ outside a sector of the left half λ-plane defined by the boundary curve of the form $\lambda(s) = \sigma(s) + i\tau(s)$ such that

$$(9) \qquad \lim_{\tau(s)\uparrow\infty} \frac{-\sigma(s)}{\tau(s)} = \tan\epsilon = \lim_{\tau(s)\downarrow-\infty} \frac{\sigma(s)}{\tau(s)}, \qquad \epsilon > 0.$$

Moreover, $\|(\lambda I - A)^{-1}\|$ is of the order $|\tau|^{-1}$ when $|\tau|$ tends to ∞ outside the above sector and lying in the left half λ-plane. Thirdly, the representation theorem of Hille holds for the semigroup satisfying (8)

$$(10) \qquad \exp(tA) = \frac{1}{2\pi i} \int_{\lambda(s)} e^{\lambda t}(\lambda I - A)^{-1}\, d\lambda, \qquad t > 0,$$

the integral being taken in the uniform operator topology along the path of integration $\lambda(s) = 2^{-1}\sigma(s) + i\tau(s)$.

3. Unique continuation theorem of the diffusion equations

Consider a diffusion equation

$$(11) \qquad \frac{\partial u}{\partial t} = Au, \qquad t > 0,$$

where the differential operator

$$(12) \qquad A = a^{ij}(x)\frac{\partial^2}{\partial x_i \partial x_j} + b^i(x)\frac{\partial}{\partial x_i} + c(x)$$

is strongly elliptic in a connected region G of an m-dimensional Euclidean space E^m. For the sake of simplicity of the exposition we assume that $G = E^m$. We assume that the real-valued coefficients a, b, and c are C^∞ in E^m and that

(13) $a^{ij}(x)$ *and its first and second partials,* $b^i(x)$ *and its first partials, and* $c(x)$ *are, in absolute values, all bounded on* E^m *by a positive constant* γ *and* δ *such that*

$$(14) \qquad \gamma \sum_{j=1}^{m} \xi_j^2 \geqq a^{ij}(x)\xi_i\xi_j \geqq \delta \sum_{j=1}^{m} \xi_j^2 \qquad on \qquad E^m$$

for every real vector $(\xi_1, \xi_2, \cdots, \xi_m)$.

Let $H_1 = H_1(E^m)$ be the space of complex-valued C^∞ functions $f(x) = f(x_1, x_2, \cdots, x_m)$ in E^m for which

$$(15) \qquad \|f\|_1 = \left(\int_{E^m} |f(x)|^2\, dx + \sum_{j=1}^{m} \int_{E^m} |f_{x_i}(x)|^2\, dx\right)^{1/2} < \infty.$$

Then the completion of H_1 by the norm

$$(16) \qquad ||f|| = \left(\int_{E^m} |f(x)|^2 \, dx \right)^{1/2}$$

is the space $L_2(E^m) = L_2$.

LEMMA 1.   *Let us consider A as an operator defined on $\{f; f \in H_1, Af \in H_1\} \subseteq L_2$ into L_2. Then the smallest closed extension $\hat{A}$, in L_2, of A is the infinitesimal generator of a semigroup $T_t = \exp(tA)$ of type $S(1, \beta)$.*

For the proof, see Yosida [21] and [23]. See also Phillips [15]. The proof is based upon the Milgram-Lax theorem [7].

LEMMA 2.   *A satisfies the condition (8).*

PROOF.   (Yosida [23]. Compare Phillips [16].) Let $\sigma > 0$ be sufficiently large. Then we obtain, by partial integration, and by making use of the inequality $|\epsilon\kappa| \leq 2^{-1}(|\epsilon|^2 + |\kappa|^2)$,

$$(17) \qquad \text{Re} \{[(\sigma + \sqrt{-1}\,\tau)I - A]w, w\}$$

$$= \sigma||w||^2 + \text{Re} \left\{ \int_{E^m} a^{ij} \frac{\partial w}{\partial x_i} \frac{\partial \bar{w}}{\partial x_j} \, dx + \int_{E^m} \frac{\partial a^{ij}}{\partial x_i} \frac{\partial w}{\partial x_j} \bar{w} \, dx \right.$$

$$\left. - \int_{E^m} b^i \frac{\partial w}{\partial x_i} \bar{w} \, dx - \int_{E^m} cw\bar{w} \, dx \right\}$$

$$\geq (\sigma - \delta - \eta_0)||w||^2 + (\delta - m\eta\nu)||w||_1^2,$$

where $(\delta - m\eta\nu) > 0$ and $\eta_0 = m\eta(m\nu^{-1} - \nu + m^{-1}) > 0$ for sufficiently small $\nu > 0$. Similarly we have

$$(18) \qquad |\text{Im} \{[(\sigma + \sqrt{-1}\,\tau)I - A]w, w\}|$$

$$\geq |\tau| \cdot ||w||^2 - m\eta\{||w||_1^2 + m||w||^2\} = (|\tau| - m^2\eta)||w||^2 - m\eta||w||_1^2.$$

If we assume that there exists $w \in H_1$, with $w \neq 0$, and sufficiently large $|\tau|$ such that $m\eta||w||_1^2 \geq 2^{-1}(|\tau| - m^2\eta)||w||^2$, then

$$(19) \qquad \text{Re} \{[(\sigma + \sqrt{-1}\,\tau) - A]w, w\} \geq (\delta - m\eta\nu) \frac{|\tau| - m^2\eta}{2m\eta} ||w||^2.$$

Therefore, by virtue of the Schwarz inequality,

$$(20) \qquad ||[(\sigma + \sqrt{-1}\,\tau) - A]w|| \cdot ||w|| \geq |\{[(\sigma + \sqrt{-1}\,\tau) - A]w, w\}|,$$

we see that A satisfies the latter condition in (8). The former condition in (8) is clear since we have proved, in lemma 1, that $\hat{A}$ is the infinitesimal generator of a semigroup of type $S(1, \beta)$.

LEMMA 3.   *For any $f \in L$, $u(t, x) = \exp(t\hat{A})f(x)$ is C^∞ in $t > 0$ and in $x \in E^m$ and satisfies the Cauchy problem*

$$\frac{\partial u}{\partial t} = Au, \qquad\qquad t > 0,$$

$$(21)$$

$$L_2 - \lim_{t \downarrow 0} u(t, x) = f(x).$$

PROOF.   If we apply, in the sense of the distribution of Schwartz, the elliptic differential operators $(\partial^2/\partial t^2 + A)$ any number of times to $u(t, x)$, then the result is locally square integrable in the product space $(0 < t < \infty) \times E^m$. Thus, by the Weyl-Schwartz theorem, $u(t, x)$ is equivalent to a function which is C^∞ in $(0 < t < \infty) \times E^m$. Hence the lemma is proved.

Now, by the time-like analyticity, as proved in lemma 2, of the semigroup $T_t = \exp(t\hat{A}) \in H(\theta, M, \beta)$, it is easy to see that the above solution $u(t, x)$ satisfies the *time-like unique continuation theorem*, Yosida [23]: if, for a certain $t_0 > 0$, $u(t_0, x) = 0$ on an open domain G of E^m, then $u(t, x) = 0$ for all $t > 0$ and all $x \in G$. Hence, by applying the space-like *unique continuation theorem* of Mizohata [13], we see that $u(t, x)$ satisfies the *unique continuation theorem*, Itô and Yamabe [6]: if, for a certain $t_0 > 0$, $u(t_0, x) = 0$ on an open domain G of E^m, then $u(t, x) = 0$ for all $t > 0$ and all $x \in E^m$.

REMARK.   By virtue of (7$'$) and (7), we see that $u(t, x) = \exp(tA)f(x)$ is an analytic vector in the sense of Nelson [14], that is, for any fixed $t > 0$,

$$(22) \qquad \sum_{n=1}^{\infty} \epsilon^n (n!)^{-1} \|A^n \exp(t_0 A)f\| < \infty$$

for sufficiently small $\epsilon > 0$. Hence, if we assume that the coefficients of the differential operator A are real analytic in x, then $u(t, x)$ is real analytic in x. Therefore, in this case, the unique continuation theorem for $u(t, x)$ may be proved without appealing to Mizohata's result. This observation is due to Komatsu [10], [11].

4. Fundamental solutions of temporally inhomogeneous evolution equations

Consider an equation of evolution

$$(23) \qquad \frac{du}{dt} = A(t)u(t), \qquad\qquad a \leqq t \leqq b,$$

where $u(t) \in X$ and $A(t)$ is a linear operator in X. Such an equation was investigated by Kato [8] under the assumption that $A(t) \in S(1, 0)$, and recently by Tanabe [19] under the assumption that $A(t) \in H(\theta, M, \beta)$. A family of bounded linear operators $U(t, s)$ in X is called a fundamental solution of (23) if it satisfies the following conditions:

(24)   *$U(t, x)$ is defined for $t \geqq s$ and is strongly continuous there,*

$$(25) \qquad\qquad U(t, t) = I,$$

(26)   *for every $x \in D[A(s)]$, $U(t, s)x$ belongs to the domain $D[A(t)]$ and*

is strongly differentiable in t such that $\dfrac{d}{dt} U(t, s)x = A(t)U(t, s)x$.

If a fundamental solution $U(t, s)$ exists and if the uniqueness theorem for the Cauchy problem of (23) holds, then every solution $u(t)$ of (23) is expressed as

$u(t) = U(t, s)u(s)$. The uniqueness theorem was proved by Kato [8] under the sole solution that $A(t) \in S(1, 0)$.

Tanabe proved that (23) has a fundamental solution under the following conditions:

(27) $\exp\left[\tau A(t)\right] \in S(\theta, M, \beta)$ *for* $t \in [a, b], \tau > 0,$ *and that the constants* $\theta, M,$ *and* β *are independent of* t,

(28)   *the domain* $D[A(t)] = D$ *is independent of* t,

(29)   *there exists a bounded linear operator* A_0 *which maps* X *onto* D *in a one-to-one manner and such that*

(30) $$B(t) = A(t)A_0$$

is uniformly Lipschitz continuous in t *in the uniform topology of operators and is strongly continuously differentiable in* t.

Tanabe's construction of the fundamental solution may be written as

$$U(t, s) = \exp\left[(t - s)A(s)\right] + W(t, s),$$

$$W(t, s) = \int_s^t \exp\left[(t - r)A(r)\right]R(r, s)\, dr,$$

(31) $$R(t, s) = \sum_{m=1}^{\infty} R_m(t, s),$$

$$R_1(t, s) = \begin{cases} [A(t) - A(s)]\exp\left[(t - s)A(s)\right], & t > s, \\ 0, & t = s, \end{cases}$$

$$R_m(t, s) = \int_s^t R_1(t, r)R_{m-1}(r, s)\, dr, \qquad m = 2, 3, \cdots.$$

He proved that every operator appearing in the above formulas is strongly continuous in t and s, with $s < t$, and that every integral and series converge. The crucial points in his proof are revealed in the following lemmas.

LEMMA 4. $\exp\left[(t - s)A(s)\right]$ *is strongly differentiable in* s *and* t *and* $(\partial/\partial t + \partial/\partial s)\exp\left[(t - s)A(s)\right]$ *is uniformly bounded in* $a \leq s < t \leq b$.

LEMMA 5.   *There exist positive constants* $K_1, K_2,$ *and* $\rho, 0 < \rho < 1,$ *such that, when* $a \leq s < \tau < t \leq b,$ *we have*

(32) $$\|R(t, s) - R(\tau, s)\| \leq K_1(t - \tau)(t - s)^{-1} + K_2(t - s)^\rho(t - \tau)^{1-\rho}.$$

Tanabe applied his result to the integration of the temporally inhomogeneous diffusion equation

(33) $$\frac{\partial u}{\partial t} = a^{ij}(t, x)\frac{\partial^2 u}{\partial x_i \partial x_j} + b^i(t, x)\frac{\partial u}{\partial x_i} + c(t, x)u + f(t, x)$$

in a bounded domain of E^m and for $a \leq t \leq b$.

5. Analyticity of the fundamental solution

Komatsu [10] gave an important remark to Tanabe's result. Let Δ be a convex complex neighborhood of the real segment $[a, b]$. Suppose that $A(t)$ is defined on Δ and satisfies the conditions

(34) $\qquad A(t) \in S(\theta, M, \beta)$ *for* $t \in \Delta$, *where* θ, M, *and* β *are independent of* t,

(35) $\qquad D[A(t)] = D$ *is independent of* $t \in \Delta$,

(36) $\qquad$ *there exists a bounded linear operator* A_0 *which maps* X *onto* D *in a one-to-one manner and such that* $B(t) = A(t)A_0$ *is analytic in* t *for* $t \in \Delta$.

Under these conditions, Komatsu [11] proved that the fundamental solution $U(t, s)$ of (25) constructed as in (33) is analytic in t and s if $|\arg (t - s)| < \theta$. His proof is based upon the following

LEMMA 6. *We write* $t >_\theta s$ *when* $|\arg (t - s)| < \theta$. *Let* $P(t, s)$ *and* $Q(t, s)$ *be bounded linear operators in* X *defined for* $t >_\theta s$ *with* t *and* $s \in \Delta$. *If they are uniformly bounded and analytic there, then*

$$(37) \qquad \int_s^t P(t, r)Q(r, s)\, dr$$

is uniformly bounded and analytic in t *and* $s \in \Delta$ *when* $t >_\theta s$.

This result may be applied, as in the case of temporally homogeneous equations discussed in section 3, to the unique continuation theorem of diffusion equations. This is Komatsu's proof of the extension of Shirota [17], [18] of the unique continuation theorem of Itô and Yamabe [6].

6. Fractional powers of infinitesimal generators and the analyticity of the semigroups generated by them

Let $T = \exp (tA)$, with $t > 0$, be a semigroup of type $S(M, 0)$. A fractional power of A

$$(38) \qquad\qquad -(-A)^\alpha, \qquad\qquad 0 < \alpha < 1,$$

was defined by Bochner [3] and Phillips [15] as the infinitesimal generator of the semigroup

$$(39) \qquad \hat{T}_t x = \hat{T}_{t,\alpha} x = \int_0^\infty T_\lambda x\, d\gamma_{t,\alpha}(\lambda),$$

where the measure $d\gamma_{t,\alpha}(\lambda) \geqq 0$ is defined through the Laplace integral

$$(40) \qquad \exp (-ta^\alpha) = \int_0^\infty \exp (-\lambda a)\, d\gamma_{t,\alpha}(\lambda), \qquad t, a > 0; 0 < \alpha < 1.$$

We (Kato [9], Yosida [24], and Balakrishnan [1]) can prove that the semigroup T_t is of type $S(\theta, M, \beta)$. To this purpose, invert the Laplace integral (40). Then we see that the measure $d\gamma_{t,\alpha}(\lambda)$ has the density $f_{t,\alpha}(\lambda)$ given by

$$(41) \qquad f_{t,\alpha}(\lambda) = (2\pi i)^{-1} \int_{\sigma-i\infty}^{\sigma+i\infty} \exp (z\lambda - z^\alpha t)\, dz \qquad \text{for any} \quad \sigma > 0,$$

so that we have

$$(39') \qquad \hat{T}_t x = \hat{T}_{t,\alpha} x = \int_0^\infty f_{t,\alpha}(\lambda) T_\lambda x \, d\lambda.$$

Take any θ with $\pi/2 \leq \theta \leq \pi$. Then we obtain

$$(41') \qquad f_{t,\alpha}(\lambda) = \pi^{-1} \int_0^\infty \exp\left(\lambda r \cos\theta - t r^\alpha \cos\alpha\theta\right)$$

$$[\sin\left(\lambda r \sin\theta - t r^\alpha \sin\alpha\theta + \theta\right)] \, dr$$

by deforming the path of integration in (41) to the union of two paths

$$(42) \qquad \begin{aligned} r \exp(-i\theta), & \qquad \infty > r > 0, \\ r \exp(i\theta), & \qquad 0 < r < \infty. \end{aligned}$$

Taking $\theta = \theta_\alpha = \pi/(1 + \alpha)$ in (41') and differentiating (39') with respect to t, we obtain

$$(43) \qquad \frac{d}{dt} \hat{T}_t x = \int_0^\infty T_\lambda x \, d\lambda \left(\int_0^\infty \exp\left[(\lambda r + t r^\alpha) \cos\theta_\alpha\right] \right.$$

$$\left. \{\sin\left[(\lambda r - t r^\alpha) \sin\theta_\alpha\right]\} r^\infty \, dr \right).$$

This formal differentiation is justified, since the right side reduces, upon changing the variables of integration, to

$$(44) \qquad (t\pi)^{-1} \int_0^\infty T_{\nu t^{1/\alpha}} \, d\nu \left(\int_0^\infty \exp\left[(s\nu + s^\alpha) \cos\theta_\alpha\right] \{\sin\left[(s\nu - s^\alpha) \sin\theta_\alpha\right]\} s^\alpha \, ds \right)$$

which is, by $\cos\theta_\alpha < 0$ and $\|T_t\| \leq M$, uniformly convergent in $t \geq t_0$ for any fixed $t_0 > 0$. We have incidentally proved that $\hat{T}_t$ satisfies (7), and thus $\hat{T}_t$ belongs to the class $S(\theta, M, \beta)$.

From (43) it is easy to deduce the following formulas which were proved earlier by Balakrishnan [1] (see Krasnoselski and Sobolevski [12]),

$$(45) \qquad -(-A)^\alpha x = [-\Gamma(-\alpha)]^{-1} \int_0^\infty \lambda^{-\alpha-1}(T_\lambda - I)x \, d\lambda, \qquad x \in D(A),$$

$$= \pi^{-1} \sin\alpha\pi \int_0^\infty (\lambda I - A)^{-1} A x \, d\lambda, \qquad x \in D(A).$$

See Yosida [24] and Kato [8].

7. An application of the fractional power operators to the reduced Cauchy problem

Let A be the infinitesimal generator of a semigroup $T_t = \exp(tA) \in S(M, 0)$. Then for each $u_0 \in D(A)$, we have that $u(t) = \hat{T}_{t,1/2} u_0$ is a solution of

$$(46) \qquad \frac{d^2 u}{dt^2} + A u = 0, \qquad t \geq 0,$$

and satisfies the conditions

$$(47) \qquad\qquad s - \lim_{t \downarrow 0} u(t) = u_0, \qquad\qquad \sup_t \|u(t)\| < \infty,$$

and

$$(48) \qquad\qquad s - \lim_{t \downarrow 0} \frac{d}{dt} u(t) = s - \lim_{t \downarrow 0} \frac{d}{dt} \hat{T}_{t,1/2} u_0 = B u_0,$$

where $B = -(-A)^{1/2}$. According to Balakrishnan [2], the solution of (46) is uniquely determined by condition (47). Hence, for a solution $u(t)$ of (46) and (47), the initial condition for the first partial derivative $du(0)/dt$ is determined by (48), and other values cannot be prescribed. In this sense, the Cauchy problem for (46) and (47) is reduced. For a general definition of the reduced Cauchy problem introduced by Hille see [5].

We follow Balakrishnan's proof. Put $\hat{T}_{t,1/2} = T_{1/2}(t)$. Let $v(t)$ be a twice strongly and continuously differentiable solution of (46) satisfying the condition

$$(47') \qquad\qquad s - \lim_{t \downarrow 0} v(t) = u_0 \in D(A), \qquad\qquad \sup_t \|v(t)\| < \infty.$$

Put $w(t) = T_{1/2}(1/n)v(t)$. If we can prove, for all positive integer n, that $dw(0)/dt = Bw(0)$, then we obtain $dv(0)/dt = Bv(0)$ by letting $n \to \infty$.

To prove this fact we first observe that $w(t)$ is a solution of (46), and by (47'), $\|dw(t)/dt\|$ is of exponential growth at $t \to \infty$. Thus we see, by putting $w_0 = w(0)$ and $w_1 = dw(0)/dt$, that

$$(49) \qquad\qquad \frac{dw(t)}{dt} + Bw(t) = T_{1/2}(t)Bw_0 + T_{1/2}(t)w_1,$$

because both sides satisfy the Cauchy problem

$$(50) \qquad\qquad \frac{du}{dt} = Bu, \qquad u(0) = Bw_0 + w_1$$

and are of exponential growth at $t \to \infty$. Hence we have

$$(51) \qquad\qquad \frac{d}{dt}\left[T_{1/2}(t)w(t)\right] = T_{1/2}(2t)w_1 + T_{1/2}(2t)Bw_0,$$

so that

$$(52) \qquad\qquad T_{1/2}(t)w(t) = T_{1/2}(2t)w_0 + \frac{1}{2}\int_0^{2t} T_{1/2}(s)(w_1 - Bw_0)\,ds.$$

Hence

$$(53) \qquad\qquad T_{1/2}(t)Bw(t) = T_{1/2}(2t)Bw_0 + \frac{1}{2}\left[T_{1/2}(2t) - I\right](w_1 - Bw_0).$$

Since $T_{1/2}(t)$ satisfies (7), $Bw(t) = BT_{1/2}(1/n)v(t)$ is bounded in t by the assumption that $v(t)$ is bounded in t. On the other hand, because of the time-like analyticity of $T_{1/2}(t)$, zero does not belong to the point spectrum of $T_{1/2}(t)$ for any

$t > 0$. We write $T_{1/2}(-t)$ for the inverse of $T_{1/2}(t)$. Hence we see, by applying $T_{1/2}(-t)$ to (53),

$$(54) \qquad \sup_t \|T_{1/2}(-t)(w_1 - Bw_0)\| < \infty.$$

From this we can prove that $z = (w_1 - Bw_0) = 0$. To this purpose we put

$$(55) \qquad F(\lambda) = \int_0^\infty e^{\lambda t} T_{1/2}(-t) z \, dt, \qquad\qquad \mathrm{Re}\,(\lambda) < \infty.$$

Then $\lambda F(\lambda)$ is bounded when $\mathrm{Re}\,(\lambda) < 0$, $|\mathrm{Im}\,(\lambda)|/|\mathrm{Re}\,(\lambda)| < c$ for any positive constant c. On the other hand, it is easily verified that

$$(56) \qquad -F(\lambda) = (\lambda I - B)^{-1}$$

for λ with $\mathrm{Re}\,(\lambda) < 0$ and in the resolvent set of B. Hence $-\lambda F(\lambda)$ is the analytical extension of $\lambda(\lambda I - B)^{-1}z$ into the left half λ-plane. Moreover, as was indicated in section 2, $\lambda(\lambda I - B)^{-1}$ is bounded in a sector of the form $-\pi/2 - \epsilon \leqq \arg \lambda \leqq \pi/2 + \epsilon$ for $\epsilon > 0$. Hence, by Liouville's theorem, $-\lambda F(\lambda)$ must reduce to a constant vector. Thus, by $s - \lim_{\lambda \downarrow 0} \lambda(\lambda I - B)^{-1}z = z$, we obtain $\lambda(\lambda I - B)^{-1}z = z$. Hence $Bz = 0$ and so $T_{1/2}(t)z = z$. Therefore, by (52),

$$(57) \qquad T_{1/2}(t)w(t) = T_{1/2}(2t)w_0 + t(w_1 - Bw_0),$$

and hence the boundedness of $w(t)$ implies that $w_1 = Bw_0$.

REFERENCES

[1] A. V. Balakrishnan, "An operational calculus for infinitesimal generators of semigroups," *Trans. Amer. Math. Soc.*, Vol. 91 (1959), pp. 330–353.

[2] ———, "Fractional powers of closed operators and the semigroups generated by them," to be published.

[3] S. Bochner, "Diffusion equations and stochastic processes," *Proc. Nat. Acad. Sci. U.S.A.*, Vol. 35 (1949), pp. 369–370.

[4] W. Feller, "On a generalization of M. Riesz' potentials and the semigroups generated by them," *Comm. Sém. Math. Univ. Lund*, Vol. 21 Suppl. (1952), pp. 72–81.

[5] E. Hille and R. S. Phillips, "Functional analysis and semigroups," *Amer. Math. Soc. Colloq. Publ.*, Providence, 1957.

[6] S. Itô and H. Yamabe, "A unique continuation theorem for solutions of a parabolic equation," *J. Math. Soc. Japan*, Vol. 10 (1958), pp. 314–321.

[7] P. D. Lax and A. Milgram, "Parabolic equations," *Contributions to the Theory of Partial Differential Equations*, Annals of Mathematics Studies, No. 33, Princeton, Princeton University Press, 1954, pp. 167–190.

[8] T. Kato, "Integration of the equation of evolution in a Banach space," *J. Math. Soc. Japan*, Vol. 5 (1953), pp. 208–234.

[9] ———, "Note on fractional powers of linear operators," *Proc. Japan Acad.*, Vol. 36 (1960), pp. 94–96.

[10] H. Komatsu, "A characterization of real analytic functions," *Proc. Japan Acad.*, Vol. 36 (1960), pp. 90–93.

[11] ———, "Abstract analyticity in time and unique continuation property of solutions of a parabolic equation," to be published.

[12] M. A. Krasnoselski and P. E. Sobolevski, "Fractional power operators defined on Banach spaces," *Dokl. Akad. Nauk SSSR*, Vol. 129 (1959), pp. 499–502. (In Russian.)

[13] S. MIZOHATA, "Unicité du prolongement des solutions pour quelques opérateurs différentiels paraboliques," *Mem. Coll. Sci. Univ. Kyoto, Ser. A., Math.*, Vol. 31 (1958), pp. 219–239.

[14] E. NELSON, "Analytic vectors," *Ann. of Math.*, Vol. 70 (1959), pp. 572–615.

[15] R. S. PHILLIPS, "On the generation of semigroups of linear operators," *Pacific J. Math.*, Vol. 2 (1952), pp. 343–389.

[16] ———, "On the integration of the diffusion equation with boundary conditions," to be published.

[17] T. SHIROTA, "A unique continuation theorem of a parabolic differential equation," *Proc. Japan Acad.*, Vol. 35 (1959), pp. 455–460.

[18] ———, "A remark on my paper 'A unique continuation theorem of a parabolic differential equation'," *Proc. Japan Acad.*, Vol. 36 (1960), pp. 133–135.

[19] H. TANABE, "A class of the equations of evolution in a Banach space," *Osaka Math. J.*, Vol. 11 (1959), pp. 121–145.

[20] K. YOSIDA, "Semigroup theory and the integration problem of diffusion equations," *Proc. Internat. Congress Math.*, Vol. 1 (1954), pp. 405–420.

[21] ———, "An operator-theoretical integration of wave equations," *J. Math. Soc. Japan*, Vol. 8 (1956), pp. 79–82.

[22] ———, "On the differentiability of semigroups of linear operators," *Proc. Japan Acad.*, Vol. 34 (1958), pp. 337–340.

[23] ———, "An abstract analyticity in time for solutions of diffusion equations," *Proc. Japan Acad.*, Vol. 35 (1959), pp. 109–113.

[24] ———, "Fractional powers of infinitesimal generators and the analyticity of the semigroups generated by them," *Proc. Japan Acad.*, Vol. 36 (1960), pp. 86–89.

On the integration of the equation of evolution

J. Fac. Sci. Univ. Tokyo **I. 9** (1963) 397–402

We are concerned with the integration of the so-called equation of evolution in a Banach space X:

$$(1) \qquad dx(t)/dt = A(t)x(t), \qquad a \leq t \leq b.$$

Here the unknown $x(t)$ is an element of X depending on a real parameter t, while $A(t)$ is a given, in general unbounded, linear operator in X depending also upon t. If $A(t) = A$ is independent of t in such a way that A is the infinitesimal generator of a semi-group of class (C_0), then the solution of the initial value problem:

$$(2) \qquad dx(t)/dt = Ax(t), \quad a \leq t \leq b, \quad x(a) = x_0 \in D(A),$$

where $D(A)$ is the domain of A, is given by $x(t) = \exp((t-a)A)x_0 = \text{strong-lim}_{n \to +\infty} \exp((t-a)A_n)x_0$, where $A_n = A(I - n^{-1}A)^{-1}$. We are thus lead to the following *problem*:

Let $A(t)$ satisfy the conditions:

(3) $A(t)$ is, for $a \leq t \leq b$, a closed linear operator with dense domain $D(A(t)) \subseteq X$ and range $R(A(t)) \subseteq X$ such that, for $\lambda \geq 0$, the resolvent $(\lambda I - A(t))^{-1}$ exists with the estimate $\|\lambda(\lambda I - A(t))^{-1}\| \leq 1 + \lambda^{-1}M$ for $\lambda \geq 1$, where M is independent of λ and t,

(4) The strong derivative $dA(t)^{-1}/dt = B(t)$ exists and is strongly continuous in t, $a \leq t \leq b$.

We now consider the following initial value problem

(5) $dx_n(t)/dt = A_n(t)x_n(t), \; a \leq t \leq b, \; x_n(a) = A(a)^{-1}y$, where y is an arbitrary element of X and $A_n(t) = A(t)(I - n^{-1}A(t))^{-1} = n(J_n(t) - I), \; J_n(t) = (I - n^{-1}A(t))^{-1}$.

Because of (4), $J_n(t)$ is strongly continuously differentiable with respect to t, and so $J_n(t)$ is uniformly continuous in t in the sense of the operator norm. Thus problem (5) has a uniquely determined solution $x_n(t)$ which may be obtained, for instance, by successive approximation starting with the first approximation $\exp((t-a)A_n(a))A(a)^{-1}y$. We are thus lead to the *problem* of finding conditions concerning $A(t)$ under which the sequence $x_n(t)$ converges, strongly or weakly, to the solution $x(t)$ of the initial value problem

---

1) Presented at the Conference on Analysis held at Yale University in June, 1962 honoring Professor Einar Hille.

$(1)'$ $$dx(t)/dt = A(t)x(t), \quad a \leq t \leq b, \quad x(a) = A(a)^{-1}y.$$

The purpose of the present note is to give partial answers to this problem, based upon a uniqueness lemma and the reflexivity of X.

LEMMA. The solution of $(1)'$, if it exists, must satisfy the estimate

(6) $$\|x(t)\| \leq \|x(a)\| \exp((t-a)M).$$

PROOF. We follow an argument which is essentially due to T. Kato [1]. Since $x(t)$ is in $D(A(t))$, we have, for $\delta > 0$,

$$\begin{aligned}
x(t+\delta) &= x(t) + \delta A(t)x(t) + o(\delta) \\
&= (I + \delta A(t))(I - \delta A(t))(I - \delta A(t))^{-1}x(t) + o(\delta) \\
&= (I - \delta A(t))^{-1}x(t) - \delta^2 A(t)(I - \delta A(t))^{-1}A(t)x(t) + o(\delta) \\
&= (I - \delta A(t))^{-1}x(t) - \delta((I - \delta A(t))^{-1} - I)A(t)x(t) + o(\delta).
\end{aligned}$$

Because of (3), we have strong-$\lim_{\delta \downarrow 0} (I - \delta A(t))^{-1}z = z$ for any $z \in X$. Thus, by the estimate in (3), we have

$$\|x(t+\delta)\| \leq (1 + \delta M)\|x(t)\| + o(\delta),$$

and so $d^+\|x(t)\|/dt \leq M\|x(t)\|$, proving the lemma.

THEOREM 1. We assume that X is a reflexive Banach space and, besides (3) and (4), that

(7) $A(t)B(t) = A(t)[dA(t)^{-1}/dt]$ is strongly continuous in t.

Then there exists a uniquely determined solution of $(1)'$ which is obtained as $x(t)$ = weak-limit$_{n \to \infty}$ $x_n(t)$.

PROOF. Since $A_n(t) = n(J_n(t) - I)$ and $\|J_n(t)\| \leq 1 + n^{-1}M$, we have, for $\delta > 0$,
$$\|(I - \delta A_n(t))x\| = \|(I - \delta n(J_n(t) - I))x\| \geq (1 + n\delta)\|x\| - \delta n(1 + n^{-1}M)\|x\| = (1 - \delta M)\|x\|.$$
Hence, for $0 < \delta < M^{-1}$, the bounded inverse $(I - \delta A_n(t))^{-1}$ exists with the estimate $\|(I - \delta A_n(t))^{-1}\| \leq (1 - \delta M)^{-1}$. Therefore, by the Lemma, we see that the solution $x_n(t)$ of (5) satisfies

(8) $$\|x_n(t)\| \leq \|x_n(a)\| \exp((t-a)M'),$$

where M' is independent of n and t. Thus $x_n(t)$ is uniquely determined by the initial condition $x_n(a) = A(a)^{-1}y$ and so we may put

$9)$ $\quad x_n(t) = U_n(t, a)x_n(a) = U_n(t, a)A(a)^{-1}y$, where $\|U_n(t, a)\| \leq \exp((t-a)M')$.

We next put

(10) $$y_n(t) = A_n(t)U_n(t, a)A(a)^{-1}y = A_n(t)x_n(t) = dx_n(t)/dt.$$

We see, by (4), that $A_n(t)$ is strongly continuously differentiable in t and

$$(11) \qquad dA_n(t)/dt = -A_n(t)B(t)A_n(t) ,$$

because we have

$$(12) \qquad \delta^{-1}(A_n(t+\delta) - A_n(t)) = -A_n(t+\delta)\{[A_n(t+\delta)^{-1} - A_n(t)^{-1}]/\delta\}A_n(t) ,$$

where

$$A_n(t)^{-1} = A(t)^{-1} - n^{-1}I .$$

Thus $y_n(t)$ satisfies the initial value problem

$$(13) \qquad dy_n(t)/dt = A_n(t)y_n(t) - A_n(t)B(t)y_n(t) , \quad a \leqq t \leqq b , \quad y_n(a) = J_n(a)y .$$

We see by (7), that $A_n(t)B(t) = J_n(t)A(t)B(t)$ is strongly continuous in t, and so is uniformly bounded in t and n by (3). Thus

$$(14) \qquad \|A_n(t)B(t)\| \leqq C$$

for $a \leqq t \leqq b$ and $n=1, 2, \cdots$. We have, for $\delta > 0$, $y_n(t+\delta) = y_n(t) + \delta A_n(t)y_n(t) - \delta A_n(t)B(t)y_n(t) + o(\delta)$. Hence, as in the proof of the Lemma, we obtain $d^+\|y_n(t)\|/dt \leqq (M'+C)\|y_n(t)\|$, which proves that

$$(15) \qquad \begin{aligned} \|y_n(t)\| &\leqq \|y_n(a)\| \exp((t-a)(M'+C)) \\ &= \|J_n(a)y\| \exp((t-a)(M'+C)) \\ &\leqq (1+n^{-1}M) \exp((t-a)(M'+C))\|y\| . \end{aligned}$$

Hence, by $dx_n(t)/dt = y_n(t)$, we see that $x_n(t)$ is bounded and strongly continuous in t, uniformly in t and n. Thus by the reflexivity of X, there exists a subsequence $\{n'\}$ of $\{n\}$ such that weak-lim $x_{n'}(t) = x(t)$ exists for all t, $a \leqq t \leqq b$, simultaneously.

We next prove that

$$(16) \quad x(t) \in D(A(t)) \text{ and } A(t)x(t) = \underset{n' \to \infty}{\text{weak-lim}} A_{n'}(t)x_{n'}(t) \text{ is bounded and strongly mea-}$$
surable in t.

In the truth we can prove that $A(t)x(t)$ is strongly continuous in t. In the proof below, we shall often use the fact that, for the dual operators, we have

$$(17) \qquad \begin{aligned} J_n(t)' &= ((I - n^{-1}A(t))^{-1})' = (I' - n^{-1}A(t)')^{-1} , \\ (A(t)^{-1})' &= (A(t)')^{-1} . \end{aligned}$$

Since X is reflexive, $D(A(t)')$ is strongly dense in the dual space X'. Thus, by $\|J_n(t)'\| \leqq 1 + n^{-1}M$, which follows from (3), we see that $\underset{n \to \infty}{\text{strong-lim}} J_n(t)'f' = f'$ for every $f' \in X'$. Hence, by $\underset{n' \to \infty}{\text{weak-lim}} x_{n'}(t) = x(t)$, we get

$$(18) \qquad \text{weak-lim } J_{n'}(t)x_{n'}(t) = x(t) .$$

We have

Kôsaku Yosida

$$\langle A(t)J_n(t)x_n(t), (A(t)^{-1})'f'\rangle = \langle J_n(t)x_n(t), f'\rangle ,$$

and so, by (17) and (18),

$$(19) \qquad \lim \langle A(t)J_{n'}(t)x_{n'}(t), (A(t)^{-1})'f'\rangle = \langle x(t), f'\rangle .$$

On the other hand, the range $(A(t)^{-1})'X'$ is strongly dense in X'. For, otherwise, the reflexivity of X asserts the existence of a $w \in X$, $w \neq 0$, such that $0 = \langle w, (A(t)^{-1})'f'\rangle = \langle A(t)^{-1}w, f'\rangle$, proving that $A(t)^{-1} = 0$, i.e., $w = 0$. Hence, by the boundedness in t and n of $y_n(t) = A_n(t)x_n(t)$, we see that weak-$\lim_{n'\to\infty} A_{n'}(t)x_{n'}(t) = u(t)$ must exist. We have, by (19) $\langle u(t), (A(t)^{-1})'f'\rangle = \langle x(t), f'\rangle$, that is, $A(t)^{-1}u(t) = x(t)$. Since the strong continuity in t of $A_n(t)x_n(t)$ implies that the closure of the set $\{A_n(t)x_n(t), a \leq t \leq b, n = 1, 2, \cdots\}$ is separable, the weak measurability of $u(t) = \text{weak-}\lim_{n'\to\infty} x_{n'}(t)$ in t implies, by a theorem of Pettis, its strong measurability. To prove that $u(t)$ is strongly continuous in t, we integrate (13), obtaining

$$(20) \qquad y_n(t) = U_n(t, a)y_n(a) - \int_a^t U_n(t, s)y_n(s)x_{n'}(s)ds .$$

As in the proof of the strong continuity in t of $x_n(t) = U_n(t, a)x_n(a)$, we see that $U_n(t, a)y_n(a)$ is strongly continuous in t, uniformly in t and n, if $A_n(a)y_n(a) = A(a)J_n(a)J_n(a)y$ is bounded in n. For, $U_n(t, a)y_n(a)$ is a solution of $dz(t)/dt = A_n(t)z(t)$. $A_n(a)y_n(a) = A(a)J_n(a)J_n(a)y$ is bounded in n if y is the domain $D(A(a))$ of $A(a)$, which is dense in X. Thus, by virtue of the lemma, we see that, for any $y \in X$, the first term on the right of (20) is strongly continuous in t, uniformly in t and n. The second term on the right is also strongly continuous in t, uniformly in t and n, because of the estimate (9) and the boundedness of $y_n(t)$ in t and n. Thus $y_n(t)$ is strongly continuous in t, uniformly in t and n. This proves the strong continuity in t of $u(t) = A(t)x(t) = \text{weak-}\lim_{n'\to\infty} y_{n'}(t)$. Therefore, by letting $n' \to \infty$ in the relation

$$x_{n'}(t) - x_{n'}(a) = x_{n'}(t) - A(a)^{-1}y = \int_a^t A_{n'}(s)x_{n'}(s)ds ,$$

we obtain

$$(21) \qquad x(t) - A(a)^{-1}y = \int_a^t A(s)x(s)ds .$$

Hence, by the strong continuity in t of $u(t) = A(t)x(t)$, we see that (1) is satisfied. Since the solution $x(t)$ is uniquely determined by the initial value $x(a)$, we see that the original sequence $\{x_n(t)\}$ itself must converge weakly to $x(t)$.

REMARK. Set $A(t)dA(t)^{-1}/dt = C(t)$, and let $W(t)$ be the solution of $dW(t)/dt = -C(t)W(t)$, with $W(a) = I$. Then $d(A(t)^{-1}W(t))/dt = A(t)^{-1}C(t)W(t) - A(t)^{-1}C(t)W(t)$

$=0$, and so $A(a)^{-1}W(a)=A(a)^{-1}=A(t)^{-1}W(t)$. From this we easily see that the domain $D(A(t)) =$ the range $R(A(t)^{-1})$ is independent of t. These observations are due to T. Kato. The case of the time dependent domain $D(A(t))$ was recently discussed by H. Tanabe [2] by extending his earlier methods. I shall give a theorem which is easily obtained by the same idea as Theorem 1, and which covers a case which was not handled by Tanabe's methods.

THEOREM 2. Let X be a Hilbert space, and let $A(t)$ satisfy, beside (3)–(4), further conditions:

(22) $A(t)$ is self-adjoint and $(A(t)x, x)\leqq-\|x\|^2$ for and $x\in D(A(t))$,

(23) there exists a constant α with $2^{-1}\leqq\alpha\leqq1$ such that $(-A(t))^\alpha B(t)$ is bounded in t.

Here the fractional power $(-A(t))^\alpha$ is defined, by virtue of the spectral resolution $A(t)=\int_{-\infty}^{-1}\lambda dE_t(\lambda)$, through

$$(24) \qquad (-A(t))^\alpha=\int_{-\infty}^{-1}(-\lambda)^\alpha dE_t(\lambda)\,.$$

Then the sequence $\{x_n(t)\}$ converges weakly to the uniquely determined strongly continuous solution of the integral equation (21).

PROOF. The above proof of the Theorem 1 shows that we have only to prove that $y_n(t)$ is uniformly bounded in t and n. We have

$$(25) \qquad \frac{d}{dt}\|y_n(t)\|^2=2\mathrm{Re}\left(\frac{dy_n(t)}{dt}, y_n(t)\right)$$
$$=2\mathrm{Re}(A_n(t)y_n(t), y_n(t))-2\mathrm{Re}(A_n(t)B(t)y_n(t), y_n(t))\,.$$

Since $-A_n(t)=-A(t)(I-n^{-1}A(t))^{-1}$ is self-adjoint and non-negative with $-A(t)$, we have, by Schwarz inequality,

$$|\mathrm{Re}(A_n(t)B(t)x, x)|\leqq|((-A_n(t))^\alpha B(t), (-A_n(t))^{1-\alpha}x)|$$
$$\leqq\|(-A_n(t))^\alpha B(t)x\|\cdot\|(-A_n(t))^{1-\alpha}x\|$$
$$\leqq2^{-1}\varepsilon\|(-A_n(t))^\alpha B(t)x\|^2$$
$$+2^{-1}\varepsilon^{-1}\|(-A_n(t))^{1-\alpha}x\|^2 \text{ for any } \varepsilon>0.$$

Since $A_n(t)=A(t)J_n(t)=\int_{-\infty}^{-1}\lambda(1-n^{-1}\lambda)^{-1}dE_t(\lambda)$, the spectrum of $-A_n(t)$ lies in the interval $[(1+n^{-1})^{-1}, \infty)$. Hence, by $2^{-1}\leqq a\leqq1$, we see that

$$2^{-1}\varepsilon^{-1}\|(-A_n(t))^{1-\alpha}x\|^2\leqq\|(-A_n(t))^{1/2}x\|^2=(-A_n(t)x, x)$$

for all t and n, by taking $\varepsilon>0$ appropriately.

Therefore we see that the right hand side of (25) is, by (22)–(23), smaller than

$2^{-1}\varepsilon^{-1}\|(-A_n(t))^\alpha B(t)y_n(t)\| = 2^{-1}\varepsilon^{-1}\|J_n(t)^\alpha(-A(t))^\alpha B(t)y_n(t)\| \leq K\|y_n(t))\|^2$, where K is independent of t and n. Hence $\|y_n(t)\| \leq \|y_n(a)\| \cdot \exp((t-a)K^{1/2})$ and so $y_n(t)$ is bounded in t and n.

REMARK. Let $X = L^2(0,1)$, and $A(t)$ be the multiplication operator $A(t)x(s) = -(1+|t-s|^{-\beta})x(s)$, $x(s)\varepsilon L^2(0,1)$. If $\beta \geq 2$, then, by taking $\alpha \geq 1/2$ in such a way that $\beta(1-\alpha)-1\geq 0$, we see that the conditions of the Theorem is satisfied. Similar case was treated by J. Tanabe [1] recently by his method of integration. His method cannot be applied to the case $\beta = 1$. In the latter case, we have $B(t) = (1+|t-s|)^{-2}$ or $-(1+|t-s|)^{-2}$ according as $t > s$ or $t < s$. Thus the right hand side of (25) is ≤ 0, and so $\|y_n(t)\| \leq \|y_n(a)\|$. Hence our Theorem 2 applies to this case also.

University of Tokyo

Reference

[1] T. Kato: Integration of the equation of evolution in a Banach space, J. Math. Soc. Japan, **5** (1953), 208–234.

[2] H. Tanabe: Evolutional equations of parabolic type, Proc. Japan Acad., **37** (1961), 610–613.

(Received October 23, 1962)

Holomorphic semi-groups in a locally convex linear topological space

Osaka Math. J. **15** (1963) 51–57

The purpose of the present note is to show that the analytical theory of holomorphic semi-groups in a Banach space, given in a preceding note[1]. can be extended to locally convex linear topological spaces. The result may thus be applied to the "abstract Cauchy problem" in such spaces.

Let X be a *locally convex, sequentially complete linear topological space*, and $L(X, X)$ be the set of all continuous linear operators defined on X into X. Let $T_t \in L(X, X)$, $t \geq 0$, satisfy the conditions:

(i) $T_t T_s = T_{t+s}(t, s \geq 0)$, $T_0 = I =$ the identity operator,

(ii) $\lim_{t \to t_0} T_t x = T_{t_0} x$ for all $t_0 \geq 0$ and $x \in X$,

(iii) $\{T_t\}$ is *equi-continuous* in $t \geq 0$ in the sense that, for any continuous semi-norm $p(x)$ on X, there exists a continuous semi-norm $q(x)$ on X such that $p(T_t x) \leq q(x)$ for all $t \geq 0$ and all $x \in X$.

Such a system $\{T_t\}$ is said to constitute an *equi-contituous semi-group of class* (C_0). *The infinitesimal generator* A of T_t is defined by

(iv) $Ax = (D_t T_t x)_{t=0} = \lim_{t \downarrow 0} t^{-1}(T_t - I)x$, i. e., the domain $D(A)$ of A is the set of those $x \in X$ for which the right hand limit exists, and when $x \in D(A)$ we have $Ax = \lim_{t \downarrow 0} t^{-1}(T_t - I)x$.

As in the case where X is a Banach space and $\sup_{t \geq 0} \| T_t \| < \infty$, such A is characterized by the following properties:

(v) A is a closed linear operator with dense domain $D(A)$, i. e., $D(A)^a = X$[2],

---

1) K. Yosida: *On the differentiability of semi-groups of linear operators*, Proc. Japan Acad. **34** (1958), 337–340. Cf. also E. Hille–R. S.Phillips: Functional Analysis and Semi-groups, Providence (1957).

2) M^a denotes the closure of $M \subseteq X$.

(vi)   the resolvent $R(\lambda\,;A)=(\lambda I-A)^{-1}\in L(X,\,X)$ exists for $Re(\lambda)>0$ and the system of linear operators $\{(\lambda R(\lambda\,;A))^n\}$ is equi-continuous in $\lambda\geq 1$ and in $n=1,\,2,\,\cdots$

Moreover, the resolvent $R(\lambda\,;A)$ is obtained from the original group by

(vii) $\displaystyle (\lambda R(\lambda\,;A))^n x = \frac{\lambda^n}{(n-1)!}\int_0^\infty e^{-\lambda t}t^{n-1}T_t x\,dt$ for $Re(\lambda)>0$ and $x\in X$.

After these preliminaries, we are ready to discuss those semi-groups T_t which can, as functions of the parameter t, be continued holomorphically into a sector of the complex plane containing the positive t-axes.

Lemma. *Suppose that, for all $t>0$, $T_t X\subseteq D(A)$, Then, for any $x\in X$, $T_t x$ is infinitely differentiable in $t>0$ and we have*

(1) $$T_t^{(n)}x = (T_{t/n}')^n x \quad \text{for all} \quad t>0,$$

where $T_t'=D_t T_t$, $T_t''=D_t T_t'$, $\cdots$, $T_t^{(n)}=D_t T_t^{(n-1)}$

Proof. It $t>t_0>0$, then $T_t'x=AT_t x=T_{t-t_0}AT_{t_0}x$ by the commutativity of A and T_s, which is an easy consequence from (i) and (iv). Thus $T_t'X\subseteq T_{t-t_0}X\subseteq D(A)$ when $t>0$, and so $T_t''x$ exists for all $t>0$ and $x\in X$. Since A is a closed linear operator, we have $T_t''x=D_t(AT_t)x = A\cdot\lim_{n\uparrow\infty} n(T_{t+(1/n)}-T_t)x = A(AT_t)x = AT_{t/2}AT_{t/2}x = (T_{t/2}')^2 x.$ Repeating the argument, we obtain (1).

Theorem. *For an equi-continuous semi-group T_t of class (C_0) in a locally convex, sequentially complete linear topological space X, the following three conditions are mutually equivalent.*

(I)   *For all $t>0$, $T_t X\subseteq D(A)$ and there exists a positive constant $C\leq 1$ such that the family of operators $\{(CtT_t')^n\}$ is equi-continuous in $n=1,\,2,\,\cdots$ and $0<t\leq 1$.*

(II)   *T_t admits a holomorphic extension T_λ given by*

(2) $$T_\lambda x = \sum_{n=0}^\infty (\lambda-t)^n T_t^{(n)}x/n! \quad \text{for} \quad |\arg\lambda|<\mathrm{Tan}^{-1}(Ce^{-1}),$$

in such a way that

(3)   *the family of operators $\{e^{-\lambda}T_\lambda\}$ is equi-continuous in λ for $|\arg\lambda|<\mathrm{Tan}^{-1}(2^{-1}Ce^{-1})$.*

(III)   *Let A be the infinitesimal generator of T. Then there exists a positive constant C_1 such that the family of operators $\{(C_1\lambda R(\lambda\,;A))^n\}$ is equi-continuous in $n=1,\,2,\,\cdots$ and λ with $Re(\lambda)\geq 1+\varepsilon$, where $\varepsilon>0$.*

Proof. *The implication* (I)→(II). Let p be any continuous semi-norm on X. Then, by hypothesis, there exists a continuous semi-norm q on X such that $p((t T'_t)^n x) \leq C^{-n} q(x)$ for $1 \geq t \geq 0$, $n \geq 0$ and $x \in X$. Hence, by (1), we obtain, for any $t > 0$,

$$p((\lambda - t)^n T_t^{(n)} x / n!) \leq \frac{|\lambda - t|^n}{t^n} \frac{n^n}{n!} \frac{1}{C^n} \, p\left(\left(\frac{t}{n} \, C T'_{t/n}\right)^n x\right)$$

$$\leq \left(\frac{|\lambda - t|}{t} \, C^{-1} e\right)^n q(x), \quad \text{whenever } 0 < t/n \leq 1.$$

Thus the right side of (2) surely converges for $\|\arg \lambda\| < \mathrm{Tan}^{-1}(Ce^{-1})$, and so, by the sequential completeness of X, $T_\lambda x$ is well defined and is holomorphic in λ for $|\arg \lambda| < \mathrm{Tan}^{-1}(Ce^{-1})$. Next put $S_t = e^{-t} T_t$. Then $S'_t = -e^{-t} T_t + e^{-t} T'_t$ and so, by $0 \leq t e^{-t} \leq 1$ $(0 \leq t)$ and (I), we easily see that $\{(2^{-1} C t S'_t)^n\}$ is equi-continuous in $t > 0$ and $n \geq 0$, in virtue of the equi-continuity of $\{T_t\}$. The equi-continuous semi-group S_t of class (C_0) satisfies the condition that $S_t X \subseteq D(A - I) = D(A)$, where $(A - I)$ is the infinitesimal generator of S_t. Therefore, by the same reasoning as applied to T_t above, we can prove that the holomorphic extension $e^{-\lambda} T_\lambda$ of $S_t = e^{-t} T_t$ satisfies the estimate (3).

By the way, we can prove the following

Corollary (due to E. Hille). *If, in particular, X is a complex B-space and* $\overline{\lim\limits_{t \downarrow 0}} \|t T'_t\| < e^{-1}$,[1] *then* $X = D(A)$.

Proof. For a fixed $t > 0$, we have $\overline{\lim\limits_{n \to \infty}} \|(t/n) T'_{t/n}\| < e^{-1}$, and so the series

$$\sum_{n=0}^{\infty} (\lambda - t)^n T_t x / n! = \sum \frac{(\lambda - t)^n}{t^n} \frac{n^n}{n!} \left(\frac{t}{n} \, T'_{t/n}\right)^n x$$

converge in some circle

$$\{\lambda \,;\, |\lambda - t| / t < 1 + \delta \quad \text{with a} \quad \delta > 0\}$$

of the complex λ-plane. This circle surely contains $\lambda = 0$ in its interior.

The implication (II)→(III). We have, by (vii),

$$(4) \qquad (\lambda R(\lambda \,;\, A))^n x = \frac{\lambda^{n+1}}{n!} \int_0^\infty e^{-\lambda t} t^n T_t x \, dt \quad \text{for} \quad Re(\lambda) > 0, \qquad x \in X.$$

Hence

$$((\sigma + 1 + i\tau) R(\sigma + 1 + i\tau \,;\, A))^{n+1} x = \frac{(\sigma + 1 + i\tau)^{n+1}}{n!} \int_0^\infty e^{-(\sigma + i\tau)t} t^n S_t x \, dt, \qquad \sigma > 0,$$

Let $\tau < 0$. Since the integrand is holomorphic, we can deform, by the estimate (3) and Cauchy's integral theorem, the path of integration: $0 \leq t < \infty$ to the ray: $re^{i\theta}$ $(0 \leq r < \infty)$ contained in the sector $0 < \arg \lambda < \mathrm{Tan}^{-1}(2^- Ce^{-1})$ of the complex λ-plane. We thus obtain

$$((\sigma+1+i\tau)R(\sigma+1+i\tau\,;\,A))^{n+1}x = \frac{(\sigma+1+i\tau)^{n+1}}{n!} \times$$

$$\int_0^\infty e^{-(\sigma+i\tau)re^{i\theta}} r^n\, e^{in\theta} S_{re^{i\theta}} x\, e^{i\theta} dr\,,$$

and so, by (3),

$$p((\sigma+1+i\tau)R(\sigma+1+i\tau\,;\,A))^{n+1}x)$$

$$\leq \frac{|(\sigma+1+i\tau)|^{n+1}}{n!} \int_0^\infty e^{(-\sigma\cos\theta+\tau\sin\theta)r} r^n p(S_{re^{i\theta}})\, dr$$

$$\leq q'(x)\, \frac{|\sigma+1+i\tau|^{n+1}}{|\tau\sin\theta-\sigma\cos\theta|^{n+1}}\,,$$

where q' is a continuous semi-norm on X. A similar estimate is obtained for the case $\tau > 0$ also. Hence, combined with (vi), we have proved (III).

The implication (III) $\to$ (I). For any continuous semi-norm p on X, there exists a continuous semi-norm q on X such that

$$p((C_1\lambda R(\lambda\,;\,A))^n x) \leq q(x) \quad \text{whenever} \quad Re(\lambda) \geq 1+\varepsilon,\ \varepsilon > 0 \quad \text{and} \quad n \geq 0.$$

Hence, if $Re(\lambda_0) \geq 1+\varepsilon$, we have

$$p(((\lambda-\lambda_0)R(\lambda_0\,;\,A))^n x) \leq \frac{|\lambda-\lambda_0|^n}{(C_1|\lambda_0|)^n}\, q(x) \qquad (n = 0, 1, 2, \cdots)\,.$$

Thus, if $|\lambda-\lambda_0|/C_1|\lambda_0| < 1$, the resolvent $R(\lambda\,;\,A)$ exists and is given by

$$R(\lambda\,:\,A)x = \sum_{n=0}^\infty (\lambda_0-\lambda)^n R(\lambda_0\,;\,A)^{n+1} x \quad \text{such that}$$

$$p(R(\lambda\,;\,A)x) \leq (1-C_1^{-1}|\lambda_0|^{-1}|\lambda-\lambda_0|)^{-1} q(R(\lambda_0\,;\,A)x)\,.$$

Therefore, by (III) there exists an angle θ_0 with $\pi/2 < \theta_0 < \pi$ such that $R(\lambda\,;\,A)$ exists and satisfies the estimate

$$(5) \qquad p(R(\lambda\,;\,A)x) \leq \frac{1}{|\lambda|}\, q'(x)$$

with a continuous semi-norm q' on X in the sectors $\pi/2 \leq \arg \lambda \leq \theta_0$ and $-\theta_0 \leq \arg \lambda \leq -\pi/2$ and also for $Re(\lambda) \geq 0$, when $|\lambda|$ is sufficiently large. Hence the integral

$$(6) \qquad \hat{T}_t x = (2\pi i)^{-1} \int_{C_2} e^{\lambda t} R(\lambda \,;\, A) x \, d\lambda \qquad (t > 0,\ x \in X)$$

converges if we take the path of integration $C_2 = \lambda(\sigma)$, $-\infty < \sigma < \infty$, in such a way that $\lim\limits_{|\sigma| \uparrow \infty} |\lambda(\sigma)| = \infty$ and, for some $\varepsilon > 0$,

$$\pi/2 + \varepsilon \leqq \arg \lambda(\sigma) \leqq \theta_0 \quad \text{and} \quad -\theta_0 \leqq \arg \lambda(\sigma) \leqq -\pi/2 - \varepsilon$$

when $\sigma \uparrow +\infty$ and $\sigma \downarrow -\infty$, respectively; for not large $|\sigma|$, $\lambda(\sigma)$ lies in the right half plane of the complex λ-plane.

We shall show that $\hat{T}_t$ coincides with the semi-group T_t itself[3]. We first show that $\lim\limits_{t \downarrow 0} \hat{T}_t x = x$ for all $x \in D(A)$. Let x_0 be any element $\in D(A)$, and choose any complex number λ_0 to the right of the contour C_2 of integration, and denote $(\lambda_0 I - A) x_0 = y_0$. Then, by the resolvent equation,

$$\hat{T}_t x_0 = \hat{T}_t R(\lambda_0 \,;\, A) y_0 = (2\pi i)^{-1} \int_{C_2} e^{\lambda t} R(\lambda \,;\, A) R(\lambda_0 \,;\, A) y_0 \, d\lambda$$

$$= (2\pi i)^{-1} \int_{C_2} e^{\lambda t} (\lambda_0 - \lambda)^{-1} R(\lambda \,;\, A) y_0 \, d\lambda$$

$$- (2\pi i)^{-1} \int_{C_2} e^{\lambda t} (\lambda_0 - \lambda)^{-1} R(\lambda_0 \,;\, A) y_0 \, d\lambda \,.$$

The second integral on the right is equal to zero, as may be seen by shifting the path of integration to the left. Hence

$$\hat{T}_t x_0 = (2\pi i)^{-1} \int_{C_2} e^{\lambda t} (\lambda_0 - \lambda)^{-1} R(\lambda \,;\, A) y_0 \, d\lambda, \quad y_0 = (\lambda_0 I - A) x_0 \,.$$

Because of the estimate (5), the passage to the limit $t \downarrow 0$ under the integral sign is justified, and so

$$\lim_{t \downarrow 0} \hat{T}_t x_0 = (2\pi i)^{-1} \int_{C_2} (\lambda_0 - \lambda)^{-1} R(\lambda \,;\, A) y_0 \, d\lambda, \quad y_0 = (\lambda_0 I - A) x_0 \,.$$

To evaluate the right hand integral, we make a closed contour out of the original path of integration C_2 by adjoining the arc of the circle $|\lambda| = r$ which is to the right of the path C_2, and throwing away that portion of the original path C_2 which lies outside the circle $|\lambda| = r$. The value of the integral along the new arc and the discarded arc tends to zero as $r \downarrow \infty$, in virtue of (5). Hence the value of the integral is equal to the residue inside the new closed contour, that is, the value

---

3)   Adapted from P. D. Lax and A. N. Milgram: Parabolic equations, Contributions to the Theory of Partial Differential Equations, Princeton (1954).

$R(\lambda_0 ; A)y_0 = x_0$. We have thus proved $\lim_{t \downarrow 0} \hat{T}_t x_0 = x_0$ when $x_0 \in D(A)$.

We next show that $\hat{T}'x = A\hat{T}_t x$ for $t > 0$ and $x \in X$. We have $R(\lambda ; A)X = D(A)$ and $AR(\lambda ; A) = \lambda R(\lambda ; A) - I$, so that, by the convergence factor $e^{\lambda t}$, the integral $(2\pi i)^{-1} \int_{C_2} e^{\lambda t} AR(\lambda ; A) x \, d\lambda$ has a sense. This integral is equal to $A\hat{T}_t x$, as may be seen by approximating the integral (6) by Riemann sum and using the fact that A is *closed*: $\lim_{n \to \infty} x_n = x$ and $\lim_{n \to \infty} Ax_n = y$ imply $x \in D(A)$ and $Ax = y$. Therefore

$$A\hat{T}_t x = (2\pi i)^{-1} \int_{C_2} e^{\lambda t} AR(\lambda ; A) x \, d\lambda, \quad t > 0.$$

On the other hand, by differentiating (6) under the integral sign, we obtain

$$(8) \qquad \hat{T}'_t x = (2\pi i)^{-1} \int_{C_2} e^{\lambda t} \lambda R(\lambda ; A) x \, d\lambda, \quad t > 0.$$

In fact, the difference of these two integrals is $(2\pi i)^{-1} \int_{C_2} e^{\lambda t} x \, d\lambda$, and the value of the last integral is zero, as may be seen by shifting the path of integration to the left.

Thus we have proved that $\hat{x}(t) = \hat{T}_t x_0$, $x_0 \in D(A)$, satisfies i) $\lim_{t \downarrow 0} \hat{x}(t) = x_0$, ii) $d\hat{x}(t)/dt = A\hat{x}(t)$ for $t > 0$, and iii) $\{\hat{x}(t)\}$ is bounded when $t \uparrow \infty$, as may be seen from (6). On the other hand, since $x_0 \in D(A)$ and since $\{T_t\}$ is equi-continuous in $t \geq 0$, we see that $x(t) = T_t x_0$ also satisfies $\lim_{t \downarrow t_0} x(t) = x_0$, $dx(t)/dt = Ax(t)$ for $t \geq 0$, and $\{x(t)\}$ is bounded when $t \geq 0$. Let us put $\hat{x}(t) - x(t) = y(t)$. Then $\lim_{t \downarrow 0} y(t) = 0$, $dy(t)/dt = Ay(t)$ for $t > 0$ and $\{y(t)\}$ is bounded when $t \uparrow \infty$. Hence we may consider the Laplace transform

$$L(\lambda ; y) = \int_0^\infty e^{-\lambda t} y(t) \, dt, \quad Re(\lambda) > 0.$$

We have

$$\int_\alpha^\beta e^{-\lambda t} y'(t) = \int_\alpha^\beta e^{-\lambda t} Ay(t) \, dt = A \int_\alpha^\beta e^{-\lambda t} y(t) \, dt, \quad 0 \leq \alpha < \beta < \infty,$$

by approximating the integral by Riemann sum and using the fact that A is closed. By partial integration, we obtain

$$\int_\alpha^\beta e^{-\lambda t} y'(t) \, dt = e^{-\lambda \beta} y(\beta) - e^{-\lambda \alpha} y(\alpha) + \lambda \int_\alpha^\beta e^{-\lambda t} y(t) \, dt$$

which tends to $\lambda L(\lambda ; y)$ as $\alpha \downarrow 0$, $\beta \uparrow \infty$. For, $y(0) = 0$ and $\{y(\beta)\}$ is bounded as $\beta \uparrow \infty$. Thus again, by using the closure property of A, we obtain

$$AL(\lambda\,;\,y) = \lambda L(\lambda\,;\,y),\quad Re(\lambda) > 0.$$

Since the inverse $(\lambda I - A)^{-1}$ exists for $Re(\lambda) > 0$, we must have $L(\lambda\,;\,y) = 0$ when $Re(\lambda) > 0$. Thus, for any continuous linear functional $f \in X'$, the dual space of, we have

$$\int_0^\infty e^{-\lambda t} f(y(t))\, dt = 0 \quad \text{when} \quad Re(\lambda) > 0.$$

We set $\lambda = \sigma + i\tau$ and put

$$g_\sigma(t) = e^{-\sigma t} f(y(t)) \quad \text{or} \quad = 0 \quad \text{according as} \quad t \geq 0 \quad \text{or} \quad t < 0.$$

Then, the above equality shows that the Fourier transform

$$(2\pi)^{-1} \int_{-\infty}^\infty e^{-i\tau t} g_\sigma(t)\, dt \quad \text{vanishes identically in} \quad \tau,\ -\infty < \tau < \infty,$$

so that, by Fourier's integral theorem, $g_\sigma(t) = 0$ identically. Thus $f(y(t)) = 0$ and so we must have $y(t) = 0$ identically, in virtue of Hahn-Banach's theorem.

Therefore $\hat{T}_t x = T_t x$ for all $t > 0$ and $x \in D(A)$. $D(A)$ being dense in X and $\hat{T}_t$, T_t both belong to $L(X, X)$, we easily conclude that $\hat{T}_t x = T_t x$ for all $x \in X$ and $t > 0$. Hence, by defining $\hat{T}_0 = I$, we have $\hat{T}_t = T_t$ for all $t \geq 0$. Hence, by (7). $T_t' x = (2\pi i)^{-1} \int_{C_2} e^{\lambda t} \lambda R(\lambda\,;\,A) x\, dt$, $t > 0$, and so, by (1) and (5), we obtain

$$(T_{t/n}')^n x = T_t^{(n)} x = (2\pi i)^{-1} \int_{C_2} e^{\lambda t} \lambda^n R(\lambda\,;\,A) x\, d\lambda.$$

Hence

$$(t\, T_t')^n x = (2\pi i)^{-1} \int_{C_2} e^{n\lambda t} (t\lambda)^n R(\lambda\,;\,A) x\, d\lambda.$$

Therefore, by (III),

$$p((t\, T_t')^n x) \leq (2\pi)^{-1} q(x) \int_{C_2} |e^{n\lambda t}|\, |t^n|\, |\lambda|^{n-1} d|\lambda|.$$

The last integral is majoraized by C_3^n with some positive constant C_3, when $1 \geq t > 0$. Hence we have proved (I).

University of Tokyo

(Received March 6, 1963)

A perturbation theorem for semi-groups of linear operators

Proc. Japan Acad. **41** (1965) 645–647

Let A be the infinitesimal generator of a contraction semi-group T_t of class (C_0) on the Banach space X.[1] A is thus a closed linear operator with the domain $D(A)$ and the range $R(A)$ both in X such that: i) $D(A)$ is dense in X, and ii) the resolvent $(\lambda I - A)^{-1}$ exists as a bounded linear operator on X into X satisfying the estimate $\| \lambda(\lambda I - A)^{-1} \| \leq 1$ for all $\lambda > 0$. Let B likewise be the infinitesimal generator of another contraction semi-group of class (C_0) on X. Then the condition $D(B) \geq D(A)$ implies, by the closed graph theorem, that there exist positive constants a and b such that

$$\| Bx \| \leq a \| Ax \| + b \| x \| \quad \text{for} \quad x \in D(A).$$

As an important remark to Theorem 2 in H. F. Trotter [1] (Cf. T. Kato [1]), E. Nelson [1] proved that $(A+B)$ with the domain $D(A+B) = D(A)$ is the infinitesimal generator of a contraction semi-group of class (C_0) if we can take $a < 1/2$.

The purpose of the present note is to propose a sufficient condition in order that Nelson's hypothesis be satisfied. We shall prove

Theorem. Let $0 < \alpha < 1$. Let $\hat{A}_\alpha = -(-A)^\alpha$ be the fractional power of A, and let us assume that $D(B) \geq D(\hat{A}_\alpha)$. Then $(A+B)$ with the domain $D(A+B) = D(A)$ is the infinitesimal generator of a contraction semi-group of class (C_0) on X.

Corollary. Assume, furthermore, that A generates a holomorphic semi-group, then $(A+B)$ with the domain $D(A+B) = D(A)$ also generates a holomorphic semi-group.

Remark 1.[2] The fractional power $\hat{A}_\alpha$ is defined as the infinitesimal generator of the semi-group of class (C_0):

$$(1) \qquad \hat{T}_{t,\alpha} x = \int_0^\infty f_{t,\alpha}(s) T_s x \, ds \quad (t > 0, \, x \in X),$$

where

$$(2) \qquad f_{t,\alpha}(s) = \frac{1}{2\pi i} \int_{\sigma - i\infty}^{\sigma + i\infty} e^{zs - tz^\alpha} dz \, (\sigma > 0, \, t > 0, \, s \geq 0),$$

the branch of z^α being taken so that $Re(z^\alpha) > 0$ for $Re(z) > 0$. According to V. Balakrishnan [1], we have $D(\hat{A}_\alpha) \geq D(A)$ and

$$(3) \qquad (-A)^\alpha x = \frac{\sin \pi \alpha}{\pi} \int_0^\infty \lambda^{\alpha-1} (\lambda I - A)^{-1} (-Ax) d\lambda \quad \text{for} \quad x \in D(A).$$

---

1) See, e.g., E, Hille–R. S. Phillips [1] or K. Yosida [1].
2) See, e.g., K. Yosida [1].

Remark 2.[3]　A contraction semi-group T_t of class (C_0) is called a holomorphic semi-group if there exist positive numbers C_1 and C_2 $(<C_1)$ such that T_t admits holomorphic extension T_λ given by Taylor's expansion

$$(4)\qquad T_\lambda x = \sum_{n=0}^{\infty} (\lambda - t)^n T_t^{(n)} x/n! \text{ for } x \in X \text{ and } |\arg \lambda| < \tan^{-1} C_1$$

satisfying the estimate

$$(5)\qquad \| e^{-\lambda} T_\lambda \| \leq C_2 \text{ for } |\arg \lambda| < \tan^{-1} C_2,$$

the existence of the strong derivatives $T_t^{(n)} x$ with respect to t being assumed for all $x \in X$. It is known that a contraction semi-group T_t of class (C_0) is a holomorphic semi-group if and only if, for fixed $\sigma_0 > 0$, we have the estimate

$$(6)\qquad \varlimsup_{|\tau| \uparrow \infty} \| \tau((\sigma_0 + i\tau)I - A)^{-1} \| < \infty.$$

Proof of the Theorem.　Being infinitesimal generators of contraction semi-groups of class (C_0), A and B are both dissipative in the sense of G. Lumer and R. S. Phillips [1]. That is, if we take, for $x \in X$, a continuous linear functional $\varphi = \varphi_x$ defined on X such that

$$\| \varphi_x \| = 1 \text{ and } \langle x, \varphi_x \rangle = \| x \|,$$

then we have

$$Re\langle Ax, \varphi_x \rangle \leq 0 \, (x \in D(A)) \text{ and } Re\langle Bx, \varphi_x \rangle \leq 0 \, (x \in D(B)).$$

Thus $Re\langle (\lambda I - A - B)x, \varphi_x \rangle \geq \lambda \| x \|$ and hence $\| (\lambda I - A - B)x \| \geq \lambda \| x \|$ for all $x \in D(A)$ and $\lambda > 0$. Therefore

(7)　the inverse $(\lambda I - A - B)^{-1}$ exists with the estimate

$$\| (\lambda I - A - B)^{-1} f \| \leq \lambda^{-1} \| f \| \text{ for all } f \in R(\lambda I - A - B).$$

Since $D(A+B) = D(A)$ is dense in X, we have only to prove that $R(\lambda I - A - B) = X$ for some $\lambda > 0$. For, then it is easy to show that $(\lambda I - A - B)^{-1}$ is everywhere defined on X with the estimate

$$\| \lambda(\lambda \tau - A - B)^{-1} \| \leq 1 \text{ for all } \lambda > 0.$$

Now formula (3) may be written as

$$(-A)^\alpha x = \frac{\sin \pi \alpha}{\pi} \Big\{ \int_0^\delta \lambda^{\alpha - 1} [I - \lambda(\lambda I - A)^{-1}] x \, d\lambda$$

$$+ \int_\delta^\infty \lambda^{\alpha - 2} \lambda(\lambda I - A)^{-1}(-Ax) d\lambda \Big\}.$$

Hence, by ii), we obtain

$$(8)\qquad \| \hat{A}_\alpha x \| \leq a \| Ax \| + b \| x \| \text{ for all } x \in D(A),$$

where a can be taken arbitrarily small by taking b appropriately.[4]

---

3) See, e.g., K. Yosida [1].

4) Mr. K. Masuda kindly called the author's attention that (8) may be replaced by a sharper one:

$$(8)'\qquad \| \hat{A}_\alpha x \| \leq \frac{\sin \pi \alpha}{\pi} \frac{2^{1-\alpha}}{\alpha(1 - \alpha)} \| Ax \|^\alpha \| x \|^{1-\alpha} \text{ for all } x \in D(A),$$

which is obtained by considering the minimum with respect to δ of

$$\frac{\sin \pi \alpha}{\pi} \Big\{ \| x \| \int_0^\delta 2 \cdot \lambda^{\alpha - 1} d\lambda + \| Ax \| \int_\delta^\infty \lambda^{\alpha - 2} d\lambda \Big\}.$$

We also have, by $D(B) \geqq D(\hat{A}_\alpha)$ and the closed graph theorem, that there exists a positive constant c such that

$$(9) \qquad \| Bx \| \leqq c(\| \hat{A}_\alpha x \| + \| x \|) \text{ for all } x \in D(\hat{A}_\alpha).$$

Thus, by (8) and (9),

$$(10) \qquad \| Bx \| \leqq ac \| Ax \| + c(b+1) \| x \| \text{ for all } x \in D(A).$$

Since $D(A) \leqq D(\hat{A}_\alpha) \leqq D(B)$, the operator $B(\lambda I - A)^{-1}$ with $\lambda > 0$ is everywhere defined on X and so, by (10) and the closed graph theorem, we obtain

$$(11) \qquad \| B(\lambda I - A)^{-1} \| \leqq ac \| A(\lambda I - A)^{-1} \| + c(b+1) \| (\lambda I - A)^{-1} \|.$$

Hence, by ii) and $A(\lambda I - A)^{-1} = \lambda(\lambda I - A)^{-1} - I$, we see that $\| B(\lambda I - A)^{-1} \| < 1$ for $2ac + \lambda^{-1}c(b+1) < 1$. This proves that, for $2ac < 1$ and for sufficiently large $\lambda > 0$, the inverse $(I - B(\lambda I - A)^{-1})^{-1} = \sum_{m=0}^{\infty} (B(\lambda I - A)^{-1})^m$ exists as a bounded linear operator on X into X so that $R((I - B(\lambda I - A)^{-1}) = X$. Therefore, by $R(\lambda I - A) = X$, we see that $R(\lambda I - A - B) = R((I - B(\lambda I - A)^{-1})(\lambda I - A)) = X$.

Proof of the Corollary. Since A and $(A+B)$ both generate contraction semi-groups of class (C_0), we know that $((\sigma_0 + i\tau)I - A - B)^{-1}$ and $((\sigma_0 + i\tau)I - A)^{-1}$ both exist as bounded linear operators on X into X with the estimates

$$\| \sigma_0((\sigma_0 + i\tau)I - A - B)^{-1} \| \leqq 1, \ \| \sigma_0((\sigma_0 + i\tau)I - A)^{-1} \| \leqq 1$$

and

$$\lim_{|\tau| \uparrow \infty} |\tau| \| ((\sigma_0 + i\tau)I - A)^{-1} \| < \infty.$$

Hence, as in the proof of Theorem, we prove that for $2ac < 1$ and for sufficiently large $|\tau|$, the estimate

$$\| \tau((\sigma_0 + i\tau)I - A - B)^{-1} \| \leqq \| (I - B((\sigma_0 + i\tau)I - A)^{-1})^{-1} \|$$
$$\cdot \| \tau((\sigma_0 + i\tau)I - A)^{-1} \|$$

so that $|\tau| \cdot \| ((\sigma_0 + i\tau)I - A - B)^{-1} \|$ is bounded as $|\tau| \uparrow \infty$. This proves that $(A+B)$ generates a holomorphic semi-group.

References

V. Balakrishnan [1]: Fractional powers of closed operators and the semi-group generated by them. Pacific J. Math., **10**, 419–437 (1960).

E. Hille and R. S. Phillips [1]: Functional Analysis and Semi-groups. Providence (1957).

T. Kato [1]: Fundamental properties of Hamiltonian operators of Schrödinger type. Trans. Amer. Math. Soc., **7**, 195–211 (1951).

G. Lumer and R. S. Phillips [1]: Dissipative operators in a Banach space. Pacific J. Math., **11**, 679–698 (1961).

E. Nelson [1]: Feynman integrals and the Schrödinger equation. J. Math. Physics, **1**, 332–343 (1964).

E. F. Trotter [1]: On the product of semigroups of operators. Proc. Amer. Math. Soc., **10**, 545–551 (1959).

K. Yosida [1]: Functional Analysis. Springer (1965).

Time dependent evolution equations in a locally convex space

Math. Ann. **162** (1965) 83–86

Let X be a locally convex linear topological space (see, e.g., G. Köthe [2] and K. Yosida [3]), and denote by $L(X, X)$ the totality of continuous linear operators defined on X into X. Consider an abstract Cauchy problem for the evolution equation in X:

$$(1) \qquad \frac{dx(t)}{dt} = A(t)\, x(t),\ x(0) = y \in X \quad \text{and} \quad x(t) \in X,\ 0 \leq t \leq 1,$$

where d/dt denotes the differentiation in the topology of X, and $A(t)$ is, for each t, a linear operator whose domain $D(A(t))$ and range $R(A(t))$ both belong to X. We do not assume that $D(A(t)) = X$ nor that $A(t)$ is a continuous linear operator.

The purpose of the present note is to devise a simplified presentation, adapted to the locally convex space X, of the existence proof of solutions of (1) due to T. Kato [1] which was designed for a Banach space.

We shall assume the following five conditions:

(2) X is sequentially complete.

(3) $D(A(t))$ is independent of t and it is dense in X.

(4) For every $\lambda > 0$ and $t\, (0 \leq t \leq 1)$, the resolvent $(\lambda I - A(t))^{-1}$ exists as an operator $\in L(X, X)$. There exists a fundamental family P of continuous semi-norms p on X such that, for every $p \in P$, there exists a positive constant M satisfying

$$p\big((I - \lambda A(t_n))^{-1} (I - \lambda A(t_{n-1}))^{-1} \ldots (I - \lambda A(t_1))^{-1}x\big) \leq M\,p(x)\,,$$

M being independent of $\lambda > 0$, $t's$ with $0 \leq t_1 \leq t_2 \leq \cdots \leq t_n \leq 1$, n and $x \in X$.

(5) $\qquad A(s)^{-1} \in L(X, X) \quad \text{and} \quad A(t)\, A(s)^{-1} \in L(X, X) \quad (0 \leq s, t \leq 1)\,.$

(6) For every $x \in X$, $(t-s)^{-1} C(t,s)x = (t-s)^{-1}(A(t)\, A(s)^{-1} - I)x$ is bounded and uniformly continuous in t and s, $t \neq s$, and $\displaystyle \lim_{k \to \infty} k\, C\left(t, t - \frac{1}{k}\right) x = C(t)x$ exists uniformly in t, where $C(t) \in L(X, X)$. We have, moreover, $p((t-s)^{-1} C(t,s)x) \leq N\,p(x)$ with a positive constant N which is independent of $x \in X$, t and s.

---

*) The author expresses his sincere thanks to Professors K. Jörgens and E. T. Poulsen who kindly read the manuscript and pointed out careless mistakes the author overlooked originally. Thus, in particular, condition (4) below is rewritten thanks to their kind suggestion.

6*

Remark. We have, by (4),

(4)' $p((I - n^{-1}A(s))^{-m}x) \leq M p(x)$ for all positive integers n, m and $s (0 \leq s \leq 1)$.

$D(A(s))$ being dense in X by (3), we know by (2) and (4)' that $A(s)$ is the infinitesimal generator of an equi-continuous semi-group $\{\exp(tA(s)); t \geq 0\}$ of class (C_0). That is, we have the results stated in (7), (8) and (9) below (see, e.g., K. Yosida [3]):

(7) $\exp(tA(s))x = \lim\limits_{n \to \infty} \exp\big(tA(s)(I - n^{-1}A(s))^{-1}\big)x = \lim\limits_{n \to \infty}(I - n^{-1}tA(s))^{-n}x$
uniformly in $t, 0 \leq t \leq 1$,
$\exp(t_1 A(s)) \exp(t_2 A(s)) = \exp((t_1 + t_2)A(s))$, and
$\lim\limits_{t \to t_0} \exp(tA(s))x = \exp(t_0 A(s))x$.

(9) $\dfrac{d \exp(tA(s))y}{dt} = A(s) \exp(tA(s))y = \exp(tA(s))A(s)y$ for $y \in D(A(s))$.

We shall have

Theorem. *For every positive integer k and $0 \leq s \leq t \leq 1$, define the operator $U_k(t, s) \in L(X, X)$ through[1]*

(10)
$$U_k(t,s) = \exp\left((t-s)A\left(\frac{i-1}{k}\right)\right) \text{ for } \frac{i-1}{k} \leq s \leq t \leq \frac{i}{k} \ (1 \leq i \leq k),$$
$$U_k(t,r) = U_k(t,s) U_k(s,r) \text{ for } 0 \leq r \leq s \leq t \leq 1.$$

Then, for every $x \in X$ and $0 \leq s \leq t \leq 1$,

(11) $U(t, s)x = \lim\limits_{k \to \infty} U_k(t, s)x$ *exists uniformly in t and s.*

If, in particular, $y \in D(A(0))$, then the Cauchy problem (1) is solved by $x(t) = U(t, 0)y$ which satisfies the estimate $p(x(t)) \leq M p(y)$.

Proof. By (4)', (7) and (10), we have

(12) $\qquad p(U_k(t, s)x) \leq M p(x) \quad (k = 1, 2, \ldots; 0 \leq s \leq t \leq 1; x \in X)$.

We also can prove that, for $W_k(t, 0) = A(t) U_k(t, 0) A(0)^{-1}$,

(13) $\qquad p(W_k(t, 0)x) \leq (1 + k^{-1}N) M \cdot \exp(tMN) \cdot p(x)$.

In fact we have, by (10) and $A(s)^{-1} \exp((t-s)A(s)) = \exp((t-s)A(s))A(s)^{-1}$,

$$W_k(t,0) = A(t)A\left(\frac{[kt]}{k}\right)^{-1} U_k\left(t, \frac{[kt]}{k}\right) A\left(\frac{[kt]}{k}\right) A\left(\frac{[kt]-1}{k}\right)^{-1} U_k\left(\frac{[kt]}{k}, \frac{[kt]-1}{k}\right).$$

$$\ldots A\left(\frac{2}{k}\right) A\left(\frac{1}{k}\right)^{-1} U_k\left(\frac{2}{k}, \frac{1}{k}\right) A\left(\frac{1}{k}\right) A(0)^{-1} U_k\left(\frac{1}{k}, 0\right) .$$

---

[1] T. Kato [1] uses the approximation of the form

$$\exp((t - t_k) A(t_k)) \exp((t_k - t_{k-1}) A(t_{k-1})) \ldots \exp((t_1 - t_0) A(t_0))$$

$$(s = t_0 < t_1 < \cdots < t_k < t)$$

instead of our $U_k(t, s)$.

Expanding the right side, we obtain

$$(14)\quad W_k(t,0) = \left(I + C\left(t,\frac{[kt]}{k}\right)\right)\left\{U_k(t,0) + \sum_{ks=1}^{[kt]-1} U_k(t,s)\, C\left(s,s-\frac{1}{k}\right) U_k(s,0) + \right.$$
$$\left. + \sum_{ku=1}^{[kt]-1} U_k(t,u)\, C\left(u,u-\frac{1}{k}\right) \sum_{ks=1}^{[ku]-1} U_k(u,s)\, C\left(s,s-\frac{1}{k}\right) U_k(s,0) + \cdots\right\},$$

that is,[2])

$$W_k(t,0) = \left(I + C\left(t,\frac{[kt]}{k}\right)\right)\left\{U_k(t,0) + W_k^{(1)}(t,0) + W_k^{(2)}(t,0) + \cdots\right\},$$

$$(14)'\quad W_k^{(1)}(t,0) = \sum_{ks=1}^{[kt]-1} U_k(t,s)\, C\left(s,s-\frac{1}{k}\right) U_k(s,0)\,,$$

$$W_k^{(m+1)}(t,0) = \sum_{ks=1}^{[kt]-1} U_k(t,s)\, C\left(s,s-\frac{1}{k}\right) W_k^{(m)}(s,0)\quad (m=1,2,\ldots,[kt]-1)\quad.$$

We have $p\left(C\left(s,s-\frac{1}{k}\right)x\right) \leqq k^{-1}N\,p(x)$ by (6). Hence, by (12),

$$(15)\quad p(W_k^{(1)}(t,0)x) \leqq t\cdot M\,N\,M\cdot p(x),\; p(W_k^{(2)}(t,0)x) \leqq \frac{t^2}{2}M\,N\cdot M\,N\,M\cdot p(x),\ldots,$$
$$p(W_k^{(m)}(t,0)x) \leqq \frac{t^m}{m!}\,M\,(N\,M)^m\cdot p(x),\ldots$$

so that we obtain (13).

In particular, we see that $U_k(t,0)\,y \in D(A(t))$ for $y \in D(A(0))$, and analogously that $U_k(t,s)\,y \in D(A(t))$ for $y \in D(A(0))$. Hence, by (9),

$$U_k(t,s)y = \exp\left(\left(t-\frac{[kt]}{k}\right) A\left(\frac{[kt]}{k}\right)\right) U_k\left(\frac{[kt]}{k},s\right) y$$

is differentiable at $t \neq \frac{i}{k}\; (i = 0, 1, \ldots, k)$:

$$(16)\quad \frac{d\,U_k(t,s)y}{dt} = A\left(\frac{[kt]}{k}\right) U_k(t,s)y,\; y \in D(A(0))\,.$$

Similarly we prove the differentiability at $s \neq \frac{i}{k}\; (i = 0, 1, \ldots, k)$:

$$(16)'\quad \frac{d\,U_k(t,s)y}{ds} = -U_k(t,s)\, A\left(\frac{[ks]}{k}\right) y,\; y \in D(A(0))\,.$$

By (3), (4), (5), (6), (7) and (12), these derivatives are bounded and are continuous in t and in s except at $t = \frac{i}{k}$ and $s = \frac{i}{k}$. By the fact that $U_k(s,0)\,A(0)^{-1}x \in D(A(s))$ proved above, we see, remembering (16) and (16)',

$$(U_k(t,0) - U_n(t,0))\,A(0)^{-1}x = [U_n(t,s)\,U_k(s,0)\,A(0)^{-1}x]_{s=0}^{s=t}$$

$$= \int_0^t \frac{d}{ds}\{U_n(t,s)\,U_k(s,0)\,A(0)^{-1}x\}\,ds$$

$$= \int_0^t U_n(t,s)\left\{A\left(\frac{[ks]}{k}\right) - A\left(\frac{[ns]}{n}\right)\right\} A\left(\frac{[ks]}{k}\right)^{-1} A\left(\frac{[ks]}{k}\right) U_k(s,0)\,A(0)^{-1}x\,ds$$

$$= -\int_0^t U_n(t,s)\, C\left(\frac{[ns]}{n},\frac{[ks]}{k}\right) A\left(\frac{[ks]}{k}\right) A(s)^{-1} W_k(s,0)x\,ds\,.$$

---

[2]) This key formula was obtained in T. Kato [1] by solving a Volterra type equation.

Thus, we see, by (2), (12), (13) and (6), that

$$\lim_{k \to \infty} U_k(t, 0) A(0)^{-1}x \quad \text{exists uniformly in } t, 0 \leq t \leq 1 .$$

Hence, by (3) and (12), we prove that, for every $x \in X$,

$$\lim_{k \to \infty} U_k(t, 0)x = U(t, 0)x \quad \text{exists uniformly in } t, 0 \leq t \leq 1 .$$

Similarly we prove that, for every $x \in X$,

$$(17) \qquad \lim_{k \to \infty} U_k(t, s)x = U(t, s)x \quad \text{exists uniformly in } t \text{ and } s .$$

Therefore, by (7), $U(t, s)x$ is uniformly continuous in t and s.

Finally we easily prove, remembering (14)', (15), (17) and (6), the bounded convergence of $\{W_k(t, 0)x; k = 1, 2, \ldots\}$:

$$\lim_{k \to \infty} W_k(t, 0)x = W(t, 0)x = U(t, 0)x + W^{(1)}(t, 0)x + W^{(2)}(t, 0)x + \cdots,$$

$$W^{(1)}(t, 0)x = \int_0^t U(t, s) C(s) U(s, 0)x \, ds ,$$

$$W^{(m+1)}(t, 0)x = \int_0^t U(t, s) C(s) W^{(m)}(s, 0)x \, ds \qquad (m = 1, 2, \ldots) .$$

Therefore, if $y \in D(A(0))$, the limits

$$\lim_{k \to \infty} U_k(t, 0)y = U(t, 0)y \quad \text{and} \quad \lim_{k \to \infty} A(t) U_k(t, 0)y = W(t, 0) A(0)y$$

both exist boundedly and uniformly in t, and $W(t, 0) A(0)y$ is uniformly continuous in t. $A(t)$ is a closed linear operator by (5), and so we have proved that, for every $y \in D(A(0))$,

$$U(t, 0)y \quad \text{is in } D(A(t)), \text{ and}$$

$$(19) \quad A(t) U(t, 0)y = \lim_{k \to \infty} A(t) A\left(\frac{[kt]}{k}\right)^{-1} A\left(\frac{[kt]}{k}\right) U_k(t, 0)y = W(t, 0) A(0)y$$

boundedly and uniformly in $t, 0 \leq t \leq 1$.

Hence, by letting $k \to \infty$ in

$$U_k(t, 0)y - y = \int_0^t \frac{d}{ds} U_k(s, 0)y \, ds = \int_0^t A\left(\frac{[ks]}{k}\right) U_k(s, 0)y \, ds ,$$

we see that $x(t) = U(t, 0)y$ is a solution of the Cauchy problem (1).

References

[1] Kato, T.: Integration of the equation of evolution in a Banach space, J. Math. Soc. Japan 5, 208—234 (1953).
[2] Köthe, G.: Topologische Lineare Räume, 1. Berlin-Göttingen-Heidelberg: Springer 1960.
[3] Yosida, K.: Functional Analysis. Berlin-Heidelberg-New York: Springer 1965.

(Received June 28, 1965)

Druck: Brühlsche Universitätsdruckerei Gießen

VI. Diffusion Equations

Comments (Seizô Itô)

The series of papers commented on in this section is, in short, a very important contribution to the theory on integration of forward and backward diffusion equations in a Riemannian space. In most of these papers, the proof of main theorems is based on the celebrated Hille-Yosida theorem on one-parameter semi-groups of bounded linear operators in a Banach space (see T. Kato's comment on [57]) for the operator theoretical part and an elegant construction of a suitable parametrix for a given diffusion operator. The outline of the main results of those papers are mentioned below.

Let R be an orientable Riemannian space of dimension ≥ 2 with metric $ds^2 = g_{ij}(x)dx^i dx^j$. The forward diffusion equation, often called the Fokker-Planck equation, is of the form

$$(1) \qquad \frac{\partial f(t, x)}{\partial t} = Af(t, x)$$

where A is the elliptic differential operator given by

$$(2) \qquad Af = \frac{1}{g(x)^{1/2}} \frac{\partial^2}{\partial x^i \partial x^j} \{g(x)^{1/2} b^{ij}(x)f\}$$

$$- \frac{1}{g(x)^{1/2}} \frac{\partial}{\partial x^i} \{g(x)^{1/2} a^i(x)f\} \quad (g(x) = \det[g_{ij}(x)]),$$

and the backward diffusion equation is of the form

$$(3) \qquad \frac{\partial f(t, x)}{\partial t} = A'f(t, x)$$

where the operator A' is the formal adjoint of A:

$$(4) \qquad A'f = b^{ij}(x)\frac{\partial^2 f}{\partial x^i \partial x^j} + a^i(x)\frac{\partial f}{\partial x^i}.$$

In the paper [60] the following result is obtained. Under the assumption that the Riemannian space R is compact, the equation (1) with the initial condition

421

$f(0, x) = f(x) \in L^1(R)$ has a solution $f(t, x) = (U_t f)(x)$, where $\{U_t; t \geq 0\}$ is a strongly continuous one-parameter semi-group of positive linear operators in $L^1(R)$ preserving the norm of positive elements of $L^1(R)$ such that

$$\text{strong-lim}_{\delta \downarrow 0}\ \delta^{-1}(U_{t+\delta} - U_t)f = \tilde{A}U_t f,$$

$\tilde{A}$ being the closed extension of A. An analogous result is obtained in [67] for the equation (3) by making use of the resolvent of the semi-group generated by A' in the Banach space of continuous functions with the usual norm.

The equation (1) in a connected compact region $G\ (\subset R)$ with smooth boundary ∂G is treated in [66]. By means of Green's formula, the integral of $h \cdot Af - f \cdot A'h$ over G is transformed into an integral over ∂G, which will be denoted by I. It is shown that (1) has a unique solution in $L^1(G)$ if and only if

$$(*)\quad \begin{array}{l}\text{for no } m > 0 \text{ there exists a bounded solution } h \text{ of } A'h = mh \text{ which satisfies}\\ \text{the boundary condition specifying the integral } I \text{ to vanish identically.}\end{array}$$

This condition $(*)$ is verified in the case of Laplacian in a subdomain of euclidean space R^n by using the result of [64]; also [64] deals with the case of an exterior domain.

The paper [70] is essentially a continuation of [67]; the elliptic differential operator A_1 considered in [70] is of the form $A_1 f = A'f + c(x)f$ where A' is defined by (4). The existence of the resolvent $(I - m^{-1}A_1)^{-1}$ follows as in [67], and it is proved that, for m sufficiently large, the resolvent may be expressed as an integral transformation with a kernel $p_m(x, y)$. The kernel is considered as the Laplace transform of a transition probability, and the result has an immediate application to the diffusion process generated by A_1. In the special case of a symmetric operator $(A_1 = A_1')$, a bilinear expansion for the transition probability is obtained.

In the paper [65], a diffusion equation concerning differential forms on R (closed or open) is treated, and an ergodic theorem associated with the solution of the equation is proved; the result is considered as a stochastic interpretation and proof of Hodge's theory of harmonic integrals. Let H' be the set of C^∞ exterior differential forms α on R vanishing outside a compact set, and H be the Hilbert space obtained by completing H' under the norm $\|\alpha\| = (\alpha, \alpha)^{1/2}$ where $(\alpha, \beta) = \int_R \alpha\beta^*$, β^* being the adjoint form of β. The Laplacian $\Delta = d\delta + \delta d$ is defined as usual, and it is proved that, starting with any $\alpha \in H'$, one may obtain its harmonic part by a stochastic process. Denote by $\tilde{\Delta}$ the Friedrichs extension of Δ which is self-adjoint and non-positive definite. It is proved that $\tilde{\Delta}$ generates a strongly continuous semi-group $T_t\ (t \geq 0)$ such that, for any $\alpha \in D(\tilde{\Delta})$, $T_t\alpha$ satisfies the diffusion equation

$$\frac{\partial T_t\alpha}{\partial t} = \text{strong-lim}_{\delta \to 0} \frac{T_{t+\delta}\alpha - T_t\alpha}{\delta} = \tilde{\Delta}T_t\alpha \quad (t \geq 0)$$

with the initial condition: $\text{strong-lim}_{t \downarrow 0} T_t\alpha = \alpha$. The ergodic theorem proved in this paper asserts the existence of $\text{strong-lim}_{t \to \infty} T_t\alpha = T_\infty\alpha$, here $\Delta T_\infty\alpha = 0$ and

$(\alpha - T_\infty\alpha, \beta) = 0$ for every β such that β and $\Delta\beta$ belong to H and $\Delta\beta = 0$. The ellipticity of Δ is used in proving infinite differentiability of the various forms involved.

In the paper [71], the initial-value problem for both the forward equation (1) and the backward equation (3) is treated in the case where the space R is compact. Yosida proved not only the existence and the uniqueness of solution of (1), but also the following fact: the solution $f(t, x)$ with the initial value $f(x)$ satisfies

$$(5) \qquad \int_R |f(t, x)|dx \leqq \int_R |f(x)| \, dx$$

and the positivity of $f(x)$ on R implies that $f(t, x) \geqq 0$ for any $t > 0$ and any $x \in R$ and that the equality holds in (5). Similar results are obtained for the equation (3); in particular, the solution $f(t, x)$ of (3) with the initial value $f(x)$ satisfies that

$$(6) \qquad \min_x f(x) \leqq f(t, x) \leqq \max_x f(x).$$

$\left(\max_x f(t, x) = \max_x f(x) \text{ does not hold in general even if } f(x) \geqq 0 \text{ on } R. \right)$ The above-mentioned results concerning the equation (1) are extended to the case where R is a connected domain with smooth boundary and a certain boundary condition is assigned.

In the papers [73], [74] and [75], the results in [71] are extended to the case where the space R is non-compact, the coefficients a^i and b^{ij} in (2) and (4) depend on both t and x, and the term $c(t, x)f$ is added to the right-hand sides of (2) and (4). A fundamental solution is constructed in [73] by making use of a parametrix. In [74], a weak solution (a solution in the sense of Schwartz distribution) of the equation (1) is first obtained by means of operator-theoretical treatment, and then it is shown by making use of a parametrix given in [73] that the weak solution is equivalent to a genuine solution. The paper [75] shows that the existence proof in the preceding paper [74] may be modified so as to yield the existence of a solution admitting of a kernel representation of the form

$$f(t, s, x) = \int_R P(t, s, x, y)f(y) \, dy;$$

the kernel function $P(t, s, x, y)$ can be so chosen that it satisfies the equation $\partial P/\partial t = A_{t,x}P$.

An abstract analyticity in time for solutions of a diffusion equation discussed in [83] is also an interesting result. Considered in this paper is the equation

$$(7) \qquad \frac{\partial u(t, x)}{\partial t} = Au(t, x) \qquad (t > 0, x \in G \subset E^m)$$

where $A = a^{ij}(x)\partial^2/\partial x^i\partial x^j + b^i(x)\partial/\partial x^i + c(x)$ is an elliptic differential operator whose coefficients satisfy certain smoothness and boundedness conditions in G. It is proved that the operator A (considered in a suitable manner in $L^2(G)$) generates an analytic semi-group $\{T_t\}$ of linear operators. By making use of this fact, the following results are shown: (i) for any $f \in L^2(G)$, $u(t, x) = (T_t f)(x)$ is a solution of (7) such that $\|u(t, \cdot) - f\|_{L^2} \to 0$ as $t \downarrow 0$; (ii) if, for a fixed $t_0 > 0$, $u(t_0, x) \equiv 0$ on an open set $G_0 \subset G$, then $u(t, x) = 0$ for all $t > 0$ and all $x \in G$. It is also mentioned that $u(t, x)$ satisfies the so-called *unique continuation property*, namely the conclusion of (ii) can be replaced by "$u(t, x) = 0$ for all $t > 0$ and all $x \in G$."

Integration of Fokker-Planck's equation in a compact Riemannian space

Ark. Mat. 1 (1950) 71–75

1. The result. Let R be an n-dimensional compact Riemannian space with the metric $ds^2 = g_{ij}(x)\,dx^i\,dx^j$, and consider a *temporally homogeneous* MARKOFF *process* on R for which $P(t, x, y)$, $t > 0$, is the *transition probability* that the point x be transferred to y after the elapse of t units of time. We assume that $P(t, x, y)$ is continuous in (t, x, y) and hence satisfies SMOLUCHOVSKI's equation

$$(1.1) \qquad P(t + s, x, y) = \int_R P(t, x, z)\,P(s, z, y)\,dz \qquad (t, s > 0),$$

where the volume measure

$$dz = \sqrt{g(z)}\,dz^1 \ldots dz^n, \quad g(z) = \det\,[g_{ij}(z)],$$

and the *probability hypothesis*

$$(1.2) \qquad P(t, x, y) \geqq 0, \quad \int_R P(t, x, y)\,dy = 1.$$

The "continuity" of the transition process $P(t, x, y)$ may be defined as follows.[1] Let $L^1(R)$ be the Banach space of functions $f(x)$ integrable with respect to dx over R. There exist functions $f(x)$ dense in $L^1(R)$ for which the so called FOKKER-PLANCK *equation* holds:

$$(1.3) \quad \frac{\partial}{\partial t} f(t, x) = A \cdot f(t, x)\,(t \geqq 0),$$

$$f(t, x) = \int_R f(y)\,P(t, y, x)\,dy \quad (t > 0),$$

$$f(0, x) = f(x),$$

where the operator A is defined by

---

[1] A. KOLMOGOROFF: Zur Theorie der stetigen zufälligen Prozesse, Math. Ann. 108 (1933). W. FELLER: Zur Theorie der stochastischen Prozesse, Math. Ann., 113 (1937).

$$(1.4) \quad (A f)(x) = \frac{1}{\sqrt{g(x)}} \frac{\partial}{\partial x^i} \left[-\sqrt{g(x)}\, a^i(x)\, f(x) \right] +$$

$$+ \frac{1}{\sqrt{g(x)}} \frac{\partial^2}{\partial x^i \partial x^j} \left[\sqrt{g(x)}\, b^{ij}(x)\, f(x) \right]$$

with a positive definite quadratic form $b^{ij}(x)\, \xi_i \xi_j$.

The purpose of the present note is to show that, under certain continuity assumptions on $a(x)$ and $b(x)$, we may integrate (1.3) in the following manner: *there exists one and only one one-parameter semi-group* $\{U_t\}$, $0 \leq t < \infty$, *of linear operators on* $L^1(R)$ *to* $L^1(R)$ *such that:*

$$(1.5) \qquad U_t U_s = U_{t+s} \quad (t, s \geq 0), \ U_0 = I = \text{the identity;}$$

$$(1.6) \qquad \text{strong} \lim_{t \to t_0} U_t f = U_{t_0} f \ \text{ for } f \in L^1(R);$$

$$(1.7) \qquad \text{strong} \lim_{\delta \to 0} \frac{1}{\delta} [U_{t+\delta} - U_t] f = \tilde{A} U_t f \ \text{ for } f \ \text{ in a dense set in } L^1(R), \ \tilde{A}$$

denoting the closed extension of the operator A;

$(1.8) \qquad$ *if* $f(x) \in L^1(R)$ *is non-negative, then* $f(t, x) = (U_t f)(x)$ *is also non-negative and* $\| U_t f \|_1 = \int_R |f(t, x)|\, dx = \int_R |f(x)|\, dx = \| f \|_1$ (U_t *may be called a transition operator on* $L^1(R)$ *to* $L^1(R)$).

The method of proof is based upon the theory of semi-groups of linear operators due to E. HILLE[1] and the author[2] according to which *the operator* U_t *satisfying* (1.5)—(1.7) *is unique and may be given by*

$$(1.9) \qquad U_t f = \text{strong} \lim_{n \to \infty} \exp [t \tilde{A} (I - n^{-1} \tilde{A})^{-1}] f =$$

$$= \text{strong} \lim_{n \to \infty} \exp [t\, n\, (I - n^{-1} \tilde{A})^{-1} - I] f,$$

if $I_n = (I - n^{-1} \tilde{A})^{-1}$ *exists and is of norm* ≤ 1 $(n = 1, 2, \ldots)$.[3] The condition (1.8) is implied by the fact that the I_n are transition operators since

$$\exp (t A I_n) f = \exp [n t (I_n - I)] f = \exp (-n t) \exp (n t I_n) f.$$

These results may be considered as an extension of the case in which R is the surface of the three-sphere.[4]

---

[1] Functional analysis and semi-groups, New York (1948) Chap. 12.

[2] On the differentiability and the representation of one-parameter semi-groups of linear operators, Journ. Math. Soc. Japan, Vol. 1, No. 1 (1948), 15—21.

[3] Cf. E. HILLE, loc. cit., pp. 403—407, where the case of the n-dimensional euclidean space R and of constant $a(x)$, $b(x)$ is treated. Cf. also K. YOSIDA: An operator theoretical treatment of temporally homogeneous Markoff process, to appear in the Journ. Math. Soc. Japan, where the case of one-dimensional euclidean space R and non-constant $a(x)$, $b(x)$ is treated.

[4] K. YOSIDA: Brownian motion on the surface of the 3-sphere, to appear in the Ann. of Math.

2. Hypotheses and proof. The operator A may be written as

$$(2.1) \qquad (A f)(x) = \frac{1}{\sqrt{g(x)}} \frac{\partial}{\partial x^i}\left[\sqrt{g(x)}\, b^{ij}(x) \frac{\partial f}{\partial x^j}\right] + c^i(x) \frac{\partial f}{\partial x^i} + e(x) f(x).$$

We assume that

$(2.2) \qquad g(x),\ b(x)$ *and their first derivatives,* $c(x)$ *and* $e(x)$ *are all continuous in* R;

$(2.3) \qquad b^{ij}(x)\, \xi_i\, \xi_j \geqq a^{-1} \sum_i \xi_i^2$ *with a positive constant* a.

Then the formally self-adjoint operator

$$(2.4) \qquad H = \frac{1}{\sqrt{g(x)}} \frac{\partial}{\partial x^i}\left[\sqrt{g(x)}\, b^{ij}(x) \frac{\partial}{\partial x^j}\right]$$

has a *hypermaximal extension* $\tilde{H}$ and, since R is without boundary,

$$(2.5) \qquad (H f, f) = -\int_R b^{ij}(x) \frac{\partial f}{\partial x^i} \frac{\partial f}{\partial x^j} d x \leqq 0.$$

Our process of integration may be carried out in two steps.

The first step: I_n *exists as an operator on* $L^2(R)$ *to* $L^2(R)$ *for large* n.
P r o o f. Since H satisfies (2.5), $(I - n^{-1} \tilde{H})^{-1}$ exists with the norm

$$\|(I - n^{-1} \tilde{H})^{-1}\|_2 \leqq 1$$

as operator on $L^2(R)$ to $L^2(R)$ $(n = 1, 2, 3, \ldots)$. Hence the range

$$\{h;\ h = (I - n^{-1} H) f,\ f \in \text{domain } D(H) \text{ of } H\}$$

is dense in $L^2(R)$. When $h = (I - n^{-1} H) f$, we have, by (2.4) since

$$\|(I - n^{-1} \tilde{H})^{-1}\|_2 \leqq 1,$$

$$\left\|n^{-1} c^i(x) \frac{\partial f}{\partial x^i}\right\|_2 \leqq \left\{n^{-2} \int_R \sum_i [c^i(x)]^2 d x \cdot \int_R \sum_i \left(\frac{\partial f}{\partial x^i}\right)^2 d x\right\}^{\frac{1}{2}}$$

$$\leqq \left\{a\, n^{-2} \int_R \sum_i [c^i(x)]^2 d x \cdot \int_R b^{ij}(x) \frac{\partial f}{\partial x^i} \frac{\partial f}{\partial x^j} d x\right\}^{\frac{1}{2}}$$

$$= \left\{a\, n^{-1} \int_R \sum_i [c^i(x)]^2 d x\right\}^{\frac{1}{2}} \cdot [-n^{-1}(H f, f)]^{\frac{1}{2}},$$

where

$$0 \leqq - n^{-1}(Hf, f) = - (f, f) + (h, f) \leqq \| I - n^{-1}\tilde{H})^{-1} h \|_2^2$$

$$+ \| h \|_2 \| (I - n^{-1}\tilde{H})^{-1} h \|_2 \leqq 2 \| h \|_2^2$$

so that

$$\left\| n^{-1} c^i(x) \frac{\partial}{\partial x^i} (I - n^{-1}\tilde{H})^{-1} h \right\|_2 \leqq \left[2 a n^{-1} \int_R \sum_i [c^i(x)]^2 \, dx \right]^{\frac{1}{2}} \| h \|_2 .$$

We have also

$$\| n^{-1} e (I - n^{-1}\tilde{H})^{-1} h \|_2 \leqq n^{-1} \sup | e(x) | \cdot \| h \|_2 .$$

Therefore, since $\{ h; h = (I - n^{-1}H)f, f \in D(H) \}$ is dense in $L^2(R)$,

$$(2.6) \quad \| n^{-1} \tilde{K} \|_2 \leqq 1, \quad (Kh)(x) = c^i(x) \frac{\partial}{\partial x^i} (I - n^{-1}\tilde{H})^{-1} h(x) +$$

$$+ e(x)(I - n^{-1}\tilde{H})^{-1} h(x) ,$$

provided

$$(2.7) \qquad \left\{ 2 a n^{-1} \int_R \sum_i [c^i(x)]^2 \, dx \right\}^{\frac{1}{2}} + n^{-1} \sup | e(x) | < 1 .$$

Hence $(I - n^{-1}\tilde{K})^{-1}$ exists and the range $\{ g; g = (I - n^{-1}K)h, h \in D(K) \}$ is dense in $L^2(R)$ if (2.7) is satisfied. Therefore

$$\left\{ g; g = (I - n^{-1}H)f - n^{-1} c^i \frac{\partial f}{\partial x^i} - n^{-1} e f, f \in D(H) \right\}$$

$$= \{ g; g = (I - n^{-1}A)f, f \in D(H) \}$$

is dense in $L^2(R)$. Since $\| n^{-1}\tilde{K} \|_2 \leqq 1$ and $\| (I - n^{-1}\tilde{H})^{-1} \|_2 \leqq 1$, the solution f of

$$g = (I - n^{-1}A)f \quad \text{with} \quad (I - n^{-1}H)f = h, \ (I - n^{-1}K)h = g$$

satisfies

$$\| f \|_2 \leqq \| (I - n^{-1}\tilde{H})^{-1} h \|_2 \leqq \| h \|_2 \leqq \| (I - n^{-1}\tilde{K})^{-1} g \|_2 \leqq$$

$$\leqq \| (I - n^{-1}\tilde{K})^{-1} \|_2 \cdot \| g \|_2 .$$

Thus $(I - n^{-1}\tilde{A})^{-1}$ exists as a bounded operator on $L^2(R)$ to $L^2(R)$.

The second step: *Let $f(x)$ have continuous second order derivatives and put* $f(x) - n^{-1}(Af)(x) = g(x)$, *then, if* $n > \sup | e(x) |$, *we have*

$$f(x_0) \geqq [1 - n^{-1} e(x_0)]^{-1} g(x_0) \ \text{ for some } x_0 .$$

Proof. Let $f(x)$ reach its minimum at $x = x_0$. Since $\dfrac{\partial f}{\partial x^i}(x_0) = 0$, we have

$$f(x_0) - n^{-1} b^{ij}(x_0) \frac{\partial^2 f}{\partial x^i \partial x^j}(x_0) - n^{-1} e(x_0) f(x_0) = g(x_0),$$

that is,

$$f(x_0)\left[1 - n^{-1} e(x_0)\right] \geqq g(x_0).$$

Moreover we have

$$\int_R f(x)\, dx = \int_R f(x)\, dx - n^{-1} \int_R (A f)(x)\, dx = \int_R g(x)\, dx$$

since R is without boundary. Thus if the sequence $\{g_n(x)\}$ tends L^2-strongly to a non-negative function $g_\infty(x) \in L^2(R) < L^1(R)$, the corresponding functions $f_n(x)$ must tend to an almost everywhere non-negative function

$$f_\infty(x) \in L^2(R) < L^1(R) \quad \text{and} \quad \|f_\infty\|_1 = \|g_\infty\|_1.$$

Since $L^2(R)$ is L^1-dense in $L^1(R)$. I_n exists as a transition operator on $L^1(R)$ to $L^1(R)$ if n is sufficiently large. By formula (1.9), U_t exists as a transition operator satisfying (1.5)—(1.8). Thus the integration process has been carried through.

Mathematical Institute, Nagoya University. January 19, 1949.

Tryckt den 24 maj 1949

Uppsala 1949. Almqvist & Wiksells Boktryckeri AB

A theorem of Liouville's type for meson equation

Proc. Japan Acad. **27** (1951) 214 215

(Comm. by K. Kunugi, m.j.a., May 16, 1951.)

In connection with the stochastic integrability of Fokker-Planck's equation[1], the author encountered with the following

Problem. Let R be a connected domain with smooth boundaries ∂R of an n-dimensional euclid space $R_n (n \geq 2)$. Does there exist, for $m > 0$, a bounded solution $h(x)$ other than 0 of the meson equation

$$(1) \qquad \varDelta h(x) = m\, h(x) \text{ in } R$$

with the boundary condition

$$(2) \qquad \frac{\partial h}{\partial n} = O \text{ on } \partial R,\ n \text{ denoting outer normal?}$$

The purpose of the present note is to give an answer to this problem of Liouville's type. It reads as follows:

Let the boundaries ∂R lie entirely in the bounded part of R_n, and let $m(x)$ be a continuous function such that

$$(3) \qquad \inf_{x \in R} m(x) = m > 0.$$

Then the solution $h(x)$ of

$$(1)' \qquad \varDelta h(x) = m(x)\, h(x) \text{ in } R$$

satisfying the boundary condition (2) together with the order relation

$$(4) \qquad h(x) = O\ (exp\ (\alpha|x|)), \text{ where } 2\alpha < \sqrt{2m},$$

must vanish identically. Here $|x|$ denotes the distance of the point x from the origin of R.

Proof. 1st case (R is a bounded domain). By (2) and the Green's integral theorem, we have

$$m \int_R h(x)^2\, dx \leq \int_R h(x)\, \varDelta h(x)\, dx \leq \int_{\partial R} h(x)\, \frac{\partial h}{\partial n}\, dS = O.$$

Here dx and dS respectively denote the volume element of R and the hypersurface element of ∂R. Thus $h(x)$ must $\equiv O$.

---

1) K. Yosida: Integration of Fokker-Planck's equation with a boundary condition, Journal of the Mathematical Society of Japan, Takagi's Congratulation Number. The result obtained below implies the existence of the Brownian motion in R with ∂R as reflecting barrier.

2nd case (R extends to infinity). Let r_0 be so large that the sphere K_{r_0} of radius r_0 with the origin as its centre contains ∂R entirely. Applying Green's integral theorem for the domain D_r bounded by the hypersurface ∂K_r of K_r and by ∂R, we have, by (2),

$$(5) \qquad m\int_{D_r} h(x)^2\,dx \leq \int_{D_r} h(x)\,\varDelta h(x)\,dx \leq \int_{\partial K_r} h(x)\frac{\partial h}{\partial r}\,dS, \quad \text{for } r>r_0.$$

If we put

$$(6) \qquad\qquad F(r) = \int_{D_r} h(x)^2\,dx,$$

we have

$$F'(r) = \int_{\partial K_r} h(x)^2\,dS, \quad F''(r) = \int_{\partial K_r}\frac{\partial h^2}{\partial r}\,dS + \int_{\partial K_r} h(x)^2\,\frac{dS}{dr}.$$

Thus, by (5),

$$2m\,F(r) \leq F''(r) \quad \text{for } r > r_0.$$

Multiplying by $F'(r)$ and integrating from r_0 to r, we obtain

$$2m\,(F(r)^2 - F(r_0)^2) \leq F'(r)^2 - F'(r_0)^2.$$

If there exists $r_1 > r_0$ such that $F(r_1) > F(r_0)$, we have

$$\int_{r_1}^{r}\frac{dF}{\sqrt{2m(F^2 - F(r_0)^2) + F'(r_0)^2}} \geq \int_{r_1}^{r} dr.$$

Hence $\lim\limits_{r\to\infty} F(r) = \infty$ and

(7) $F(r)$ is of order not smaller than exp $(\sqrt{2m}\,r)$ as $r\to\infty$.

This contradicts to the assumption (4) and the definition (6) of $F(r)$. Thus $F(r) = F(r_0)$ for all $r > r_0$. Hence $h(x)$ must $= 0$ for $|x| > r_0$. Therefore we obtain $h(x) = 0$ in R by the same argument as in the 1st case.

An ergodic theorem associated with harmonic integrals

Proc. Japan Acad. **27** (1951) 540–543

(Comm. by K. Kunugi, m.j.a., Nov. 12, 1951.)

Let R be an n-dimensional $(n \geq 2)$, infinitely differentiable, orientable Riemann space, closed or open. We consider the totality H' of the infinitely differentiable exterior differential p-forms α on R which vanish outside compact sets of R. Let $d\alpha$ and $\delta\alpha = (-1)^{np+n}(d\alpha^*)^*$ be the exterior differential and codifferential of $\alpha \in H'$, so that we have $(d\alpha, \beta) = -(\alpha, \delta\beta)$. Here $(\alpha, \beta) = \int_R \alpha\beta^*$ and β^* is the adjoint form of β. Let H be the Hilbert space obtained as the completion of the pre-Hilbert space H', metrized by the norm $\| \alpha \| = (\alpha, \alpha)^{1/2}$. We denote by H'' the linear subspace of H consisting of the totality of the infinitely differentiable p-forms α such that α and $\Delta\alpha$, $\Delta = d\delta + \delta d$ (the Laplacian), both belong to H. Surely we have $H' \subseteq H'' \subseteq H$.

Suggested by an interesting paper by A. N. Milgram and P. C. Rosenbloom[1], we will prove the following ergodic theorem as a stochastic interpretation and proof of Hodge's theory of harmonic integrals[2].

The ergodic theorem. Let us consider Δ as a symmetric, non-positive definite operator defined on $H' \subseteq H$ with values in $H' \subseteq H$. Let $\tilde{\Delta}$ be the Friedrichs-Freudenthal's[3] self-adjoint, non-positive definite extension of Δ; $\tilde{\Delta}$ is the adjoint Δ' of Δ restricted to the domain $D(\tilde{\Delta})$ which is the intersection of the domain $D(\Delta')$ of Δ' with the completion H_1 of H' by the new norm $\| \alpha \|_1 = (\alpha - \Delta\alpha, \alpha)^{1/2}$. If we consider the diffusion equation

$$(1) \qquad \frac{\partial T_t\alpha}{\partial t} = \text{strong} \lim_{\delta \to 0} \frac{T_{t+\delta}\alpha - T_t\alpha}{\delta} = \tilde{\Delta}T_t\alpha, \quad t \geq 0,$$

with the initial condition

---

1) Harmonic forms and heat conduction, Proc. Nat. Acad. Sci., **37** (1951), 180–184.

2) W. V. D. Hodge: The theory and applications of harmonic integrals, Cambridge (1941). Cf. P. Bidal–G. de Rham: Le formes différentielles harmoniques, Comm. Math. Helv., **19** (1946), 1–49. K. Kodaira: Harmonic fields in Riemannian manifolds, Ann. of Math., **50** (1949), 581–665, G. de Rham–K. Kodaira: Harmonic integrals, Princeton (1950).

3) K. Friedrichs: Spektraltheorie halbbeschränkter Operatoren, Math. Ann., 109 (1934), 465–487. H. Freudenthal: Über Friedrichsche Fortsetzung halbbeschränkter Hermitescher Operatoren, Proc. Acad. Amsterdam, **39** (1936), 832–833.

$$(2) \qquad \text{strong } \lim_{t \to 0} T_t \alpha = \alpha \in D(\tilde{\varDelta}),$$

then there exists $T_\infty \alpha \in H''$ such that

$$(3) \qquad \text{strong } \lim_{t \to \infty} T_t \alpha = T_\infty \alpha \quad \text{and}$$

$$\varDelta T_\infty \alpha = 0, \ (\alpha - T_\infty \alpha, \beta) = 0 \ \text{for any } \beta \in H'' \text{ with } \varDelta \beta = 0.$$

Supplement. i) If R is closed, then we may take, in (1), $\tilde{\varDelta}$ as the smallest closed extension $(\varDelta')'$ of $\varDelta$. Moreover, when $\alpha \in H''$,

$$(4) \qquad \int_0^\infty T_t(\alpha - T_\infty \alpha)dt \ \text{may be considered to belong to } H'' \text{ and}$$

$$T_\infty \alpha - \alpha = \varDelta \int_0^\infty T_t(\alpha - T_\infty \alpha)dt.$$

ii) When R is the euclidean n-space, we may take, in (1), $\tilde{\varDelta}$ as the smallest closed extension $(\varDelta')'$ of $\varDelta$.

Proof of the ergodic theorem. Since $\varDelta$ is self-adjoint with its spectra lying on the half line $(-\infty, 0)$, we may apply the semi-group theory[4]. Thus there exists a one-parameter semi-group of linear operators $T_t (t \geq 0)$:

$$(5) \qquad T_t \alpha = \text{strong } \lim_{m \to \infty} (I - m^{-1} t \tilde{\varDelta})^{-m} \alpha, \ \alpha \in H,$$

$$(I = \text{the identity operator})$$

such that

$$(6) \quad T_0 = I, \ T_t T_s = T_{t+s}, \ \text{strong } \lim_{t \to 0} T_t \alpha = \alpha \ \text{for } \alpha \in H, \ \| T_t \| \leq 1$$

and

$$(1)' \quad \frac{\partial T_t \alpha}{\partial t} = \text{strong } \lim_{\delta \to 0} \frac{T_{t+\delta} - T_t}{\delta} \alpha \ \text{exists only for } \alpha \in D(\tilde{\varDelta}) \text{ and}$$

$$= \tilde{\varDelta} T_t \alpha,$$

$$(1)'' \quad T_t \alpha - \alpha = \int_0^t \tilde{\varDelta} T_t \alpha dt \quad \text{for } \alpha \in D(\tilde{\varDelta}).$$

Let

$$(7) \qquad \tilde{\varDelta} = \int_{-\infty}^0 \lambda dE_\lambda$$

be the spectral resolution of $\tilde{\varDelta}$. Then, by (5), we have

$$(8) \qquad T_t \alpha = \int_{-\infty}^0 \exp(\lambda t) dE_\lambda \alpha, \ \alpha \in H.$$

Hence, by

$$\| T_t \alpha - T_{t'} \alpha \|^2 = \int_{-\infty}^0 |\exp(\lambda t) - \exp(\lambda t')|^2 d \| E_\lambda \alpha \|^2,$$

we easily see that

---

4) E. Hille Functional analysis and semi-groups, New York (1948). K. Yosida: On the differentiability and the representation of one-parameter semi-group of linear operators, J. Math. Soc. Japan, **1** (1948), 15-21.

5) L. Schwartz: Théorie des distributions, Paris (1950).

(9)   strong $\lim_{t\to\infty} T_t \alpha = T_\infty \alpha$ exists and $T_t T'_\infty \alpha = T_\infty \alpha$ for $\alpha \in H$.

Therefore, by (1)$'$, we have

(10) $\tilde{\varDelta} T_\infty \alpha = 0$ for $\alpha \in H$.

Consider the "distribution"[5] $S(\gamma) = (\gamma, T_\infty \alpha)$, $\gamma \in H'$. Then, by (10), S satisfies the differential equation in the sense of the "distribution":

$$\varDelta S = 0 .$$

Since $\varDelta$ is "elliptic", there exists[6] infinitely differentiable β such that $\varDelta \beta = 0$ and $S(\gamma) = (\gamma, \beta)$ for every $\gamma \in H'$. Hence we may consider that $T_\infty \alpha$ is in H'' and

(10)$'$ $\varDelta T_\infty \alpha = 0 .$

Thus we see that

(11) $T_\infty = E_0 - E_{0-0}$, viz. T_∞ is the projection operator upon the closed linear subspace spanned by the solutions $\beta \in H''$ of $\varDelta \beta = 0$.

Hence, starting from any $\alpha \in H$, we may obtain its harmonic part $T_\infty \alpha$ by a stochastic procedure (9). For we have

(12) $\varDelta \beta = 0\,(\beta \in H'')$ implies $T_\infty \beta = \beta$ and hence

$$(\alpha - T_\infty \alpha, \beta) = (\alpha, \beta) - (T_\infty \alpha, \beta) = (\alpha, \beta) - (\alpha, T_\infty \beta)$$
$$= (\alpha, \beta) - (\alpha, \beta) = 0 .$$

The proof of the supplement. i) We have

(13) $\| (I - \varDelta)\alpha \|^2 \geq \| \alpha \|^2 - (\varDelta \alpha, \alpha) \geq \| \alpha \|^2.$

Moreover, the range $\{\beta;\ \beta = (I - \varDelta)\alpha,\ \alpha,\ \alpha \in H'\}$ is strongly dense in H. For, if otherwise, there would exist $\tilde{\gamma} \in H$ such that $\tilde{\gamma} \neq 0$ and $((I - \varDelta)\alpha, \tilde{\gamma}) = 0$ for every $\gamma \in H'$. By the "ellipticity" of the operator $(I - \varDelta)$, we see that $\tilde{\gamma}$ may be considered to be infinitely differentiable. Thus, by (13), $\tilde{\gamma}$ must equal to 0, contrary to $\tilde{\gamma} \neq 0$. Hence the resolvent $(I - \tilde{\varDelta})^{-1}$ exists as a bounded self-adjoint operator defined on H. We have, from (13),

(14) $\| (I - \varDelta)\alpha \| \leq 1$ implies $\| \alpha \| \leq 1$ and $0 \leq -(\varDelta \alpha, \alpha) \leq 1.$

By virtue of the compactness of R, we see, by extending F. Rellich's argument[7], that the resolvent $(I - \tilde{\varDelta})^{-1}$ is completely continuous. Hence the spectra of form a discrete set which accumulates only at $-\infty$. Hence

(7)$'$ $\tilde{\varDelta} = \sum_{i=1}^{\infty} \lambda_i P_i$, $O = \lambda_1 > \lambda_2 > \cdots$, $\lim_{i\to\infty} \lambda_i = -\infty$, where P_i is the projection on the closed linear subspace spanned by the solutions $\beta \in H''$ of $\varDelta \beta = \lambda_i \beta_i .$

---

6) L. Schwartz: ibid.

7) Ein Satz über mittlere Konvergenz, Göttingen Nachrichten (1930), 15–21.

Here we have again made use of the "ellipticity" of the operator $(\lambda_i I - \Delta)$. Thus we have

$$(8)' \qquad T_t \alpha = \sum_{i=1}^{\infty} \exp(\lambda_i t) P_i \alpha, \quad \alpha \in H.$$

By $P_1 = T_\infty$, $P_i P_j = 0 \ (i \neq j)$, we have

$$T_t(\alpha - T_\infty \alpha) = \sum_{i=2}^{\infty} \exp(\lambda_i t) P_i \alpha.$$

Hence the integral (in the sense of S. Bochner)

$$\int_0^t T_t(\alpha - T_\infty \alpha) dt$$

converges strongly, when $t \uparrow \infty$, to an element $\int_0^\infty T_t(\alpha - T_\infty \alpha) dt \in H$. Thus, by $(1)''$ and the closure of the operator $\tilde{\Delta}$, we obtain

$$(15) \qquad T_\infty \alpha - \alpha = \tilde{\Delta} \int_0^\infty T_t(\alpha - T_\infty \alpha) dt \quad \text{for} \quad \alpha \in H''.$$

Since $T_\infty \alpha - \alpha$ is infinitely differentiable and since the operator Δ is "elliptic", $\int_0^\infty T_t(\alpha - T_\infty \alpha) dt$ may be considered to be infinitely differentiable and so we have (4).

ii) It would be sufficient, as above, to prove that the range $\{\beta; \beta = (I - m^{-1}\Delta)\alpha, \ \alpha \in H'\}$ is, for $m > 0$, strongly dense in H. Let us assume the contrary. Then there must exist $\tilde{\gamma} \in H$ such that $\tilde{\gamma} \neq 0$ and $((I - m^{-1}\Delta)\alpha, \tilde{\gamma}) = 0$ for every $\alpha \in H'$. By the "ellipticity" of the operator $(I - m^{-1}\Delta)$, there must exist infinitely differentiable γ which satisfies $(I - m^{-1}\Delta)\gamma = 0$ and $(\alpha, \tilde{\gamma}) = (\alpha, \gamma)$ for every $\alpha \in H'$. Such γ is, in the case of the ordinary Laplacian Δ in the euclidean n-space $(n \geq 2)$[8], identically zero, contrary to $\tilde{\gamma} \neq 0$.[9]

8) See K. Yosida: A theorem of Liouville's type for meson equation, Proc. Japan Acad., **27** (1951), 214-215.

9) At this juncture, the author wishes to express his hearty thanks to J. Igusa who kindly pointed out that, in the general case of open space R, the smallest closed extension$(\Delta')'$ may have positive spectra. The original manuscript is revised according to his criticism.

Integration of Fokker-Planck's equation with a boundary condition

J. Math. Soc. Japan 3 (1951) 69–73

1. Introduction. We consider Fokker-Planck's equation[1]

$$(1) \qquad \frac{\partial f(t, x)}{\partial t} = Af(t, x), \ t \geq 0,$$

$$(Af)(x) = \frac{1}{\sqrt{g(x)}} \frac{\partial^2}{\partial x^i \partial x^j} \left(\sqrt{g(x)}\, b^{ij}(x) f(x) \right)$$

$$+ \frac{1}{\sqrt{g(x)}} \frac{\partial}{\partial x^i} \left(- \sqrt{g(x)} a^i(x) f(x) \right)$$

in a connected region R of an n-dimensional orientable Riemannian space with the metric $ds^2 = g_{ij}(x) dx^i dx^j$. As usual, the volume element in R is defined by $dx = \sqrt{g(x)} dx^1 dx^2 \cdots dx^n$, $g(x) = det(g_{ij}(x))$. We assume that the contravariant tensor $b^{ij}(x)$ be such that $b^{ij}(x)\xi_i\xi_j > 0$ in R (for $\sum_i \xi_i^2 > 0$). The $a^i(x)$ obeys, by the coordinate change $x \to \bar{x}$, the transformation rule

$$(2) \qquad \bar{a}^i(\bar{x}) = \frac{\partial \bar{x}^i}{\partial x^k} a^k(x) + \frac{\partial^2 \bar{x}^i}{\partial x^k x^s} b^{ks}(x).$$

These properties of the coefficients $a^i(x)$ and $b^{ij}(x)$ are connected with the probabilistic meaning of the equation (1).

We assume that $g_{ij}(x)$, $a^i(x)$ and $b^{ij}(x)$ are infinitely differentiable functions of the coordinates $x = (x^1, x^2, \cdots, x^n)$. The purpose of the present note is to consider a certain natural boundary condition on the boundary ∂R of R for the probability density $f(t, x)$ at the time moment $t > 0$ and to discuss, for this boundary condition, the stochastic integrability (in the sense to be explained in §3) of the equation (1). As in the previous papers, our treatment and the method of proof relies upon the theory of semi-group of linear operators,[2] which is, so to speak, an operator-theo-

---

1) A. Kolmogoroff: Zur Theorie der stetigen zufälligen Prozess, Math. Ann., **108** (1933), 149–160. K. Yosida: An extension of Fokker-Planck's equation, Proc. Japan Acad., **25** (1949), (9), 1–3.

2) E. Hille: Functional Analysis and Semi-groups, New York (1948). K. Yosida: On the differentiability and the representation of one-parameter semi-group of linear operators, Journ. Math. Soc. Japan, 1 (1949), 1, 15–21, and K. Yosida: An operator-theoretical treatment of temporally homogeneous Markoff process, ibid., 1 (1949), 1, 244–235.

retical adaptation of the Laplace transform method in partial differential equations.

2. Green's formula and the boundary condition. Let A' be the formally adjoint operator of A:

$$(3) \qquad (A'h)(x) = b^{ij}(x)\frac{\partial^2 h}{\partial x^i \partial x^j} + a^i(x)\frac{\partial h}{\partial x^i}.$$

By partial integration, we obtain the Green's formula:

$$(4) \qquad \int_G (h(x)(Af)(x) - f(x)(A'h)(x))\,dx =$$

$$\int_{\partial G} \sqrt{g(x)}\,b^{ij}(x)\,\left(h(x)\frac{\partial f}{\partial x^j} - f(x)\frac{\partial h}{\partial x^j}\right)\Pi_i(x)\,dS +$$

$$\int_{\partial G}\left(\frac{\partial \sqrt{g(x)}\,b^{ij}(x)}{\partial x^j} - \sqrt{g(x)}\,a^i(x)\right)\Pi_i(x)f(x)h(x)\,dS,$$

where $\Pi_i(x)$ is $\cos(n, x^i)$, n being outer normal at the point x of the boundary ∂G of the connected domain $G \subseteq R$, and dS denotes hypersurface area on ∂G. If $b^{ij}(x)\Pi_i(x)\Pi_j(x) > 0$ at $x \in \partial G$, we may define the outer transversal direction ν at x by

$$(5) \qquad \frac{dx^i}{\sqrt{g(x)}\,b^{ij}(x)\,\Pi_j(x)} = d\nu \quad (i = 1, 2, \cdots, n)$$

so that

$$(6) \qquad \sqrt{g(x)}\,b^{ij}(x)\left(h(x)\frac{\partial f}{\partial x^j} - f(x)\frac{\partial h}{\partial x^j}\right)\Pi_i(x)\,dS =$$

$$\left(h(x)\frac{\partial f}{\partial \nu} - f(x)\frac{\partial h}{\partial \nu}\right)dS.$$

We consider A to be an additive operator defined for the totality $D(A)$ of infinitely differentiable functions $f(x)$ on R which vanish outside some compact set (depending upon $f(x)$) and which also satisfy the boundary condition on ∂R:

$$(7) \qquad \sqrt{g(x)}\,b^{ij}(x)\frac{\partial f}{\partial x^j}\Pi_i(x) + \left(\frac{\partial \sqrt{g(x)}\,b^{ij}(x)}{\partial x^j}\right.$$

$$\left. - \sqrt{g(x)}\,a^i(x)\right)\Pi_i(x)f(x) = 0.$$

$D(A)$ is surely dense in the Banach space $L_1(R)$ of integrable (with respect to dx) functions on R, metrized by the norm $\|f\| = \int_R |f(x)|\,dx$. Thus A

may be considered as an additive operator defined for $D(A) \subseteq L_1(R)$ with values in $L_1(R)$.

3. Stochastic integration of (1) with the boundary condition (7). We first prove

Lemma 1. Let $f(x) \in D(A)$ be positive (negative) in a connected domain $G \subseteq R$ such that $f(x)$ vanishes on $\partial G - \partial R$, viz. on the part of ∂G not contained in ∂R. Then we have, for any positive number m, the inequality

$$(8) \qquad \int_G (f(x) - m^{-1}(Af)(x)) dx \geq \int_G f(x) dx > 0 \, (\leq \int_G f(x) dx < 0).$$

Proof. By (4)–(7), we have

$$\int_G (Af)(x) dx = \int_{G-\partial R} \frac{\partial f}{\partial \nu} dS \leq 0 \quad (\geq 0)$$

Corollary. For any $f(x) \in D(A)$, we have

$$(9) \qquad \|f - m^{-1}Af\| \geq \|f\| \text{ for } m > 0, \text{ and for } f \in D(A).$$

Proof. Let $h(x)$ be $= 1, -1$ or 0 according as $f(x)$ is > 0, < 0 or $= 0$. Since the conjugate space $L_1(R)^*$ of $L_1(R)$ is the space of all the essentially bounded measurable functions $k(x)$ with the norm $\|k\|^* = $ essential sup $|k(x)|$, we have, by the above lemma,

$$\|f - m^{-1}Af\| \geq \int_R h(x)(f(x) - m^{-1}(Af)(x)) dx = \int_R |f(x)| dx$$

$$- m^{-1} \sum_i \int_{P_i} (Af)(x) dx + m^{-1} \sum_j \int_{N_j} (Af)(x) dx,$$

where $P(N)$ is connected domain in which $f(x) > 0 (< 0)$ such that $f(x)$ vanishes on the boundary $\partial P(\partial N)$. Q. E. D.

Thus, there exists the bounded additive inverse

$$(10) \qquad (I - m^{-1}\tilde{A})^{-1},$$

where I and $\tilde{A}$ respectively denote identity operator and the smallest closed extension of A. Thus we have the

Lemma 2. The resolvent I_m ($=$ everywhere defined inverse $(I - m^{-1}\tilde{A})^{-1}$) exists if and only if the range $\{(I - m^{-1}A)f; f \in D(A)\}$ of the operator $(I - m^{-1}A)$ is dense in $L_1(R)$. Moreover, if the resolvent I_m exists, it is a transition operator, viz.

$$(11) \quad f(x) \geq 0 \text{ and } f \in L_1(R) \text{ imply } (I_m f)(x) \geq 0 \text{ and } \int_R (I_m f)(x) dx = \int_R f(x) dx.$$

Proof. The first part of the lemma is evident. Let the resolvent I_m

438

exist, For any $g(x)\geq 0$ of $L_1(R)$, there exists a sequence $\{f_k(x)\}\subseteq D(A)$ such that

$$\text{strong } \lim_{k\to\infty} f_k=f \text{ exists and strong } \lim_{k\to\infty} (f-m^{-1}Af_k)=f-m^{-1}\tilde{A}f=g.$$

On the other hand, by the boundary condition (7), we have

$$\int_R (f_k(x)-m^{-1}(Af_k)(x))dx=\int_R f_k(x)dx.$$

Hence, in the limit $k\to\infty$, we have

$$\int_R g(x)dx=\int_R f(x)dx \text{ and } \int_R |g(x)|\,dx\geq \int_R |f(x)|\,dx \quad \text{(by (9))}.$$

Therefore $f(x)\geq 0$ almost everywhere and $\|g\|=\|f\|$. Q. E. D.

By the semi-group theory, there exists a one-parameter semi-group of transition operators T_t satisfying the conditions:

$$(12) \qquad T_t T_s=T_{t+s}, \ (s, \ t\geq 0), \ T_0=\text{the identity,}$$

$$\text{strong } \lim_{t\to t_0} T_t f=T_{t_0}f, \ f\epsilon L_1(R),$$

$$\text{strong } \lim_{\delta\to 0} \frac{T_{t+\delta}-T_t}{\delta}f=\tilde{A}T_t f \text{ for } f \text{ in the domain } D(\tilde{A}) \text{ of } \tilde{A},$$

if and only if the resolvents I_m (for $m>0$) exist as transition operators. This T_t is, in fact, defined by

$$(13) \qquad T_t f=\text{strong } \lim_{m\to\infty} (I-m^{-1}t\,\tilde{A})^{-m}f, \ f\epsilon L_1(R).$$

The existence of this semi-group is just the stochastic integrability mentioned in the introduction. That $T_t f(t>0)$, $f\epsilon L_1(R)$, satisfies the boundary condition (7) in a limiting sense may be seem from (13) and the definition of the smallest closed extention $\tilde{A}$ of A.

4. The theorem. Since the conjugate space of $L_1(R)$ is the space of essentially bounded measurable functions, we see that the stochastic integrability of (1) is equivalent to the non-existence of bounded measurable function $h(x)$ such that

$$(14) \qquad \|h\|^*>0 \text{ and } \int_R h(x)(f(x)-m^{-1}(Af)(x))dx=0 \text{ for all } f\epsilon D(A).$$

Thus, if we define the distribution $H(f)$ in the sense of L. Schwartz:

$$(15) \qquad H(f)=\int_R h(x)f(x)dx,$$

we see that H satisfies the elliptic differential equation (in the sense of the distribution)[3]

$$(16) \qquad A'H=mH.$$

---

3) L. Schwartz: Théorie des distributions, 1, Paris (1950).

Hence,[4] if $n \geq 2$, there exists a function $\tilde{h}(x)$ infinitely differentiable in R such that

$$(17) \qquad (A'\tilde{h})(x) = m\tilde{h}(x) \text{ in } R, \quad H(f) = \int_R \tilde{h}(x) f(x) dx.$$

Surely $\tilde{h}(x)$ is equal to $h(x)$ almost everywhere, and so does not vanish identically. Let $\{R_k\}$ be a monotone increasing sequence of connected domains $\subseteq R$ such that the boundary ∂R_k tends, as $k \to \infty$, to the boundary ∂R very smoothly. Then we have, by the Green's formula (4), (14) and (17),

$$(18) \quad \int_{\partial R_k} \sqrt{g(x)}\, b^{ij}(x) \left(\tilde{h}(x) \frac{\partial f}{\partial x^j} - f(x) \frac{\partial \tilde{h}}{\partial x^j} \right) \Pi_i(x) dS +$$

$$\int_{\partial R_k} \left(\frac{\partial \sqrt{g(x)} b^{ij}(x)}{\partial x^j} - \sqrt{g(x)}\, a^i(x) \right) \Pi_i(x) f(x) \tilde{h}(x) dS = 0, f \epsilon D(A).$$

By the boundedness of $\tilde{h}(x)$ and the boundary condition of $f(x) \epsilon D(A)$, we have

$$(19) \qquad \lim_{k \to \infty} \int_{\partial R_k} \sqrt{g(x)}\, b^{ij}(x) f(x) \frac{\partial \tilde{h}}{\partial x^j} \Pi_i(x) dS = 0, \ f \epsilon D(A).$$

Therefore we have the

Theorem. Let the dimension n of R be ≥ 2. Then the stochastic integrability of (1) with the boundary condition (7) is equivalent to the non-existence, for $m > 0$, of bounded solution $\tilde{h}(x) \not\equiv 0$ of

$$(20) \qquad (A'\tilde{h})(x) = m\tilde{h}(x) \text{ in } R$$

satisfying the boundary candition (19).

Remark. The above condition of the stochastic integrability is satisfied in the case of compact Riemannian spach R, as is shown by the following argument. At a maximizing (minimizing) point x_0 of $\tilde{h}(x)$ we have $(A'\tilde{h})(x_0) \leq (\geq) 0$, so that a continuous solution $\tilde{h}(x)$ of (20) cannot have either a positive maximum or a negative minimum. Other applications of the Theorem to concrete examples will be published elsewhere.

Nagoya University

---

4) L. Schwartz: loc. cit. Cf. also an early paper by K. Kodaira: Harmonic tensor fields in Riemannian manifolds, Ann. of Math., **50** (1949), 587–655.

5) Cf. K. Yosida: Integration of Fokker-Planck's equation in a compact Riemannian space, Arkiv för Matematik, **1**, (1949), 9, 1–3.

Integrability of the backward diffusion equation in a compact Riemannian space

Nagoya Math. J. **3** (1951) 1–4

1. Introduction. Let R be an orientable, compact Riemannian space with the metric $ds^2 = g_{ij}(x)dx^i dx^j$, and consider the *backward diffusion equation*

$$\text{(1)} \qquad \frac{\partial f(t, x)}{\partial t} = A \cdot f(t, x), \qquad t \geqq 0,$$

$$(Af)(x) = b^{ij}(x)\frac{\partial^2 f}{\partial x^i \partial x^j} + a^i(x)\frac{\partial f}{\partial x^i}.$$

Here $b^{ij}(x)$ is a contravariant tensor such that the quadratic form $b^{ij}(x)\xi_i\xi_j$ is >0 for $\sum_i \xi_i^2 > 0$, and $a^i(x)$ changes, by a coordinate transformation $x \to \bar{x}$, as follows:

$$\text{(2)} \qquad \bar{a}^i(\bar{x}) = \frac{\partial \bar{x}^i}{\partial x^k}a^k(x) + \frac{\partial^2 \bar{x}^i}{\partial x^k \partial x^s}b^{ks}(x).$$

These conditions for the coefficients $a^i(x)$ and $b^{ij}(x)$ are connected with the probabilistic interpretation of the equation (1).[1] In preceding notes,[2] the author treated the stochastic integrability of the *forward diffusion equation* (*Fokker-Planck's equation*)

$$\text{(3)} \qquad \frac{\partial f(t, x)}{\partial t} = A' \cdot f(t, x), \qquad t \geqq 0,$$

$$(A'f)(x) = \frac{1}{\sqrt{g(x)}}\frac{\partial^2}{\partial x^i \partial x^j}(\sqrt{g(x)}b^{ij}(x)f(x))$$

$$+ \frac{1}{\sqrt{g(x)}}\frac{\partial}{\partial x^i}(-\sqrt{g(x)}a^i(x)f(x)), \quad g(x) = \det(g_{ij}(x))$$

Received March 15, 1951.

[1] A. Kolmogoroff: Zur Theorie der stetigen zufälligen Prozesse, Math. Ann., **108** (1933), 149–160. K. Yosida: An extension of Fokker-Planck's equation, Proc. Japan Acad., **25** (1949), (9), 1–3.

[2] K. Yosida: Integration of Fokker-Planck's equation in a compact Riemannian space, Arkiv för Matematik, 1 (1949), 9, 71–75. K. Yosida: Integration of Foker-Planck's equation with boundary condition, Journ. Math. Soc. Japan, (1951), Takagi's Congratulation volume.

by the semi-group theory.[3] The purpose of the present note is to consider the *stochastic integrability* (to be explained below) of (1), also by the semi-group theory. The result[4] may be considered, in a certain sense, a dual of the result in the preceding notes referred to above.

2. The theorems. For the sake of simplicity, we assume that R is an analytic manifold and that $a^j(x)$ and $b^{ij}(x)$ are holomorphic functions of the coordinates $x=(x^1, x^2, \ldots, x^n)$. We consider A as an additive operator whose domain $D(A)$ is the totality of infinitely differentiable functions defined in R, with values in the Banach space $C(R)$ of the totality of continuous functions $f(x)$ defined in R and metrized by the norm $\|f\| = \max_{x \in R} |f(x)|$. The following two simple lemmas are essential for our arguments.

LEMMA 1. For any $f \in D(A)$ and for any positive number m, we have

$$(4) \qquad \max_{x \in R} h(x) \geqq f(x) \geqq \min_{x \in R} h(x), \quad h(x) = f(x) - m^{-1}(Af)(x).$$

Proof. Let $f(x)$ reach its maximum (minimum) at $x_0(x_1)$. Then we have

$$h(x_0) = f(x_0) - m^{-1}(Af)(x_0) \geqq f(x_0) \quad (h(x_1) = f(x_1) - m^{-1}(Af)(x_1) \leqq f(x_1)).$$

COROLLARY. For any $f \in D(A)$ and for any positive number, we have

$$(5) \qquad \|f - m^{-1}Af\| \geqq \|f\|.$$

LEMMA 2. Let $\{f_n\} \subseteq D(A)$ and $\{Af_n\}$ converge, as $n \to \infty$, strongly[5] to 0 and h respectively. Then we have $h(x) \equiv 0$.

Proof. For any $k(x) \in D(A)$, we have $(dx = \sqrt{g(x)} dx^1 dx^2 \ldots dx^n$, $g(x) = \det(g_{ij}(x))$

$$\int_R h(x)k(x)dx = \lim_{n \to \infty} \int_R (Af_n)k(x)dx = \lim_{n \to \infty} \int_R f_n(x)(A'k)(x)dx = 0,$$

and thus $h(x)$ must $\equiv 0$.

COROLLARY. The smallest closed extension $\bar{A}$ of the operator A exists. $\bar{A}$ is defined as follows:

(6) $\qquad \bar{A}f$ is defined and $= h$ if there exists $\{f_n\} \subseteq D(A)$ such that $\{f_n\}$ and $\{Af_n\}$ converge, as $n \to \infty$, strongly to f and h.

---

[3] E. Hille: Functional Analysis and Semi-groups, New York (1948). K. Yosida: On the differentiability and the representation of one-parameter semi-group of linear operators, Journ. Math. Soc. Japan, **1** (1949), 1, 15–21, and K. Yosida: An operator-theoretical treatment of temporally homogeneous Markoff process, ibid., **1** (1949), 1, 224–235.

[4] Cf. another approach by K. Itô: Stochastic differential equations on a differentiable manifold, Nagoya Math. J., **1** (1950), 35–48.

[5] By the topology defined by the norm $\|f\|$, viz. by the uniform convergence on R.

From these two lemmas we have the

THEOREM 1. *The inverse $I_m = (-m^{-1}\overline{A})^{-1}$ exists (I=the identity operator) for $m>0$ and I_m is a positive, contraction operator, leaving the constant functions invariant:*

(7)   if $h(x)$ in the domain $D(\overline{A})$ of $\overline{A}$ be non-negative, then $(I_m h)(x)$ is also non-negative and $\|I_m h\| \leq \|h\|$; $I_m \cdot 1 = 1$.

By the semi-group theory, the coincidence of the domains $D(I_m)$ of I_m with $C(R)$ is the necessary and sufficient condition for the existence of the one-parameter semi-group of linear operators in $C(R)$ with the properties:

(8) $T_t T_s = T_{t+s}$ $(t, s \geqq 0)$, $T_0 = I$;

T_t are positive, contraction operators, leaving constant functions invariant; strong $\lim_{t \to t_0} T_t f = T_{t_0} f$, $f \in C(R)$;

strong $\lim_{\delta \to 0} \dfrac{T_{t+\delta} - T_t}{\delta} f = \overline{A} T_t f = T_t \overline{A} f$ for f in $D(\overline{A})$, which is surely dense in $C(R)$.

This T_t is, in fact, defined by

(9) $T_t f = $ strong $\lim_{m \to \infty} (I - t\, m^{-1}\overline{A})^{-m} f$; $f \in C(R)$.

The existance of this semi-group T_t may be considered as the *stochastic integrability* of (1). The coincidence of the domains $D(I_m)$ with $C(R)$ is equivalent to the denseness of the ranges $R(I-m^{-1}A) = \{f - m^{-1}Af\,;\ f \in D(A)\}$, $m>0$, in $C(R)$. Hence we have the

COROLLARY. (1) is stochastically integrable if and only if positive numbers m do not belong to the residual spectra of the operator A.

When the dimension n of the space R is $\geqq 2$, we have the

THEOREM 2. *The backward diffusion equation* (1) *is stochastically integrable if the compact space R is of dimension $\geqq 2$.*

Proof. Let the range $R(I-m^{-1}A)$ be not dense in $C(R)$. Then there exists a measure φ, countably additive for Borel sets of R such that

(10) $0 <$ total variation of φ in $R < \infty$,

(11) $\displaystyle\int_R (f(x) - m^{-1}(Af)(x))\varphi(dx) = 0$ for $f \in D(A)$,

since the conjugate space of $C(R)$ is the space of measures, countably additive for Borel sets and of bounded total variations. If we define the *distribution* (in the sense of L. Schwartz[6]) by

(12) $\displaystyle H(f) = \int_R f(x)\varphi(dx)$, $f \in D(A)$,

---

[6] L. Schwartz: Théorie des distributions, 1, Paris (1950).

H satisfies, by (11), the *differential equation* (*in the sense of the distribution*)

(13) $A'H=mH.$

By the *elliptic character* of the differential operator A', there must exist[7] an infinitely differentiable function $h(x)$ such that

(14) $(A'h)(x)=mh(x), \qquad H(f)=\int_R f(x)h(x)dx.$

By (10), we have

(15) $h(x)\not\equiv 0.$

Let $k(x)$ be $=1$, -1 or $=0$ according as $h(x)>0$, <0 or $=0$. Then we have

(16) $$0=\int_R |h(x)-m^{-1}(A'h)(x)|dx\geqq \int_R (h(x)-m^{-1}(A'h)(x))k(x)dx$$

$$=\int_R |h(x)|dx-m^{-1}\sum_i \int_{P_i}(A'h)(x)dx+m^{-1}\sum_j \int_{N_i}(A'h)(x)dx,$$

where $P(N)$ are connected domains in which $h(x)>0$ (<0) such that $h(x)=0$ on their boundaries $\partial P (\partial N)$. By the integral theorem of Green's type we have

(17) $$\int_P (A'h)(x)dx=\int_{\partial P}\frac{\partial h}{\partial n}dS,$$

where n and dS respectively denote outer normal and positive measure on ∂P.

Hence we have $\int_P (A'h)(x)dx\leqq 0$, and similarly $\int_N (A'h)(x)dx\geqq 0$. Therefore we obtain, from (14)$-$(15), a contradiction $0\geqq\int_R |h(x)|dx>0.$

Mathematical Institute,
Nagoya University

---

7) L. Schwartz: loc. cif., p. 136,

On the existence of the resolvent kernel for elliptic differential operator in a compact Riemann space

Nagoya Math. J. **4** (1952) 63–72

§1. Introduction.

We consider the differential operator

$$(1.1) \qquad (Af)(x) = b^{ij}(x)\frac{\partial^2 f}{\partial x^i \partial x^j} + a^i(x)\frac{\partial f}{\partial x^i} + c(x)f(x)$$

in an n-dimensional $(n \geqq 2)$, orientiable, compact Riemann space R with the metric $ds^2 = g_{ij}(x)dx^i dx^j$. Here $b^{ij}(x)$ is a contravariant tensor such that the quadratic form $b^{ij}(x)\xi_i\xi_j$ is > 0 for $\sum_{i=1}^{n}\xi_i^2 > 0$, and $a^i(x)$ changes, by the coordinates transformation $x \to \bar{x}$, as follows:

$$(1.2) \qquad \bar{a}^i(\bar{x}) = \frac{\partial \bar{x}^i}{\partial x^k}a^k(x) + \frac{\partial^2 \bar{x}^i}{\partial x^j \partial x^s}b^{js}(x).$$

These transformation rules for the coefficients are connected with the fact that the value of $(Af)(x)$ is independent of the local coordinates $(x^1, \ldots, x^n)$.

For the sake of simplicity, we assume that R is an infinitely differentiable manifold and that $g_{ij}(x)$, $b^{ij}(x)$, $a^i(x)$, $c(x)$ are infinitely differentiable functions of the local coordinates $(x^1, \ldots, x^n)$. We consider A as an additive operator whose domain $D(A)$ is the totality of real-valued infinitely differentiable functions on R, with values in the Banach space $C(R)$ of the totality of real-valued continuous functions $f(x)$ on R, metrized by the norm $\|f\| = \max_{x \in R} |f(x)|$. As in a preceding note,[1] we may prove (§2) the folowing existence theorem:

Let us consider $D(A)$ as a linear subspace of $C(R)$ and let $\tilde{A}$ be the smallest closed extension of the operator A. Then, if

$$(1.3) \qquad m > \max_{x}|c(x)|,$$

the operator $(I - m^{-1}\tilde{A})$ $(I =$ the identity operator) admits a bounded linear inverse, the resolvent $I_m = (I - m^{-1}\tilde{A})^{-1}$ defined on $C(R)$.

Received September 25, 1951.

[1] K. Yosida: Integrability of the backward diffusion equation in a compact Riemannian space, Nagoya Math. Journal, Vol. 3, 1–4 (1951). At this juncture, the author wishes to correct the errata in the cited paper. $(-m^{-1}\tilde{A})$ on page 3, line 2 must be corrected as $(I - m^{-1}\tilde{A})$. $D(A)$ and A on page 3, line 5 must be corrected as $D(I_m)$ and I_m respecteively.

The purpose of the present note is to show that this resolvent may, for sufficiently large m, be represented as an integral operator of the form

$$(1.4) \qquad (I_m f)(x) = \int_{R} p_m(x, y) f(y) dy, \qquad dy = \sqrt{g(x)} dx^1 \ldots dx^n,$$

$$g(x) = \det(g_{ij}(x)),$$

with a measurable kernel $p_m(x, y)$. The result will be applied to the explicit expression for the transition probability of the stochastic process defined by the diffusion equation

$$(1.5) \qquad \frac{\partial f}{\partial t} = Af \qquad (t \geqq 0).$$

§2. The existence of the resolvent I_m. We will prepare lammas.

LEMMA 1. Let m satisfy (1.3) and let $((I - m^{-1}A)f)(x) = g(x)$ for $f \in D(A)$. Then we have

$$(2.1) \qquad \max_{x} g(x) \geqq (1 - m^{-1}\|c\|) \max_{x} f(x) \quad \text{for} \quad \max_{x} f(x) \geqq 0$$

$$\geqq (1 - m^{-1}(\min_{x} c(x)) \max_{x} f(x) \quad \text{for} \quad \max_{x} f(x) \leqq 0,$$

$$(2.1) \qquad \min_{x} g(x) \leqq (1 - m^{-1}\|c\|) \min_{x} f(x) \quad \text{for} \quad \min_{x} f(x) \leqq 0$$

$$\leqq (1 - m^{-1}(\min_{x} c(x)) \min_{x} f(x) \quad \text{for} \quad \min_{x} f(x) \geqq 0.$$

Proof. Let $f(x)$ reach its maximum and minimum at $x = x_1$ and x_2. Then we have, by

$$b^{ij}(x) \frac{\partial^2 f}{\partial x^i \partial x^j} \leqq 0 \ (\text{at } x = x_1), \qquad b^{ij}(x) \frac{\partial^2 f}{\partial x^i \partial x^j} \geqq 0 \ (\text{at } x = x_2),$$

the inequalities

$$f(x_1) - m^{-1}c(x_1)f(x_1) \leqq g(x_1), \qquad f(x_2) - m^{-1}c(x_2)f(x_2) \geqq g(x_2).$$

LEMMA 2. The smallest closed extension $\tilde{A}$ of A exists. It is defined as follows: $\tilde{A}f = e$ if there exists $\{f_k\} \subseteq D(A)$ such that the strong $\lim_{k \to \infty} f_k = f$, strong $\lim_{k \to \infty} Af_k = e$. Here strong lim means the lim defined by the norm of $C(R)$.

Proof. By the integral theorem of Green, we have

$$(2.2) \qquad \int_{R} (Af_k)(x)h(x)dx = \int_{R} f_k(x)(A'h)(x)dx, \qquad h \in D(A), \quad \text{where}$$

$$(2.3) \qquad (A'h)(x) = \frac{1}{\sqrt{g(x)}} \frac{\partial^2}{\partial x^i \partial x^j} (\sqrt{g(x)} b^{ij}(x) h(x))$$

$$- \frac{1}{\sqrt{g(x)}} \frac{\partial}{\partial x^i} (\sqrt{g(x)} a^i(x) h(x))$$

$$+ c(x)h(x) = (A_1 h)(x) + c(x)h(x).$$

Thus, if strong $\lim_{k\to\infty} f_k = 0$, we would have

$$\int_R e(x)h(x)dx = \lim_{k\to\infty}\int_R (Af_k)(x)h(x)dx = \lim_{k\to\infty}\int_R f_k(x)(A'h)(x)dx = 0.$$

Hence we must have $e(x) \equiv 0$ for strong $\lim_{k\to\infty} f_k = 0$. Therefore $\widetilde{A}f$ is a one-valued function of f, independent of the sequence $\{f_k\}$ which defines f.

LEMMA 3. The range $\{(I - m^{-1}A)f;\ f\in D(A)\}$ is strongly dense in $C(R)$.

Proof. If otherwise, there would exists a measure $\mu(E)$, countably additive for Borel set E of R, such that

(2.4) $\qquad\qquad$ the total variation of μ on R is $\neq 0$,

(2.5) $\qquad\qquad \int_R ((I - m^{-1}A)f)(x)\mu(dx) = 0 \quad$ for $\quad f\in D(A)$.

Since the operator $(I - m^{-1}A)$ is elliptic, there must exist[2] infinitely differentiable function $h(x)$ such that

(2.6) $\qquad\qquad \mu(E) = \int_E h(x)dx, \qquad ((I - m^{-1}A')h)(x) = 0.$

Let[3] $k(x)$ be $=1$, $=-1$ or $=0$ according as $h(x) > 0$, <0 or $=0$. Then we have

$$0 = \int_R k(x)((I - m^{-1}A')h)(x)dx \geqq \int_R (1 - m^{-1}\|c\|)\,|\,h(x)\,|\,dx$$

$$- m^{-1}\sum_i\int_{P_i}(A_1h)(x)dx + m^{-1}\sum_j\int_{N_j}(A_1h)(x)dx,$$

where $P(N)$ are connected domains in which $h(x) > 0$ (<0) such that $h(x)$ vanishes on the boundaries $\partial P(\partial N)$. We have, by Green's integral theorem,

$$\int_{P_i}(A_1h)(x)dx = \int_{\partial P_i}\frac{\partial h}{\partial n}dS,$$

where n and dS denote outer normal and positive measure on ∂P respectively. Hence $\int_{P_i}(A_1h)(x) \leqq 0$. Similarly we have $\int_{N_j}(A_1h)(x)dx \geqq 0$. Thus we must have $h(x) \equiv 0$ and hence $\mu(E) = \int_E h(x)dx = 0$, contrary to (2.4).   Q.E.D.

We have incidentally proved the following lemma, which plays an important role in §4 below.

LEMMA 4. For any $h\in D(A)$, we have, for sufficiently large m,

[2] L. Schwartz: Théorie des distributions, Paris (1950).

[3] Cf. K. Yosida: Integration of Fokker-Planck's equation with a boundary condition, Journal of the Math. Soc. of Japan, Vol. 3, No. 1, 69–73 (1951).

$$(2.7) \qquad \int_R |((I-m^{-1}A)h)(x)|\,dx \geqq \frac{1}{2}\int_R |h(x)|\,dx .$$

By the above three lemmas 1, 2 and 3, we see that, for $m > \|c\|$, the resolvent

$$(2.8) \qquad I_m = (I-m^{-1}\tilde{A})^{-1}$$

exists as a bounded linear operator on $C(R)$. Moreover, by lemma 1, the operator I_m is positive :

$$(2.9) \qquad g(x) \geqq 0 \text{ on } R \text{ implies } f(x) = ((I-m^{-1}\tilde{A})^{-1}g)(x) \geqq 0 \text{ on } R.$$

Hence, for fixed $x_0 \in R$, $(I_m g)(x_0)$ is a bounded linear functional on $C(R)$ and thus

$$(2.10) \qquad (I_m g)(x_0) = \int_R P_m(x_0,\, dy) g(y) ,$$

where $P_m(x_0, E)$ is a non-negative set function, countably additive for Borel set E. $P_m(x, E)$ is also Borel measurable in x for fixed E.

We will show (§4) that, for sufficiently large m,

$$(2.11) \qquad P_m(x,\, E) = \int_E P_m(x,\, y)dy, \quad \text{with a measurable density } P_m(x,\, y)$$

$$\text{satisfying certain regularity conditions (see (4.12) below).}$$

To this purpose, we need a parametrix in the large, viz. almost Green's function of the operator $(I-m^{-1}A')$. This will be introduced in the next §.

§3. The parametrix in the large. We adopt a new metric

$$(3.1) \qquad dr^2 = b_{ij}(x)dx^i dx^j ,$$

where $(b_{ij}(x))$ is the inverse matrix of the matrix $(b^{ij}(x))$. We also assume that the local coordinates $(x^1, \ldots, x^n)$ are a normal coordinates in the vicinity of the point $P = (0, \ldots, 0)$. Thus the adjoint poerator A' of A is of the form $(b(x) = \det (b_{ij}(x)))$:

$$(3.2) \qquad (A'f)(x) = \frac{1}{\sqrt{b(x)}} \frac{\partial^2}{\partial x^i \partial x^j} (\sqrt{b(x)}\, b^{ij}(x)f(x))$$

$$- \frac{1}{\sqrt{b(x)}} \frac{\partial}{\partial x^2} (\sqrt{b(x)}\, a^i(x)f(x))$$

$$+ c(x)f(x)$$

$$= (\Delta f)(x) + e^i(\dot{x})\frac{\partial f}{\partial x^i} + k(x)f(x), \quad \text{where}$$

$$(\Delta f)(x) = b^{ij}(x)\left[\frac{\partial^2 f}{\partial x^i \partial x^j} - \frac{\partial f}{\partial x^\alpha}\left\{ \begin{matrix} \alpha \\ ij \end{matrix} \right\} \right] \quad \text{(the Laplacian)} ,$$

$$\left\{ \begin{matrix} \alpha \\ ij \end{matrix} \right\} = \frac{1}{2}b^{\alpha\mu}\left[\frac{\partial b_{\mu i}}{\partial x^j} + \frac{\partial b_{j\mu}}{\partial x^i} - \frac{\partial b_{ij}}{\partial x^\mu} \right] .$$

Let $\Gamma = r^2$ be the square of the geodesic distance of the point $Q = (x^1, \ldots, x^n)$

from the point $P=(0,\ldots,0)$. We have the well-known identity

$$(3.3) \qquad \Gamma = \Gamma_{PQ} = r^2 = r_{PQ}^2 = b_{\alpha\beta}(0)x^\alpha x^\beta,$$
$$b_{\alpha\sigma}(x)x^\sigma = b_{\alpha\sigma}(0)x^\sigma,$$
$$\left\{{k \atop ij}\right\}x^i x^j = 0.$$

Let $\Phi(\Gamma)$ be a function of $\Gamma = \Gamma_{PQ}^2$. Then, from

$$\frac{\partial\Phi}{\partial x^\alpha} = \frac{d\Phi}{d\Gamma}\frac{\partial\Gamma}{\partial x^\alpha}, \quad \frac{\partial^2\Phi}{\partial x^\alpha \partial x^\beta} = \frac{d^2\Phi}{d\Gamma^2}\frac{\partial\Gamma}{\partial x^\alpha}\frac{\partial\Gamma}{\partial x^\beta} + \frac{d\Phi}{d\Gamma}\frac{\partial^2\Gamma}{\partial x^\alpha \partial x^\beta},$$

we obtain

$$(3.4) \quad (A'\Phi)(x) = \frac{d^2\Phi}{d\Gamma^2}b^{\alpha\beta}(x)\frac{\partial\Gamma}{\partial x^\alpha}\frac{\partial\Gamma}{\partial x^\beta} + \frac{d\Phi}{d\Gamma}\Delta\Gamma + \frac{d\Phi}{d\Gamma}e^\alpha(x)\frac{\partial\Gamma}{\partial x^\alpha} + k(x)\Phi(\Gamma).$$

The coefficients in this equation may be simplified as follows.[4] From (3.3)

$$b^{\alpha\beta}\frac{\partial\Gamma}{\partial x^\alpha}\frac{\partial\Gamma}{\partial x^\beta} = 4b^{\alpha\beta}b_{\alpha\sigma}(0)x^\sigma b_{\beta\tau}(0)x^\tau = 4b^{\alpha\beta}b_{\alpha\sigma}x^\sigma b_{\beta\tau}(0)x^\tau = 4\Gamma.$$

From (3.3) and the definition of the Laplacian in (3.2),

$$\Delta\Gamma = 2b^{\alpha\beta}b_{\alpha\beta}(0) - 2b^{\alpha\beta}x^\sigma\frac{\partial b_{\alpha\sigma}}{\partial x^\beta} + b^{\alpha\beta}x^\sigma\frac{\partial b_{\alpha\beta}}{\partial x^\sigma} = 2n + x^\sigma\frac{\partial \log b}{\partial x^\sigma}.$$

The last equality may be obtained by differentiating the 2nd identity of (3.3) with respect x^β and summing on the indices α and β:

$$b^{\alpha\beta}x^\sigma\frac{\partial b_{\alpha\sigma}}{\partial x^\beta} = -n + b^{\alpha\beta}b_{\alpha\beta}(0).$$

Therefore we have

$$(3.5) \quad (A'\Phi)(x) = 4\Gamma\frac{d^2\Phi}{d\Gamma^2} + \left[2n + x^\sigma\frac{\partial \log b}{\partial x^\sigma} + 2e^\alpha b_{\alpha\beta}(0)x^\beta\right]\frac{d\Phi}{d\Gamma} + k\Phi.$$

Thus, by taking

$$(3.6) \qquad \Phi_m(\Gamma_{PQ}) = -\frac{m}{2\pi}\log r_{PQ}, \quad (n=2),$$

$$= \frac{m}{N}r_{PQ}^{2-n}, \quad N = (n-2)2(\pi)^{n/2}/\Gamma(n/2), \quad (n \geq 3),$$

we have

$$(3.7) \quad (A'\Phi_m)(x) = -\frac{m}{2\pi}\left\{\frac{1}{2}\left(x^\sigma\frac{\partial \log b}{\partial x^\sigma} + 2e^\alpha b_{\alpha\beta}(0)x^\beta\right)r^{-2} + k\log r\right\}, \quad (n=2),$$

$$= \frac{m}{N}\left\{\left(\frac{2-n}{2}\right)\left(x^\sigma\frac{\partial \log b}{\partial x^\sigma} + 2e^\alpha b_{\alpha\beta}(0)x^\beta\right)r^{-n} + kr^{2-n}\right\}, \quad (n \geq 3).$$

---

[4] We follow T. Y. Thomas and E. W. Titt: On the elementary solution of the general linear differential equation of the second order with analytic coefficients, Journal de Math., tome **18**, 217-248 (1939).

Hence (3.6) is a parametrix in the large of the operator $(I - m^{-1}A')$ in the following sense. By the integral theorem of Green $(dx = \sqrt{b(x)}\, dx^1 \ldots dx^n)$, we obtain

$$\int_D h(x)((I - m^{-1}A)f)(x)dx - \int_D f(x)((I - m^{-1}A')h(x)dx$$

$$= m^{-1}\int_D (f(x)(A'h)(x) - h(x)(Af)(x))dx$$

$$= -m^{-1}\int_{\partial D}\left\{f\frac{\partial h}{\partial \nu} - h\frac{\partial f}{\partial \nu} + Lfh\right\}dS,$$

where ν is the inner transversal direction defined by

$$\frac{dx^j}{\sqrt{b(x)}\, b^{ij}(x)\cos(n,\, x^i)} = d\nu \quad (n \text{ denotes the inner normal}),$$

and dS is the hypersurface element on the boundary ∂D which surrounds the point $P = (0, \ldots, 0)$, and L is a function continuous for $P = (0, \ldots, 0)$. If we take $\Phi_m(\Gamma_{PQ})$ for $h(x)$ and the geodesic sphere of radius δ and $P = (0, \ldots, 0)$ as centre for ∂D, we obtain, in the limit,

$$(3.8) \qquad \lim_{\delta \downarrow 0} -m^{-1}\int_{\partial D} = \text{the value at } P \text{ of the function } f.$$

This we prove, in view of (3.6), by taking the local coordinates in such a way that $b_{ij}(0) = \delta_{ij}$—the geodesic coordinates at P. In this way, we have

$$(3.9) \qquad \int_R K_m(k,\, y)((I - m^{-1}A)f)(y)dy = f(x) + \int_R L_m(x,\, y)f(y)dy,$$

where

$$(3.10) \qquad K_m(x,\, y) = \Phi_m(r_{x,\,y}), \quad r_{x,\,y} = \text{the geodesic distance of } x \text{ and } y,$$

and

$$(3.11) \qquad L_m(x,\, y) = ((I - m^{-1}A')K_m(x,\, y) \text{ is infinitely differentiable for } x \neq y$$
$$\text{and is, in the vicinity of } x = y, \text{ of the order}$$
$$\begin{cases} r_{x,\,y}^{-1}, & (n = 2), \\ r_{x,\,y}^{1-n}, & (n \geqq 3). \end{cases}$$

§ 4. The integral representation of the resolvent I_m.

We have, from (3.9),

$$(4.1) \quad (I_m g)(x) + \int_R L_m(x,\, y)(I_m g)(y)dy = \int_R K_m(x,\, y)g(y)dy \quad \text{for } g \in C(R).$$

This may be written as

$$(4.1)' \qquad I_m g + L_m I_m g = K_m g.$$

Hence we have

$$I_m g + L_m(K_m g - L_m I_m g) = K_m g, \text{ that is,}$$
$$I_m g - L_m^{(2)} I_m g = (K_m - L_m K_m)g, \quad \text{where}$$

(4.2)
$$(L_m^{(2)} g)(x) = \int_R \left\{ \int_R L_m(x, z) L_m(z, y) dz \right\} g(y) dy,$$

$$(L_m K_m g)(x) = \int_R \left\{ \int_R L_m(x, z) K_m(z, y) dz \right\} g(y) dy.$$

Thus we obtain

$$I_m g - L_m^{(2)}(L_m^{(2)} I_m g + K_m - L_m K_m g) = K_m g - L_m K_m g, \text{ that is,}$$
$$I_m g - L_m^{(4)} I_m g = (K_m - L_m K_m + L_m^{(2)} K_m - L_m^{(3)} K_m)g.$$

Repeating the process, we obtain the integral equation of the form

(4.3)
$$I_m g - L_m^{(k)} I_m g = (K_m - L_m K_m + \ldots)g.$$

Because of (3.10) and (3.11), we may take k so large that

(4.4) $M_m(x, y) = L_m^{(k)}(x, y)$ is continuous in (x, y) and
$N_m(x, y) = (K_m - L_m K_m + \ldots)(x, y)$ is continuous for $x \neq y$ and
has the same order of singularity, for $x = y$, as $K_m(x, x)$.

We have thus proved that $(I_m g)(x)$ must satisfy the integral equation

(4.5)
$$(I_m g)(x) - \int_R M_m(x, y)(I_m g)(y) dy = \int_R N_m(x, y) g(y) dy.$$

By the continuity of the kernel $M_m(x, y)$, we may apply the classical theory of Fredholm to (4.5). Thus there exist a continuous kernel $Q_m(x, y)$ and k' functionals $c_1(g), c_2(g), \ldots, c_{k'}(g)$ such that

(4.6) $$(I_m g)(x) = \int_R N_m(x, y) g(y) dy$$
$$+ \int_R Q_m(x, z) dz \left\{ \int_R N_m(z, y) g(y) dy \right\} + \sum_{i=1}^{k'} c_i(g) \varphi_i(x),$$

where $\varphi_1(x), \varphi_2(x), \ldots, \varphi_{k'}(x)$ form the linearly independent base of the solutions of the homogenous equations

(4.7)
$$\int_R M_m(x, y) \varphi(y) dy = \varphi(x).$$

Because of the lemmas 1–3, $(I_m g)(x)$ may, for fixed x, be considered as a bounded linear functional of $g \in C(R)$. Hence we have

(4.8)
$$c_i(g) = \int_R \mu_i(dy) g(y),$$

where μ_i are regular measures, countably additive for Borel sets E. These measures must, for sufficiently large m, be absolutely continuous with respect to the measure dy, and with bounded measurable densities:

$$(4.9) \qquad \mu_i(E) = \int_E \nu_i(y)dy, \text{ essential supremum } |\nu_i(y)| < \infty .$$

This we see from the lemma 4, viz. from

$$(4.10) \qquad \lim_{s \to \infty} \int_R |h_s(x)|dx = 0 \quad \text{if} \quad \lim_{s \to \infty} \int_R |((I - m^{-1}A)h_s)(x)|dx = 0 .$$

Summing up, we have obtained the result: for sufficiently large m,

$$(4.11) \qquad (I_m g)(x) = \int_R p_m(x, y)g(y)dy, \quad g \in C(R) ,$$

with a kernel $p_m(x, y)$ enjoying the conditions:

$$(4.12) \qquad p_m(x, y) \text{ is measurable in } (x, y),$$
$$p_m(x, y) \text{ is continuous in } x \text{ for fixed } y \neq x,$$
$$p_m(x, y) \text{ is, for } x = y, \text{ of the same order as } K_m(x, y), \text{ viz.}$$
$$p_m(x, y) = \begin{cases} O(\log r_{x,y}), & n = 2 \\ O(r_{x,y}^{2-n}), & n \geqq 3. \end{cases}$$

§5. An application to the stochastic processes. *We will consider the special case of a symmetric operator A:*

$$(5.1) \qquad\qquad\qquad A = A' .$$

Since the singularity of the resolvent kernel $p_m(x, y)$ is given by (4.12), we see that its k-th iterated kernel $p_m^{(k)}(x, y)$ is, for sufficiently large k, a bounded measurable function of (x, y). Thus, by Hilbert-Schmidt's expansion theorem, the Fourier series of the kernel $p_m^{(k)}(x, y)$ are absolutely and uniformly convergent on the product space $R \times R$. By virtue of this fact, we may prove[5] that the series

$$(5.2) \qquad\qquad \sum_{i=1}^{\infty} \frac{\psi_i(x)\psi_i(y)}{(1 - m^{-1}\lambda_i)^k}$$

are, for sufficiently large k, absolutely and uniformly convergent on $R \times R$. Here $\{\psi_i(x)\}$ is a complete system of normal orthogonal eigenfunctions of the differential operator $A: \psi_i(x)$ belonging to the eigenvalue λ_i.

Proof. Let $\psi(x)$ be any eigenfunction of the operator I_m:

$$(5.3) \qquad\qquad (I - m^{-1}\widetilde{A})^{-1}\psi = \mu\psi .$$

We define, by the function $\psi(x)$, a distribution in the sense of Laurent Schwartz:[6]

$$(5.4) \qquad\qquad \Phi(f) = \int_R \psi(x)f(x)dx, \quad f \in D(A) .$$

---

[5] The same result is proved in other ways by K. Kodaira (unpublished) and by S. Minakshsundarum and A. Pleijel: Some properties of the eigenfunctions of the Laplace-operator on Riemannian manifolds, Canadian Journal of Math., Vol. 1, 242-256 (1950).

[6] Schwartz: ibid.

By virtue of (5.3), $\varnothing$ satisfies the differential equation in the sense of the distribution:

$$(5.5) \qquad (I - m^{-1}A)\varnothing = \mu^{-1}\varnothing .$$

Since $(I - m^{-1}A)$ is elliptic, there exists[7] an infinitely differentiable function $\varphi(x)$ such that

$$(5.6) \qquad ((I - m^{-1}A)\varphi)(x) = \mu^{-1}\varphi(x), \quad \varphi(x) = \psi(x)$$

almost everywhere with respect to the measure dx.

Therefore we may assume $\psi(x)$ to be an eigenfunction of the differential operator A, belonging to the eigenvalue $m(1 - \mu^{-1})$:

$$(5.7) \qquad (A\psi)(x) = m(1 - \mu^{-1})\psi(x) .$$

It is easy to see that, conversely, any eigenfunction of (5.7), belonging to the eigenvalue λ, is also an eigenfunction of $(I - m^{-1}\widetilde{A})^{-1}$, viz. of the kernel $p_m(x, y)$, belonging to the eigenvalue $(1 - m^{-1}\lambda)^{-1}$.

Therefore, by the absolute and uniform convergence of the Fourier series of the kernel $p_m^{(k)}(x, y)$, we see that the Fourier series (5.2) converge absolutely and uniformly on $R \times R$.

If we assume the negativity of the eigenvalues λ of A, which is surely satisfied for the operator (5.11), we have

$$(5.8) \qquad (1 - m^{-1}t\lambda_i)^m \leqq \exp(-\lambda_i t) \quad \text{for} \quad t > 0 .$$

Thus, by (5.2), the series

$$(5.9) \qquad \sum_{i=1}^{\infty} \exp(\lambda_i t)\psi_i(x)\psi_i(y) = P(t, x, y)$$

are, for $t > 0$, absolutely and uniformly convergent on $R \times R$.

Let us assume further that

$$(5.10) \qquad \int_R dx = 1$$

and

$$(5.11) \qquad (Af)(x) = \frac{1}{\sqrt{g(x)}} \frac{\partial}{\partial x^i}\left(\sqrt{g(x)}\, b^{ij}(x)\frac{\partial f}{\partial x^j}\right).$$

Then we may prove the probability condition

$$(5.12) \qquad P(t, x, y) \geqq 0, \quad \int_R P(t, x, y)dy = 1.$$

Proof. The last equality is proved by the orthonormality of $\{\psi_i(x)\}$ and the fact that we may take $\psi_1(x) \equiv 1$.

---

[7] Schwartz: ibid.

The proof of $P(t, x, y) \geqq 0.$[8] We have, for

$$(5.13) \qquad f(t, x) = \int_R P(t, x, y) f(y) dy, \quad f(x) = \sum_{i=1}^{s} c_i \psi_i(x),$$

the diffusion equation

$$(5.14) \qquad \frac{\partial f(t, x)}{\partial t} = A_x f(t, x) \ (t > 0), \ \text{strong} \lim_{t \downarrow 0} f(t, x) = f(x).$$

Hence we have, for

$$(5.15) \qquad g_\varepsilon(t, x) = \exp(-\varepsilon t) f(t, x),$$

the differential equation

$$(5.16) \qquad \frac{\partial g_\varepsilon(t, x)}{\partial t} = A_x g_\varepsilon(t, x) - \varepsilon g_\varepsilon(t, x), \quad g_\varepsilon(0, x) = f(0, x) = f(x).$$

Let ε be > 0 and let $g_\varepsilon(t, x)$ reach its minimum at the point (t_1, x_1). Then we have

$$(5.17) \qquad g_\varepsilon(t_1, x_1) \geqq \min_x f(x) \quad \text{when} \cdot t_1 = 0$$
$$\geqq 0 \quad \text{when} \quad t_1 = \infty \quad \text{or when} \quad 0 < t_1 < \infty.$$

The first two inequalities are evident. For, we have

$$g_\varepsilon(0, x_1) = f(0, x_1) \geqq \min_x f(x) \quad \text{and} \quad g_\varepsilon(\infty, x_1) = 0.$$

Let $0 < t_1 < \infty$. Then, from

$$\frac{\partial g_\varepsilon(t_1, x_1)}{\partial t} = 0, \ (A_x g_\varepsilon)(t_1, x_1) \geqq 0 \quad \text{and} \quad (5.16),$$

we obtain $g_\varepsilon(t_1, x_1) \geqq 0$. Thus we have (5.17) and hence, by letting $\varepsilon \downarrow 0$,

$$(5.18) \qquad f(t, x) \geqq \min(0, \min_x f(x)).$$

Therefore, by the denseness of $f(x)$ in $C(R)$, we must have $P(t, x, y) \geqq 0$.
$$\text{Q.E.D.}$$

We have thus proved that, under the conditions (5.10) and (5.11), the series $P(t, x, y)$ give the explicite expression for the transition probability of the temporally homogeneous Markoff process, defined by the diffusion equation (5.14).

In concluding the paper, the author wishes to express his hearty thanks to Dr. Tosio Kato for his friendly criticism of the original manuscript.

Mathematical Institute,
Nagoya University

[8] Cf. K. Yosida: Brownian motion on the surface of the 3-sphere, Ann. of Math. Statistics, Vol. **20**, 292–296 (1949).

On the integration of diffusion equations in Riemannian spaces

Proc. Amer. Math. Soc. **3** (1952) 864–873

1. Introduction. Let R be a connected domain of an infinitely differentiable, orientable, m-dimensional ($m \geq 2$) Riemannian space with the metric $ds^2 = g_{ij}(x)dx^i dx^j$. Under a certain "continuity condition" of Lindeberg's type, the temporally homogeneous stochastic process in R is governed by a pair of equations:[1]

$$(1.1) \qquad \frac{\partial f(x, t)}{\partial t} = b^{ij}(x) \frac{\partial^2 f(x, t)}{\partial x^i \partial x^j} + a^i(x) \frac{\partial f(x, t)}{\partial x^i}, \qquad t \geq 0,$$

$$(1.2) \qquad \begin{aligned} \frac{\partial h(x, t)}{\partial t} &= \frac{1}{(g(x))^{1/2}} \frac{\partial^2}{\partial x^i \partial x^j} ((g(x))^{1/2} b^{ij}(x) h(x, t)) \\ &\quad - \frac{1}{(g(x))^{1/2}} \frac{\partial}{\partial x^i} ((g(x))^{1/2} a^i(x) h(x, t)), \qquad t \geq 0. \end{aligned}$$

These are called the "backward diffusion equation" and the "forward diffusion equation" respectively, the latter being sometimes called the Fokker-Planck's equation. In these equations, the symmetric contravariant tensor $b^{ij}(x)$ is assumed to be such that the quadratic form $b^{ij}(x)\xi_i\xi_j$ is, for $\sum_i \xi_i^2 > 0$, greater than 0 in R and $a^i(x)$ is assumed to obey, in the coordinate change $x \to \bar{x}$, the transformation rule

$$(1.3) \qquad \bar{a}^i(\bar{x}) = \frac{\partial \bar{x}^i}{\partial x^k} a^k(x) + \frac{\partial^2 \bar{x}^i}{\partial x^k \partial x^s} b^{ks}(x).$$

Hence the two elliptic differential operators on the right-hand sides of (1.1) and (1.2) are formally adjoint to each other and they have a meaning independent of the local coordinates $(x^1, \cdots, x^m)$. We assume that the coefficients $g_{ij}(x)$, $a^i(x)$, and $b^{ij}(x)$ are infinitely differentiable functions of the local coordinates $(x^1, \cdots, x^m)$.

The purpose of the present note is to prove the following three theorems.

THEOREM 1. *Let R be a compact Riemannian space. Then, to any function $f(x)$, infinitely differentiable in R, there corresponds a uniquely determined solution $f(x, t)$ of (1.1) satisfying the conditions:*

Received by the editors March 28, 1952.

[1] A. Kolmogoroff, *Zur Theorie der stetigen zufälligen Prozesse*, Math. Ann. vol. 108 (1933) pp. 149–160. W. Feller, *Zur Theorie der stochastischen Prozesse*, Math. Ann. vol. 113 (1937) pp. 113–160.

(1.4) $$\lim_{t \downarrow 0} f(x, t) = f(x) \text{ uniformly in } x,$$

(1.5) $$\min_{x} f(x) \leq f(x, t) \leq \max_{x} f(x), \text{ and}$$

$$\max_{x} f(x, t) = \max_{x} f(x) \text{ when } f(x) \text{ is non-negative.}$$

THEOREM 2. *Let R be a compact Riemannian space. Then, to any function $h(x)$, infinitely differentiable in R, there corresponds a uniquely determined solution $h(x, t)$ of (1.2) satisfying the conditions:*

(1.6) $\lim_{t \downarrow 0} \int_{R} |h(x, t) - h(x)| dx = 0$, where $dx = (g(x))^{1/2} dx^1 \cdots dx^m$, $g(x) = \det(g_{ij}(x))$,

(1.7) $\int_{R} |h(x, t)| dx \leq \int_{R} |h(x)| dx$,

and $h(x, t)$ is non-negative with $\int_{R} h(x, t) dx = \int_{R} h(x) dx$ when $h(x)$ is non-negative.

If R is a connected domain with the smooth boundary ∂R, Theorem 2 is extended to the following theorem.

THEOREM 2′. *Let D be the totality of infinitely differentiable functions $h(x)$ in R with compact carriers (supports in the terminology of L. Schwartz) satisfying the boundary condition on ∂R:*

(1.8)
$$(g(x))^{1/2} b^{ij}(x) \frac{\partial h}{\partial x^i} \cos(n, x^j)$$
$$+ \left(\frac{\partial (g(x))^{1/2} b^{ij}(x)}{\partial x^i} - (g(x))^{1/2} a^j(x) \right) \cos(n, x^j) h(x) = 0,$$

(n denotes the outer normal). Then, to every $h(x) \in D$, there corresponds a uniquely determined solution $h(x, t)$ of (1.2) satisfying the conditions (1.6)–(1.7) if and only if the following hypothesis is satisfied:

THE HYPOTHESIS. *Let $\{R_k\}$ be a monotone increasing sequence of connected domains $\subseteq R$ such that the boundary ∂R_k tends smoothly, as $k \to \infty$, to the boundary ∂R. Then, for any $m > 0$, the equation*

(1.9) $$b^{ij}(x) \frac{\partial^2 f(x)}{\partial x^i \partial x^j} + a^i(x) \frac{\partial f(x)}{\partial x^i} = mf(x)$$

does not admit a bounded solution $f(x) \not\equiv 0$ satisfying the boundary condition

(1.10) $$\lim_{k \to \infty} \int_{\partial R_k} (g(x))^{1/2} b^{ij}(x) h(x) \frac{\partial f}{\partial x^i} \cos(n, x^j) dS = 0$$

for all $h(x) \in D$,

dS denoting the hypersurface element of ∂R_k.

These theorems may be proved by refining, with the aid of parametrix considerations, the operator-theoretical integration of the diffusion equations given in the preceding notes.[2] It is to be remarked that our construction of the parametrix for the general diffusion equation is carried out by an elementary calculation without appealing to the theory of integral equations nor to the power series expansion. It is an extension of the construction due to S. Minakshisundaram and Å. Pleijel.[3]

2. **The construction of the parametrix.** Let $e(x)$ be an infinitely differentiable function and let

$$(2.1) \qquad A = A_x = b^{ij}(x) \frac{\partial^2}{\partial x^i \partial x^j} + a^i(x) \frac{\partial}{\partial x^i} + e(x).$$

Let $\Gamma = \Gamma(P, Q) = r(P, Q)^2$ be the square of the smallest distance of two points P and Q of R according to the new metric $dr^2 = b_{ij}(x)dx^i dx^j$, where $(b_{ij}(x)) = (b^{ij}(x))^{-1}$. We have then the lemma.

LEMMA 1. *For any positive integer k, we may construct the parametrix*

$$H_k(P, Q, t - \tau)$$

$$(2.2) \qquad = (t - \tau)^{-m/2} \exp\left(-\frac{\Gamma(P, Q)}{4(t - \tau)}\right) \sum_{i=0}^{k} u_i(P, Q)(t - \tau)^i, \, t > \tau,$$

such that

(2.3) $u_i(P, Q)$ $(i = 0, 1, \cdots, k)$ *are infinitely differentiable in the vicinity of $Q = P$ and $u_0(P, P) = 1$,*

$$\left(-\frac{\partial}{\partial \tau} - A_Q\right) H_k(P, Q, t - \tau)$$

$$(2.4)$$

$$= (t - \tau)^{k-m/2} \exp\left(-\frac{\Gamma(P, Q)}{4(t - \tau)}\right) c_k(P, Q),$$

where $c_k(P, Q)$ is infinitely differentiable in the vicinity of $Q = P$.

---

[2] K. Yosida, *Integration of Fokker-Planck's equation in a compact Riemannian space*, Arkiv. för Matematik vol. 1 (1949) pp. 71–75 (to be referred to as [I]). K. Yosida, *Integration of Fokker-Planck's equation with a boundary condition*, Journal of the Mathematical Society of Japan vol. 3 (1951) pp. 69–73 (to be referred to as [II]). K. Yosida, *Integrability of the backward diffusion equation in a compact Riemannian space*, Nagoya Math. J. vol. 3 (1951) pp. 1–4 (to be referred to as [III]).

[3] *Some properties of the eigenfunctions of the Laplace operator on Riemannian manifolds*, Canadian Journal of Mathematics vol. 1 (1949) pp. 242–256.

PROOF. We introduce the normal coordinates y^σ of $Q = (x^1, \cdots, x^m)$ in the vicinity of P:

$$(2.5) \qquad y^\sigma = (\Gamma(P, Q))^{1/2} \left(\frac{dx^\sigma}{dr}\right)_{Q=P}.$$

Let

$$(2.6) \qquad dr^2 = \beta_{ij}(y)dy^i dy^j.$$

We have the well known formulae

$$(2.7) \qquad \Gamma(P, Q) = \beta_{ij}(0)y^i y^j, \qquad \beta_{ij}(y)y^j = \beta_{ij}(0)y^j.$$

By virtue of (2.7), the operator

$$(2.8) \qquad A = A_y = \beta^{ij}(y) \frac{\partial^2}{\partial y^i \partial y^j} + \alpha^i(y) \frac{\partial}{\partial y^i} + e(y),$$

$$(\beta^{ij}(y)) = (\beta_{ij}(y))^{-1},$$

when applied to the function $f(\Gamma, y)$, where Γ is considered as a function of y, may be written as follows:

$$(2.9) \qquad A_y f = 4\Gamma \frac{\partial^2 f}{\partial \Gamma^2} + 4y^\sigma \frac{\partial^2 f}{\partial \Gamma \partial y^\sigma} + M \frac{\partial f}{\partial \Gamma} + N(f),$$

where

$$M = \beta^{ij} \frac{\partial^2 \Gamma}{\partial y^i \partial y^j} + \alpha^i \frac{\partial \Gamma}{\partial y^i} = 2m + O(y),$$

$$N(f) = \beta^{ij} \frac{\partial^2 f}{\partial y^i \partial y^j} + \alpha^i \frac{\partial f}{\partial y^i} + ef.$$

The differentiations in $A_y f$ and $N(f)$ are to be performed as if Γ and y are independent variables. Hence we have

$$-A_y H_k = \sum_{i=0}^{k} -\frac{\Gamma}{4}(t - \tau)^{i-2-m/2} \exp\left(-\frac{\Gamma}{4(t-\tau)}\right)$$

$$+ \sum_{i=0}^{k} (t-\tau)^{i-1-m/2} \exp\left(-\frac{\Gamma}{4(t-\tau)}\right) \left\{ y^\sigma \frac{\partial u_i}{\partial y^\sigma} \right.$$

$$\left. + \frac{M}{4} u_i - N(u_{i-1}) \right\}$$

$$+ (t-\tau)^{k-m/2} \exp\left(-\frac{\Gamma}{4(t-\tau)}\right) N(u_k),$$

where $u_{-1} \equiv 0$ and hence $N(u_{-1}) \equiv 0$. Therefore, since

$$-\frac{\partial}{\partial \tau} H_k = \sum_{i=0}^{k} (t - \tau)^{i-1-m/2} \exp\left(-\frac{\Gamma}{4(t-\tau)}\right) \cdot u_i\left(\frac{-m}{2} + i + \frac{\Gamma}{4(t-\tau)}\right),$$

we obtain Lemma 1 if the u_i are successively so chosen that

$$(2.10) \qquad y^\sigma \frac{\partial u_i}{\partial y^\sigma} + \left(\frac{-m}{2} + i + \frac{M}{4}\right) u_i = N(u_{i-1}),$$

$u_i(P, Q)$ being infinitely differentiable in the vicinity of $Q = P$ with $u_{-1} \equiv 0$ and $u_0(P, P) = 1$. Such u_i may be determined, in view of the order relation

$$(2.11) \qquad\qquad M = 2m + O(y).$$

For this purpose, put $y^\sigma = \eta^\sigma s$ and transform (2.10) into ordinary differential equations in s containing the parameters η. The equations are integrated by

$$(2.12) \qquad
\begin{aligned}
u_0(P, Q) &= \exp\left(-\int_0^s s^{-1}\left(\frac{-m}{2} + \frac{M}{4}\right) ds,\right. \\
u_i(P, Q) &= u_0 s^{-i} \int_0^s s^{i-1} u_0^{-1} N(u_{i-1}) ds \qquad (i = 1, \cdots, k).
\end{aligned}$$

REMARK. By (2.9) and (2.10), we have

$$(2.13) \qquad
\begin{aligned}
A_Q u_0(P, Q) \Gamma(P, Q)^{(2-m)/2} &= N(u_0(P, Q)) \Gamma(P, Q)^{(2-m)/2}, \quad m \geqq 3, \\
A_Q u_0(P, Q) \log \Gamma(P, Q) &= N(u_0(P, Q)) \log \Gamma(P, Q), \quad m = 2.
\end{aligned}$$

We have thus obtained the parametrix for the elliptic differential operator A.

3. **The proof of Theorem 2′.** Let $e \equiv 0$ in the operator A and let

$$(3.1) \qquad
\begin{aligned}
(A'h)(x) &= \frac{1}{(g(x))^{1/2}} \frac{\partial^2}{\partial x^i \partial x^j} \left((g(x))^{1/2} b^{ij}(x) h(x)\right) \\
&\quad + \frac{1}{(g(x))^{1/2}} \frac{\partial}{\partial x^i} \left(-(g(x))^{1/2} a^i(x) h(x)\right)
\end{aligned}$$

be the formally adjoint operator of A. Let $L_1(R)$ be the Banach space of the totality of the functions $h(x)$ integrable with respect to dx in

R, metrized by the norm $\|h\| = \int_R |h(x)| \, dx$. D is a dense linear subset of $L_1(R)$. We take the operator A' to be an additive operator defined on $D \subseteq L_1(R)$ into $L_1(R)$, and let $\tilde{A}'$ be the smallest closed extension of the operator A'. Then[4] there exists a uniquely determined one-parameter semi-group[5] of linear operators T_t on $L_1(R)$ into $L_1(R)$ satisfying the conditions

$$T_t T_s = T_{t+s} \ (t, s) \geq 0), \ T_0 = \text{the identity},$$

$$\text{strong } \lim_{t \to t_0} T_t h = T_{t_0} h \quad \text{for } h \in L_1(R),$$

(3.2)

$(T_t h)(x)$ is non-negative and $\|T_t h\| = \|h\|$ when $h(x)$ is non-negative,

$$\partial_t T_t h = \text{strong } \lim_{\delta \to 0} \frac{T_{t+\delta} h - T_t h}{\delta} = \tilde{A}' T_t h \quad \text{for } h \text{ in the domain}$$

$\tilde{D}'$ of the operator $\tilde{A}'$,

if and only if the hypothesis of Theorem 2′ is satisfied.

We thus have to show that the function $h(x, t) = (T_t h)(x)$ is equivalent (in the sense to be explained below) to a function which is continuously differentiable once in t and twice in x.

For this purpose we prepare two lemmas.

LEMMA 2. *Let $f(x, t)$ be an infinitely differentiable function which vanishes outside a compact coordinate neighborhood of P. Then*

$$\int_R h(y, t) f(y, t) \, dy$$

$$(3.3) \qquad = \int_R h(y, 0) f(y, 0) \, dy$$

$$+ \int_0^t d\tau \int_R \left\{ \partial_\tau h(y, \tau) f(y, \tau) + h(y, \tau) \frac{\partial f(y, \tau)}{\partial \tau} \right\} dy.$$

PROOF. By (3.2), $h(y, \tau)$ is strongly differentiable in τ with the differential quotient $\tilde{A}_y' h(y, \tau)$. Hence $h(y, \tau) f(y, \tau)$ is weakly differ-

---

[4] See [II]. Cf. also [I]. The hypothesis is surely satisfied when R is a connected domain in m-dimensional Euclidean space whose boundary lies entirely in the compact part of the space and when, moreover, the operator A is the usual Laplacian. See K. Yosida, *A theorem of Liouville's type for meson equation*, Proc. Imp. Acad. Tokyo vol. 27 (1951) pp. 214–215.

[5] E. Hille, *Functional analysis and semi-groups*, New York, 1948. Cf. also K. Yosida, *On the differentiability and the representation of one-parameter semi-group of linear operators*, Journal of the Mathematical Society of Japan vol. 1 (1948) pp. 15–21.

entiable in τ with the differential quotient

$$f(y, \tau)\tilde{A}'_y h(y, \tau) + \frac{\partial f(y, \tau)}{\partial \tau} h(y, \tau)$$

which is strongly continuous in τ. Thus, by integration, we obtain (3.3).

LEMMA 3. *We have*

$$(3.4) \quad \int_R h(y, t)f(y, t)dy = \int_R h(y, 0)f(y, 0)dy$$

$$= \int_0^t d\tau \int_R \left\{ h(y, \tau)\left(\frac{\partial f(y, \tau)}{\partial \tau} + A_y f(y, \tau)\right)\right\} dy.$$

PROOF. The right-hand side is, by (3.2) and (3.3), equal to

$$-\int_0^t d\tau \int_R \left\{ h(y, \tau)\left(-\frac{\partial f(y, \tau)}{\partial \tau} - A_y f(y, \tau)\right)\right.$$

$$\left. - f(y, \tau)(\partial_\tau h(y, \tau) - \tilde{A}'_y h(y, \tau))\right\} dy$$

$$= \left[\int_R h(y, \tau)f(y, \tau)dy \right]_0^t$$

$$+ \int_0^t d\tau \int_R \left\{ h(y, \tau)A_y f(y, \tau) - f(y, \tau)\tilde{A}'_y h(y, \tau)\right\} dy.$$

That the second integral on the right-hand side is equal to zero may be seen by the following argument. Let $\{h_k(y, \tau)\} \subseteq D$ be such that strong $\lim_{k \to \infty} h_k(y, \tau) = h(y, \tau)$, strong $\lim_{k \to \infty} A'_y h_k(y, \tau) = \tilde{A}'_y h(y, \tau)$. The existence of such a sequence $\{h_k(y, \tau)\}$ is assured by the fact that $h(y, \tau) = (T_\tau h)(y)$ is in the domain $\tilde{D}'$ of the operator $\tilde{A}'$ which is the smallest closed extension of the operator A' with the domain D. We have thus

$$\int_R \left\{ h(y, \tau)A_y f(y, \tau) - f(y, \tau)\tilde{A}'_y h(y, \tau)\right\} dy$$

$$= \lim_{k \to \infty} \int_R \left\{ h_k(y, \tau)A_y f(y, \tau) - f(y, \tau)A'_y h_k(y, \tau)\right\} dy.$$

The right-hand integrals are equal to zero as may be seen by Green's integral theorem and the fact that $f(y, \tau)$ vanishes outside a compact coordinate neighborhood of P.

After these preliminaries, we may give:

PROOF OF THE THEOREM 2'. We shall apply (3.4) to the case when

$$(3.5) \qquad f(y, \tau) = f(Q, \tau) = H_k(P, Q, t + \epsilon - \tau)\delta(P, Q)\delta(P_0, P).$$

Here P_0 is an arbitrary point of R, fixed in the following argument, ϵ is a positive constant, and

$$(3.6) \qquad\qquad \delta(P, Q) = \alpha(r(P, Q)),$$

where $\alpha(r)$ denotes an infinitely differentiable function of r satisfying

$$\alpha(r) = \begin{cases} 1 & \text{for} \quad r \leqq 2^{-1}\eta, \\ \text{between 0 and 1} & \text{for} \quad 2^{-1}\eta < r < \eta, \\ 0 & \text{for} \quad r \geqq \eta. \end{cases}$$

We assume that the positive constant η is chosen so small that

(3.7) the points Q satisfying $\delta(P_0, P)\delta(P, Q) > 0$ are contained in a compact coordinate neighborhood of P_0.

We have thus

$$\int_R h(Q, t)H_k(P, Q, \epsilon)\delta(P_0, P)\delta(P, Q)dy$$

$$(3.8) \qquad\qquad - \int_R h(Q, 0)H_k(P, Q, t + \epsilon)\delta(P_0, P)\delta(P, Q)dy$$

$$= - \int_0^t d\tau \int_R h(Q, \tau)K_k(P, Q, t + \epsilon - \tau)dy,$$

where

$$(3.9) \qquad \begin{aligned} &K_k(P, Q, t + \epsilon - \tau) \\ &\quad = \left(- \frac{\partial}{\partial \tau} - A_Q\right)(H_k(P, Q, t + \epsilon - \tau)\delta(P_0, P)\delta(P, Q)). \end{aligned}$$

Let k be taken so large that

$$(3.10) \qquad\qquad \frac{-m}{2} + k \geqq 2.$$

Then, in view of Lemma 1, $K_k(P, Q, t+\epsilon-\tau)$ is, for $r(P_0, P) \leqq 2^{-1}\eta$, devoid of the singularity even if $(t+\epsilon-\tau) = 0$. We shall next show that the first term on the left-hand side of (3.8) tends, as $\epsilon \downarrow 0$, strongly to $h(P, t)$ in the vicinity of P_0. This may be proved as follows. We have, by (3.6) and (3.7),

$$\int_R \delta(P_0,\, P)dP \left| \int_R h(Q,\, t)H_k(P,\, Q,\, \epsilon)\delta(P,\, Q)dy \right.$$

$$\left. - h(P,\, t)\int_R H_k(P,\, Q,\, \epsilon)\delta(P,\, Q)dy \right|$$

$$\leqq C \int_{r(P_0,Q)\leqq 2\eta} \left(\int_{r(P_0,P)\leqq \eta} |h(Q,\, t) - h(P,\, t)|\, dP \right) |H_k(P,\, Q,\, \epsilon)|\, dy$$

$$\leqq C_1 \int \cdots \int \left(\int |h(z + \epsilon^{1/2}\xi,\, t) - h(z,\, t)|\, dz \right)$$

$$\cdot \exp\left(- \sum_i (\xi^i)^2 \right) d\xi^1 \cdots d\xi^m.$$

Here $(z^1, \cdots, z^m)$ and $(z^1+y^1, \cdots, z^m+y^m)$ are the coordinates of the points P and Q respectively in the coordinate neighborhood of P_0, defined in (3.7), and C and C_1 denote suitable positive constants. The inner integral on the right-hand side converges, when $\epsilon \downarrow 0$, to zero boundedly by Lebesgue's theorem.

Therefore, there exists a sequence $\{\epsilon_i\}$ with $\epsilon_i \downarrow 0$ such that

$$h(P,\, t) \lim_{i \to \infty} \int_R H_k(P,\, Q,\, \epsilon_i)\delta(P,\, Q)dy$$

(3.11)
$$= \int_R h(Q,\, 0)H_k(P,\, Q,\, t)\delta(P,\, Q)dy$$

$$- \int_0^t d\tau \int_R h(Q,\, \tau)K_k(P,\, Q,\, t - \tau)dy$$

almost everywhere with respect to P in the vicinity of P_0. Hence, by (3.10)–(3.11), $h(P,\, t)$ may be considered to be continuously differentiable once in $t>0$ and twice in P in the vicinity of P_0 if

(3.12) $\quad \lim\limits_{\epsilon \downarrow 0} \int_R H_k(P,\, Q,\, \epsilon)\delta(P,\, Q)dy$ is positive and twice continuously

differentiable in P in the vicinity of P_0.

The proof of (3.12) may be obtained as follows. We have

$$\lim_{\epsilon \downarrow 0} \int_R H_k(P,\, Q,\, \epsilon)\delta(P,\, Q)dy$$

$$= \lim_{\epsilon \downarrow 0} \int_{r(P,Q)\leqq \zeta} \epsilon^{-m/2} \exp\left(- \frac{\Gamma(P,\, Q)}{4\epsilon} \right) dy$$

for any positive constant ζ. Hence, by putting

$$(3.13) \qquad ds^2 = \gamma_{ij}(y)dy^idy^j, \qquad y^i = \epsilon^{1/2}\xi^i,$$

we obtain, in view of the arbitrariness of ζ,

$$
\begin{aligned}
(3.14) \quad & \lim_{\epsilon \downarrow 0} \int_R H_k(P, Q, \epsilon)\delta(P, Q)dy \\
&= \lim_{\epsilon \downarrow 0} \int \cdots \int_{-\zeta \leq \epsilon^{1/2}\xi^i \leq \zeta} \exp\left(-\beta_{ij}(0)\xi^i\xi^j\right)(\gamma(0))^{1/2}d\xi^1 \cdots d\xi^m \\
&= \pi^{m/2}(\gamma(0))^{1/2}(\beta(0))^{1/2} = \pi^{m/2}(g(P))^{1/2}/(b(P))^{1/2},
\end{aligned}
$$

where

$$g(P) = \det\,(g_{ij}(P)) \quad \text{and} \quad b(P) = \det\,(b_{ij}(P)).$$

4. The proof of Theorem 1. Let $C(R)$ be the Banach space of the totality of continuous functions $f(x)$ in R, metrized by the norm $\|f\|$ $= \max_x |f(x)|$. D is a dense linear subset of $C(R)$. Let $\epsilon \equiv 0$ in the operator A. We consider this A to be an additive operator defined on $D \subseteq C(R)$ into $C(R)$, and define the smallest closed extension $\tilde{A}$ of A. Then[6] there exists a uniquely determined one-parameter semi-group of linear operators S_t on $C(R)$ into $C(R)$ satisfying the conditions:

$$
\begin{aligned}
& S_t S_s = S_{t+s} \quad (t, s \geq 0), \qquad S_0 = \text{the identity } I \\
& \text{strong } \lim_{t \to t_0} S_t f = S_{t_0} f \quad \text{for} \quad f \in C(R),
\end{aligned}
$$

$$
\begin{aligned}
(4.1) \quad & (S_t f)(x) \text{ is non-negative and } \max_x (S_t f)(x) = \max_x f(x) \text{ if } f(x) \\
& \text{is non-negative} \\
& \partial_t S f = \text{strong } \lim_{\delta \to 0} \frac{S_{t+\delta}f - S_t f}{\delta} = \tilde{A}S_t f \text{ for } f \text{ in the domain } \tilde{D} \text{ of}
\end{aligned}
$$

the operator A.

Therefore, as in §3, we may prove that $f(x, t) = (S_t f)(x)$ may be considered to be continuously differentiable once in t and twice in x.

Nagoya University

---

[6] See [III]

On the fundamental solution of the parabolic equation in a Riemannian space

Osaka Math. J. **5** (1953) 65–74

1. Introduction. Let $\boldsymbol{R}$ be a connected domain of an infinitely differentiable, m-dimensional Riemannian space with the metric $ds^2 = g_{ij}(x)dx\,dx^j$. We consider the general parabolic equation

$$(1.1) \qquad L_{tx}f = \frac{\partial f(t,\,x)}{\partial t} - A_{tx}f(t,\,x),$$

where

$$(1.2) \qquad A_{tx}f(t,\,x) = g(x)^{-1/2}\frac{\partial^2}{\partial x^i\,\partial x^j}\left(g(x)^{1/2}a^{ij}(t,\,x)f(t,\,x)\right)$$

$$-\,g(x)^{-1/2}\frac{\partial}{\partial x^i}\left(g(x)^{1/2}b^i(t,\,x)f(t,\,x)\right)+c(t,\,x)f(t,\,x),$$

$$g(x) = \det\left(g_{ij}(x)\right).$$

The operator A_{tx} is assumed to be elliptic in x in the sense that

$$(1.3) \qquad a^{ij}(t,\,x)\xi_i\xi_j > 0 \quad \text{for} \quad \sum_i (\xi_i)^2 > 0.$$

Since the value of $A_{tx}f(t,\,x)$ must be independent of the local coordinates $(x^1,\,\ldots,\,x^m)$, we must have, by the coordinates change $x \to \bar{x}$, the transformation rule

$$(1.4) \qquad \bar{a}^{ij}(t,\,\bar{x}) = \frac{\partial \bar{x}^i}{\partial x^k}\frac{\partial \bar{x}^j}{\partial x^s}\,a^{ks}(t,\,x),$$

$$(1.4) \qquad \bar{b}^i(t,\,\bar{x}) = \frac{\partial \bar{x}^i}{\partial x^k}\,b^k(t,\,x)+\frac{\partial^2 \bar{x}^i}{\partial x^k\,\partial x^s}\,a^{ks}(t,\,x).$$

For the sake of simplicity, we assume that the coefficients $a^{ij}(t,\,x)$, $b^i(t,\,x)$, $c(t,\,x)$ and $g_{ij}(x)$ are infinitely differentiable function of the local coordinates $(x^1,\,\ldots,\,x^m)$. The purpose of the present note is to construct, under a certain HYPOTHESIS, which is surely satisfied for compact Riemannian space $\boldsymbol{R}$, the fundamental solution

$$(1.6) \qquad P(s,\,t,\,y,\,x), \qquad (\,s < t \quad \text{and} \quad y,\,x \in \boldsymbol{R})$$

of (1.1) with the following four properties:

i) For $s < t$,

$$(1.7) \qquad L'_{sy}P = -\frac{\partial P}{\partial s} - A'_{sy}P = 0, \qquad L_{tx}P = \frac{\partial P}{\partial t} - A_{tx}P = 0.$$

where

$$(1.8) \qquad A'_{sy}h(s, y) = a^{ij}(s, y)\frac{\partial^2 h(s, y)}{\partial y^i \partial y^j} + b^i(s, y)\frac{\partial h(s, y)}{\partial y^i} + c(s, y)h(s, y).$$

ii) When $t \downarrow s_0$, $s \uparrow s_0$ and distance $(x, x_0) \to 0$, distance $(y, x_0) \to 0$, the function $P(s, t, y, x)$ exhibits the principal singularity

$$(1.9) \qquad \pi^{-m/2}(a(s_0, x_0)/g(x_0))^{1/2}(t-s)^{-m/2}\exp\left(-a_{ij}(s_0, x_0)(x^i - y^i)(x^j - y^j)\right.$$
$$\left. \times 4^{-1}(t-s)^{-1}\right), \qquad \text{where}$$
$$a(s, x) = \det(a_{ij}(s, x)), \ (a_{ij}(s, x)) = (a^{ij}(s, x))^{-1}.$$

iii) We have

$$(1.10) \qquad P(s, t, y, x) \text{ is, for any } \varepsilon > 0, \text{ bounded in } y(x) \text{ for fixed } x(y)$$
when s and $t(>s+\varepsilon)$ are bounded.

$$(1.11) \qquad \int_R |P(s, t, y, x)|\,dx, \text{ where } dx = g(x)^{1/2}dx^1 \cdots dx^m, \text{ is bounded in}$$
y when s and $t(>s)$ are bounded.

$$(1.11)' \qquad \int_R P(s, t, y, x)\,dx = 1 \quad \text{when} \quad c(t, x) \equiv 0.$$

iv) The Chapman-Kolmogoroff's equation holds, viz.

$$(1.12) \qquad P(s, t, y, x) = \int_R P(s, u, y, z)P(u, t, z, x)\,dz, \quad s < u < t.$$

An Application to the Stochastic Processes. Let $c(t, x) \equiv 0$. When R is a compact Riemannian space, we have, besides i)-iv), the condition

$$(1.13) \qquad P(s, t, y, x) \text{ is everywhere non-negative.}$$

Thus, in such a case, $P(s, t, y, x)$ may be considered as the transition probability governed by the corresponding pair of Kolmogoroff's equations.

The following construction of P is based upon a construction[1] of a " fairly regular " parametrix for the adjoint equation of (1.1). Mr. Seizô Itô kindly discussed the manuscript and remarked that, when R is an Euclidean space, the fundamental solution for (1.1) was constructed by F. G. Dressel[2] starting with an entirely different parametrix.

---

1) Cf. K. Yosida: On the integration of diffusion equations in Riemannian spaces, the Proc. Amer. Math. Soc. 3, 1952, 864-873.

2) The fundamental solution of the parabolic equations, Duke Math. J., 7 (1940), 186-203.

His method is an extension of W. Feller's paper[3] for the case $m = 1$. Mr. Itô also has succeeded in constructing the fundamental solution for the differentiable manifold R by extending Feller-Dressel's method. See the immediately following paper by Mr. Itô.

2. The Parametrix for the Adjoint Equation of (1.1). Let, according to the new metric $dr(\tau)^2 = a_{ij}(\tau, x)dx^i dx^j$,

$$(2.1) \qquad \Gamma = \Gamma(\tau, y, x) = r(\tau, y, x)^2$$

be the square of the smallest distance of y and x of R. Then we have the

Lemma. *Let the positive integer k be* $>(2+m/2)$. *We may construct a parametrix for the adjoint equation of* (1.1)

$$(2.2) \quad H_1(\tau, t, y, x) = (t-\tau)^{-m/2} \exp\left(-\frac{\Gamma(\tau, y, x)}{4(t-\tau)}\right)\sum_{i=0}^{k} u_i(\tau, y, x)(t-\tau)^i, \quad t>\tau,$$

such that

$(2.3) \qquad u_i(\tau, y, x)$ *are infinitely differentiable in the vicinity of* $y = x$ *and* $u_0(\tau, x, x) = 1$,

$(2.4) \qquad L'_{\tau y}H_1(\tau, t, y, x) = (t-\tau)^{k-m/2} \exp\left(-\frac{\Gamma(\tau, y, x)}{4(t-\tau)}\right) c_k(\tau, y, x)$, *where*

$\qquad\qquad c_k(\tau, y, x)$ *is infinitely differentiable in the vicinity of* $y = x$.

Proof. We regard the point z on the geodesic (according to the new metric $dr(\tau)^2 = a_{ij}(\tau, x)dx^i dx^j$) joining x and y as a function of $r = r(\tau, x, z)$. We have then the well-known identities[4]

$$(2.5) \qquad L(\tau, z, \dot{z}) = a_{ij}(\tau, z)\dot{z}^i\dot{z}^j = 1, \quad \dot{z}^i = \frac{dz^i}{dr},$$

$$\frac{\partial\Gamma(\tau, y, x)}{\partial y^i} = r(\tau, y, x)\frac{\partial L(\tau, y, \dot{y})}{\partial \dot{y}^i} = 2r(\tau, y, x)a_{ij}(\tau, y)\dot{y}^j.$$

Hence we have the important identity

$$(2.6) \qquad \Gamma(\tau, y, x) = r(\tau, y, x)^2 2^{-1}\dot{y}^k \frac{\partial L(\tau, y, \dot{y})}{\partial \dot{y}^k}$$

$$= r(\tau, y, x)^2 4^{-1}a^{jk}(\tau, y)\frac{\partial L(\tau, y, \dot{y})}{\partial \dot{y}^j}\frac{\partial L(\tau, y, \dot{y})}{\partial \dot{y}^k}$$

$$= 4^{-1}a^{jk}(\tau, y)\frac{\partial\Gamma(\tau, y, x)}{\partial y^j}\frac{\partial\Gamma(\tau, y, x)}{\partial y^k}.$$

---

3) W. Feller: Zur Theorie der stochastischen Prozesse, Math. Ann. **113** (1936), 113-160.

4) See, for example, M. Riesz: L'intégrale de Riemann-Liouville et le problème de Cauchy, Acta Math, **81** (1948), p. 171.

Thus the operator $A'_{\tau y}$, when applied to a function $F(\Gamma, y)$, where Γ being looked as a function of y, may be written as

$$(2.7) \qquad A'_{\tau y}F = 4\Gamma \frac{\partial^2 F}{\partial \Gamma^2} + M \frac{\partial F}{\partial \Gamma} + 2\left(a^{\sigma j}\frac{\partial \Gamma}{\partial y^j}\right)\frac{\partial^2 F}{\partial \Gamma \partial y^\sigma} + N(F), \quad \text{where}$$

$$M = a^{ij}\frac{\partial^2 \Gamma}{\partial y^i \partial y^j} + b^i \frac{\partial \Gamma}{\partial y^i} = 2m + \sum_i 0(x^i - y^i),$$

$$N(F) = a^{ij}\frac{\partial^2 F}{\partial y^i \partial y^j} + b^i \frac{\partial F}{\partial y^i} + cF.$$

Here the differentiation must be performed as if Γ and y are independent variables. Hence we have

$$-A'_{\tau y}H_1(\tau, t, y, x) = \sum_{i=0}^{k} -\frac{\tau}{4}(t-\tau)^{i-2-m/2}\exp\left(-\frac{\Gamma}{4(t-\tau)}\right)u_i$$

$$+ \sum_{i=0}^{k}(t-\tau)^{i-1-m/2}\exp\left(-\frac{\Gamma}{4(t-\tau)}\right)\left\{\frac{1}{2}\left(a^{\sigma j}\frac{\partial \Gamma}{\partial y^j}\right)\frac{\partial u_i}{\partial y^\sigma} + \frac{M}{2}u_i - N(u_{i-1})\right.$$

$$-(t-\tau)^{k-m/2}\exp\left(-\frac{\Gamma}{4(t-\tau)}\right)N(u_k),$$

where $u_{-1} \equiv 0$ and hence $N(u_{-1}) \equiv 0$. Therefore, by

$$-\frac{\partial}{\partial \tau}H_1 = \sum_{i=0}^{k}(t-\tau)^{i-1-m/2}\exp\left(-\frac{\Gamma}{4(t-\tau)}\right)u_i\left(-\frac{m}{2}+i+\frac{1}{4}\frac{\partial \Gamma}{\partial \tau}+\frac{\Gamma}{4(t-\tau)}\right)$$

$$-\sum_{i=0}^{k}(t-\tau)^{i-m/2}\exp\left(-\frac{\Gamma}{4(t-\tau)}\right)\frac{\partial u_i}{\partial \tau},$$

we obtain the lemma if u_i are successively so determined that

$$(2.8) \qquad \frac{1}{2}\left(a^{\sigma j}\frac{\partial \Gamma}{\partial y^j}\right)\frac{\partial u_i}{\partial y^\sigma} + \left(-\frac{m}{2}+i+\frac{M}{4}+\frac{1}{4}\frac{\partial \Gamma}{\partial \tau}\right)u_i = N(u_{i-1}) + \frac{\partial u_{i-1}}{\partial \tau},$$

where u_i are infinitely differentiable in the vicinity of $y = x$ and $u_{-1} \equiv 0$, $u_0(\tau, x, x) = 1$.

To this purpose, we introduce the normal coordinates of y around x according to the new metric $dr(\tau)^2 = a_{ij}(\tau, y)dy^i dy^j$

$$(2.9) \qquad \eta^i = r(\tau, y, x)\left(\frac{dy^i}{dr}\right)_{r=0} = r\xi^i.$$

Then we have, by (2.5),

$$\frac{1}{2}a^{\sigma j}\frac{\partial \Gamma}{\partial y^j}\frac{\partial u_i}{\partial y^\sigma} = r\xi^\sigma \frac{\partial u_i}{\partial \eta^\sigma}.$$

We have also the order relations

$$M = 2m + 0(r),$$

$$\frac{\partial \Gamma}{\partial \tau} = 0\,(r)\,.$$

Hence the equations (2.8) are transformed into ordinary differential equations in r containing the parameters ξ

$$(2.10)\qquad r\frac{du_i}{dr}+\left(-\frac{m}{2}+i+\frac{M}{4}+\frac{1}{4}\frac{\partial\Gamma}{\partial\tau}\right)u_i=N(u_{i-1})+\frac{\partial u_{i-1}}{\partial\tau}\,.$$

By $u_{-1}\equiv 0$ and $u_0(\tau,y,x)=1$, these equations may be integrated as

$$(2.11)\qquad u_0(\tau,y,x)=\exp\left(-\int_0^r\rho^{-1}\left(-\frac{m}{2}+\frac{M}{4}+\frac{1}{4}\frac{\partial\Gamma}{\partial\tau}\right)\right)d\rho\,,$$

$$u_i(\tau,y,x)=u_0r^{-i}\int_0^r\rho^{i-1}u_0^{-1}\left(N(u_{i-1})+\frac{\partial u_{i-1}}{\partial\tau}\right)d\rho\,,\quad (i=1,2,3,\dots,k).$$

2. The Fundamental Solution of the Adjoint Equation of (1.1). We assume the following

Hypothesis. There exists a positive constant η with the properties: Let $\delta(S)$ be infinitely differentiable and ≥ 0 for $S\geq 0$ such that

$$(3.1)\qquad \delta(S)=1 \quad\text{for}\quad 0\leq S\leq\eta \quad\text{and}\quad \delta(S)=0 \quad\text{for}\quad S\geq 2\eta\,.$$

Let $S(x,y)$ denote the distance of x and y according to the original metric $ds^2=g_{ij}(x)dx^idx^j$. Then

i) the function

$$(3.2)\qquad H(s,t,y,x)=\pi^{-m/2}(a(t,x)/g(x))^{1/2}H_1(s,t,y,x)\delta(S(y,x))$$

is defined everywhere and the integral

$$(3.3)\qquad \int_R|H(s,t,y,x)|\,dx \text{ is bounded in } y \text{ when } s \text{ and } t(>s) \text{ are bounded.}$$

ii) The function

$$(3.4)\qquad K(s,t,y,x)=L'_{sy}H(s,t,y,x) \text{ is bounded everywhere when } s \text{ and }$$
$$t(>s) \text{ are bounded.}$$

iii) The integral

$$(3.5)\qquad \int_{S(x,y)\leq 2\eta} dy \qquad \text{is bounded in } x\,.$$

The above HYPOTHESIS is surely satisfied when R is a compact Riemannian space. In the general case, the HYPOTHESIS will impose conditions upon the coefficients $g_{ij}(x)$, $a^{ij}(t,x)$, $b^i(t,x)$ and $c(t,x)$.

Theorem 1. *Let the Hypothesis[5]. be satisfied. Then the function*

(3.6) $$P(s, t, y, x) = H(s, t, y, x) - \int_s^t d\tau \int_{\boldsymbol{R}} H(s, \tau, y, z)Q(\tau, t, z, x)dz,$$

where

(3.7) $$Q(s, t, y, x) = \sum_{n=1}^{\infty} (-1)^{n+1} K_n(s, t, y, x),$$

$$K_1 = K, \quad K_n(s, t, y, x) = \int_s^t d\tau \int_{\boldsymbol{R}} K(s, \tau, y, z)K_{n-1}(\tau, t, z, x)dz,$$

satisfies $L'_{sy}P(s, t, y, x) = 0$, (1.10) *and* (1.11).

Proof. We obtain, by the integral formula due to Dirichlet

$$\int_s^t d\tau \int_s^\tau M(\sigma, \tau)N(\tau, t)d\sigma = \int_s^t d\sigma \int_\sigma^t M(\sigma, \tau)N(\tau, t)d\tau,$$

the associative law

(3.8) $$(K \otimes L) \otimes M = K \otimes (L \otimes M)$$

for the "convolution"

(3.9) $$(L \otimes M)(s, t, y, x) = \int_s^t d\tau \int_{\boldsymbol{R}} L(s, \tau, y, z)M(\tau, t, z, x)dz.$$

Let us, by (3.4)-(3.5), put

$$\sup_{\substack{s_0 \leq s < t \leq t_0 \\ x, y \in \boldsymbol{R}}} |K(s, t, y, x)| = N, \quad \sup_x \int_{S(x, y) \leq 2\eta} dy = A.$$

Then since $K(s, t, y, x)$ vanishes for $S(y, x) \geq 2\eta$ independently of s and t, we have, for $s_0 \leq s < t \leq t_0$,

(3.10) $$\sup_{x, y \in \boldsymbol{R}} |K_n(s, t, y, x)| \leq N^n A^{n-1}(t-s)^{n-1}/(n-1)!,$$

$$\sup_y \int_{\boldsymbol{R}} |K_n(s, t, y, x)| dx \leq N^n A^n (t-s)^{n-1}/(n-1)!.$$

This proves the convergence of (3.7). Thus we have, by (3.3)-(3.8),

(3.11) $$P = H - H \otimes Q = H - P \otimes K.$$

We have also (1.9)-(1.11) by applying Fubini's theorem.

The proof of $L'_{sy}P(s, t, y, x) = 0$ may be obtained as follows. We first prove the fundamental limit theorem

---

5) It is to be noted that our HYPOTHESIS is independent of the choice of the local coordinates, whereas Mr. Itô's conditions are dependent upon the local coordinates since his conditions are referred to the "canonical coordinates system".

$$(3.12) \quad f(x) = \lim_{s \uparrow t_0,\, t \downarrow t_0} \int_R f(y) H(s, t, y, x) dy = \lim_{s \uparrow t_0,\, t \downarrow t_0} \int_R f(y) H(s, t, x, y) dy$$

for any continuous function $f(y)$.

This may be proved as in the note referred to 1). Thus, if we know

$$(3.13) \quad \lim_{\varepsilon \downarrow 0} L'_{sy} \int_s^{s+\varepsilon} d\tau \int_R H(s, \tau, y, z) Q(\tau, t, z, x) dz = 0 ,$$

we have

$$
\begin{aligned}
L'_{sy} P &= L'_{sy} H + \lim_{\varepsilon \downarrow 0} \frac{\partial}{\partial s} \int_{s+\varepsilon}^t d\tau \int_R H(s, \tau, y, z) Q(\tau, t, z, x) dz \\
&\quad - \lim_{\varepsilon \downarrow 0} \int_{s+\varepsilon}^t d\tau \, L'_{sy} \int_R H(s, \tau, y, z) Q(\tau, t, z, x) dz \\
&= L'_{sy} H - \lim_{\varepsilon \downarrow 0} \int_R H(s, s+\varepsilon, y, z) Q(s+\varepsilon, t, z, x) dz \\
&\quad - \lim_{\varepsilon \downarrow 0} \int_{s+\varepsilon}^t d\tau \int_R Q(\tau, t, z, x) L'_{sy} H(s, \tau, y, z) dz \\
&= K - Q - K \otimes Q = 0 .
\end{aligned}
$$

The proof of (3.13) may be obtained by changing z into the normal coordinates ζ around y according to the metric $dr(s)^2 = a_{ij}(s, z) dz^i dz^j$ and then changing the coordinates ζ and τ into ξ and κ :

$$\zeta^i = (\tau - s)^{1/2} \xi^i, \quad (\tau - s)^{1/2} = \kappa .$$

4. The Fundamental Formula and the Identity of the Fundamental Solution of (1.1) with the that of the Adjoint Equation of (1.1). Starting with the parametrix $H^*(s, t, y, x)$ for L_{tx}, we may construct the fundamental solution $P^*(s, t, y, x)$ for L_{tx} with the same properties as those given in (1.9)-(1.11). Of course we must impose the HYPO-THESIS for H^* similar to that for H. We may prove the identity

$$(4.1) \quad P^*(s, t, y, x) = P(s, t, y, x) .$$

Proof. We will make use of the FUNDAMENTAL FORMULA

$$
\begin{aligned}
(4.2) \quad &\int_R h(t, y) f(t, y) dy - \int_R h(s, y) f(s, y) dy \\
&= \int_s^t d\tau \int_R \{ h(\tau, y) L_{\tau y} f(\tau, y) - f(\tau, y) L'_{\tau y} h(\tau, y) \} dy ,
\end{aligned}
$$

if $h(\tau, y)$ and $f(\tau, y)$ are continuously differentiable once in τ and twice in y and if, moreover, $h(\tau, y)$ vanishes outside a compact set of y which is independent of τ. This may be proved by

$$\left[\int_{R} h(\tau, y)f(\tau, y)dy\right]_{s}^{t} = \int_{s}^{t} d\tau \frac{d}{d\tau}\int_{R} h(\tau, y)f(\tau, y)dy$$

and

$$\int_{R} \{f(\tau, y)A'_{\tau y}h(\tau, y) - h(\tau, y)A_{\tau y}f(\tau, y)\}\,dy = 0\,,$$

the latter being proved by Green's integral theorem and the vanishing of $h(\tau, y)$ outside a compact set of y independently of τ.

Now let $t_2 < s < t < t_1$, and apply (4.2) to

$$h(\tau, y) = H(\tau, t_1, y, z),\ f(\tau, y) = P^*(t_2, \tau, x, y).$$

Thus

$$\int_{R} H(t, t_1, y, z)P^*(t_2, t, x, y)dy - \int_{R} H(s, t_1, y, z)P^*(t_2, s, x, y)dy$$

$$= \int_{s}^{t} d\tau \int_{R} H(\tau, t, y, z)L_{\tau y}P^*(t_2, \tau, x, y)dy$$

$$- \int_{s}^{t} d\tau \int_{R} P^*(t_2, \tau, x, y)L'_{\tau y}H(\tau, t_1, y, z)dy$$

$$= - \int_{s}^{t} d\tau \int_{R} P^*(t_2, \tau, x, y)K(\tau, t_1, y, z)dy\,.$$

By letting $t_1 \downarrow t$ and remembering (3.12), we obtain

$$P^*(t_2, t, x, z) - \int_{R} H(s, t, y, z)P^*(t_2, s, x, y)dy$$

$$= - \int_{s}^{t} d\tau \int_{R} P^*(t_2, \tau, x, y)K(\tau, t, y, z)dy\,.$$

Next, by letting $t_2 \uparrow s$ and remembering the limit theorem

$$(3.12)'\quad f(x)\left(= \lim_{s \uparrow t_0, t \downarrow t_0}\int_{R} f(y)P^*(s, t, y, x)dy\right) = \lim_{s \uparrow t_0, t \downarrow t_0}\int_{R} f(y)P^*(s, t, x, y)dy$$

for any (integrable and) bounded continuous function $f(y)$,

which may be proved as (3.12), we obtain

$$(4.3)\quad P^*(s, t, x, z) - H(s, t, x, z) = - \int_{s}^{t} d\tau \int_{R} P^*(s, \tau, x, y)K(\tau, t, y, z)dy\,,$$

viz.

$$(4.3)'\qquad\qquad\qquad P^* = H - P^*\otimes K\,.$$

Therefore the continuous kernel

$$S(s, t, y, x) = P(s, t, y, x) - P^*(s, t, y, x)$$

satisfies the conditions

$$S = -S\otimes K\,,$$

$$\sup_{y} \int_{R} |S(s, t, y, x)| dx \text{ is bounded if } s \text{ and } t(>s) \text{ are bounded.}$$

Hence

$$S = -S \otimes K_n \quad (n = 1, 2, \ldots),$$

and thus, by (3.10), we must have $S(s, t, y, x) \equiv 0$.

5. The Uniqueness Lemmas and their Application to the Proof of (1.11)′, (1.12) and (1.13).

The Uniqueness Lemma 1. *Let $f(t, x)$ be a continuous (for $t \geq s$) solution of $L_{tx}f = 0$, $t > s$, such that*

$$f(s, x) = 0 , \quad x \in \boldsymbol{R} ,$$

$$\int_{R} |f(t, x)| dx \text{ is bounded for bounded } t(>s) .$$

Then we must have $f(t, x) \equiv 0$.

Proof. By applying the same argument as was used in the proof of (4.3), we obtain, for $\varepsilon > 0$,

$$\int_{R} f(t, y)H(t, t+\varepsilon, y, x)dy = - \int_{s}^{t} d\tau \int_{R} f(\tau, y)K(\tau, t+\varepsilon, y, x)dy .$$

Hence, by letting $\varepsilon \downarrow 0$ and remembering (3.12), we have

$$f(t, x) = - \int_{s}^{t} d\tau \int_{R} f(\tau, y)K(\tau, t, y, x)dy .$$

Thus we obtain $f(t, x) \equiv 0$ by the same argument as was used in the proof of $P = P^*$.

Similarly we obtain the

Uniqueness Lemma 2. *Let $h(s, y)$ be a continuous (for $s \geq t$) solution of $L'_{sy}h = 0$, $s < t$, such that*

$$h(t, y) = 0, \quad y \in \boldsymbol{R}$$

$$\sup_{y} |h(s, y)| \text{ is bounded if } s(<t) \text{ is bounded.}$$

Then $h(s, y) \equiv 0$.

We are now able to prove (1.11)′, (1.12) and (1.13).

The proof of (1.12). Let $s < u < t$, and consider

$$T(s, t, y, x) = \int_{R} P(s, u, y, z)P(u, t, z, x)dz .$$

It is easy to see from (3.6), (3.10) and (1.11), that

$$L_{tx}T = \int_R P(s, u, y, z)L_{tx}P(u, t, z, x)dz = 0, \quad t > u.$$

Moreover $T(s, u, y, x) = P(s, u, y, x)$, by

$$(3.12)'' \quad f(x)\left(= \lim_{s \uparrow t_0, t \downarrow t_0} \int_R f(y)P(s, t, y, x)dy\right) = \lim_{s \uparrow t_0, t \downarrow t_0} \int_R f(y)P(s, t, x, y)dy$$

for any (integrable and) bounded continuous function $f(y)$, which may be proved as (3.12). Thus we obtain

$$T(s, t, y, x) = P(s, t, y, x) \quad \text{for} \quad t > u$$

by the uniqueness lemma 1. Similarly we may prove

$$T(s, t, y, x) = P(s, t, y, x) \quad \text{for} \quad s < u.$$

The proof of (1.11)'. The function

$$p(s, t, y) = \int_R P(s, t, y, x)dx$$

is bounded when s and $t(> s)$ are bounded and satisfies

$$-\frac{\partial p}{\partial s} - a^{ij}(s, y)\frac{\partial^2 p}{\partial y^i \partial y^j} - b(s, y)\frac{\partial p}{\partial y^i} = 0,$$

$$\lim_{s \to t} p(s, t, y) = 1 \quad \text{(by } (3.12)'').$$

Hence, by the uniqueness lemma 2, we have

$$p(s, t, y) \equiv 1.$$

The proof of (1.13). Let $f(x)$ be non-negative and continuous. It is sufficient to prove the non-negativity of

$$F(\varepsilon, s, t, y) = \exp(\varepsilon s)\int_R P(s, t, y, x)f(x)dx$$

for any $\varepsilon < 0$ and for any such $f(x)$. We have, by $L'_{sy}P = 0$,

$$(5.1) \qquad -\frac{\partial F}{\partial s} - a^{ij}(s, y)\frac{\partial^2 F}{\partial y^i \partial y^j} - b^i(s, y)\frac{\partial F}{\partial y^i} - \varepsilon F = 0, \quad s < t.$$

$F(\varepsilon, t, t, y)$ is non-negative by $(3.12)''$. Let $F(\varepsilon, s, t, y)$ be, for fixed ε and t, negative somewhere and let $F(\varepsilon, s_0, t, y_0) < 0$. Then $\hat{F}(s, y) = F(\varepsilon, s, t, y)$ must, in the product space

$$\{s ; s_0 \leq s \leq t\} \times R,$$

reach its negative minimum at a certain point (s_1, y_1), $s_0 \leq s_1 < t$. We have, at (s_1, y_1),

$$\frac{\partial F}{\partial s} \geq 0, \quad a^{ij}\frac{\partial^2 F}{\partial y^i \partial y^j} \geq 0, \quad b^i\frac{\partial F}{\partial y^i} = 0, \quad \varepsilon F > 0,$$

contrary to (5.1).

(Received March 6, 1953)

On the integration of the temporally inhomogeneous diffusion equation in a Riemannian space

Proc. Japan Acad. **30** (1954) 19–23

(Comm. by K. Kᴜɴᴜɢɪ, ᴍ.ᴊ.ᴀ., Jan. 12, 1954)

1. Introduction. Let R be a connected domain of an m-dimensional, orientable C^∞ Riemann space with the metric $ds^2 = g_{ij}(x)dx^i\,dx^j$. We consider the forward diffusion equation in R

$$(1.1) \qquad E_{tx}f = \frac{\partial f(t,\,x)}{\partial t} - A_{tx}f(t,\,x) = 0,\, t > s ,$$

where

$$(1.2) \quad A_{tx}f(t,\,x) = g(x)^{-1/2}\frac{\partial^2}{\partial x^i \partial x^j}(g(x)^{1/2}a^{ij}(t,\,x)f(t,\,x))$$

$$- g(x)^{-1/2}\frac{\partial}{\partial x^i}(g(x)^{1/2}b^i(t,\,x)f(t,\,x)) + c(t,\,x)f(t,\,x) ,$$

$$g(x) = \det(g_{ij}(x)) .$$

The associated backward diffusion equation is defined by

$$(1.3) \qquad E_{sy}^* h = -\frac{\partial h(s,\,y)}{\partial s} - A_{sy}^* h(s,\,y) = 0,\, s < t ,$$

where A_{sy}^* is the formal adjoint of A_{ty}:

$$(1.4) \quad A_{sy}^* h(s,y) = a^{ij}(s,y)\frac{\partial^2 h(s,y)}{\partial y^i \partial y^j} + b^i(s,y)\frac{\partial h(s,y)}{\partial y^i} + c(s,y)h(s,y).$$

The operator $A_t = A_{tx}$ is assumed to be elliptic in x in the sense that

$$(1.5) \qquad a^{ij}(t,x)\xi_i\xi_j > 0 \text{ for } \sum_i(\xi_i)^2 > 0 .$$

Since the value of $A_{tx}f(t,x)$ should be independent of the local coordinates $(x^1, \ldots, x^m)$, we must have, by the coordinates change $x \to \bar{x}$, the transformation rule

$$(1.6) \qquad \bar{a}^{ij}(t,\bar{x}) = \frac{\partial x^i}{\partial x^k}\frac{\partial x^j}{\partial x^n}a^{kn}(t,x) ,$$

$$\bar{b}^i(t,\bar{x}) = \frac{\partial \bar{x}^i}{\partial x^k}b^k(t,x) + \frac{\partial^2 \bar{x}^i}{\partial x^k \partial x^n}a^{kn}(t,x) .$$

For the sake of simplicity, we assume that the coefficients $a^{ij}(t,x)$, $b^i(t,x)$, $c(t,x)$ and $g_{ij}(x)$ are C^∞ functions of (t,x).

The purpose of the present note is to give a sketch of a method[1]

---

1) Another method was proposed by Tosio Kato: (Integration of the equation of evolution in a Banach space, J. Math. Soc. Japan, **5**, 208–234 (1953)). His method is much general and elegant. However, it may not be easy to apply his method to the concrete equation such as (1.1), since he assumes that the domain $D(\bar{A}_t)$ of the closed extention $\bar{A}_t$ of A_t is independent of t.

of the integration in the function space $L = L_1(R)$[2] of (1.1) with the initial condition

(1.7) $$f(s, x) = f(x) .$$

Thus we firstly consider an approximate equation

(1.1)′ $$E_{tx}^{(n)} f^{(n)} = \frac{\partial f^{(n)}(t, x)}{\partial t} - A_{tx}^{(n)} f^{(n)}(t, x) = 0, \ t \geqq s \ (n = 1, 2, \ldots),$$

$$f^{(n)}(s, x) = f(x) ,$$

where $A_t^{(n)} = A_{tx}^{(n)}$ is a bounded operator in L which converges, in the sense to be explained below, to $A_t = A_{tx}$ as $n \to \infty$. Secondly it will be shown that there exists a subsequence $\{f^{(n')}(t, x)\}$ of the solutions $\{f^{(n)}(t, x)\}$ of (1.1)′ which converges, in the sense of the "distribution",[3] to a solution T_t of

(1.1)″ $$\frac{\partial (T_t, \varphi)}{\partial t} - (T_t, A_t^* \varphi) = 0, \ t > s ,$$

$$(T_s, \varphi) = (f, \varphi) = \int_R f(x)\varphi(x)dx .$$

Here (T_t, φ) is the value of the distribution T_t at φ, $\varphi(x)$ denoting a C^∞ function whose carrier is compact and is contained in an open domain of R. The totality of such functions $\varphi(x)$ will be denoted by $D(R)$. Finally we will show, by a parametrix consideration, that this T_t is defined by a "genuine" solution of (1.1). (See the Theorem below in 3.)

2. The Construction of the Distribution T_t. Let D be a set of C^∞ functions $f(x)$ with compact carriers such that D is L-dense in L. We regard $A_t = A_{tx}$ as an additive operator defined on $D \subseteqq L$ to L. Let $\bar{A}_t$ be the smallest closed extension of the operator A_t, and we will make the following:

Hypothesis.[4] Let, for all sufficiently large integer n (independently of t), the resolvents

(2.1) $$I_t^{(n)} = (I - n^{-1} \bar{A}_t)^{-1}$$

exist as bounded operators on L to L such that

(2.2) $I_t^{(n)} f(x)$ is non-negative and $\int_R I_t^{(n)} f(x)dx = \int_R f(x)dx$

if $f(x) \in L$ is non-negative,

(2.3) $I_t^{(n)} f$ is strongly continuous in t.

The Hypothesis implies that

---

2) The function space of the Borel measurable functions $f(x)$ which are integrable with respect to the measure $dx = g(x)^{1/2} dx^1 \ldots dx^m$. The norm of f is hence given by $\|f\| = \int_R |f(x)| dx$. It is to be noted that our method of integration of (1.1) may, with slight modifications, be extended to the case of the function space $L_p(R)$, $1 \leqq p \leqq \infty$.

3) L. Schwartz : Théorie des distributions, I et II, Paris (1950 et 1951).

4) Cf. K. Yosida: On the integration of diffusion equations in Riemannian spaces, Proc. Amer. Math. Soc., **3**, 864–873 (1952).

$$(2.4) \qquad \| I_t^{(n)} \| \leq 1$$

and

$$(2.5) \qquad A_t^{(n)} = \overline{A}_t I_t^{(n)} = n(I_t^{(n)} - I)$$

satisfies

$$(2.6) \qquad \text{strong} \lim_{n \to \infty} A_t^{(n)} f = \overline{A}_t f \text{ for } f \text{ in the domain } D(\overline{A}_t) \text{ of } \overline{A}_t .$$

We first prove the

Lemma 1. For any $f \in L$, there exists a solution $f_{ts}^{(n)} = f^{(n)}(t, s, x) \in L$ of

$$(2.7) \qquad D_t f_{ts}^{(n)} = \text{strong} \lim_{\delta \to 0} \delta^{-1}(f_{t+\delta, s}^{(n)} - f_{ts}^{(n)}) = A_t^{(n)} f_{ts}^{(n)}, \quad t \geq s,$$

$$\text{strong} \lim_{t \to s} f_{ts}^{(n)} = f,$$

satisfying

$$(2.8) \qquad f^{(n)}(t, s, x) \text{ is non-negative and } \int_R f^{(n)}(t, s, x)dx = \int_R f(x)dx$$

if $f(x)$ is non-negative.

Proof. Putting

$$P^{(n)}(t, s) = \exp\left((t-s)A_s^{(n)}\right) = \sum_{k=0}^{\infty} (k!)^{-1}(n(t-s)(I_s^{(n)} - I))^k,$$

we have the bounded operator

$$W^{(n)}(t, s) = E_t^{(n)} P^{(n)}(t, s) = (A_t^{(n)} - A_s^{(n)})P^{(n)}(t, s).$$

Then the solution $f_{ts}^{(n)}$ of (2.7) may be defined by

$$(2.9) \qquad f_{ts}^{(n)} = P^{(n)}(t, s)f - \int_s^t P^{(n)}(t, \tau)Q^{(n)}(\tau, s)f \, d\tau,$$

where

$$(2.10) \qquad Q^{(n)}(t, s) = \sum_{m=1}^{\infty} (-1)^{m+1} W_m^{(n)}(t, s),$$

$$W_m^{(n)}(t, s) = \int_s^t W^{(n)}(t, \tau) W_{m-1}^{(n)}(\tau, s)d\tau, \quad W_1^{(n)} = W^{(n)}.$$

We next prove

$$(2.11) \qquad \| f_{ts}^{(n)} \| \leq \| f \| .$$

For this purpose, we start, by (2.5) and (2.7),

$$(2.12) \qquad f_{t+\delta, s}^{(n)} = f_{ts}^{(n)} + \delta(n(I_t^{(n)} - I)f_{ts}^{(n)}) + o(\delta), \quad \delta > 0 .$$

Then, by (2.4),

$$\| f_{t+\delta, s}^{(n)} \| \leq (1 - \delta n) \| f_{ts}^{(n)} \| + \delta n \| f_{ts}^{(n)} \| + o(\delta) \leq \| f_{ts}^{(n)} \| + o(\delta),$$

and hence

$$\frac{d^+ \| f_{ts}^{(n)} \|}{dt} \leq 0$$

which proves (2.11). We next assume that $f^{(n)}(t, s, x)$ to be non-negative. Then, by (2.2) and (2.12),

$$\int_R f^{(n)}(t+\delta, s, x)dx = \int_R f^{(n)}(t, s, x)dx + \int_R \delta n(I_t^{(n)} - I)f^{(n)}(t, s, x)dx$$

$$+ \int_R o(\delta)dx \geq \int_R f^{(n)}(t, s, x)dx + o(\delta),$$

which implies

$$(2.13) \qquad \frac{d^+}{dt} \int_R f^{(n)}(t, s, x)dx \geq 0 .$$

Hence, if $f(x)$ is non-negative, we have, by (2.11) and (2.13),

$$\| f_{ts}^{(n)} \| = \int_R | f^{(n)}(t, s, x) | \, dx \leq \| f \| = \int_R | f(x) | \, dx$$
$$= \int_R f(x)dx \leq \int_R f^{(n)}(t, s, x)dx .$$

Therefore $f^{(n)}(t, s, x)$ must be non-negative (almost everywhere) with $f(x)$.

Lemma 2. There exists a subsequence $\{f_{ts}^{(n')}\}$ of $\{f_{ts}^{(n)}\}$ such that

$$(2.14) \qquad \lim_{n' \to \infty} (f_{ts}^{(n')}, \varphi) = (T_t , \varphi), \quad \varphi \in D(R) ,$$

where (T_t , φ) satisfies $(1.1)''$. Actually, T_t is a distribution defined by a measure ρ_{ts} :

$$(2.15) \qquad (T_t , \varphi) = \int_R \varphi(x)d\rho_{ts}(x) .$$

Proof. Integrating (2.7), we have

$$(f_{ts}^{(n)}, \varphi) - (f, \varphi) = \int_s^t (A_\tau^{(n)} f_{\tau s}^{(n)}, \varphi)d\tau = \int_s^t (I_\tau^{(n)} f_{\tau s}^{(n)}, A_\tau^* \varphi)d\tau ,$$

and hence, by (2.4) and (2.11),

$$(2.16) \qquad | (f_{t_1, s}^{(n)} - f_{ts}^{(n)}, \varphi) | \leq \| f \| \cdot | \int_t^{t_1} \max_x | A_{\tau x}^* \varphi(x) | \, d\tau | .$$

By virtue of (2.11) and (2.16), we may choose a subsequence $\{n'\}$ of $\{n\}$ such that (2.14) holds for a distribution T_t which satisfies

$$(2.17) \qquad (T_t , \varphi) \text{ is continuous in } t.$$

We see that (2.15) also holds good by (2.8). We have also

$$(2.18) \qquad \lim_{n' \to \infty} (I_\tau^{(n')} f_{\tau s}^{(n')}, A_\tau^* \varphi) = (T_\tau , A_\tau^* \varphi) \text{ boundedly in } \tau, \ s \leq \tau \leq t ,$$

since, by (2.11),

$$(I_\tau^{(n)} f_{\tau s}^{(n)}, \varphi) - (I_\tau^{(n)} f_{\tau s}^{(n)}, (I - n^{-1} A_\tau^*)\varphi) = n^{-1}(I_\tau^{(n)} f_{\tau s}^{(n)}, A_\tau^* \varphi) ,$$
$$(I_\tau^{(n)} f_{\tau s}^{(n)}, (I - n^{-1} A_\tau^*)\varphi) = ((I - n^{-1} \overline{A}_\tau)(I - n^{-1} \overline{A}_\tau)^{-1} f_{\tau s}^{(n)}, \varphi) = (f_{\tau s}^{(n)}, \varphi),$$
$$n^{-1} | (I_\tau^{(n)} f_{\tau s}^{(n)}, A_\tau^* \varphi) | \leq n^{-1} \| f \| \cdot \max_x | A_{\tau x}^* \varphi(x) | .$$

Thus T_t satisfies $(1.1)''$.

3. **The Theorem.** Let x_0 be any point of R and let $U(x_0)$ be a sufficiently small neighbourhood of x_0. Let $V(x_0)$ be any neighbourhood of x_0 such that its closure is contained in $U(x_0)$. We may construct[5] a parametrix $H(x, y, t, s)$ for the equation (1.3) such that

$(3.1) \qquad H(x, y, t, s)$ is, for $t > s$, C^∞ in (x, y, t, s),

$(3.2) \qquad E_{sx}^* H(x, y, t, s) = K(x, y, t, s)$ is C^∞ in (x, y, t, s) even when $t = s$,

$(3.3) \qquad H(x, y, t, s) \equiv 0$ if x or y is outside of $U(x)$,

$(3.4) \qquad f(x) = \lim_{t \downarrow t_0, s \uparrow t_0} \int_R f(y)H(x, y, t, s)dy = \lim_{t \downarrow t_0, s \uparrow t_0} \int_R f(y) H(y, x, t, s)dy$

for any $x \in V(x_0)$ and for any continuous function $f(y)$.

---

5) Cf. K. Yosida: On the fundamental solution of the parabolic equation in a Riemannian space, Osaka Math. J., **5**, 65–74 (1953).

We have thus, for any $\varepsilon > 0$,

$$(f^{(n)}(t, s, x), H(x, y, t+\varepsilon, t)) - (f^{(n)}(s, s, x), H(x, y, t+\varepsilon, s))$$

$$= \int_s^t \frac{d}{d\tau} (f^{(n)}(\tau, s, x), H(x, y, t+\varepsilon, \tau)) d\tau$$

$$= \int_s^t (E_{\tau x}^{(n)} f^{(n)}(\tau, s, x), H(x, y, t+\varepsilon, \tau)) d\tau$$

$$+ \int_s^t (f^{(n)}(\tau, s, x), -E_{\tau x}^* H(x, y, t+\varepsilon, \tau)) d\tau$$

$$+ \int_s^t \{ (\overline{A}_{\tau x} I_\tau^{(n)} f^{(n)}(\tau, s, x), H(x, y, t+\varepsilon, \tau))$$

$$- (I_\tau^{(n)} f^{(n)}(\tau, s, x), A_{\tau x}^* H(x, y, t+\varepsilon, \tau)) \} d\tau$$

$$+ \int_s^t (I_\tau^{(n)} f^{(n)}(\tau, s, x) - f^{(n)}(\tau, s, x), A_{\tau x}^* H(x, y, t+\varepsilon, \tau)) d\tau .$$

On the right hand side, the second term vanishes[6] in virtue of the Green's integral theorem and (3.3). And the third term tends, as $n = n' \to \infty$, to zero. This we see by (2.14) and (2.18).

Therefore we have, for any $\varphi(y) \in D(R)$,

$$\int_R d\rho_{ts}(x)(H(x, y, t+\varepsilon, t), \varphi(y)) = \int_R f(x)(H(x, y, t+\varepsilon, s), \varphi(y)) dx$$

$$+ \int_s^t \{ \int_R d\rho_{\tau s}(x)(K(x, y, t+\varepsilon, \tau), \varphi(y)) \} d\tau.$$

By letting $\varepsilon \downarrow 0$ and remembering (3.4), we obtain

(3.5) $$\int_R d\rho_{ts}(x)\varphi(x) = \int_R f(x)(H(x, y, t, s), \varphi(y)) dx$$

$$+ \int_s^t \{ \int_R d\rho_{\tau s}(x) [\int_R K(x, y, t, \tau)\varphi(y) dy] \} d\tau .$$

The measure $\int_A d\rho_{ts}(x)$ is thus absolutely continuous with respect to the measure $\int_A dx$, and the density $f(t, s, x)$ satisfies, by (3.5),

(3.6) $$\int_R f(t, s, x)\varphi(x) dx = \int_R f(x)(H(x, y, t, s), \varphi(y)) dx$$

$$+ \int_s^t \{ \int_R d\rho_{\tau s}(x) [\int_R K(x, y, t, \tau), \varphi(y)] \} d\tau.$$

Hence $f(t, s, x)$, which satisfies

(3.7) $$(T_t, \varphi) = \int_R f(t, s, x)\varphi(x) dx ,$$

is equivalent to

(3.8) $$\int_R f(y)H(y, x, t, s) dy + \int_s^t \{ \int_R d\rho_{\tau s}(y) K(y, x, t, \tau) \} d\tau .$$

This is surely continuously differentiable once in t and twice in x for $t > s$ and for $x \in V(x_0)$.

We have thus proved the following:

Theorem. Let the Hypothesis be satisfied. Then for any $f \in L$, there exists a solution $f_{ts} = f(t, s, x) \in L$ of (1.1) with the initial condition

(3.9) $$\lim_{t \downarrow s} f(t, s, x) = f(x) \text{ almost everywhere.}$$

The uniqueness of this solution may be proved by the known argument. Moreover, $f(t, s, x)$ is non-negative with $f(x)$.

---

6) By a similar argument as in the paper by K. Yosida. Cf. 4), p. 870.

On the integration of the temporally inhomogeneous diffusion equation in a Riemannian space. II

Proc. Japan Acad. **30** (1954) 273–275

(Comm. by K. Kunugi, m.j.a., April 12, 1954)

1. Introduction. In a preceding note with the same title,[1] the author devised an existence proof of the solution for the Cauchy's problem of the temporally inhomogeneous diffusion equation with C^∞ coefficients:

$$(1.1) \qquad \frac{\partial f(t, s, x)}{\partial t} - A_{tx} f(t, s, x) = 0, \quad t > s,$$

$$\lim_{t \to s} f(t, s, x) = f(x) \in L_1(R)^{[2]} \text{ almost everywhere,}$$

$$A_{tx} f(t, s, x) = g(x)^{-1/2} \frac{\partial^2}{\partial x^i \partial x^j} (g(x)^{1/2} a^{ij}(t, x) f(t, s, x))$$

$$- g(x)^{-1/2} \frac{\partial}{\partial x^i} (g(x)^{1/2} b^i(t, x) f(t, s, x)) + c(t, x) f(t, s, x),$$

in a connected domain R of an m-dimensional, orientable C^∞ Riemannian space with the metric $ds^2 = g_{ij}(x) dx^i dx^j$.

The purpose of the present note is to show that the existence proof in [I] may be modified so as to yield the existence proof of the solution admitting the kernel representation

$$(1.2) \qquad f(t, s, x) = \int_R P(t, s, x, y) f(y) dy \quad \text{for every } f(x) \in L_1(R).$$

2. The Proof of the Kernel Representation. Let D denote a linear set of C^∞ functions with compact carriers such that D is $L_1(R)$-dense in $L_1(R)$. We regard $A_t = A_{tx}$ as an additive operator on $D \subseteq L_1(R)$ to $L_1(R)$, and let $\overline{A}_t$ be the smallest closed extension of A_t. We assume that D is so chosen that the following Hypothesis is satisfied.

Hypothesis: Let, for sufficiently large integer n (independently of t), the resolvents

$$(2.1) \qquad I_t^{(n)} = (I - n^{-1} \overline{A}_t)^{-1}$$

---

1) Proc. Japan Acad., **30** (1954), No. 1, 19–23. This note will be referred to as [I]. At this juncture, the author wants to give the following corrigenda to [I]: (i) On page 23, lines 9–10, "the second term" and "the third term" should be read respectively as "the third term" and "the fourth term". (ii) On page 23, line 24, "is equivallent to" should be read as "equivalent, when $x \in V(x_0)$, to".

2) The Banach space of Borel measurable functions which are integrable, with respect to the measure $dx = g(x)^{1/2} dx^1 \ldots dx^m$, over R.

exist as bounded operators on $L_1(R)$ to $L_1(R)$ such that

(2.2) $I_t^{(n)}f(x)$ is non-negative and $\int_R I_t^{(n)}f(x)dx = \int_R f(x)dx$ when $f(x) \in L_1(R)$ is non-negative,

(2.3) $I_t^{(n)}f(x)$ is strongly continuous in t for every $f(x) \in L_1(R)$.

Then the following results were proved in [I] actually: Consider the approximate equation of (1.1) in $L_1(R)$

(2.4) strong $\lim_{\delta \to 0} \delta^{-1}\{f^{(n)}(t+\delta,s,x) - f^{(n)}(t,s,x)\} = \overline{A}_t I_t^{(n)} f^{(n)}(t,s,x)$, $t \geq s$,

strong $\lim_{t \to s} f^{(n)}(t,s,x) = f(x) \in L_1(R)$.

Then we may choose a subsequence $\{n'\}$ of $\{n\}$ such that $\{f^{(n')}(t,s,x)\}$ converges, for every $t > s$, in the sense of the "distribution", to a solution of (1.1) satisfying the conditions below.

(2.5) $\int_R |f(t,s,x)|\,dx \leq \int_R |f(x)|\,dx$ and $f(t,s,x)$ is non-negative when $f(x)$ is non-negative,

(2.6) for any $x_0 \in R$ and for any sufficiently small vicinity $U(x_0)$, there exist a vicinity $V(x_0) \subseteq U(x_0)$ and kernels $H(x,y,t,\tau)$, $K(x,y,t,\tau)$ such that (i) $H(x,y,t,\tau)$ and $K(x,y,t,\tau)$ are C^∞ for $t > \tau$ and $t \geq \tau$ respectively, (ii) $H(x,y,t,\tau)$ and $K(x,y,t,\tau)$ both vanish if x or y is outside of $U(x_0)$, and (iii) when $x \in V(x_0)$ we have the representation

$$f(t,s,x) = \int_R H(y,x,t,s)\,f(y)dy + \int_s^t \left\{\int_R K(y,x,t,\tau)f(\tau,s,y)dy\right\}d\tau.$$

We may modify the choice of the subsequence $\{n'\}$ as follows. Firstly we remark that the mapping $f(x) \to f^{(n)}(t,s,x)$ is a bounded linear mapping on $L_1(R)$ to $L_1(R)$. This we see by the boundedness and the strong continuity in t of the operator $\overline{A}_t I_t^{(n)} = n(I_t^{(n)} - I)$. Actually it was proved in [I] that

(2.7) $\|f^{(n)}(t,s,x)\| \leq \|f(x)\|$.

Thus, since $L_1(R)$ is separable, we may choose, by a diagonal method, a subsequence $\{n''\}$ of $\{n\}$ such that $\{f^{(n'')}(t,s,x)\}$ converges, for every $t > s$ and for every $f(x) \in L_1(R)$ simultaneously, in the sense of the "distribution" to a solution $f(t,s,x)$ of (1.1) satisfying (2.5) and (2.6). Hence, for any triple (t,s,x) with $t > s$ and $x \in V(x_0)$, the value $f(t,s,x)$ of this solution may be considered as an additive functional $F(f) = F_{t,s,x}(f)$ of $f(x) \in L_1(R)$. Moreover, we have, by (2.5) and (2.6),

$$|F_{t,s,x}(f)| \leq \int_R |f(y)|\,dy \cdot \max_y |H(x,y,t,s)|$$

$$+ (t-s)\int_R |f(y)|\,dy \cdot \max_{y,s \leq \tau \leq t} |K(x,y,t,\tau)|.$$

Therefore there exists a bounded measurable function in y depending upon (t, s, x), say $P(t, s, x, y)$, such that (1.2) holds good.

Remark 1. Comparing (1.2) with (2.6) and remembering (2.5), we see that the kernel $P(t, s, x, y)$ may be considered to be measurable in (t, s, x, y) for $t > s$. Thus, again by (1.2) and (2.6), $P(t, s, x, y)$ is equivalent, when $x \in V(x_0)$, to the function

$$H(x,y,t,s) + \int_s^t \left\{ \int_R K(z, x, t, \tau) P(\tau, s, z, y) dz \right\} d\tau.$$

Hence we see that $P(t, s, x, y)$ satisfies the equation

$$(2.8) \qquad \frac{\partial P}{\partial t} - A_{tx} P = 0, \quad t > s. [3]$$

Remark 2. The original sequence $\{f^{(n)}(t, s, x)\}$ itself converges in the sense of the "distribution" to the solution $f(t, s, x)$ of (1.1) satisfying (1.2) and (2.5)–(2.6), if it is assured that the solution of (1.1) satisfying (2.5) is unique. A condition for the uniqueness was given in another paper.[4] However, the author ·is not so far able to prove that whether the above Hypothesis assures this uniqueness or not.

---

3) Cf. K. Yosida: On the fundamental solution of the parabolic equation in a Riemannian space, Osaka Math. J., **5** (1953), No. 1, 65–74. See also S. Itô: The fundamental solution of the parabolic equation in a differentiable manifold, ibid., 75–92.

4) See the papers referred to in 3).

An abstract analyticity in time for solutions of a diffusion equation

Proc. Japan Acad. **35** (1959) 109–113

(Comm. by Z. SUETUNA, M.J.A., March 12, 1959)

1. *Introduction and the result.* Consider an equation of evolution

$$(1.1) \qquad \frac{\partial u}{\partial t} = Au, \qquad t > 0,$$

where the differential operator

$$(1.2) \qquad A = a^{ij}(x) \frac{\partial^2}{\partial x_i \partial x_j} + b^i(x) \frac{\partial}{\partial x_i} + c(x)$$

is elliptic in a connected domain G of an m-dimensional euclidean space E^m. Under certain conditions upon the coefficients a, b and c of A, we can specify a linear subspace D of $L_2(G)$ with the following three properties.

(i) The functions $\in D$ are C^∞ in G, and D is $L_2(G)$-dense in $L_2(G)$ such that $Af \in L_2(G)$ for $f \in D$.

(ii) If we consider A as an operator on $D \subseteq L_2(G)$ into $L_2(G)$, then A admits, in $L_2(G)$, the smallest closed extension $\hat{A}$.

(iii) $\hat{A}$ is the infinitesimal generator of a semi-group T_t of normal type in $L_2(G)$ such that, for any $f \in L_2(G)$, $u(t, x) = (T_t f)(x)$ is a solution of (1.1) with the initial condition

$$(1.1)' \qquad L_2(G)\text{-}\lim_{t \downarrow 0} u(t, x) = f(x)$$

satisfying the " forward and backward unique continuation property ":
(1.3)   If, for a fixed $t_0 > 0$, $u(t_0, x) \equiv 0$ on an open set $G_0 \subseteq G$, then
$u(t, x) = 0$ for every $t > 0$ and every $x \in G_0$.

The proof of (1.3) is based upon the fact that $T_t f$ is an $L_2(G)$-valued abstract analytic function of t in a certain sector of the complex plane which contains the positive t-axis in its interior and with $t = 0$ as its vertex. Such abstract analyticity in time is implied by the estimate (2.11) below of the resolvent of $\hat{A}$.[1]

Our result (1.3) gives a partial answer to a conjecture proposed by S. Ito and H. Yamabe [2]. Actually, our solution $u(t, x) = (T_t f)(x)$ enjoys the "unique continuation property":
(1.3)'   If, for a fixed $t_0 > 0$, $u(t_0, x) \equiv 0$ on an open set $G_0 \subseteq G$, then $u(t, x) = 0$ for every $t > 0$ and every $x \in G$.

---

1) This estimate was given in the author's lecture at Yale University in the fall of 1958.

483

This may be proved by combining (1.3) with the "space-like unique continuation theorem for solutions of parabolic equations" obtained recently by S. Mizohata [3]. Thus we obtain another proof of the unique continuation theorem of S. Ito and H. Yamabe [2].[2]

2. *The proof of the result.* For the sake of simplicity of exposition, we shall be concerned with the case[3] $G = E^m$. We assume that the real-valued coefficients a, b and c are C^∞ in E^m and that

(2.1) $a^{ij}(x)$ and its first and second partials, $b^i(x)$ and its first partials and $c(x)$ are, in absolute values, all bounded on E^m by a positive constant β.

Thus the strict ellipticity of A implies the existence of two positive constants γ and δ such that

$$(2.2) \qquad \gamma \sum_{j=1}^{m} \xi_j^2 \geq a^{ij}(x)\xi_i\xi_j \geq \delta \sum_{j=1}^{m} \xi_j^2 \qquad \text{on } E^m$$

for any real vector $(\xi_1, \xi_2, \cdots, \xi_m)$.

Let $H_1 = H_1(E^m)$ be the space of complex-valued C^∞ functions $f(x) = f(x_1, \cdots, x_m)$ in E^m for which

$$(2.3) \qquad \|f\|_1 = \left(\int_{E^m} |f(x)|^2 dx + \sum_{j=1}^{m} \int_{E^m} |f_{x_j}(x)|^2 dx \right)^{1/2} < \infty,$$

and let $\widehat{H}_1 = L_2(E^m) = L_2$ be the completion of H_1 with respect to the norm

$$(2.4) \qquad \|f\| = \left(\int_{E^m} |f(x)|^2 dx \right)^{1/2}.$$

We denote by RH_1 (and RL_2) the totality of real-valued functions belonging to H_1 (and to L_2).

Lemma. There exist two positive constants α_0 and β_0 such that, for any $f \in RH_1$, the equation

$$(2.5) \qquad \alpha u - Au = f, \quad \alpha > \max(\alpha_0, \delta + \beta_0),$$

admits a uniquely determined solution $u(x) = u_f(x) \in RH_1$, and we have the estimate

$$(2.6) \qquad \|u_f\| \leq (\alpha - \delta - \beta_0)^{-1} \|f\|.$$

Proof. The existence of the solution $u_f \in RH_1$ for sufficiently large α is proved in K. Yosida [4]. If we denote by (f, g) the inner product $\int_{E^m} f(x)g(x)dx$, then for any $u \in RH_1$,

$$(2.7) \qquad \|(\alpha I - A)u\| \cdot \|u\| \geq |((\alpha I - A)u, u)|$$

by Schwarz inequality. By partial integration, we have (see K. Yosida [4])

---

2) For, these two authors treat the case where $\hat{A}$ is self-adjoint with its spectrum lying on negative real axis, and the estimate (2.11) is clear for such operator $\hat{A}$.

3) If G is a bounded domain of E^m, the method of the following proof may be modified so as to apply to the case where A is an elliptic differential operator of $2n$-order $(n > 1)$.

$$((\alpha I - A)u,\, u) = \alpha \|u\|^2 + \int_{E^m} a^{ij} \frac{\partial u}{\partial x_i} \frac{\partial u}{\partial x_j}\, dx + \int_{E^m} \frac{\partial a^{ij}}{\partial x_i} \frac{\partial u}{\partial x_j}\, u\, dx$$

(2.8)

$$- \int_{E^m} b^i \frac{\partial u}{\partial x_i}\, u\, dx - \int_{E^m} c\, u\, u\, dx.$$

Hence we have, by (2.1)–(2.2) and the inequality $|\varepsilon\eta| \leq 2^{-1}(|\varepsilon|^2 + |\eta|^2)$,

$$((\alpha I - A)u,\, u) \geq \alpha \|u\|^2 + \delta(\|u\|_1^2 - \|u\|^2)$$

(2.9)
$$- m\beta[\nu(\|u\|_1^2 - \|u\|^2) + \nu^{-1}m\|u\|^2 + m^{-1}\|u\|^2]$$
$$= [\alpha - \delta - m\beta(m\nu^{-1} - \nu + m^{-1})]\|u\|^2 + (\delta - m\beta\nu)\|u\|_1^2$$

for any $\nu > 0$. Thus we have (2.6) from (2.7), by taking $\nu > 0$ so small that $(\delta - m\beta\nu) > 0$ and $\beta_0 = m\beta(m\nu^{-1} - \nu + m^{-1}) > 0$.

Corollary. Let us consider A as an operator defined on $\{f;\, f \in RH_1,$ $Af \in RH_1\} \subseteq RL_2$ into RL_2. Then the smallest closed extension $\widetilde{A}$, in RL_2, of A satisfies the condition that, for $\alpha > \max(\alpha_0,\, \delta + \beta_0)$, the inverse $(\alpha I - \widetilde{A})^{-1}$ exists as a bounded linear operator defined on RL_2 into RL_2 with the estimate

(2.10)
$$\|(\alpha I - \widetilde{A})^{-1}\| \leq (\alpha - \delta - \beta_0)^{-1}.$$

Theorem 1. If we consider A as an operator on $\{f;\, f \in H_1,\, Af \in H_1\}$ $\subseteq L_2$ into L_2, then the smallest closed extension $\widehat{A}$, in L_2, of A is the infinitesimal generator of a semi-group T_t in L_2 which is strongly continuous in t, $\|T_t\| \leq \exp((\delta + \beta_0)t)$ and such that

(2.11)
$$\varlimsup_{|\tau| \uparrow \infty} |\tau| \cdot \|((\alpha + \sqrt{-1}\tau)I - A)^{-1}\| < \infty.$$

Proof. By the lemma and the reality of the coefficients of A, we see that the range $(\alpha I - A) \cdot H_1$ is, for $\alpha > \max(\alpha_0,\, \delta + \beta_0)$, L_2-dense in L_2. Moreover we have, for $(u + \sqrt{-1}v) \in H_1$,

$$\|(\alpha I - A)(u + \sqrt{-1}v)\|^2 = \|(\alpha I - A)u\|^2 + \|(\alpha I - A)v\|^2$$
$$\geq (\alpha - \delta - \beta_0)^2 \|u\|^2 + (\alpha - \delta - \beta_0)^2 \|v\|^2.$$

Thus $(\alpha I - \widehat{A})^{-1}$ is a bounded linear operator on L_2 into L_2 satisfying

(2.12)
$$\|(\alpha I - \widehat{A})^{-1}\| \leq (\alpha - \delta - \beta_0)^{-1}.$$

Hence the first part of the theorem is proved (see E. Hille-R. S. Phillips [1] or K. Yosida [5]). We have to show that (2.11) holds good. We have, for $w \in H_1$, $\alpha > \max(\alpha_0,\, \delta + \beta_0)$,

$$\|((\alpha + \sqrt{-1}\tau)I - A)w\| \cdot \|w\| \geq |(((\alpha + \sqrt{-1}\tau)I - A)w,\, w)|.$$

As in (2.9), we obtain

$$|\text{Real Part }(((\alpha + \sqrt{-1}\tau)I - A)w,\, w)|$$

$$= \alpha \|w\|^2 + \text{Real Part}\left(\int_{E^m} a^{ij} \frac{\partial w}{\partial x_i} \frac{\partial \overline{w}}{\partial x_j}\, dx + \int_{E^m} \frac{\partial a^{ij}}{\partial x_i} \frac{\partial w}{\partial x_j}\, \overline{w}\, dx \right.$$

$$\left. - \int_{E^m} b^i \frac{\partial w}{\partial x_i}\, \overline{w}\, dx - \int_{E^m} c\, w\, \overline{w}\, dx \right)$$

$$\geq (\alpha - \delta - \beta_0)\|w\|^2 + (\delta - m\beta\nu)\|w\|_1^2.$$

Similarly we have

$$|\text{Imaginary Part }(((\alpha+\sqrt{-1}\tau)I-A)w,\,w)|$$
$$\geqq||\tau|\cdot||w||^2-m\beta\{||w||_1^2+m||w||^2\}|=|(|\tau|-m^2\beta)||w||^2-m\beta||w||_1^2|.$$

If we assume that there exists $w\in H_1$, $||w||\neq 0$, such that

$$|\text{Imaginary Part }(((\alpha+\sqrt{-1}\tau)I-A)w,\,w)|\leqq 2^{-1}(|\tau|-m^2\beta)||w||^2$$

for sufficiently large τ (or for sufficiently large $-\tau$), then, for such large τ (or $-\tau$),

$$m\beta||w||_1^2\geqq 2^{-1}(|\tau|-m^2\beta)||w||^2.$$

Hence, for such large τ (or $-\tau$),

$$|\text{Real Part }(((\alpha+\sqrt{-1}\,\tau)I-A)w,\,w)|\geqq(\delta-m\beta\nu)\frac{(|\tau|-m^2\beta)}{2m\beta}||w||^2.$$

Thus (2.11) is proved.

Theorem 2. The semi-group T_t is, for $t>0$, strongly differentiable in t any number of times. Actually, if we denote by $T_t^{(k)}$ the k-th strong derivative of T_t with respect to t, then there exists a positive constant ε such that, for any $t>0$, the sequence of operators

$$\sum_{k=0}^{n}(k!)^{-1}(\lambda-t)^k T_t^{(k)}$$

is, as $n\uparrow\infty$, convergent in the sense of the norm of operators when

$$(2.12)\qquad\qquad |\lambda-t|<\varepsilon t.$$

Proof. See K. Yosida [6].[4]

Corollary. For any $f\in L_2$, $u(t,\,x)=(Tf)(x)$ is infinitely differentiable in $t>0$ and $x\in E^m$ and satisfies the Cauchy problem $(1.1)-(1.1)'$.

Proof. If we apply, in the sense of the distribution of L. Schwartz, the elliptic differential operator

$$\left(\frac{\partial^2}{\partial t^2}+A\right)$$

any number of times to $u(t,\,x)$, then the result is locally square integrable in the product space $(0<t<\infty)\times E^m$. Thus $u(t,\,x)$ is equivalent to a function which is C^∞ in $(0<t<\infty)\times E^m$. See, for the details, K. Yosida [4].

Proof of 1.3. Since $T_t^{(k)}=A^k T_t$, we have, by Theorem 2,

$$\lim_{n\to\infty}||T_{t_0+h}f-\sum_{k=0}^{n}(k!)^{-1}h^k A^k T_{t_0}f||=0$$

for sufficiently small h. Hence there exists a sequence $\{n'\}$ of natural numbers such that

$$u(t_0+h,\,x)=\lim_{n'\to\infty}\sum_{k=0}^{n'}(k!)^{-1}h^k A^k u(t_0,\,x)\qquad\text{for almost all }x\in E^m.$$

By the hypothesis in (1.3), we have $A^k u(t_0,\,x)\equiv 0$ in G_0, and hence $u(t_0+h,\,x)\equiv 0$ in G_0. Repeating the process we see that $u(t,\,x)=0$ for every $t>0$ and every $x\in G_0$.

---

4) The "if" part of Theorem 2 in K. Yosida [6] must be corrected as: if $\lim_{|\tau|\uparrow\infty}\log|\tau|\cdot||R(1+i\tau,\,A)||=0$, then T_t' exists for every $t>0$.

References

[1] E. Hille and R. S. Phillips: Functional Analysis and Semi-groups, New York (1958).
[2] S. Ito and H. Yamabe: A unique continuation theorem for solutions of a parabolic equations, J. Math. Soc. Japan, **10**, no. 3, 314–321 (1958).
[3] S. Mizohata: Unicité du prolongement des solutions pour quelques opérateurs différentiels paraboliques, Mem. Colleg. Sci. Univ. Kyoto, sect. A, **31**, no. 3, 219–239 (1958).
[4] K. Yosida: An operator-theoretical integration of wave equations, J. Math. Soc. Japan, **8**, no. 1, 79–92 (1956).
[5] K. Yosida: On Semi-group Theory and Its Application to Cauchy's Problem in Partial Differential Equations, Bombay (1957).
[6] K. Yosida: On the differentiability of semi-groups of linear operators, Proc. Japan Acad., **34**, 337–340 (1958).

VII. Markov Processes

Comments (Shinzo Watanabe)

Yosida's works on Markov processes are devoted primarily to applications of semigroup theory to the problem of characterizing and constructing Markov processes. A foundation of the modern theory of Markov processes was laid down by a celebrated paper of Kolmogorov (Math. Ann. 104, 1931) in which he established basic analytical methods in the study of Markov processes in continuous time parameter by introducing, among others, the Kolmogorov differential equations for transition probabilities. Now these deep connections between Markov processes and differential equations can be better described and studied by means of analytical theory of semigroups. The leit-motif in Yosida's works was the study of local characteristics of Markov processes in terms of the infinitesimal generators of semigroups intrinsically associated with the processes.

In [58], the class of spatially homogeneous Markov processes (equivalently, the class of temporally homogeneous additive processes or processes with stationary independent increments) is determined in terms of the infinitesimal generators of associated semigroups acting on the L_1-space on the line and this gives an operator-theoretical counterpart of the Lévy-Khinchin characterization of the infinitely divisible distributions. Also, starting from the Fokker-Planck equation on the line and following the recipe of the Hille-Yosida theory, one can achieve the integration of this equation by constructing a semigroup and thereby a Markov process on the line. In this paper, the importance in the Markov process theory of the Yosida approximation

$$A_n = AI_n = n(I_n - I), \qquad I_n = (I - n^{-1}A)^{-1}, \quad (A: \text{the generator})$$

was emphasized. Actually, the Markov process generated by A_n is a pure jump Markov process quite similar to a compound Poisson process and the Yosida approximation is an approximation of any Markov process by such simple Markov processes.

The problems discussed in [58] were further pursued and developed in subsequent papers. In [59], a characterization of continuous Markov processes (i.e. diffusions) on the sphere invariant under the rotation was discussed and it was shown that such processes are essentially the Brownian motion on the sphere by noting that the invariant second order differential operators are essentially the Laplace-Beltrami operator. The construction problem was solved by using the eigenfunction expansion. The invariant diffusions on homogeneous Riemannian spaces were further studied in [69].

In [61], a general form of the time dependent infinitesimal generator for C^2 functions in the domain was obtained for transition operators of a (not necessarily temporally homogeneous) Markov process on Euclidean space. This is a space dependent generalization of the form obtained in [58] as an operator theoretical counterpart of the Lévy-Kchinchin characterization of infinitely divisible laws. It should be remarked that the same idea was used by A. D. Wentzell to obtain the so-called Wentell's boundary conditions for diffusion processes (Th. Prob. its Appl. 4, 1959).

One of the most successful application of the Hille-Yosida theory to Markov processes is Feller's characterization of one-dimensional diffusion processes. In other words Feller was able to determine, for the case of one-dimensional space, the most general class of operators A satisfying the local property and the local minimum principle. Here the local minimum principle is that if $f(x)$ has a local minimum at x_0 then $Af(x_0) \geq 0$. In [78], an extension of this result to higher dimensional spaces is discussed. But additional assumptions are required. The difficulty comes from the fact that the above local properties are topological ones and do not refer to the differentiable structure of the space. Yosida's treatise on Feller's theory can be found in his book; *Functional Analysis*, Springer, 1965, Chapter XIII, 6.

In [63], it was remarked that a Markov process can be constructed from flows (one-parameter group of transformations) in such a way that its infinitesimal generator is a sum of the squares of vector fields generating flows. In [77], a one-to-one correspondence was established between a Markov semigroup and its 1-resolvent. The 1-resolvent kernel is called here the generating parametrix of the associated Markov process. The 1-resolvent is a typical potential operator of Markov process and, in later years, Yosida's interest was directed to potential operators of Markov processes. In [94], the notion of holomorphic Markov processes was introduced in connection with the notion of holomorphic semigroups, and a general class of one-dimensional diffusion processes was shown to be holomorphic when its transition semigroup acts on the space of continuous functions.

Yosida's works on Markov processes are restricted to the analytical treatments and the sample paths are considered only as a heuristic background. It should be pointed out that Yosida's analytical results on Markov processes are now given probabilistic interpretations by means of the sample path description of Markov processes.

Finally, we will review some of major developments of analytical treatments of Markov processes after Yosida's work.

A convergence theorem of semigroups which is fundamental in limit theorems of Markov processes was obtained by H. F. Trotter (Pacific J. Math. 8, 1958).

The characterization problem of invariant Markov processes was solved, among others, in the case of Brownian motions on Lie groups by K. Ito (Proc. Japan

Acad. 26, 1950) and in the general case on Lie groups and their quotient spaces by G. A. Hunt (TAMS. 81, 1956).

Among many construction problems of Markov processes, the case of diffusion processes with Wentzell's boundary conditions has been studied extensively. In the analytical construction by the semigroup theory, a fundamental route was laid down by K. Sato and T. Ueno (J. Math. Kyoto Univ. 4, 1965) and, following it, a general class of diffusions was constructed by J. M. Bony, Ph. Courrège and P. Priouret (Ann. Inst. Fourier 18, 1968), K. Taira (Academic Press, 1988) and others.

The symmetry of transition semigroups is closely related to the time reversibility of Markov processes. The theory of symmetric Markov processes has been extensively developed mainly by M. Fukushima (North-Holland/Kodansha, 1980) based on the use of Dirichlet forms. The method of Dirichlet forms has often proved to be most useful to characterize and construct symmetric Markov processes.

An operator-theoretical treatment of temporally homogeneous Markoff process

J. Math. Soc. Japan 1 No. 3 (1949) 244–253

(Received April 1, 1948)

1. *Introduction.* Let $\{U_t\}$, $0 \leq t < \infty$, be a one-parameter semi-group of linear (=everywhere defined additive, continuous) operators from a complex Banach space E to E:

(1.1) $\qquad U_t U_s = U_{t+s}.\quad U_0 = I \quad$ (=the identity operator).

(1.2) $\qquad \sup_t \| U_t \| \leq 1,$

(1.3) $\qquad \lim_{t \to t_0} U_t \, x = U_{t_0} \, x, \; 0 \leq t_0 < \infty \;$ (lim=strong limit).

In a preceding note[1], the author obtained the following results. i) If D is the totality of x for which

(1.4) $\qquad$ weak $\lim_{h \downarrow 0} h^{-1} (U_h - I) x = Ax$

exists, then D coincides with the totality of x for which

(1.4)′ $\qquad \lim_{h \downarrow 0} h^{-1}(U_h - I) x = Ax$

exists and D is dense in E. The differential quotient operator (d.q.o.) A is a closed additive operator from D to E with the properties:

(1.5) $\qquad U_t x - x = \int_0^t U_s \, Ax \, ds \quad$ for $x \in D,$

(1.6) $\quad$ for any positive integer n, $I_n = (I - n^{-1} A)^{-1}$ exists and $\| I_n \| \leq 1$, $A I_n = n(I_n - I)$, $\lim_{n \to \infty} A I_n x = Ax \quad$ for $x \in D,$

(1.7) $\qquad I_n x = \int_0^\infty n \exp \, (-nt) U_t x \, dt \quad$ and $\lim_{n \to \infty} I_n x = x \quad$ for $x \in E.$

(1.8) $\quad U_t x = \lim_{n \to \infty} \exp \, (t A I_n) \, x, \; x \in E,$ uniformly in t for any finite interval of t.[2]

ii) Let conversely A be an additive operator from a dense linear subset D of E such that (1.6) is satisfied for any positive integer n, then there

---

1) On the differentiability and the representation of the one-parameter semi-group of linear operators, the Journal of the Math. Soc. of Japan, 1 (1948).

2) We may obtain, similarly as (1.8), another representation of U_t :
(1.8)′ $\qquad U_t \, x = \lim_{n \to \infty} (I - n^{-1} t \, A)^{-n} \, x.$

exists a uniquely determined one-parameter semi-group $\{U_t\}$ which satisfies $(1.1) - (1.5)$. This $\{U_t\}$ is given by (1.8).

The purpose of the present note is to give, as an application of these results, a characterisation of the *temporally homogeneous Markoff process*. By virtue of the composition rules for the d.q.o.'s and the differentiability theorem $(1.4)'$, we may determine the explicit form of the d.q.o. A in the special case where the Markoff process is not only temporally but also spatially homogeneous. The result may be considered as an operator-theoretical interpretation of the infinitely divisible law.[3] The results in **2** are also applied to the integration of the Fokker-Planck's equation.

2. *A characterisation of the d.q.o. of the temporally homogeneous Markoff process.* Let E be an *abstract-L-space*[4] and let, for $t \geq 0$,

(2.1) U_t be a positive operator ($U_t x \geq 0$ for $x \geq 0$) isometric on positive elements ($\|U_t\, x\| = \|\,x\,\|$ for $x \geq 0$.)

Such operator may be called a *transition operator*, and the semi-group $\{U_t\}$ may be considered as an abstract form of the temporally homogeneous Markoff process. In this case,

(2.2) $I_n = (I - n^{-1}\, A)^{-1} (= \int_0^\infty n\, \exp(-nt)\, U_t\, dt)$ is, for each $n = 1, 2, \ldots\ldots,$

a transition operator.

Conversely it is easy to see that if (2.2) is satisfied for large n then
$$U_t x = \lim_{n \to \infty} \exp\,(tAI_n)\,x = \lim_{n \to \infty} \exp(tn\,(I_n - I))\,x$$
$$= \lim_{n \to \infty} \exp\,(-nt)\,\exp\,(tnI_n)\,x$$
is also a transition operator. Thus we may construct all the temporally homogeneous Markoff process satisfying the continuity condition (1.3).

Let, in particular, E be the space $L_1(-\infty,\ \infty)$ and let $x \geq 0$ mean $x(t) \geq 0$ almost everywhere on $(-\infty,\ \infty)$. Then the additive operators

(2.3) $(Ax)(s) = \gamma x'(s)$ ($\gamma\ real \neq 0$),
$$= \sigma x''(s)\qquad (\sigma > 0),$$
$$= \lambda(x(s-u) - x(s))\quad (\lambda > 0,\ u \neq 0).$$

satisfy (1.6) aud (2.2). For the proof see the examples below.

Example 1. Let us consider the translation:
(2.4) $(U_t x)(s) = x(s+t),\quad x(s) \in L_1(-\infty,\ \infty).$
We have

---

3) P. Lévy: Théorie de l'addition des variables aléatoires, Paris (1937).

K. Yosida

$$y_n(s) = (I_n x)(s) = \int_s^\infty n \, \exp\,(n(s-k)) \; x(k) \; dk,$$

and hence, if $x(s)$ is continuous,

$$y_n'(s) - n y_n(s) = -n \; x(s).$$

Thus

$$(2.5) \qquad (A I_n x)(s) = (A y_n)(s) = (n(I_n - I)x)(s) = y_n'(s)$$

$$= \frac{d}{ds} \int_s^\infty n \, \exp\,(+n(s-k)) \; x(k)dk$$

and hence the operator $A = A_T$ is the differential operator $\left(\dfrac{d}{ds}\right)$
We have, by (1.8) and (1.8)' two expansions of Taylor's type[5].

Example 2.   Consider the integral

$$(2.6) \quad (U_t x)(s) = \frac{1}{\sqrt{\pi t}} \int_{-\infty}^\infty exp\left(-\frac{(s-k)^2}{t}\right) x(k)dk, \; x(k) \in L_1(-\infty,\infty)$$

corresponding to the *Gaussian distribution*.   We have

$$y_n(s) = (I_n x)(s) = \int_0^\infty x(k) \, dk \int_{-0}^\infty \frac{1}{\sqrt{\pi t}} \; n \, \exp\,\left(-nt - \frac{(s-k)^2}{t}\right) dt$$

$$= \int_{-\infty}^\infty \sqrt{n} \, \exp\,(-2\sqrt{n}\,|s-k|) \; x(k)dk,$$

and hence, if $x(s)$ is continuous,

$$y_n''(s) = 4n \; y_n(s) - 4nx(s).$$

Thus

$$(2.7) \qquad (A I_n \, x)(s) = (A y_n)(s) = (n(I_n - I)x)(s) = 4^{-1} \, y_n''(s)$$

$$= \frac{1}{4} \, \frac{d^2}{ds^2} \int_{-\infty}^\infty \sqrt{n}\, \exp\,(-2\sqrt{n}\,|s-k|) \; x(k) \; dk.$$

Therefore $A = A_G$ is the differential operator $\left(\dfrac{1}{4}\,\dfrac{d^2}{ds^2}\right)$, and we have, by

(1.8) and (1.8)', two expansions, the first of which improves Eddington's
formal expansion.[6]

4)   G. Birkhoff: Lattice Theory, New York (1940).   S. Kakutani; Concrete representation of abstract (L)-spaces and the mean ergodic theorem, Ann, of Math., **42** (1941).

5)   Cf. N. Dunford and I. E. Segal: Semi-groups of operators and the Weierstrass theorem, Bullet. Amer. Math. Soc., **52** (1946).

6)   A. A. Eddington: On a formula for correcting statistics for the effect of a known probable error of observations, Monthly Notice R. Astr. Soc., **73** (1914).

Example 3. Let $\lambda > 0$, $u \neq 0$ and consider

$$(2.8) \qquad (U_t\, x)\,(s) = \exp\,(-\lambda\, t) \sum_{k=0}^{\infty} \frac{(\lambda t)^k}{k!}\; x\,(s-ku),\ x\,(s) \in L_1(-\infty,\infty)$$

corresponding to the *Poisson distribution.* We have

$$y_n\,(s) = (I_n x)\,(s) = \int_0^{\infty} n\,\exp(-\,(n+\lambda)t)\sum_{k=0}^{\infty}\frac{(\lambda\, t)^k}{k!}\; x\,(s-ku)\ dt$$

$$= \sum_{k=0}^{\infty}\frac{n\lambda^k}{(n+\lambda)^{k+1}}\; x\,(s-ku)\,,$$

and therefore, when $n \to \infty$.

$$(2.9) \quad (Ay_n)\,(s) = (n(I_n - I)x)\,(s) \longrightarrow (Ax)\,(s) = \lambda\ (x(s-u) - x(s))\,.$$

$A = A_P$ is thus the difference operator.

Example 4. Let A be a linear operator defined on the abstract-L-space E satisfying the condition:

(2.10) $P = k^{-1}\,(A+kI)$ is a transition operator for a certain positive number k.

In this case we may show that $(I-n^{-1}A)^{-1}$ satisfies, for $n > 0$, (2.2) and (1.6).

Proof We have

$$I-n^{-1}A = (1+\sigma)(I- \frac{\sigma}{1+\sigma}\ P),\ \sigma = n^{-1}k > 0,$$

and hence, by $\left\| \frac{\sigma}{1+\sigma}\ P \right\| = \frac{\sigma}{1+\sigma} < 1 = \| I \|,$

$$(2.11) \qquad (I-n^{-1}\,A)^{-1} = (1+\sigma)^{-1}\Big\{I + \sum_{m=1}^{\infty}(\frac{\sigma}{1+\sigma})^m P^m\Big\}$$

exists. It is easy to see that this expansion defines a transition operator with P. $\hspace{4cm}$ (Q.E.D.).

The example 3 surely satisfies (2.10). The criterion (2.10) may also be used to deduce

Kolmogoroff's theorem.[7] Let E be the space of the n-dimensional complex vectors $x = (\xi_1, \xi_2,\ldots\ldots\xi_n)$ with the norm $\| x \| = \sum_{i=1}^{n} |\xi_i|$, and let $x \geq 0$ mean $\xi_i \geq 0$ $(i=1,2,\ldots\ldots,n)$. Then the transition operator P on E is represented by the matrix (p_{ij}) satisfying the condition:

---

7) A. Kolmogoroff: Die analytische Methoden in der Wahrscheinlichkeitsrechnung, Math. Ann., **104** (1931).

$$(2.12) \qquad p_{ij} \geq 0, \quad \sum_{j=1}^{n} p_{ij} = 1.$$

In this case, the d.q. matrix $A = (a_{ij})$ of the one-parameter semi-group of transition matrices $U(t) = (u_{ij}(t))$ is characterised by

$$(2.13) \qquad \sum_{j=1}^{n} a_{ij} = 0, \quad a_{ij} \geq 0 \; (i \neq j), \quad a_{ii} \leq 0.$$

3. *Composition of the d. q. o. 's.* We shall give two lemmas which enable us to construct another d.q.o. of the temporally homogeneous Markoff process from, for example, the d.q.o.'s in **2**.

Lemma 1. Let the intersection D of the domains of two additive operators A_1 and A_2 satisfying (1:6) and (2.2) be dense in E. Let, moreover, A_1 and A_2 be commutative in the sense that

$$(3.1) \qquad A_1 A_2 \, x = A_2 A_1 \, x,$$

if either $A_1 A_2 x$ or $A_2 A_1 x$ is well defined. Then

$$(3.2) \qquad A = A_1 + A_2$$

also satisfies (1.6) and (2.2).

Proof. Put

$$(3.3) \qquad I_{1n} = (I - n^{-1} A_1)^{-1}, \quad I_{2n} = (I - n^{-1} A)^{-1}.$$

Then

$$(3.4) \qquad A^{(n)} = A_1 \, I_{1n} + A_2 \, I_{2n} = n(I_{1n} - I) + n(I_{2n} - I)$$

satisfies (2.10) and hence (1.6) and (2.2) too. Thus the semi-group

$$(3.5) \qquad U_t^{(n)} = \exp \, (t A^{(n)})$$

constitutes a temporally homogeneous Markoff process satisfying (1.3). By (3.1), $A^{(m)}$ is commutative with $U_t^{(n)}$ Hence we have

$$\left\| (U_t^{(m)} - U_t^{(n)}) x \right\| = \left\| \int_0^t \frac{d}{ds} \left(\exp((t-s) A^{(n)}) U_s^{(m)} x \right) ds \right\|$$

$$= \left\| \int_0^t \left(\exp((t-s) A^{(n)}) U_s^{(m)} (A^{(n)} - A^{(m)}) \right) x \, ds \right\|$$

$$\leq \int_0^t \left\| (A^{(m)} - A^{(n)}) x \right\| \, ds$$

by $\| \exp(t A^{(n)}) \| \leq 1$. Therefore, by (1.6),

$$U_t \, y = \lim_{n \to \infty} U_t^{(n)} \, y \qquad (y \in D)$$

exists uniformly in t for any finite interval of t. Since D is dense in E

and since $\|U_t^{(n)}\| \leq 1$, we see that $U_t\, y$ exists for all $y \in E$ and satisfies (1.1) — (1.3). Surely U_t is a transition operator with $U_t^{(n)}$. We have, from (3.5)

$$U_t^{(n)}x - x = \int_0^t U_s^{(n)}\, A^{(n)}\, x\, ds, \quad x \in E.$$

Hence, by letting $n \to \infty$,

$$U^t x - x = \int_0^t U_s\, Ax\, ds, \quad x \in D,$$

in virtue of (1.6). Therefore $A = A_1 + A_2$ is the d.q.o. of U_t.

Similarly we may prove the

Lemma 2. Let $\{A^{(n)}\}$ be a sequence of mutually commutative linear operators satisfying (1.6) and (2.2), and let

$$(3.6) \qquad \lim_{n \to \infty} A^{(n)}x = Ax$$

exist for a dense linear subset D of E. Then

$$(3.7) \qquad U_t x = \lim_{n \to \infty} \exp\,(tA^{(n)})x$$

exists uniformly in t for any finite interval of t. Thus $\{U_t\}$ defines a temporally homogeneous Markoff process satisfying (1.3) whose d.q.o. is given by A.

4. *Temporally and spatially homogeneous Markoff process.* As an application of the above results, we shall give an operator-theoretical interpretation of the infinitely divisible law, to the effect that the examples given in 2 exhaust, in a certain sense, the d. q. o. A of the temporally and spatially homogeneous Markoff process.

Let U_t be defined by

$$(4.1) \qquad (U_t x)\,(s) = \int_{-\infty}^{\infty} x(s-u)\,d_u\, F(t,u)\,, \quad x(s) \in L_1(-\infty,\ \infty)\,,$$

where $F(t,u)$ is, for any $t \geq 0$, a distribution function of u. Then, for any $x(s)$ from the domain of the d.q.o. A,

$$(4.2) \qquad (Ax)\,(s) = \underset{n \to \infty}{\text{strong}}\ \text{limit}\ n\left(\int_{-\infty}^{\infty} x(s-u)\,d_u\, F(n^{-1},u) - x(s)\right).$$

Hence, by the Fourier transformation,

$$(4.3)\quad X(\lambda) \int_{-\infty}^{\infty} (\exp(i\lambda u) - 1)n\, d_u F(n^{-1}u)\,, \quad \left(X(\lambda) = \frac{1}{\sqrt{2\pi}}\right.$$

$$\left.\int_{-\infty}^{\infty} \exp\,(i\,\lambda\, s)\,x(s)\,ds\right)$$

K. YOSIDA

converges, as $n \to \infty$, uniformly in λ. Since the domain of A is dense in $L_1(-\infty, \infty)$, it is easy to see that

$$(4.4) \qquad \lim_{n \to \infty} \int_{-\infty}^{\infty} (\exp\,(i\,\lambda\,u) - 1) n\,d_u\,F(n^{-1},u)$$

exists uniformly in any finite interval of λ. We put

$$(4.5) \qquad G_n(u) = \int_0^u n\,\frac{u^2}{1+u^2}\,d_u\,F(n^{-1},u)\;.$$

Then, following after A. Khintchine's argument[8], we may prove that the suquence $\{G_n(u)\}$ of the monotone increasing functions contains a subsequence $\{G_{n'}(u)\}$ such that

$$(4.6) \qquad \text{a bounded } \lim_{n' \to \infty} G_{n'}(u) = G(u) \text{ exists,}$$

$$(4.7) \qquad \lim_{a \to \infty} \int_{|v| > a} dG_{n'}(u) = 0 \quad \text{uniformly in } n',$$

$$(4.8) \qquad \text{a finite } \lim_{n' \to \infty} \int_{-\infty}^{\infty} u^{-1}\,dG_{n'}(u) = \gamma \quad \text{exists.}$$

Thus, by $G(0) = 0$, we have, for continuous function $x(s) \in L_1(-\infty,\infty)$ whose continuous second derivative $x''(s)$ is also contained in $L_1(-\infty,\infty)$,

$$(4.9) \qquad \text{weak } \lim_{h \downarrow 0} (h^{-1}(U_h - I)x)(s) = (Ax)(s)$$

$$= -\gamma x'(s) + \sigma x''(s)$$

$$+ \lim_{\varepsilon \downarrow 0} \int_{|u| > \varepsilon} \left(x(s-u) - x(s) + \frac{ux'(s)}{1+u^2}\right)\frac{1+u^2}{u^2}\,dG(u),$$

where

$$(4.10) \qquad \sigma = \lim_{\varepsilon \downarrow 0} (G(\varepsilon) - G(-\varepsilon))\;.$$

Conversely we see, by the two lemmas in **3**, that the operator A defined by (4.9) is the d.q.o. of a temporally and spatially homogeneous Maikoff process. Here we make use of the fact that the operators (2.7) are all the d.q.o.'s of the temporally and spatially homogeneous Markoff processes.

5. *On the integration of the Fokker-Planck's equation.*[9] Consider

8) A. Khintchine: Déduction nouvelle d'une formule de P. Lévy, Bullet. de l'université d'état à Moscow, Sect. A, 1 (1937).

9) Cf. W. Feller: Zur Theorie der stochastischen Prozesse, Math. Ann., 113 (1936). K. Itô: On stochastic processes (Ⅱ), to appear in the Mem. of the Am. Math. Soc. Our method of integration may be extended to the Fokker-Planck's equation in homogeneous Riemannian spaces. For example, we may determine the "Brownian motion" on the surface of the sphere. The details will be published elsewhere. Here I express my hearty thanks to Dr. K. Itô for his friendly criticism during the preparation of the present note.

$$(5.1) \qquad \frac{\partial y(s,t)}{\partial t} = \frac{\partial (a(s)y(s,t))}{\partial s} + \frac{\partial^2 (b(s)y(s,t))}{\partial s^2}$$

$$= \delta(s)y(s,t) + \gamma(s)\frac{\partial y(s,t)}{\partial s} + \frac{\beta(s)}{4}\frac{\partial^2 y(s,t)}{\partial s^2}$$

where $t \geq 0$, $-\infty < s < \infty$, with positive $b(s)$ and

$$(5.2) \qquad \delta(s) = a'(s) + b''(s), \quad \gamma(s) = a(s) + 2b'(s), \quad \beta(s) = 4b(s).$$

If we assume

(5.3) $\quad \delta(s)$, $\gamma(s)$, $\beta(s)^{-1}$ and $\beta'(s)$ are all bounded and continuous,

$$(5.4) \quad \bar{s} = \int_0^s \frac{ds}{\sqrt{\beta(s)}} \to \infty \ as \ s \to \infty, \ and \ \bar{s} = \int_0^s \frac{ds}{\sqrt{\beta(s)}} \to -\infty \ as \ s \to -\infty,$$

then the additive Operator A defined by

$$(5.5) \qquad (Ay)(s) = (a(s)y(s))' + (b(s)y(s))''$$

$$= \delta(s)y(s) + \gamma(s)y'(s) + 4^{-1}\beta(s)y''(s)$$

in $L_1(-\infty, \infty)$ is the d.q.o. of a temporally homogeous Markoff process.

Proof. We have only to show that A satisfies (1.6) and (2.2) for large n.

The above example 2 suggests us that the solution $y_n(s)$ of

$$(5.6) \qquad y_n(s) - n\,(Ay_n)\,(s) = x(s)$$

will be given by the integral equation

$$(5.7) \quad y_n(s) = \int_{-\infty}^{\infty} \sqrt{n} \ \exp\left(-2\sqrt{n}\left|\int_k^s \frac{du}{\sqrt{\beta(u)}}\right|\right)\left(\frac{8\gamma(k) - \beta'(k)}{8n} y_n'(k)\right.$$

$$+ \frac{\delta(k)}{n} y_n\,(k) + x(k)\bigg)\frac{dk}{\sqrt{\beta(k)}}\ .$$

That (5.7) admits solution $y_n(s)$ for continuous $x(s) \in L_1(-\infty, \infty)$ will be seen as follows. Put

$$(5.8) \qquad C = \sup_s \left(8^{-1}|8\gamma(s) - \beta'(s)|, |\delta(s)|, \frac{1}{\sqrt{\beta(s)}}\right).$$

Then, for the successive approximations

$$y_{n1}(s) = \int_{-\infty}^{\infty} \sqrt{n} \ \exp\left(-2\sqrt{n}\left|\int_k^s \frac{du}{\sqrt{\beta(u)}}\right|\right)x(k)\frac{dk}{\sqrt{\beta(k)}},$$

$$y_{n,m}(s) = \int_{-\infty}^{\infty} \sqrt{n} \ \exp\left(-2\sqrt{n}\left|\int_k^s \frac{du}{\sqrt{\beta(u)}}\right|\right)\left(\frac{8\gamma(k) - \beta'(k)}{8n}\right.$$

$$y'_{n,m-1}(k) + \frac{\delta(k)}{n}\, y_{n,m-1}(k))\frac{dk}{\sqrt{\beta(k)}}$$

we easily obtain

$$\sup_s |y_{n,1}(s)| \leq \sup_s |x(s)|, \quad \sup_s |y'_{n,1}(s)| \leq 2\sqrt{n}\,C \sup_s |x(s)|,$$

$$\sup_s |y_{n,m}(s)| \leq n^{-1}C\,(\sup_s |y'_{n,m-1}(s)| + \sup_s |y_{n,m-1}(s)|,$$

$$\sup_s |y'_{n,m}(s)| \leq \frac{2\sqrt{n}\,C^2}{n}(\sup_s |y'_{n,m-1}(s)| + \sup_s |y_{n,m-1}(s)|).$$

Hence, for large n, the two series

$$\sum_{m=1}^{\infty} y_{n,m}(s), \quad \sum_{m=1}^{\infty} y'_{n,m}(s)$$

are uniformly and absolutely convergent to bounded continuous functions. Therefore

$$(5.9) \qquad\qquad y_n(s) = \sum_{m=1}^{\infty} y_{n,m}(s)$$

satisfies (5.7) and hence (6.6).

That this $y_n(s)$ belongs to $L_1(-\infty, \infty)$ with $x(s)$ will be seen as follows. If $\int_{-\infty}^{\infty} x(s)|\,\overline{ds} < \infty$, then

$$\int_{-\infty}^{\infty}|y_{n,1}(s)|\,\overline{ds} \leq \int_{-\infty}^{\infty}|x(s)|\,\overline{ds} \quad \int_{-\infty}^{\infty}|y'_{n1}(s)|\,\overline{ds} \leq 2\sqrt{n}\,C\int_{-\infty}^{\infty}|x(s)|\,\overline{ds,})$$

$$\int_{-\infty}^{\infty}|y_{n,m}(s)|\,\overline{ds} \leq n^{-1}C(\int_{-\infty}^{\infty}|y'_{n,m-1}(s)|\,\overline{ds} + \int_{-\infty}^{\infty}|y_{n,m-1}(s)|\,\overline{ds}),$$

$$\int_{-\infty}^{\infty}|y'_{n,m}(s)|\,\overline{ds} \leq \frac{2\sqrt{n}\,C^2}{n}(\int_{-\infty}^{\infty}|y'_{n,m-1}(s)|\,\overline{ds} + \int_{-\infty}^{\infty}|y_{n,m-1}(s)|\,\overline{ds}).$$

Hence we easily see that $y_n(s) = \sum_{m=1}^{\infty} y_{n,m}(s) \in L_1(-\infty, \infty)$. Moreover, the above inequalities show that

$$y_n(s) - y_{n,1}(s) = \sum_{m=2}^{\infty} y_{n,m}(s)$$

converges to zero strongly in $L_1(-\infty, \infty)$. Since the strong $\lim_{n\to\infty} y_{n1} = x$, we see that strong $\lim_{n\to\infty} y_n = x$.

On the other hand we see that, for large n,

$$y(s) - n^{-1}(Ay)(s) \geq 0$$

implies $y(s) \geq 0$, because such $y(s)$ cannot have negative minimum by the positivity of $b(s)$. Thus if $x(s) \geq 0$ satisfies $\int_{-\infty}^{\infty} |x(s)| \, ds < \infty$ and $\int_{-\infty}^{\infty} |x(s)|$

$ds < \infty$, then the solution $y_n(s) \in L_1 \ (-\infty, \infty)$ obtained above of (5.6) satisfies

$$\int_{-\infty}^{\infty} |y_n(s)| \, ds = \int_{-\infty}^{\infty} y_n(s) \, ds = \int_{-\infty}^{\infty} |x(s)| \, ds ,$$

because

$$\int_{-\infty}^{\infty} (A y_n)(s) ds = [a(s) y_n(s)]_{-\infty}^{\infty} + [(b(s) y_n(s))']_{-\infty}^{\infty} = 0.$$

Since $x(s) \in L_1 \ (-\infty, \infty)$ satisfying $\int_{-\infty}^{\infty} |x(s)| \, \bar{ds} < \infty$ are dense in L_1

$(-\infty, \infty)$, the above results shows that $I_n = (I - n^{-1}A)^{-1}$ exists and satisfies (2.2). Since $AI_n x = I_n Ax$ if x is in the domain D of A, we have (1.6) by strong $\lim y_n = x$, viz. $\lim I_n x = x$.

Thus we may integrate the original equation (5.1) by virtue of (1.8) or $(1.8)'$.

Mathematical Institute,

Nagoya University, Nagoya, Japan.

Added during the proof. On reading the manuscript of the present note, Prof. E. Hille kindly remarked me that essentially the same results as stated in **1** was already obtained by him by a different method. See his book: Functional Analysis and Semi-groups, New York (1948). He also kindly sent to me his manuscript " On the integration problem for Fokker-Planck's equation in theory of stochastic processes " which, replacing my analysis by a simpler argument, extends the results in **5**. After the present note was presented to the M. S. of Japan, I published two notes concerning the integration of F—P equation: Brownian motion on the surface of the 2–sphere, Ann. of Math. Stat., **20**, No. 2 (1949) ; Integration of Fokker-Planck's equation in a compact Riemannian space, Arkiv för Math. **1**, No. 9 (1949).

501

Brownian motion on the surface of the 3-sphere

Ann. Math. Statist. **20** (1949) 292–296

1. Introduction. Let S be a n-dimensional compact riemann space with the metric $ds^2 = g_{ij}(x)\, dx^i dx^j$ such that the totality G of the isometric transformations of S onto S constitutes a Lie group transitive on S. Consider a temporally homogeneous Markoff process by which $P(t, x, y)$, $t > 0$, is the transition probability that a point x is transferred to y after the elapse of t-unit time. We assume that $P(t, x, y)$ is a Baire function in (t, x, y) and continuous in t, then P satisfies Smoluchouski's equation

$$(1.1) \qquad P(t + s, x, y) = \int_S P(t, x, z)P(s, z, y)\, dz \qquad (t, s > 0),$$

dz being the G-invariant measure $\sqrt{g(x)}dx^1\, dx^2 \cdots dx^n$, $g(x) = \det(g_{ij}(x))$, and

$$(1.2) \qquad P(t, x, y) \geqq 0,$$

$$(1.3) \qquad \int_S P(t, x, y)\, dy = 1.$$

The spatial homogeneity of the transition process may be defined by

$$(1.4) \qquad P(t, Tx, Ty) = P(t, x, y) \qquad \text{for } T \,\epsilon\, G.$$

The "continuity" of the transition process may be defined, following after A. Kolmogoroff and W. Feller,[1] as follows. Let $L_1(S)$ be the function space of integrable (with respect to dx) functions $f(x)$ on S, then, for those $f(x)$ which are dense in $L_1(S)$,

$$\frac{\partial f(t, x)}{\partial t} = A \cdot f(t, x), \qquad (t \geqq 0);$$

$$(1.5)$$

$$f(t, x) = \int_S f(y)P(t, y, x)\, dy, \qquad (t > 0), \qquad f(0, x) = f(x),$$

where, with non-negative $b^{ii}(x)$

$$(1.6) \qquad (Af)(x) = \frac{1}{\sqrt{g(x)}} \frac{\partial}{\partial x^i} \left(- \sqrt{g(x)}\, a^i(x)f(x) \right)$$

$$+ \frac{1}{\sqrt{g(x)}} \frac{\partial^2}{\partial x^i \partial x^j} \left(\sqrt{g(x)}\, b^{ij}(x)f(x) \right).$$

---

[1] A. Kolmogoroff, "Zur Theorie der stetigen zufälligen Prozesse," *Math. Annalen*, Vol. 108 (1933); W. Feller, "Zur Theorie der stochastischen Prozesse," *Math. Annalen*, Vol. 113 (1937).

The temporally and spatially homogeneous "continuous" Markoff process may, if it exists, be called a Brownian motion on the homogeneous space S. The purpose of the present note is to show that, under some derivability hypothesis concerning $a^i(x)$ and $b^{ij}(x)$, there exists one and (essentially) only one Brownian motion on the surface of the 3-sphere S^3.

I here express my hearty thanks to Dr. Kiyosi Itô who proposed to me the problem and discussed and much improved the manuscript.

2. The defining equation for the Brownian motion. The spatial homogeneity (1.4) is equivalent to the fact that A is commutative with every operator $\tilde{T}$ defined by

$$(2.1) \qquad (\tilde{T}f)(x) = f(Tx), \qquad T \in G,$$

because we have

$$\int_S f(y)P(t, y, Tx)\, dy = \int_S f(Ty)P(t, Ty, Tx)\, dTy = \int_S f(Ty)P(t, y, x)\, dy.$$

The condition (2.1) is equivalent to

$$(2.2) \qquad XA = AX \text{ for any infinitesimal operator } X = \xi^k(x)\frac{\partial}{\partial x^k}$$

induced on S by the infinitesimal operator of the Lie group G. Thus, assuming the derivability of $a^i(x)$ and $b^{ij}(x)$ of necessary orders, we obtain from (2.2) the conditions:

$$(2.3) \qquad \xi^k(x)\frac{\partial}{\partial x^k}\left(\frac{1}{\sqrt{g(x)}}\frac{\partial G^i(x)}{\partial x^i}\right) = 0,$$

$$\left(G^i(x) = -\sqrt{g(x)}\, a^i(x) + \frac{\partial \sqrt{g(x)}b^{ij}(x)}{\partial x^j}\right),$$

$$(2.4) \qquad \frac{1}{\sqrt{g(x)}}H^i(x)\frac{\partial \xi^k(x)}{\partial x^i} + b^{ij}(x)\frac{\partial^2 \xi^k(x)}{\partial x^i \partial x^j} = \xi^i(x)\frac{\partial}{\partial x^i}\left(\frac{1}{\sqrt{g(x)}}H^k(x)\right),$$

$$\left(H^i(x) = G^i(x) + \frac{\partial}{\partial x^j}(\sqrt{g(x)}\, b^{ij}(x))\right),$$

$$(2.5) \qquad b^{ij}(x)\frac{\partial \xi^k(x)}{\partial x^j} + b^{kj}(x)\frac{\partial \xi^i(x)}{\partial x^j} = \xi^j(x)\frac{\partial b^{ik}(x)}{\partial x^j}.$$

Now for the surface of the 3-sphere S^3,

$$ds^2 = d\theta^2 + \sin^2\theta \cdot d\varphi^2, \qquad g(\theta, \varphi) = \sin^2\theta,$$

and the infinitesimal operators

$$X_x = \sin \varphi\, \frac{\partial}{\partial \theta} + \frac{\cos\theta \cos\varphi}{\sin\theta}\frac{\partial}{\partial \varphi},$$

$$X_y = \cos \varphi\, \frac{\partial}{\partial \theta} - \frac{\cos\theta \sin\varphi}{\sin\theta}\frac{\partial}{\partial \varphi},$$

$$X_z = \frac{\partial}{\partial \varphi}$$

respectively correspond to the rotations about the x-, y- and z-axis.

From (2.5) we see that, by taking $X = X_z$,

(2.6) $\qquad\qquad\qquad b^{ij}(\theta, \varphi)$ is independent of φ.

By taking $X = X_z$ in (2.4) we see that H^k is independent of φ. Hence, by (2.6),

(2.7) $\qquad\qquad\qquad a^i(\theta, \varphi)$ is independent of φ.

Thus, by taking $k = 1$, $X = X_z$ we obtain from (2.4),

$$\frac{1}{\sin\theta} H^2(\theta) \cos\varphi - b^{22}(\theta) \sin\varphi = \sin\varphi \frac{d}{d\theta}\left(\frac{1}{\sin\theta} H^1(\theta)\right)$$

and thus

(2.8) $\qquad H^2(\theta) = 0, \qquad b^{22}(\theta) + \frac{d}{d\theta}\left(\frac{1}{\sin\theta} H^1(\theta)\right) = 0.$

Hence, by taking $k = 2$, $X = X_x$ or $X = X_y$, we obtain from (2.4)

$$\frac{-H^1(\theta) \cos\varphi}{\sin^3\theta} + 2b^{11}(\theta) \frac{\cos\theta \cos\varphi}{\sin^3\theta} + 2b^{12}(\theta) \frac{\sin\varphi}{\sin\theta} - b^{22}(\theta) \frac{\cos\theta \cos\varphi}{\sin\theta} = 0,$$

$$\frac{H^1(\theta) \sin\varphi}{\sin^3\theta} - 2b^{11}(\theta) \frac{\cos\theta \sin\varphi}{\sin^3\theta} + 2b^{12}(\theta) \frac{\cos\varphi}{\sin^2\theta} + b^{22}(\theta) \frac{\cos\theta \sin\varphi}{\sin\theta} = 0.$$

From these two equations we obtain

(2.9) $\qquad b^{12}(\theta) = 0, \qquad \frac{H^1(\theta)}{\sin^3\theta} - 2b^{11}(\theta) \frac{\cos\theta}{\sin^3\theta} + b^{22}(\theta) \frac{\cos\theta}{\sin\theta} = 0.$

By taking $i = 2$, $k = 1$, $X = X_x$, we obtain from (2.5), (2.9)

$$b^{22}(\theta) \cos\varphi + b^{11}(\theta) \frac{d}{d\theta}\left(\frac{\cos\theta \cos\varphi}{\sin\theta}\right) = 0$$

and hence

(2.10) $\qquad\qquad\qquad b^{22}(\theta) = \frac{b^{11}(\theta)}{\sin^2\theta}.$

Similarly by taking $i = 1$, $k = 1$, $X = X_x$ we obtain from (2.5)

$$b^{12}(\theta) \cos\varphi + b^{12}(\theta) \cos\varphi = \sin\varphi \frac{db^{11}(\theta)}{d\theta}$$

and hence by (2.9), (2.10)

(2.11) $\qquad\qquad b^{11}(\theta) = \text{constant } C, \qquad b^{22}(\theta) = \frac{C}{\sin^2\theta}.$

Thus we obtain from (2.4)

$$H^1(\theta) = -a^1(\theta) \sin \theta + 2C \cos \theta, \qquad H^2(\theta) = -\sin \theta \cdot a^2(\theta)$$

and thus, by (2.8),

$$(2.12) \qquad\qquad a^2(\theta) = 0.$$

Substituting (2.11) in (2.9) we obtain

$$(2.13) \qquad\qquad a^1(\theta) = \frac{C \cos \theta}{\sin \theta}.$$

Therefore since $b^{11}(\theta)$ and $b^{22}(\theta)$ are non-negative, A is (essentially) equal to the Laplace operator

$$(2.14) \qquad \Lambda = \frac{1}{\sin \theta} \frac{\partial}{\partial \theta} \sin \theta \frac{\partial}{\partial \theta} + \frac{1}{\sin^2 \theta} \frac{\partial^2}{\partial \varphi^2}.$$

Thus we may obtain $P(t, x, y)$ by integrating the equation

$$(2.15) \qquad \frac{\partial f(t; \theta, \varphi)}{\partial t} = \Lambda \cdot f(t; \theta, \varphi), \qquad (t \geqq 0),$$

and by putting

$$(2.16) \qquad f(t; \theta, \varphi) = f(t, x) = \int_{S^2} f(y) P(t, y, x) \, dy.$$

3. Integration of the equation (2.15)–(2.16). Consider the Laplacian (real) spherical harmonics

$$(3.1) \qquad Y_k^{(m)}(\theta, \varphi) = Y_k^{(m)}(x), \qquad (-k \leq m \leq k; k = 0, 1, \cdots).$$

They constitute an orthonormal function system complete for continuous functions on S^2, and we have

$$(3.2) \qquad \Lambda \cdot Y_k^{(m)}(\theta, \varphi) = -k(k+1) Y_k^{(m)}(\theta, \varphi).$$

Since, as is well-known,

$$(3.3) \qquad Y_k^{(m)}(T^{-1}x) = \sum_{n=-k}^{k} u_{nm}^{(k)}(T) Y_k^{(n)}(x)$$

by an irreducible orthogonal representation $(u_{nm}^{(k)}(T))$ of the rotation group G, we have

$$(3.4) \qquad \max_x |Y_k^{(m)}(x)|^2 \leq (2k+1) \min_x \sum_{n=-k}^{k} |Y_k^{(n)}(x)|^2,$$

by applying the Schwarz inequality and the transitivity of the group G on S^2. The right hand member satisfies, by the orthonormality

$$(3.5) \qquad (2k+1)^2 / (\text{area of } S^3).$$

Therefore the double series (for $t > 0$)

$$(3.6) \quad P(t; \theta, \varphi; \theta', \varphi') = \sum_{k=0}^{\infty} \sum_{m=-k}^{k} \exp\left(-k(k+1)t\right) Y_k^{(m)}(\theta, \varphi) Y_k^{(m)}(\theta', \varphi')$$

is absolutely and uniformly convergent on S^3. We will show that this P is the required (unique) Brownian motion on S^3.

The proof may be given in three steps. i) We see by (3.2) and (3.6), that $\int_{S_3} f(y) P(t, y, x)\, dx$ satisfies (2.15) if

$$f(x) \sim \sum_{k=0}^{\infty} \sum_{m=-k}^{k} d_k^{(m)} Y_k^{(m)}(x), \qquad \sum_{k=0}^{\infty} \sum_{m=-k}^{k} \exp\left(-k(k+1)t\right) k(k+1)\, d_k^{(m)} Y_k^{(m)}(x)$$

are both absolutely and uniformly convergent. By the completeness of $\{Y_k^{(m)}(x)\}$, such $f(x)$ are dense in $L_1(S)$.

ii) Because of (3.3) we see that (3.6) satisfies the spacial homogeneity (1.4).

iii) (1.3) is obvious by the orthonormality of $\{Y_k^{(m)}(x)\}$ and the constancy on S^3 of $Y_0^{(0)}(x)$. Next, for the solution $f(t, x)$ of (2.15)–(2.16), let $f(x) = f(0, x)$ be non-negative on S^3, then $g_\epsilon(t, x) = \exp(-\epsilon t) f(t, x)$, $(\epsilon > 0)$, satisfies

$$\frac{\partial g_\epsilon(t, x)}{\partial t} = \Lambda \cdot g_\epsilon(t, x) - \epsilon g_\epsilon(t, x), \qquad (t > 0),$$

$$g_\epsilon(0, x) = f(x) \geqq 0 \qquad\qquad \text{(on } S^3\text{)}.$$

Thus $g_\epsilon(t, x) \geqq 0$ on S^3, since $g_\epsilon(t, x)$ cannot have a negative minimum on the product space $[t_1, t_2] \times S^3$, for any $t_2 > t_1 > 0$. For at such minimizing point we must have

$$\frac{\partial g_\epsilon}{\partial t} = 0, \qquad \frac{\partial g_\epsilon}{\partial \theta} = 0, \qquad \frac{\partial g_\epsilon}{\partial \varphi} = 0, \qquad \frac{\partial^2 g_\epsilon}{\partial \theta^2} \geqq 0, \qquad \frac{\partial^2 g_\epsilon}{\partial \varphi^2} \geqq 0.$$

Therefore, since $\epsilon > 0$, $t_2 > t_1 > 0$ were arbitrary, we conclude that $f(t, x) \geqq 0$ on S^3 for $t > 0$ if $f(x) = 0$ on S^3. This proves (1.2). The same argument simultaneously shows us that the solution P of (2.15)–(2.16) and (1.2)–(1.3) is unique.

━━━◆━━━

An extension of Fokker-Planck's equation

Proc. Japan Acad. **25** (1949) 1–3

(Comm. by T. TAKAGI, M. J. A., Oct. 12, 1949.)

Let the possible states of a stochastic system be represented by the points $x = (x_1, \ldots, x_n)$ of the n-dimensional Riemannian space R. We denote by $P(s, x, t, E)$, $s \leq t$, the transition probability that the state x at the time moment s is transferred into the Borel set $E \subseteq R$ at the later time moment t. The function P will satisfy the probability conditions

$$(1) \qquad P(s, x, t, E) \geq 0, \quad P(s, x, t, R) = 1,$$

$$(2) \qquad P(s, x, s, E) = 1 \text{ or } = 0 \text{ according as } x \,\epsilon\, E \text{ or } x \,\bar{\epsilon}\, E,$$

and the Chapman-Smoluchouski's equation

$$(3) \qquad P(s, x, t, E) = \int_R P(s, x, u, dz) P(u, z, t, E), \quad s \leq u \leq t.$$

Let $C(R)$ be the Banach space of real-valued bounded continuous functions $f\,x)$ on R with the norm $\|f\| = \sup |f(x)|$. We assume that

$$(4) \qquad (U_{st}f)(x) = \int_R P(s, x, t, dy) f(y)$$

defines a system of linear operators $\{U_{st}\}$ on $C(R)$ in $C(R)$. Then

$$(5) \qquad (U_{st}f)(x) \text{ is non-negative with } f(x) \text{ and } \|U_{st}\| = 1,$$

$$(6) \qquad U_{ss} = I \text{ (the identity)}, \quad U_{su}U_{ut}f = U_{st}f.$$

In the special case of the temporal homogeneity

$$(7) \qquad U_{su} = T_{u-s},$$

the strong continuity in t of T_t implies the strong differentiability of $T_t f$ for those f which are strongly dense in $C(R)$[1]:

$$(8) \qquad \frac{dT_t f}{dt} = \text{strong} \lim_{\varDelta \downarrow 0} \frac{T_{t+\varDelta} - T_t}{\varDelta} f = A T_t f = T_t A f, \quad A f = \left(\frac{dT_t f}{dt}\right)_{t=0}.$$

In the general case, a formal extenssion of the above equation will be

$$(9) \qquad \frac{\partial U_{st} f}{\partial s} = -A_s U_{st} f.$$

It may be called as Fokker-Planck's equation corresponding to the stochastic process $P(s, x, t, E)$.

The purpose of the present note is to give a possible form of the un-

---

1) E. Hille: Functional Analysis and Semi-groups, New York (1948). K. Yosida: On the differentiability and the representation of one-parameter semi-group of linear operators, Journal of the Math. Soc. of Japan, Vol. 1. No. 1 (1948).

bounded operator As as an extension of the form given by A. Kolmogoroff[1] and W. Feller.[2] It has a certain connection with the infinitely divisible law of P. Lévy,[3] and it reads as follows.

Theorem. Let there exists a sequence $\{m\}$ of positive integers such that

$$(10) \qquad (A_sf)(x)=a \text{ finite } \lim_{m\to\infty} m\left[\int_R P(s, x, s+m^{-1}, dy)f(y)-f(x)\right]$$

exists if $f(x)$ and its 1st and 2nd-derivatives are bounded and continuous in R,

$$(11) \qquad \lim_{a\to\infty} m \int_{d(x,y)\geqq a} P(s, x, s+m^{-1}, dy)=0$$

uniformly in m, $(d(x, y)=$ the geodesic
distance of x and $y)$.
Then we have

$$(12) \qquad (A_sf)(x)=\sum_{j=1}^{n} a_j(s, x)\frac{\partial f}{\partial x_j}+\sum_{j,k=1}^{n} b_{jk}(s, x)\frac{\partial^2 f}{\partial x_j\partial x_k}$$

$$+\lim_{\varepsilon\downarrow 0}\int_{d(x,y)\geqq\varepsilon}\left\{f(y)-f(x)-\frac{\rho(y, x)}{1+d(y, x)^2}\sum_{j=1}^{n}(y_j-x_j)\frac{\partial f}{\partial x_j}\right\}\frac{1+d(y, x)^2}{d(y, x)^2}G(s, x, dy)$$

where i) $G(s, x, E)$ is a countably additive non-negative set function in E and $G(s, x, R)<\infty$, ii) $\rho(x, y)$ is continuous in (x, y) such that $\rho(x, y)$ is 1 or 0 according as $d(x, y)\leqq\delta/2$ or $\geqq\delta\,\delta>0)$, iii) the quadratic form $\sum_{j,k=1}^{n}b_{jk}(s, x)\xi_j\xi_k$ is non-negative definite.

Proof. From (10) and (11) we see that

$$(13) \qquad G_m(s, x, E)=m\int_E \frac{d(y, x)^2}{1+d(y, x)^2} P(s, x, s+m^{-1}, dy)$$

satisfies

$$(14) \qquad \lim_{a\to\infty}\int_{d(x,y)\leqq a} G_m(s, x, dy)=0 \qquad \text{uniformly in } m,$$

$$(15) \qquad G_m(s, x, E) \text{ is uniformly bounded in } E \text{ and in } m.$$

Hence, for any fixed (s, x), there exists a subsequence $\{m'\}$ such that, if

$$g(x) \in C(R),$$

---

1)　Math. Ann., *104* (1931) and *108* (1933).

2)　Math. Ann., *113* (1936).

3)　See K. Yosida: An operator-theoretical treatment of temporally homogeneous Markoff process, to appear in the Journal of the Math. Soc. of Japan. A formula analogus to (13) below was also obtained by K. Itô in connection with his theory of stochastic differential equations, to appear soon elsewhere. P. Lévy: Théorie de l'addition des variable aléatoires, Paris (1937), Chapitre 7.

(16) a finite $\displaystyle\lim_{m'\to\infty}\int_R g(y)G_{m'}(s, x, dy)$ exists and $=\displaystyle\int_R g(y)G(s, x, dy)$ with $G(s, x, E)$ satisfying the above i).

Now

(17)
$$m\left[\int_R P(s, x, s+m^{-1}, dy)f(y)-f(x)\right]$$

$$=\int_R\left\{\left[f(y)-f(x)-\frac{\rho(x, y)}{1+d(y, x)^2}\sum_{j=1}^{n}(y_j-x_j)\frac{\partial f}{\partial x_j}\right]\frac{1+d(y, x)^2}{d(y, x)^2}\right\}G_m(s, x, dy)$$

$$+\int_R\frac{\rho(x, y)}{d(y, x)^2}\sum_{j=1}^{n}(y_j-x_j)\frac{f}{\partial x_j}G_m(s, x, dy)).$$

We have, for sufficiently small $d(y, x)$,

$$\{\ \}=\sum_{j=1}^{n}(y_j-x_j)\frac{\partial f}{\partial x_j}+\sum_{k, j=1}^{n}(y_j-x_j)(y_k-x_k)\left(\frac{\partial^2 f}{\partial X_j\partial X_k}\right)\frac{1+d(y, x)^2}{d(y, x)^2}$$

where $X_j=x_j+\theta(y_j-x_j)$, $0<\theta<1$. Thus $\{\ \}$ is bounded and continuous in y. Hence, by (16) the first term on the right side of (17) tends, as $m'\to\infty$, to $\displaystyle\int_R\{\ \}G(s, x, dy)$. Therefore, by (10),

(18) a finite $\displaystyle\lim\int_R\frac{\rho(y, x)}{d(y, x)^2}\sum_{j=1}^{n}(y_j-x_j)\frac{\partial f}{\partial x_j}G_{m'}(s, x, dy)=\sum_{j=1}^{n}a_j(s, x)\frac{\partial f}{\partial x_j}$

exists and hence we have (12), by taking

(19) $\displaystyle b_{jk}(s, x)=\lim_{\varepsilon\to 0}\ \lim_{m'\to\infty}\int_{d(y,x)\leq\varepsilon} m'(y_j-x_j)(y_k-x_k)P(s, x, s+m'^{-1}, dy).$

Stochastic processes built from flows

Proc. Japan Acad. **26** No. 8 (1950) 1–3

(Comm. by T. TAKAGI, M.J.A., Oct. 12, 1950.)

By virtue of the theory of semi-groups due to E. Hille[1] and the present author[2], we may construct stochastic processes in a separable measure space R from flows in R.

1. A flow in R is a one-parameter group $\{F_t\}$ of equi-measure transformations in R which is continuous in the sense that $f(F_t \cdot x)$, $f(x) \in L_p(R)$ $(1 \leq p < \infty)$, is strongly continuous in t. Thus the flow $\{F_t\}$ induces a one-parameter group $\{T_t\}$ of linear operators in $L_p(R)$:

$$(1.1) \qquad (T_t f)\ (x) = f(F_t \cdot x),\ f \in L_p(R),$$

$$(1.2) \qquad T_t T_s = T_{t+s},\ T_0 = I \quad \text{(the identity)},$$

$$(1.3) \qquad \text{strong} \lim_{t \to t_0} T_t f = T_{t_0} f.$$

Each T_t is a transition operator in $L_p(R)$:

$$(1.4) \quad f(x) \geq 0 \text{ implies } (T_t f)\ (x) \geq 0 \text{ and } \int_R f(x)\, dx = \int_R (T_t f)\ (x)\, dx.$$

By the semi-group theory, $\{T_t\}$ admits infinitesimal generator A:

$$(1.5) \quad \begin{cases} Af = \text{strong} \lim_{t \downarrow 0} \dfrac{T_t - I}{t} f \text{ for those } f \text{ which are dense in } L_p(R), \\[2mm] T_t f = \exp(tA)f = \text{strong} \lim_{n \to \infty} \exp\left(nt[(I - n^{-1}A)^{-1} - I]\right) f,\ -\infty < t < \infty. \end{cases}$$

Since $(I - n^{-1}A)^{-1}$ exists as a transition operator for $n \geq 0$[3],

$$(1.6) \qquad (I - n^{-1}A^2)^{-1} = (I - \sqrt{n^{-1}}A)^{-1}\,(I + \sqrt{n^{-1}}A)^{-1}$$

exists as a transition operator. Hence A^2 is the infinitesimal generator of a one-parameter semi-group $\{S_t\}$ of transition operators:

$$(1.7) \qquad S_t f = \exp(tA^2)f,\ 0 \leq t < \infty.$$

Thus the Fokker-Planck's equation in a Riemannian space R:

---

* The following result was, under somewhat more restricted conditions and without proof, reported in May 1950 to the Conference in Probability of the International Congress of Mathematicians. Similar result with an interesting formulation was also obtained by Dr. Kiyosi Itô, by virtue of his theory of stochastic differential equations. See his paper in the same issue of this Proceedings.

1) Functional Analysis and Semi-groups, New York (1948).

2) On the differentiability and the representation of one-parameter semi-group of linear operators, J. Math. Soc. Japan, 1 (1948).

3) Since $\{T_t\}$ is a group (not only a semi-group).

510

$$(1.8) \qquad \partial f(t,\ x)/\partial t = A^2 f(t,\ x),\ f(o,\ x) = f(x) \in L_1(R),\ t \geqq 0$$

is integrable stochastically if

$$(1.9) \qquad A = p^i(x)\frac{\partial}{\partial x^i}$$

is the infinitesimal transformation of a one-parameter Lie group of equi-measure transformations in R.

    2. Let a Riemannian space R admit several flows:

$$(2.1) \qquad \exp\ (tA_k),\ -\infty < t < \infty,\ (k = 1,\ 2,\ldots,\ m),$$

$$A_k = p^{ki}(x)\frac{\partial}{\partial x^i}\ .$$

Then, if the matrix (h^{ij}) is symmetric and positive definite, the operator

$$(2.2) \qquad C = h^{ij}\, A_i A_j$$

is the infinitesimal generator of a one-parameter semi-group o transition operators:

$$(2.3) \qquad \exp\ (tC),\ t \geq 0.$$

    Proof. C is, as an operator in $L_2(R)$, symmetric and negativ definite :

$$(2.4) \qquad \begin{cases} (Cf,\ g) = (f,\ Cg)\ \text{and}\ (Cf,\ f) \leq 0\ \text{for twice (continuously} \\ \text{differentiable functions}\ f,g\ \text{which are} \equiv 0\ \text{outside a compac} \\ \text{set of}\ R.^{4)} \end{cases}$$

Hence[5] C admits self-adjoint extension $\tilde{C}$ which is also negativ definite. Thus, for $n > 0$, $(I - n^{-1}\tilde{C})^{-1}$ exists as linear operator i $L_2(R)$ of norm ≤ 1. Therefore, for any $g \in L_1(R) \cap L_2(R)$, there ex ists uniquely determined $f \in L_2(R)$ such that

$$(2.5) \qquad (I - n^{-1}\tilde{C})\, f = g.$$

Let a compact set R_1 be so chosen that

$$(2.6) \qquad \int_{R - R_1} |\, g(x)\, |\, dx < \varepsilon.$$

Then we may find $f_\varepsilon(x)$ with the properties :

$$(2.7) \qquad f_\varepsilon(x) \equiv 0\ \text{for}\ x \in R - R_1 ,$$

$$\int_R |\, g(x) - g_\varepsilon(x)\, |\ dx < 2\varepsilon\ \text{where}\ g_\varepsilon = (I - n^{-1}C)f_\varepsilon .$$

This we see from (2.6) and the vanishing of $g_\varepsilon(x)$ in $R - R_1$. Therefor

---

4) Since the divergences of the vectors $(p^{k1}\,(x),\ p^{k2}\,(x),\ldots,\ p^{km}(x)\,)$ vanish.

5) Cf. II. Freudenthal: Über die Friedrichssche Fortsetzung halbbesch änkter Hermitescher Operatoren, **Proc. Amsterdam Acad.**, **39** (1936).

(2.8) $\quad \begin{cases} \text{the range } \{(I-n^{-1}C)f \,; \, f \in L_1(R) \cap (\text{the domain of } C)\} \text{ is dense} \\ \text{in } L_1(R). \end{cases}$

On the other hand C is of the form

$$(2.9) \qquad C = b^{ij}(x)\frac{\partial^2}{\partial x^i\,\partial x^j} + a^i(x)\frac{\partial}{\partial x^i}\,,$$

where $b^{ij}(x)\xi_i\xi_j \geq 0$. Hence we have, from

(2.10) $\qquad (I-n^{-1}C)f_\varepsilon = g_\varepsilon \; (f_\varepsilon = g_\varepsilon = 0 \text{ outside the compact set } R_1)$,

(2.11) $\qquad \min f_\varepsilon(x) = f_\varepsilon(x_0) \geq g_\varepsilon(x_0)$

and

$$(2.12) \qquad \int_R f_\varepsilon(x)\,dx = \int_R f_\varepsilon(x)\,dx - n^{-1}\int_R (Cf_\varepsilon)(x)\,dx = \int_R g_\varepsilon(x)\,dx^{7)}\,.$$

From (2.10), (2.11) add (2.12) we see that, if $n > 0$, $(I-n^{-1}\bar C)^{-1}$ exists as a transition operator in $L_1(R)^{7)}$. Here $\bar C$ denotes a closed extension of C. Thus $\exp(t\bar C)$ is, for $t \geq 0$, a one-parameter semi-group of transition operators.

3. Let, in particular, the group G of motions of R be a compact semi-simple Lie group with the infinitesimal transformations $X_1, X_2,\ldots, X_m$ transitive on R. Then the so-called Casimir operator

(3.1) $\quad C = h^{ij}X_iX_j$, where $(h^{ij}) = (h_{ij})^{-1}$, $h_{ij} = c_{i\rho}^{\sigma} c_{j\sigma}^{\rho}$, $[X_i, X_j] = c_{ij}^k X_k$

is commutative with every X_i. By the compactness of G, (h^{ij}) is a (truly) positive definite symmetric matrix. Therefore, by the result in 2, C is the infinitesimal geperator of a one-parameter semi-group $\{\exp(tC)\}$, $t \geq 0$, of transition operators. Moreover, by the commutativity

(3.2) $\qquad [C, X_i] = 0 \qquad (i = 1, 2,\ldots, m)$,

$\{\exp(tC)\}$ defines a temporally and spatially homogeneous " continuous " stochastic process in $R^{8)}$ ——a Brownian motion in the homogeneous space R.

---

6) By the same reason as stated in 4).

7) Cf. K. Yosida : Integration of Fokker-Planck's equation in a compact Riemannian space, Arkiv för Matematik, 1, No. 2 (1949).

8) Cf. K. Yosida : Brownian motion on the surface of the 3-sphere, Ann. of Math. Statistics, 20, No. 2 (1949).

On Brownian motion in a homogeneous Riemannian space

Pacific J. Math. **2** (1952) 263–270

1. **Introduction.** Let R be an n-dimensional, orientable, infinitely differentiable Riemannian space such that the group G of isometric transformations S^* of R onto R constitutes a Lie group transitive on R. Consider a temporally homogeneous Markoff process in R, and let $P(t, x, E)$ be the transition probability that the point $x \in R$ is, by this process, transferred into a Borel set $E \subseteq R$ after the lapse of t units of time, $t > 0$. We assume that $P(t, x, E)$ is, for fixed (t, x), countably additive for Borel sets E and, for fixed (t, E), Borel measurable in x. Then we must have the probability condition

$$(1.1) \qquad P(t, x, E) \geq 0, \; P(t, x, R) = 1,$$

and Smoluchowski's equation

$$(1.2) \qquad P(t + s, x, E) = \int_R P(t, x, dy) \, P(s, y, E) \qquad (t, s > 0).$$

We further assume that the process is spatially homogeneous:

$$(1.3) \qquad P(t, x, E) = P(t, S^*x, S^*E) \qquad \text{for every } S^* \in G.$$

The purpose of the present note is to prove the following:

THEOREM 1. *Let x_0 be any point of R and assume that the Lie subgroup $\{S^* \in G; S^*x_0 = x_0\}$ of G is compact* [1]. *Let us denote by $d(x, y)$ the distance of two points $x, y \in R$. Then the continuity condition:*

$$(1.4) \qquad \lim_{t \to 0+} \frac{1}{t} \int_{d(x, y) > \epsilon} P(t, x, dy) = 0 \qquad \text{for any} \quad \epsilon > 0,$$

implies the condition of Lindeberg's type:

---

[1] At first, this condition was overlooked. Mr. Seizô Itô kindly remarked that this condition is necessary for the convergence of the integral (2.11) below.

Received August 30, 1951.

Pacific J. Math. **2** (1952), 263–270

$$(1.5) \qquad \overline{\lim_{t \to 0+}} \ \frac{1}{t} \ \int_R \frac{d(x, y)^2}{1 + d(x, y)^2} \ P(t, x, dy) < \infty.$$

From this theorem we may deduce:

THEOREM 2. *The finite limits* $(x = (x^1, x^2, \cdots, x^n))$

$$(1.6) \qquad a^i(x) = \lim_{t \to 0+} \frac{1}{t} \int_{d(x, y) \leq \epsilon} (y^i - x^i) \, P(t, x, dy),$$

$$(1.7) \qquad b^{ij}(x) = \lim_{t \to 0+} \frac{1}{t} \int_{d(x, y) \leq \epsilon} (y^i - x^i)(y^j - x^j) \, P(t, x, dy)$$

exist, independently of the sufficiently small $\epsilon > 0$. *Moreover, if a real-valued function* $f_0(x)$ *be such that* $f_0(x)$, $\partial f_0 / \partial x^i$, $\partial^2 f_0 / \partial x^i \partial x^j$ *are bounded and uniformly continuous on* R, *then*

$$(1.8) \qquad \lim_{t \to 0+} \frac{1}{t} \left(\int_R f_0(y) \, P(t, x, dy) - f_0(x) \right) = a^i(x) \frac{\partial f_0}{\partial x^i} + b^{ij}(x) \frac{\partial^2 f_0}{\partial x^i \partial x^j}$$

REMARK. In the literature [4; 2; 6], (1.8) is derived by assuming the condition of Lindeberg's type:

$$(1.5)' \qquad \lim_{t \to 0+} \left(\int_R d(x, y)^3 \, P(t, x, dy) / \int_R d(x, y)^2 \, P(t, x, dy) \right) = 0$$

and some differentiability hypothesis concerning $P(t, x, E)$. Considering the Brownian motion on the real line, Seizô Itô raised the question whether, under the condition of the spatial homogeneity (1.3), "the (almost sure) continuity of the sample motions of the temporally homogeneous Markoff process" which is equivalent to the continuity condition (1.4), would be sufficient to derive Theorem 2. And he proved Theorem 2 in the special case where $R = G$ and G is a maximally almost periodic Lie group. The present note gives an extension of his result to general homogeneous space, without the hypothesis of the maximal almost periodicity of the group G of motions of the space R. Thus we may define the Brownian motions in a homogeneous Riemannian space R as temporally and spatially homogeneous Markoff processes satisfying the condition (1.4) of continuity.

2. **Preliminaries.** Let us denote by $C(R)$ the totality of real-valued bounded functions $f(x)$ on R which are uniformly continuous on R. The space $C(R)$ is a Banach space by the norm

$$(2.1) \qquad ||f|| = \sup_x |f(x)|.$$

We define, for any $f \in C(R)$,

$$(2.2) \qquad (T_t f)(x) = \int_R P(t, x, dy) f(y);$$

then we have, by (1.1),

$$(2.3) \qquad \sup_x |(T_t f)(x)| \leq \sup_x |f(x)|.$$

We have, by (1.3),

$$(2.3) \quad (T_t f)(S^* x) = \int_R P(t, S^* x, dy) f(y) = \int_R P[t, S^* x, d(S^* y)] f(S^* y)$$

$$= \int_R P(t, x, dy) f(S^* y),$$

and hence the commutativity

$$(2.4) \qquad T_t S = S T_t,$$

where S is defined by

$$(2.5) \qquad (Sf)(x) = f(S^* x), \qquad\qquad S^* \in G.$$

Thus, if $S^* \in G$ be such that $S^* x = x'$, we have

$$(2.6) \qquad (T_t f)(x) - (T_t f)(x') = (T_t f)(x) - (ST_t f)(x) = T_t(f - Sf)(x).$$

By the uniform continuity of $f(x)$, and by (2.3) and (2.6), we see that $(T_t f)(x)$ is bounded and uniformly continuous in x. Hence T_t defines a bounded linear transformation on $C(R)$ into $C(R)$ such that

$$(2.7) \qquad ||T_t|| = \sup_{||f|| = 1} ||T_t f|| = 1.$$

We have, from (1.2),

$$(2.8) \qquad T_{t+s} = T_t T_s \qquad\qquad (t, s > 0).$$

We have also, from (1.1),

$$(T_t f)(x) - f(x) = \int_R P(t, x, dy)[f(y) - f(x)]$$

$$= \int_{d(x, y) \leq \epsilon} P(t, x, dy)[f(y) - f(x)] + \int_{d(x, y) \geq \epsilon} P(t, x, dy)[f(y) - f(x)].$$

Thus, in view of conditions (1.4) and (1.1), and the uniform continuity of $f(x)$, we have

$$(2.9) \qquad \lim_{t \to 0+} (T_t f)(x) = f(x) \qquad \text{boundedly in } x.$$

Hence T_t is weakly continuous in t, and therefore, by (2.8) and N. Dunford's theorem [1], T_t is strongly continuous in t and

$$(2.9)' \qquad \text{strong } \lim_{t \to 0+} T_t f = f; \text{ that is, } \lim_{t \to 0+} || T_t f - f || = 0 .$$

Therefore we may apply the theory [3; 5] of one-parameter semigroups of bounded linear operators to the semigroup $\{T_t\}$. In particular, we have the result:

(2.10)   strong $\lim_{t \to 0+} (T_t f - f)/t = A f$ exists, for those f which constitute a linear subset $D(A)$ of $C(R)$ which is dense in $C(R)$. Moreover, A is a closed linear operator defined on $D(A) \subseteq C(R)$ with values in $C(R)$.

LEMMA. *Let $g(x) \in C(R)$ vanish outside a compact set. Then the convolution*

$$(2.11) \qquad (f \otimes g)(x) = \int_G f(S_y^* x) \, g(S_y^* x_0) \, dy$$

belongs to $D(A)$ if f belongs to $D(A)$. Here S_y^* is a general element of G, dy is a right invariant Haar measure of G, and x_0 is any fixed point of R.

Proof. The integral may be approximated by the Riemann sum

$$(2.12) \qquad \sum_{i=1}^{m} f(S_{y_i}^* x) \, c_i$$

uniformly in x. This we see by the uniform continuity of $f(x)$ and the fact that $g(x)$ vanishes outside a compact set. We know, from (2.4), that A is commutative with every S_y:

$$(2.13) \qquad f \in D(A) \text{ implies } S_y f \in D(A) \text{ and } S_y A f = A S_y f .$$

Hence (2.12) belongs to $D(A)$, and we have

$$(2.14) \qquad A \left(\sum_{i=1}^{m} f(S_{y_i}^* x) \, c_i \right) = A \left(\sum_{i=1}^{m} (S_{y_i} f)(x) \, c_i \right) = \sum_{i=1}^{m} (S_{y_i} h)(x) \, c_i ,$$

where $h = Af$. Therefore, since $h \in C(R)$, we see that (2.14) converges, when

$m \longrightarrow \infty$, to a function $\in C(R)$ uniformly in x. Since A is a closed operator, we must have $(f \otimes g)(x) \in D(A)$.

COROLLARY 1. *The convolution $(f \otimes g)(x)$ is infinitely differentiable if $g(x)$ is infinitely differentiable.*

Proof. It is possible, for sufficiently small $d(x, x_0)$, to choose $S^*(x) \in G$ such that

$$(2.15) \qquad S^*(x)x = x_0 \text{ and } S^*(x)x_0 \text{ depends analytically on } x^1, \cdots, x^n.$$

This we see from the fact that the set $\{S_y^* \in G; \ S_y^* x = x_0\}$ forms an analytic submanifold of G; it is one of the cosets of G with respect to the Lie subgroup $\{S_y^* \in G; \ S_y^* x_0 = x_0\}$. Hence, by the right invariance of dy, we have

$$(2.16) \qquad (f \otimes g)(x) = \int_G f(S_y^* S^*(x)x)\,(g(S_y^* S^*(x)x_0)\,dy$$

$$= \int_G f(S_y^* x_0)\,g(S_y^* S^*(x)x_0)\,dy.$$

The right side is infinitely differentiable in the vicinity of x_0, and

$$(2.17) \qquad \frac{\partial^{q_1 + \cdots + q_n} (f \otimes g)(x)}{\partial(x^1)^{q_1} \cdots \partial(x^n)^{q_n}}$$

$$= \int_G f(S_y^* x_0)\,\frac{\partial^{q_1 + \cdots + q_n} g(S_y^* S^*(x)x_0)}{\partial(x^1)^{q_1} \cdots \partial(x^n)^{q_n}}\,dy$$

belongs to $C(R)$.

COROLLARY 2. (i) *There exist infinitely differentiable functions $F^1(x)$, $F^2(x), \cdots, F^n(x) \in D(A)$ such that the Jacobian*

$$(2.18) \qquad \frac{\partial(F^1(x), \cdots, F^n(x))}{\partial(x^1, \cdots, x^n)} \quad \text{does not vanish at } x = x_0.$$

(ii) *There exists an infinitely differentiable function $F_0(x) \in D(A)$ such that*

$$(2.19) \qquad (x^i - x_0^i)(x^j - x_0^j)\,\frac{\partial^2 F}{\partial x_0^i\,\partial x_0^j} \geq \sum_{i=1}^{n} (x^i - x_0^i)^2.$$

Proof. In (2.16), f belongs to $D(A)$, which is dense in $C(R)$; and $g(x) \in C(R)$

is arbitrary except that $g(x)$ must vanish outside a compact set. Thus, by taking $F(x) = (f \otimes g)(x)$ suitably, we may prove (i) and (ii).

3. **Proof of Theorem 1.** Because of their functional independence, we may take $F^1(x), \cdots, F^n(x)$ as local coordinates of the points x which satisfy $d(x, x_0) < \epsilon$ for sufficiently small $\epsilon > 0$. Since $F^i(x) \in D(A)$,

$$(3.1) \quad \text{a finite } \lim_{t \to 0+} \frac{1}{t} \int_R (F^i(x) - F^i(x_0)) P(t, x_0, dx) \text{ exists} \quad (i = 1, \cdots, n).$$

Because of (1.4), this limit is equal to

$$(3.1)' \qquad \lim_{t \to 0+} \frac{1}{t} \int_{d(x, x_0) \leq \epsilon} (F^i(x) - F^i(x_0)) P(t, x, dx),$$

independently of the positive constant ϵ. We shall denote these new local coordinates $F^1(x), F^2(x), \cdots, F^n(x)$ by the letters $x^1, x^2, \cdots, x^n$. Then

$$(3.1)'' \qquad \lim_{t \to 0+} \frac{1}{t} \int_{d(x, x_0) \leq \epsilon} (x^i - x_0^i) P(t, x, dx) = a^i(x_0) \text{ exists}$$

$$(i = 1, \cdots, n),$$

independently of $\epsilon > 0$. The function $F_0(x)$ belongs to $D(A)$; hence, by (1.4),

$$(3.2) \quad (AF_0)(x_0) = \lim_{t \to 0+} \frac{1}{t} \int_{d(x, x_0) \leq \epsilon} (F_0(x) - F_0(x_0)) P(t, x_0, dx),$$

independently of $\epsilon > 0$. This limit is equal to

$$\lim_{t \to 0+} \left[\frac{1}{t} \int_{d(x, x_0) \leq \epsilon} (x^i - x_0^i) \frac{\partial F_0}{\partial x_0^i} P(t, x_0, dx) \right.$$

$$\left. + \frac{1}{t} \int_{d(x, x_0) \leq \epsilon} (x^i - x_0^i)(x^j - x_0^j) \left(\frac{\partial^2 F_0}{\partial x^i \partial x^j} \right)_{x = x_0 + \theta(x - x_0)} P(t, x_0, dx), \right]$$

$$0 < \theta < 1.$$

The first term in [ ] has the limit

$$a^i(x_0) \frac{\partial F_0}{\partial x_0^i},$$

and hence the second term has a limit. Thus, by virtue of (1.1) and (2.19),

$$(3.3) \qquad \overline{\lim_{t \to 0+}} \frac{1}{t} \int_{d(x,\, x_0) \leq \epsilon} \sum_{i=1}^{n} (x^i - x_0^i)^2 \, P(t, x_0, dx) < \infty.$$

Hence, by (1.1) and Schwarz's inequality,

$$(3.4) \qquad \frac{1}{t} \int_{d(x,\, x_0) \leq \epsilon} (x^i - x_0^i)(x^j - x_0^j) \, P(t, x_0, dx) \text{ is bounded in } t > 0.$$

Therefore, by (1.4), we obtain (1.5).

4. Proof of Theorem 2. Since $c_{ij}(\epsilon)$ is of order ϵ, we have

$$(4.1) \qquad \frac{(T_t f_0)(x_0) - f_0(x_0)}{t} = \frac{1}{t} \int_{d(x,\, x_0) \leq \epsilon} (x^i - x_0^i) \, P(t, x_0, dx) \, \frac{\partial f_0}{\partial x_0^i}$$

$$+ \frac{1}{t} \int_{d(x,\, x_0) \leq \epsilon} (x^i - x_0^i)(x^j - x_0^j) \, P(t, x_0, dx) \, \frac{\partial^2 f_0}{\partial x_0^i \, \partial x_0^j}$$

$$+ \frac{1}{t} \int_{d(x,\, x_0) \leq \epsilon} (x^i - x_0^i)(x^j - x_0^j) \, c_{ij}(\epsilon) \, P(t, x_0, dx)$$

$$+ \frac{1}{t} \int_{d(x,\, x_0) \geq \epsilon} (f_0(x) - f_0(x_0)) \, P(t, x_0, dx)$$

$$= I_1(t, \epsilon) + I_2(t, \epsilon) + I_3(t, \epsilon) + I_4(t, \epsilon).$$

Now

$$(4.2) \qquad \lim_{t \to 0+} I_1(t, \epsilon) = a^i(x_0) \, \frac{\partial f_0}{\partial x_0^i} \text{ by } (3.1)''; \quad \lim_{\epsilon \to 0+} I_3(t, \epsilon) = 0 \text{ by } (3.4);$$

$$\lim_{t \to 0+} I_4(t, \epsilon) = 0 \text{ by } (1.4).$$

On the other hand, by (1.4) and (3.4), the finite limits

$$\overline{\lim_{t \to 0+}} \frac{1}{t} \int_{d(x,\, x_0) \leq \epsilon} (x^i - x_0^i)(x^j - x_0^j) \, P(t, x_0, dx) = b_1^{ij}(x_0),$$

$$\underline{\lim_{t \to 0+}} \frac{1}{t} \int_{d(x,\, x_0) \leq \epsilon} (x^i - x_0^i)(x^j - x_0^j) \, P(t, x_0, dx) = b_2^{ij}(x_0)$$

exist and are independent of $\epsilon > 0$. Let us, in place of $f_0(x)$, take $F_0(x)$ of the form $(f \otimes g)(x)$. We may choose $F_0(x)$ such that $\partial F_0/\partial x_0^i\, \partial x_0^j$ assumes values arbitrarily near to given constants $\alpha_{ij}\,(i, j = 1, \cdots, n)$. Thus, by (4.1) and (4.2) and the fact $F_0(x) \in D(A)$, we see that $b_1^{ij}(x_0)$ must be equal to $b_2^{ij}(x_0)$. Hence (1.7) is proved.

Therefore, by (1.4), (3.1)$''$, and (4.2), we obtain (1.8).

References

1. Nelson Dunford, *On one-parameter groups of linear transformations*, Ann. of Math. 39 (1938), 569-573.

2. W. Feller, *Zur Theorie der stochastischen Prozesse*, Math. Ann. **113** (1936), 113-160.

3. E. Hille, *Functional analysis and semi-groups*, American Mathematical Society, New York, 1948.

4. A. Kolmogoroff, *Zur Theorie der stetigen zufälligen Prozesse*, Math. Ann. 108 (1933), 149-160.

5. K. Yosida, *On the differentiability and the representation of one-parameter semi-groups of linear operators*, J. Math. Soc. Japan 1 (1948), 15-21.

6. ———, *An extension of Fokker-Planck's equation*, Proc. Japan Acad. 25 (1949), No. 9, 1-3.

MATHEMATICAL INSTITUTE,

NAGOYA UNIVERSITY.

On the generating parametrix of the stochastic processes

Proc. Nat. Acad. Sci. U.S.A. **41** (1955) 240–244

Communicated by Marston Morse, January 24, 1955

1. Let X be a locally compact space, and let $\mathfrak{B}$ be the Borel field generated from the open sets of X. A temporally homogeneous discrete Markoff process (d.M.p.) in X is defined by the transition probability $G(x, E)$ satisfying the conditions

$$G(x, E) \geqq 0, \ G(x, X) = 1, \tag{1.1}$$

$$G(x, E) \text{ is countably additive in } E \ \epsilon \ \mathfrak{B} \text{ and } \mathfrak{B}\text{-measurable in } x. \tag{1.2}$$

A temporally homogeneous Markoff process (M.p.) in X is defined by a one-parameter family of transition probabilities $P(t, x, E)$, $t > 0$, satisfying the conditions

$$P(t, x, E) \geqq 0, \ P(t, x, X) = 1, \tag{1.1'}$$

$$P(t, x, E) \text{ is countably additive in } E \ \epsilon \ \mathfrak{B} \text{ and } \mathfrak{B}\text{-measurable in } x, \tag{1.2'}$$

$$P(t + s, x, E) = \int_X P(t, x, dy) P(s, y, E). \tag{1.3}$$

We assume[1] the "continuity condition"

$$\lim_{t \downarrow 0} P(t, x, X - V(x)) = 0 \text{ for any vicinity } V(x) \text{ of } x. \tag{1.4}$$

Hence we see, by (1.3), that $P(t, x, E)$ is measurable in t. The integral

$$G(x, E) = \int_0^\infty \exp(-t) P(t, x, E) \, dt \tag{1.5}$$

defines a d.M.p.

The purpose of the present note is to show that (see, for the exact formulation, the theorem in sec. 2) a certain class $\mathfrak{P}$ of $P(t, x, E)$ is related in one-one manner with a certain class $\mathfrak{G}$ of $G(x, E)$ in such a way that

$$\mathfrak{G} \ni G(x, E) = \int_0^\infty \exp(-t) P(t, x, E) \, dt \quad \text{if } P(t, x, E) \text{ is } \epsilon \ \mathfrak{P}, \tag{1.6}$$

$$\mathfrak{P} \ni P(t, x, E) = \lim_{\alpha \downarrow 0} \exp\left(t(G(x, E) - \delta(x, E))/(\alpha\delta(x, E) + \right.$$

$$(1 - \alpha)G(x, E))) \text{ if } G(x, E) \text{ is } \epsilon \ \mathfrak{G} \text{ and if } \delta(x, E) \text{ denotes the}$$
$$\text{Dirac measure:} \tag{1.7}$$

$$\delta(x, E) = 1 \text{ or } 0 \text{ according as } x \ \epsilon \ E \text{ or not.} \tag{1.8}$$

We shall call $G(x, E)$ of the class $\mathfrak{G}$ as "the generating parametrix" of the M.p. $P(t, x, E)$. The construction and the properties of the M. processes $\epsilon \ \mathfrak{P}$ are thus reduced to those of the d.M. processes $\epsilon\mathfrak{G}$. As an application, it will be shown in section 3 that we may decompose X (under a M.p. $P(t, x, E)$) into ergodic parts and dissipative part.

2. Let $C(X)$ be a Banach space consisting of a certain linear set of bounded continuous real-valued functions $f(x)$ with the norm

$$\|f\| = \sup_x |f(x)|, \tag{2.1}$$

such that $C(X)$ contains all the continuous functions whose carriers are compact and are contained in the interior of X.

Let $\mathfrak{P}$ be the class of M. processes $P(t, x, E)$ in X satisfying $(1.1)'$, (1.4) and the condition

$(P_t f) (x) = \int_X P(t, x, dy)f(y)$ defines a linear operator on $C(X)$ into $C(X)$. $\tag{2.2}$

Let $\mathfrak{G}$ be the class of d.M. processes $G(x, E)$ in X satisfying (1.1), (1.2) and the conditions

$(Gf) (x) = \int_X G(x, dy)f(y)$ defines a one-one mapping of $C(X)$ onto a strongly dense linear subspace of $C(X)$; $\tag{2.3}$

for any $\alpha, 0 < \alpha < 1$, the linear operator $\alpha I + (1 - \alpha)G$ ($I =$ the identity operator) defines a one-one mapping of $C(X)$ onto $C(X)$ such that the linear operator $G_\alpha = G(\alpha I + (1 - \alpha)G)^{-1}$ on $C(X)$ into $C(X)$ is defined by a d.M.p. $G_\alpha(x, E)$ through $(G_\alpha f) (x) = \int_X G_\alpha(x, dy)f(y)$. $\tag{2.4}$

Then we have the

THEOREM. For any $P(t, x, E) \epsilon \mathfrak{P}$, $G(x, E) = \int_0^\infty \exp(-t)P(t, x, E)\, dt$ belongs to $\mathfrak{G}$. Conversely, any $G(x, E) \epsilon \mathfrak{G}$ defines $P(t, x, E) \epsilon \mathfrak{P}$ by the linear operator

$P_t f = \text{strong} \lim_{\alpha \downarrow 0} \exp(t(G - I)(\alpha I + (1 - \alpha)G)^{-1})$ (uniformly in any bounded interval of t) $\tag{2.5}$

in such a way that

$(P_t f) (x) = \int_X P(t, x, dy)f(y)$ and $G(x, E) = \int_0^\infty \exp(-t)P(t, x, E)\, dt.$

The proof may be reduced to the theory of semigroup of linear operators[2] through the

LEMMA. Let $\{P_t\}$, $t > 0$, be a semigroup

$$P_t P_s = P_{t+s} \tag{2.6}$$

of bounded linear operators on a Banach space S into S such that

$$\text{weak} \lim_{t \downarrow 0} P_t f = f \text{ for every } f \epsilon S. \tag{2.7}$$

Then we have

$$\text{strong} \lim_{t \to 0} P_t f = f \text{ for every } f \epsilon S. \tag{2.8}$$

Proof: For any $f_0 \in S$, the strong closure $\tilde{S}_0$ of the linear subspace S_0 spanned by $\{P_t f_0; t > 0\}$ is mapped into $\tilde{S}_0$ by every P_t. This we see by (2.6). By the equivalence of the strong closure with the weak closure in the Banach space, we see, by (2.6)–(2.7), that $\tilde{S}_0$ is strongly separable and that $f_0 \in \tilde{S}_0$. P_t is weakly measurable on $\tilde{S}_0$ by (2.7). Also, by (2.7), $\|P_t\|$ is bounded in t. Hence, by a theorem of N. Dunford,[3] we see that P_t is, for $t > 0$, strongly continuous in t on $\tilde{S}_0$. Thus we see, by (2.6)–(2.7), that strong $\lim_{t \to 0} P_t f_0 = f_0$.

Proof of the Theorem: Let $P(t, x, E)$ be $\in \mathfrak{P}$, and let $f \in C(X)$. For any $x \in X$ and for any $\epsilon > 0$, let the vicinity $V(x_0)$ of x_0 be chosen such that $x \in V(x_0)$ implies $|f(x) - f(x_0)| < \epsilon$. Then we have

$$(P_t f)\,(x_0) - f(x_0) = \int_{V(x_0)} P(t, x_0, dx)\,(f(x) - f(x_0)) + \int_{x - V(x_0)} P(t, x_0, dx)\,(f(x) - f(x_0)).$$

Thus, by (1.4), we see that

$$\lim_{t \to 0}\,(P_t f)\,(x_0) = f(x_0) \text{ boundedly in } x_0. \tag{2.9}$$

Hence weak $\lim_{t \downarrow 0} P_t f = f$. We have, by (1.3), (2.6) also. Thus we have, by the lemma,

$$\text{strong } \lim_{t \downarrow 0} P_t f = f \text{ for every } f \in C(X). \tag{2.10}$$

Therefore, we may apply the semigroup theory to $\{P_t\}$. Let A be the infinitesimal generator of P_t, viz., let

$$Af = \text{strong } \lim_{t \downarrow 0} t^{-1}(P_t - I)f \text{ if the right-hand limit exists.} \tag{2.11}$$

Such A is characterized by the following properties:

the domain $D(A)$ of A is strongly dense in $C(X)$, $\tag{2.12}$

for any $\alpha > 0$, the resolvent $G_\alpha = (I - \alpha A)^{-1}$ exists as a bounded
linear operator on $C(X)$ into $C(X)$ such that $\tag{2.13}$

$G_\alpha.1 = 1$ and $(G_\alpha f)\,(x)$ is nonnegative for nonnegative function $f(x)$. $\tag{2.14}$

Moreover, we have

$$G_\alpha = \alpha^{-1} \int_0^\infty \exp\,(-\alpha^{-1}t)P_t\,dt. \tag{2.15}$$

Thus we have

$$AG_\alpha = \alpha^{-1}(G_\alpha - I) \tag{2.16}$$

and hence

$$(I - \alpha A)G_1 = G_1 - \alpha(G_1 - I) = \alpha I + (1 - \alpha)G_1 \tag{2.17}$$

gives a one-one linear mapping of $C(X)$ onto $C(X)$. Hence

$$(\alpha I + (1 - \alpha)G_1)^{-1} = (I - A)G_\alpha,$$

and thus

$$G_\alpha = G \, (\alpha I + (1 - \alpha)G)^{-1}, \text{ where } G = G_1. \tag{2.18}$$

We have thus proved the first half of the theorem. The latter half of the theorem is proved also by the semigroup theory. In fact, the conditions (2.3)–(2.4) imply the existence of the uniquely determined semigroup

$$\begin{aligned}
P_t f &= \text{strong} \lim_{\alpha \downarrow 0} \exp\,(tAG_\alpha) = \text{strong} \lim_{\alpha \downarrow 0} \exp\,(tAG(\alpha I + (1 - \alpha)G)^{-1}) \\
&= \text{strong} \lim_{\alpha \to 0} \exp\,(t(G - I)\,(\alpha I + (1 - \alpha)G)^{-1}
\end{aligned} \tag{2.19}$$

strongly continuous in t with the infinitesimal generator

$$A = I - G^{-1}. \tag{2.21}$$

That this P_t is defined by a M.p. may be seen from

$$\begin{aligned}
\exp\,(t(G - I)\,(\alpha I + (1 - \alpha)G)^{-1}) &= \exp\,(tAG_\alpha) \\
&= \exp\,(t\alpha^{-1}(G_\alpha - I)) \\
&= \exp\,(-\alpha^{-1}t)\,\exp\,(t\alpha^{-1}G_\alpha)
\end{aligned} \tag{2.22}$$

and (2.14). The continuity condition (1.4) of the corresponding $P(t, x, E)$ may be proved by the strong continuity (2.11) of P_t.

Remark: We have, by (2.16),

$$AG = G - I, \text{ where } G = G_1. \tag{2.16$'$}$$

Thus $G(x, E)$ is the "parametrix" of the infinitesimal generator A of P_t. Hence we call $G(x, E)$ as the "generating parametrix" of the Markoff process $P(t, x, E)$.

3. A nonnegative, normalized measure $\varphi(E)$, countably additive for $E \, \epsilon \, \mathfrak{B}$, is called an "invariant measure" of the process $P(t, x, E)$ if

$$\varphi(E) = \int_X \varphi(dx)P(t, x, E) \text{ for every } t > 0 \text{ and for every } E \, \epsilon \, \mathfrak{B}. \tag{3.1}$$

It is easy to see, by the theorem above in section 2, that $\varphi(E)$ is invariant for $P(t, x, E)$ if and only if $\varphi(E)$ is invariant for the corresponding $G(x, E)$:

$$\varphi(E) = \int_X \varphi(dx)G(x, E) \text{ for every } E \, \epsilon \, \mathfrak{B}. \tag{3.2}$$

Let us consider the special case when X is separable and when the operator G maps continuous functions with compact carriers into continuous functions with compact carriers. In another paper[4] the present author proved the following extension of the results due to J. von Neumann,[5] N. Kryloff and N. Bogoliouboff,[6] and M. Beboutoff:[7]

X is decomposable into a dissipative part D and ergodic parts E_λ in such a way that

$$\varphi(D) = 0 \text{ for every invariant measure } \varphi, \tag{3.3}$$

$$G(x, E_\lambda) = 1 \text{ for every } x \, \epsilon \, E_\lambda, \tag{3.4}$$

$G(x, E)$, considered (by (3.4)) as a d.M.p. in E_λ, has one and only one invariant measure φ_λ. (3.5)

It is easy to see, by the theorem in section 2, that the above decomposition of X with respect to the d.M.p. $G(x, E)$ gives also the decomposition of X into a dissipative part and ergodic parts with respect to the corresponding M.p. $P(t, x, E)$ for every t simultaneously.

[1] Recently, W. Feller has published interesting results concerning the general form of the infinitesimal generator of $P(t, x, E)$ for the case, on real line X, where the "Lindeberg's condition" $\lim_{t \downarrow 0} t^{-1} P(t, x, X - V(x)) = 0$ is satisfied. See, for example, his paper "The General Diffusion Operator and Positivity Preserving Semi-groups in One Dimension," *Ann. Math.*, **60**, No. 3, 417–436, 1954. In this connection see also K. Yosida, "On Brownian Motion in a Homogeneous Riemannian Space," *Pacific J. Math.*, **2**, 263–270, 1952.

[2] E. Hille, *Functional Analysis and Semi-groups*, (New York, 1948), and K. Yosida, "On the Differentiability and the Representation of One Parameter Semi-group of Linear Operators," *J. Math. Soc. Japan*, **1**, 15–21, 1948.

[3] "On One-Parameter Groups of Linear Transformations," *Ann. Math.*, **39**, 569–573, 1938.

[4] "Simple Markoff Process with a Locally Compact Phase Space," *Mathematica Japonica*, **1**, No. 3, 1–5, 1948.

[5] "Zur Operatorenmethode in der klassischen Mechanik," *Ann. Math.*, **33**, 587–642, 1932.

[6] "La Théorie générale de la mesure dans son application à l'étude des systèmes dynamiques de la mécanique non linéaire," *Ann. Math.*, **38**, 65–113, 1947.

[7] "Simple Markoff Process with a Compact State Space," *Rec. Math.*, **10**, 22–30, 1942.

A characterization of the second order elliptic differential operators

Proc. Japan Acad. 31 (1955) 406–409

(Comm. by Z. SUETUNA, M.J.A., July 12, 1955)

§ 1. **Introduction and the theorem.** We consider the elliptic differential operator

$$(1.1) \qquad (Ef)(x) = a^{ij}(x)\, \frac{\partial^2 f}{\partial x^i \partial x^j} + b^i(x)\, \frac{\partial f}{\partial x^i}\ (a^{ij}(x)\xi_i\xi_j \geqq 0)$$

as a linear operator from real-valued continuous functions $f(x)$ of the point $x=(x^1,\ldots,x^m)$ of an m-dimensional C^∞ manifold R to real-valued continuous functions $(Ef)(x)$ of the point x. As was stressed by W. Feller,[1] it enjoys two important properties:

(1.2) *The local property* which means that the value $(Ef)(x_0)$ is determined by the values of $f(x)$ in any small neighbourhood of x_0.

(1.3) *The property* (E) which says that if $f(x)$ has a local minimum at x_0 then $(Ef)(x_0) \geqq 0$.

W. Feller[2] has determined, for the case of one-dimensional R, the most general class of operators with these two characteristic properties. According to him such operator E can, under the condition $E \cdot 1 = 0$, be represented by means of repeated differentiation

$$(1.4) \qquad\qquad D_v D_u f$$

with respect to monotone non-decreasing functions u and v (of these two functions, u is a continuous function).

It is desirable to extend Feller's result to the case of higher dimensional R. The purpose of the present note is to give, by refining the method in a preceding note,[3] a partial answer to this problem. Our result gives a characterization of, so to speak, "the smallest closed extension" A of the differential operator E.

Let us formulate the situation precisely. We assume that (i) R *is a homogeneous Riemann space*, viz. R is a C^∞ Riemann space

---

1) W. Feller: The general diffusion operator and positivity preserving semi-groups in one dimension, Ann. Math., **60**, No. 3, 417–436 (1954). W. Feller: On second order differential operators, Ann. Math., **61**, No. 1, 90–105 (1955).

2) See the reference referred to in 1).

3) K. Yosida: On Brownian motion in a homogeneous Riemannian space, Pacific Journ. Math., **2**, No. 2, 263–270 (1952). In this paper, the function space $C(R)$ can be corrected to be the Banach space which is the closure, by the norm defined by the maximum of the absolute value of the function, of the totality of continuous functions on R with compact supports.

such that the (Lie) group G of isometries of R is transitive on R, and we further assume that (ii) *those isometries leaving a definite point x_0 invariant constitute a compact (Lie) subgroup of G.* Let A be a linear operator whose domain and range are real-valued functions on R. Following after W. Feller, we say that *f belongs to the domain of A at x_0,* in symbols, $f \in D(A:x_0)$, if both f and Af are defined and continuous in some neighbourhood of x_0. This domain $D(A:x_0)$ is supposed to be linear, that is, $f_i \in D(A:x_0)$ implies $\sum_i c_i f_i \in D(A:x_0)$. The operator A is assumed to have (iii) *the local property* and (iv) *the property (E) for those f which are contained in the intersection $D(A:x_0) \cap C^\infty((x_0))$.*[4] Let, moreover, (v) $A \cdot 1 = 0$.

We shall say that (vi) *A is locally closed at x_0* if, whenever a sequence $\{f_k(x)\} \subseteq D(A:x_0)$ satisfies

(1.5) $\lim\limits_{k\to\infty} f_k(x) = f_\infty(x),\ \lim\limits_{k\to\infty} (Af_k)(x) = h_\infty(x)$ uniformly in a neighbourhood of x_0, then $f_\infty \in D(A:x_0)$ and $(Af_\infty)(x_0) = h_\infty(x_0)$.

We also say that (vii) *A is regular at x_0* if $D(A:x_0)$ includes functions arbitrarily and uniformly near to any continuous function which is $\neq 0$ at x_0 and 0 outside any small neighbourhood of x_0.

We may state our theorem as follows.

Theorem. Let the conditions (i)–(vii) be satisfied. Let us further assume that $(viii)$ if $f \in D(A:x_0)$ then $f_\alpha(x) = f(T_\alpha x) \in D(A:x_0)$ for T_α in sufficiently small neighbourhood of the identity T_{α_0} of the group G and $(Af_\alpha)(x)$ is continuous in the two variables (x, α) in some neighbourhood of (x_0, α_0). Then $D(A:x_0) \cap C^\infty((x_0))$ includes functions whose successive derivatives at x_0 are arbitrarily near to the corresponding derivatives at x_0 of any function $\in C^\infty((x_0))$, and, moreover, $(Af)(x_0)$ is, for $f \in D(A:x_0) \cap C^\infty((x_0))$, given by

(1.6) $(Af)(x_0) = a^{ij}(x_0)\dfrac{\partial^2 f}{\partial x_0^i \partial x_0^j} + b^i(x_0)\dfrac{\partial f}{\partial x_0^i}$ *with* $a^{ij}(x_0)\xi_i\xi_j \geq 0$.

§ 2. The proof of the theorem

Lemma. Let $f \in D(A:x_0)$ and let a continuous function $g(x)$ vanish outside a sufficiently small neighbourhood of x_0. Then the "convolution"

(2.1) $$(f \otimes g)(x) = \int_G f(T_\alpha x) g(T_\alpha x_0) d\alpha$$

belongs to $D(A:x_0)$. Here $d\alpha$ is a right invariant Haar measure of G.

Proof. The integral may be approximated by the Riemann sum

---

4) $C^\infty((x_0))$ denotes the totality of functions which are C^∞ in some neighbourhood of x_0. Thus $D(A:x_0) \cap C^\infty((x_0))$ might be void except constant functions. However, see the theorem below.

(2.2)
$$\sum_i f(T_{\alpha_i} x) c_i$$

uniformly in x in a sufficiently small neighbourhood of x_0. This we see by the local uniform continuity of f, the condition (ii) and the fact that $g(x)$ vanishes outside a sufficiently small neighbourhood of x_0. By the condition (viii) we see that (2.2) and

$$A\left(\sum_i f_{\alpha_i} c_i\right)(x) = \sum_i (A f_{\alpha_i})(x) c_i$$

both converges uniformly in a sufficiently small neighbourhood of x_0. Thus we see, by (vi), that (2.1) belongs to $D(A : x_0)$.

Corollary. There exist functions $f^1(x), \ldots, f^m(x) \in D(A : x_0) \cap C^\infty((x_0))$ such that the Jacobian

(2.3) $\qquad \partial(f^1(x), \ldots, f^m(x))/\partial(x^1, \ldots, x^m)$ does not vanish at $x = x_0$.

There exists, for any system of m^2 constants α_{ij} with $\alpha_{ij} = \alpha_{ji}$, a function $f_0(x) \in D(A : x_0) \cap C^\infty((x_0))$ such that the values $\partial^2 f_0/\partial x_0^i \partial x_0^j$ are arbitrarily near to the values α_{ij} respectively. In particular, there exists a function $f_0(x) \in D(A : x_0) \cap C^\infty((x_0))$ such that

(2.4) $\qquad (x^i - x_0^i)(x^j - x_0^j)(\partial^2 f_0/\partial x_0^i \partial x_0^j) \geq 2 \sum_i (x^i - x_0^i)$.

Proof. It is possible, for x in a sufficiently small neighbourhood of x_0, to choose $T(x) \in G$ such that

(2.5) $\qquad T(x)x = x_0$ and $T(x)x_0$ is a C^∞ function of $x = (x^1, \ldots, x^m)$.

This we see from the fact that the set $\{T_\alpha \in G : T_\alpha x = x_0\}$ constitutes an analytic submanifold of G; it is one of the coset of G with respect to the Lie subgroup $\{T_\alpha \in G : T_\alpha x_0 = x_0\}$. Hence, by the right invariance of $d\alpha$, we have

(2.6)
$$(f \otimes g)(x) = \int_G f(T_\alpha T(x)x) g(T_\alpha T(x)x_0) d\alpha$$

$$= \int_G f(T_\alpha x_0) g(T_\alpha T(x)x_0) d\alpha.$$

Thus, if g is a C^∞ function, we have

(2.7)
$$\partial^{q_1 + \cdots + q_m}(f \otimes g)(x)/\partial(x^1)^{q_1} \ldots \partial(x^m)^{q_m}$$

$$= \int_G d\alpha f(T_\alpha x_0) \partial^{q_1 + \cdots + q_m} g(T_\alpha T(x)x_0)/\partial(x^1)^{q_1} \ldots \partial(x^m)^{q_m}.$$

Hence, by making use of the regularity of A at x_0, we obtain the corollary. We have only to choose, in $(f \otimes g)(x) \in D(A : x_0)$, f and g appropriately.

Proof of the theorem.[5] We may take, by (2.3), the functions $f^1(x), \ldots, f^m(x)$ from $D(A : x_0)$ as the local coordinates of the point x in a small neighbourhood of x_0. We shall denote these new local coordinates $f^1(x), \ldots, f^m(x)$ by the letters $x^1, \ldots, x^m$. Thus let

(2.8) $\qquad (Ax^i)(x_0) = b^i(x_0) \qquad (i = 1, 2, \ldots, m)$.

The function $f_0(x)$ in the corollary belongs to $D(A : x_0)$. Hence, by the linearity of $D(A : x_0)$, (2.8), and the condition (v), we see that

---

5) Cf. the paper referred to in 3).

$$(2.9) \qquad \hat{f}_0(x) = f_0(x) - f_0(x_0) - (x^i - x_0^i)(\partial f_0/\partial x_0^i)$$
$$= (x^i - x_0^i)(x^j - x^j)(\partial^2 f_0/\partial X^i \partial X^j)_{X = x_0 + 0(x - x_0)}$$

belongs to $D(A : x_0)$.

Let us consider the subadditive[6] functional

$(2.10) \quad \bar{A}(g) = \inf (Af)(x_0)$ for those $f \in D(A : x_0) \cap C^\infty((x_0))$ which satisfies the conditions: $f(x_0) = g(x_0)$ and $f(x) \geq g(x)$ in a neighbourhood of x_0.

It is easy to see, by the property (E) in (iv), that

$(2.11) \quad f \in D(A : x_0) \cap C^\infty((x_0))$ implies $(Af)(x_0) = A(f) = \underline{A}(f)$ where

$$\underline{A}(g) = -\bar{A}(-g).$$

We have, by (iii)–(iv) and (2.4),

$$(2.12) \qquad 0 \leq \bar{A}\left(\sum_i (x^i - x_0^i)^2\right) \leq (A\hat{f}_0)(x_0) = (Af_0)(x_0) - b^i(x_0)(\partial f_0/\partial x_0^i).$$

When $\partial^2 f_0/\partial x_0^i \partial x_0^j \neq 0$, we have

$$(2.13) \qquad (\partial^2 f_0/\partial X^i \partial X^j)_{X = x_0 + 0(x - x_0)} = (1 + \varepsilon_{ij}(x))(\partial^2 f_0/\partial x_0^i \partial x_0^j)$$
$$\text{where } \lim_{x \to x_0} \varepsilon_{ij}(x) = 0.$$

Let $\varepsilon > 0$. Then, by (2.13) and (iii)–(iv),

$$\bar{A}((x^i - x_0^i)(x^j - x_0^j)(\partial^2 f_0/\partial x_0^i \partial x_0^j)) - \varepsilon \bar{A}\left(\sum_i (x^i - x_0^i)^2\right)$$
$$\leq (A\hat{f}_0)(x_0)$$
$$\leq \bar{A}((x^i - x_0^i)(x^j - x_0^j)(\partial^2 f_0/\partial x_0^i \partial x_0^j)) + \varepsilon \bar{A}\left(\sum_i (x^i - x_0^i)^2\right).$$

Hence, by (2.12), we see that

$$(2.14) \qquad (A\hat{f}_0)(x_0) = \bar{A}((x^i - x_0^i)(x^j - x_0^j)(\partial^2 f_0/\partial x_0^i \partial x_0^j)).$$

Similarly we obtain

$$(2.14)' \qquad (A\hat{f}_0)(x_0) = \underline{A}((x^i - x_0^i)(x^j - x_0^j)(\partial^2 f_0/\partial x_0^i \partial x_0^j)).$$

We may choose, by the corollary, $f_0 \in D(A : x_0)$ in such a way that $\partial^2 f_0/\partial x_0^i \partial x_0^j$ are arbitrarily near to the given constants $\alpha_{ij}(= \alpha_{ji})$ respectively. Thus we see, by the similar argument by which we have obtained (2.14) and (2.14)', that every quadratic form $\alpha_{ij}(x^i - x_0^i)(x^j - x_0^j)$ belongs to the domain of the linear functional $\tilde{A}$ defined by

$(2.15) \quad \tilde{A}(g) = \bar{A}(g) = \underline{A}(g)$ if the latter two values are equal.

We have thus

$$(2.16) \qquad \tilde{A}(\alpha_{ij}(x^i - x_0^i)(x^j - x_0^j)) = \alpha_{ij}\tilde{A}((x^i - x_0^i)(x^j - x_0^j)).$$

By repeating again the similar arguments, we see that any $f \in C^\infty((x_0))$ belongs to the domain of the linear functional $\tilde{A}$ and

$$\tilde{A}(f) = \tilde{A}(f(x_0) + (x^i - x_0^i)(\partial f/\partial x_0^i) + (x^j - x_0^j)(x^i - x_0^i)(\partial^2 f/\partial x_0^j \partial x_0^i))$$
$$= 0 + b^i(x_0)(\partial f/\partial x_0^i) + a^{ji}(x_0)(\partial^2 f/\partial x_0^j \partial x_0^i), \quad \text{where}$$
$$a^{ji}(x_0) = \tilde{A}((x^j - x_0^j)(x^i - x_0^i)).$$

Hence, by (2.11), we have the theorem. That $a^{ji}(x_0)\xi_j\xi_i$ is ≥ 0 follows from the non-negativity of $(x^j - x_0^j)(x^i - x_0^i)\xi_j\xi_i$ and (iv).

---

6) $\bar{A}(g+h) \leq \bar{A}(g) + \bar{A}(h)$ and $\bar{A}(\alpha g) = \alpha\bar{A}(g)$ for $\alpha \geq 0$.

On holomorphic Markov processes

Proc. Japan Acad. **42** (1966) 313–317

(Comm. by Zyoiti Suetuna, m.j.a., April 12, 1966)

Under appropriate regularity conditions, a temporally homogeneous Markov process is associated with a contraction semi-group $\{T_t; t \geq 0\}$ of class (C_0) [1] in a suitable Banach space X. In certain cases where X are complex Banach spaces, T_t admits a holomorphic extension T_λ given by strongly convergent Taylor series for all $x \in X$:

$$(1) \qquad T_\lambda x = \sum_{n=0}^{\infty} \frac{(\lambda-t)^n}{n!} T_t^{(n)} x \quad \text{for} \quad \frac{|\lambda-t|}{t} \leq \text{some positive constant } C,$$

the existence of the n-th strong derivative $T_t^{(n)}x$ in x of $T_t x$ being assumed for any $t>0$ and any $x \in X$ ($n=1, 2, \cdots$). Such is the case of the semi-group

$$(2) \qquad (T_t f)(x) = (2\pi t)^{-\frac{1}{2}} \int_{-\infty}^{\infty} e^{-|x-y|^2/2t} f(y)dy \qquad (t>0),$$
$$= f(x) \qquad\qquad\qquad (t=0)$$

in the Banach space $C[-\infty, \infty]$ of bounded uniformly continuous, complex valued functions $f(x)$ on $(-\infty, \infty)$ endowed with the maximum norm. Suggested by this example, we shall call a Markov process a *holomorphic Markov process* if the associated semi-group T_t admits a holomorphic extension T_t of the form given in (1). This notion seems to be of some interest. For instance, we can prove

Proposition. *Let a semi-group T_t with the infinitesimal generator A be associated with a holomorphic Markov process through*

$$(3) \qquad (T_t f)(x) = \int P(t, x, dy) f(y), \qquad f \in X$$

where $P(t, x, dy)$ is the transition probability of this process. Suppose that $T_{t_0} f_0 = 0$ for some $t_0 > 0$ and $f_0 \in X$. Then $f_0 = 0$.

Proof. We have $A^n T_{t_0} f_0 = T_{t_0}^{(n)} f_0 = 0$ ($n=0, 1, \cdots$) by the linearity of A. Hence, by Taylor expansion (1), we see that $T_\lambda f_0 = 0$ for $|\lambda-t|/t \leq C$. Repeating the argument, we easily see that $T_t f_0 = 0$ for all $t>0$ and so $f_0 = s - \lim_{t \downarrow 0} T_t f_0 = 0$.

There are abundant examples of holomorphic Markov processes. In fact, the fractional power [2] $\hat{A}_\alpha$ ($0 < \alpha < 1$) of the infinitesimal generator A of a contraction semi-group T_t of class (C_0) generates a constraction semi-group $\hat{T}_{t,\alpha}$ of class (C_0) which admits a holomorphic extension $\hat{T}_{\lambda,\alpha}$ of the similar form given in (1). Moreover, since

$$(4)\qquad \hat{T}_{t,\alpha}x=\int_0^\infty f_{t,\alpha}(s)\,T_s x\,ds \quad \text{with a function } f_{t,\alpha}(s)\geqq 0 \text{ satisfying}$$

$$\int_0^\infty f_{t,\alpha}(s)ds=1,$$

we see that $\hat{T}_{t,\alpha}$ is associated with a holomorphic Markov process if T_t is associated with a Markov process.

The purpose of the present note is to devise another method for the construction of holomorphic Markov processes. It is based upon

Theorem. *Let B be the infinitesimal generator of an equi-continuous group of class (C_0) in a complex Banach space X. Then $A=B^2$ is the infinitesimal generator of an equi-continuous semi-group of class (C_0) which is also a holomorphic semi-group [3] characterized by any one of the following three conditions:*

(I) For all $t>0$, $T_t X\subseteq D(A)$, the domain of A, and there exists a positive constant C_1 such that the family of operators $\{(C_1 t T_t')^n;\ 0<t\leqq 1,\ n=0,1,\cdots\}$ is equi-continuous.

(II) T_t admits a holomorphic extension T_λ of the form given in (1) such that the family of operators $\{e^{-\lambda}T_\lambda;\ |\arg\lambda|\leqq\tan(k^{-1}C_1)$ with some fixed $k>0\}$ is equi-continuous.

(III) There exists a positive constant C_2 such that the family of operators $\{(C_2\lambda(\lambda I-A)^{-1})^n;\ Re(\lambda)\geqq 1$ and $n=0,1,\cdots\}$ is equi-continuous.

Proof. Since B generates an equi-continuous group of class (C_0), $D(B)$ is dense in X and the resolvents $(\sqrt{\lambda}\,I\pm B)^{-1}$ both exist as bounded linear operators on X into X for $Re(\sqrt{\lambda})>0$ satisfying the condition

$$(5)\qquad \{(Re(\sqrt{\lambda})(\sqrt{\lambda}\,I\pm B)^{-1})^n;\ Re(\sqrt{\lambda})>0 \text{ and } n=0,1,\cdots\}$$
$$\text{is equi-continuous.}$$

Thus, by

$$(6)\qquad (\lambda I-A)^{-1}=(\sqrt{\lambda}\,I-B)^{-1}(\sqrt{\lambda}\,I+B)^{-1}(Re(\lambda)>0)$$

we see that $D(A)=$ the range of $(\lambda I-A)^{-1}$ is dense in X with $D(B)$. (6) implies also that

$$(7)\qquad \{(\lambda(\lambda I-A)^{-1})^n$$
$$=(\sqrt{\lambda}\,(\sqrt{\lambda}\,I-B)^{-1}\cdot\sqrt{\lambda}\,(\sqrt{\lambda}\,I+B)^{-1})^n;\ \lambda>0,\ n=0,1,2,\cdots\}$$
$$\text{is equi-continuous.}$$

Hence A generates an equi-continuous semi-group of class (C_0). Moreover,

$$\left\{\left(\left(\sqrt{1+\tau^2}\,\cos^2\left(\frac{1}{2}\tan^{-1}\tau\right)((1+i\tau)I-A)^{-1}\right)\right)^n\right\}$$
$$=\{(Re(\sqrt{1+i\tau})(\sqrt{1+i\tau}I-B)^{-1}\,Re(\sqrt{1+i\tau})(\sqrt{1+i\tau}\,I+B)^{-1})^n\}$$

is equi-continuous in $-\infty<\tau<\infty$ and in $n=0, 1, \cdots$. Hence, by (III), the operator A generates a holomorphic semi-group.

　　An example of holomorphic Markov processes. Let $X=C[-\infty, \infty]$ and consider the operator

$$(8)\qquad A=a^2(x)\frac{d^2}{dx^2}+b(x)\frac{d}{dx}+q(x).$$

Suppose that $a(x)$, $a'(x)$, $b(x)$, and $q(x)$ are uniformly continuous, bounded real-valued functions in $(-\infty, \infty)$ satisfying conditions

(9)　$q(x)\leqq0$ and $0<\delta\leqq a(x)$ in $(-\infty, \infty)$, where δ is a positive constant.

Then A generates a contraction holomorphic semi-group T_t in X which is *positive*, i.e., $f(x)\geqq0$ in $(-\infty, \infty)$ implies $(T_t f)(x)\geqq0$ in $(-\infty, \infty)$. Thus T_t is associated with a holomorphic Markov process.

　　Proof. A may be written as

$$(8)'\qquad A=\left(a(x)\frac{d}{dx}\right)^2+p(x)\frac{d}{dx}-\varepsilon\frac{d}{dx}+q(x),\quad\text{where}$$

$$p(x)=b(x)-a(x)a'(x)+\varepsilon\ \text{and}\ \varepsilon>2\sup_{-\infty<x<\infty}|\,[b(x)-a(x)a'(x)]\,|.$$

We shall prove (i): $E=\left(a\dfrac{d}{dx}\right)^2$ generates a contraction positive holomorphic semi-group in X, (ii): $p\dfrac{d}{dx}$ and $-\varepsilon\dfrac{d}{dx}$ both generate contraction positive semi-groups of class (C_0) in X, (iii): $q(x)$ generates a contraction positive semi-group of class (C_0) in X, (iv): for $1>\alpha>\dfrac{1}{2}$, the domain $D\left(p\dfrac{d}{dx}\right)$ contains the domain $D(\hat{E}_\alpha)$, where $\hat{E}_\alpha$ is the fractional power operator of E and (v): for $1>\alpha>\dfrac{1}{2}$, the domain $D\left(-\varepsilon\dfrac{d}{dx}\right)$ contains the domain $D(\hat{F}_\alpha)$, where $\hat{F}_\alpha$ is the fractional power operator of $F=E+p\dfrac{d}{dx}$ with the domain $D(F)=D(E)$.

　　Then, by a theorem proved in a preceding note in these Proceedings [4], $F=\left(E+p\dfrac{d}{dx}\right)$ generates a contraction holomorphic semi-group in X. By H. F. Trotter's product formula [5], we have

$$e^{t\left(E+p\frac{d}{dx}\right)}=s-\lim_{n\to\infty}(e^{\frac{t}{n}E}\cdot e^{\frac{t}{n}p\frac{d}{dx}})^n$$

so that the semi-group $e^{t\left(E+p\frac{d}{dx}\right)}$ generated by F is positive by the positivity of semi-groups e^{tE} and $e^{tp\frac{d}{dx}}$. Similarly, by (v), $F-\varepsilon\dfrac{d}{dx}$ with the domain $D\left(F-\varepsilon\dfrac{d}{dx}\right)=D(F)$ generates a positive contraction holomorphic semi-group in X. The multiplication operator q is a bounded operator in X which generates a positive contraction semi-group of class (C_0) in X by (iii). Hence, by a similar argument as above, $A = F - \varepsilon \dfrac{d}{dx} + q$ with the domain $D(A) = D(F)$ generates a positive contraction holomorphic semi-group in X.

The proof of (i) through (iv) is given as follows.

(i): $B = a \dfrac{d}{dx}$ generates a positive contraction group of class (C_0) in X of translations

$$f(x(y)) \to f(x(y \pm t)), \quad \text{where} \quad y(x) = \int_0^x a(s)^{-1} ds.$$

Hence the resolvents $(\sqrt{\lambda}\, I \pm B)^{-1}$ are positive operators in X for $\lambda > 0$ and so $(\lambda I - E)^{-1} = (\sqrt{\lambda}\, I - B)^{-1}(\sqrt{\lambda}\, I + B)^{-1}$ is a positive operator in X. Thus, remembering the Theorem and the representation $e^{tE} = s - \lim\limits_{n \to \infty} \left(I - \dfrac{t}{n} E \right)^{-n}$, we have proved (i).

(ii): As in (i), we prove that $p \dfrac{d}{dx}$ and $-\varepsilon \dfrac{d}{dx}$ both generate positive contraction group of class (C_0) in X.

(iv): The resolvent of $\hat{E}_\alpha$ is given by T. Kato's formula [6]

$$(10) \qquad (\lambda I - \hat{E}_\alpha)^{-1} = \frac{\sin \alpha \pi}{\pi} \int_0^\infty (rI - E)^{-1} \frac{r^\alpha}{\lambda^2 - 2\lambda r^\alpha \cos \alpha \pi + r^{2\alpha}} dr.$$

We have

$$B(rI - E)^{-1} = B(\sqrt{r}\, I - B)^{-1}(\sqrt{r}\, I + B)^{-1}$$
$$= (\sqrt{r}\,(\sqrt{r}\, I - B)^{-1} - I)(\sqrt{r}\, I + B)^{-1}$$

and so, by $\|(\sqrt{r}\, I + B)^{-1}\| \leq r^{-1/2}$, we see that the right side of

$$B(\lambda I - \hat{E}_\alpha)^{-1} = \frac{\sin \alpha \pi}{\pi} \int_0^\infty B(rI - E)^{-1} \frac{r^\alpha}{\lambda^2 - 2\lambda r^\alpha \cos \alpha \pi + r^{2\alpha}} dr$$

converges when $1 > \alpha > \dfrac{1}{2}$. This proves (iv).

(v): Remembering

$$B(rI - F)^{-1} = B(I - F)^{-1}(I - F)(rI - F)^{-1}$$
$$= B(I - F) \cdot [(rI - F)^{-1} + I - r(rI - F)^{-1}]$$
$$= O(r^{-1}) \qquad \text{for small } r$$

and

$$B(rI - F)^{-1} = B(rI - E)^{-1} \left\{ I - p \frac{d}{dx}(rI - E)^{-1} \right\}^{-1}$$

$$= O(r^{-1/2}) \qquad \text{for large } r,$$

we prove (v) as in (iv).

Remark. That $b(x)$ in (8) may change sign on $(-\infty, \infty)$ was suggested, thanks to a conversation with Professor S. Ito and Dr. H. Tanaka.

References

[1]　E. Hille-R. S. Phillips: Functional Analysis and Semi-groups. Providence (1957); K. Yosida: Functional Analysis. Springer (1965).

[2]　K. Yosida: Loc. cit.

[3]　——: Loc. cit.

[4]　——: A perturbation theorem for semi-groups of linear operators. Proc. Japan Acad., **41** (8), 645-647 (1965).

[5]　H. F. Trotter: On the product of semi-groups of operators. Proc. Amer. Math. Soc., **10**, 545-551 (1959).

[6]　See, for instance, K. Yosida: Loc. cit.

VIII. Hyperbolic Equations

Comments (S. T. Kuroda and Sigeru Mizohata)

The theory of semi-groups of linear operators and some related methods of operator-theoretical integration were quickly applied to Cauchy problems for partial differential equations by Yosida himself, P. D. Lax, J.-L. Lions, and others. Among these, Yosida's papers [72], [80] and [79] are concerned with the Cauchy problem for the wave equation $\partial^2 u/\partial t^2(t, x) = Au(t, x)$, where A is a second order elliptic operator with coefficients depending, in [72] and [79], on x and, in [80], on x and t. "Elliptic" means that the associated quadratic form is uniformly bounded and uniformly strictly positive. In all of these papers an L^2-solution is constructed by an operator-theoretical means and the differentiability of the obtained solution is proved afterwards.

In [72], A is assumed to be formally self-adjoint and an L^2 solution is constructed in terms of the sine and the cosine of $(-A)^{-1/2}$. A major part of the paper is concerned with an elaborate construction of a parametrix, suggested by M. Riesz' work and applicable also to a non-self-adjoint operator, by means of which the differentiability of the L^2-solution is established.

In [79] Yosida proved the solvability of the Cauchy problem for the wave equation $\partial^2 u/\partial t^2(t, x) = Au(t, x)$ in the Euclidean space E^m as seemingly the first substantial application of the semi-group theory to the wave equation. As a key to the proof, Yosida presents a basic L^2-resolvent estimate for the second order elliptic operator A in E^m with a detailed proof. The estimate is then applied to the first order system corresponding to the wave equation to construct the L^2-solution. The differentiability is proved by the ellipticity argument. This resolvent estimate also provides a key step in [83]. As a due generalization of [79], J. L. Lions (J. Math. Soc. Japan, 1957) showed that an analogous method can be applied to general mixed initial-boundary value problems with higher order elliptic operators.

In [80] it is assumed, among others, that the generator $\mathscr{A}_t$, depending on t, of the system satisfies the resolvent estimate uniformly with respect to t. As in [74], the main idea in the construction of an L^2-solution is first to solve the approximate equation whose generator is, at each fixed t, the Yosida approximation of $\mathscr{A}_t$.

Before Yosida's work, generally speaking, the construction of solutions relied partially on the Cauchy-Kowalewsky theorem, where the coefficients of the equation were approximated by real-analytic ones. By using operator theory Yosida showed that this rather unnatural way of argument is not necessary. Afterwards in 1959 S. Mizohata (J. Math. Soc. Japan) showed that Yosida's method is applicable to the existence theorem for general hyperbolic systems by using a priori estimates of solutions. Furthermore, T. Kato started in 1970 (J. Fac. Sci. Univ. Tokyo, Sect. IA) a systematic study based on the semi-group theory and has developed a powerful theory of evolution equations of "hyperbolic" type.

On Cauchy's problem in the large for wave equations

Proc. Japan Acad. **28** (1952) 396–403

(Comm. by Z. SUETUNA, M.J.A., Oct. 13, 1952.)

§ 1. *Introduction.* Let R be a connected domain of an orientable, m-dimensional Riemannian space with the metric $ds^2 = g_{ij}(x)dx^i dx^j$. We consider the wave equation

$$(1.1) \qquad \frac{\partial^2 u(x,t)}{\partial t^2} = A_x u(x,t), \quad -\infty < t < \infty,$$

with Cauchy's data

$$(1.2) \qquad u(x,0) = f(x), \quad \frac{\partial u(x,0)}{\partial t} = h(x).$$

Here the differential operator $A = A_x$ defined by

$$(1.3) \qquad A_x f(x) = b^{ij}(x)\frac{\partial^2 f(x)}{\partial x^i \partial x^j} + a^i(x)\frac{\partial f(x)}{\partial x^i} + e(x)f(x)$$

is *elliptic* in the sense that the quadratic form $b^{ij}(x)\xi_i\xi_j$ is > 0 for $\sum_i (\xi_i)^2 > 0$. Since the value of $A_x f(x)$ must be independent of the local coordinates $(x^1, \ldots, x^m)$ of the point x, the coefficients $a^i(x)$ and $b^{ij}(x)$ must be transformed, by the coordinates change $x \to \bar{x}$, respectively into

$$(1.4) \qquad \bar{a}^i(\bar{x}) = \frac{\partial \bar{x}^i}{\partial x^k}a^k(x) + \frac{\partial^2 \bar{x}^i}{\partial x^k \partial x^s}b^{ks}(x) \text{ and } \bar{b}^{ij}(\bar{x}) = \frac{\partial \bar{x}^i}{\partial x^k}\frac{\partial \bar{x}^j}{\partial x^s}b^{ks}(x).$$

For the sake of simplicity, we assume that $g_{ij}(x)$, $b^{ij}(x)$, $a^i(x)$ and $e(x)$ are infinitely differentiable functions of the local coordinates $(x^1, \ldots, x^m)$.

Since we are concerned with *the existence in the large of the integral* of (1.1)–(1.2), it will perhaps be necessary to rely upon operator-theoretical method[1]. We here assume that the operator A_x is, as in the case of Laplacian, *formally self-adjoint* and *non-positive definite*, viz.

$$(1.5) \qquad \int_R (A_x f(x))h(x)dx = \int_R f(x)(A_x h(x))dx \text{ and } \int_R (A_x f(x))f(x)dx \leq 0$$
$$(dx = \sqrt{g(x)}\ dx^1 \ldots dx^m, \quad g(x) = det(g_{ij}(x))),$$

if $f(x)$ and $h(x)$ are twice continuously differentiable such that $f(x)$ vanishes outside a compact set contained in the interior of R. Then we may integrate, by virtue of the Hilbert space technique, an operator-theoretical variant of (1.1)–(1.2) It will next be shown, by a parametrix consideration, that this operator-theoretical integral is, for sufficiently smooth initial data (1.2), equivalent to the ordinary integral of the genuine differential equation (1.1)–(1.2). It is to be noted that the Lemma 2 below, which is of the type of Poisson's equation, may be of use in other problems relating to the elliptic differential operator.

§2. *An operator-theoretical integration.* Let L be the linear space of twice continuously differentiable real-valued functions $f(x)$ vanishing outside compact set and satisfying a certain linear boundary condition on the boundary ∂R of R. It is assumed that the boundary condition is chosen in such a way that we have

$$(2.1) \qquad \int_R (A_x f(x))h(x)dx = \int_R f(x)(A_x h(x))dx \text{ and}$$

$$(2.2) \qquad \int_R (A_x f(x))f(x)dx \leq 0 \qquad \text{for } f, h \in L.$$

Such boundary condition is possible because of the assumption (1.5). L is a pre-Hilbert space by the norm

$$(2.3) \qquad ||f|| = (\int_R f(x)^2 dx)^{1/2} = (f, f)^{1/2},$$

such that the completion L^a of this linear normed space L is a real Hilbert space $L_2(R)$.

We consider $A = A_x$ to be an additive operator defined on $L \subseteq L^a$ into L^a. Let $\tilde{A}$ be a non-positive definite self-adjoint extension of A. Such $\tilde{A}$ may be defined as follows[2]: Let L' be the completion of the linear space L by the new metric

$$(2.4) \qquad ||f||' = ((-Af, f) + (f, f))^{1/2}.$$

Because of (2.2), we may identify L' with a linear subspace of L^a. Then

(2.5) $\tilde{A}$ is the contraction of the adjoint operator A^* of A restricted to the domain $D(\tilde{A}) = L' \wedge D(A^*)$, where $D(A^*)$ is the domain of A^*.

We have, by (2.1),

$$(2.6) \qquad L \subseteq D(\tilde{A}).$$

Let (2.7):

$$-\tilde{A} = \int_0^\infty \lambda dE(\lambda)$$

be the spectral resolution of $-\tilde{A}$ and let

$$(2.8) \qquad (-\tilde{A})^{1/2} = \int_0^\infty \lambda^{1/2} dE(\lambda)$$

be the positive square root of the operator $-\tilde{A}$. Surely we have

(2.9)   the domain $D((-\tilde{A})^{1/2})$ of $(-\tilde{A})^{1/2} \supseteq D(\tilde{A})$, and hence, by (2.6),

$$(2.6)' \qquad L \subseteq D(\tilde{A}) \subseteq D((\tilde{A})^{1/2}).$$

Let us consider, for f and $h \in L$,

$$(2.10) \quad \tilde{u}(x, t) = (\cos(-\tilde{A})^{1/2}t)f(x) + (\sin((-\tilde{A})^{1/2}t)/(-\tilde{A})^{1/2})h(x)$$

$$= \int_0^\infty \cos(\lambda^{1/2}t)dE(\lambda)f(x) + \int_0^\infty (\sin(\lambda^{1/2}t)/\lambda^{1/2})dE(\lambda)h(x).$$

The convergence of the right hand integral is clear. We see, by (2.6)′, that $\tilde{u}(x, t)$ satisfies the operator-theoretical differential equation

(2.11) $\qquad \partial_t\partial_t u(x, t) = \tilde{A}_x \tilde{u}(x, t),$ where

$$\partial_t\tilde{u}(x, t) = \text{strong} \lim_{\delta\to 0} \delta^{-1}(\tilde{u}(x, t+\delta) - \tilde{u}(x, t)).$$

We have also (2.12): $\qquad \tilde{u}(x, 0) = f(x), \quad \partial_t\tilde{u}(x, 0) = h(x).$
Therefore we have:

Theorem 1. (2.10) is an operator-theoretical solution of Stokes' type of the operator-theoretical variant (2.11)–(2.12) of (1.1)–(1.2).

Let D be the subset of L consisting of all the infinitely differentiable functions $f(x)$ such that $f(x) \in$ the domain $D(\tilde{A}_x^q)$ of the operator $\tilde{A}_x^q$ for every $q > 0$. Such is the case for infinitely differentiable function $f(x)$ when $f(x)$ vanishes outside a compact set contained in the interior of R. From the definition (2.10) and (2.6)′, we see that (2.13): if f and h are both in D, the function $\tilde{u}(x, t)$ given by (2.10) is in the domain $D(\tilde{A}_x^q)$ for any $q > 0$. We will show, in §4, that such $\tilde{u}(x, t)$ is equal (x, t)-almost everywhere to a function $u(x, t)$ which is infinitely differentiable in (x, t), so that $u(x, t)$ is an ordinary integral of the genuine differential equation (1.1)–(1.2).

§3. *The parametrix for the iterated elliptic operator.* The hypothesis of the formal self-adjointness of the operator $A = A_x$ is not needed in this §. Thus let

(3.1) $\qquad A'_x f(x) = b^{ij}(x)\dfrac{\partial^2 f}{\partial x^i \partial x^j} + c^i(x)\dfrac{\partial f}{\partial x^i} + p(x)f(x)$

be the formally adjoint operator of A_x. We will construct a parametrix for the iterated elliptic operator (3.2): $A_x'^{q-1}$. To this purpose, let $\Gamma(P, Q) = r(P, Q)^2$ be the square of the smallest distance between the two points $P = (x'^1, \dots, x'^m)$ and $Q = (x^1, \dots, x^m)$ of R according to the new metric (3.3): $dr^2 = b_{ij}(x)dx^i dx^j$, where $(b_{ij}(x)) = (b^{ij}(x))^{-1}$. We have then:

Lemma 1[3]. Let the dimension m be odd. For any positive integer n and for any even $\alpha \geq 0$, we may construct a parametrix $W_\alpha(P, Q)$ for the operator $A' = A'_x$:

(3.4) $\quad W_\alpha(P, Q) = \sum\limits_{k=0}^{n} \Gamma(P, Q)^{(\alpha+2k-m)/2} V_k(P, Q)/K_m(\alpha)L_m(\alpha+2k),$

where $\quad K_m(\alpha) = 2^{\alpha/2}\Gamma(\alpha/2),\ L_m(\alpha+2k) = 2^{(\alpha+2k)/2}\Gamma((\alpha+2k+2-m)/2)$
and $V_k(P, Q)$ *are infinitely differentiable in the vicinity of $Q = P$ and $V_0(P, P) = 1$*

so that (3.5): $\quad A'_x W_{\alpha+2}(P, Q) = W_\alpha(P, Q)$

$$+ \Gamma(P, Q)^{(\alpha+2+2n-m)/2}A'_x W_\alpha(P, Q)/K_m(\alpha+2)L_m(\alpha+2+2n).$$

Proof. We introduce the normal coordinates y of $Q = (x^1, \dots, x^m)$ in the vicinity of P:

(3.6) $\qquad y^\sigma = (\Gamma(P, Q))^{1/2}\left(\dfrac{dx^\sigma}{dr}\right)_{r=0}.$

Let   (3.7): $dr^2 = \beta_{ij}(y)dy^i dy^j.$

We have the well-known formulae

(3.8) $\Gamma(P, Q) = \beta_{ij}(0)y^i y^j, \quad \beta_{ij}(y)y^j = \beta_{ij}(0)y^j.$

By virtue of (3.8), the operator

(3.9) $A' = A'_y = \beta^{ij}(y)\dfrac{\partial^2}{\partial y^i \partial y^j} + \alpha^i(y)\dfrac{\partial}{\partial y^i} + \gamma(y)$

$$((\beta^{ij}(y)) = (\beta_{ij}(y))^{-1}),$$

when applied to the function of the form $f(\Gamma(P, Q), y)$, may be written as follows:

(3.10) $A'_y f = 4\Gamma\dfrac{\partial^2 f}{\partial \Gamma^2} + 4y^\sigma\dfrac{\partial^2 f}{\partial \Gamma \partial y^\sigma} + M\dfrac{\partial f}{\partial \Gamma} + N(f),$ where

$$M = \beta^{ij}\dfrac{\partial^2 \Gamma}{\partial y^i \partial y^j} + \alpha^i\dfrac{\partial \Gamma}{\partial y^i} = 2m + 0(y), \quad N(f) = \beta^{ij}\dfrac{\partial^2 f}{\partial y^i \partial y^j} + \alpha^i\dfrac{\partial f}{\partial y^i} + \gamma f.$$

The differentiation in A'_y and in $N(f)$ are to be performed as if Γ and y^σ are independent variables.   Hence, by

(3.11) $\alpha/K_m(\alpha+2) = 1/K_m(\alpha), \quad (\alpha+2-m)/L_m(\alpha+2) = 1/L_m(\alpha),$

we obtain $A'_y W_{\alpha+2}(P, Q) = \displaystyle\sum_{k=0}^{n} \dfrac{\Gamma(P, Q)^{(\alpha+2k-m)/2}}{K_m(\alpha+2)L_m(\alpha+2k)}$

$$\times \left\{ 2y^\sigma\dfrac{\partial V_k}{\partial y^\sigma} + \left(\dfrac{M}{2} + 2k - m + \alpha\right)V_k + A'_y V_{k-1}(P,Q)\right\}$$

$$+ \dfrac{\Gamma(P, Q)^{(\alpha+2+2n-m)/2}}{K_m(\alpha+2)L_m(\alpha+2+2n)}A'_y V_n$$

$$= W_\alpha(P, Q) + \dfrac{\Gamma(P, Q)^{(\alpha+2+2n-m)/2}}{K_m(\alpha+2)L_m(\alpha+2+2n)}A'_y V_n,$$

if $V_k(P, Q)$ may be so determined that $V_k(P, Q)$ are infinitely differentiable in the vicinity of $Q = P$, $V_{-1}(P, Q) \equiv 0$, $V_0(P, P) = 1$ and

(3.12) $2y^\sigma\dfrac{\partial V_k}{\partial y^\sigma} + (\dfrac{M}{2} + 2k - m)V_k(P, Q) + A'_y V_{k-1}(P, Q)$

$$= 0, \quad (k = 0, 1, \ldots, n).$$

Such $V_k(P, Q)$ exist by virtue of the order relation

(3.13) $M = 2m + 0(y).$

   Proof. By putting $y^\sigma = r\eta^\sigma$, (3.12) is reduced to the ordinary differential equation in r containing the parameters η :

(3.12)' $2r\dfrac{dV_k(P, r\eta)}{dr} + (\dfrac{M(r\eta)}{2} + 2k - m)V_k(P, r\eta) = -A'_y V_{k-1}(P, r\eta).$

Hence, by $V_{-1}(P, Q) \equiv 0$ and $V_0(P, P) = 1$, we obtain

(3.14) $V_0 = \exp\left(-\displaystyle\int_0^r (2t)^{-1}(\dfrac{M}{2} - m)dt\right),$

$$V_k = -V_0 r^{-k}\int_0^r t^{k-1}V_0^{-1}A'_y V_{k-1}dt.$$

   Corollary.

(3.15) $\cdot$ $A'^{q-i}_y W_{2q}(P, Q) = W_{2i}(P, Q) + 0(\Gamma(P, Q)^{(2i+2+2n-m)/2}),$

$A'^q_y W_{2q}(P, Q) = 0(\Gamma(P, Q)^{(2+2n-m)/2})$ for $P \neq Q.$

Next let P_0 be any inner point of R and consider, for sufficiently small $\varepsilon > 0$,

(3.16) $\quad U_\alpha(P, Q) = W_\alpha(P, Q)\delta(\Gamma(P, Q))\delta(\Gamma(P_0, P))$, where $\delta(x) \gtrless$ is infinitely differentiable in $x \geq 0$ such that $\delta(x) = 1$ or 0 according as $x \leq \varepsilon$ or $x \geq 2\varepsilon$.

Thus, in a certain vicinity of P_0,

(3.17) $\quad A_y'^{q-1} U_{2q}(P, Q) = U_{2i}(P, Q) + 0(\Gamma(P, Q)^{(2i+2+2n-m)/2})$,

$\qquad A_y'^q U_{2q}(P, Q) = 0(\Gamma(P, Q)^{(2+2n-m)/2})$ for $P \neq Q$.

After these preliminaries, we may prove an analogue of Poisson's equation, viz.

Lemma 2. Let the dimension m be odd and ≥ 2, and let $k(Q)$ be $\in L$. Then we have, for $2n \geq m$,

(3.18) $\quad C(P)k(P) = \displaystyle\int_R (A_y'^{q-1} U_{2q}(P, Q))(A_y k(Q))dQ$, *where $C(P)$ is infinitely differentiable and $\neq 0$ in a certain vicinity of P_0.*

Proof. We have, by Green's integral theorem and (3.17),

$$\int_R (A_y'^{q-1} U_{2q}(P, Q)(A_y k(Q))dQ$$

$$= \lim_{\kappa \to 0} \int_{R-\{Q; \Gamma(P, Q) \leq \kappa\}} (A_y'^{q-1} U_{2q}(P, Q))(A_y k(Q))dQ$$

$$= \lim_{\kappa \to 0} \int_{R-\{Q; \Gamma(P, Q) \leq \kappa\}} (A_y'(A_y'^{q-1} U_{2q}(P, Q))k(Q)dQ$$

$$+ \lim_{\kappa \to 0} \int_{\Gamma(P, Q)=\kappa} \left\{ \frac{A_y'^{q-1} U_{2q}(P, Q)}{\partial \nu} k(Q) - (A_y'^{q-1} U_{2q}(P, Q)) \frac{\partial k(Q)}{\partial \nu} \right\} dS$$

where ν is the transversal direction defined by

(3.19) $\quad \dfrac{\partial \nu}{\partial y^i} = (\sqrt{g(y)}\, \beta^{ij}(y) \cos(r, y^j))^{-1}, \quad (i = 1, 2, \ldots, m)$

and dS is the hypersurface element on $\Gamma(P, Q) = \kappa$.

We have, from (3.17),

$A_y'^q U_{2q}(P, Q) = 0(\Gamma(P, Q)^{(2+2n-m)/2})$ for $P \neq Q$,

$A_y'^{q-1} U_{2q}(P, Q) = (4\Gamma((4-m)/2))^{-1}\Gamma(P, Q)^{(2-m)/2} + 0(\Gamma(P, Q)^{(2+2n-m)/2})$.

Hence we have, when $\Gamma(P, Q) = \kappa$ tends to zero

$$\frac{\partial A_y'^{q-1} U_{2q}(P, Q)}{\partial \nu} \doteqdot (8\Gamma((4-m)/2)^{-1}(2-m)\Gamma^{-m/2}\frac{\partial \Gamma}{\partial y^i}\sqrt{g(y)}\,\beta^{ij}(y)\cos(r, y^j)$$

$$= (4\Gamma(4-m)/2)^{-1}(2-m)\Gamma^{-m/2}\beta_{ik}(0)y^k\sqrt{g(y)}\,\beta^{ij}(y)\cos(r, y^j) \quad \text{(by (3.8))}$$

$$= (4\Gamma((4-m)/2)^{-1}(2-m)y^j\Gamma^{-m/2}\sqrt{g(y)}\,\cos(r, y^j) \quad \text{(by (3.8))}$$

$$= (4\Gamma((4-m)/2)^{-1}(2-m)\Gamma^{(1-m)/2}\sqrt{g(r\eta)}\,\sum_{j=1}^m (\eta^j)^2 \quad \text{(by putting } y^j = r\eta^j).$$

Therefore we have $\displaystyle\int_R (A_y'^{q-1} U_{2q}(P, Q))(A_y k(Q))dQ$

$$= \lim_{\kappa \to 0} \int_{\beta_{ij}(P)\eta^i\eta^j = 1} (4\Gamma((4-m)/2)^{-1}(2-m)\kappa^{(1-m)/2}\sqrt{g(\sqrt{\kappa}\,\eta)}\,\sum_{j=1}^m (\eta^j)^2 dS_{\sqrt{\kappa}\,\eta}$$

$$= (4\Gamma(4-m)/2)^{-1}(2-m)\sqrt{g(P)} \int_{\beta_{ij}(P)\eta^i\eta^j = 1} \sum_{j=1}^m (\eta^j)^2 dS_\eta.$$

This proves (3.16).

§ 4. *The differentiability of the operator-theoretical solution* $\tilde{u}(Q, t)$. We first remark that we are dealing with the case $A'=A$. We will prepare two lemmas.

Lemma 3. For fixed t, there exists a sequence of functions $\{k_i(Q)\}\subseteqq L$ such that

$$(4.1) \quad \text{strong } \lim_{i\to\infty} k_i(Q)=\tilde{u}(Q, t),$$

$$\lim_{i\to\infty}\int_R w(Q)(A_y k_i(Q))dQ=\int_R w(Q)(\tilde{A}_y\tilde{u}(Q, t))dQ \text{ for every } w(Q)\in L.$$

Proof. By $\tilde{u}(Q, t)\in D(\tilde{A}_y)$ and the definition (2.5) of $\tilde{A}$, there exists a sequence of functions $\{k_i(Q)\}\subseteqq L$ such that strong $\lim_{i\to\infty} k_i(Q)$ $=\tilde{u}(Q, t)$. We have, for any $w(Q)\in L$,

$$\lim_{i\to\infty}\int_R w(Q)(A_y k_i(Q))dQ= \lim_{i\to\infty}\int_R (A_y w(Q))k_i(Q)dQ$$

$$=\int_R (A_y w(Q))\tilde{u}(Q, t)dQ=\int_R w(Q)(\tilde{A}_y\tilde{u}(Q, t))dQ$$

by (2.1) and by the definition (2.5) of $\tilde{A}$.

Lemma 4. We have, for $w(Q)\in L$ and for $1\leq i\leq q$,

$$(4.2) \qquad \int_R w(P)(A_y^{q-i}U_{2q}(P, Q))dP\in L.$$

Proof. By (3.16), we see that the integral vanishes outside a compact coordinate neighbourhood of P_0. Moreover, by (3.4), (3.15), (3.16) and (3.17), we see that the integral is twice continuously differentiable in Q ($Q.E.D.$).

We have, by (3.18),

$$C(P)k_i(P)=\int_R (A_y^{q-1}U_{2q}(P, Q))(A_y k_i(Q))dQ$$

in a certain vicinity of P_0. Let $w(Q)\in L$ vanish outside this vicinity. Letting $i\to\infty$ in

$$\int_R w(P)C(P)k_i(P)dP =\int_R w(P)dP\left\{\int_R (A_y^{q-1}U_{2q}(P, Q))(A_y k_i(Q))dQ\right\} ,$$

we obtain, by the Lemma 3 and Lemma 4,

$$(4.3) \quad \tilde{u}(P, t)=C(P)^{-1}\int_R (A_y^{q-1}U_{2q}(P, Q))(\tilde{A}_y\tilde{u}(Q, t))dQ \quad \text{almost every-}$$

where in P in a certain vicinity of P_0.

The function $\tilde{u}(Q, t)$ belongs to $D(\tilde{A}_y^p)$ for every $p>0$. Thus we see, by the Lemma 3, that there exists a sequence of functions $\{k_i(Q)\}\subseteqq L$ such that

$$(4.4) \quad \text{strong } \lim_{i\to\infty} k_i(Q)=\tilde{A}_y\tilde{u}(Q, t) ,$$

$$\lim_{i\to\infty}\int_R w(Q)(A_y k_i(Q))dQ =\int_R w(Q)(\tilde{A}_y^2\tilde{u}(Q, t))dQ \text{ for every } w(Q)\in L.$$

Hence we have

$$(4.5) \quad \int_R (A_y^{q-1}U_{2q}(P, Q))(\tilde{A}_y\tilde{u}(Q, t))dQ = \lim_{i\to\infty}\int_R (A_y^{q-1}U_{2q}(P, Q))k_i(Q)dQ$$

almost everywhere in P. Also, by Green's integral theorem,

$$\int_R (A_y^{q-1}U_{2q}(P, Q))k_i(Q)dQ$$

$$= \lim_{\kappa\to 0}\int_{R-\{Q\,;\,\Gamma(P,\,Q)\leq\kappa\}} (A_y^{q-1}U_{2q}(P, Q))k_i(Q)dQ$$

$$= \lim_{\kappa\to 0}\int_{R-\{Q\,;\,\Gamma(P,\,Q)\leq\kappa\}} (A_y^{q-2}U_{2q}(P, Q))(A_y k_i(Q))dQ$$

$$- \lim_{\kappa\to 0}\int_{\Gamma(P,\,Q)=\kappa}\left\{\frac{\partial A_y^{q-2}U_{2q}(P,Q)}{\partial\nu}k_i(Q) - (A_y^{q-2}U_{2q}(P,Q))\frac{\partial k_i(Q)}{\partial\nu}\right\}dS$$

$$= \int_R (A_y^{q-2}U_{2q}(P, Q))(A_y k_i(Q))dQ.$$

The last equality may be obtained, as in the proof of (3.18), from the order relation (3.17):

$$A_y^{q-2}U_{2q}(P, Q) = 0(\Gamma(P, Q)^{(4-m)/2}).$$

Hence, for any $w(P)\in L$, we have

$$\int_R w(P)dP\left\{\int_R (A_y^{q-1}U_{2q}(P, Q))k_i(Q)dQ\right\}$$

$$= \int_R w(P)dP\left\{\int_R (A_y^{q-2}U_{2q}(P, Q))(A_y k_i(Q))dQ\right\}.$$

Thus, by letting $i\to\infty$, we obtain, from (4.4), (4.5) and the Lemma 4,

$$\int_R (A_y^{q-1}U_{2q}(P, Q))(\tilde{A}_y\tilde{u}(Q, t))dQ = \int_R (A_y^{q-2}U_{2q}(P, Q))(\tilde{A}_y^2\tilde{u}(Q, t))dQ$$

almost everywhere in P. Repeating the process, we obtain, from (4.3),

Theorem 2. Let the dimension m be odd and ≥ 2, and let $2n \geq m$ in the definition of $U_{2q}(P, Q)$. Then, for the initial data f and h in D, we have

$$(4.6) \quad \breve{u}(P, t) = C(P)^{-1}\int_R U_{2q}(P, Q)(\tilde{A}_y^q\tilde{u}(Q, t))dQ \quad \textit{almost everywhere in}$$

P in a certain vicinity of P_0.

Corollary. $\tilde{u}(Q, t)$ is, for fixed t, equal almost everywhere to a function $u(P, t)$ which is infinitely differentiable in P in a certain vicinity of P_0 such that

$$(4.6)' \quad u(P, t) = C(P)^{-1}\int_R U_{2q}(P, Q)(\tilde{A}_y^q\tilde{u}(Q, t))dQ.$$

Proof. We see that, if $q\geq m$,

$$u(P, t) = C(P)^{-1}\int_R U_{2q}(P, Q)(\tilde{A}_y^q\tilde{u}(Q, t))dQ$$

is, by (3.17), q times continuously differentiable in P. As q may be taken arbitrarily large, the Corollary is proved.

In the above, we have assumed that the dimention m be odd and ≥ 2. Let us consider the case in which m does not satisfy this condition. In such a case, let $m'>m$ be odd and ≥ 2. We consider the function

$$\hat{u}(\hat{Q}, t)=u(y^1 \ldots, y^m,\ t)\ \text{exp}\ (-(y^{m+1})^2-\ldots-(y^{m'})^2)$$

of m' independent variables $y^1, \ldots,\ y^m,\ y^{m+1}, \ldots,\ y^{m'}$. By introducing the operator

$$(4.7) \qquad A^{(1)}=A+\frac{\partial^2}{\partial(y^{m+1})^2}+\ldots+\frac{\partial^2}{\partial(y^{m'})^2}$$

in place of the operator $A=A_y$, we see, as above, that $(4.6)'$ holds good for $u(\hat{Q}, t)$ in this case also. *Proof.* $\tilde{A}^{(1)q}\hat{u}(\hat{Q}, t)$ belongs, for fixed t, to the product Hilbert space

$$L^a \times L_2(-\infty<y^{m+1}<\infty, \ldots, -\infty<y^{m'}<\infty)$$

and hence we may apply the proof of the Theorem 2 above[1].

Next since $u(Q, t)$ belong to $D(\tilde{A})_y^p)$ for every $p>0$, it is easy to see, by (2.10) that

$$(4.8) \qquad (\partial_t\partial_t)^r \tilde{A}_y^q u(Q, t)=\tilde{A}_y^{q+r}u(Q, t) \text{ for every } r\geq 0.$$

Thus we see, by $(4.6)'$, that $u(P, t)$ is, for fixed P, infinitely differentiable in t.

Moreover, since $u(Q, t)$ is infinitely differentiable in Q, we have

$$(4.9) \qquad \tilde{A}_y^{q+r}u(Q, t)=A_y^{q+r}u(Q, t) \text{ almost everywhere in } Q.$$

For, we have, by the definition (2.5) of $\tilde{A}$,

$$\int_R w(Q)(\tilde{A}_y^{q+r}u(Q, t)dQ=\int_R (A_y^{q+r}w(Q))u(Q, t)dQ=\int_R w(Q)(A_y^{q+r}u(Q, t))dQ,$$

when $w(Q)$ is infinitely differentiable and vanishes outside a compact set contained in the interior of R.

Therefore, in view of (2.11), we have proved finally the

Theorem 3. When f and h are in D, the function $\tilde{u}(x, t)$ given by (2.10) is (x, t)-almost everywhere equal to an infinitely differentiable function $u(x, t)$ satisfying (1.1)-(1.2).

---

1) Cf. K. Yoshida: On the integration of diffusion equations in Rimannian spaces, to appear in the Proc. Amer. Math. Soc.

2) See K. Friedrichs: Spektraltheorie halbbeschränkter Operatoren, Math. Ann. **109** (1934), 456-487. H. Fruedenthal: Über die Friedrichssche Fortsetzung halbbeschränkter Hermitescher Operatoren, Proc. Amsterdam Acad. **39** (1936), 832-833.

3) Suggested by M. Riesz: L'intégrale de Riemann-Liouville et le problème de Cauchy, Acta Math. **81** (1948), 1-223. Cf. L. Schwartz: Théorie des distributions, I (1950), p. 47.

4) This argument may be called a method of descent. Cf. J. Hadamard: Le problème de Cauchy et les équations aux dérivees partielles linéaires hyperboliques, (1932) p. 287.

An operator-theoretical integration of the wave equation

J. Math. Soc. Japan **8** (1956) 79–92

(Received Jan. 30, 1956)

§ 1. Introduction and the theorem. We consider the Cauchy problem for the wave equation in m-dimensional euclidean space E^m:

$$(1.1) \qquad \frac{\partial^2 u(t, x)}{\partial t^2} = Au(t, x), \quad u(0, x) = f(x), \quad u_t(0, x) = g(x),$$

$$A = a^{ij}(x)\frac{\partial^2}{\partial x_i \partial x_j} + b^i(x)\frac{\partial}{\partial x_i} + c(x), \quad x = (x_1, \cdots, x_m).$$

The problem is equivalent to the matricial equation

$$(1.1)' \qquad \frac{\partial}{\partial t}\begin{pmatrix} u(t, x) \\ v(t, x) \end{pmatrix} = \begin{pmatrix} 0 & I \\ A & 0 \end{pmatrix}\begin{pmatrix} u(t, x) \\ v(t, x) \end{pmatrix}, \quad \begin{pmatrix} u(0, x) \\ v(0, x) \end{pmatrix} = \begin{pmatrix} f(x) \\ g(x) \end{pmatrix},$$

and we may apply the theory of semi-group of linear operators[1] to the integration in the large of (1.1), by considering, in a suitable Banach space, the "resolvent equation"

$$(1.2) \qquad \left(\begin{pmatrix} I & 0 \\ 0 & I \end{pmatrix} - n^{-1}\begin{pmatrix} 0 & I \\ A & 0 \end{pmatrix}\right)\begin{pmatrix} u \\ v \end{pmatrix} = \begin{pmatrix} f \\ g \end{pmatrix} \quad \text{for large } |n|, \quad (n = \text{integer})$$

and obtaining the estimate

$$(1.3) \qquad \left\|\begin{pmatrix} u \\ v \end{pmatrix}\right\| \leq (1 + |n^{-1}|\beta)\left\|\begin{pmatrix} f \\ g \end{pmatrix}\right\|$$

where β is a positive constant independent of n, f and g. The irrelevance of the sign of n implies that

$$(1.4) \qquad \begin{pmatrix} 0 & I \\ A & 0 \end{pmatrix}$$

---

1) E. Hille: Functional Analysis and Semi-groups, New York (1948).

K. Yosida: On the differentiability and the representation of one-parameter semi-group of linear operators, J. Math. Soc. Japan, 1 (1948), 15–21.

generates a group $\{T_t\}_{-\infty < t < \infty}$ such that

$$(1.5) \qquad T_t \begin{pmatrix} f(x) \\ g(x) \end{pmatrix} = \begin{pmatrix} u(t, x) \\ v(t, x) \end{pmatrix}$$

yields a solution of $(1.1)'$ when the initial functions $\{f(x), g(x)\}$ are prescribed appropriately.

In this way we may prove the solvability of the Cauchy problem in the large of (1.1) without appealing to the classical Cauchy-Kowalewski existence theorem or to the Laplace-Fourier transform theory[2]. Our result reads as the

THEOREM. *Let* (i) *the coefficients* $a^{ij}(x)$, $b^i(x)$, $c(x)$ *are real-valued* C^∞ *functions and let*

$$(1.6) \qquad \max \left(\sup_x |a^{ij}(x)|, \ \sup_x |b^i(x)|, \ \sup_x |c(x)|, \ \sup_x \left| \frac{\partial a^{ji}(x)}{\partial x_k} \right|, \right.$$

$$\left. \sup_x \left| \frac{\partial b^i(x)}{\partial x_j} \right|, \ \sup_x \left| \frac{\partial^2 a^{ij}(x)}{\partial x_k \partial x_s} \right| \right) = \eta < \infty$$

Let (ii), *moreover, there exist positive constants* λ *and* μ *such that*

$$(1.7) \qquad \mu \sum_i \xi_i^2 \geq a^{ij}(x)\xi_i \xi_j \geq \lambda \sum_i \xi_i^2 .$$

Then there exists a positive constant β *such that, for sufficiently small positive constant* α_0, *the Cauchy problem for* (1.1) *is solvable in the following sense: For any pair* $\{f(x), g(x)\}$ *of* C^∞ *functions such that* $(A^k f)(x)$, $(A^k g)(x)$ *and their first partial derivatives are square integrable over* E^m *(for all* $k = 0, 1, \cdots$*), the equation* (1.1) *admits a* C^∞ *solution* $u(t, x)$ *satisfying the estimate*

$$(1.8) \quad ((u - \alpha_0 A u, u) + \alpha_0 (u_t, u_t))^{1/2} \leq \exp(\beta |t|) ((f - \alpha_0 A f, f) + \alpha_0 (g, g))^{1/2} ,$$

(h, k) *denoting, as usual, the inner product*

$$(1.9) \qquad (h, k) = \int_{E^m} h(x)k(x)dx, \ dx = dx_1 dx_2 \cdots dx_m .$$

---

2) Cf. J. Schauder: Der Anfangswertproblem einer quasi-linearen hyperbolischen Differentialgleichungen, Fund. Math. 24 (1935), 213–246, and J. Leray: Symbolic Calculus with Several Variables, Projections and Boundary Value Problems for Differential Equations, Princeton (1952). The two authors ingeneously make use of the Cauchy-Kowalewski existence theorem in their treatment.

Before proceeding to the proof of the theorem, we must prepare some lemmas concerning the elliptic differential operators A and its formal adjoint A^*:

$$(1.10) \quad (A^*f)(x) = \frac{\partial^2}{\partial x_i \partial x_j}(a^{ij}(x)f(x)) - \frac{\partial}{\partial x_i}(b^i(x)f(x)) + c(x)f(x).$$

§2. Lemma 1 (concerning the partial integration). Let H be the space of real-valued C^∞ functions $f(x)$ for which

$$(2.1) \qquad \|f\|_1 = \left(\int_{E^m} f^2 dx + \int_{E^m} \sum_i \left(\frac{\partial f}{\partial x_i} \right)^2 dx \right)^{1/2},$$

and let H_1 be the completion of H with respect to the norm $\|f\|_1$. Let similarly H_0 be the completion of H with respect to the norm

$$(2.2) \qquad \|f\|_0 = \left(\int_{E^m} f^2 dx \right)^{1/2}.$$

We have thus introduced two real Hilbert spaces H_1 and H_0, and H and H_1 are $\|\ \|_0$-dense in H_0.

LEMMA 1. *Let $f, g \in H_0$. and let $Af \in H_0$. Then we have*

$$(2.3) \qquad (Af, g) = -\int_{E^m} a^{ij} \frac{\partial f}{\partial x_i} \frac{\partial g}{\partial x_j} dx - \int_{E^m} \frac{\partial a^{ij}}{\partial x_j} \frac{\partial f}{\partial x_i} g\, dx$$

$$+ \int_{E^m} b^i \frac{\partial f}{\partial x_i} g\, dx + \int_{E^m} cfg\, dx,$$

viz. we may, in (Af, g), partially integrate the terms containing the second order derivatives as if the integrated terms are nought.

PROOF. By (1.6), $Af \in H_0$ and the fact that f and g both belong to H_1 we see that $a^{ij} \dfrac{\partial^2 f}{\partial x_i \partial x_j} \cdot g$ is integrable over E^m. We have, by Fubini's theorem,

$$\int_{E^m} a^{ij} \frac{\partial^2 f}{\partial x_i \partial x_j} \cdot g\, dx = \lim_{\substack{\delta_1 \to \infty \\ \varepsilon_1 \to -\infty}} \int_{-\infty}^{\infty} \cdots \int dx_2 \cdots dx_m \int_{\varepsilon_1}^{\delta_1} a^{ij} \frac{\partial^2 f}{\partial x_i \partial x_j} \cdot g\, dx_1,$$

and

$$\int_{\varepsilon_1}^{\delta_1} a^{ij} \frac{\partial^2 f}{\partial x_i \partial x_1} \cdot g\, dx_1 = \left[a^{1j} \frac{\partial f}{\partial x_j} g \right]_{x_1 = \varepsilon_1}^{x_1 = \delta_1} + \left\{ \int_{\varepsilon_1}^{\delta_1} - a^{1j} \frac{\partial f}{\partial x} \frac{gx}{\partial x_1} dx_1 \right.$$

$$-\int_{\epsilon_1}^{\delta_1} \frac{\partial a^{1j}}{\partial x_1} \frac{\partial f}{\partial x_1} \cdot g \, dx_1 \bigg] + \int_{\epsilon_1}^{\delta_1} \sum_{i,j \neq 2} a^{ij} \frac{\partial^2 f}{\partial x_i \partial x_j} g \, dx_1$$

$$= \kappa_1(\delta_1, \epsilon_1, x_2, \cdots, x_m) + \kappa_2(\delta_1, \epsilon_1, x_2, \cdots, x_m) + \kappa_3(\delta_1, \epsilon_1, x_2, \cdots, x_m) .$$

By (1.6) and Schwarz inequality we have

$$\left| \int_{-\infty}^{\infty} \cdots \int dx_2 \cdots dx_m \, \kappa_1(\delta_1, \epsilon_1, x_2, \cdots, x_m) \right| \leq$$

$$\eta \sum_i \left(\int_{-\infty}^{\infty} \cdots \int \left| \frac{\partial f(\delta_1, x_2, \cdots, x_m)}{\partial x_j} \right|^2 dx_2 \cdots dx_m \times \int_{-\infty}^{\infty} \cdots \int g(\delta_1, x_2, \cdots, x_m)^2 \, dx_2 \cdots dx_m \right)^{1/2}$$

$+$similar terms pertaining to ϵ_1 instead of δ_1.

We have, by Fubini's theorem,

$$\int_{E^m} g^2 \, dx = \int_{-\infty}^{\infty} dx_1 \int_{-\infty}^{\infty} \cdots \int g(x_1, \cdots, x_m)^2 \, dx_2 \cdots dx_m .$$

Hence we see that

$$\lim \int_{-\infty}^{\infty} \cdots \int g(x_1, \cdots, x_m)^2 \, dx_2 \cdots dx_m = 0$$

when x_1 tends to ∞ (or $-\infty$) without taking the values of x_1 which form a set of finite measure. The same reasoning applies also when we replace g by $\partial f / \partial x_j$. Therefore there exist two sequences $\{\delta_1^{(k)}\}$ and $\{\epsilon_1^{(k)}\}$ such that

$$(2.4) \qquad \lim_{\substack{\delta_1^{(k)} \to +\infty \\ \epsilon_1^{(k)} \to -\infty}} \int_{-\infty}^{\infty} \cdots \int \kappa_1(\delta_1^{(k)}, \epsilon_1^{(k)}, x_2, \cdots, x_m) \, dx_2 \cdots dx_m = 0 .$$

On the other hand, we see, remembering (1.6) and the fact that $f, g, \partial g / \partial x_1$ and $\partial f / \partial x_j$ belongs to H_0, that

$$\lim_{\substack{\delta_1 \to +\infty \\ \epsilon_1 \to -\infty}} \int_{-\infty}^{\infty} \cdots \int \kappa_2(\delta_1, \epsilon_1, x_2, \cdots, x_m) \, dx_2 \cdots dx_m$$

$$= \int_{E^m} \left\{ -a^{1j} \frac{\partial f}{\partial x_j} \frac{\partial g}{\partial x_1} - \frac{\partial a^{1j}}{\partial x_1} \frac{\partial f}{\partial x_j} g \right\} dx = \kappa_2$$

exists and is finite. Therefore

$$\int_{E^m} a^{ij} \frac{\partial^2 f}{\partial x_i \partial x_j} g\, dx = \kappa_2 + \lim_{\substack{\delta_1^{(k)} \to \infty \\ \varepsilon_1^{(k)} \to -\infty}} \int_{-\infty}^{\infty} \cdots \int \kappa_3(\delta_1^{(k)}, \varepsilon_1^{(k)}, x_2, \cdots, x_m)\, dx_2 \cdots dx_m.$$

Thus

$$\int_{\varepsilon_1^{(k)}}^{\delta_1^{(k)}} \sum_{i,j \neq 1} a^{ij} \frac{\partial^2 f}{\partial x_i \partial x_j} g\, dx_1$$

is integrable over the domain defined by $-\infty < x_i < \infty$ $(i=2,\cdots,m)$. Hence

$$\kappa_3 = \lim_{\substack{\delta_1^{(k)} \to \infty \\ \varepsilon_1^{(k)} \to -\infty}} \int_{-\infty}^{\infty} \cdots \int \kappa_3(\delta_1^{(k)}, \varepsilon_1^{(k)}, x_2, \cdots, x_m)\, dx_2 \cdots dx_m$$

$$= \lim_{\substack{\delta_1^{(k)} \to \infty \\ \varepsilon_1^{(k)} \to -\infty}} \lim_{\substack{\delta_2 \to \infty \\ \varepsilon_2 \to -\infty}} \int_{-\infty}^{\infty} \cdots \int dx_3 \cdots dx_m \left\{ \int_{\varepsilon_2}^{\delta_2} dx_2 \int_{\varepsilon_1^{(k)}}^{\delta_1^{(k)}} \sum_{i,j \neq 1} a^{ij} \frac{\partial^2 f}{\partial x_i \partial x_j} g\, dx_1 \right\}.$$

However

$$\left\{ \quad \right\} = \int_{\varepsilon_1^{(k)}}^{\delta_1^{(k)}} dx_1 \int_{\varepsilon_2}^{\delta_2} \sum_{i,j \neq 1} a^{ij} \frac{\partial^2 f}{\partial x_i \partial g_j} g\, dx_2$$

$$= \int_{\varepsilon_1^{(k)}}^{\delta_1^{(k)}} dx_1 \left[\left[a^{2j} \frac{\partial f}{\partial x_j} g \right]_{x_2 = \varepsilon_2}^{x_2 = \delta_2} + \int_{\varepsilon_2}^{\delta_2} -a^{2j} \frac{\partial f}{\partial x_j} \frac{\partial g}{\partial x_2}\, dx_2 - \int_{\varepsilon_2}^{\delta_2} \frac{\partial a^{2j}}{\partial x_2} \frac{\partial f}{\partial x_j} g dx_2 \right]$$

$$+ \int_{\varepsilon_1^{(k)}}^{\delta_1^{(k)}} \int_{\varepsilon_2}^{\delta_2} \sum_{i,j \neq 1,2} a^{ij} \frac{\partial^2 f}{\partial x_i \partial x_j} g\, dx_1\, dx_2.$$

By (1.6) and Schwarz inequality, we have

$$\left| \int_{-\infty}^{\infty} \cdots \int dx_3 \cdots dx_m \int_{\varepsilon_1^{(k)}}^{\delta_1^{(k)}} a^{2j} \frac{\partial f}{\partial x_j} g\, dx_1 \right|$$

$$\leq \eta \sum_j \left(\int_{-\infty}^{\infty} \cdots \int \left(\frac{\partial f}{\partial x_j} \right)^2 dx_1\, dx_3 \cdots dx_m \times \int_{-\infty}^{\infty} \cdots \int g^2\, dx_1\, dx_3 \cdots dx_m \right)^{1/2}$$

Since $\partial f / \partial x_j$ and g both belong to H_0 we see, as in the case of (2.4), that there exist two sequences $\{\delta_2^{(l)}\}$ and $\{\varepsilon_2^{(l)}\}$ such that

$$\lim_{\substack{\delta_3^{(l)}\to\infty \\ \varepsilon_2^{(l)}\to-\infty}} \int_{-\infty}^{\infty}\cdots\int dx_3\cdots dx_m\int_{\varepsilon_1}^{\delta_1}\left[a^{2j}\frac{\partial f}{\partial x_j}\,g\right]_{x_2=\varepsilon_2^{(l)}}^{x_2=\delta_2^{(l)}}dx_1=0$$

uniformly with respect to δ_1 and ε_1. We have also, by (1.6) and the fact that $\partial f/\partial x_j$, $\partial g/\partial x_2$ all belong to H_0,

$$\lim_{\substack{\delta_1^{(k)}\to\infty \\ \varepsilon_1^{(k)}\to-\infty}}\lim_{\substack{\delta_2\to\infty \\ \varepsilon_2\to-\infty}}\int_{-\infty}^{\infty}\cdots\int dx_3\cdots dx_m\int_{\varepsilon_1^{(k)}}^{\delta_1^{(k)}}dx_1\int_{\varepsilon_2}^{\delta_2}\left(-a^{2j}\frac{\partial f}{\partial x_j}\frac{\partial g}{\partial x_2}-\frac{\partial a^{2j}}{\partial x_2}\frac{\partial f}{\partial x_j}\,g\right)dx_2$$

$$=\int_{-\infty}^{\infty}\cdots\int\left(-a^{2j}\frac{\partial f}{\partial x_j}\frac{\partial g}{\partial x_2}-\frac{\partial a^{2j}}{\partial x_2}\frac{\partial f}{\partial x_j}\,g\right)dx_1\cdots dx_m.$$

Therefore

$$\int_{E^m}a^{ij}\frac{\partial^2 f}{\partial x_i\partial x_j}\,g\,dx=-\left(\int_{E^m}\sum_{i\text{ or }j=1,2}a^{ij}\frac{\partial f}{\partial x_i}\frac{\partial g}{\partial x_j}\,dx\right)$$

$$-\left(\int_{E^m}\sum_{i\text{ or }j=1,2}\frac{\partial a^{ij}}{\partial x_i}\frac{\partial f}{\partial x_j}\,g\,dx\right)$$

$$+\lim_{\substack{\delta_1^{(k)}\to\infty \\ \varepsilon_1^{(k)}\to-\infty}}\lim_{\substack{\delta_2^{(l)}\to\infty \\ \varepsilon_2^{(l)}\to-\infty}}\int_{-\infty}^{\infty}\cdots\int dx_3\cdots dx_m\int_{\varepsilon_1^{(k)}}^{\delta_1^{(k)}}\int_{\varepsilon_2^{(l)}}^{\delta_2^{(l)}}\sum_{i,j\neq 1,2}a^{ij}\frac{\partial^2 f}{\partial x_i\partial x_j}\,g\,dx_1\,dx_2.$$

Repeating the same argument we obtain (2.3).

REMARK. If $f,g\in H$ and if $A^*f\in H_0$, we may, in (A^*f,g), partially integrate the terms containing the second order derivatives as if the integrated terms are nought:

$$(2.3)'\quad (A^*f,g)=-\int_{E^m}a^{ij}\frac{\partial f}{\partial x_i}\frac{\partial g}{\partial x_j}\,dx-\int_{E^m}\frac{\partial a^{ij}}{\partial x_i}f\frac{\partial g}{\partial x_i}\,dx$$

$$-\int_{E^m}b^i f\frac{\partial g}{\partial x_i}\,dx+\int_{E^m}cfg\,dx.$$

COROLLARY. *There exist a positive constant κ and, for sufficiently small $\alpha>0$, positive constants γ and δ such that,*

$$(2.5)\qquad \delta\|f\|_1^2\leq \begin{matrix}(f-\alpha Af,f)\\[4pt](f-\alpha A^*f,f)\end{matrix}\leq(1+\alpha\gamma)\,\|f\|_1^2\quad \begin{matrix}\text{if }f\in H\quad\text{and}\quad Af\in H_0,\\[4pt]\text{if }f\in H\quad\text{and}\quad A^*f\in H_0,\end{matrix}$$

$$(2.6) \quad \begin{aligned} |(f-\alpha Af, g)| & \qquad \text{if } f, g \in H \text{ and } Af \in H_0, \\ & \leq (1+\alpha\gamma)\,\|f\|_1 \cdot \|g\|_1 \\ |(f-\alpha A^*f, g)| & \qquad \text{if } f, g \in H \text{ and } A^*f \in H_0, \end{aligned}$$

$$(2.7) \quad |(Af, g)-(f, Ag)| \leq \kappa \|f\|_1 \cdot \|g\|_0 \qquad \text{if } f, g \in H \text{ and } Af, Ag \in H_0.$$

PROOF. (2.5)–(2.6) may be proved by (1.6)–(1.7) and (2.3)–(2.3)′ remembering the inequality

$$2\alpha\,|ab| \leq \alpha\,(\epsilon|a|^2 + \epsilon^{-1}|b|^2), \qquad (\alpha \text{ and } \epsilon > 0),$$

since f and g both belong to H_1. Similarly we obtain (2.7) from

$$(Af, g)-(f, Ag) = -\int_{E^m} \left(2\frac{\partial a^{ij}}{\partial x_i}\frac{\partial f}{\partial x_j}g \right.$$
$$\left. + \frac{\partial^2 a^{ij}}{\partial x_i \partial x_j}fg - 2b^i\frac{\partial f}{\partial x_i}g - \frac{\partial b^i}{\partial x_i}fg \right) dx.$$

The right hand side is obtained by (2.3) and the expression obtained from (2.3) corresponding to (f, Ag) in which we have partially integrated the terms containing the factors like $f \times (\partial g/\partial x_j)$.

§ 3. Lemma 2 (concerning the existence of solutions of $u - n^{-2}Au = f$).

We invoke to Milgram-Lax theorem[3] for the proof of the Lemma 2 below. For the sake of completeness, we here give the full statement of the theorem together with its proof.

MILGRAM-LAX THEOREM. *Let a bilinear functional $B(u, v)$ defined on the Hilbert space H_1 satisfy the conditions:*

$$(3.1) \quad |B(u, v)| \leq \gamma'\|u\|_1 \cdot \|v\|_1, \qquad 0 < \gamma' < \infty,$$

$$(3.2) \quad \delta'\|u\|_1^2 \leq B(u, u), \qquad 0 < \delta' < \infty,$$

Then, to any $v \in H_1$ there corresponds a uniquely determined $v^* = Sv \in H_1$ such that

$$(3.3) \quad (u, v)_1 = B(u, Sv) \quad \text{for all } u \in H_1 \; ((u, v)_1 \text{ denotes the inner product in } H_1),$$

$$(3.4) \quad \delta'\|Sv\|_1 \leq \|v\|_1$$

---

3) P. D. Lax and A. N. Milgram: Parabolic Equations in " Contributions to the Theory of Partial Differential Equations ", Princeton (1954), 167–190.

Proof. Let $\{v, v^*\}$ be a pair of elements of H_1 for which we have $(u, v)_1 = B(u, v^*)$ for every $u \in H_1$. v^* is determined uniquely by v, since $B(u, v^*) = 0$ for all $u \in H_1$ implies

$$\delta' \|v^*\|_1^2 \leq B(v^*, v^*) = 0$$

Moreover, the operator $S(v^* = Sv)$ is continuous and (3.4) holds good since

$$\delta' \|Sv\|_1^2 \leq B(Sv, Sv) = (Sv, v)_1 \leq \|Sv\|_1 \cdot \|v\|_1 .$$

Thus the domain $D(S)$ of the operator S is a closed linear subspace of H_1. Assume $D(S) \neq H_1$. Then there exists $w_0 \in H_1$ such that

$$(3.5) \qquad (w_0, v)_1 = 0 \quad \text{for every} \quad v \in D(S) \quad \text{and} \quad \|w_0\| \neq 0 .$$

We consider the linear functional $F(z) = B(z, w_0)$ on H_1. It is a bounded functional since

$$|F(z)| = |B(z, w_0)| \leq \gamma' \|z\|_1 \cdot \|w_0\|_1 ,$$

and hence, by Riesz theorem, there exists $w_0' \in H_1$ such that $F(z) = B(z, w_0) = (z, w_0')_1$. Therefore $w_0' \in D(S)$ and $Sw_0' = w_0$. This is a contradiction, because of (3.5) and (3.2):

$$\delta' \|w_0\|_1^2 \leq B(w_0, w_0) = (w_0, w_0')_1 = 0 .$$

Therefore $D(S) = H_1$ and the theorem is proved.

Lemma 2. *Let a positive number α_0 be chosen so small that the Corollary of the Lemma 1 is valid for $0 < \alpha \leq \alpha_0$. Then, for any function $f(x) \in H$, the equation*

$$(3.6) \qquad u - \alpha A u = f \qquad (0 < \alpha \leq \alpha_0)$$

admits a uniquely determined solution $u_f(x) \in H$.

Proof. Let us define a bilinear functional

$$\hat{B}(u, v) = (u - \alpha A^* u, v)$$

for functions $u, v \in H$ satisfying $A^* u \in H_0$. From the Corollary of the Lemma 1, we have

$$(3.7) \qquad |\hat{B}(u, v)| \leq (1 + \alpha \gamma) \|u\|_1 \cdot \|v\|_1 , \quad \delta \|u\|_1^2 \leq \hat{B}(u, u) .$$

Hence, by continuity, $\hat{B}(u, v)$ may be extended to the bilinear functional $B(u, v)$ defined on H_1 satisfying

$$(3.7)' \qquad |B(u, v)| \leq (1 + \alpha\gamma) \|u\|_1 \cdot \|v\|_1 , \quad \delta \|u\|_1^2 \leq B(u, u) .$$

Consider the linear functional $F(u) = (u, f)$ defined on H_1. It is a bounded functional since

$$|(u, f)| \leq \|u\|_0 \cdot \|f\|_0 \leq \|u\|_1 \cdot \|f\|_1 ,$$

and hence, by Riesz theorem, there exists a uniquely determined $v = v(f) \in H_1$ such that $(u, f) = (u, v(f))_1$. Thus, by Milgram-Lax theorem, we have

$$(3.8) \qquad (u, f) = B(u, Sv(f)) \quad \text{for all} \quad u \in H_1 .$$

Let u run over C^∞ functions with compact supports, and let $v_n \in H$ be such that

$$\lim_{n \to \infty} \|v_n - Sv(f)\|_1 = 0 .$$

Then

$$B(u, Sv(f)) = \lim_{n \to \infty} B(u, v_n) = \lim_{n \to \infty} \hat{B}(u, v_n) = \lim_{n \to \infty} (u - \alpha A^* u, v_n)$$

$$= (u - \alpha A^* u, Sv(f)) ,$$

since the norm $\| \ \|_1$ is larger than the norm $\| \ \|_0$. Hence

$$(3.8)' \qquad (u, f) = (u - \alpha A^* u, Sv(f)) .$$

$f(x)$ being any C^∞ function with compact support and $(I - \alpha A^*)$ being an elliptic differential operator with C^∞ coefficients, we see, by L. Schwartz theorem[4], that $u_f = Sv(f) \in H_1$ is a C^∞ solution of (3.6).

The proof of the uniqueness of the solution of (3.6). Let a function $u \in H$ satisfy

$$u - \alpha A u = 0 ,$$

Then Au belongs to H and hence to H_0. Thus, by the Corollary of the Lemma 1, we obtain

$$0 = (u - \alpha A u, u) \geq \delta \|u\|_1^2 , \quad \text{viz.} \quad u = 0 .$$

§4. Proof of the Theorem.

We first prove the

LEMMA 3. *Let the integer n be such that $|n^{-1}|$ is sufficiently small.*

---

4) L. Schwartz: Théorie des Distributions, Paris (1950), 136.

Then, for any pair $\{f, g\}$ of elements $\in H$ such that $Af \in H_0$, the resolvent equation

$$(1.2) \qquad \left(\begin{pmatrix} I & 0 \\ 0 & I \end{pmatrix} - n^{-1} \begin{pmatrix} 0 & I \\ A & 0 \end{pmatrix} \right) \begin{pmatrix} u \\ v \end{pmatrix} = \begin{pmatrix} f \\ g \end{pmatrix}$$

admits a uniquely determined solutions $\{u, v\}$, u and $v \in H$, satisfying

$$(4.1) \quad ((u - \alpha_0 Au, u) + \alpha_0(v, v))^{1/2} \leqq (1 + \beta |n^{-1}|) ((f - \alpha_0 Af, f) + \alpha_0(g, g))^{1/2},$$

with a positive constont β independent of n and $\{f, g\}$.

PROOF. Let $u_1 \in H$ and $v_2 \in H$ respectively be the solutions of

$$u_1 - n^{-2} Au_1 = f \quad \text{and} \quad v_2 - n^{-2} Av_2 = g.$$

The existence of such solutions was proved in the Lemma 2. Then

$$u = u_1 + n^{-1} v_2, \quad v = n^{-1} Au_1 + v_2$$

satisfies (1.2).

The proof of (4.1). We first remark that

$$Au = n(v - g) \in H \subseteqq H_0 \quad \text{and hence} \quad Av = n(Au - Af) \in H_0.$$

Therefore we may apply the Corollary of the Lemma 1. Thus, by (1.2),

$$(f - \alpha_0 Af, f) = (u - n^{-1}v - \alpha_0 A(u - n^{-1}v), u - n^{-1}v)$$

$$= (u - \alpha_0 Au, u) - 2n^{-1}(u, v) + \alpha_0 n^{-1}(Au, v) + \alpha_0 n^{-1}(Av, u)$$

$$+ n^{-2}(v - \alpha_0 Av, v)$$

and

$$\alpha_0(g, g) = \alpha_0(v - n^{-1}Au, v - n^{-1}Au)$$

$$= \alpha_0(v, v) - \alpha_0 n^{-1}(v, Au) - \alpha_0 n^{-1}(Au, v) + \alpha_0 n^{-2}(Au, Au)$$

imply that there exists a positive constant β satisfying

$$(f - \alpha_0 Af, f) + \alpha_0(g, g) \geqq (u - \alpha_0 Au, u) + \alpha_0(v, v)$$

$$- \alpha_0 |n^{-1}| \, |(Av, u) - (Au, v)| - 2|n^{-1}| \, |(u, v)|$$

$$\geqq (1 + \beta |n^{-1}|)^{-2} ((u - \alpha_0 Au, u) + \alpha_0(v, v))$$

for sufficiently large $|n|$.

The above estimate for the solutions $\{u, v\}$ belonging to H shows that the solutions are uniquely determined by $\{f, g\}$. Q. E. D.

An operator-theoretical integration of the wave equation.

The product space $H_1 \otimes H_0$ of vectors

$$(4.2) \qquad \binom{u}{v} = \{u, v\}', \quad \text{where} \quad u \in H_1 \quad \text{and} \quad v \in H_0,$$

is a Banach space by the norm

$$(4.3) \qquad \left\| \binom{u}{v} \right\| = \| \{u, v\}' \| = ((u - \alpha_0 Au, u) + \alpha_0 (v, v))^{1/2}.$$

Let the domain $D(\mathfrak{A})$ of the operator

$$(4.4) \qquad \mathfrak{A} = \begin{pmatrix} 0 & I \\ A & 0 \end{pmatrix}$$

be the vectors $\{u, v\}' \in H_1 \otimes H_0$ such that

$$u, v \in H \quad \text{and} \quad A(u - n^{-1}v) \in H_0, \quad v - n^{-1}Au \in H.$$

Then, the Lemma 3 shows that the range of the additive operator $\begin{pmatrix} I & 0 \\ 0 & I \end{pmatrix} - n^{-1}\mathfrak{A}$ coincides with the set of vectors $\{f, g\}'$ in the Lemma 3. Moreover, it is easy to see that the set of such vectors $\{f, g\}'$ is $\| \ \|$-dense in the Banach space $H_1 \otimes H_0$. Hence we have the

COROLLARY. *The smallest closed extension* $\overline{\mathfrak{A}}$ *of the operator* $\mathfrak{A}$ *is such that the operator*

$$(4.5) \qquad \mathfrak{J} - n^{-1}\overline{\mathfrak{A}} = \begin{pmatrix} I & 0 \\ 0 & I \end{pmatrix} - n^{-1}\overline{\mathfrak{A}}, \qquad (n = integer),$$

admits, for sufficiently large $|n|$, *everywhere (in* $H_1 \otimes H_0$*) defined inverse* $\mathfrak{J}_n = (\mathfrak{J} - n^{-1}\overline{\mathfrak{A}})^{-1}$ *satisfying*

$$(4.6) \qquad \| \mathfrak{J}_n \| \leqq (1 + \beta |n^{-1}|).$$

Hence, by the semi-group theory[1] and the irrelevance of the sign of n, there exists a uniquely determined group T_t:

$$(4.7) \qquad T_t \binom{f}{g} = \text{strong} \lim_{n \to \infty} \exp (t \overline{\mathfrak{A}} \mathfrak{J}_n) \binom{f}{g}$$

of linear bounded operators T_t on $H_1 \otimes H_0$ into $H_1 \otimes H_0$ such that

$$(4.8) \qquad T_t T_s = T_{t+s} (-\infty < t, s < \infty), \quad T_0 = \text{the identity I},$$

$$(4.9) \qquad \| T_t \| \leqq \exp (\beta |t|), \quad \text{strong} \lim_{t \to t_0} T_t \binom{f}{g} = T_{t_0} \binom{f}{g},$$

K. Yosida

(4.10)   if $\begin{pmatrix} f \\ g \end{pmatrix}$ is in the domain of the "infinitesimal generator" $\overline{\mathfrak{A}}$,

we have strong $\lim_{h \to 0} h^{-1}(T_{t+h}-T_t)\begin{pmatrix} f \\ g \end{pmatrix} = \overline{\mathfrak{A}}T_t\begin{pmatrix} f \\ g \end{pmatrix} = T_t\overline{\mathfrak{A}}\begin{pmatrix} f \\ g \end{pmatrix}$

Now, by the assumption of the Theorem,

(4.11) $A^k f \in H$ and $A^k g \in H$ $(k=0, 1, \cdots)$.

Hence we see that

(4.12) $\overline{\mathfrak{A}}^k\begin{pmatrix} f \\ g \end{pmatrix} = \mathfrak{A}^k\begin{pmatrix} f \\ g \end{pmatrix} \in H_1 \otimes H_0$ $(k=0, 1, \cdots)$,

viz. the vector $\{f, g\}'$ is in the domain of the every power of $\overline{\mathfrak{A}}$. Therefore, by (4.10), the vectors

(4.13) $\begin{pmatrix} u(t, x) \\ v(t, x) \end{pmatrix} = T_t\begin{pmatrix} f(x) \\ g(x) \end{pmatrix}$

are in the domain of every power of $\overline{\mathfrak{A}}$ and

$$\overline{\mathfrak{A}}^k\begin{pmatrix} u(t, x) \\ v(t, x) \end{pmatrix}$$

belongs to $H_1 \otimes H_0$. Therefore, the "distribution"

(4.14) $U_t \cdot \varphi = \int_{E^m} u(t, x)\,\varphi(x)\,dx$ (the testing functions φ run over

C^∞ functions with compact supports)

is such that, for every $k=0, 1, \cdots$, the "distribution"

(4.15) $A^k U_t$

is the "distribution" defined by a function which is locally summable (in the truth, this function belongs to H_0). A being an elliptic differential operator with C^∞ coefficients, we see, by a theorem due to L. Schwartz[5], that $u(t, x)$ is a C^∞ function in x.

Thus $u(t, x)$ is, for fixed t, not only belongs to H_1 but also belongs

---

5) L. Schwartz: Théorie des Distributions, II, Paris (1951), 47. Actually, the theorem is proved for the case when $A=$ the Laplacian. However, since the proof is based upon the fact that the parametrix of the iterated Laplacian Δ^k becomes more smooth as k becomes large, the theorem may be extended to general elliptic differential operator A with C^∞ coefficients.

556

to H. Hence the value at $(t, x) = (t, x_1, \cdots, x_m)$ of $u(t, x)$ is determined without ambiguity. We also see, from (4.7), that this function $u(t, x)$ is measurable in (t, x). And, by the estimate (4.9), we see that the function $u(t, x)$ is locally summable in $t-x$ space. To this function we may apply every power of A and hence every power of

$$(4.16) \qquad \partial_{t^2} = \text{the strong second order derivative with respect to } t,$$

and

$$(4.17) \qquad (\partial_{t^2})^k\, u(t, x) = A^k u(t, x) \qquad (k = 0, 1, \cdots).$$

This we see by (4.10) and the fact that (4.12) holds good for our initial functions $\{f, g\}'$. Thus the "distribution"

$$(4.18) \qquad U\psi = \int_{Em}\int_{-\infty}^{\infty} u(t, x)\, \psi(t, x)\, dx dt \text{ (the testing functions } \psi \text{ run over}$$

C^∞ functions in (t, x) with compact supports)

is such that, for any $k = 0, 1, \cdots$, the "distribution"

$$(4.19) \qquad \left(\frac{\partial^2}{\partial t^2} + A\right)^k U = (2A)^k U$$

is a "distribution" defined by a locally summable function in (t, x). The operator

$$\left(\frac{\partial^2}{\partial t^2} + A\right)$$

being elliptic in (t, x), we see, again by making use of Schwartz theorem[5], that $u(t, x)$ is a C^∞ function in (t, x). Thus it is easy to see that $u(t, x)$ is a C^∞ solution of (1.1).

Finally the inequality (1.8) is identical with the estimate $\|T_t\| \leq \exp(\beta|t|)$ in (4.9).

REMARK 1. We may prove

$$(1.8)' \qquad ((A^k u - \alpha_0 A^{k+1} u,\ A^k u) + \alpha_0 (A^k u_t,\ A^k u_t))^{1/2}$$

$$\leq \exp(\beta|t|)((A^k f - \alpha_0 A^{k+1} f,\ A^k f) + \alpha_0(A^k g,\ A^k g))^{1/2},\ (k = 0, 1, \cdots),$$

since $(A^k u)(t, x)$ is the solution of the original wave equation (1.1) with the initial condition

$$(A^k u)(0, x) = (A^k f)(x),\ (A^k u)(0, x) = (A^k g)(x),$$

K. Yosida

to be obtained by our method.

REMARK 2. The above obtained solution $u(t, x)$ together with $v(t, x) = u_t(t, x)$ satisfy, by (4.10) and (4.9),

$$(4.20) \qquad \left\| h^{-1} \begin{pmatrix} u(t+h, x) - u(t, x) \\ v(t+h, x) - v(t, x) \end{pmatrix} - \begin{pmatrix} 0 & I \\ A & 0 \end{pmatrix} \begin{pmatrix} u(t, x) \\ v(t, x) \end{pmatrix} \right\| \to 0 \quad \text{as } h \to 0,$$

$$\left\| \begin{pmatrix} u(t, x) \\ v(t, x) \end{pmatrix} - \begin{pmatrix} f(x) \\ g(x) \end{pmatrix} \right\| \to 0 \quad \text{as } t \to 0,$$

$$\left\| \begin{pmatrix} u(t, x) \\ v(t, x) \end{pmatrix} \right\| \leq \exp(\beta|t|) \left\| \begin{pmatrix} f(x) \\ g(x) \end{pmatrix} \right\|.$$

As was proved by E. Hille[6], such solution is unique since the resolvent $\mathfrak{J}_n = (\mathfrak{J} - n^{-1}\overline{\mathfrak{A}})^{-1}$ exists and satisfies (4.6) for sufficiently large $|n|$, n denoting integers.

Department of Mathematics,
Tokyo University.

---

6)   A note on Cauchy's problem, Ann. Soc. Polonaise de Math., 25 (1952), 59.

An operator-theoretical integration of the temporally inhomogeneous wave equation

J. Fac. Sci. Univ. Tokyo I. 7 (1957) 463–466

1. Introduction. We consider the Cauchy problem for the wave equation in a connected domain G of the m-dimensional euclidean space E^m:

$$(1.1) \qquad \frac{\partial^2 u(t, x)}{\partial t^2} = A_t u(t, x), \ u(0, x) = f(x), \ \frac{\partial u(0, x)}{\partial t} = g(x),$$

$$A_t = a_{ij}(t, x) \frac{\partial^2}{\partial x_i \partial x_k} + b_i(t, x) \frac{\partial}{\partial x_i} + c(t, x), \ x = (x_1, \cdots, x_m).$$

We assume that the coefficients $a_{ij}(t, x)$, $b_i(t, x)$ and $c(t, x)$ are C^∞ functions of (t, x) for $x \in G$ and $t_1 < t < t_2$, where $-\infty \leq t_1 < 0 < t_2 \leq \infty$. We also assume that A_t is elliptic in the sense that the quadratic form $a_{ij}(t, x)\, \xi_i\, \xi_j$ is >0 for $\sum\limits_{i=1}^{m} \xi_i^2 > 0$.

The problem (1.1) is equivalent to the matricial equation

$$(1.1)' \qquad \frac{\partial}{\partial t} \begin{pmatrix} u(t, x) \\ v(t, x) \end{pmatrix} = \mathfrak{A}_t \begin{pmatrix} u(t, x) \\ v(t, x) \end{pmatrix} = \begin{pmatrix} 0 & I \\ A_t & 0 \end{pmatrix} \begin{pmatrix} u(t, x) \\ v(t, x) \end{pmatrix}, \ \begin{pmatrix} u(0, x) \\ v(0, x) \end{pmatrix} = \begin{pmatrix} f(x) \\ g(x) \end{pmatrix}.$$

The purpose of the present note is to give a sketch of a method of integration in the large of (1.1)'. It is simple and is based upon the same idea which was applied to the integration in the large of the temporally inhomogeneous diffusion equation.[1] We firstly consider an approximate equation

$$(1.2) \qquad \frac{\partial}{\partial t} \begin{pmatrix} u^{(n)}(t, x) \\ v^{(n)}(t, x) \end{pmatrix} = \mathfrak{A}_t(\mathfrak{I} - n^{-1}\mathfrak{A}_t)^{-1} \begin{pmatrix} u^{(n)}(t, x) \\ v^{(n)}(t, x) \end{pmatrix}, \ \begin{pmatrix} u^{(n)}(t, x) \\ v^{(n)}(t, x) \end{pmatrix} = \begin{pmatrix} f(x) \\ g(x) \end{pmatrix},$$

$$\mathfrak{I} = \begin{pmatrix} I & 0 \\ 0 & I \end{pmatrix},$$

---

1) K. Yosida: Integration of temporally inhomogeneous diffusion equation in a Riemannian space, Proc. Japan Acad., **30** (1954), 19–23 and 273–275. Another ingeneous method of integration of the temporally inhomogeneous wave equation was developed by P. Lax: On Cauchy's problem for hyperbolic equations and the differentiability of solutions of elliptic equations, Comm. of Pure and Applied Math., 8 (1956), 615–633. Prof. J. L. Lions kindly communicated to the author that he has devised a method of integration applicable to a wide class of the temporally inhomogeneous wave equations by adapting Galerkin's procedure. He also communicated to the author that the method of the present paper can be extended to the case where A_t is an elliptic differential operator of order higher than 2, provided the highest order terms are of constant coefficients.

the integration of which is easy since the operator

$$(1.3) \qquad \mathfrak{A}_t^{(n)} = \mathfrak{A}_t\, (\mathfrak{J} - n^{-1}\mathfrak{A}_t)^{-1}$$

would be a bounded linear operator in a suitable Banach space. We then let n tend to ∞ and obtain the genuine solution of $(1.1)'$ if the initial functions $f(x)$ and $g(x)$ are prescribed appropriately.

2. The approximate equation. Let k be an integer $\geqq 0$. The totality of the pairs $\begin{pmatrix} f(x) \\ g(x) \end{pmatrix}$ of C^∞ functions with compact supports contained in the interior of G constitutes a pre-Hilbert space $H_k \times H_{k-1}$ by the norm

$$(2.1) \qquad \left\| \begin{pmatrix} f \\ g \end{pmatrix} \right\|_k = \left(\sum_{|n| \leq k} \int_G |D^{(n)} f(x)|^2 \, dx + \sum_{|n| \leq k-1} \int_G |D^{(n)} g(x)|^2 \, dx \right)^{1/2}, \quad \text{where}$$

$$D^{(n)} = \partial^{n_1 + \cdots + n_m} / \partial x_1^{n_1} \cdots \partial x_m^{n_m}, \quad |n| = \sum n_i, \quad dx = dx_1\, dx_2 \cdots dx_m.$$

Let $\mathfrak{H}_k$ be the Hilbert space obtained as the completion of $H_k \times H_{k-1}$, and we consider $\mathfrak{A}_t$ as an additive operator defined on $H_k \times H_{k-1} \leqq \mathfrak{A}_k$ into $H_k \times H_{k-1} \leqq \mathfrak{H}_k$. Let $\overline{\mathfrak{A}_t}$ be the smallest closed extension of $\mathfrak{A}_t$. We assume the following Hypothesis is satisfied.

Hypothesis. Let the positive integer k be sufficiently large, to be specified later. Let, for sufficiently large positive integer $|n|$ which is independent of t with $t_1 < t < t_2$, the resolvent

$$(2.2) \qquad \mathfrak{J}_t^{(n)} = (\mathfrak{J} - n^{-1}\overline{\mathfrak{A}_t})^{-1}$$

exists as a bounded linear operator on $\mathfrak{H}_k$ into $\mathfrak{H}_k$ such that

$(2.3) \qquad ||\mathfrak{J}_t^{(n)}||_k \leqq (1 + \beta\, |n^{-1}|)$ with positive constant β which is independent of n and t,

$(2.4) \qquad \mathfrak{J}_t^{(n)} U$ is strongly continuous in t for every $U \in \mathfrak{H}_k$.

Remark. The existence of the resolvents $\mathfrak{J}_t^{(n)}$ with the property (2.3) may be verified by applying Milgram-Lax theorem, if we assume, for a certain $\alpha > 0$, the norm $(((I - \alpha\, A_t)^k f, f) + ((I - \alpha\, A_t)^{k-1} g, g))^{1/2}$ is equivalent to the norm $\left\| \begin{pmatrix} f \\ g \end{pmatrix} \right\|_k$. Cf. the preceding paper[2].

The Hypothesis implies that

$$(2.5) \qquad \mathfrak{A}_t^{(n)} = \overline{\mathfrak{A}_t}\,(\mathfrak{J} - n^{-1}\overline{\mathfrak{A}_t})^{-1} = n\,(\mathfrak{J}_t^{(n)} - \mathfrak{J})$$

---

2) K. Yosida: An operator-theoretical integration of the wave equation, J. Math. Soc. Japan, **8**, No. 1 (1956), 79–92. See also J. L. Lions: Une remarque sur les applications du Théorème de Hille-Yosida, to appear in the J. Math. Soc. Japan. It is to be noted here that Prof. P. Lax kindly communicated to the author that he had applied the semi-group theory to the mixed boundary value problem of the wave equation $u_{tt} = Au + Mu_t$, M denoting any first order operator (Abstract No. 180, Bull. Amer. Math. Soc., **58**, no. 2 (1952), 182). According to him, his proof of the differentiability in t of the operator solution was entirely different from mine.

is a bounded linear operator on $\mathfrak{H}_k$ into $\mathfrak{H}_k$ strongly continuous in t for $t_1 < t < t_2$. Thus, by (2.4), we may obtain, by the method of successive approximation, the solution $U_t^{(n)} \in \mathfrak{H}_k$ of the approximate equation (where $nt > 0$)

$$(2.6) \qquad D_t U_t^{(n)} = \text{strong} \lim_{\delta \to 0} \delta^{-1}(U_{t+\delta}^{(n)} - U_t^{(n)}) = \mathfrak{A}_t^{(n)} U_t^{(n)}, \quad U_0^{(n)} = F \in \mathfrak{H}_k.$$

For the approximate solutions, we may prove the fundamental

Lemma.

$$(2.7) \qquad ||U_t^{(n)}||_k \leq \exp(\beta|t|) \cdot ||F||_k, \quad \text{for } nt > 0.$$

Proof. We have, by (2.5) and (2.6),

$$U_{t+\delta}^{(n)} - U_t^{(n)} = \delta n(I_t^{(n)} - I) U_t^{(n)} + o(\delta).$$

Let $t > 0$ and let $\delta > 0$ be sufficiently small. Then we have, by (2.3) and $n > 0$,

$$||U_{t+\delta}^{(n)}||_k \leq (1 - n\delta)||U_t^{(n)}||_k + \delta n(1 + \beta n^{-1})||U_t^{(n)}||_k + o(\delta) = (1 + \beta\delta)||U_t^{(n)}||_k + o(\delta).$$

Hence we obtain

$$(2.8) \qquad \frac{d^+||U_t^{(n)}||_k}{dt} \leq \beta ||U_t^{(n)}||_k, \quad t \geq 0.$$

Similarly we obtain, for $n < 0$,

$$(2.9) \qquad \frac{d^-||U_t^{(n)}||_k}{dt} \geq -\beta ||U_t^{(n)}||_k, \quad t \leq 0.$$

Therefore, by the initial condition $U_0^{(n)} = F$, we have (2.7).

3. The integration of (1.1)′. Let $D(G)$ be the totality of pairs of C^∞ functions whose supports are compact and are contained in the interior of G. Then we have, by (2.6) and partial integration,

$$(3.1) \qquad (U_t^{(n)}, \Phi)_0 = (F, \Phi)_0 + \int_0^t (\mathfrak{A}_\tau^{(n)} U_\tau^{(n)}, \Phi)_0 \, d\tau$$

$$= (F, \Phi)_0 + \int_0^t (I_\tau^{(n)} U_\tau^{(n)}, \mathfrak{A}_\tau^* \Phi) \, d\tau, \quad \Phi = \begin{pmatrix} \varphi(x) \\ \psi(x) \end{pmatrix} \in D(G),$$

where

$$(3.2) \qquad (F, \Phi)_0 = \int_G f(x)\varphi(x)dx + \int_G g(x)\psi(x)dx,$$

and

$$(3.3) \qquad \mathfrak{A}_t^* = \begin{pmatrix} 0 & A_t^* \\ I & 0 \end{pmatrix}, \quad A_t^* = \text{the formal adjoint of } A_t.$$

Thus, by (2.7), we see that $(U_t^{(n)}, \Phi)_0$ is, for fixed $\Phi \in D(G)$, equi-continuous in t for variable n with $nt > 0$. Therefore we see, by making use of (2.7) again, that there exists, for all t simultaneously, a subsequence $\{U_t^{(n')}\}$ of $\{U_t^{(n)}\}$ with the property

$$(3.4) \qquad \text{weak} \lim_{|n'| \to \infty} U_t^{(n')} = U_t \text{ exists such that } ||U_t||_k \leq \exp(\beta|t|) \, ||F||_k.$$

We next prove that

$$(3.5) \qquad \lim_{|n'| \to \infty} (\mathfrak{J}_t^{(n')} U_t^{(n')}, \mathfrak{A}_t^* \Phi)_0 = (U_t, \mathfrak{A}_t^* \Phi)_0, \quad \Phi \in D(G),$$

boundedly in t for any compact interval of t. This may be proved by (3.4),

$$|(\mathfrak{J}_t^{(n)} U_t^{(n)}, \mathfrak{A}_t^* \Phi)_0 - (\mathfrak{J}_t^{(n)} U_t^{(n)}, (\mathfrak{J} - n^{-1} \mathfrak{A}_t^*) \mathfrak{A}_t^* \Phi)_0|$$

$$\leq n^{-1} |(\mathfrak{J}_t^{(n)} U_t^{(n)}, \mathfrak{A}_t^{*2} \Phi)_0| \leq |n^{-1}|(1 + \beta |n^{-1}|) \exp(\beta |t|) ||F||_k \cdot ||\mathfrak{A}_t^{*2} \Phi||_k$$

and

$$(\mathfrak{J}_t^{(n)} U_t^{(n)}, (\mathfrak{J} - n^{-1} \mathfrak{A}_t^*) \mathfrak{A}_t^* \Phi)_0 = (U_t^{(n)}, \mathfrak{A}_t^* \Phi)_0 .$$

Hence we have

$$(3.6) \qquad (U_t, \Phi)_0 = (F, \Phi)_0 + \int_0^t (U_\tau, \mathfrak{A}_\tau^* \Phi)_0 \, d\tau .$$

On the other hand, since $U_t = \begin{pmatrix} u(t, x) \\ v(t, x) \end{pmatrix} \in \mathfrak{H}_k$, the Soboleff lemma shows that for fixed t,

$$(3.7) \qquad u(t, x) \in C^{k-[m/2]-1} \quad \text{and} \quad v(t, x) \in C^{k-1-[m/2]-1} \quad \text{in } G .$$

Hence, by partial integration,

$$(3.8) \qquad (U_t, \Phi)_0 = (F, \Phi)_0 + \int_0^t (\mathfrak{A}_\tau U_\tau, \Phi)_0 \, d\tau , \quad \Phi \in D(G) .$$

Thus we see, by (2.7), that, for sufficiently large k, U_t is, in the parameter t, differentiable number of times locally and weakly in G such that the differential quotients being locally integrable in (t, x). Hence, again by making use of Soboleff lemma, we see, for sufficiently large k, that $U_t = U(t, x)$ is C^1 in t and C^2 in x. Thus (3.8) shows that $U(t, x)$ is a genuine solution of (1.1)'.

Department of Mathematics, Tokyo University.

IX. Potential Theory

Comments (Shinzo Watanabe)

Yosida's mathematical works since 1965 were devoted mainly to operator theoretical treatments of potential theory. These works were inspired, as he himself stated, by the fheory of G. A. Hunt (Illinois J. Math. 1 and 2, 1957–58), a highlight in the modern investigation of the relationship between Markov processes and potential theory which has been extensively studied following the pioneering works by P. Lévy, S. Kakutani and J. L. Doob. Yosida was mainly interested in clarifying, from the operator theoretical point of view, the analytical mechanism in those parts of Hunt's theory which are concerned with potential operators of Markov processes. However, Yosida's potential operators can be defined in a general setting of abstract Banach spaces, cf. [100], [101].

Yosida's definition of the potential operator V associated to a strongly continuous equi-bounded semigroup T_t of linear operators on a Banach space B is as follows. Let $J_\lambda f = (\lambda - A)^{-1} f = \int_0^\infty e^{-\lambda t} T_t f dt$, $\lambda > 0$, $f \in B$, be the resolvent where A is the infinitesimal generator with domain $D(A)$. Then, the potential operator V is defined, when and only when $D(V) := \{f \in B | {}^3 s\text{-}\lim_{\lambda \downarrow 0} J_\lambda f\}$ is dense in B, by $Vf = s\text{-}\lim_{\lambda \downarrow 0} J_\lambda f$ for $f \in D(V)$. Among others, a main result on potential operators is as follows, cf. [91], [95] and [97].

(i) If the potential operator V can be defined, then $D(V) = R(A)$ (the range of A), $A: D(A) \to R(A)$ is one-to-one, and $V = A^{-1}$.

(ii) The following conditions are equivalent:

    (a) The potential operator V can be defined.

    (b) $R(A)$ is dense in B.

    (c) $s\text{-}\lim_{\lambda \downarrow 0} \lambda J_\lambda f = 0$ for all $f \in B$.

    (d) $w\text{-}\lim_{\lambda \downarrow 0} \lambda J_\lambda f = 0$ for all $f \in B$.

In the proof, a key is the Abelian ergodic theorem or the ergodic theorem of Hille-type: $R(I - \lambda J_\lambda)$ is independent of $\lambda > 0$ and its closure in B is equal to $\{f \in B | s\text{-}\lim_{\lambda \downarrow 0} \lambda J_\lambda f = 0\}$.

The potential operators were studied in [95], [96] and [99] in the following setting which seems to be the most important case: $B = C_\infty(X)$ is the closure with respect to the uniform norm of the space $C_0(X)$ of all continuous functions with compact support on a locally compact but non-compact separable Hausdorff space X and T_t is a strongly continuous contraction semigroup of positive operators on B, (i.e. a Markov process semigroup or a Feller semigroup). Among many results in these papers, the following one in [95] is, perhaps, the most important in the sense that it proved one of the fundamental results of Hunt by a masterful operator theoretical method. It is stated as follows: If V is a positive linear operator from $C_0(X)$ into $C_\infty(X)$ with dense range satisfying the *principle of positive maximum*, then there exists a unique Feller semigroup T_t on $C_\infty(X)$ such that $Vf = \int_0^\infty T_t f \, dt$ for $f \in C_0(X)$. Here the principle of positive maximum is that, for every $f \in C_0(x)$, $\sup\{Vf(x)|x \in X, f(x) > 0\} = \sup\{Vf(x)|x \in X\}$, provided that the latter supremum is positive. A crucial point in the proof is to show that $(\lambda V + I) \cdot C_0(X)$ is dense in $C_\infty(X)$ for any $\lambda > 0$ and once this can be shown, the resolvent J_λ can be obtained as the continuous extension to the whole space of the operator $\hat{J}_\lambda$ defined by $\hat{J}_\lambda(\lambda Vf + f) = Vf$, $f \in C_0(X)$. In [96] one can find an important remark that such an operator V is always pre-closed in B, so that its closure coincides with the potential operator in Yosida's sense associated to the semigroup T_t.

It should be remarked that Yosida's potential operator V associated with a Feller semigroup, when it exists, does not necessarily satisfy $C_0(X) \subset D(V)$ so that it is not necessarily given by a positive kernel. Actually, Yosida's potential operator can be defined for many recurrent Markov processes and, in [98], it was shown that the potential operator can be defined for Brownian motions in any dimension, including the recurrent cases of dimensions 1 and 2. Yosida's works left an important problem of describing potential operators precisely in concrete cases, that is, the problem of finding their explicit expressions and determining their domains or cores. Remarkable results in this problem were obtained by K. Sato (Proc. 6th Berkeley Symp. Math. Stat. Probab. Vol. III, 1972).

Positive pseudo-resolvents and potentials

Proc. Japan Acad. 41 (1965) 1–5

(Comm. by Zyoiti Suetuna, m.j.a., Jan. 12, 1965)

1. Introduction. Let Ω be a set, and denote by X a Banach space of real-valued bounded functions $f(x)$ defined on Ω and normed by $\|f\| = \sup_{x \in \Omega} |f(x)|$. We assume that X is closed with respect to the lattice operations $(f \wedge g)(x) = \min(f(x), g(x))$ and $(f \vee g)(x) = \max(f(x), g(x))$. For any linear subspace Y of X, we shall denote by Y^+ the totality of functions $f \in Y$ which are ≥ 0 on Ω, in symbol $f \geq 0$. We also use the notation $f^+ = f \vee 0$ and $f^- = (-f) \vee 0$.

We denote by $L(X, X)$ the totality of continuous linear operators defined on X into X. A family $\{J_\lambda; \lambda > 0\}$ of operators $\in L(X, X)$ is called a *pseudo-resolvent* if it satisfies the *resolvent equation*

$$(1) \qquad J_\lambda - J_\mu = (\mu - \lambda) J_\lambda J_\mu.$$

Suggested by the case of the resolvent $J_\lambda = (\lambda I - A)^{-1}$ of the infinitesimal generator A of a semi-group $\{T_t; t \geq 0\}$ of operators $\in L(X, X)$ of class $(C_0)^{1)}$ mapping X^+ into X^+, we shall assume conditions:

(2) J_λ is *positive*, in symbol $J_\lambda \geq 0$, that is, $f \geq 0$ implies $J_\lambda f \geq 0$ for all $\lambda > 0$.

$$(3) \qquad \|\lambda J_\lambda\| \leq 1 \quad \text{for all } \lambda > 0.$$

Then, an element $f \in X$ is called *superharmonic* (or *subharmonic*) if $\lambda J_\lambda f \leq f$ (or $\lambda J_\lambda f \geq f$) for all $\lambda > 0$, and an element $f \in X$ is called a *potential* if there exists a $g \in X$ such that $f = s\text{-}\lim_{\lambda \downarrow 0} J_\lambda g$, where s-lim denotes the strong limit in X, i.e., uniform limit on Ω.

We shall be concerned with the *potential operator* V defined by

(4) $Vf = s\text{-}\lim_{\lambda \downarrow 0} J_\lambda f$ (when $s\text{-}\lim_{\lambda \downarrow 0} J_\lambda f^+$ and $s\text{-}\lim_{\lambda \downarrow 0} J_\lambda f^-$ both exist).

Our main results are stated in the following two theorems.

Theorem 1. Let J_λ satisfy (1) and (2). Then $V \geq 0$ and we have:

(5) Let $f \in X^+$, $g \in X^+$ and $\lambda > 0$, and define $V_\lambda = V + \lambda^{-1} I$. If $(V_\lambda f)(x) \leq (Vg)(x)$ on the support (f), we must have $V_\lambda f \leq Vg$. (*the principle of majoration*).

Theorem 2. Let J_λ satisfy (1), (2) and (3). If the *range* $R(V)$ of the potential operator V is dense in X, then $R(V_\lambda)$ is also dense in X and the *null space* $N(V) = \{f; Vf = 0\}$ consists of the zero vector only. Moreover, J_λ is the resolvent of a linear operator A with dense domain $D(A)$ defined through the *Poisson equation* $AVf = -f$.

Remark. Two special cases of X are important for concrete

---

1) See, e.g., K. Yosida: Functional Analysis, Springer, to appear soon.

application. *The first case*: Ω is a locally compact Hausdorff space and X is the totality of real-valued continuous functions defined on Ω which *tend to zero at infinity*; X is the closure with respect to the norm $\|f\| = \sup_{x \in \Omega} |f(x)|$ of the space $C_0(\Omega)$ of continuous functions with compact support defined on Ω. *The second case*: Ω is a σ-additive family of subsets of a set, and X is the Banach space of σ-additive measures defined on Ω and normed by the total variation of the measure. The first case was discussed by G. A. Hunt[2] in the view to characterize the operator $\tilde{V}$ defined through

$$(4)' \qquad (\tilde{V}f)(x) = \int_0^\infty (T_t f)(x)\,dt,$$

where $\{T_t; t \geq 0\}$ is the semi-group associated with a Markov process in a locally compact space Ω. *In the first case*, we can prove, under condition(5), an analogue of Hunt's research:

Theorem 3. If $V \geq 0$ and if $R(V)$ is dense in X, we have:

$(5)_1$ Let $f \in C_0(\Omega)^+$ and $g \in X^+$, and let $(Vf)(x) \leq (Vg)(x)$ on the support (f). Then $Vf \leq Vg$.

If furthermore, $V(C_0(\Omega)^+)$ is dense in X^+, then we obtain:

$(5)_2$ Let $f \in C_0(\Omega)$ and let $(Vf)(x_0) = \max_{x \in \Omega} (Vf)(x)$. Then $f(x_0) \geq 0$.

Theorem 4. Let V be a closed linear operator whose domain $D(V)$ and range $R(V)$ both belong to $X = C_0(\Omega)^a$ in such a way that V satisfies (5) and further conditions:

$(6) \qquad\qquad\qquad V \geq 0,$

(7) Vf is defined if and only if Vf^+ and Vf^- are both defined.

$(8) \qquad\qquad\qquad N(V) = \{0\}.$

$(9) \qquad\qquad\qquad C_0(\Omega) \subseteqq D(V).$

(10) $V(C_0(\Omega)^+)$ is dense in X^+ and $V_\lambda(C_0(\Omega))$ is dense in X for $\lambda > 0$. Then, for the operator A defined through the Poisson equation $AVf = -f$ and for $\lambda > 0$, the resolvent $J_\lambda = (\lambda I - A)^{-1}$ exists as an operator $\in L(X, X)$ such that (1), (2), (3) and (4) hold.

2. Proof of the theorems. We shall rely upon a lemma which is a special case of the so-called *Abelian ergodic theorem*.[3]

Lemma. Under condition (1), we have

$(11) \qquad\qquad\qquad J_\lambda J_\mu = J_\mu J_\lambda.$

Under conditions (1) and (3), we have:

(12) $R(J_\lambda)$ is independent of λ, and its closure $R(J_\lambda)^a$ coincides with $\{f;\ s\text{-}\lim_{\lambda \to \infty} \lambda J_\lambda f = f\}$.

2) Markoff processes and potentials, I, II and III, Illinois J. of Math., **1**, 44-93, 316-469 (1957) and **2**, 151-213 (1958). Further researches are given, e.g., in Séminaire du Potentiels, dirigés par M. Brelot, G. Choquet et J. Deny, Fac. Sci. Paris (1950-).

3) K. Yosida: Ergodic theorems for pseudo-resolvents. Proc. Japan Acad., **37**, 422-423 (1961). Cf. E. Hille-R. S. Phillips, Functional Analysis and Semi-groups, Providence (1957).

(12)' $R(I-\lambda J_\lambda)$ is independent of λ and its closure $R(I-\lambda J_\lambda)^a$ coincides with $\{f; s\text{-}\lim_{\lambda\downarrow 0}\lambda J_\lambda f=0\}$.

Proof. See the reference cited in the footnotes 1) and 3).

Proof of Theorem 1. The operator V defined through (4) satisfies

(13) $Vf=\lambda J_\lambda Vf+J_\lambda f=V\lambda J_\lambda f+J_\lambda f$ for $f\in D(V)$,

because of (1), (4) and (11). Thus, if $f\geq 0$ belongs to the domain $D(V)$, then Vf is superharmonic by $J_\lambda f\geq 0$.

Next we show that, if $f\in X^+$ be such that $\mu J_\mu f\leq f$ for all μ with $0<\mu\leq\lambda$, then

(14) $\lim_{\mu\to 0}(J_\mu(\lambda I-\lambda^2 J_\lambda)f)(x)=(\lambda J_\lambda f)(x)-f_h(x)$, where

$$f_h(x)=\lim_{\mu\downarrow 0}(\mu J_\mu f)(x)^{4)}.$$

To prove this, we first observe that $\lambda(I-\lambda J_\lambda)f$ is ≥ 0. We have, by (1),

$$J_\mu(\lambda f-\lambda^2 J_\lambda f)=\lambda J_\mu f-\frac{\lambda^2}{\lambda-\mu}(J_\mu-J_\lambda)f=\frac{-\lambda}{\lambda-\mu}\mu J_\mu f+\frac{\lambda^2}{\lambda-\mu}J_\lambda f.$$

We also have, by (1),

$$(I+(\mu-\lambda)J_\lambda)(I-\mu J_\mu)f=(I-\lambda J_\lambda)f.$$

Hence, if $0<\mu\leq\lambda$, the condition $\mu J_\mu f\leq f$ (for $0<\mu\leq\lambda$) implies that $0\leq\mu J_\mu f\leq\lambda J_\lambda f$. Thus $\lim_{\mu\downarrow 0}(\mu J_\mu f)(x)=f_h(x)$ exists and so we obtain (14).

We are now able to prove (5). Put $v(x)=\min((V_\lambda f)(x),(Vg)(x))$. Then $v\geq 0$ by $f\geq 0$, $g\geq 0$ and $V\geq 0$, and we have

(15) $\mu J_\mu v\leq v$ for $0<\mu\leq\lambda$.

We have only to show that $\mu J_\mu V_\lambda f\leq V_\lambda f$. But we obtain

$$\mu J_\mu V_\lambda f=\mu J_\mu Vf+\frac{\mu}{\lambda}J_\mu f=\mu J_\mu Vf+J_\mu f+\left(\frac{\mu}{\lambda}-1\right)J_\mu f$$

$$=Vf+\left(\frac{\mu}{\lambda}-1\right)J_\mu f\leq Vf\leq V_\lambda f.$$

Thus $w=\lambda(I-\lambda J_\lambda)v\geq 0$ and we have, by (14),

(16) $\lim_{\mu\downarrow 0}(J_\mu w)(x)=(\lambda J_\lambda v)(x)-v_h(x)$, where $v_h(x)=\lim_{\mu\downarrow 0}(\mu J_\mu v)(x)\geq 0$.

Hence, by (13) and the positivity of J_μ, we obtain

(17) $\lim_{\mu\downarrow 0}(J_\mu w)(x)\leq(\lambda J_\lambda v)(x)=v(x)-\lambda^{-1}w(x)$

$$\leq(\lambda J_\lambda V_\lambda f)(x)=(Vf)(x)$$
$$=(V_\lambda f)(x)-\lambda^{-1}f(x).$$

We have $(V_\lambda f)(x)=v(x)$ on the support (f) by hypothesis. Hence, by (17), $f(x)\leq w(x)$ on the support (f), and so, by $f\geq 0$ and $w\geq 0$, we must have $f\leq w$. Therefore, by (17) and the positivity of J_μ, we obtain

$(V_\lambda f)(x)\leq\lim_{\mu\downarrow 0}(J_\mu w)(x)+\lambda^{-1}w(x)\leq v(x)\leq(Vg)(x)$, that is, $V_\lambda f\leq Vg$.

Proof of Theorem 2. By (13), we see that $R(V)^a=X$ implies $R(V_\lambda)^a=X$ and $R(J_\lambda)^a=X$. $R(J_\lambda)^a=X$ implies, by (12), that $N(J_\lambda)=$

---

4) Originally, the author tacitly concluded that $s\text{-}\lim_{\lambda\downarrow 0}\lambda J_\lambda f=f_h$ exists. This was pointed out by Mr. D. Fujiwara.

{0} which, in turn, implies the existence of the inverse J_λ^{-1}. By (1), it is easy to see that $(\lambda I - J_\lambda^{-1})$ is independent of λ so that $J_\lambda = (\lambda I - A)^{-1}$ where $A = \lambda I - J_\lambda^{-1}$. Moreover, $D(A) = R(J_\lambda) \supseteq R(V)$ is dense in X. By (13), $Vf = 0$ implies $J_\lambda f = 0$ so that $N(V) = \{0\}$ if $R(V)^a = X$. Finally, we have, by (13) and $J_\lambda = (\lambda I - A)^{-1}$,

$$(\lambda I - A)Vf = \lambda Vf + f, \text{ that is, } AVf = -f.$$

Proof of Theorem 3. We first prove $(5)_1$. Since $R(V)$ is dense in X, there exists an $h \geq 0$ such that $f(x) \leq (Vh)(x)$ on the support (f) which is compact by hypothesis. For any $\varepsilon > 0$, take $\lambda > 0$ such that $\lambda^{-1} < \varepsilon$. Then $(V_\lambda f)(x) \leq (Vg)(x) + \lambda^{-1} f(x) \leq (V(g + \varepsilon h))(x)$ on the support (f). Hence, by (5), $Vf \leq V_\lambda f \leq V(g + \varepsilon h)$. Letting $\varepsilon \downarrow 0$, we obtain $Vf \leq Vg$.

Proof of $(5)_2$. Since $Vf \in X = C_0(\Omega)^a$, we must have $(Vf)(x_0) \geq 0$. Let us tentatively assume that $(Vf)(x_0) > 0$. Then we can show that $f(y) \geq 0$ at some point $y \in E = \{x; (Vf)(x_0) = (Vf)(x)\}$. Since $V \geq 0$, the condition $(Vf)(x_0) > 0$ implies that f^+ is not equal to zero. For any point $y \in E \cap \text{support}(f^+)$, we have surely $f(y) \geq 0$. If $E \cap \text{support}(f^+)$ is void, then there exists an $\varepsilon > 0$ such that $(Vf)(x_0) > \varepsilon$ and that $(Vf)(x) \leq (Vf)(x_0) - \varepsilon$ on the support (f^+). Since $V(C_0(\Omega)^+)$ is dense in X^+ by hypothesis, there exists an $h \geq 0$ such that $(Vh)(x_0) = (Vf)(x_0) - \varepsilon$ and $(Vh)(x) \geq (Vf)(x_0) - \varepsilon$ on the support (f^+). Hence $(Vf^+)(x) \leq (Vf^-)(x) + (Vh)(x)$ on the support (f^+), and so, by $(5)_1$, we must have $Vf \leq Vh$. Thus we have a contradiction $(Vf)(x_0) \leq (Vh)(x_0) = (Vf)(x_0) - \varepsilon$. We now turn to the general case $(Vf)(x_0) \geq 0$, and take any compact set $\hat{E}$ of Ω containing x_0 as an interior point. Since $V(C_0(\Omega)^+)$ is dense in X^+, there exists a $g \in C_0(\Omega)^+$ such that $(Vg)(x_0) > \max_{x \in \Omega - \hat{E}} (Vg)(x)$. Then, for any $\varepsilon > 0$, the function $(V(f + \varepsilon g))(x)$ takes its positive maximum at a point $\in \hat{E}$ and not at points outside $\hat{E}$. Hence, as proved above, there must exist at least one point $y \in \hat{E}$ such that $f(y) + \varepsilon g(y) \geq 0$. Therefore, we obtain $f(x_0) \geq 0$ by letting $\varepsilon \downarrow 0$.

Proof of Theorem 4. By (8), we can define the operator A through $AVf = -f$. Thus

(18) $$(\lambda I - A)Vf = \lambda Vf + f.$$

We first prove that the condition $V_\lambda f = 0$ with $\lambda > 0$ implies that $f = 0$. For, then $(V_\lambda f^+)(x) \leq (Vf^-)(x)$ on the support (f^+) and so $Vf^+ \leq V_\lambda f^+ \leq Vf^-$ by (5). Similarly we obtain $Vf^+ \geq Vf^-$ and hence $Vf = 0$ so that $f = 0$ by (8).

Therefore the inverse $J_\lambda = (\lambda I - A)^{-1}$ exists as an operator which maps $(\lambda Vf + f)$ onto Vf. We can prove that

(19) $J_\lambda = (\lambda I - A)^{-1}$ is positive.

Let $h \geq 0$ be $\in D(J_\lambda)$. Then $J_\lambda h = g = Vf$ with $f \in D(V)$ and

(20) $$h=(\lambda I-A)J_\lambda h=\lambda g-Ag=\lambda Vf+f.$$

Since $h\geq0$, we have $(\lambda Vf^+)(x)\geq(\lambda Vf^{-1})(x)+f^-(x)$ on the support (f^-) and so, by (5), $\lambda Vf^+\geq\lambda Vf^-+f^-$, that is, $J_\lambda h=Vf\geq\lambda^{-1}f^-\geq0$.

Since A is a closed linear operator with V, we see that $D(J_\lambda)^a\supseteq V_\lambda(C_0(\Omega))^a=X$ implies, by $V\geq0$, that $\lambda>0$ is in the resolvent set of A and $J_\lambda=(\lambda I-A)^{-1}\in L(X,X)$.

We next show that (3) is true. Let $h\in V_\lambda(C_0(\Omega))$. Then, by (20), we can show that $\min\limits_{x\in\Omega} h(x)\leq(\lambda Vf)(x)\leq\max\limits_{x\in\Omega} h(x)$. In fact, let $(Vf)(x_0)=\max\limits_{x\in\Omega}(Vf)(x)$. Then, by $(5)_2$, we have $f(x_0)\geq0$ so that $(\lambda Vf)(x)\leq h(x_0)\leq\max\limits_{x\in\Omega} h(x)$. Similarly we obtain $(\lambda Vf)(x)\geq\min\limits_{x\in\Omega} h(x)$. Thus we have proved (3).

We finally prove that $Vf=s\text{-}\lim\limits_{\mu\downarrow0} J_\mu f$ for $f\in D(V)$. We have, by (18),

(21) $$Vf=\lambda J_\lambda Vf+J_\lambda f.$$

We also have, by $J_\lambda=(\lambda I-A)^{-1}$,

(22) $$(I-\lambda J_\lambda)f=-J_\lambda Af\quad\text{for }f\in D(A).$$

On the other hand, the range $R(A)=D(V)\supseteq C_0(\Omega)$ is dense in X and the range $R(J_\lambda)=D(A)=R(V)$ is dense in X. Thus we see that (22) implies that $R(I-\lambda J_\lambda)^a=X$. Hence, by (12)', $s\text{-}\lim\limits_{\lambda\downarrow0}\lambda J_\lambda f=0$ for every $f\in X$. Therefore, by (21), we obtain $Vf=s\text{-}\lim\limits_{\lambda\downarrow0} J_\lambda f$ for every $f\in D(V)$.

Positive resolvents and potentials (An operator-theoretical treatment of Hunt's theory of potentials)

Z. Wahrscheinlichkeitstheorie und verw. Geb. **8** (1967) 210–218

1. Introduction

Let X be a separable, locally compact, non compact Hausdorff space, and B the completion with respect to the maximum norm of the space $C_0(X)$, the space of real-valued continuous functions $f(x)$ with compact supports defined in X. G. A. HUNT [3] introduced the notion of *potential operator* V which is defined as a positive linear operator from $C_0(X)$ into B satisfying the following three conditions[1]:

(α) For $f \in C_0(X)^+$ and $g \in C_0(X)^+$, the inequality $Vf \leq Vg$ holds everywhere if it holds on the support of f.

(β) There exists a sequence of functions $h_n \in C_0(X)^+$ such that $V h_n$ increases everywhere to 1 as $n \to \infty$.

(γ) The range of V is dense in B.

HUNT [3] proved[2] by (α) and (β) the so-called "complete maximum principle" for V:

(δ) Let a be a positive constant, f and g functions in $C_0(X)^+$. Then $a + Vg$ majorizes Vf if it does so on the support of f. By virtue of (α) through (δ), he proved that there exists a uniquely determined contraction semi-group[3] $\{T_t; t \geq 0\}$ of class (C_0) of bounded positive linear operators T_t on B into B such that

$$(1) \qquad Vf = s - \lim_{\lambda \downarrow 0} \int_0^\infty e^{-\lambda t} T_t f \, dt \quad \text{for all} \quad f \in C_0(X).$$

The purpose of the present paper is to discuss, from an operator-theoretical point of view, the analytical mechanism in HUNT's theory of potentials. Our analysis is expressed step by step by the following four theorems[4].

---

[1] For any subset M of B, we denote by M^+ the totality of functions $f \in M$ which are ≥ 0 everywhere in X, in symbol $f \geq 0$. An operator T is called *positive*, in symbol $T \geq 0$, if $f \geq 0$ implies $Tf \geq 0$. For any function $f \in B$, we denote by f^+ and f^- the functions $f^+(x) = \max(f(x), 0)$ and $f^-(x) = -\min(f(x), 0)$, respectively.

[2] Cf. J. DENY [1], G. LION [4] and P. A. MEYER [5] and the references cited there.

[3] For general theory of analytical semi-groups, see K. YOSIDA [6] and E. HILLE-R. PHILLIPS [2].

[4] For a linear operator V, we shall denote by $D(V)$, $R(V)$ and $N(V)$ the *domain*, the *range* and the *null space* of V, respectively.

Theorem 1. *Let $\{J_\lambda; \lambda > 0\}$ be a family of bounded linear operators on B into B such that the "resolvent equation" holds*[5]:

$$(2) \qquad J_\lambda - J_\mu = (\mu - \lambda) J_\lambda J_\mu .$$

Let, moreover, J_λ be positive. Then, the operator V defined through

$$(3) \qquad V f = s - \lim_{\lambda \downarrow 0} J_\lambda f$$

satisfies the "principle of majoration":

(4) *Let f and g be both $\in D(V)^+$. If, for $V_\lambda = V + \lambda^{-1} I$ with a positive constant λ, the inequality $(V_\lambda f)(x) \leqq (Vg)(x)$ holds on the support of f, then we must have $V_\lambda f \leqq Vg$ everywhere.*

Theorem 2. *If the positive pseudo-resolvent J_λ is the resolvent of a linear operator with dense domain A, i.e. $J_\lambda = (\lambda I - A)^{-1}$, and if $D(V)$ is dense in B, then $N(V)$ consists of 0 only and $R(V)$ is dense in B.*

Theorem 3. *Let a positive linear operator V on $D(V) \subseteqq B$ into B satisfy the principle of majoration (4). Let us further assume*

$$(5) \qquad D(V) \supseteqq C_0(X) ,$$

$$(6) \qquad \text{the image} \quad V \cdot C_0(X) \quad \text{is dense in } B,$$

and HUNT's *condition (β). Then V satisfies the "principle of positive maximum"*[6]:

(7) $\qquad\qquad$ Let $\quad f \in C_0(X) , \quad$ and let

$$P = \left\{ x_0 ; (Vf)(x_0) = \sup_x (Vf)(x) \right\}$$

and

$$N = \left\{ x_1 ; (Vf)(x_1) = \inf_x (Vf)(x) \right\} .$$

Then $f(x) \geqq 0$ for $x \in P$, and $f(x) \leqq 0$ for $x \in N$.

Theorem 4. *Let a positive linear operator V on $D(V) \subseteqq B$ into B satisfy conditions (5), (6) and (7). Then there exists, for all positive constants λ, a uniquely determined resolvent $J_\lambda = (\lambda I - A)^{-1}$ of a closed linear operator A satisfying the following three conditions:*

$$(8) \qquad A V f = -f \quad \text{when} \quad f \in C_0(X) ,$$

$$(9) \qquad J_\lambda = (\lambda I - A)^{-1} \quad \text{is positive and} \quad \| \lambda J_\lambda \| \leqq 1$$

and

$$(10) \qquad V f = s - \lim_{\lambda \downarrow 0} (\lambda I - A)^{-1} f \quad \text{when} \quad f \in C_0(X) .$$

---

[5] Such J_λ is called the pseudo-resolvent by E. HILLE.

[6] As far as the author knows, such statement of the "principle of positive maximum" was firstly given by J. DENY [1]. In other references, the principle is stated as

$$f(x_0) \geqq 0 \quad \text{if} \quad (Vf)(x_0) = \sup_{x \in X}(Vf)(x) \quad \text{is} \quad > 0 .$$

Furthermore, we have:

(11)   *the inverse V^{-1} exists and $A = -V^{-1}$ if and only if V is a closed linear operator, and $(\lambda I - A)^{-1} = V(\lambda V + I)^{-1}$ in this case.*

Remark. Since $D(A)$ is dense in B by (5) and (8), we see from (9) that A in Theorem 4 is the infinitesimal generator of a positive contraction semi-group T_t of class (C_0) such that (1) holds.

Comments. As it should be, the key to the whole theory of potentials is the "principle of positive maximum" (7). Hunt [3] derived (7) from (γ) and (δ), i.e., from (α), (β) and (γ). In the proof of Theorem 3, we derive (7) from (β), (γ) and (4). Since the constant function 1 is outside our function space B, it seems preferable not to rely upon the "principle of complete maximum" (δ); instead, the author should like to propose the "principle of majoration" (4) which, as Theorem 1 shows, is intimately connected with Hunt's setting (1) of the potential theory. Theorem 2 says that the denseness of the *domain $D(V)$* of V implies that of the *range $R(V)$* of V, and, moreover, the fact that the *null space $N(V)$* of V consists of 0 only, that is, the existence of the inverse V^{-1} of V. It is to be noted that, in the proof of Theorem 4, the "ergodic theorems of the Hille type" in K. Yosida [6] plays an important role.

2. Proof of Theorems

Proof of Theorem 1[7]. By (2), (3) and the commutativity $J_\lambda J_\mu = J_\mu J_\lambda$ implied by (2), we have

(12) $$V f = \lambda J_\lambda V f + J_\lambda f = V \lambda J_\lambda f + J_\lambda f \quad \text{for} \quad f \in D(V).$$

Thus, if $f \in D(V)^+$ then $\lambda J_\lambda V f \leq V f$ by the positivity of V, i.e., $V f$ is "superharmonic"[8]. We next prove:

(13)   if $f \in B^+$ be such that $\mu J_\mu f \leq f$ for all constants μ with $0 < \mu \leq \lambda$, then

$$\lim_{\mu \downarrow 0} (J_\mu (\lambda I - \lambda^2 J_\lambda) f)(x) = (\lambda J_\lambda f)(x) - f_h(x),$$

where

$$f_h(x) = \lim_{\mu \downarrow 0} (\mu J_\mu f)(x).$$

In fact, we have, by (2) and the commutativity $J_\lambda J_\mu = J_\mu J_\lambda$,

$$J_\mu(\lambda f - \lambda^2 J_\lambda f) = \lambda J_\mu f - \frac{\lambda^2}{\lambda - \mu}(J_\mu - J_\lambda) f = \frac{-\lambda}{\lambda - \mu} \mu J_\mu f + \frac{\lambda^2}{\lambda - \mu} J_\lambda f$$

and

$$(I + (\mu - \lambda) J_\lambda)(I - \mu J_\mu) f = (I - \lambda J_\lambda) f.$$

---

[7] This theorem was stated and proved in K. Yosida [7]. We reproduce it here for the sake of comprehension of our operator-theoretical setting of the potential theory.

[8] Condition (β) thus means that the function 1 is the supremum of "superharmonic functions" $V f \leq 1$ such that $f \geq 0$.

Since $\mu \le \lambda$ and $J_\mu \ge 0$, the latter equation and the assumption $\mu J_\mu f \le f$ (for $0 < \mu \le \lambda$) imply that $0 \le \mu J_\mu f \le \lambda J_\lambda f$. Thus

$$f_h(x) = \lim_{\mu \downarrow 0} (\mu J_\mu f)(x)$$

exists and so we obtain (13).

We are now able to prove the principle of majoration. Put $v(x) = \min((V_\lambda f)(x),$ $(Vg)(x))$. Then $v \ge 0$ by $f \ge 0, g \ge 0$ and $V \ge 0$. We shall prove

$$(14) \qquad \mu J_\mu v \le v \quad \text{for} \quad 0 < \mu \le \lambda.$$

We have only to show that $\mu J_\mu V_\lambda f \le V_\lambda f$. This is clear from

$$\mu J_\mu V_\lambda f = \mu J_\mu V f + \frac{\mu}{\lambda} J_\mu f = \mu J_\mu V f + J_\mu f + \left(\frac{\mu}{\lambda} - 1\right) J_\mu f$$
$$= V f + \left(\frac{\mu}{\lambda} - 1\right) J_\mu f \le V f \le V_\lambda f.$$

Thus $w = \lambda(I - \lambda J_\lambda)v \ge 0$ and we have, by (13),

$$(15) \qquad \lim_{\mu \downarrow 0} (J_\mu w)(x) = (\lambda J_\lambda v)(x) - v_h(x),$$

where

$$v_h(x) = \lim_{\mu \downarrow 0} (\mu J_\mu v)(x) \ge 0.$$

Hence, by $v \le V_\lambda f$ and the positivity of J_λ, we obtain

$$\lim_{\mu \downarrow 0} (J_\mu w)(x) \le (\lambda J_\lambda v)(x) = v(x) - \lambda^{-1} w(x)$$
$$(16) \qquad\qquad \le (\lambda J_\lambda V_\lambda f)(x) = (V f)(x)$$
$$= (V_\lambda f)(x) - \lambda^{-1} f(x).$$

We have $(V_\lambda f)(x) = v(x)$ on the support of f by hypothesis. Hence, by (12), $f(x) \le w(x)$ on the support of f, and so, by $f \ge 0$ and $w \ge 0$, we must have $f \le w$. Therefore, by (16) and the positivity of J_μ, we obtain

$$(V_\lambda f)(x) \le \lim_{\mu \downarrow 0} (J_\mu w)(x) + \lambda^{-1} w(x) \le v(x) \le (Vg)(x),$$

that is, $V_\lambda f \le Vg$.

Proof of Theorem 2. $D(V)$ being dense in B, we can define the dual operator V^* of V. V^* is a linear mapping from the dual space B^* of B into B^*. Since J_λ is a bounded linear operator, we have (see, e.g., K. YOSIDA [6], p. 195) $(J_\lambda V)^* = V^* J_\lambda^*$ and $(V J_\lambda)^*$ is an extension of $J_\lambda^* V^*$. Hence, by (12), we obtain

$$V^* v^* = V^* \lambda J_\lambda^* v^* + J_\lambda^* v^* = \lambda J_\lambda^* V^* v^* + J_\lambda^* v^* \quad \text{for} \quad v^* \in D(V^*).$$

Thus $V^* v^* = 0$ implies $J_\lambda^* v^* = 0$ and so $v^* = 0$, because J_λ^* is the resolvent of the operator A^* (see, e.g., K. YOSIDA [6], p. 223). Hence, by the Hahn-Banach extension theorem, we see that $R(V)$ must be dense in B. That $N(V) = \{0\}$ is also clear by (12), since $V f = 0$ implies $J_\lambda f = 0$.

Proof of Theorem 3. Since $V f \in B$, we surely have

$$\sup_{x \in X} (V f)(x) \ge 0.$$

We first consider the case when this supremum is strictly positive. Then, by the positivity of V, f cannot be ≤ 0, that is, we must have $f^+ \neq 0$. Let

$$a = \sup_{f(x) > 0} (Vf)(x) .$$

Then, denoting by a^+ the max $(a, 0)$, we have $Vf^+ \leq a^+ + Vf^-$ on the support of f^+. Let $b > a^+$ and let $\varepsilon > 0$ be such that

$$b > a^+ + \varepsilon \cdot \sup_{x \in X} f^+(x) .$$

Consider the sequence of functions $V(f^- + b h_n)$, where the functions h_n are the ones mentioned in (β). For some value of n, according to Dini's theorem, we have $Vf^+ + \varepsilon f^+ \leq V(f^- + b h_n)$ on the support of f^+ which is compact by hypothesis. Thus, by the principle of majoration (4), this inequality holds everywhere, and so $Vf + \varepsilon f^+ \leq b V h_n \leq b$. Letting $b \downarrow a^+$ and $\varepsilon \downarrow 0$, we obtain $Vf \leq a^+$. Hence $a^+ = a > 0$ and

$$\sup_{x \in X} (Vf)(x) = \sup_{f(x) > 0} (Vf)(x) .$$

We next consider the general case

$$\sup_{x \in X} (Vf)(x) = (Vf)(x_0) \geq 0 ,$$

and take a compact neighbourhood M of the point x_0. By (6), there exists a $g \in C_0(X)$ such that

$$(Vg)(x_0) > \sup_{x \in X - M} (Vg)(x) .$$

The right hand supremum is ≥ 0 by $Vg \in B$. Hence, for any positive constant ε,

$$\sup_{x \in X} (V(f + \varepsilon g))(x)$$

is strictly positive, and this supremum is attained on M and not on $(X - M)$. Therefore, by what we have proved already, we have $f(y) + \varepsilon g(y) \geq 0$ at some point y of M. Letting M tend to x_0 and $\varepsilon \downarrow 0$, we obtain $f(x_0) \geq 0$.

Proof of Theorem 4. The crucial point in the proof of Theorem 4 is the fact:

(17)   For any positive constant λ, the image $V_\lambda \cdot C_0(X)$ is dense in B.

To prove this we recall first that X is separable by the assumption. Thus by (5) we can find a sequence of functions $f_n \in C_0(X)^+ \subseteq D(V)$ such that

$$\bigcup_{n=1}^{\infty} \{x ; f_n(x) > 0\} = X .$$

Choose a sequence of positive numbers α_n such that

$$b = \sum_n \alpha_n f_n$$

and

$$\sum_n \alpha_n (V f_n)$$

both strongly converge in B. Following G. A. Hunt [3], we put

$$a_n(x) = \min(n \cdot b(x), 1) .$$

Then,

(18). $\quad a_n(x) > 0$ everywhere and $a_n(x)$ converges increasingly to 1 as $n \to \infty$.

It is easy to see that the mapping

$$(19) \qquad\qquad f \to V a_n f \qquad (f \in C_0(X))$$

can be extended by continuity to a bounded linear positive operator V_n on B into B. Hence, for any positive number λ with $\|\lambda V_n\| < 1$, the bounded linear inverse $(\lambda V_n + I)^{-1} = \sum_{k=0}^{\infty} (-\lambda V_n)^k$ exists and so

$$(20) \qquad\qquad J_{\lambda}^{(n)} = V_n(\lambda V_n + I)^{-1}$$

is a bounded linear operator on B into B. We shall prove that

$$(21) \qquad\qquad J_{\lambda}^{(n)} \text{ is positive and} \quad \|\lambda J_{\lambda}^{(n)}\| < 1 .$$

In fact, we have

$$(22) \qquad\qquad J_{\lambda}^{(n)}(\lambda V_n + I)f = V_n f \quad \text{for all} \quad f \in B .$$

Applying the principle of positive maximum to $V_n f = V a_n f$ for the case $f \in C_0(X)$, we obtain $f(x_0) \geq 0$ at points x_0 where $(V_n f)(x)$ reaches the supremum value, and also $f(x_1) \leq 0$ at points x_1 where $(V_n f)(x)$ reaches the infimum value. Hence, for every $f \in C_0(X)$,

$$(23) \qquad \inf_{x \in X} (\lambda V_n f + f)(x) \leq \lambda(V_n f)(x) \leq \sup_{x \in X} (\lambda V_n f + f)(x) \quad \text{in } X .$$

Since V_n is a bounded linear operator on B and since $C_0(X)$ is dense in B, we see that (23) holds for every $f \in B$, proving that (21) is true.

We next show that, for all positive constant μ, the bounded linear operator $(\mu V_n + I)^{-1}$ is defined such that $J_{\mu}^{(n)} = V_n(\mu V_n + I)^{-1}$ is positive and satisfies the estimate $\|\mu J_{\mu}^{(n)}\| \leq 1$. To this purpose, let $\|\lambda V_n\| < 1$ and take any positive number μ satisfying $\|(\mu - \lambda)J_{\lambda}^{(n)}\| \leq |(\mu - \lambda)\lambda^{-1}| < 1$. Then $(I + (\mu - \lambda)J_{\lambda}^{(n)})$ admits bounded linear inverse, and hence, by

$$(24) \qquad \begin{aligned} \mu V_n + I &= (I + (\mu - \lambda) V_n(\lambda V_n + I)^{-1})(\lambda V_n + I) \\ &= (I + (\mu - \lambda)J_{\lambda}^{(n)})(\lambda V_n + I) , \end{aligned}$$

we see that $(\mu V_n + I)^{-1}$ exists as a bounded linear operator on B into B. Then as in the case for $J_{\lambda}^{(n)}$, we see that $J_{\mu}^{(n)} = V_n(\mu V_n + I)^{-1}$ also satisfies

$$(23') \qquad \begin{aligned} \inf_{x \in X} (\mu V_n f + f)(x) &\leq \mu J_{\mu}^{(n)}(\mu V_n f + f)(x) = \mu(V_n f)(x) \\ &= \sup_{x \in X} (\mu V_n f + f)(x) \quad \text{in } X, \end{aligned}$$

and hence $J_{\mu}^{(n)}$ is positive with the estimate $\|\mu J_{\mu}^{(n)}\| \leq 1$. Repeating the process, we see that, for every positive constant μ,

$(21')$ $\quad J_{\mu}^{(n)} = V_n(\mu V_n + I)^{-1}$ is a bounded linear positive operator with the

estimate $\|\mu J_{\mu}^{(n)}\| \leq 1$.

We are now able to prove (17). For any $g \in C_0(X)^+$, we have, by (21'),

$$f_n = J_\mu^{(n)} g = V_n g - \mu J_\mu^{(n)} V_n g \geqq 0 .$$

Since $f_n = (\mu V_n + I)^{-1} V_n g$, we have $V_n g = \mu V_n f_n + f_n$ so that, by

$$0 < a_n(x) \leqq 1 ,$$

$$0 \leqq V_n g - (\mu V a_n f_n + a_n f_n) = (1 - a_n) f_n \leqq (1 - a_n) V_n g .$$

Since g is with compact support, we have $V_n g = V g$ for sufficiently large n by (18) and Dini's theorem. Hence $(1 - a_n) V_n g$ converges strongly to 0 as $n \to \infty$. V_n being a bounded linear operator and $V \cdot C_0(X)$ being dense in B by hypothesis, we have proved that, for any $h \in B$ and $\varepsilon > 0$, there exists an $f \in C_0(X)$ such that

$$\| h - (\mu V f + f) \| < \varepsilon .$$

This proves (17).

As in (23), we obtain, for any $f \in C_0(X)$ and $\lambda > 0$,

$$(25) \qquad \inf_{x \in X} (\lambda V f + f)(x) \leqq \lambda(V f)(x) \leqq \sup_{x \in X} (\lambda V f + f)(x) \qquad \text{in } X .$$

Thus we see that the mapping

$$(26) \qquad \hat{J}_\lambda : (\lambda V f + f) \to V f \qquad (f \in C_0(X))$$

is well defined and it is a positive linear operator satisfying the estimate

$$(27) \qquad \| \lambda \hat{J}_\lambda g \| \leqq \| g \| \quad \text{when} \quad g = \lambda V f + f \quad \text{with} \quad f \in C_0(X) .$$

Since $V_\lambda \cdot C_0(X)$ is dense in B by (17), we see that the operator $\hat{J}_\lambda$ can be extended by continuity to a bounded linear operator J_λ such that

$$(28) \qquad J_\lambda \text{ is positive and } \quad \| \lambda J_\lambda \| \leqq 1 \quad \text{for all positive constant } \lambda .$$

We shall show that J_λ is a resolvent. In fact, we have, by (26),

$$J_\lambda(\lambda V f + f) - J_\mu(\lambda V f + f) = V f - J_\mu\left(\frac{\lambda}{\mu}(\mu V f + f) + \left(1 - \frac{\lambda}{\mu}\right) f\right)$$

$$= V f - \frac{\lambda}{\mu} V f - \left(1 - \frac{\lambda}{\mu}\right) J_\mu f = \left(1 - \frac{\lambda}{\mu}\right)(V f - J_\mu f)$$

and

$$(\mu - \lambda) J_\mu J_\lambda(\lambda V f + f) = (\mu - \lambda) J_\mu V f = (\mu - \lambda) J_\mu\left(V f + \frac{1}{\mu} f - \frac{1}{\mu} f\right)$$

$$= (\mu - \lambda) \frac{1}{\mu} V f - (\mu - \lambda) \frac{1}{\mu} J_\mu f = \left(1 - \frac{\lambda}{\mu}\right)(V f - J_\mu f)$$

when $f \in C_0(X)$. Thus the bounded linear operators $(J_\lambda - J_\mu)$ and $(\mu - \lambda) J_\mu J_\lambda$ coincide on a dense subset $V_\lambda \cdot C_0(X)$ of B, and so $(J_\lambda - J_\mu) = (\mu - \lambda) J_\mu J_\lambda$, that is, J_λ is a pseudo-resolvent. Hence, by $\| \lambda J_\lambda \| \leqq 1$, we can apply the "ergodic theorems of the Hille type" in K. Yosida [6]:

$$(29) \qquad \text{the closure in } B \text{ of } R(J_\lambda) = \left\{ f \in B; \, s - \lim_{\mu \to \infty} \mu J_\mu f = f \right\},$$

$$(30) \qquad \text{the closure in } B \text{ of } R(I - \lambda J_\lambda) = \left\{ f \in B; \, s - \lim_{\mu \downarrow 0} \mu J_\mu f = 0 \right\}.$$

We see, from the resolvent equation, that the null space $N(J_\lambda)$ is independent of λ. Hence, by (29) and the denseness of $R(J_\lambda)$ implied by (6) and (26), we see that $N(J_\lambda) = \{0\}$. Thus, by Theorem 1 in K. YOSIDA [6], p. 216, we see that J_λ is the resolvent of a linear operator A, and the range $R(J_\lambda)$ coincides with the domain $D(A)$ of A. Hence $D(A)$ is dense in B. J_λ being a bounded linear transformation, the operator A must be closed. We have thus proved

(31) $J_\lambda = (\lambda I - A)^{-1}$, where the domain $D(A)$ of the closed linear operator A is dense in B.

By (26), we have

$$(32) \qquad Vf = \hat{J}_\lambda(\lambda Vf + f) = \lambda J_\lambda Vf + J_\lambda f \quad \text{when} \quad f \in C_0(X),$$

and hence

$$(\lambda I - A)J_\lambda(\lambda Vf + f) = \lambda Vf + f = (\lambda I - A)Vf = \lambda Vf - A Vf \quad \text{when} \quad f \in C_0(X),$$

that is,

$$(33) \qquad A Vf = -f \quad \text{when} \quad f \in C_0(X).$$

We are now able to prove

$$(34) \qquad Vf = s - \lim_{\lambda \downarrow 0} J_\lambda f = s - \lim_{\lambda \downarrow 0} (\lambda I - A)^{-1} f \quad \text{when} \quad f \in C_0(X).$$

In fact, we have, by $J_\lambda = (\lambda I - A)^{-1}$,

$$(35) \qquad (I - \lambda J_\lambda)f = -J_\lambda A f \quad \text{when} \quad f \in D(A).$$

As proved above in (33), $R(A)$ is dense in B, and also $R(J_\lambda) \supseteq V \cdot C_0(X)$ is dense in B by (6). Hence the range $R(I - \lambda J_\lambda)$ is dense in B. Therefore, by (30), we must have

$$s - \lim_{\lambda \downarrow 0} \lambda J_\lambda g = 0$$

for every $g \in B$. This proves (34) by (32).

We finally shall prove (11). By (31), $A = \lambda I - J_\lambda^{-1}$ is closed with J_λ^{-1}. Thus the equation $A = -V^{-1}$ implies that V must be closed. Next let V be closed. Then, for any $g \in B$, there exists, by (17), a sequence of functions $f_n \in C_0(X)$ such that

$$s - \lim_{n \to \infty} (\lambda Vf_n + f_n) = g.$$

Thus, by the continuity of the operator J_λ, we have

$$s - \lim_{n \to \infty} J_\lambda(\lambda Vf_n + f_n) = s - \lim_{n \to \infty} \hat{J}_\lambda(\lambda Vf_n + f_n) = s - \lim_{n \to \infty} Vf_n = J_\lambda g.$$

Hence, by the closure property of V, we must have

$$J_\lambda g = Vf = s - \lim_{n \to \infty} Vf_n \quad \text{and} \quad g = s - \lim_{n \to \infty} (\lambda Vf_n + f_n) = \lambda Vf + f.$$

That f is uniquely determined by g is clear from these equations. Hence

$$(36) \qquad J_\lambda = (\lambda I - A)^{-1} = V(\lambda V + I)^{-1}.$$

We have incidentally proved that $(\lambda V + I)$ is a closed linear operator. Thus the inverse $(\lambda V + I)^{-1}$ is an everywhere defined closed linear operator, and so, by the closed graph theorem, the inverse $(\lambda V + I)^{-1}$ is a bounded linear operator on B into B.

We have

$$(32') \qquad V f = \lambda J_\lambda V f + J_\lambda f \quad \text{when} \quad f \in D(V),$$

and hence

$$(\lambda I - A) J_\lambda (\lambda V f + f) = \lambda V f + f = (\lambda I - A) V f = \lambda V f - A V f \quad \text{when} \quad f \in D(V),$$

that is,

$$(33') \qquad A V f = -f \quad \text{when} \quad f \in D(V).$$

Hence $N(V) = \{0\}$ and so V^{-1} exists. Therefore, by (36), we obtain

$$(\lambda I - A) = J_\lambda^{-1} = (\lambda V + I) V^{-1} = \lambda I + V_\lambda^{-1},$$

which proves that $A = -V^{-1}$. $\hfill$ Q. E. D. [9]

References

1. Deny, J.: Les principes fondamentaux de la théorie du potentiel. Séminaire de Théorie du Potentiel dirigé par M. Brelot, G. Choquet et J. Deny (1960/61).
2. Hille, E., and R. S. Phillips: Functional analysis and semi-groups. Providence (1957).
3. Hunt, G. A.: Markoff processes and potentials, II. Illinois J. Math. 1, 316—369 (1957).
4. Lion, G.: Construction du semi-groupe associé a un noyau de Hunt. Séminaire de Théorie du Potentiel dirigé par M. Brelot, G. Choquet et J. Deny (1960/61).
5. Meyer, P. A.: Probability and potentials. New York: Blaisdell Publishing Company 1966.
6. Yosida, K.: Functional analysis. Berlin-Heidelberg-New York: Springer 1966.
7. — Positive pseudo-resolvents and potentials. Proc. Japan Acad. 41, 1—5 (1965).

Department of Mathematics
University of Tokyo
Honjo, Tokyo

---

[9] In the literature (see G. Lion [4] and P. A. Meyer [5]), J_λ is defined as

$$s - \lim_{n \to \infty} J_\lambda^{(n)} = J_\lambda$$

by making use of the fact that for $f \in C_0(X)$, the sequence $J_\lambda^{(n)} a_n^{-1} f$ decreases as n increases. This monotone property is proved by the principle of positive maximum. However, even if f is in $C_0(X)$, it seems not certaint that $J_\lambda^{(n)} a_n^{-1} f$ is with compact support and hence the principle of positive maximum might not be applied directly to

$$V(a_n J_\lambda^{(n)} a_n^{-1} f - a_m J_\lambda^{(m)} a_m^{-1} f) = \lambda J_\lambda^{(m)} a_m^{-1} f - \lambda J_\lambda^{(n)} a_n^{-1} f.$$

Therefore, G. Lion [4] and P. A. Meyer [5] have to extend the principle of positive maximum to functions outside B.

On the pre-closedness of the potential operator

(*with T. Watanabe and H. Tanaka*)

J. Math. Soc. Japan **20** (1968) 419–421

(Received Oct. 30, 1967)

§1. Introduction. Let X be a separable, locally compact, non-compact Hausdorff space, and B be the completion with respect to the maximum norm of the space $C_0(X)$ of real-valued continuous functions with compact supports defined in X. G. A. Hunt [1] introduced the notion of the potential operator V as a positive linear operator on $D(V) \subsetneqq B$ with $D(V) \supseteqq C_0(X)$ into B satisfying the "principle of positive maximum"[1]:

(1)　For any $f \in C_0(X)$, we have $\sup\limits_{f(x)>0} (Vf)(x) = \sup\limits_{x \in X} (Vf)(x)$ if the latter supremum is positive.

The fundamental result of Hunt reads as follows:

THEOREM. *Let V satisfy* (1) *and the condition that*

(2)　$V \cdot C_0(X)$ *is dense in* B.

Then, there exists a uniquely determined semi-group $\{T_t ; t \geqq 0\}$ *of class* (C_0) *of positive contraction linear operators* T_t *on* B *into* B *such that*

(3)　$AVf = -f$, $f \in C_0(X)$, *for the infinitesimal generator* A *of* T_t.

An operator-theoretical proof of this theorem was given in K. Yosida [2], showing that the resolvent $J_\lambda = (\lambda I - A)^{-1}$, $\lambda > 0$, of A is the continuous extension to the whole space B of the operator $\hat{J}_\lambda$ defined by

(4)　　　　　　　　$\lambda Vf + f \to Vf$, 　 $f \in C_0(X)$,

with an additional remark that

(5)　V^{-1} exists and $V^{-1} = -A$ if and only if V is closed.

The purpose of the present note is to show that *the restriction $V|C_0(X)$ of V to $C_0(X)$ is pre-closed so that its smallest closed extention, which shall be*

---

1)　This principle, sometimes called as the "weak principle of positive maximum", is proved on page 220 of [2] in the course of the proof of:

(1)′　For any $f \in C_0(X)$, the condition $(Vf)(x_0) = \sup\limits_{x \in X} (Vf)(x)$ implies $f(x_0) \geqq 0$.

It is also proved on the same page that (1)′ is a consequence of (1) and (2).

denoted by the same letter V, satisfies the true Poisson equation for the potential of functions:

$$(5)' \qquad V = -A^{-1}.$$

The closure property of V is important, since, as in [2], we can prove:

(6)   for any $\lambda > 0$, the inverse $(\lambda V + I)^{-1}$ exists as a continuous linear operator on B into B so that $J_\lambda = (\lambda I - A)^{-1} = V(\lambda V + I)^{-1}$. In particular, for any $g \in B$, there exists a uniquely determined f in the domain $D(V)$ of V with $\lambda V f + f = g$.

Thus, applying the closed range theorem and its corollary in K. Yosida [3] to the closed linear operator $(\lambda V + I)$, we obtain:

(7)   for any $\lambda > 0$, the inverse $(\lambda V^* + I^*)^{-1}$, of the dual operator $(\lambda V^* + I^*)$ of $(\lambda V + I)$ exists as a continuous linear operator on the dual space B^* of B into B^*. In particular, for any measure $\gamma \in B^*$, there exists a uniquely determined measure $\varphi \in D(V^*)$ such that $\lambda V^* \varphi + \varphi = \gamma$.

Moreover, since the domain $D(A)$ of the infinitesimal generator A is dense in B, we have $(A^*)^{-1} = (A^{-1})^*$ by the denseness in B of the range $R(A) = D(V)$. For the proof, see p. 224 in K. Yosida [3]. Therefore, by $(5)'$, we have *the true Poisson equation for the potential of measures:*

$$(5)'' \qquad V^* = -(A^*)^{-1}.$$

§2. Proofs of the pre-closedness of the operator $V \mid C_0(X)$.

We have to prove $g = 0$ from $f_n \in C_0(X)$, $s - \lim\limits_{n \to \infty} f_n = 0$ and $s - \lim\limits_{n \to \infty} V f_n = g$.

The first proof (by Tanaka).   From (33) in [2], we have $AVf = -f$. Thus, by the closure property of the operator A, we have $Ag = 0$. By $J_\lambda = (\lambda I - A)^{-1}$, we have $(I - \lambda J_\lambda)g = -J_\lambda A g = 0$ so that $g = s - \lim\limits_{\lambda \downarrow 0} \lambda J_\lambda g$. That the latter $s - \lim$ is $= 0$ is proved in the paragraph following (35) of [2].

The second proof (by Yoshida).   By (32) in [2], we have $Vf = \lambda J_\lambda V f + J_\lambda f$ so that $g = s - \lim\limits_{n \to \infty} V f_n = \lambda J_\lambda g$. Hence $g = 0$ as in the first proof.

The third proof (by Watanabe).   It is straightforward in the sense that it only makes use of the principle of positive maximum. It reads as follows.

Suppose $g(x_0) > 0$ for some $x_0 \in X$. Let K be any compact set of X such that $K \ni x_0$. Let $h \in C_0(X)$ be such that $h(x) = 1$ on K. Then $\| V(f_n h) \| \leq \| f_n \| \cdot \| V h \|$. Hence $s - \lim\limits_{n \to \infty} V(1-h)f_n = g$. Choose n so large that

$$(8) \qquad |(V(1-h)f_n)(x) - g(x)| < \frac{1}{4} g(x_0) \qquad \text{for all } x.$$

Then

$$(V(1-h)f_n)(x_0) \geqq \frac{3}{4} g(x_0) > 0, \text{ and support } ((1-h)f_n) \subsetneq X-K.$$

By the principle of positive maximum, there exists thus a point $x_1 \in X-K$ such that $(V(1-h)f_n)(x_1) \geqq (V(1-h)f_n)(x_0)$ and so, by (8), $|g(x_1)| \geqq \frac{1}{2} g(x_0)$. Since K was arbitrary, this contradicts to the fact that $g(x) \in B$ tends to zero at infinity. Therefore g must be $\leqq 0$ everywhere. In the same way, we can prove $g \geqq 0$ everywhere. This prove $g = 0$.

University of Tokyo (Yosida, Tanaka)
and
University of Osaka (Watanabe)

References

[1] G. A. Hunt, Markov processes and potentials, II, Illinois J. Math., 1 (1957), 316–369.

[2] K. Yosida, Positive resolvents and potentials, Z. für Wahrscheinlichkeitstheorie und Verw. Gebiete, 8 (1967), 210–218.

[3] K. Yosida, Functional Analysis, Springer-Verlag, Berlin-Heidelberg-New York, 1966.

The existence of the potential operator associated with an equicontinuous semi-group of class (C_0)

Studia Math. **31** (1968) 531–533

Hunt [2] introduced the notion of potential operators V associated with transient Markov processes in a separable, locally compact, non-compact Hausdorff space. The present author gave an operator-theoretical treatment of Hunt's theory of potentials (see [4] and [5]). This treatment suggests us to give an *abstract definition of the potential operator which may be applied to transient as well as to some recurrent Markov processes.*

Let X be a locally convex, sequentially complete, linear topological Hausdorff space. Let a family $\{T_t; t \geqslant 0\}$ of continuous linear operators T_t on X into X satisfy the following three conditions:

(1) $\quad T_t T_s = T_{t+s}, \quad T_0 = I = \text{the identity (the semigroup property)};$

(2) $\quad$ for any continuous seminorm $p(x)$ on X, there exists a continuous seminorm $q(x)$ on X such that $p(T_t x) \leqslant q(x)$ for all $t \geqslant 0$ and $x \in X$ (the equicontinuity);

(3) $\quad \lim T_t x = T_{t_0} x$ for every $t_0 \geqslant 0$ and $x \in X$ (the class (C_0) property).

Thus $\{T_t; t \geqslant 0\}$ is an equicontinuous semigroup of class (C_0) in X (see [3]). We can prove the following existence theorem:

THEOREM. *The infinitesimal generator A of T_t defined through*

$$(4) \qquad Ax = \lim_{h \downarrow 0} h^{-1}(T_t x - x)$$

admits a densely defined inverse A^{-1} if and only if

$$(5) \qquad \lim_{\lambda \downarrow 0} \int_0^\infty \lambda e^{-\lambda t} T_t x \, dt = 0 \quad \text{for all } x \in X.$$

Moreover, (5) is a consequence of an apparently weaker condition

$$(5') \qquad \operatorname{weak-lim}_{\lambda \downarrow 0} \int_0^\infty \lambda e^{-\lambda t} T_t x \, dt = 0 \quad \text{for all } x \in X.$$

By virtue of this Theorem, we may give the

Definition. In case when (5) is satisfied, we shall call

$$(6) \qquad V = -A^{-1}$$

the *potential operator associated with the semigroup* $\{T_t; t \geqslant 0\}$.

Remark 1. The definition of the potential operator

$$(6') \qquad Vx = \lim_{\lambda \downarrow 0} (\lambda I - A)^{-1} x$$

given in [4] well fits to (6).

Remark 2. Consider the case where X is the completion, with respect to the maximum norm, of the space $C_0(R^1)$ of real-valued continuous functions $x(\xi)$ with compact supports in the whole real line R^1. Then the semigroup

$$(7) \qquad (T_t x)(\xi) = (2 \pi t)^{-1/2} \int_{-\infty}^{\infty} e^{-|\xi - \eta|^2/2t} x(\eta) d\eta, \qquad x(\xi) \epsilon X,$$

satisfies condition (5), although the Brownian motion in R^1 is recurrent.

Proof of the Theorem. It is known (see, e.g., [3]) that

$$(8) \qquad \lambda(\lambda I - A)^{-1} x = \int_{0}^{\infty} \lambda e^{-\lambda t} T_t x \, dt$$

and

$$(9) \qquad p\big(\lambda(\lambda I - A)^{-1} x\big) \leqslant q(x) \qquad \text{for all } \lambda > 0 \text{ and } x \epsilon X.$$

Hence the condition $Ax = 0$ is equivalent to $\lambda(\lambda I - A)^{-1} x = x$ (for all $\lambda > 0$). Thus condition (5) implies, by (8), the existence of the inverse A^{-1}.

On the other hand, we have

$$(10) \qquad A(\lambda I - A)^{-1} x = \big(\lambda(\lambda I - a)^{-1} - I\big) x$$

and hence

$$(11) \qquad \text{the range } R(A) = \text{the range } R\big(I - \lambda(\lambda I - A)^{-1}\big).$$

By virtue of condition (9), we can apply the ergodic theorem of the Hille type in [3] (cf. Theorem 18. 6. 2. in [1]) to the effect that

$$(12) \qquad R\big(I - \lambda(\lambda I - A)^{-1}\big) \text{ is independent of } \lambda > 0 \text{ and its closure in}$$
X is equal to the set $\{x \epsilon X;\ \lim_{\lambda \downarrow 0} \lambda(\lambda I - A)^{-1} x = 0\}$.

Hence, condition (5) is equivalent to the condition that the range $R(A)$ is dense in X. That condition (5) may be replaced by a weaker condition $(5')$ is proved in the ergodic theorem of the Hille type mentioned above.

Therefore the Theorem is proved.

Added in proof. Professor H. Komatsu called the author's attention that another proof of the above Theorem may be obtained by applying Theorem 3.1 in his paper *Fractional powers of operators,* Pacific J. Math. 19 (1966), p. 285.

References

[1] E. Hille and R. S. Phillips, *Functional analysis and semi-groups,* Providence 1957.

[2] G. H. Hunt, *Markov processes and potentials, II,* Illinois J. Math. 1 (1957), p. 316-369.

[3] K. Yosida, *Functional analysis,* 1965.

[4] — *Positive resolvents and potentials,* Z. Wahrscheinlichkeitstheorie und verwandte Gebiete 8 (1967), p. 210-218.

[5] — and T. Watanabe and H. Tanaka, *Pre-closedness of potential operators,* to appear in J. Math. Soc. Japan.

DEPARTMENT OF MATHEMATICS, UNIVERSITY OF TOKYO

Reçu par la Rédaction le 6. 1. 1968

On the potential operators associated with Brownian motions

J. Analyse Math. **23** (1970) 461–465

Let X be a locally convex, sequentially complete, linear topological Hausdorff space, and $\{T_t; t \geq 0\}$ be an equi-continuous semi-group of class (C_0) of continuous linear operators T_t on X into X (see [4] and [1]):

(1) $T_t T_s = T_{s+t}$, $T_0 = I =$ the identity operator (the semi-group property);

(2)   for any continuous semi-norm $p(x)$ on X, there exists a continuous semi-norm $q(x)$ on X such that $p(T_t f) \leq q(f)$ for all $t \geq 0$ and all $f \in X$ (the equi-continuity);

(3) $\lim_{t \to t_0} T_t f = T_{t_0} f$ for all $t_0 \geq 0$ and all $f \in X$ (the class (C_0) property).

In a preceding paper [7], the present author proved the following

Theorem.   *The infinitesimal generator A of T_t defined through*

$$(4) \qquad Af = \lim_{h \downarrow 0} h^{-1}(T_h - I)f$$

admits a densely defined inverse A^{-1} if and only if

$$(5) \qquad \lim_{\lambda \downarrow 0} \int_0^\infty \lambda e^{-\lambda t} T_t f\, dt = \lim_{\lambda \downarrow 0} \lambda(\lambda I - A)^{-1} f = 0 \text{ for every } f \in X.$$

By virtue of this theorem and suggested by [5] and [6], the author defined in [7] the *potential operator V associated with the semi-group* $\{T_t; t \geq 0\}$ through

$$(6) \qquad V = -A^{-1} \text{ if and only if } A \text{ satisfies (5).}$$

The purpose of the present note is to show explicitly that the above definition of the *potential operator V based upon the Theorem is relevantly applicable to transient as well as to recurrent Brownian motions.*

We first remark that

$$(6)' \qquad Vf = -A^{-1}f = \lim_{\lambda \downarrow 0} (\lambda I - A)^{-1}f \text{ for } f \in D(A^{-1}) = D(V)$$

since, by (5),

$$-A^{-1}f - (\lambda I - A)^{-1}f = [-I - (\lambda I - A)^{-1}A]A^{-1}f = \lambda(\lambda I - A)^{-1}A^{-1}f$$

tends to 0 as $\lambda \downarrow 0$.

Next let X be the Banach space $C_\infty(R^d)$ obtained as the completion of the space $C_0(R^d)$, the space of real-valued continuous functions on R^d with compact supports, pertaining to the maximum norm

$$\|f\| = \sup_{x \in R^d} |f(x)|.$$

Let us define through

$$(7) \qquad (T_t^{(d)}f)(x) = (2\pi t)^{-d/2} \int_{R^d} e^{-|x-y|^2/2t}f(y)dy, \quad f \in C_\infty(R^d)$$

$$(dy = dy_1 dy_2 \cdots dy_d)$$

the semi-group associated with the d-dimensional Brownian motion. It is known that, for $d \geq 3$, the Brownian motion is *transient* so that we can make use of Hunt's definition (see [2]) of the potential operator

$$(8) \qquad (V^{(d)}f)(x) = \int_0^\infty (T_t^{(d)}f)(x)\,dt$$

without appealing to our definition (6)'. However, for $d \leq 2$, the Brownian motion is *recurrent* and Hunt's definition (8) diverges for non-negative functions $f(x) \not\equiv 0$ from $C_\infty(R^d)$. We shall show that our definition (6)' is pertinently fit also to such recurrent cases.

In fact we have (see [3], p. 146)

$$(9) \quad \int_0^\infty e^{-\lambda t}(2\pi t)^{-d/2}e^{-|x-y|^2/2t}dt = 2(2\pi)^{-d/2}\left(\frac{\sqrt{2\lambda}}{|x-y|}\right)^{\frac{d}{2}-1} K_{\frac{d}{2}-1}\left(\sqrt{2\lambda}|x-y|\right),$$

where $K_\alpha(z)$ is the modified Bessel function so that, in particular,

$$(10) \quad K_0(z) = -\pi \int_0^\pi e^{z\cos\theta}[\log(2z\cdot\sin^2\theta) + \gamma]d\theta, \quad \gamma = \text{the Euler constant},$$

$$K_{\frac{1}{2}-1}(z) = (\pi/2z)^{1/2}e^{-z}.$$

On the other hand, it is known from the semi-group theory (see [1] and [4]) that

$$(11) \qquad \left\| \lambda(\lambda I - A^{(d)})^{-1} \right\| \leqq 1 \quad \text{for all } \lambda > 0,$$

where $A^{(d)}$ is the infinitesimal generator of the semi-group $T_t^{(d)}$. Hence, in virtue of (9) and (10), we can easily verify the condition

$$\operatorname*{s\text{-}lim}_{\lambda\downarrow 0} \lambda(\lambda I - A^{(d)})^{-1}f = 0 \quad \text{for all } f\in C_\infty(R^d) \quad \text{with } d \leqq 2$$

by verifying it for functions $f\in C_0(R^d)$ only. Thus we can apply definition (6)′ to the recurrent Brownian motion with $d = 1$ and $d = 2$.

To this purpose, we recall that $A^{(d)}$ is known to be equal to half the d-dimensional Laplacian so that by $\int_{R^d} (T_t^{(d)}g)(x)\,dx = \int_{R^d} g(x)\,dx$

$$(12) \qquad \int_{R^d} f(y)dy = 0 \quad \text{whenever } f = A^{(d)}g \in D((A^{(d)})^{-1}) \cap C_0(R^d).$$

Now, in virtue of(9), (10) and (6)′, we obtain, for $f\in D((A^{(2)})^{-1}) \cap C_0(R^2)$

satisfying $\int_{R^2} f(y)dy = 0,$

$$(V^{(2)}f)(x) = (-(A^{(2)})^{-1}f)(x)$$

$$= \text{s-}\lim_{\lambda \downarrow 0} \int_{R^2} f(y)dy \int_0^{\pi} (-\pi)^2 e^{\sqrt{2\lambda}|x-y|\cos\theta}[\log(2\sqrt{2\lambda}\cdot\sin^2\theta) + \log|x-y| + \gamma]d\theta$$

$$= -\pi^{-1} \int_{R^2} \log|x-y|\cdot f(y)dy$$

$$+ \text{s-}\lim_{\lambda \downarrow 0}\left[-\pi^2\sqrt{2\lambda}\log\sqrt{2\lambda}\int_{R^2} dy\left\{\sqrt{2\lambda}^{-1}\int_0^{\pi} (e^{\sqrt{2\lambda}|x-y|\cos\theta}f(y) - f(y))d\theta\right\}\right.$$

$$= -\pi^{-1} \int_{R^2} \log|x-y|\cdot f(y)dy.$$

In this way, we obtain a semi-group theoretical interpretation of the logarithmic potential.

Similarly, we obtain, for $f \in D((A^{(1)})^{-1}) \cap C_0(R^1)$ satisfying $\int_{R^1} f(y)dy = 0$,

$$(V^{(1)}f)(x) = (-(A^{(1)})^{-1}f)(x)$$

$$= \text{s-}\lim_{\lambda \downarrow 0} (2\lambda)^{-1/2} \int_{R^1} (e^{-\sqrt{2\lambda}|x-y|}f(y) - f(y))dy$$

$$= -\int_{R^1} |x-y|\cdot f(y)dy.$$

This is a semi-group theoretical interpretation of the one-dimensional Newtonian potential.

REFERENCES

1. E. Hille and R. S. Phillips, Functional Analysis and Semi-groups, Providence (1957).

2. G. H. Hunt, Markov processes and potentials, II, *Illinois J. of Math.*, **1** (1957), 316–369.

3. A. Erdélyi, Tables of Integral Transforms, I, New York (1954).

4. K. Yosida, Functional Analysis, Springer (1965), the 2nd edition (1967).

5. K. Yosida, Positive resolvent and potentials, *Z. für Wahrscheinlichkeitstheorie und verwandte Gebiete*, **8** (1967), 210–218.

6. K. Yosida, T. Watanabe and H. Tanaka, Pre-closedness of potential operators, *J. Math. Soc. of Japan*, **20**, Nos 1–2 (1968), 419–421.

7. K. Yosida, The existence of the potential operator associated with an equi-continuous semi-group of class (C_0), in *Stud. Math.*, **31** (1968), 531–533.

DEPARTMENT OF MATHEMATICS
KYOTO UNIVERSITY

On the pre-closedness of Hunt's potential operators and its applications

Proc. Intern. Conf. Functional Analysis and Related Topics (Tokyo, 1969), Univ. Tokyo Press (1970) 324–331

§ 1. Introduction.

Let X be a locally compact, non-compact, separable Hausdorff space, and $C_\infty(X)$ be the completion with respect to the maximum norm of the space $C_0(X)$ of real-valued continuous functions with compact support defined in X. $C_\infty(X)$ is thus a Banach space with respect to the maximum norm. G. A. Hunt [2] introduced the notion of the *potential operator* V as a positive linear operator on the domain $D(V) \subseteqq C_\infty(X)$ into $C_\infty(X)$ satisfying the following three conditions:

(1)
$$D(V) \supseteqq C_0(X);$$

(2)
$$V \cdot C_0(X) \text{ is strongly dense in } C_\infty(X);$$

(3) For any $f \in C_0(X)$, we have $\sup_{f(x)>0}(Vf)(x) = \sup_{x \in X}(Vf)(x)$ if the latter supremum is positive (the so-called *principle of positive maximum*).

THE FUNDAMENTAL RESULT OF HUNT states: There exists a uniquely determined semi-group $\{T_t ; t \geqq 0\}$ of class (C_0) of positive contraction operators T_t on $C_\infty(X)$ into $C_\infty(X)$ in such a way that Poisson's equation holds for $f \in C_0(X)$, i. e.,

(4)
$$AVf = -f, \text{ where } A \text{ is the infinitesimal generator of } T_t.$$

The present speaker (see K. Yosida [8]) gave an operator-theoretical proof of the above result by showing that for $\lambda > 0$ the resolvent

(5)
$$J_\lambda = (\lambda I - A)^{-1}$$

of A given by

(5)'
$$(\lambda I - A)^{-1}f = \int_0^\infty e^{-\lambda t} T_t f \, dt \qquad (f \in C_\infty(X))$$

is the continuous extension to the whole space $C_\infty(X)$ of the operator $\hat{J}_\lambda$ defined through

(6)
$$J_\lambda(\lambda Vf + f) = Vf, \qquad f \in C_0(X).$$

The crucial points in this proof are on the one hand the fact that

(7)   for any $\lambda > 0$, $(\lambda V + I) \cdot C_0(X)$ is strongly dense in $C_\infty(X)$

and on the other hand THE ABELIAN ERGODIC THEOREM or THE ERGODIC THEOREM OF THE HILLE TYPE (K. Yosida [7]):

Let a family $\{J_\lambda; \lambda > 0\}$ of bounded linear operators J_λ on a Banach space B into B satisfy

(8) $$J_\lambda - J_\mu = (\mu - \lambda) J_\lambda J_\mu \quad \text{(the resolvent equation)}$$

and

(9) $$\sup_{\lambda > 0} \| \lambda J_\lambda \| < \infty .$$

Then we have

(10)   the strong closure in B of the range $R(J_\lambda)$ is independent
of λ and is equal to $\{ f \in B ; \operatorname*{s\text{-}lim}_{\lambda \uparrow \infty} \lambda J_\lambda f = f \}$

and

(11)   the strong closure in B of the range $R(I - \lambda J_\lambda)$ is independent
of λ and is equal to $\{ f \in B ; \operatorname*{s\text{-}lim}_{\lambda \downarrow 0} \lambda J_\lambda f = 0 \} .$

The merit of the operator-theoretical treatment is that we could obtain several comments of interest pertinent to Hunt's theory of potentials. These comments shall be stated as Propositions together with their Corollaries in this lecture. The most important among them is the fact expressed in

PROPOSITION 1.   *The restriction* $V|C_0(X)$ *of the operator* V *to* $C_0(X)$ *is pre-closed (or closable), that is,*

(12)   *if* $f_n \in C_0(X)$ *be such that* $\operatorname*{s\text{-}lim}_{n \to \infty} f_n = 0$ *and*

$\operatorname*{s\text{-}lim}_{n \to \infty} V f_n = g,$ *then* g *must be* $= 0$.

This result was proved in K. Yosida-T. Watanabe-H. Tanaka [9]. It implies, combined with Theorem 4 in K. Yosida [8], the following

COROLLARY 1.   *We shall denote the smallest closed extension of* $V|C_0(X)$ *by* U. *Then the inverse* U^{-1} *of* U *exists and we have the true Poisson's equation:*

(13) $$U^{-1} = -A, \text{ that is, } U = -A^{-1} ,$$

where A *is the infinitesimal generator stated in the fundamental result of Hunt as given above.*

REMARK.   As is pointed out to the author by A. Yamada, formula (11) in K. Yosida [8] must be corrected as the inverse V^{-1} exists and $A = -V^{-1}$ if and only if V is the smallest closed extension in $C_\infty(X)$ of $V|C_0(X)$, and $(\lambda I - A)^{-1} = V(\lambda V + I)^{-1}$ in this case.

COROLLARY 2. *U is defined through*

(14) $$Uf = \text{s-lim}_{\lambda \downarrow 0} (\lambda I - A)^{-1} f.$$

PROPOSITION 2. *The principle of positive maximum expressed in (3) can be deduced from the following three properties of V:*

(15) *V is a positive linear operator on $D(V) \subseteqq C_\infty(X)$ with $D(V) \supseteqq C_0(X)$ into $C_\infty(X)$ such that $V \cdot C_0(X)$ is strongly dense in $C_\infty(X)$;*

(16) *There exists a sequence of non-negative functions $h_n \in C_0(X)$ such that Vh_n increases everywhere to 1 as $n \to \infty$;*

(17) *Let f and g be both non-negative and $\in D(V)$, and let $V_\lambda = V + \lambda^{-1}I$ with a positive constant λ. If the inequality $(V_\lambda f)(x) \leqq (Vg)(x)$ holds on the support of f, then we must have $V_\lambda f \leqq Vg$ everywhere.*

REMARK. The speaker called (17) as the *principle of majoration* and gave the proof of Proposition 2 in K. Yosida [8]. The following Proposition explains the origin of (17).

PROPOSITION 3. *Let a family $\{J_\lambda; \lambda > 0\}$ of positive continuous linear operators J_λ on $C_\infty(X)$ into $C_\infty(X)$ satisfy (8). Then the operator V defined through*

(14)′ $$Vf = \text{s-lim}_{\lambda \downarrow 0} J_\lambda f$$

satisfies the principle of majoration (17).

REMARK. Recently A. Yamada [6] gave a simplification of the speaker's original proof of Proposition 3 in K. Yosida [8].

PROPOSITION 4. *Let the operator V be defined through (14)′ where J_λ is the resolvent of a linear operator A with the domain $D(A)$ strongly dense in $C_\infty(X)$, i. e., $J_\lambda = (\lambda I - A)^{-1}$. If the domain $D(V)$ of V is strongly dense in $C_\infty(X)$, then the inverse V^{-1} exists and the range $R(V)$ of V is strongly dense in $C_\infty(X)$.*

REMARK. This Proposition explains the scope of (15). The original proof in K. Yosida [8] is based upon the Hahn-Banach extension theorem. A straightforward proof is obtained recently by A. Yamada [6].

In this lecture, the speaker shall reproduce the proof of Proposition 1 and its Corollaries in § 2. It will be shown in § 3, that Hunt's theory of potentials is restricted solely to *transient Markov processes*. Corollary 1 suggests us to propose a definition of the potential operator which is applicable to transient as well as to a well defined subclass of *recurrent Markov processes*. This new definition is useful in yielding the Newtonian potentials for the Brownian motion in Euclidean space of every dimension; as is well known, Brownian motions in one and two dimensions are recurrent Markov processes, while those of dimension greater than or equal to 3 are transient Markov processes.

§ 2. The pre-closedness of Hunt's potential operators.

PROOF OF (12). We shall prove (12) following the argument due to T. Watanabe (see K. Yosida-T. Watanabe-H. Tanaka [9]). Assume the contrary and let $g(x_0) > 0$ for some $x_0 \in X$. Let K be any compact set of X containing x_0. Let a non-negative function $h(x) \in C_0(X)$ be such that $h(x) = 1$ on a neighborhood of K. Then $\sup_x |(V(f_n h))(x)| \leq \|f_n\| \cdot \|Vh\|$ by the positivity and the linearity of the operator V. Hence s-$\lim_{n \to \infty} V((1-h)f_n) = g$. Choose n so large that

$$(18) \qquad |(V((1-h)f_n))(x) - g(x)| < \frac{1}{4} g(x_0) \qquad \text{for all } x \in X.$$

Then

$$(V((1-h)f_n))(x_0) > \frac{3}{4} g(x_0) > 0 \quad \text{and support } ((1-h)f_n) \subseteq (X-K).$$

By the principle of positive maximum, there exists thus a point $x_1 \in (X-K)$ such that $(V((1-h)f_n))(x_1) \geq (V((1-h)f_n))(x_0)$ and so, by (18), $|g(x_1)| > 2^{-1}g(x_0)$. Since K was arbitrary, this contradicts to the fact that $g \in C_\infty(X)$ tends to zero at infinity. Thus g must be ≤ 0 everywhere. In the same way, g must be ≥ 0 everywhere and hence g must be $= 0$.

PROOF OF (13). By (6) we have $J_\lambda(\lambda V f + f) = V f, f \in C_0(X)$. Hence, by the continuity of the operator J_λ, we obtain,

$$(19) \qquad J_\lambda(\lambda U f + f) = U f \qquad \text{for all } f \in D(U).$$

Thus, by $J_\lambda = (\lambda I - A)^{-1}$,

$$\lambda U f + f = (\lambda I - A) J_\lambda (\lambda U f + f) = (\lambda I - A) U f = \lambda U f - A U f \qquad \text{if } f \in D(U)$$

and so

$$(20) \qquad A U f = -f \quad \text{whenever} \quad f \in D(U).$$

Thus the inverse U^{-1} exists. Hence, by (19) and $J_\lambda = (\lambda I - A)^{-1}$, the inverse $(\lambda U + I)^{-1}$ exists and

$$(\lambda I - A) = J_\lambda^{-1} = (\lambda U + I) U^{-1} = \lambda I + U^{-1}, \qquad \text{that is, } A = -U^{-1}.$$

PROOF OF (14). We have

$$(21) \qquad A(\lambda I - A)^{-1} f = -f + \lambda(\lambda I - A)^{-1} f \qquad \text{for all } f \in C_\infty(X).$$

Hence if s-$\lim_{\lambda \downarrow 0} (\lambda I - A)^{-1} f = g$ exists for a certain $f \in C_\infty(X)$, the closure property of the infinitesimal generator A implies that $Ag = -f$, i. e., $g = A^{-1}Ag = -A^{-1}f = U f$ by (13). This proves s-$\lim_{\lambda \downarrow 0} (\lambda I - A)^{-1} f = U f$. On the other hand, let $f \in D(U)$ so that $U f = -A^{-1}f$ by (13). By virtue of (21), we see that the range $R(A)$ of A coincides with the range $R(I - \lambda(\lambda I - A)^{-1})$. However, $R(A) = D(U)$

must be strongly dense in $C_\infty(X)$ since $D(U) \supseteq C_0(X)$ by the definition of U. Thus, by the ergodic theorem (11), we must have

$$(22) \qquad \text{s-}\lim_{\lambda \downarrow 0} \lambda(\lambda I - A)^{-1}g = 0 \qquad \text{for every } g \in C_\infty(X).$$

Therefore, by (21) and

$$(23) \qquad -A^{-1}f - (\lambda I - A)^{-1}f = -A^{-1}f - A(\lambda I - A)^{-1}A^{-1}f = -\lambda(\lambda I - A)^{-1}A^{-1}f,$$

we obtain

$$(24) \qquad Uf = -A^{-1}f = \text{s-}\lim_{\lambda \downarrow 0} (\lambda I - A)^{-1}f.$$

This completes the proof of (14).

§ 3. The existence of the potential operator associated with an equi-continuous semi-group of class (C_0).

We first show

PROPOSITION 5. *Hunt's potential operator V satisfying* (1), (2) *and* (3) *is given by*

$$(25) \qquad (Vf)(x) = \int_0^\infty (T_t f)(x)dt \quad \text{whenever} \quad f \in C_0(X),$$

where $\{T_t; t \geq 0\}$ is the semi-group of positive contraction linear operators T_t on $C_\infty(X)$ into $C_\infty(X)$ stated in The Fundamental Result of Hunt in the Introduction.

PROOF. By (24) and (5)′, we have

$$(26) \qquad Uf = Vf = \text{s-}\lim_{\lambda \downarrow 0} \int_0^\infty e^{-\lambda t}T_t f\, dt \quad \text{whenever } f \in C_0(X) \subseteq D(V) \cap D(U).$$

Thus when $f \in C_0(X)$ is non-negative we have

$$(Vf)(x) \geq \lim_{\lambda \downarrow 0} \int_0^M e^{-\lambda t}(T_t f)(x)dt = \int_0^M (T_t f)(x)dt \qquad \text{for every } M > 0.$$

Hence it is easy to see that (25) is true for non-negative $f \in C_0(X)$, and thus (25) is true.

Therefore Hunt's definition of the potential operator satisfying (1), (2) and (3) is related solely to *transient Markov processes*; it cannot be applied to *recurrent Markov processes*. The speaker should like to propose the following definition which is suggested by (13) and is applicable to transient as well as to a definite subclass of recurrent Markov processes.

DEFINITION. Let $\{T_t; t \geq 0\}$ be an equi-continuous semi-group of class (C_0) of bounded linear operators T_t on $C_\infty(X)$ into $C_\infty(X)$, and let A be its infinitesimal generator. Then T_t is called to admit the *potential operator U* if and only if

(27)   the inverse A^{-1} exists with its domain $D(A^{-1})=R(A)$ strongly dense in $C_\infty(X)$,

and in case (27) the potential operator U is defined by (24).

PROPOSITION 6.   (27) *is equivalent to* (22).

We shall sketch the proof which was given in K. Yosida [10]. The condition $Af=0$ is equivalent to $\lambda(\lambda I-A)^{-1}f=f$, and so (22) implies the existence of A^{-1}. As proved in §2, $D(A^{-1})=R(A)=R(I-\lambda(\lambda I-A)^{-1})$. Thus, by the ergodic theorem (11), (22) implies (27).

COROLLARY.   (27) *is equivalent to*

$$(28) \qquad \operatorname*{s\text{-}lim}_{\lambda\downarrow0}\int_0^\infty e^{-s}T_{\frac{s}{\lambda}}\,f\,ds=0 \qquad for\ every\ f\subset C_0(X).$$

PROOF.   By (5)$'$, we have $\lambda(\lambda I-A)^{-1}f=\int_0^\infty e^{-s}T_{\frac{s}{\lambda}}\,f\,ds$.

REMARK.   It is proved in K. Yosida [7] that, in the ergodic theorem (10) and (11), s-lim $\lambda J_\lambda f=f$ and s-lim $\lambda J_\lambda f=0$ can be replaced by w-lim $\lambda J_\lambda f=f$ and w-lim $\lambda J_\lambda f=0$, respectively. Hence, in (22) and (28), we may replace s-lim by w-lim.

Next we shall verify the scope of the above Definition of the potential operator together with (24) to the special case of the semi-group $\{T_t^{(n)}; t\geq0\}$ associated with the Brownian motion in n-dimensional Euclidean space $X=R^n$:

$$(29) \qquad (T_t^{(n)}f)(x)=(2\pi t)^{-\frac{n}{2}}\int_{R^n}e^{-|x-y|^2/2t}f(y)dy,\ f\in C_\infty(R^n);$$

$$where\ dy=dy_1\cdots dy_n.$$

We know (see A. Erdélyi [4]) that

$$(30) \qquad \int_0^\infty e^{-\lambda t}(2\pi t)^{-\frac{n}{2}}e^{-|x-y|^2/2t}\,dt=2(2\pi)^{-\frac{n}{2}}\left(\frac{\sqrt{2\lambda}}{|x-y|}\right)^{\frac{n}{2}-1}K_{\frac{n}{2}-1}(\sqrt{2\lambda}\,|x-y|),$$

where $K_\alpha(z)$ is the modified Bessel function. Hence, in particular,

$$(31) \qquad K_0(z)=-\frac{1}{\pi}\int_0^\pi e^{z\cos\theta}(\log(2z\cdot\sin^2\theta)+\gamma)d\theta,\ \gamma=\text{the Euler constant},$$

$$K_{\frac{1}{2}-1}(z)=(\pi/2z)^{-\frac{1}{2}}e^{-z}.$$

On the other hand, we have, by the semi-group theory, the inequality

$$(32) \qquad \sup_{\lambda>0}\|\lambda(\lambda I-A^{(n)})^{-1}\|\leq1,$$

where $A^{(n)}$ is the infinitesimal generator of $T_t^{(n)}$. Hence, in virtue of (30) and (31), we can easily verify the condition

$$(22)' \qquad \operatorname*{s\text{-}lim}_{\lambda\downarrow0}\lambda(\lambda I-A^{(n)})^{-1}f=0 \qquad for\ every\ f\in C_\infty(R^n)\ (n=1\ and\ 2)$$

by verifying (22)′ for $f \in C_0(R^n)$ only.

Since $A^{(1)}$ is half the Laplacian $2^{-1}d^2/dx^2$, we easily see that $f \in D((A^{(1)})^{-1})$ implies that $\int_{-\infty}^{\infty} f(y)dy = 0$. Suggested by this fact, we shall consider the potential

$$(-(A^{(1)})^{-1}f)(x) \quad \text{for} \quad f \in C_0(R^1) \cap D((A^{(1)})^{-1}) \quad \text{with} \quad \int_{-\infty}^{\infty} f(y)dy = 0,$$

and obtain, by (30), (31) and (24),

$$(-(A^{(1)})^{-1}f)(x) = \underset{\lambda \downarrow 0}{\text{s-lim}}\ (2\lambda)^{-\frac{1}{2}} \int_{-\infty}^{\infty} e^{-\sqrt{2\lambda}|x-y|} f(y)dy$$

$$= \underset{\lambda \downarrow 0}{\text{s-lim}}\ (2\lambda)^{-\frac{1}{2}} \int_{-\infty}^{\infty} (e^{-\sqrt{2\lambda}|x-y|} f(y) - f(y))dy$$

$$= -\int_{-\infty}^{\infty} |x-y| \cdot f(y)dy.$$

This is precisely the one-dimensional Newtonian potential.

Similarly, we obtain the logarithmic potential (see K. Yosida [11])

$$(-(A^{(2)})^{-1}f)(x)$$

$$= \underset{\lambda \downarrow 0}{\text{s-lim}} \int_{R^2} f(y)dy \int_0^{\pi} (-\pi^{-2}) e^{\sqrt{2\lambda}|x-y|\cos\theta} [\log(2\sqrt{2\lambda}\ \sin^2\theta) + \log|x-y| + \gamma]d\theta$$

$$= -\pi^{-1} \int_{R^2} \log|x-y| f(y)dy \quad \text{when} \quad f \in C_0(R^2) \cap D((A^{(2)})^{-1}) \quad \text{with} \quad \int_{R^2} f(y)\,dy = 0.$$

References

[1] E. Hille and R.S. Phillips, Functional Analysis and Semi-groups, Providence, 1957.

[2] G.A. Hunt, Markoff processes and potentials, II, Illinois J. Math., 1 (1957), 316–369.

[3] G. Lion, Familles d'opérateurs et frontière en théorie du potentiel, Ann. Inst. Fourier, Grenoble, 16 (1966), 389–453.

[4] A. Erdélyi, Tables of Integral Transforms, 1, New York, 1954.

[5] P.-A. Meyer, Probability and Potentials, Blaisdell, 1966.

[6] A. Yamada, On the correspondence between potential operators and semi-groups associated with Markov processes, Z. Wahrscheinlichkeitstheorie und Verw. Gebiete, to appear soon.

[7] K. Yosida, Functional Analysis, 2nd ed., Springer, 1968.

[8] K. Yosida, Positive resolvents and potentials, Z. Wahrscheinlichkeitstheorie und Verw. Gebiete, 8 (1967), 210–218.

[9] K. Yosida, T. Watanabe and H. Tanaka, On the pre-closedness of the potential

operator, J. Math. Soc. Japan, 20 (1968), 419–421.

[10] K. Yosida, The existence of the potential operator associated with an equi-continuous semi-group of class (C_0), Studia Math., 31 (1968), 531–533.

[11] K. Yosida, On the potential operators associated with Brownian motions, J. Analyse Math., to appear.

RESEARCH INSTITUTE FOR MATHEMATICAL SCIENCES
KYOTO UNIVERSITY
KITASHIRAKAWA, KYOTO
606 JAPAN

On the existence and a characterization of abstract potential operators

Troisième Colloque sur l'Analyse Fonctionnelle (Liège, 1970) 129–136

1. The existence theorem

Let X be a Banach space, and $\{J_\lambda; \lambda>0\}$ be a family of bounded linear operators defined on X with values in X satisfying the *resolvent equation*

$$(1) \qquad J_\lambda - J_\mu = (\mu-\lambda)J_\lambda J_\mu.$$

Such J_λ is called, by E. Hille (E. Hille-R.S. Phillips [1]), a *pseudo-resolvent* in X. The dual operator J_λ^* of J_λ is a pseudo-resolvent in the dual space X^* of X. As a refinement with extension of the results in preceding papers (K. Yosida [2]-[5]), we first prove:

Theorem 1. We assume that

$$(2) \qquad \sup_{\lambda>0} \|\lambda J_\lambda\| \leqq 1 .$$

Then the following five conditions are mutually equivalent ([2]).

$$(3) \qquad \begin{cases} w\text{--}\lim_{\lambda\uparrow\infty} \lambda J_\lambda x = x & \text{for all} \quad x \in X , \\[2mm] w\text{--}\lim_{\lambda\downarrow 0} \lambda J_\lambda x = 0 & \text{for all} \quad x \in X; \end{cases}$$

([1]) The results in this lecture were announced in two preprints: K. YOSIDA: On the existence of abstract potential eporators and the principle of majoration associated with them, Research Institute for Mathematical Sciences, Kyoto University, RIMS-63 (1970, April) and K. YOSIDA: A characterization of abstract potential operators, *ibid.*, 65 (1970, May).

([2]) w-lim and s-lim respectively mean limit in the sense of weak topology and strong topology of X. w^*-lim means lim in the sense of w^*-topology in X^*, i.e., the topology of the convergence of functionals $\in X^*$ at every point of X. Strongly dense (s-dense), weakly dense (w-dense) and w^*-dense should be understood according to the respective topologies.

$$
(4) \quad
\begin{cases}
s\text{-}\lim_{\lambda \uparrow \infty} \lambda J_\lambda x = x & \text{for all } x \in X, \\[2ex]
s\text{-}\lim_{\lambda \downarrow 0} \lambda J_\lambda x = 0 & \text{for all } x \in X;
\end{cases}
$$

$$
(5) \quad
\begin{cases}
w^*\text{-}\lim_{\lambda \uparrow \infty} \lambda J_\lambda^* f^* = f^* & \text{for all } f^* \in X^*, \\[2ex]
w^*\text{-}\lim_{\lambda \downarrow 0} \lambda J_\lambda^* f^* = 0 & \text{for all } f^* \in X^*;
\end{cases}
$$

(6)   there exists a closed linear operator A with domain $D(A)$ and range $R(A)$ both s-dense in X such that $J_\lambda = (\lambda I - A)^{-1}$ and the inverse A^{-1} exists in such a way that $V = -A^{-1} = s\text{-}\lim_{\lambda \downarrow 0} J_\lambda$;

(7)   there exists a closed linear operator A^* with domain $D(A^*)$ and range $R(A^*)$ both w^*-dense in X^* such that $J_\lambda^* = (\lambda I^* - A^*)^{-1}$ and the inverse $(A^*)^{-1}$ exists in such a way that $V^* = (-A^*)^{-1} = w^*\text{-}\lim_{\lambda \downarrow 0} J_\lambda^*$.

Remark 1. Actually, A^* is the dual operator of A as will be seen from $A^* = \lambda I^* - (J_\lambda^*)^{-1} = (\lambda I - J_\lambda^{-1})^*$. See the proof of Theorem 1.

Remark 2. A is thus the infinitesimal generator of a uniquely determined contraction semi-group $\{T_t; t \geq 0\}$ of class (C_0) of bounded linear operators T_t on X into X, and the second condition in (4) was introduced by the present speaker as the existence criterion for the operator $V = -A^{-1} = s\text{-}\lim_{\lambda \downarrow 0} J_\lambda$ with strongly dense domain. Suggested by G. Hunt's theory (G. Hunt [1]) of potentials, this operator V shall be called the *abstract potential operator* associated with the semi-group T_t.

Proof of Theorem 1. (4) implies (3). w-lim in (3) can be strengthened to s-lim by the abelian ergodic theorem or the ergodic theorem of the Hille type in Yosida [1]. Hence (3) implies (4). (4) implies (5) and (5) implies (3).

The proof of the implication $(4) \rightarrow (6)$. By (1), the null space $N(J_\lambda) = \{x \in X; J_\lambda x = 0\}$ is independent of λ. Hence the first condition in (4) implies that the null space $N(J_\lambda)$ consists of zero vector only. By virtue of (1),

$$
J_\lambda J_\mu (\lambda I - J_\lambda^{-1} - \mu I + J_\mu^{-1}) = (\lambda - \mu) J_\lambda J_\mu - J_\lambda J_\mu (J_\lambda^{-1} J_\mu^{-1}) = 0,
$$

so that $\lambda I - J_\lambda^{-1}$ is independent of λ and we have $J_\lambda = (\lambda I - A)^{-1}$ with $A = \lambda I - J_\lambda^{-1}$, proving that $D(A) = R(J_\lambda)$. Therefore the first condition in (4) implies that the strong closure of $D(A)$ is equal to X. Similarly, by using $AJ_\lambda x = -x + \lambda J_\lambda x$ and the second condition in (4), we prove that the strong closure of $R(A)$ is equal to X. Since $Ax_0 = 0$ implies $\lambda J_\lambda x_0 = x_0$, the second condition in (4) implies the existence of the inverse A^{-1}. Let $x \in D(V) = D(A^{-1})$. Then, by the second condition in (4), $J_\lambda x - Vx = (J_\lambda A + I)A^{-1}x = \lambda J_\lambda A^{-1}x$ tends strongly to 0 as $\lambda \downarrow 0$, i.e., $s\text{-}\lim_{\lambda \downarrow 0} J_\lambda x = Vx$. Conversely, let $s\text{-}\lim_{\lambda \downarrow 0} J_\lambda x = z$. Then, by $AJ_\lambda x = -x + \lambda J_\lambda x$ and the closure property of A, the second condition in (4) implies that $Az = -x$ so that $Vx = -A^{-1}x = z$. Hence $x \in D(V)$ and $Vx = s\text{-}\lim_{\lambda \downarrow 0} J_\lambda x$. This proves that the operator V is defined by $V = s\text{-}\lim_{\lambda \downarrow 0} J_\lambda$. We have thus proved the implication (4) → (6).

The implication (5) → (7) will be proved similarly.

The proof of the implication (6) → (4). Let $x \in D(A)$. Then $\lambda^{-1}(\lambda J_\lambda Ax) = -x + \lambda J_\lambda x$. Hence, by (2), we have $s\text{-}\lim_{\lambda \uparrow \infty} \lambda J_\lambda x = x$ for $x \in D(A)$. $D(A)$ being strongly dense in X, we see, again using (2), that the first condition in (4) holds good. Next we have, by (1),

$$\lambda J_\lambda (I - \mu J_\mu) = (1 - \mu(\mu - \lambda)^{-1})\lambda J_\lambda - \lambda(\lambda - \mu)^{-1}\mu J_\mu,$$

and so, for any $y = AJ_\mu x = -x + \mu J_\mu x$, we have $s\text{-}\lim_{\lambda \downarrow 0} \lambda J_\lambda y = 0$ by (2). $R(A)$ being strongly dense in X, we have proved the second condition in (4) by (2).

The proof of the implication (7) → (4). The inverse J_λ^{-1} exists. For, if there would exist an $x_0 \neq 0$ such that $J_\lambda x_0 = 0$, then $J_\lambda^* f^*(x_0) = 0$ for all $f^* \in X^*$, contrary to the w^*-denseness of $D(A^*) = R(J_\lambda^*)$. Next we prove that $R(J_\lambda)$ is s-dense in X. For, if otherwise, there would exist a non-zero functional $f_0^* \in X^*$ such that $f_0^*(J_\lambda x) = 0$ for all $x \in X$ so that $J_\lambda^* f_0^* = 0$, contrary to the existence of the inverse $(J_0^*)^{-1}$. Therefore, by R. S. Phillips' theorem (see, e.g., Theorem 1 in K. Yosida [1], p. 224), we have $(J_\lambda^*)^{-1} = (J_\lambda^{-1})^*$. Hence A^* is the dual operator of $A = (\lambda I - J_\lambda^{-1})$ and $J_\lambda = (\lambda I - A)^{-1}$. The s-denseness of $R(J_\lambda)$ implies that $D(A) = R(J_\lambda)$ is s-dense in X. Similarly to the proof of the s-denseness of $R(J_\lambda)$, we can prove the s-denseness of $R(A)$ by virtue of the existence of the inverse $(A^*)^{-1}$. Therefore, $D(A)$ and $R(A)$ are both s-dense in X and so, by what we have proved already, we see that (7) implies (4).

The proof of the implication (6) → (7) is performed by the following steps: (6) → (4), (4) → (5) and (5) → (7).

2. A characterization of abstract potential operators

To give another characterization of abstract potential operators $V = s\text{-}\lim_{\lambda \downarrow 0} J_\lambda$, we shall prove two theorems below:

Theorem 2. The abstract potential operator V and its dual operator

V^* satisfy the following inequalities:

(8) $\qquad \|\lambda Vx + x\| \geqq \|\lambda Vx\|$ for all $x \in D(V)$ and all $\lambda > 0$,

(9) $\qquad \|\lambda V^* f^* + f^*\| \geqq \|\lambda V^* f^*\|$ for all $f^* \in D(V^*)$ and all $\lambda > 0$.

Proof. Letting $\mu \downarrow 0$ in (1), we obtain

(10) $\quad J_\lambda x - Vx = -\lambda J_\lambda Vx$ for all $x \in D(V)$ and all $\lambda > 0$.

As proved in Theorem 1, (4) implies that $V^* f^* = w^*\text{-}\lim_{\lambda \downarrow 0} J_\lambda^* f^*$ for $f^* \in D(V^*)$. Hence we obtain

(11) $\quad J_\lambda^* f^* - V^* f^* = -\lambda J_\lambda^* V^* f^*$ for all $f^* \in D(V^*)$ and all $\lambda > 0$

from

(12) $\quad J_\lambda^* - J_\mu^* = (\mu - \lambda) J_\lambda^* J_\mu^*$

which is a consequence from the equation equivalent to (1):

(1)′ $\qquad\qquad\qquad J_\lambda - J_\mu = (\mu - \lambda) J_\mu J_\lambda.$

Therefore we obtain (8) and (9) by (10), (11) and (2).

Corollary. The inverse $(\lambda V + I)^{-1}$ and the inverse $(\lambda V^* + I^*)^{-1}$ both exist as bounded linear operators.

Proof. The existence of $(\lambda V + I)^{-1}$ and $(\lambda V^* + I^*)^{-1}$ are clear from (8) and (9), respectively. We shall prove

(13) $\qquad\qquad\qquad R(\lambda V + I) = X$ for all $\lambda > 0$.

The existence of the inverse $(\lambda V^* + I^*)^{-1}$ implies, by the Hahn-Banach theorem, that the range $R(\lambda V + I)$ is s-dense in X. Hence, for any $y \in X$, there exists a sequence $\{x_n\} \subseteqq D(V)$ such that $s\text{-}\lim_{n \to \infty} (\lambda Vx_n + x_n) = y$. Thus, by (8), we have

$$\|\lambda V(x_n - x_m) + (x_n - x_m)\| \geqq \|\lambda V(x_n - x_m)\|$$

and so $s\text{-}\lim_{n \to \infty} Vx_n = z$ exists, proving that $s\text{-}\lim_{\mu \to \infty} x_n = x$ exists. Since an abstract potential operator is closed by definition, the operator V must be closed. Hence we must have $Vx = z$, i.e., $y = Vx + x$. This proves (13), and thus $(\lambda V + I)^{-1}$ is a bounded linear operator by the closed graph theorem. Hence the closed range theorem (see, e.g., K. Yosida [1]) implies that

(14) $\qquad\qquad\qquad R(\lambda V^* + I^*) = X^*$ for all $\lambda > 0$,

and so $(\lambda V^* + I^*)^{-1}$ is a bounded linear operator on X^* into X^*.

Theorem 3. Let a closed linear operator V with domain $D(V)$ and range $R(V)$ both s-dense in X be such that V and its dual operator V^* satisfy inequalities (8) and (9), respectively. Then there exists a pseudo-resolvent $\hat{J}_\lambda$ with $\sup \|\lambda \hat{J}_\lambda\| \leq 1$ such that

$$s\text{-}\lim_{\lambda \uparrow \infty} \lambda \hat{J}_\lambda x = x \quad \text{and} \quad s\text{-}\lim_{\lambda \downarrow 0} \lambda \hat{J}_\lambda x = 0 \quad \text{for all } x \in X,$$

and, moreover, the inverse V^{-1} exists in such a way that

$$V = s\text{-}\lim_{\lambda \downarrow 0} (\lambda I + V^{-1})^{-1} = s\text{-}\lim_{\lambda \downarrow 0} \hat{J}_\lambda.$$

Hence V is an abstract potential operator.

Proof. We first remark that the Corollary above holds good for V and V^*, and hence $(\lambda V + I)^{-1}$ is a bounded linear operator on X into X. Moreover, the linear operator $\hat{J}_\lambda$ defined through

(15) $$\hat{J}_\lambda(\lambda V x + x) = V x \quad \text{(for all } x \in D(V) \text{ and } \lambda > 0)$$

is a bounded linear operator on X into X such that

(16) $$\hat{J}_\lambda = V(\lambda V + I)^{-1} \quad \text{and} \quad \sup_{\lambda > 0} \|\lambda \hat{J}_\lambda\| \leq 1.$$

We can prove, by (15), that $\hat{J}_\lambda$ is a pseudo-resolvent. In fact, we have

$$\hat{J}_\lambda(\lambda V x + x) - \hat{J}_\mu(\lambda V x + x) = V x - \hat{J}_\mu\left(\frac{\lambda}{\mu}(\mu V x + x) + \left(1 - \frac{\lambda}{\mu}\right)x\right)$$

$$= V x - \frac{\lambda}{\mu} V x - \left(1 - \frac{\lambda}{\mu}\right)\hat{J}_\mu x = \left(1 - \frac{\lambda}{\mu}\right)(V x - \hat{J}_\mu x)$$

and

$$(\mu - \lambda)\hat{J}_\mu \hat{J}_\lambda(\lambda V x + x) = (\mu - \lambda)\hat{J}_\mu V x = (\mu - \lambda)\hat{J}_\mu\left(V x + \frac{1}{\mu}x - \frac{1}{\mu}x\right)$$

$$= (\mu - \lambda)\frac{1}{\mu} V x - (\mu - \lambda)\frac{1}{\mu}\hat{J}_\mu x = \left(1 - \frac{\lambda}{\mu}\right)(V x - \hat{J}_\mu x).$$

Therefore, by (16), we can apply the ergodic theorem of the Hille type in K. Yosida [1] to the effect that:

(17) the closure in X of $R(\hat{J}_\lambda) = \{x \in X; \ s\text{-}\lim_{\mu \uparrow \infty} \mu \hat{J}_\mu x = x\}$

and

(18) the closure in X of $R(I - \lambda \hat{J}_\lambda) = \{x \in X; \ s\text{-}\lim_{\mu \downarrow 0} \mu \hat{J}_\mu x = 0\}.$

We see, from the resolvent equation for $\hat{J}_\lambda$, that the null space $N(\hat{J}_\lambda)$ of $\hat{J}_\lambda$ is independent of λ. The range $R(V)$ being s-dense in X, we see by (15) that the range $R(\hat{J}_\lambda) = R(V)$ is s-dense in X. Thus, by (17), $N(\hat{J}_\lambda)$ consists of the zero vector only and so $\hat{J}_\lambda$ is the resolvent of a linear operator A:

$$(19) \quad \hat{J}_\lambda = (\lambda I - A)^{-1}, \qquad A = \lambda I - \hat{J}_\lambda^{-1}.$$

Since $D(A) = R(\hat{J}_\lambda) = R(V)$ is s-dense in X, we have, by (17),

$$(20) \qquad s\text{-}\lim_{\mu \uparrow \infty} \mu \hat{J}_\mu x = x \quad \text{for all } x \in X.$$

We next have, by (19) and (15),

$$(\lambda I - A)\hat{J}_\lambda (\lambda V x + x) = \lambda V x + x = (\lambda I - A) V x = \lambda V x - A V x,$$

that is,

$$(21) \qquad -A V x = x \quad \text{whenever} \quad x \in D(V).$$

Hence $R(A) \supseteq D(V)$ is s-dense in X. Thus, since $R(\hat{J}_\lambda) = R(V)$ is s-dense in X, $R(I - \lambda \hat{J}_\lambda) \supseteq R(-\hat{J}_\lambda A)$ must be s-dense in X. Therefore, by (18), we obtain

$$(22) \qquad s\text{-}\lim_{\mu \downarrow 0} \mu \hat{J}_\mu x = 0 \quad \text{for all } x \in X.$$

Hence, by (20) and (22), we can apply Theorem 1 to the effect that

$$(23) \qquad s\text{-}\lim_{\lambda \downarrow 0} (\lambda I - A)^{-1} = s\text{-}\lim_{\lambda \downarrow 0} \hat{J}_\lambda \quad \text{is equal to } (-A)^{-1}.$$

That $V = -A^{-1}$ may be proved following the arguments given in K. Yosida [2]. Firstly the inverse V^{-1} exists, since $Vx = 0$ implies $x = 0$ by (21). Thus, by (16) and (19), we obtain

$$\lambda I - A = \hat{J}_\lambda^{-1} = (\lambda V + I) V^{-1} = \lambda I + V^{-1},$$

which proves that $-A = V^{-1}$.

3. Comments and remarks

Following G. Lumer-R.S. Phillips [1] (see, e.g., K. Yosida [1], p. 250), we can associate, to each pair $\{x, y\}$ of vectors of X, a *semi-scalar product* $[x, y]$ in such a way that

$$(24) \qquad [x+y, z] = [x, z] + [y, z], \quad [\lambda x, y] = \lambda [x, y],$$
$$[x, x] = \|x\|^2 \quad \text{and} \quad |[x, y]| \leq \|x\| \cdot \|y\|.$$

The operator V is called *accretive* (with respect to $[x, y]$) if

$$(25) \qquad \operatorname{Re} [x, Vx] \geq 0 \quad \text{whenever} \quad x \in D(V).$$

We can prove:

$$(26) \qquad V \text{ satisfies (8) if } V \text{ is accretive.}$$

For, we have, by (24),
$$\|\lambda Vx\|^2 = [\lambda Vx, \lambda Vx] \leqq \mathrm{Re}\ \{[\lambda Vx, \lambda Vx] + [x, \lambda Vx]\}$$
$$= \mathrm{Re}\ [\lambda Vx + x, \lambda Vx] \leqq \|\lambda Vx + x\| \cdot \|\lambda Vx\|.$$

We can also prove:

(27)   An abstract potential operator V must be accretive.

The proof reads as follows. Let $\{T_t; t \geqq 0\}$ be the contraction semi-group of class (C_0) whose infinitesimal generator A is given by $A = -V^{-1}$. We have, by $\|T_t\| \leqq 1$,

$$\mathrm{Re}\ [T_t x - x, x] = \mathrm{Re}\ [T_t x, x] - \|x\|^2$$
$$\leqq \|T_t x\| \cdot \|x\| - \|x\|^2 \leqq 0.$$

Hence, for $x \in D(A)$, we obtain

$$\mathrm{Re}\ [Ax, x] = \lim_{t \downarrow 0} \mathrm{Re}\ [t^{-1}(T_t x - x), x] \leqq 0$$

so that, for $x_0 \in D(V) = D(-A^{-1})$,

$$\mathrm{Re}\ [AVx_0, Vx_0] = \mathrm{Re}\ [-x_0, Vx_0] \leqq 0, \text{ that is, } \mathrm{Re}[x_0, Vx_0] \geqq 0.$$

In terms of the notion of the accretive operator, we can prove:

Theorem 3′. A closed linear operator V with domain $D(V)$ and range $R(V)$ both s-dense in X is an abstract potential operator if V and its dual operator V^* are both accretive.

Proof. V and V^* satisfy (8) and (9), respectively.

Corollary. A self-adjoint positive definite operator V in a Hilbert space X is an abstract potential operator.

Remark. We have (see L. Schwartz [1], p. 281)

$$(28) \quad \int_{R^2} \int_{R^2} \log \frac{1}{|x-y|} f(x)f(y) \mathrm{d}x \mathrm{d}y \geqq 0$$

whenever a real-valued continuous function $f(x)$ with compact support satisfies $\int_{R^2} f(x)\mathrm{d}x = 0$. Thus the logarithmic potential

$$(29) \quad (\varDelta^{-1} f)(x) = \frac{1}{2\pi} \int_{R^2} \log \frac{1}{|x-y|} f(y) \mathrm{d}y$$

well fits to the context of the above Theorem. The condition $\int_{R^2} f(x)\mathrm{d}x = 0$ is the property satisfied by the range of the Laplacian $\varDelta$ applied to C^2 functions with compact support.

REFERENCES

HILLE, E. and PHILLIPS, R.S., [1]: *Functional Analysis and Semi-groups*, Providence (1957).

HUNT, G.A., [1]: Markoff processes and potentials, *Illinois J. of Math.*, I (1957), 336-369.

LUMER, G. and PHILLIPS, R.S., [1]: Dissipative operators in a Banach space, *Pacific J. of Math.*, 11 (1961), 679-698.

MEYER, P.A., [1]: *Probability and Potentials*, Blaisdell Publ. Company, (1966).

SATO, K., [1]: Positive pseudo-resolvents in Banach lattices, to appear in *J. Fac. Sci. Univ. of Tokyo*.

SCHWARTZ, L., [1]: *Théorie des Distributions*, Hermann (1966).

YAMADA, A., [1]: On the correspondence between potential operators and semi-groups associated with Markoff processes, to appear in *Z. Wahrscheinlichkeitstheorie und Verw. Gebiete*.

YOSIDA, K., [1]: *Functional Analysis*, 2nd ed., Springer (1968).

YOSIDA, K., [2]: Positive resolvents and potentials, *Z. Wahrscheinlichkeitstheorie und Verw. Gebiete*, 8 (1967), 210-218.

YOSIDA, K., WATANABE, T. and TANAKA, H., [3]: On the pre-closedness of the potential operators, *J. Math. Soc. Japan,* 20 (1968), 419-421.

YOSIDA, K., [4]: The existence of the potential operator associated with an equi-continuous semi-group of class (C_0), *Studia Math.*, 31 (1968), 531-533.

YOSIDA, K., [5]: The pre-closedness of Hunt's potential operators and its applications, *Proc. Internat. Conference on Functional Analysis and Related Topics*, Tokyo, 1969.

Abstract potential operators on Hilbert space

Publ. Res. Inst. Math. Sci. Kyoto Univ. **8** (1972) 201–205

Let X be a (real or complex) Hilbert space. A linear operator V with its domain $D(V)$ and range $R(V)$ both strongly dense in X is called an *abstract potential operator* (see K. Yosida [2], p. 412) if the inverse V^{-1} exists in such a way that

$$(1) \qquad A = -V^{-1}$$

is the infinitesimal generator of a one-parameter semi-group of class (C_0) of linear contraction operators on X into X. The purpose of the present note is to prove the following existence theorem. (Hereafter, we shall denote by S^a the strong closure of a subset S of X.)

Theorem. *Let U be a linear operator satisfying three conditions*:

$$(2) \qquad D(U)^a = X,$$

$$(3) \qquad R(U)^a = X,$$

$$(4) \qquad U \text{ is } \underline{\text{accretive}}, \text{ that is, } Re(Uf, f) \geqq 0 \quad \text{for every } f \in D(U).$$

Then there exists at least one abstract potential operator V which is a closed linear accretive extension of U; V might coincide with U.

Proof. The proof is given in two steps. The first is to construct a maximal accretive extension V of U by virtue of R. S. Phillips' *theory of*

Received March 2, 1972.

* Dept. Math., Gakushuin Univ., 1-5-1, Mejiro, Toshima-ku, Tokyo 171, Japan.

Cayley transform (cf. B. Sz.-Nagy and C. Foias [1], p. 167). The second is to prove that this V is an abstract potential operator by making use of the *abelian ergodic theorem for pseudo-resolvents* (see K. Yosida [2], p. 215).

THE FIRST STEP. For every $\lambda > 0$ and $f \in D(V)$, we have, by (4),

$$(5) \qquad \|\lambda Uf + f\|^2 = (\lambda Uf + f, \lambda Uf + f) = \|\lambda Uf\|^2 + 2Re(\lambda Uf, f) + \|f\|^2$$

$$\geq \|\lambda Uf\|^2 + \|f\|^2 \geq \|\lambda Uf\|^2 - 2Re(\lambda Uf, f) + \|f\|^2 = \|\lambda Uf - f\|^2.$$

Hence the inverse $(\lambda U + I)^{-1}$ exists and moreover, the Cayley transform C defined through

$$(6) \qquad\qquad C \cdot (Uf + f) = (Uf - f)$$

is a contraction operator mapping $R(U + I)$ onto $R(U - I)$. Let us define a bounded linear extension $\hat{C}$ of C:

$$(7) \qquad \text{through continuity on } R(U + I)^a, \text{ and through putting } \hat{C} \cdot g = 0 \text{ on}$$
the orthogonal complement of $R(U + I)$.

This everywhere defined contraction operator $\hat{C}$ cannot admit eigenvalue one. Assume the contrary and let $\hat{C} \cdot f_0 = f_0$ with $\|f_0\| = 1$. Then its adjoint operator $\hat{C}^*$, which is also a contraction, must satisfy $\hat{C}^* \cdot f_0 = f_0$ because

$$\|\hat{C}^* \cdot f_0 - f_0\|^2 = \|\hat{C}^* \cdot f_0\|^2 - 2Re(\hat{C}^* \cdot f_0, f_0) + \|f_0\|^2$$

$$\leq \|f_0\|^2 - 2Re(f_0, \hat{C} \cdot f_0) + \|f_0\|^2 = 1 - 2 + 1 = 0.$$

Thus we obtain, by (6) and (7),

$$(f_0, (U - I)f) = (f_0, \hat{C} \cdot (U + I)f) = (\hat{C}^* \cdot f_0, (U + I)f) = (f_0, (U + I)f),$$

hence $(f_0, f) = 0$ and so $f_0 = 0$ by (2).

Therefore the inverse $(I - \hat{C})^{-1}$ exists and so we can define a linear operator V through

$$(8) \qquad\qquad V \cdot (I - \hat{C})f = (I + \hat{C})f.$$

V is an extension of U. In fact, we have, by (6), $(I-C)=I-(U-I)$ $(U+I)^{-1}=2(U+I)^{-1}$, that is, $U=(I+C)(I-C)^{-1}$, proving that V is an extension of U. Here the existence of $(I-C)^{-1}$ is assured by that of $(I-\hat{C})^{-1}$. We can prove that V is accretive. For, by putting $f=(I-\hat{C})^{-1}g$ and observing (8) and the contraction property of $\hat{C}$, we obatin

$$Re(Vg,\ g)=Re((I+\hat{C})f,\ (I-\hat{C})f)=\|f\|^2-\|\hat{C}\cdot f\|^2\geqq 0.$$

We can also prove, by (8) and the boundedness of the operator $\hat{C}$, that V is a closed linear operator. Moreover, by (8), we have $(I+V)=I+(I+\hat{C})(I-\hat{C})^{-1}=2(I-\hat{C})^{-1}$, and so we obtain the existence theorem

(9) $R(V+I)=D(I-\hat{C})=X$ (and also $R(\lambda V+I)=X$ whenever $\lambda>0$).

Hence the accretive extension V is maximal as regards its range $R(\lambda V+I)$ for $\lambda>0$.

THE SECOND STEP. We will show that V is an abstract potential operator following after the proof of Theorem 2 on p. 414–415 in K. Yosida [2].

V being accretive, we have, as in (5), $\|\lambda Vf+f\|\geqq\|\lambda Vf\|$ for every $f\in D(V)$ and $\lambda>0$. Hence, by (9), we can define a bounded linear operator

(10) $$J_\lambda=V(\lambda V+I)^{-1}$$

satisfying

(11) $$\|\lambda\ J_\lambda\|\leqq 1.$$

It is easy to see that J_λ is a *pseudo-resolvent*, i.e.,

(12) $$J_\lambda-J_\mu=(\mu-\lambda)J_\lambda\ J_\mu.$$

Therefore, by (11), we can apply the abelian ergodic theorem to the effect that

(13) $$R(J_\mu)^a=\{x\in X;\ \underset{\lambda\uparrow\infty}{s\text{-}\lim}\ \lambda J_\lambda x=x\}\qquad \text{for all }\mu>0,$$

(14) $\qquad R(I-\mu J_\mu)^a = \{x \in X; \underset{\lambda \downarrow 0}{\text{s-lim}}\, \lambda J_\lambda\, x = 0\}$ for all $\mu > 0$.

By $R(V)^a = R(U)^a = X$, we have $R(J_\mu)^a = X$ by (10) and so, by (11) and (12), the null space of J_λ consists of zero vector only, independently of $\lambda > 0$. Hence J_λ is the resolvent of a linear operator, i.e.,

(15) $\quad J_\lambda = (\lambda I - A)^{-1}$, where $A = \lambda I - J_\lambda^{-1}$ is independent of $\lambda > 0$.

We have thus $D(A)^a = R(J_\mu)^a = X$ and so, by (11), the operator A is the infinitesimal generator of a contraction semi-group of class (C_0). We can also prove that $R(A)^a = X$. For, we have, by (10) and (15),

$$(\lambda I - A) J_\lambda (\lambda V f + f) = \lambda V f + f = (\lambda I - A) V f = \lambda V f - A V f,$$

that is,

(16) $\qquad -AVf = f$ whenever $f \in D(V)$,

proving that $R(A)^a = D(V)^a = D(U)^a = X$. Thus, by (14) and $AJ_\mu = (\mu J_\mu - I)$, we obtain $\underset{\lambda \downarrow 0}{\text{s-lim}}\, \lambda J_\lambda f = 0$ for all $f \in X$. This implies that the inverse A^{-1} exists. In fact, the condition $Af_0 = 0$ is equivalent to $\lambda(\lambda I - A)^{-1} f_0 = f_0$ and hence $f_0 = \underset{\lambda \downarrow 0}{\text{s-lim}}\, \lambda J_\lambda f_0 = 0$.

Thus $-A^{-1}$ is an abstract potential operator. On the other hand, (16) shows that the inverse V^{-1} exists. Hence, by $(\lambda I - A) = J_\lambda^{-1} = (\lambda V + I)V^{-1} = \lambda I + V^{-1}$, we obtain $-A = V^{-1}$, completing the proof of our Theorem.

Remark. We shall verify (2), (3) and (4) for Newtonian and logarithmic potentials

(17) $\qquad (Uf)(y) = \int_{R^n} K_n(|y - z|) f(z) dz \qquad (n \geq 2),$

$$K_n(r) = r^{2-n} \text{ for } n \geq 3, \quad \text{and} \quad K_2(r) = \log r^{-1}.$$

The proof of $D(U)^a = R(U)^a = X = L^2(R^n)$ can be obtained by making use of the fact that, for $0 < \delta_1 < \delta_2$,

$$u_{x,\delta_1,\delta_2}(y)=(K_n(\delta_1)-K_2(\delta_2))^{-1}\int_{R^n}K_n(|y-z|)(d\nu_{x,\delta_1}(z)-d\nu_{x,\delta_2}(z))$$

is continuous in y satisfying

$$u_{x,\delta_1,\,\delta_2}(y)=1 \qquad \text{if } |y-x|\leq\delta_1,$$

$$=0 \qquad \text{if } |y-x|\geq\delta_2,$$

$$0<u_{x,\delta_1,\delta_2}(y)<1 \qquad \text{if } \delta_1<|y-x|<\delta_2.$$

Here $\nu_{x,\delta}$ is the unit measure uniformly distributed over the hypersurface of the sphere of centre x and radius δ in R^n.

The Gauss-Frostmann energy inequality

$$\int_{R^n}(Uf)(y)\cdot\overline{f(y)}\,dy\geq0 \qquad (n\geq2)$$

holds good whenever $f\in L^2(R^n)$ is of compact support satisfying $\int_{R^n}f(y)\,dy=0$. It is easy to prove that such f's constitute a strongly dense subset of $L^2(R^n)$.

ANOTHER TREATMENT OF THE SECOND STEP (*Added on* 20 *April*, 1972). As in the above proof of the non-existence of the eigenvalue 1 for the operator $\hat{C}$, we can show that $\hat{C}\cdot f_0=-f_0$ implies $\hat{C}^*\cdot f_0=-f_0$ and hence $(f_0,Uf)=0$, proving by (3) the non-existence of the eigenvalue -1 for $\hat{C}$. Thus $V=(I+\hat{C})(I-\hat{C})^{-1}$ given by (8) admits the inverse $V^{-1}=$ $=(I-\hat{C})(I+\hat{C})^{-1}$. Hence we can prove that V is an abstract potential operator without appealing to the abelian ergodic theorem.

Remark (added during the proof). On reading the pre-print, Prof. K. Sato gave interesting comments and extensions. See his paper to appear.

References

[1]   Sz.-Nagy, B. and C. Foias, *Harmonic Analysis of Operators on Hilbert Space*, North Holland Publ. Co., 1970.

[2]   Yosida, K., *Functional Analysis*, the Third Ed., Springer-Verlag, 1971.

X. Operational Calculus

Comments (Shigetake Matsuura)

Paper [103] is an elementary remark on Mikusiński's operational calculus that, without using Titchmarsh's theorem on the non-existence of zero divisors in the convolution algebra $C[0, \infty)$, one can treat some problems in operational calculus of Mikusiński, especially the initial value problems of differential equations of constant coefficients. This comes from the introduction of a suitable subring of fractions of $C[0, \infty)$ instead of the whole quotient field of $C[0, \infty)$.

Paper [106] gives a simple proof of Titchmarsh's theorem, which is fundamental in Mikusiński's operational calculus. The proof uses the Weierstrass polynomial approximation theorem and Liouville's theorem on bounded entire functions instead of the original proof in Mikusiński's book using results on moment problems.

Later we found that Richard Petersen had published already a similar idea in Danish in 1967.

Paper [107] is an application of the method of algebraic derivations of Mikusiński's operational calculus to the so-called Laplace equations (i.e. the second order ordinary differential equations whose coefficients are linear functions of the independent variable.)

Thus the use of Laplace transforms and Fuchsian theory of differential equations is not necessary. Bessel's equation is treated as an example.

In Paper [108] it is proved that the operator solution of a constant coefficient differential equation of the n-th order is really an n-times continuously differentiable function in the ordinary sense, if the non-homogeneous term is a continuous function. This means that, whatever kind of operators may appear in the solving process, the final result is an ordinary solution.

A note on Mikusiński's operational calculus

Proc. Japan Acad. **56** A (1980) 1–3

(Communicated Jan. 12, 1980)

§ 1. **Introduction.** In [ii], one of the present authors gave a simplified derivation of Mikusiński's operational calculus [i] without appealing to Titchmarsh's theorem concerning the vanishing of the convolution of two continuous functions defined on $[0, \infty)$.

The purpose of the present note is to give a further simplification of [ii] to the effect that we can derive the operational calculus directly from the ring C_H in [ii] without introducing the ring C_P in [ii]. For the sake of convenience for the reader, we shall begin with the definition of the ring C_H.

§ 2. **The ring C_H.** We denote by C the totality of complex-valued continuous functions defined on $[0, \infty)$. We denote such a function by $\{f(t)\}$ or simply by f, while $f(t)$ means the value at t of the function f. For $f, g \in C$ and $\alpha, \beta \in K$ (=the complex number field) we define

$$(1) \qquad \alpha f + \beta g = \{\alpha f(t) + \beta g(t)\} \quad \text{and} \quad fg = \left\{ \int_0^t f(t-s)g(s)ds \right\}.$$

Then C is a commutative ring with respect to the above addition and multiplication over the coefficient field K.

We shall denote by h (l in [i]) the constant function $\{1\} \in C$ so that we have

$$(2) \qquad h^n = \left\{ \frac{t^{n-1}}{(n-1)!} \right\} \qquad (n=1, 2, \cdots),$$

and

$$(3) \qquad hf = \left\{ \int_0^t f(s)ds \right\} \qquad \text{for } f \in C,$$

i.e. h behaves as an operation of integration. Then we have the following fairly trivial

Proposition 1. *For $k \in H = \{k ; k = h^n \ (n=1, 2, \cdots)\}$ and $f \in C$, the equation $kf = 0$ implies that $f = 0$, where 0 denotes $\{0\} \in C$.*

Therefore we can construct the commutative ring C_H of fractions:

$$(4) \qquad C_H = \left\{ \frac{f}{k} ; f \in C \text{ and } k \in H \right\}$$

where the equality is defined by

$$(5) \qquad \frac{f}{k} = \frac{f'}{k'} \qquad \text{if and only if } k'f = kf',$$

and the addition and multiplication are defined through

---

*) Department of Mathematics, Gakushuin University.

(6) $$\frac{f}{k}+\frac{f'}{k'}=\frac{k'f+kf'}{kk'} \quad\text{and}\quad \frac{f}{k}\cdot\frac{f'}{k'}=\frac{ff'}{kk'},$$

respectively.

By identifying $f \in C$ with $kf/k \in C$, the ring C can be isomorphically embedded as a subring of the ring C_H. For any complex number α, we define

(7) $$[\alpha]=\frac{\{\alpha\}}{h} \in C_H.$$

Then we have, for $\alpha, \beta \in K$, $f \in C$ and $k \in H$,

(7)′ $$[\alpha]+[\beta]=[\alpha+\beta], \qquad [\alpha][\beta]=[\alpha\beta],$$
$$[\alpha]f=\alpha f=\{\alpha f(t)\}, \qquad [\alpha]\frac{f}{k}=\frac{\{\alpha f(t)\}}{k}=\frac{\alpha f}{k}.$$

Hence $[\alpha]$ can be identified with the complex number α, not with $\{\alpha\}$, and we see that the effect of the multiplication by $[\alpha]$ is exactly the α-times multiple. In particular [1] may be identified with the multiplicative unit I of C_H:

(8) $$I=\frac{h^n}{h^n} \qquad (n=1, 2, \cdots).$$

We then define

(9) $$s=\frac{h^n}{h^{n+1}} \in C_H \qquad (n=0, 1, 2, \cdots ; h^0=I) \text{ so that } sh=I.$$

Proposition 2. *If both f and its derivative f' belong to C, then we have*

(10) $$f'=sf-f(0), \qquad where\ f(0)=[f(0)],$$

that is, s behaves as an operation of differentiation.

Proof. Clear from (9) and Newton's formula
$$hf'=\left\{\int_0^t f'(s)ds\right\}=\{f(t)-f(0)\}=f-[f(0)]h.$$

Corollary. *For n-times continuously differentiable function $f \in C$,*
$$f^{(n)}=s^n f-s^{n-1}f(0)-s^{n-2}f'(0)-\cdots-f^{(n-1)}(0),$$
(10)′
$$where\ f^{(j)}(0)=[f^{(j)}(0)].$$

Proposition 3. *For any $\alpha \in K$ and for any positive integer n, the element*
$$(s-\alpha)^n=(s-[\alpha])^n=\frac{(I-[\alpha]h)^n}{h^n} \in C_H$$

*admits a uniquely determined **multiplicative inverse** in C_H given by*

(11) $$\frac{I}{(s-\alpha)^n}=\left\{\frac{t^{n-1}}{(n-1)!}e^{\alpha t}\right\}=n\text{-times multiplication of } \{e^{\alpha t}\}.$$

Proof. We have $(s-\alpha)\{e^{\alpha t}\}=I$ by (10) and so (11) is easily obtained.

§3. **The operational calculus.** Consider the following Cauchy problem for linear ordinary differential equation with coefficients $\in K$:

$$(12) \qquad \begin{cases} \alpha_n y^{(n)} + \alpha_{n-1} y^{(n-1)} + \cdots + \alpha_0 y = f \in C \qquad (\alpha_n \neq 0), \\ y(0) = \gamma_0,\ y'(0) = \gamma_1,\ \cdots,\ y^{(n-1)}(0) = \gamma_{n-1}. \end{cases}$$

By (10)′, we can rewrite (12) into equation in C_H:

$$(12)' \qquad (\alpha_n s^n + \alpha_{n-1} s^{n-1} + \cdots + \alpha_0) y = f + \beta_{n-1} s^{n-1} + \beta_{n-2} s^{n-2} + \cdots + \beta_0,$$
$$\beta_m = \alpha_{m+1} \gamma_0 + \alpha_{m+2} \gamma_1 + \cdots + \alpha_n \gamma_{n-m-1} \qquad (m = 0, 1, 2, \cdots, n-1).$$

Since the polynomial ring of polynomials in s with coefficients in K is free from zero factors, we can define rational functions

$$F_1 = \frac{I}{\alpha_n s^n + \cdots + \alpha_0} \quad \text{and} \quad F_2 = \frac{\beta_{n-1} s^{n-1} + \cdots + \beta_0}{\alpha_n s^n + \cdots + \alpha_0}$$

and obtain their partial fraction expressions:

$$(13) \qquad F_1 = \sum_j \sum_{k=1}^{m_j} \frac{c_{jk} I}{(s - r_j)^k} \quad \text{and} \quad F_2 = \sum_j \sum_{k=1}^{m_j} \frac{d_{jk} I}{(s - r_j)^k},$$

where c_{jk} and d_{jk} belong to K and r_j are roots of the algebraic equation $\alpha_n z^n + \cdots + \alpha_0 = 0$ so that $\sum_j m_j = n$. As was proved in (11), F_1 and F_2 given in (13) belong to $C \subset C_H$ so that we obtain, from (12)′, the solution of (12):

$$\{y(t)\} = \sum_j \sum_{k=1}^{m_j} c_{jk} \left\{ \frac{t^{k-1}}{(k-1)!} e^{r_j t} \right\} \{f(t)\} + \sum_j \sum_{k=1}^{m_j} d_{jk} \left\{ \frac{t^{k-1}}{(k-1)!} e^{r_j t} \right\}$$

In this way, Mikusiński's operational calculus can be derived without appealing to Titchmarsh's theorem nor to the ring C_P in [ii], that is, the totality of fractions f/p of the form $f(\in C)$ over non-zero polynomial p (in t) with cofficients $\in K$.

References

[i] Jan Mikusiński: Operational Caluclus. Pergamon Press (1959).

[ii] Shûichi Okamoto: A simplified derivation of Mikusinski's operational calculus. Proc. Japan Acad., **55**A (1), 1–5 (1979).

A note on Mikusiński's proof of the Titchmarsh convolution theorem

(with S. Matsuura)

Conference in Modern Analysis and Probability (New Haven, Conn., 1982),
Contemp. Math., **26** (1984) 423–425

THE TITCHMARSH CONVOLUTION THEOREM

This theorem was discovered in 1926 by E. C. Titchmarsh [4]. It reads:

THEOREM. Let $C[0,\infty)$ denote the totality of complex-valued continuous functions $f = f(t)$ defined on the interval $[0,\infty)$. If the convolution $f \cdot g$ of
two functions f and g of $C[0,\infty)$ satisfies

$$f \cdot g(t) = \int_0^t f(t-\tau)g(\tau)d\tau \equiv 0,$$

then either $f(t) \equiv 0$ or $g(t) \equiv 0$ must hold.

The original proof of Titchmarsh is not easy since it involves deep
theorems on analytic functions. An elementary proof was given in 1951 by J.
Mikusiński [2] jointly with C. Ryll-Nardzewski by making use of the <u>Lerch</u>
<u>moment theorem</u> [1]. The Lerch theorem was derived in [2] from a theorem of E.
Phragmén [3].

The purpose of the present note is to propose a modification of the
Mikusiński proof so that we obtain a plain proof, by making use of the <u>Liouville</u>
<u>theorem</u> in analytic function theory combined with the <u>Weierstrass polynomial</u>
<u>approximation theorem</u>.^{***}

II.   PROOF OF THE THEOREM

<u>The first step</u> (the special case $f = g$). We start with

(1)
$$\int_0^t f(t-\tau)f(\tau)d\tau \equiv 0.$$

Let $T > 0$ and β be any real number. Put

(2)
$$\hat{f}(t) = e^{-i\beta t}f(t), \quad \text{where}\quad i = \sqrt{-1}.$$

---

  * Department of Mathematics, the University of Tokyo.
 ** Research Institute of Mathematical Sciences, Kyoto University.

*** After the present note was finished, Professor K. Masuda communicated to
    us another proof which makes use of the Liouville theorem combined with
    the Fourier transform.

Then, for any real number α , we obtain

(3)
$$(\int_{-T}^{T} e^{\alpha u} \hat{f}(T-u)du)^2 = \int_{-T}^{T}\int_{-T}^{T} e^{\alpha(u+v)} \hat{f}(T-u)\hat{f}(T-v)dudv$$

$$= \int_{-T}^{T} dv \int_{u=-v}^{T} e^{\alpha(u+v)} \hat{f}(T-u)\hat{f}(T-v)du$$

$$+ \int_{-T}^{T} dv \int_{-T}^{u=-v} e^{\alpha(u+v)} \hat{f}(T-u)\hat{f}(T-v)du \;=\; I_{\alpha,\beta,1} + I_{\alpha,\beta,2}$$

By the substitution

$$u = T - (t-\tau), \quad v = T - \tau$$

with

$$2T \geq u+v \geq 0, \quad T \geq u \geq -T, \quad T \geq v \geq -T$$

and

$$\partial(u,v)/\partial(t,\tau) = 1,$$

we have, by (1) and (2),

(4)
$$I_{\alpha,\beta,1} = \int_0^{2T} dt \int_0^t e^{\alpha(2T-t)} \hat{f}(t-\tau)\hat{f}(\tau)d\tau$$

$$= \int_0^{2T} e^{\alpha(2T-t)-i\beta t} (\int_0^t f(t-\tau)f(\tau)d\tau)dt = 0.$$

If $\alpha > 0$ then $e^{\alpha(u+v)} \leq 1$ in the integration domain of $I_{\alpha,\beta,2}$. Hence

(5)
$$|I_{\alpha,\beta,2}| \leq \int_{-T}^{T} dv \int_{-T}^{u=-v} |\hat{f}(T-u)\hat{f}(T-v)|du$$

$$\leq \int_{-T}^{T} \int_{-T}^{T} |f(T-u)f(T-v)|dudv = M^2 \quad \text{with} \quad 0 \leq M < \infty$$

Therefore, by (3), (4) and (5),

(6)
$$|\int_{-T}^{T} e^{(\alpha+i\beta)u} f(T-u)du| \leq M, \quad \text{whenever} \quad \alpha \geq 0.$$

Hence the Laplace integral

(7)
$$E_T(z) = \int_0^T e^{zu} f(T-u)du$$

satisfies, for $\alpha = \mathrm{Re}\, z \geq 0$,

(8)
$$|E_T(z)| \leq N, \quad \text{where} \quad N = M + \int_{-T}^{0} |f(T-u)|du.$$

It is easy to see that the entire analytic function

$$E_T(z) = \int_0^T e^{zu} f(T-u)du$$

satisfies

(9)
$$|E_T(z)| = |\int_0^T e^{zu} f(T-u)du| \leq N_1 \quad \text{for all} \quad z, \quad \text{where}$$

$$N_1 = \max(N,K) \quad \text{with} \quad K = \max_{0 \leq u \leq T} T \times |f(T-u)|.$$

Hence, by the Liouville theorem, we have $F_T(z) = a$ constant, and so, by differentiating $F_T(z)$ at $z = 0$, we obtain

$$\int_0^T u^n f(T-u)du = 0 \qquad (n = 0,1,2,\ldots).$$

Therefore, by the Weierstrass polynomial approximation theorem, we have

$$\int_0^T q(u)f(T-u)du = 0$$

for every continuous function $q(u)$. Thus $f(T-u)$ must vanish for $0 \leq u \leq T$, i.e., $f(t)$ must vanish for $0 \leq t \leq T$. This proves the Theorem for the special case $f = g$.

The second step (the general case $f \neq g$). Let us start with

(10) $$f \cdot g(t) = \int_0^t f(t-\tau)g(\tau)d\tau = 0 \quad \text{for all} \quad t \geq 0.$$

Following after [2], we put

(11) $$f_1(t) = tf(t) \quad \text{and} \quad g_1(t) = tg(t).$$

Then $f_1 \cdot g(t) + f \cdot g_1(t) \equiv 0$, because, by (1),

$$\int_0^t (t-\tau)f(t-\tau)g(\tau)d\tau + \int_0^t f(t-\tau)\tau g(\tau)d\tau = t\int_0^t (f(t-\tau)g(\tau)d\tau.$$

Hence, again by (10),

$$0 \equiv (f \cdot g_1) \cdot (f_1 \cdot g + f \cdot g_1)(t)$$

$$\equiv (f \cdot g) \cdot (f_1 \cdot g_1)(t) + (f \cdot g_1) \cdot (f \cdot g_1)(t) = (f \cdot g_1) \cdot (f \cdot g_1)(t).$$

That is, we have obtained

(12) $$\int_0^t f(t-\tau)\tau g(\tau)d\tau \equiv 0.$$

Iterating the same argument, we have

(12)' $$\int_0^t f(t-\tau)\tau^n g(\tau)d\tau \equiv 0 \qquad (n = 0,1,2,\ldots).$$

Thus, by the Weierstrass polynomial approximation theorem, we see that $\int_0^t f(t-\tau)g(\tau)q(\tau)d\tau \equiv 0$ for every continuous function $q(t)$. This proves that

$$f(t-\tau)g(\tau) \equiv 0 \quad \text{for} \quad 0 \leq \tau \leq t < \infty,$$

that is, either $f(t) \equiv 0$ or $g(t) \equiv 0$ must hold.

REFERENCES

[1]   E. Lerch, Sur un point de la théorie des fonctions génératrices d'Abel. Mathematica 27 (1903), 339–352.

[2]   J. Mikusiński, Operational Calculus, Pergamon Press (1967).

[3]   E. Phragmén, Sur une extension d'une théoréme classique de la théorie des fonctions, Acta Mathematica 28 (1904), 331–369.

[4]   E. C. Titchmarsh, Introduction to the Theory of Fourier Integrals, (1937), p. 327- .

The algebraic derivative and Laplace's differential equation

Proc. Japan Acad. **59 A** (1983) 1–4

(Communicated Jan. 12, 1983)

0. The purpose of the present note is to show that the differential equation with linear coefficients (so-called Laplace's differential equation)

$$(1) \qquad a_2 t y''(t) + (a_1 t + b_1) y'(t) + (a_0 t + b_0) y(t) = 0$$

is convertible into

$$(2) \qquad \frac{Dy}{y} = \frac{q(s)}{p(s)} = \frac{(-2a_2 + b_1)s - a_1 + b_0}{a_2 s^2 + a_1 s + a_0}.$$

Here D is "the operator of algebraic derivative" and s is "the operator of differentiation" in the operational calculus of J. Mikusiński (Pergamon Press (1959)), and fractions Dy/y and $q(s)/p(s)$ are "convolution quotients".

We shall show that, if the algebraic equation

$$(3) \qquad p(z) = a_2 z^2 + a_1 z + a_0$$

has two distinct roots z_1 and z_2 so that

$$\frac{q(z)}{p(z)} = \frac{\gamma_1}{z - z_1} + \frac{\gamma_2}{z - z_2} \qquad (\gamma_1 \text{ and } \gamma_2 \text{ are complex numbers}),$$

then the convolution quotient

$$(4) \qquad y = C(s - z_1 I)^{\gamma_1}(s - z_2 I)^{\gamma_2} \qquad (C \text{ is a non-zero constant})$$

satisfies equation (2). In this way, we can solve Bessel differential equation, Laguerre differential equation and the like algebraically, by simply making use of the general binomial expansion

$$(1 - \alpha z)^{\gamma} = \sum_{k=0}^{\infty} \binom{\gamma}{k} (-\alpha)^k z^k \qquad (\text{convergent for } |\alpha z| < 1),$$

without appeal to other analytical tools like the Laplace transform nor to the Fuchs theory of differential equations.

1. **The definition of D and of $(s - z_1 I)^{\gamma_1}$.** Let $C = C[0, \infty)$ be the totality of complex-valued continuous functions $f = \{f(t)\}$, $g = \{g(t)\}$, $\cdots$. C is a commutative ring by the sum $f + g = \{f(t) + g(t)\}$ and the (convolution) product $fg = \left\{ \int_0^t f(t - u)g(u)du \right\}$. By virtue of Titchmarsh's convolution theorem, we have $fg = 0$ $((fg)(t) \equiv 0)$ if and only if either $f = 0$ or $g = 0$. Hence the totality C/C of fractions (convolution quotients) f/g $(f, g \in C$ and $g \neq 0)$, $f_1/g_1 \cdots$ constitutes a commutative ring by

$$\frac{f}{g} + \frac{f_1}{g_1} = \frac{fg_1 + f_1 g}{g g_1}, \qquad \frac{f}{g}\,\frac{f_1}{g_1} = \frac{f f_1}{g g_1}.$$

C is a subring of C/C by identifying $f \in C$ with $fg/g \in C/C$.

We denote $h = \{1\}$ (the operator of integration), $I = h/h = g/g$ (the operator of the product unit) and $s = I/h = g/hg$ (the operator of differentiation). We have

(5) $$h^n = \Gamma(n)^{-1} t^{n-1} \quad (n = 1, 2, \cdots ; h^0 = I),$$

(6) $$\begin{cases} \text{If } \{f^{(n)}(t)\} \in C, \text{ then } f^{(n)} = s^n f - s^{n-1}[f(0)] - \cdots - [f^{(n-1)}(0)], \\ \text{where } [\alpha] = s\{\alpha\} \text{ for complex number } \alpha. \end{cases}$$

Definition of **D**. D is a mapping of C/C into C/C such that

(7) $$\begin{cases} Df = \{-tf(t)\} \quad \text{for } f \in C, \\ D\dfrac{f}{g} = \dfrac{(Df)g - f(Dg)}{g^2} \quad \text{for } \dfrac{f}{g} \in C/C \end{cases}$$

and it is not difficult to prove that

(8) $$\begin{cases} D\left(\dfrac{f}{g} + \dfrac{f_1}{g_1}\right) = D\dfrac{f}{g} + D\dfrac{f_1}{g_1}, \qquad D\left([\alpha]\dfrac{f}{g}\right) = [\alpha]\left(D\dfrac{f}{g}\right), \\ D\left(\dfrac{f}{g}\,\dfrac{f_1}{g_1}\right) = \left(D\dfrac{f}{g}\right)\dfrac{f_1}{g_1} + \dfrac{f}{g}\left(D\dfrac{f_1}{g_1}\right). \end{cases}$$

Moreover, it is not difficult to show that

(8)′ $$\begin{cases} \text{If } a = \dfrac{m}{n} \in C/C \text{ and } b = \dfrac{p}{q} \in C/C, \text{ then} \\ D\dfrac{a}{b} = \dfrac{(Da)b - a(Db)}{b^2} = D\dfrac{mq}{np}. \end{cases}$$

We have thus

(9) $$\begin{cases} Dh^n = -nh^{n+1} \text{ and } Ds^n = ns^{n-1} \ (n = 1, 2, \cdots ; s^0 = I), \text{ in} \\ \text{particular } Dh^0 = DI = 0, \ Ds^0 = DI = 0 \text{ and } Ds = I. \end{cases}$$

Proof. $Dh^n = \{-t\Gamma(n)^{-1} t^{n-1}\} = -\Gamma(n)^{-1}\Gamma(n+1)h^{n+1}$.

The hitherto formulas are proved by J. Mikusiński in his book mentioned above. We now define and prove the following.

For any complex number γ,

(10) $$\begin{cases} Dh^\gamma = -\gamma h^{\gamma+1}, \text{ where } h^\gamma = \dfrac{h^{\gamma+n}}{h^n} = \dfrac{\{\Gamma(\gamma+n)^{-1} t^{\gamma+n-1}\}}{\{\Gamma(n)^{-1} t^{n-1}\}}, \\ n \text{ being any integer} \geq 1 \text{ such that } \mathrm{Re}(\gamma+n) > 1. \end{cases}$$

Proof. Easy from (5), (8)′ and (9).

We have thus, by (8)′ and (10),

(10)′ $$Ds^\gamma = D\frac{I}{h^\gamma} = \frac{-Dh^\gamma}{h^{2\gamma}} = \gamma h^{\gamma+1-2\gamma} = \gamma s^{\gamma-1}.$$

The next formula is very important:

(11) $$\begin{cases} D(I - \alpha h)^\gamma = \gamma(I - \alpha h)^{\gamma-1}\alpha h^2, \text{ where} \\ (I - \alpha h)^\gamma = \sum_{k=0}^{\infty}\binom{\gamma}{k}(-\alpha)^k h^k. \end{cases}$$

Proof. $\displaystyle\sum_{k=0}^{\infty}\binom{\gamma}{k}(-\alpha)^k h^k = I + \left\{\sum_{k=1}^{\infty}\binom{\gamma}{k}(-\alpha)^k \Gamma(k)^{-1} t^{k-1}\right\} \in C/C$ and the infinite series $\sum_{k=1}^{\infty} \binom{\gamma}{k}(-\alpha)^k \Gamma(k)^{-1} t^{k-1}$ is, thanks to the convergent factors $\Gamma(k)^{-1}$, convergent at every t and it can be differentiated with respect to t by term-wise differentiation.

Now we have, by the above,

$$D(I-\alpha h)^\gamma = DI + \left\{ -t \sum_{k=1}^{\infty} \binom{\gamma}{k}(-\alpha)^k \Gamma(k)^{-1} t^{k-1} \right\}$$

$$= \sum_{k=1}^{\infty} \binom{\gamma}{k}(-\alpha)^{k-1} \alpha k h^{k+1}$$

$$= \gamma \sum_{k=1}^{\infty} \binom{\gamma-1}{k-1}(-\alpha)^{k-1} h^{k-1} \alpha h^2 = \gamma(I-\alpha h)^{\gamma-1} \alpha h^2.$$

As a corollary of (11), we have

(12) $D(s-\alpha I)^\gamma = \gamma(s-\alpha I)^{\gamma-1}$, where $(s-\alpha I)^\gamma = \dfrac{(I-\alpha h)^\gamma}{h^\gamma}$.

    Proof. $D\dfrac{(I-\alpha h)^\gamma}{h^\gamma} = \dfrac{(D(I-\alpha h)^\gamma)h^\gamma - (I-\alpha h)^\gamma(Dh^\gamma)}{h^{2\gamma}}$

$$= \dfrac{\gamma(I-\alpha h)^{\gamma-1}(\alpha h^2 h^\gamma + (I-\alpha h)h^{\gamma+1})}{h^{2\gamma}}$$

$$= \dfrac{\gamma(I-\alpha h)^{\gamma-1} h^{\gamma+1}}{h^{2\gamma}} = \gamma(s-\alpha I)^{\gamma-1}.$$

    2. **Proof of (4) and examples.**   Assuming that $y(t) \not\equiv 0$ is twice continuously differentiable, we can rewrite (1) by (6) and (7) as follows:

$$-\alpha_2 D(s^2 y - s[y(0)] - [y'(0)]) + (-a_1 D + b_1)(sy - [y(0)])$$
$$+ (-a_0 D + b_0)y = 0.$$

Hence, by $Ds = I$, we obtain (2) assuming the initial condition of $y(t)$:

(13) $y(0) = 0$ if $a_2 \neq b_1$.

This proves (4) by (12).

    Example 1 (Bessel differential equation).   For the equation

(1)′ $ty''(t) - (2\alpha - 1)y'(t) + ty(t) = 0,$

we have $a_2 - b_1 = 2\alpha$ and

(2)′ $\dfrac{Dy}{y} = \dfrac{-\alpha - 1/2}{s + iI} + \dfrac{-\alpha - 1/2}{s - iI}.$

Hence

(4)′ $y = C(s+iI)^{-\alpha-1/2}(s-iI)^{-\alpha-1/2} = C(s^2 + I)^{-\alpha-1/2}$

$$= C(h^2(I+h^2)^{-1})^{\alpha+1/2} = C\left(\sum_{k=0}^{\infty} \binom{-\alpha-1/2}{k} h^{2k} \right) h^{2\alpha+1}.$$

satisfies (2)′.   If $\operatorname{Re}\alpha \geq 0$, we obtain

$$\binom{-\alpha-1/2}{k} = \dfrac{(-1)^k \Gamma(2\alpha+2k+1)\Gamma(\alpha+1)}{2^{2k}\Gamma(k+1)\Gamma(2\alpha+1)\Gamma(\alpha+k+1)}.$$

Thus if $\operatorname{Re}\alpha > 1$, then

$$y = C\dfrac{\Gamma(\alpha+1)2^{2\alpha}}{\Gamma(2\alpha+1)} \sum_{k=0}^{\infty} \dfrac{(-1)^k}{\Gamma(k+1)\Gamma(\alpha+k+1)} \left(\dfrac{t}{2}\right)^{2k+2\alpha}$$

is twice continuously differentiable in t for $t \geq 0$ including $t = 0$. Thus the solution

$$(14) \qquad y_\alpha(t) = \sum_{k=0}^{\infty} \frac{(-1)^k}{\Gamma(k+1)\Gamma(\alpha+k+1)} \left(\frac{t}{2}\right)^{2k+2\alpha}$$

of $(2)'$ satisfying (13) is a solution of $(1)'$ for $t \geq 0$. This means that, when $t \geq 0$ and $\operatorname{Re}\alpha > 1$, the coefficient of $t^{2k+2\alpha-1}$ in the infinite series given by

$$(15) \qquad ty_\alpha''(t) - (2\alpha-1)y_\alpha'(t) + ty_\alpha(t)$$

must vanish as an analytic function of α $(k=0, 1, 2, \cdots)$.

Therefore, since $y_\alpha(t)$ with $\operatorname{Re}\alpha \geq 0$ is also twice continuously differentiable in $t > 0$, we see, as in the case of $\operatorname{Re}\alpha > 1$, that the formula

$$ty_\alpha''(t) - (2\alpha-1)y_\alpha'(t) + ty_\alpha(t)$$

must vanish, because the coefficients of $t^{2k+2\alpha-1}$ all vanish.

Thus we have proved that, when $\operatorname{Re}\alpha > 0$ or $\alpha=0$, $y_\alpha(t)$ given in (14) is a solution of $(1)'$ at every $t > 0$ satisfying (13). Hence we have obtained <u>Bessel function of the first kind and of order α</u> $(\operatorname{Re}\alpha > 0$ or $\alpha = 0)$:

$$(16) \qquad J_\alpha(t) = t^{-\alpha}y_\alpha(t) = \sum_{k=0}^{\infty} \frac{(-1)^k}{\Gamma(k+1)\Gamma(\alpha+k+1)} \left(\frac{t}{2}\right)^{2k+\alpha}$$

which satisfies the original Bessel equation

$$(17) \qquad t^2 J_\alpha''(t) + t J_\alpha'(t) + (t^2-\alpha^2)J_\alpha(t) = 0 \qquad \text{for } t > 0.$$

Example 2 (Laguerre differential equation). For the equation

$$(1)'' \qquad ty''(t) - (t+\alpha-1)y'(t) + (\alpha+\lambda)y(t) = 0$$

we have $a_2 - b_1 = \alpha$ and

$$(2)'' \qquad \frac{Dy}{y} = \frac{(-2+1-\alpha)s+1+\alpha+\lambda}{s^2-s} = \frac{-1-\alpha-\lambda}{s} + \frac{\lambda}{s-I}.$$

Hence, for $\operatorname{Re}\alpha > 0$ or for $\alpha = 0$,

$$(4)'' \qquad y_{\alpha,\lambda} = Cs^{-1-\alpha-\lambda}(s-I)^\lambda = Ch^{1+\alpha}(I-h)^\lambda$$

$$= C\sum_{k=0}^{\infty} \binom{\lambda}{k}(-1)^k \Gamma(k+\alpha+1)^{-1}t^{k+\alpha}$$

is a solution of $(2)''$ and $t^{-\alpha}y_{\alpha,\lambda}$ reduces to a polynomial in t if and only if $\lambda = n$ $(=0, 1, 2, \cdots)$. So we have, by taking C as $(n!)^{-1}\Gamma(n+\alpha+1)$,

$$t^{-\alpha}y_{\alpha,n} = \frac{\Gamma(n+\alpha+1)}{n!} \sum_{k=0}^{n} \binom{n}{k} \frac{(-t)^k}{\Gamma(k+\alpha+1)}$$

$$= \sum_{k=0}^{n} \binom{n+\alpha}{n-k} \frac{(-t)^k}{k!} = L_n^{(\alpha)}(t).$$

We have thus obtained the <u>n-th Laguerre polynomial of order α</u>: $L_n^{(\alpha)}(t)$. When $\operatorname{Re}\alpha > 0$ or $\alpha = 0$, $t^\alpha L_n^{(\alpha)}(t)$ is surely a solution of $(1)''$ with $\lambda = n$ for $t > 0$ and satisfies (13).

Remark. The equation of the form

$$\frac{Dy}{y} = \frac{\gamma}{(s-\alpha)^2}$$

is satisfied by $y = Ce^{\alpha t}e^{-\gamma h}$ which belongs to C/C. We omit the proof.

A simple complement to Mikusiński's operational calculus

Studia Math. **77** (1983) 95–98

Abstract. According to Mikusiński's operational calculus, any solution $y = \{y(t)\}$ of the Cauchy problem for the nth order linear ordinary differential equation with complex coefficients and with inhomogeneous term $f = \{f(t)\} \in C[0, \infty)$ must satisfy

$$(*) \qquad (a_n s^n + a_{n-1} s^{n-1} + \cdots + a_0) y = f + \beta_{n-1} s^{n-1} + \beta_{n-2} s^{n-2} + \cdots + \beta_0,$$

where a_j's and β_k's are complex numbers with $a_n \neq 0$. The entitled complement proves that $y = \{y(t)\}$ given by $(*)$ is n-times continuously differentiable in t so that $y(t)$ is in truth the unique solution of the original Cauchy problem. Cf. the subsequent paper by S. Okamoto (this volume, pp. 99–101).

The entitled "complement" will be stated in § 2. For the sake of the reader's convenience, I shall begin with a brief prerequisite from the operational calculus of J. Mikusiński [1] as exposed in a joint paper [2] of the present author.

§ 1. The prerequisite. Let $\mathscr{C}$ denote the totality of complex-valued continuous functions defined on $[0, \infty)$. We denote such a function by $\{f(t)\}$ or simply by f, while $f(t)$ means the value at t of the function f. For $f, g \in \mathscr{C}$ and $a, \beta \in K$ ($=$ the complex number field) we define

$$(1) \qquad af + \beta g = \{af(t) + \beta g(t)\} \quad \text{and} \quad fg = \left\{ \int_0^t f(t - \tau) g(\tau) d\tau \right\}.$$

Then $\mathscr{C}$ is a *commutative ring* with respect to the above *addition* and *multiplication* over the coefficient field K.

We shall denote by h the *constant function* $\{1\} \in \mathscr{C}$ so that we have

$$(2) \qquad hf = \left\{ \int_0^t f(\tau) d\tau \right\} \quad \text{for} \quad f \in \mathscr{C} \quad \text{and} \quad h^n = \left\{ \frac{t^{n-1}}{(n-1)!} \right\}.$$

For any integer $n \geq 1$ and $f \in \mathscr{C}$, we have, by (2),

$$(3) \qquad h^n f = 0 \quad \text{implies} \quad f = 0,$$

where 0 denotes $\{0\} \in \mathscr{C}$. Therefore we can define the commutative super-ring $\mathscr{C}_H$ of $\mathscr{C}$ by

(4)
$$\mathscr{C}_H = \{f/h^n;\ f \in \mathscr{C}\ \text{and}\ n = 1, 2, \ldots\},$$

where the *equality* means

(5)
$$\frac{f}{h^n} = \frac{f'}{h^{n'}} \quad \text{if and only if} \quad f h^{n'} = f' h^n$$

and the *addition* and *multiplication* are defined by

(6)
$$\frac{f}{h^n} + \frac{f'}{h^{n'}} = \frac{f h^{n'} + f' h^n}{h^n h^{n'}}, \quad \frac{f}{h^n}\,\frac{f'}{h^{n'}} = \frac{f f'}{h^n h^{n'}}.$$

We introduce

(7) $I = \dfrac{h^n}{h^n} \in \mathscr{C}_H \quad (n = 1, 2, \ldots) \quad$ and $\quad s = \dfrac{h^n}{h^{n+1}} \in \mathscr{C}_H \quad (n = 1, 2, \ldots)$

so that we have

(8) $\quad sh = hs = I, \quad$ and $\quad I$ is the *multiplicative unit* of the ring $\mathscr{C}_H$.

Then, if both f and its derivative f' belong to $\mathscr{C}$, we have

(9)
$$f' = sf - [f(0)], \quad \text{where} \quad [f(0)] = s\{f(0)\},$$

because of the *Newton formula*

(9)'
$$hf' = \left\{\int_0^t f'(\tau)\,d\tau\right\} = \{f(t) - f(0)\} = f - \{f(0)\}.$$

Formula (9) is generalized as follows:
If f is n-times continuously differentiable, we have

(9)''
$$f^{(n)} = s^n f - s^{n-1}[f(0)] - \cdots - [f^{(n-1)}(0)].$$

Hereafter, we shall write $f^{(j)}(0)$ for $[f^{(j)}(0)]$ in case there be no confusion of identifying $f^{(j)}(0)$ with $\{f^{(j)}(0)\}$. We have then

PROPOSITION. *For any* $\alpha \in K$ *and for any positive integer* n, *we have the result that*

$$(s - [\alpha])^n = (s - \alpha)^n = \frac{(h - [\alpha]h^2)^n}{h^{2n}} \in \mathscr{C}_H$$

admits a uniquely determined multiplicative inverse in $\mathscr{C}_H$ *given by*

(10)
$$\frac{I}{(s - [\alpha])^n} = (s - \alpha)^{-n} = \left\{\frac{t^{n-1}}{(n-1)!}\,e^{\alpha t}\right\} \in \mathscr{C} \subseteq \mathscr{C}_H,$$

because $(s - [\alpha])\{e^{\alpha t}\} = I$ *by* (9).

§ 2. The complement. Consider the following *Cauchy problem* for linear ordinary differential equation with coefficients belonging to K:

$$(11) \qquad \begin{aligned} &a_n y^{(n)} + a_{n-1} y^{(n-1)} + \ldots + a_0 y = f \in \mathscr{C} \qquad (a_n \neq 0), \\ &y(0) = \gamma_1, \quad y'(0) = \gamma_2, \ldots, y^{(n-1)}(0) = \gamma_{n-1}. \end{aligned}$$

By virtue of $(9)''$, this problem shall be converted into the equation in $\mathscr{C}_H$:

$$(11)' \qquad \begin{aligned} &(a_n s^n + a_{n-1} s^{n-1} + \ldots + a_0) y = f + \beta_{n-1} s^{n-1} + \beta_{n-2} s^{n-2} + \ldots + \beta_0, \\ &\beta_m = a_{m+1} \gamma_0 + a_{m+2} \gamma_1 + \ldots + a_n \gamma_{n-m-1} \qquad (m = 0, 1, 2, \ldots, n-1). \end{aligned}$$

Since the *polynomial ring of polynomials in s with coefficients belonging to K is free from zero factors,* we can define rational functions in s:

$$(12) \qquad F_1 = \frac{I}{a_n s^n + \ldots + a_0} \quad \text{and} \quad F_2 = \frac{\beta_{n-1} s^{n-1} + \ldots + \beta_0}{a_n s^n + \ldots + a_0},$$

and obtain their partial fraction decompositions

$$(12)' \qquad F_1 = \sum_j \sum_{k=1}^{m_j} c_{jk} (s - r_j)^{-k} \quad \text{and} \quad F_2 = \sum_j \sum_{k=1}^{m_j} d_{jk} (s - r_j)^{-k},$$

where r_j's are distinct roots of the algebraic polynomial

$$(13) \qquad p(z) = a_n z^n + \ldots + a_0 = a_n \prod_j (z - r_j)^{m_j} \qquad \left(\sum_j m_j = n \right).$$

By virtue of (10), we have

$(12)''$

$$F_1 = \sum_j \sum_{k=1}^{m_j} c_{jk} \left\{ \frac{t^{k-1}}{(k-1)!} e^{r_j t} \right\} \in \mathscr{C}, \qquad F_2 = \sum_j \sum_{k=1}^{m_j} d_{jk} \left\{ \frac{t^{k-1}}{(k-1)!} e^{r_j t} \right\} \in \mathscr{C}$$

so that we obtain the solution y of equation $(11)'$ given by a function of $\mathscr{C}$:

$$(14) \qquad \begin{aligned} y &= \frac{I}{p(s)} f + \frac{\beta_{n-1} s^{n-1} + \ldots + \beta_0}{p(s)} \\ &= \sum_j \sum_{k=1}^{m_j} c_{jk} \left\{ \frac{t^{k-1}}{(k-1)!} e^{r_j t} \right\} \{ f(t) \} + \sum_j \sum_{k=1}^{m_j} d_{jk} \left\{ \frac{t^{k-1}}{(k-1)!} e^{r_j t} \right\}. \end{aligned}$$

However, *it is not apparent that this function y of t is precisely the solution of (11), because it is not apparent that y is n-times continuously differentiable in t.* In fact, if f is not differentiable in t, then, e.g.,

$$w = \{ e^{rt} \} \{ f(t) \} = \left\{ \int_0^t e^{r(t-\tau)} f(\tau) \, d\tau \right\}$$

is not twice continuously differentiable.

Our *complement* says that, in spite of the above example, we can prove that *the function y given by* (14) *is n-times continuously differentiable so that it is the unique solution of* (11).

Proof. Multiplying both sides of (11)' by h^n we obtain

$$a_n y + a_{n-1} hy + \dots + a_0 h^n y = h^n f + \beta_{n-1} h + \dots + \beta_0 h^n.$$

Then $F(t) = h^n f + \beta_{n-1} h + \dots + \beta_0 h^n$ is surely n-times continuously differentiable. Thus, by $y \in \mathscr{C}'$ and by (2), we have the result:

$$y = -a_n^{-1}(a_{n-1} hy + a_{n-2} h^2 y + \dots + a_0 h^n y) + a_n^{-1}\{F(t)\}$$

is once continuously differentiable and its derivative satisfies

$$(15) \qquad y' = -a_n^{-1}(a_{n-1} hy' + a_{n-2} h^2 y' + \dots + a_0 h^n y') + a_n^{-1}\{F'(t)\} +$$
$$+ \text{ a polynomial in } t,$$

because, e.g.,

$$(h^3 y)' = h^2 y = h^2\big(hy' + y(0)\big) = h^3 y' + h^2 y(0)$$

by (9). Thus y' given by (15) is continuously differentiable in t and satisfies

$$y'' = -a_n^{-1}(a_{n-1} hy'' + a_{n-2} h^2 y'' + \dots + a_0 h^n y'') + a_n^{-1}\{F''(t)\} +$$
$$+ \text{ a polynomial in } t$$

and so forth.

References

[1] J. Mikusiński, *Sur les fondaments de calcul opératoire*, Studia Math. 11 (1949), 41–70.

[2] K. Yosida and S. Okamoto, *A note on Mikusiński's operational calculus*, Proc. Japan Acad. 56 A (1) (1980), 1–3.

Received April 15, 1982 (1750)

Bibliography of Kôsaku Yosida

Papers

*[1] On the asymptotic property of the differential equation $y'' + H(x)y = f(x, y, y')$. Japan. J. Math. 9 (1932), (145–152)

*[2] On the asymptotic property of the differential equation $y'' + H(x)y = f(x, y, y')$, II. Japan. J. Math. 9 (1932), (227–230)

[3] A remark to a theorem due to Halphen. Japan. J. Math. 9 (1932), (231–232)

[4] A generalisation of a Malmquist's theorem. Japan. J. Math. 9 (1932), (253–256)

*[5] Some remarks on the theory of Fredholm's integral equations. Proc. Phys.-Math. Soc. Japan 14 (1932), (381–384)

[6] On the distribution of α-points of solutions for linear differential equation of the second order. Proc. Imp. Acad. Tokyo 8 (1932), (335–336)

[7] A note on Riccati's equation. Proc. Phys.-Math. Soc. Japan 15 (1933), (227–232)

[8] On the characteristic function of a transcendental meromorphic solution of an algebraic differential equation of the first order and of the first degree. Proc. Phys.-Math. Soc. Japan 15 (1933), (337–338)

[9] On algebroid solutions of ordinary differential equations. Japan. J. Math. 10 (1933), (199–208)

[10] On a class of meromorphic functions. Proc. Phys.-Math. Soc. Japan 16 (1934), (227–235)

*[11] (With T. Shimizu and S. Kakutani) On meromorphic function I. Proc. Phys.-Math. Soc. Japan 17 (1935), (1–10)

[12] A theorem concerning the derivatives of meromorphic functions. Proc. Phys.-Math. Soc. Japan 17 (1935), (170–173)

[13] On the groups of rationality for linear differential equations. Proc. Phys.-Math. Soc. Japan 17 (1935), (498–510)

[14] On the group embedded in the metrical complete ring. Japan. J. Math. 13 (1936), (7–26)

[15] On the group embedded in the metrical complete ring. II. Japan. J. Math. 13 (1936), (459–472)

---

* These do not appear in this collection.

[16] A note on the continuous representation of topological groups. Proc. Imp. Acad. Tokyo 12 (1936), (329–331)

[17] A theorem concerning the semi-simple Lie groups. Tohoku Math. J. 43 (1937), (81–84)

[18] A problem concerning the second fundamental theorem of Lie. Proc. Imp. Acad. Tokyo 13 (1937), (152–155)

[19] A remark on a theorem of B. L. van der Waerden. Tohoku Math. J. 43 (1937), (411–413)

[20] On the exponential-formula in the metrical complete ring. Proc. Imp. Acad. Tokyo 13 (1937), (301–304)

[21] A note on the differentiability of the topological group. Proc. Phys.-Math. Soc. Japan 20 (1938), (6–10)

[22] A characterisation of the adjoint representations of the semi-simple Lie-rings. Japan. J. Math. 14 (1938), (169–173)

[23] On the fundamental theorem of the tensor calculus. Proc. Imp. Acad. Tokyo 14 (1938), (211–213)

[24] Abstract integral equations and the homogeneous stochastic process. Proc. Imp. Acad. Tokyo 14 (1938), (286–291)

[25] Mean ergodic theorem in Banach spaces. Proc. Imp. Acad. Tokyo 14 (1938), (292–294)

[26] (With S. Kakutani) Application of mean ergodic theorem to the problems of Markoff's process. Proc. Imp. Acad. Tokyo 14 (1938), (333–339)

[27] (With Y. Mimura and S. Kakutani) Integral operator with bounded kernel. Proc. Imp. Acad. Tokyo 14 (1938), (359–362)

[28] Operator-theoretical treatment of the Markoff's process. Proc. Imp. Acad. Tokyo 14 (1938), (363–367)

[29] Quasi-completely-continuous linear functional operations. Japan. J. Math. 15 (1939), (297–301)

[30] Operator-theoretical treatment of Markoff's process. II. Proc. Imp. Acad. Tokyo 15 (1939), (127–130)

[31] (With S. Kakutani) Birkhoff's ergodic theorem and the maximal ergodic theorem. Proc. Imp. Acad. Tokyo 15 (1939), (165–168)

[32] Asymptotic almost periodicities and ergodic theorems. Proc. Imp. Acad. Tokyo 15 (1939), (255–259)

[33] (With S. Kakutani) Markoff process with an enumerable infinite number of possible states. Japan. J. Math. 16 (1940), (47–55)

[34] Ergodic theorems of Birkhoff-Khintchine's type. Japan. J. Math. 17 (1940), (31–36)

[35] The Markoff process with a stable distribution. Proc. Imp. Acad. Tokyo 16 (1940), (43–48)

[36] An abstract treatment of the individual ergodic theorem. Proc. Imp. Acad. Tokyo 16 (1940), (280–284)

[37] On the theory of spectra. Proc. Imp. Acad. Tokyo 16 (1940), (378–383)

[38] (With S. Kakutani) Operator-theoretical treatment of Markoff's process and mean ergodic theorem. Ann. Math. 42 (1941), (188–228)

[39] (With M. Fukamiya) On regularly convex sets. Proc. Imp. Acad. Tokyo 17 (1941), (49–52)

[40] On vector lattice with a unit. Proc. Imp. Acad. Tokyo 17 (1941), (121–124)

[41] Vector lattices and additive set functions. Proc. Imp. Acad. Tokyo 17 (1941), (228–232)

[42] (With M. Fukamiya) On vector lattice with a unit, II. Proc. Imp. Acad. Tokyo 17 (1941), (479–482)

[43] On the representation of the vector lattice. Proc. Imp. Acad. Tokyo 18 (1942), (339–342)

[44] (With T. Nakayama) On the semi-ordered ring and its application to the spectral theorem. Proc. Imp. Acad. Tokyo 18 (1942), (555–560)

[45] (With T. Nakayama) On the semi-ordered ring and its application to the spectral theorem, II. Proc. Imp. Acad. Tokyo 19 (1943), (144–147)

[46] On the duality theorem of non-commutative compact groups. Proc. Imp. Acad. Tokyo 19 (1943), (181–183)

[47] Normed rings and spectral theorems. Proc. Imp. Acad. Tokyo 19 (1943), (356–359)

[48] Normed rings and spectral theorems, II. Proc. Imp. Acad. Tokyo 19 (1943), (466–470)

[49] Normed rings and spectral theorems, III. Proc. Imp. Acad. Tokyo 20 (1944), (71–73)

[50] Normed rings and spectral theorems, IV. Proc. Imp. Acad. Tokyo 20 (1944), (183–185)

[51] Normed rings and spectral theorems, V. Proc. Imp. Acad. Tokyo 20 (1944), (269–273)

[52] (With T. Iwamura) Equivalence of two topologies of Abelian groups. Proc. Imp. Acad. Tokyo 20 (1944), (451–453)

[53] Normed rings and spectral theorems, VI. Proc. Imp. Acad. Tokyo 20 (1944), (580–583)

[54] On the representation of functions by Fourier integrals. Proc. Imp. Acad. Tokyo 20 (1944), (655–660)

[55] On the unitary equivalence in general Euclid space. Proc. Japan Acad. 22 (1946), (242–245)

[56] Simple Markoff process with a locally compact phase space. Math. Japonica 1 (1948), (99–103)

[57] On the differentiability and the representation of one-parameter semi-group of linear operators. J. Math. Soc. Japan 1 No. 1 (1948), (15–21)

[58] An operator-theoretical treatment of temporally homogeneous Markoff process. J. Math. Soc. Japan 1 No. 3 (1949), (244–253)

[59] Brownian motion on the surface of the 3-sphere. Ann. Math. Statist. 20 (1949), (292–296)

[60] Integration of Fokker-Planck's equation in a compact Riemannian space. Ark. Mat. 1 (1950), (71–75)

[61] An extension of Fokker-Planck's equation. Proc. Japan Acad. 25 (1949), (1–3)

[62] On Titchmarsh-Kodaira's formula concerning Weyl-Stone's eigenfunction expansion. Nagoya Math. J. 1 (1950), (49–58). Correction. ibid. 6 (1953), (187–188)

[63] Stochastic processes built from flows. Proc. Japan Acad. 26 No. 8 (1950), (1–3)

[64] A theorem of Liouville's type for meson equation. Proc. Japan Acad. 27 (1951), (214–215)

[65] An ergodic theorem associated with harmonic integrals. Proc. Japan Acad. 27 (1951), (540–543)

[66] Integration of Fokker-Planck's equation with a boundary condition. J. Math. Soc. Japan 3 (1951), (69–73)

[67] Integrability of the backward diffusion equation in a compact Riemannian space. Nagoya Math. J. 3 (1951), (1–4)

[68] (With E. Hewitt) Finitely additive measures. Trans. Amer. Math. Soc. 72 (1952), (46–66)

[69] On Brownian motion in a homogeneous Riemannian space. Pacific J. Math. 2 (1952), (263–270)

[70] On the existence of the resolvent kernel for elliptic differential operator in a compact Riemann space. Nagoya Math. J. 4 (1952), (63–72)

[71] On the integration of diffusion equations in Riemannian spaces. Proc. Amer. Math. Soc. 3 (1952), (864–873)

[72] On Cauchy's problem in the large for wave equations. Proc. Japan Acad. 28 (1952), (396–403)

[73] On the fundamental solution of the parabolic equation in a Riemannian space. Osaka Math. J. 5 (1953), (65–74)

[74] On the integration of the temporally inhomogeneous diffusion equation in a Riemannian space. Proc. Japan Acad. 30 (1954), (19–23)

[75] On the integration of the temporally inhomogeneous diffusion equation in a Riemannian space. II. Proc. Japan Acad. 30 (1954), (273–275)

*[76] Semi-group theory and the integration problem of diffusion equations. Internat. Congress of Mathematicians (Amsterdam, 1954) Vol. 1, (405–420)

[77] On the generating parametrix of the stochastic processes. Proc. Nat. Acad. Sci. U.S.A. 41 (1955), (240–244)

[78] A characterization of the second order elliptic differential operators. Proc. Japan Acad. 31 (1955), (406–409)

[79] An operator-theoretical integration of the wave equation. J. Math. Soc. Japan 8 (1956), (79–92)

[80] An operator-theoretical integration of the temporally inhomogeneous wave equation. J. Fac. Sci. Univ. Tokyo I. 7 (1957), (463–466)

[81] On the reflexivity of the space of distribution. Sci. Papers Coll. Gen. Ed. Univ. Tokyo 7 (1957), (151–155)

[82] On the differentiability of semi-groups of linear operators. Proc. Japan Acad. 34 (1958), (337–340)

[83] An abstract analyticity in time for solutions of a diffusion equation. Proc. Japan Acad. 35 (1959), (109–113)

[84] Fractional powers of infinitesimal generators and the analyticity of the semi-groups generated by them. Proc. Japan Acad. 36 (1960), (86–89)

[85] On a class of infinitesimal generators and the integration problem of evolution equations. Proc. Fourth Berkeley Sympos. on Math. Stat. and Prob., (1961), (Univ. Calif. Press) Vol. II, (623–633)

[86] Ergodic theorems for pseudo-resolvents. Proc. Japan Acad. 37 (1961), (422–425)

*[87] Abelian ergodic theorems in locally convex spaces, Ergodic Theory. Proc. Internat. Sympos. (Tulane Univ. 1961), (293–299)

[88] On the integration of the equation of evolution. J. Fac. Sci. Univ. Tokyo I. 9 (1963), (397–402)

[89] Holomorphic semi-groups in a locally convex linear topological space. Osaka Math. J. 15 (1963), (51–57)

*[90] Holomorphic semi-groups. Séminaire sur les Équations aux Dérivées Partielles (1964), (68–76)

[91] Positive pseudo-resolvents and potentials. Proc. Japan Acad. 41 (1965), (1–5)

[92] A perturbation theorem for semi-groups of linear operators. Proc. Japan Acad. 41 (1965), (645–647)

[93] Time dependent evolution equations in a locally convex space. Math. Ann. 162 (1965), (83–86)

[94] On holomorphic Markov processes. Proc. Japan Acad. 42 (1966), (313–317)

[95] Positive resolvents and potentials (An operator-theoretical treatment of Hunt's theory of potentials). Z. Wahrscheinlichkeitstheorie und verw. Geb. 8 (1967), (210–218)

[96] (With T. Watanabe and H. Tanaka) On the pre-closedness of the potential operator. J. Math. Soc. Japan 20 (1968), (419–421)

[97] The existence of the potential operator associated with an equicontinuous semi-group of class (C_0). Studia Math. 31 (1968), (531–533)

[98] On the potential operators associated with Brownian motions. J. Analyse Math. 23 (1970), (461–465)

[99] On the pre-closedness of Hunt's potential operators and its applications. Proc. Intern. Conf. Functional Analysis and Related Topics (Tokyo, 1969), Univ. Tokyo Press (1970), (324–331)

[100] On the existence and a characterization of abstract potential operators. Troisième Colloque sur l'Analyse Fonctionnelle. (Liège, 1970) (129–136)

[101] Abstract potential operators on Hilbert space. Publ. Res. Inst. Math. Sci. Kyoto Univ. 8 (1972), (201–205)

[102] A note on Malmquist's theorem on first order algebraic differential equations. Proc. Japan Acad. 53 (1977), (120–123)

[103] A note on Mikusiński's operational calculus. Proc. Japan Acad. 56. A (1980), (1–3)

[104] A note on the fundamental theorem of calculus. Proc. Japan Acad. 57. A (1981), no. 5, (241)

*[105] Some aspects of E. Hille's contribution to semi-group theory. Integral Equations Operator Theory 4 (1981, Birkhäuser), no. 3, (311–329)

[106] (With S. Matsuura) A note on Mikusiński's proof of the Titchmarsh convolution theorem. Conference in Modern Analysis and Probability (New Haven, Conn., 1982), Contemp. Math., 26, (1984) (423–425)

[107] The algebraic derivative and Laplace's differential equation. Proc. Japan Acad. 59. A (1983), (1–4)

[108] A simple complement to Mikusiński's operational calculus. Studia Math. 77 (1983), (95–98)

Books (in Japanese)

1. Theory of Continuous Group, Mathematics Lecture Series Iwanani, Tokyo 1934

2. Theory of Lie Rings, Mathematics Lecture Series IV of Osaka University, Iwanami, Tokyo, 1939

3. Linear Operators, Iwanami, Tokyo, 1943

4. Spectral Analysis, Kyoritsu, Tokyo, 1947

5. Ergodic Theorems, Chubunkan, Tokyo, 1948

6. General Theory of Mathematical Physics, Nihon Hyoron, Tokyo, 1949. Revised enlarged ed. Sangyo Tosho, Tokyo, 1974

7. Theory of Integral Equations, Iwanami, Tokyo, 1950 (2nd ed. 1978). English translation: Lectures on Differential and Integral Equations, Interscience, New York—London, 1960. French translation: Equations Différentielles et Intégrales, Dunod, Paris, 1971

8. Topological Analysis I, Iwanami, Tokyo, 1951

9. Theory of Hilbert Spaces, Kyoritsu, Tokyo, 1953

10. Methods of Solving Differential Equations, Iwanami, Tokyo, 1954 (2nd ed. 1978)

11. Handbook of Applied Mathematics (with A. Amemiya, K. Itô, T. Kato and Y. Matsushima), Maruzen, Tokyo, 1954 (2nd ed. 1967)

12. Modern Analysis, Kyoritsu, Tokyo, 1956 (2nd ed. 1958)

13. Theory of Distributions, Kyoritsu, Tokyo, 1956

14. Topological Analysis, Modern Applied Mathematics Series A4, Iwanami, Tokyo, 1957

15. Theory of Integrals and Topological Analysis (with K. Kunugui, S. Nakanishi and S. Itô), Mathematics Exercise Series 15, Kyoritsu, Tokyo, 1958

16. Foundations of Topological Analysis (with Y. Kawada and T. Iwamura), Iwanami, Tokyo, 1960

17. Exercises in Applied Mathematics I (with T. Kato), Shōkabō, Tokyo, 1961

18. Measures and Integrals, Fundamental Mathematics Series, Analysis (1) (iii), Iwanami, Tokyo, 1976

19. Functional Analysis and Differential Equations (with S. Itô, A. Orihara and T. Muramatu), Modern Mathematics Exercise Series, Iwanami, Tokyo, 1976

20. My Calculus, Introduction to Analysis, Kodansha, Tokyo, 1981

21. Operational Calculus, A Theory of Hyperfunctions, University of Tokyo Press, Tokyo, 1982. English Translation (enlarged and revised). Operational Calculus, A Theory of Hyperfunctions, Springer-Verlag, New York-Berlin-Heidelberg-Tokyo, 1984
22. Analysis I, Mathematics in the 19th Century, History of Mathematics 9 Kyoritsu, Tokyo, 1986
23. Introduction to Modern Analysis (with H. Fujita), Iwanami, Tokyo, 1991

Books (in English)

i. Lectures on Semigroup Theory and its Application to Cauchy's Problem in Partial Differential Equations, Tata Institute of Fundamental Research, Bombay, 1957
ii. Functional Analysis, Grundlehren der Mathematischen Wissenschaftern 123, Springer-Verlag, Berlin, Heidelberg, 1965 (6th ed. 1980). Russian translation: Funktsional'nyĭ Analiz, Izdat. MIR, Moscow, 1967